Flowers of Greece and the Balkans

Flowers of Greece and the Balkans

a field guide

OLEG POLUNIN

With 80 pages of illustrations in colour
from photographs taken by the author and others
28 pages of line drawings
by Barbara Everard and Ann Davies
34 pages of line drawings by Pat Halliday
and maps by John Callow

Oxford   New York
OXFORD UNIVERSITY PRESS
1987

Oxford University Press, Walton Street, Oxford OX2 6DP

Oxford New York Toronto
Delhi Bombay Calcutta Madras Karachi
Petaling Jaya Singapore Hong Kong Tokyo
Nairobi Dar es Salaam Cape Town
Melbourne Auckland

and associated companies in
Beirut Berlin Ibadan Nicosia

Oxford is a trade mark of Oxford University Press

First published 1980
First issued as an Oxford University Press paperback 1987

British Library Cataloguing in Publication Data
Polunin, Oleg
Flowers of Greece and the Balkans: a field guide.
1. Wild flowers — Balkan Peninsula — Identification
I. Title
582.13'09496   QK332
ISBN 0–19–281998–4

Library of Congress Cataloging in Publication Data
Polunin, Oleg.
Flowers of Greece and the Balkans.
Bibliography: p.   Includes indexes.
1. Wild flowers — Balkan Peninsula — Identification.
2. Wild flowers — Greece – Identification. 3. Wild
flowers — Balkan Peninsula — Pictorial works. 4. Wild
flowers — Greece — Pictorial works.   I. Title.
QK332.P64 1987   582.13'09496   86–28498
ISBN 0–19–281998–4 (pbk.)

Printed in Hong Kong

Contents

List of Terrain Colour Plates

CONTENTS

Preface

The Balkan flora is undoubtedly the richest in Europe, possessing not only the largest number of species of any comparable area in Europe, but having at the same time the largest number of endemic species which grow nowhere else in the world.

This is the first serious attempt to bring together the relevant accumulated knowledge of the flora and to make it available to the layman—knowledge which is often only found in some obscure paper, in languages which are little known in the West. Not only does it describe briefly many of the richest plant-hunting areas of the Balkans, but will, it is hoped, enable the user to identify the plants seen while travelling by means of the descriptions, keys, and numerous illustrations. It is a field guide, indicating both where the plants are to be found and what they are called scientifically and locally.

The numbers of species found in our area is so great that inevitably a selection has had to be made to keep the volume within bounds. Certain families such as the *Gramineae* and the *Cyperaceae* have not been described at all, and others like the *Polygonaceae* and *Chenopodiaceae* have been largely omitted, being difficult and 'unattractive' families which require expert knowledge for their identification. Important Balkan genera with a large number of species such as *Centaureä*, *Dianthus*, *Campanula*, *Silene*, etc. have also to be curtailed, and only those species have been described which the author considers most likely to be encountered, or which are of particular interest or attractiveness, or are in some way unique. During his travels the author has frequently been assisted by local botanists who have helped in this difficult problem of selection. But it has not been an easy matter; a Bulgarian botanist, for example, may be very enthusiastic about an enclave of Mediterranean species which are very local in his country, but which are widespread and well known further south. While on the other hand the western European naturalist will find Pontic species with their Anatolian affinities, which occur nowhere else in Europe, of great interest, yet to the Bulgarian naturalist the species may be too familiar to warrant inclusion. In the final event only the author can decide, and though his selection may not satisfy the indigenous naturalist it is hoped that he will at least satisfy those who come from the West.

As in my previous volumes—*Flowers of Europe* and *Flowers of South-West Europe*—I have kept strictly to the names and authorities used in *Flora Europaea*, vols. i–v (I have been privileged to use the manuscript of vol. v), but in certain instances, particularly in the accounts of plant-hunting, I have occasionally used scientific names which are well known locally, but which are not accepted in the major work. Certain unusual species listed in the plant-hunting accounts have not been described in this volume for want of space, and in such cases reference must be made to *Flora Europaea* for their identification. For the same reason it has not been possible to include a key to the families, or keys to the genera. These can be found in the 'mother' volume, *Flowers of Europe*. Species illustrated in the mother volume are indicated by ** in Chapter 3 of the present volume. While in Chapter 2 those species marked with * are illustrated either by photographs or drawings in this volume. Species are numbered by a progressive system used throughout in *Flowers of Europe*, *Flowers of South-West Europe*, and this volume. Thus for example in *Saxifraga* the numbers 410, 410a, 410b, 410c, 410d refer respectively to *S. caesia* of central Europe, *S. media* and *S. aretioides* of western Europe, and *S. scardica* and *S. spruneri* of south-east Europe.

Area

The Balkan Peninsula as envisaged in this volume differs somewhat from the generally accepted extent of this region. It includes not only Turkey-in-Europe but all the islands surrounding the mainland, the Dalmatian, Ionian, and Aegean Islands, as well as the eastern Aegean Islands bordering the Turkish mainland, comprising Rhodes, Samos, Chios, Lesbos, etc. These latter eastern Aegean Islands are excluded from *Flora Europaea* as not being part of Europe. The northern boundary of the Balkan Peninsula is the Danube and its tributary the Sava, while further west the boundary continues along the Kupa river to its source and thence to the Adriatic, to the south of Trieste.

Distribution of species

There is no satisfactory and precise method of indicating the distribution of species in a simple format, other than by the use of distribution maps, but space precludes this. I have in consequence used the system followed in the previous volumes, but certain local variations may require explanation. Thus while **GR.** indicates the presence of a species on mainland Greece only, **GR+Is.** means that it can be found both on mainland Greece and on some, or many, or all of the islands.

GR(south)+Euboea indicates a plant's presence *only* in southern Greece and on the island of Euboea, while **GR(except south)** signifies that it occurs throughout the mainland of Greece except in the south. Similarly, **Widespread** means that the plant is found throughout the whole of our area but gives no indication of frequency within each local region.

Conservation

Since the publication of *Flowers of South-West Europe* in 1973, three major steps forward have been taken towards the co-ordination and study of the flora of southern Europe, which in the past has been so conspicuously fragmented and nationalistic in outlook, largely because of language barriers. Not only is knowledge becoming central-ized and made more accessible to serious students, but these advances are already stimulating a more intensive study of species and vegeta-tion, and have heightened our awareness of the necessity of conserva-tion on a European as well as at the national level.

The first advance was the holding of the International Symposium on the Problems of the Balkan Flora and Vegetation. Its first meeting was held in Varna (Bulgaria) in 1973. The second symposium was held in Istanbul (Turkey) in 1978. It was at the first symposium that I was able to make contact with many of the botanists who have helped so generously in the preparation of this volume.

The second was the formation of OPTIMA, the Organization of the Phyto-taxonomic Investigation of the Mediterranean Area. It held its first very successful meeting in Iraklion (Crete) in 1975, and its second in Florence (Italy) in 1977. One of the declared aims of OPTIMA is 'To stimulate and co-ordinate research activities, exploration, and conser-vation and resource studies within the Mediterranean area'.

The third major advance was the formation of the secretariat of the International Union for Conservation of Nature's Threatened Plants Committee at the Royal Botanic Gardens, Kew. In 1976 this sec-retariat published its first list of 'Rare, threatened and endemic plants for the countries of Europe'. This has been followed by Red Data sheets giving details of the individual rare and threatened species, including their present status, distribution, habitats, and ecology as well as the conservation measures taken or proposed. At last we are beginning to have international co-operation over the problems of endangered species. Individual countries now have guide-lines for action and there are prospects of real plans for the protection of the great wealth of our European plant life.

In response to this increased awareness of the necessity of plant protection, I have in the plant-hunting accounts (Chapter 2) indicated by † those species which are included in the Threatened Plants Committee's lists. I have marked all species named in this volume which fall under the categories 'endangered, vulnerable, rare, or threatened', and I have also included the 'intermediate' species whose status for conservation has not yet been assessed. By so marking these species there is a danger that their approximate whereabouts is made known to a wider public than heretofore. But on balance it has been judged that this knowledge will be beneficial and will focus attention on the necessity of taking active steps to conserve these species. May I strongly urge naturalists and others, while plant-hunting in the field, to abide by the general rules that are laid down internationally relating to the collection, picking, and destruction of such plants, and to acquaint themselves with any national laws relating to plant-collecting, etc. of the country they visit. They should also feel it their duty to report to the conservation organization or society of the country any obvious damage or destruction observed, in the hope that action will be taken to prevent further despoliation. The following organizations are deeply concerned with conservation:

Albania
Zoological Institute
Tirana

Bulgaria
Research Co-ordinate Centre for Conservation & Reproduction of the
  Environment Committee for the Protection of Nature
2 Gagarin Str.
Sofia 1113

Bulgarian Botanical Society
Institute of Botany
Acad. G. Bonchev Str.
Sofia 1113

Greece
Hellenic Society for the Protection of Nature
9 Kydathineon Str.
Athens 119

Ministry of Agriculture, Dept. of Forests
11 Kalinovou Str.
Athens 812

Friends of the Trees of Athens
4 Fokylidou Str.
Athens

Hellenic Alpine Club
7 Karagiorgi Servias Str.
Athens 126

Turkey
General Directory of National Parks and Wildlife
Ministry of Forestry
Ankara

Yugoslavia
(Serbia)
Institute for Nature Protection
5/111 Studentski trg.
Belgrade

Association for the Conservation of Nature of Yugoslavia
Molerova ul. 35
Belgrade

(Croatia)
Socijalististcka Republika Hrvatska
Republica Zavod za Zastito Pirode
Zagreb

(Bosnia-Hercegovina)
Institute for Culture Monuments and Nature Protection: Republic
   Bosnia-Hercegovina
7 Vojvode Putnika
Sarajevo

Acknowledgements

My faith in the generosity and helpfulness of European botanists, naturalists, and plant-lovers has increased greatly during the preparation of this volume. On every hand, both in this country and in the countries whose flora I have been studying, I have been helped in innumerable ways by many people, and I would like to thank them all most sincerely for this readily given encouragement. The final responsibility is of course mine, but without their sustained support this volume could not possibly have been completed in its present form.

As a result of three major journeys in the Balkans, made in the years 1970, 1973, and 1975—the first made possible by my release from teaching duties by the Governing Body of Charterhouse—I was able to make large collections in the field which were deposited in the Herbarium at Leicester University. Professor T. G. Tutin and A. O. Chater were once again absolutely crucial to the success of this volume. Not only did they name my herbarium specimens but they enabled me to identify the majority of my colour photographs taken at the same time. Their profound and broad-based knowledge of the European flora could hardly be matched elsewhere, and I am enormously grateful for all their hard work on my behalf. I only hope they will feel that this volume has justified their efforts. I have also had generous help from the members of the staff of the Herbarium of the Royal Botanic Gardens, Kew, and consider it a privilege to be able to use the facilities of the great institution so freely.

In the preparation of the plant-hunting accounts I have been most generously assisted by many people, the majority of whom have not only prepared draft accounts for me, but have kindly checked through and made additions to my final manuscript. I am particulary grateful to the following: Sir C. Barclay, who wrote the Crete account; Professor D. Phitos, who helped with the plant-hunting accounts of south and central Greece; Professor A. Strid, who wrote the Olympus and Ossa draft accounts; Dr. P. Hartvig, who wrote the Smolikas draft account; Dr. S. Kozuharov and Dr. S. Stanev, who prepared drafts and checked most thoroughly the Bulgarian accounts; Dr. Č. Šilić, who checked the Yugoslav accounts; Professor A. Baytop, for help with Turkey-in-Europe. In addition I would like to thank Dr. W. Greuter for expert advice on Crete; Dr. G. E. Wickens for botanical checking of the manuscript at various stages; Mrs. D. Williams for acting as interpreter and translator on a number of occasions. The index of popular names has only been made possible by the work of Dr. Č. Šilić, Dr. S. Kozuharov, and Professor D. Phitos, and it is in all probability the first index of its kind that has been prepared.

Illustrations are an important feature of this volume and I would like to thank my three artists for their exacting and patient work. Mrs. Barbara Everard has prepared Plates 1–13, and Mrs. Ann Davies Plates 14–28. Miss P. Halliday has drawn the double-page spreads 1–17, relating to the plant-hunting areas. It has been a great pleasure to work with them all, and they have invariably adapted their skills to the special demands that a volume of this nature requires. The prolonged and detailed secretarial work has been most intelligently carried out by Mrs. K. Cooke. I would like to thank her especially for her perseverance and willingness to carry on despite my absences and stop–go tactics. Last but not least I would like to thank my wife who has supported me in my endeavours, has accompanied me on all my journeys with the particular responsibilities of the travel arrangements—and of course undertaken the 'feeding of the brute'. She has also undertaken the laborious task of indexing. Authors without wives cannot know how much they have missed.

Once again botanists and naturalists have most generously offered the loan of transparencies taken in many parts of the Balkans. It has not been an easy task to select, for a limited number of pages, a suitable and representative collection of species, which are either unique or characteristic of the region, yet without undue repetition of species illustrated in my other volumes.

In particular I would like to thank Dr. G. Hermjakob for the loan of a superb collection of orchid transparencies, and I only wish I could afford them more space and do them full justice in a larger format. Plates 62, 63, 64 are entirely composed of photographs taken by him in the field when at the German School in Athens. Other photographs have been generously loaned by: B. Mathew 200d, 200e, 214a, 1037, 1588f, 1623b, 1643d, 1675b, 1678c, 1678e, 1681; Č. Šilić 2a, 419d, 419e, 433a, 475, 620, 708a, (757), 757, 984a; H. W. E. van Bruggen 246a, 258a, 1075a, 1210b, 1212, 1276b, 1271, 1608, 1631; J. Sotiropoulos 384d, 384e, 411j, 1017d, 1075c, 1249e, 1420a; Sir C. Barclay 960b, 971a, 1392, 1461b, 1471, (1512); A. and T. Baytop 200b, 1588e, 1588k, 1627c, 1627d; H. J. B. Birks 479a, 769d, 1060, 1269b; S. Martin 200c, 1643e, 1675; M. G. Walters 919a, 1268b; G. Barrett 1253, 1333b.

Oleg Polunin, Godalming 1980

Signs and Abbreviations, etc.

agg.	an aggregate of two or more closely related species
ann.	annual
bienn.	biennial
c. (circa)	about, used in measurements
fl. fls.	flower, flowering, flowers
fr. frs.	fruit, fruiting, fruits
incl.	including
introd.	introduced
lv. lvs.	leaf, leaves
Med.	Mediterranean
perenn.	perennial
Pl.	plate
sp. spp.	species
subsp.	subspecies
var.	variety
×	(in text) indicates hybrid; (on colour plates) indicates degree of magnification, for example, x ⅓ = one-third life size
**	(in Chapter 3) illustrated in *Flowers of Europe*
*	(in Chapter 2) illustrated in *Flowers of Greece and the Balkans*
†	(in Chapter 2) threatened plants to be conserved
' '	inverted commas round the scientific name of a species indicate that it has not been described or named in *Flora Europaea* or is for some reason excluded from this work

Authorities
The names of the authorities first describing the species are often abbreviated. For the abbreviated forms used see *Flora Europaea*, vols. i–v.

Genera
The number of species in each genera found in the Balkans—whether native, naturalized, or commonly cultivated—is given in all cases where only a proportion of the species are keyed or described in the text.

Geographical distributions

AL.	Albania	Is.	Islands of the Adriatic, Ionian, and Aegean seas
BG.	Bulgaria		
CR.	Crete	TR.	Turkey-in-Europe
GR.	Greece	YU.	Yugoslavia
+	in addition		
( )	only in		

except . . . present elsewhere but absent in . . .

+Is. on many or all islands, but this is not used if species are only present in the nearest off-shore islands, such as Euboea

Widespread on mainland excludes all islands except the immediate off-shore islands
Examples of local geographical distributions:
GR. Greek mainland only
GR+Is. Greek mainland and many or most islands
GR (Macedonia). In Macedonia only in Greece
GR (except Macedonia). Throughout Greece but not in Macedonia
YU(north). In northern Yugoslavia only
GR+Lemnos. In the mainland of Greece and Lemnos only

1. Landform, Climate, Flora, and Vegetation

Landform

The geological history of the Balkans is perhaps the most important single factor contributing to the diversity of the present-day flora. Three fundamental geological features, or events, prevail. Firstly, the presence at the surface of a mass of relatively unchanged ancient crustal rocks. Secondly, the great Alpine upheaval and folding in Tertiary times, which threw up the main mountain ranges of the Balkans and the south Aegean region. Thirdly, the uplift and fracturing of the land masses, coupled with subsidence and flooding, which led to the formation of seas, islands, and peninsulas. This latter took place relatively recently towards the end of the Tertiary and the beginning of the Quaternary periods, when the flora had evolved to something like its present-day form, and the Balkan Peninsula took on its contemporary shape.

In consequence the following main geological areas can be distinguished. The *crystalline massifs* of the Rhodope block, and the isolated Cyclades block. The *new-fold mountains* comprising the Dinaric Mountains of Yugoslavia, the Macedonian Mountains, the Pindhos of Greece, the Stara Planina of Bulgaria, and the mountains of Crete. The *islands and coastlands* of the Adriatic, Ionian, Aegean, and Cretan seas. The *lowland plains* of the Danube basin, Morava–Struma valleys, and the central plains.

The crystalline massifs form the structural nucleus of the whole of the Balkan Peninsula. Younger rocks have folded and faulted against this harder core, and stresses and tensions continue to the present day as shown by the frequent tremors and earthquakes which occur in the Balkans. These ancient rocks were probably uplifted during the Carboniferous period, and at the same time altered by pressure and chemical action to form marbles, schists, and gneiss—the metamorphosed products of limestones, clays, and igneous rocks. Subsequent

1

Geographical Regions

erosion and dissection and more recent uplift have maintained much of the area as mountains.

The Rhodope block, or more correctly the Macedonian–Thracian Massif, is the main exposure of crystalline rocks in eastern Europe. It stretches from the Black Sea and the Istranca Mountains to the mountains of east Yugoslavia, and southwards down the east coast of Greece with the mountains of Olympus, Ossa, and Pilion to Euboea. The islands of Thasos and Samothrace and the peninsulas of Khalkidhiki—including Athos—are also relics of this ancient crust. Later Tertiary upheavals have probably created the contemporary high mountain ranges which include the Rhodope Mountains (2191 m),

Rila (2925 m), Pirin (2915 m), and Olympus (2911 m). Such isolated high mountain areas have, among other things, acted as refuges for certain ancient Tertiary plant species, and they are consequently of particular interest to botanists.

To the south lies another area formed of this ancient crustal block. It is, however, much more fragmented, and as a result of subsidence and flooding remains today above sea-level as the peninsula of Attica, south Euboea, and the islands of the Cyclades. The summits and ridges of these ancient mountains are the islands and peninsulas of the present day.

The new-fold mountains are formed largely of Secondary and Tertiary sedimentary rocks of limestones, sands, and conglomerates laid down by earlier seas and lakes which covered the area. During the Alpine earth movements 50–100 million years ago 'immense lateral pressures folded and overthrust the rocks in a great arc around the old rigid block of the north-east'.[1] The Dinaric Alps were thrown up in more or less parallel ranges, stretching like necklaces from the Julian Alps in northern Yugoslavia, to the Pindhos of Greece and the mountains of the Peloponnisos, Crete, Karpathos, Rhodes, to the Taurus Mountains of southern Turkey. The continuity is only broken in Albania and it becomes less apparent where the land has sunk below sea level. The whole area has been, and still is, a region of great instability. Numerous large faults have occurred throughout the peninsula causing steep-sided valleys and plains with barrier-like mountains encircling them. It is the most rugged and extensively mountainous area of Europe outside the Alps which lay at the centre of these great upheavals.

On the mainland the mountain systems lie generally in a north-west to south-east direction. Like ripples on a pond, ridge upon ridge curve outwards, away from the ancient crustal block, to end precipitously in high coastal ranges along the shores of the Adriatic and Ionian seas, the older rocks tending to lie towards the periphery.

Further south, in Greece, the main mountain ranges continue in a north-west to south-east direction in the Pindhos, and in the Taiyetos and Parnon ranges of the Peloponnisos. In the south-east they tend to curve eastwards in Giona, Parnassos, and Parnis; the Cretan mountains lie in an east–west direction. Only infrequently do the old crystalline rocks break through the cover of folded sedimentary rocks and become exposed in the Balkans.

During a similar period the younger rocks to the north of the Rhodope Massif buckled and up-lifted into what we know today as the Balkan Mountains or Stara Planina (2376 m). These have

3

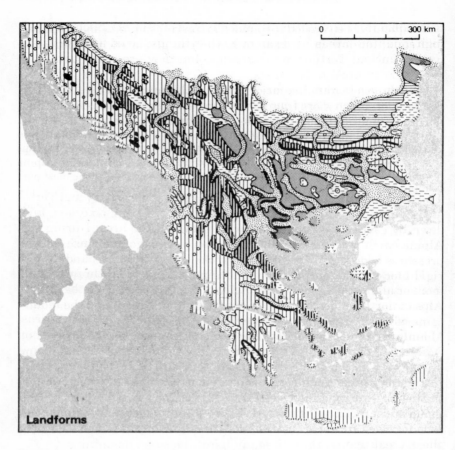

Landforms

Symbol	Description
	Hercynian massifs affected by alpine folding and forming mountains and uplands
	Alpine fold mountains, rugged in higher areas. Crystalline cores shown by heavier symbols
	Plateaux and lowlands formed by older sediment-ary rocks (Triassic – Cretaceous)
	Plateaux and lowlands, including intermontane basins, formed by younger sedimentary rocks
	Recent alluvium in river valleys, coastal plains etc.
	Volcanic rocks of recent (Tertiary) origin
	Structural faults forming prominent escarpments
	karst relief ⎫ Limestone
	Major poljes ⎭ topography
	Prominent sea cliffs

0   300 km

subsequently been eroded to rounded mountain summits and mountain plateaux, too low in altitude to have been glaciated in Quaternary times.

Within the folded areas, two major groups of rocks may be distinguished, both of which have been altered and hardened by the high temperatures and great pressures developed during folding. In the first place, the muds, sands and gravels—all shallow-water deposits—have sometimes been changed into slates, quartzites and conglomerates. The whole group of non-calcareous sedimentary rocks—shales, sandstones and marls—is known to the Alpine geologists as *Flysch*, or 'the sediments that slip'. These rocks for the most part are impermeable; streams are therefore numerous, at least during the rainy season, and the land has been easily eroded. In the second place, the calcareous sediments of the deeper clearer water have been almost entirely changed into the resistant limestones and marbles that play so great a part in the scenery . . .[2]

The older Mesozoic limestones form what is a very distinctive region of the Balkans, notably the High Karst Plateau of the Dinaric ranges, stretching from the Istra Peninsula to the Albanian border at a general level of something like 1500 m.

Following these upheavals a period of relative quiescence prevailed when the ranges were eroded and worn down to smooth contours and the valleys were filled with the sediments of less resistant rocks.

The third great period of land formation was in the late Tertiary and early Quaternary periods when the ranges were first uplifted so that some of their sedimentary rocks were raised as high as 3000 m; this was followed by great disturbances and faulting as is shown by the abrupt straight margins of many hard limestone mountains.

This shattering of the earth surface created most of the distinctive features of the present-day landscape, and not least the Aegean Sea. 'The result of the fracturing has been to divide up much of the land into blocks, subject to differential movement up and down; and the aspect of some parts of Greece has been compared to that of a badly laid pavement in which the individual blocks are tilted at different angles.'[3]

These changes resulted in the flooding of the Aegean area, followed by the fragmentation of the land masses into islands and peninsulas. This occurred over a considerable span of time, and at one period there were in all probability three land bridges connecting the Balkans with Asia Minor: that across the Bosporus to the north, through the Cyclades in the centre, and across Crete, Karpathos, and Rhodes to the south. Large water-basins separated these land bridges.

It has been estimated that Crete, Karpathos, and Rhodes have been isolated from the mainland and from each other for 10 million years, while many of the large islands near the land masses in the Aegean, Adriatic, and Ionian seas have only been isolated for less than a million years; changes in sea level are continuing at the present day.

The Quaternary Ice Ages have had relatively little direct influence on the Balkan land mass. No permanent ice caps existed, and there is little evidence of extensive glaciation. Only the highest summits of Durmitor, Olympus, Parnassos, etc. have glacial valleys and moraines to as low as 2000 m. But in the Prokletije Mountains, a range on the northern Albanian border which runs east to west to break the general trend of the Dinaric system, there is evidence of glaciation as low as 1000 m. The absence of permanent ice has undoubtedly helped in the survival of many Tertiary species of flowering plants, including those 'old' endemic species or 'palaeo-endemics' which are of such great interest in our area.

In addition, these great geological changes have had a profound effect on the recent evolution of the Balkan flora, which since late Tertiary times has been in many ways similar to the flora of the present day, with many species in common. The intervening seas have cut off the Balkan flora from that of Asia Minor with which it had many affinities, so that populations of plants which were previously able to interbreed freely have now become partially or completely isolated from each other. Thus inherited genetical material could no longer be freely interchanged—gene-flow ceased. The result was a situation similar to that on the Galápagos Islands where plants adapted and evolved largely in relation to their new environments, independent of the evolution of their nearest relatives elsewhere. Contact with the flora of the East was greatly reduced and isolation undoubtedly resulted in the disappearance of many species, while at the same time the evolution of new species was speeded up as the new environments were colonized. Contemporaneously contact with the West was if anything increased with the similarities of climate and terrain.

The population of this fragmented region was sparse; communication, particularly between the narrow coastal belt and the interior, was often difficult, while agriculture was largely limited to the bottoms of the valleys. There was also isolation of ethnic groups, often with their own agricultural and forest practices, thus helping to retain the diversity of vegetation. But these conditions are changing fast today, with improved communications, increased uniformity of living, and the universal spread of certain crops and practices.

One geological feature of great importance to the present-day landscape of the Balkans must be considered in more detail—that of the new-fold limestone mountains, often with their attendant faulting.

Limestone mountains stretch from north Dalmatia to the Peloponnisos and Crete. They are hard and slow to erode, and often persist as steep jagged escarpments, through which steep-sided gorges and canyons are cleft by the rivers draining the higher slopes.

The most extensive example of limestone mountains is the High Karst Plateau of Dalmatia. Here all the characteristic features are encountered again and again as one travels through this wild and little-populated country. Limestone is a very porous rock, yet very hard and resistant to erosion. Water is the most important erosive force, dissolving the limestone by chemical action. As it percolates down through cracks in the limestone it opens up fissures and channels, often of considerable depth, so that whole systems of underground drainage develop. During subsequent millenniums these work deeper and deeper, leaving in their wake enormous waterless caverns and grottoes and forming underground labyrinths of channels and shafts. The roofs of some of these caverns may eventually fall in, to produce great perpendicular-sided gorges, exposing the water to the surface once more. The magnificent gorges of many of the Yugoslav rivers, for example those of the Vrbas, Neretva, Tara, and Lim, are justly famous. It is often only along these gorges that communication is possible across the karst, and roads and railways have to tunnel through precipitous cliffs, or traverse narrow ledges above roaring torrents. At the same time, the purity of these rocks is such that the rivers are crystal clear, and there is little soil-making residue. Rock faces are often bare of vegetation and glaring white, but what little soil there is may collect in the hollows and support a lush vegetation, or yield narrow strips of cultivation.

One of the main features of the karst is the *planine* (plural *planina*), a mountain range of rounded, domed summits representing the worn-down edges of the limestone ridges of the Tertiary peneplain. These older ranges are in sharp contrast to the younger ridges, which have arisen as a result of faulting where the hard limestone has shifted against the softer rocks which come to form the valleys. The younger ridges have steep sides and jagged crests and are often dissected by gorges, and are known as *gore* (plural *gora*). Between the mountains lie shallow wide valleys, or steep-sided valleys, often with distinctively flat surfaces; these are known as *polje* (plural *polja*). Polja occur throughout the karst and vary in length from a few kilometres to up to 60 km and nearly 20 km in width. Many polja have deep *terra rossa* soils which are very fertile and are often the only cultivated areas in an otherwise barren and difficult terrain; other polja have shallow soils and are relatively infertile. Extreme contrasts are to be found; within a few hundred metres one may pass from rich fields and blue lakes

7

surrounded by vivid green meadows into stark limestone country, with a scanty cover of low dry scrub, where the bare ribs of the limestone show through everywhere.

Many polja are flooded during the winter and early spring, shortening the growing period and making cultivation of some crops difficult. Some have permanent lakes: Lake Shkodër is considered to be one enormous polje which has remained permanently flooded. There are no permanent surface rivers draining water off the polja, but instead water seeps away in the spring through narrow sink-holes or *ponore*, which open into caverns of great depth, and carry the water in underground drainage systems often for long distances. Ponore can also act as inlet channels and bring about the flooding of the floor of the polje during the winter months. Each polje has its own complex water system; some may remain dry throughout the year, others may have open water and marshy areas throughout the year.

The action of water is manifest throughout the karst; strong karstic features are associated with heavy rainfall. But except in certain polja water is very scarce, and one may travel far before catching a glimpse of any surface water. Occasionally water gushes out of the limestone cliffs or at the base of rock walls with a considerable force and volume, in springs known as *vrele*. Water may disappear with equal suddenness underground into holes and caverns called *jame*. Dry valleys, canyons, swallow-holes, and caverns are typical karstic features.

The *doline* (plural *dolina*) is another very characteristic declivity in the karst. Dolina are relatively small circular saucer-shaped or cone-shaped basins pitting the surface of the limestone plateau, sometimes not more than 100 metres across. The sides are often steep, and some soil, the fertile *terra rossa*, the result of limestone erosion, may collect in the bottoms. Here rich vegetable gardens, potato patches, or tiny cornfields are often cultivated. 'The patchwork pattern of these minute, walled, red fields in a setting of white sterile rocks, is as conspicuous as it is characteristic of the landscape of the dryer parts of the karst.'[4] The sides of uncultivated dolina may be clothed in shrub thickets which may serve as a local refuge for some plants which could not otherwise survive out on the open limestone.

Another typical karstic feature is the *lapiez*, or limestone pavement, where the surface of the limestone is dissected into numerous ridges. These may become so eroded that they become knife-like; between them, often several metres below, are small pockets of rich soil where plants may grow luxuriantly.

Uvale are deep irregular-shaped hollows with steep walls surrounding a level rocky floor like an amphitheatre. They are larger than dolina.

Not all the new-fold mountains are sharp and jagged. There are considerable highland areas, as in the Pindhos, where slopes are moderate and the summits are rounded, and older or softer rocks have been eroded and the debris carried down to form wide open valleys, in direct contrast to the narrow gorges through the limestone rocks. Many rivers run through both types of country alternately rushing through deep gorges or meandering over wide plains, at length to debouch on to the coastal plains often forming deltas of debris out to sea.

The islands and coastlands of the Balkan Peninsula are unique in Europe and give a very distinctive character to our area. Nowhere else is there such an abundance of small islands, or so many hundreds of kilometres of wild and miraculously unspoiled coastline.

The larger islands retain their ancient ways with thriving villages and communities of farmers and fishermen. Tiny whitewashed churches and box-like houses are scattered among the terraced olive groves and vineyards. On every hand stretches the sparkling blue sea, often whipped into a frenzy of white waves by the strong *meltemi* blowing down from the north, or the warm dry *sirocco* from the south. From the open horizons one can look beyond to countless smaller islands, or to the stark and forbidding mountains of the mainland, often capped with snow, where life is so different. There is a human scale and an intimate character in this island-world which is unlike any other place on earth.

The islands can be conveniently grouped as follows: the Dalmatian and Ionian islands of the Adriatic. The Aegean Sea encircles the central Cyclades, the more northern Sporadhes, and the northernmost Thracian Islands. Off-shore Greek islands include Kithira, Euboea, etc., while off the Turkish coast are the eastern Aegean Islands which include Lesbos, Chios, Samos, and Rhodes. Crete and Karpathos lie in the south Aegean.

The Dalmatian Islands are all that now remains above water of the outer ranges of the Dinaric Mountains. They generally run parallel to the coastal mountains. Only recently, during the late Quaternary period, have the lower valleys been flooded leaving the ridges often bare and glinting in the sun, or green and grey with pine trees or maquis; arms of the sea run in between the land masses all along the coast.

It is a young coastline with few cliffs or sandy bays. The limestone hills fall away in unbroken curves beneath the sea; tides rise and fall but a few centimetres, and in consequence of their recent separation from the mainland no endemic plants of any significance have evolved in these islands. Inland, the bare white Velebit and Biokova mountains rise abruptly from the coast with often no coastal plain at all. In winter this is a forbidding country with clouds lowering over the mountains, mists shrouding the islands, or the cold piercing *bora* blowing in gale force from the interior.

The Ionian Islands, comprising Corfu, Levkas, Cephalonia, Ithaki, and Zakinthos, are older in origin than those farther north. They lie on the continental shelf with deep water to the west in the Ionian Sea: steep cliffs, inlets, sandy bays, coves, and promontories are common, indicating an older coastline. The mountains of these islands are of limestone, and rise to altitudes of over 1000 m; Mount Ainos in Cephalonia (1620 m) is the highest. Around these limestone backbones are more recent and more fertile marls, clays, and sands where the villages and cultivation occur, provided there is enough water. Each island has small or extensive alluvial plains which are usually very fertile when brought under cultivation. The vegetation of these islands is very similar to that of the mainland coastal vegetation: Corfu has three endemic species.

The Cyclades. The numerous islands comprising the Cyclades represent all that remains above water of the ancient crustal block. This at one time formed mountainous uplands but at the present day they lie at an average depth of 150 fathoms below sea-level, with only their eroded summits above water. The peninsula of Attica and eastern Euboea are of the same origin. The Cyclades are largely composed of Cretaceous limestones (Amorgos), marble (Paros, Sifnos), schists and gneiss (Mikonos, Paros), tuffs (Thira, Milos), and igneous intrusive rocks (Naxos, Sifnos). They rise to altitudes of 1002 m in Naxos, and 944 m in Andros, but the general maximum is between 600 and 700 m on most islands. The Cyclades are separated from Ikaria and the eastern Aegean Islands by a deep channel, and to the south lies another deep depression flooded by the Cretan sea.

There are about 20 larger inhabited islands with hilly interiors, rocky valleys, and precipituous coastlines, with few fertile plains. The climate is equable and frosts are absent or rare. Winds are predominantly from the north-west but in summer fierce north-east winds, the *meltemi*, prevail. Rain occurs mostly in the winter months but some falls in the summer, thus contrasting with the mainland.

The northern **Sporadhes and Thracian Islands.** The former are largely of Cretaceous limestone lying over schist and are part of a submarine ridge. They have rounded contours and are well covered with trees and maquis, contrasting strongly with the Cyclades further south. The climate is similar to that of the mainland, but is cooler in summer with less prolonged dry periods; summer winds are from the north-east and winter winds from the south-east. Skiathos, Skopelos, and Skiros are the main islands.

The Thracian islands of Thasos and Samothrace are likewise well wooded, having climates similar to Thracian Greece. They are part of the Rhodope Massif and are largely formed of Palaeozoic and volcanic rocks; they rise to altitudes of 1203 m on Thasos and 1600 m on Samothrace. Lemnos lies further south and is composed of more recent Tertiary rocks with numbers of often steep volcanic peaks; it has a very indented coastline. It is much exposed to the winds and drought and there are no woodlands or olive groves.

In this account I have included the Sporadhes and Rhodes as well as the off-shore islands of the Turkish mainland in referring to the **eastern Aegean Islands.** Geologically these islands vary considerably. Ikaria and Samos are part of the ancient crystalline block which formed the land bridge from Attica to the Turkish mainland and Samsun Dag. They are composed of schists with a massive covering of limestone with belts of Tertiary rocks. Samos has well-developed forests and maquis particularly on the hills of the north-west. Ikaria by contrast is more or less treeless except in sheltered valleys and in cultivated areas. Lesbos and Chios are large islands which are considered to be detached portions of the western peninsula of Asia Minor. Chios is composed largely of Palaeozoic shales and mountain limestone with some Tertiary deposits. Lesbos has volcanic rocks and some serpentine and crystalline schists. Both islands come under the influence of the weather of the Turkish mainland. They are colder and wetter than the islands of the Aegean further west; the more fertile parts lie on the eastern side and it is the western coastal region which appears stark and barren by contrast.

Crete, Karpathos, and Rhodes are the part of the Tertiary mountain folds that stretched in an arc from the Peloponnisos to the southwest of Turkey. The original limestone crests have been much fractured and faulted, and with subsidence and uplift the rock structure is now complex. Some notes on the geological structure of Crete are to be found in Chapter 2.

Plains and lowlands. During the period of the Alpine folding and the formation of the main mountain ranges of the Balkans some areas experienced subsidence; these form the plains and lowlands of the present day. The Pannonian plain of Hungary and the Danube basin on the northern boundary of our area are the most extensive, while the Thracian plain of Bulgaria is of considerable importance. There are many more smaller depressions, notably the Vardar and Morava basins lying between the Rhodope block and the younger Dinaric Mountains. Many of these Balkan depressions were flooded by the northern arm of the Aegean Sea, which later receded, leaving lakes

11

from which the present-day drainage system has developed with a remarkable series of basins and gorges. Much faulting, upheaval, and subsidence have also occurred, and earthquakes and tremors still continue. Other valleys of some importance botanically are the Struma, Mesta, and Maritsa valleys which penetrate the heart of the Rhodope Mountains to the north and, in the case of the Maritsa, drains much of the Thracian plain. In eastern Greece lies the important plain of Thessaly, while on the west coast are the plains of Albania and Arta.

From the point of view of the vegetational cover and plant distribution these depressions and valleys are important in that they allow the penetration of Mediterranean air into the interior with the consequent extension of Mediterranean species inland. At the same time, particularly in the north, they act as funnels down which cold or hot air, depending on the seasons, can pour from the interior, creating a distinctive transitional vegetational zone. Valleys and plains such as these are the main communication routes to and from the interior, and likewise they act as 'migration' routes particularly for plants of disturbed ground and cultivation and for imported aliens associated with agriculture and horticulture. Most of these plains are intensely cultivated and have lost their natural vegetation except where lakes, marshes, and rivers occur locally.

Climate

The Balkan Peninsula falls within two very distinctive and contrasted climatic zones: that of the Mediterranean and that of central or continental Europe. In addition there is a transitional zone between the two climates, and the extent of this is largely determined by the presence of local mountain ranges, plains, and valleys as well as the prevailing winds.

In general the Mediterranean climate is distinguished by its hot dry summers which are rainless, contrasting strongly with the rainy but mild winter months. The central European climate is almost the reverse: rain falls throughout the year, the maximum occurring in the summer, while the Balkan winters of the interior are surprisingly cold for such southerly latitudes.

The general pressure conditions over the eastern Mediterranean are dominated by two high-pressure systems, one to the north over southern Russia, the Danube basin, and the northern Balkans, and one to the south over North Africa. Consequently the main airstreams are predominantly from the west and proceed eastward along the Mediterranean in both winter and summer.

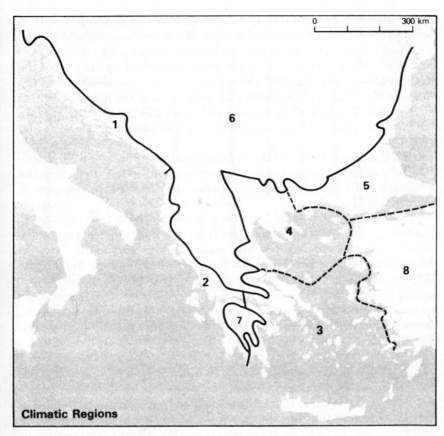

Climatic Regions

1 The Dalmatian coast
2 The coast of Albania and western Greece
3 The eastern coast of Greece and the Aegean Islands
4 The north Aegean region and islands
5 The Black Sea coast and Thrace
6 The central European region
7 The high mountains of the Peloponnisos
8 The Anatolian mountains and plateau

PLACE	TEMPERATURE Average daily temperature (max.)	Average daily temperature in °C (min.)	July average of highest temperatures	January average of lowest temperatures	RAINFALL Average annual rainfall in mm	Number of days with rain (1 mm or more)	Number of summer (Apr.–Sept.) days with rain	SUNSHINE Average annual hours of sunshine	Average number of days without sun
Athens	22·5	14·0	37·9	0·9	402	103	30	2756	20
Corfu	22·0	12·3	36·9	−1·3	1352	112	22	(2810)	—
Florina	17·3	6·3	35·9	−15·2	813	109	45	—	—
Iraklion	22·4	14·7	37·4	3·0	452	77	15	(2804)	—
Kavalla	19·7	9·7	35·4	−9·3	557	76	29	—	—
Lesbos	21·5	13·9	37·4	−0·1	739	80	18	(3058)	—
Naxos	21·2	15·7	33·8	2·4	475	80	11	(2481)	—
Salonica	20·9	11·3	38·1	−5·6	470	96	38	2624	42
Tripolis	19·5	7·8	36·9	−7·0	932	98	29	(2732)	—
Zakinthos	22·6	14·7	38·0	2·5	922	82	15	(3107)	—
Plovdiv	18·2	6·3	37·3	−18·2	492	99	48		
Sofia	15·2	6·1	33·9	−16·2	661	125	63	2046	—
Varna	17·6	8·6	36·0	−12·4	476	94	42	2253	—
Rijeka	17·8	10·7	33·6	−5·3	1548	126	59	—	—
Sarajevo	15·2	5·1	34·6	−15·4	932	157	72	1880	73
Skopje	18·4	6·0	38·1	−14·9	508	104	46	2046	—
Split	19·6	13·0	35·6	−3·7	821	113	44	2621	33

CLIMATES OF THE BALKAN PENINSULA:
YEARS 1931–60
Data from *Met. 0.856c Meteorological Office*, London, 1972.
Data in brackets from Geographical Handbook Series, Naval Intelligence Division,
Yugoslavia, i, 1944; *Greece*, iii, 1945.

The Adriatic and Aegean seas tend to bring in airstreams from a more northerly quarter; those of the Adriatic are from the north-west while those passing across the Aegean are predominantly from the north. But local weather systems may prevail, particularly where high land masses are situated close to the sea, and strong intermittent local winds may override these general surface winds for considerable periods.

During the winter months frequent depressions pass eastwards along the Mediterranean. Some come from the Atlantic through the Straits of Gibraltar, others develop in the Tyrrhenian and Adriatic seas. These depressions travel eastwards, passing across the Aegean Sea either to the south of the Greek peninsula, or across the Greek mainland, often to veer northwards via the Sea of Marmara to the Black Sea. Depressions are frequent in winter and spring, and bring characteristically unsettled rainy weather with either hot or cold air currents, depending on whether each depression passes to the north or to the south of a particular area. In Crete, for example, where the majority of depressions pass to the north of the island, many of the winds are southerly, bringing warm air from North Africa. Depressions also pass down the Danube basin to the Black Sea; they are more frequent in summer. In the eastern Mediterranean depressions are rare in summer and also the land masses of the Middle East heat up, pressure falls, and the strong northerly Etesian winds, known locally as the *meltemi*, blow down the Aegean from the north.

Rainfall

The rainfall of the Balkans, which has such a profound influence on the vegetation, is as varied as anywhere in Europe, in both its range and its seasonal effect. In general most rain falls in the western parts of the Balkans, diminishing eastwards where the total rainfall may be below 400 mm per annum in Attica, the Cyclades, and eastern Crete. The west coast of Greece receives about 1100 mm of rain in contrast to an average of about 500 mm on the east coast. The Dinaric ranges, the mountains of Albania, and the Pindhos act as effective barriers to the predominantly westerly rain-bearing depressions, and in their shadow lie very much drier areas, modified to a greater or lesser extent by the continental air masses to the north and south. There are also the two contrasting rainfall regimes: that of the Mediterranean with maximum rainfall in winter and that of central Europe with maximum rainfall in summer, so that the incidence of rain varies very considerably from place to place.

Of particular interest is the extremely heavy rainfall of the coastal

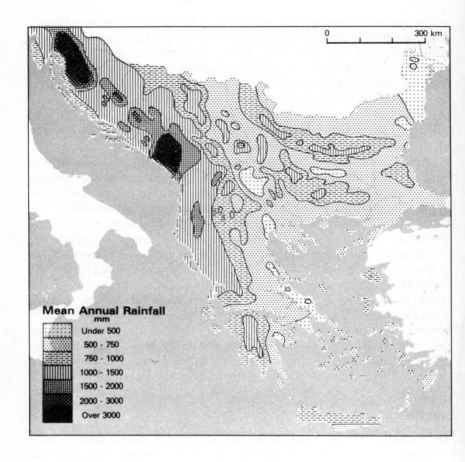

Mean Annual Rainfall
mm
Under 500
500 - 750
750 - 1000
1000 - 1500
1500 - 2000
2000 - 3000
Over 3000

mountains of Dalmatia, and further inland on the Prokletije range on the borders of Yugoslavia and Albania; here over 3000 mm of rain falls annually. In fact in southern Dalmatia, inland from Kotor, lies the wettest place in Europe, at Crkvice, with an annual total of as much as 4622 mm of rain. By contrast, in the interior to the east, in the Vardar valley, the annual rainfall is as low as 457 mm. The rainfall of the more northerly interior Balkan valleys such as the Sava and Danube have annual means of 500–700 mm. To a lesser extent the Pindhos and the mountains of the Peloponnisos, with averages of 1200 mm, act as a rain-barrier creating the drier area of the south-east of the peninsula. Relief also has a considerable local effect on the rainfall where many of the interior mountain ranges (*gora* and *planina*) have heavier rainfalls than the intervening polja which may be conspicuously drier.

16

Seasonal Rainfall

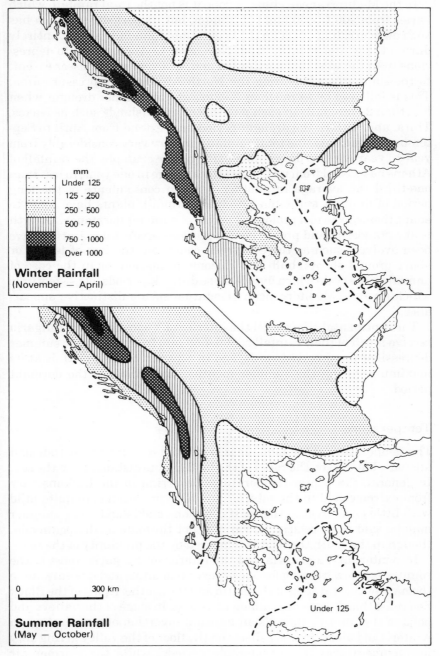

mm
Under 125
125 - 250
250 - 500
500 - 750
750 - 1000
Over 1000

Winter Rainfall
(November — April)

0   300 km

Summer Rainfall
(May — October)

Under 125

17

In the Mediterranean climatic zone it is not the total rainfall, which varies from 400 to 800 mm per annum, but its periodicity which has such an overriding effect on the vegetation. Rain falls almost entirely during the winter months, from November to April, owing to depressions from the west and north-west and, to a very much lesser extent, to the southerly winds from the North African high-pressure area. This is followed by as much as 3–5 months of summer drought when less than 25 mm falls in each month. On some islands such as Naxos, Thira, and Kithira the drought period may extend from April to September. In addition, the total rainfall can vary very considerably from year to year with no discernible pattern. For example, the rainfall of Athens has varied from up to twice the average in one year, to less than one-third the average in another year. In consequence of the long period of drought and often uncertain rainfall, plants have come to adapt themselves to a prolonged period of summer dormancy. Growth is at a standstill, and protective adaptations against dessication have been evolved, such as the evergreen condition, the loss of leaves or above-ground shoots, bulb and rhizome formation, or in the case of annuals a life cycle which avoids these difficult periods, etc.—features which give the Mediterranean vegetation such a distinctive appearance.

The continental climates of the interior of Yugoslavia and Bulgaria receive much of their rainfall of 500–659 mm from numerous summer depressions passing eastwards, and in contrast plant growth is at its maximum during the summer months, and winter is the dormant period.

Temperature

The temperature of the land mass is affected not only by latitude and altitude, but also by the relief of the land and its distance from the sea. In general, the winter temperatures occurring in the Balkans vary from extreme cold in the interior and high mountains, to quite mild with little or no frost in the coastal regions and islands. The summer may be said to be relatively warm to hot throughout the peninsula, though in the south the heat is tempered by the proximity of the sea.

In winter, in Yugoslavia and the interior of Bulgaria, most of the country is below freezing for one or several months, and it is only along the narrow coastal strip of Dalmatia and the southern part of the Black Sea coast that frost is uncommon or absent. In some of the valleys and polja of the interior there is an unusual inversion of temperature in winter. Cold air sinks and collects on the floor of the valley giving very low temperatures for considerable periods, while the warmer air

Actual Surface Temperatures

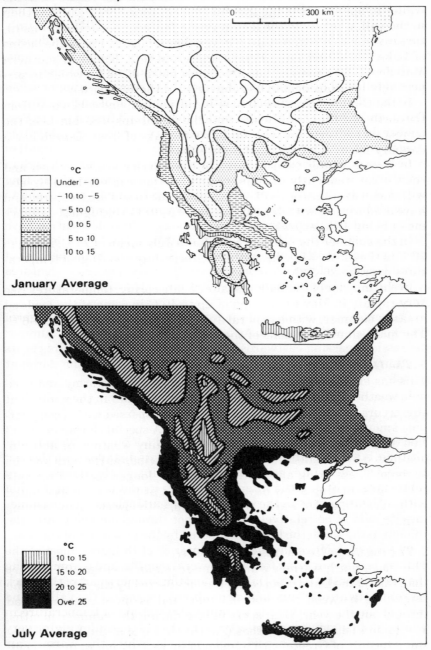

0 — 300 km

°C
Under −10
−10 to −5
−5 to 0
0 to 5
5 to 10
10 to 15

January Average

°C
10 to 15
15 to 20
20 to 25
Over 25

July Average

19

creeps up the surrounding mountainsides with the result that the winter temperatures of the summits may be considerably higher than in the valley below. In the mountains of Macedonia similar temperature inversions occur, resulting in very low temperatures in the basins of Lake Ohrid and Lake Prespa and elsewhere. Bitola in southern Macedonia, for example, has a January mean of −1°C, which is surprisingly low considering its relatively southerly position.

In the Danube basin the climate is continental, and cold winds from the north, and the *kosava* winds from the mountains of Serbia, keep the temperatures very low with up to 70–80 days of frost; Zagreb has a January mean of −0·5°C and Sofia −0·6°C.

In Greece winter temperatures in the interior are low. Frost and snow occur frequently in January and February in northern Greece, with mean averages of 1–5°C. By contrast the coasts are frost-free and have mild winters with January means of 5–10°C, the highest January mean being 12°C at Iraklion (Crete).

In the summer the Adriatic coast has a July mean of 24°C, rising to 26°C in the south, while in southern Greece July and August temperatures average 26–7°C.

In the interior of the Balkans, after a late spring, temperatures rise very rapidly in May to maxima of up to 40°C in some of the lowland areas in summer, with a general July and August average of 21–6°C. The main climatic regions of the Balkan Peninsula are:

1. *The Dalmatian coast*
This has long hot dry summers with little rain, and stormy and variable weather in the autumn. There is heavy rainfall in the winter but this occurs in relatively mild conditions interspersed with a considerable amount of sun, and bright clear skies. 'A special characteristic of this area is the particularly stormy and rainy weather in autumn, dominated by the effects of such contrasting winds as the bora and the scirocco. These bring alternating shorter or longer spells of bitingly cold winds, lasting a few days or more and at times even associated with snowfall, and of warm "muggy" days, with low cloud and driving rain.'[5]

2. *The coast of Albania and western Greece*
This has a much warmer winter climate, on an average 10°C higher than that of the Dalmatian coast. It is unaffected by the cold *bora*. The rainfall is between 500 and 900 mm, and is more evenly spread throughout the year, with some falling during the summer months. The Ionian island of Zakinthos (Zanthe) has one of the highest sunshine records of Greece with a mean of up to 3107 hours in the year;

this compares with a mean of 2753 hours of sunshine in Tripolis in the interior (Eastbourne has 1833 hours—Britain's maximum). The climate is generally more 'Atlantic' and there are few periods of summer drought.

3. *The eastern coast of Greece and the Aegean Islands*

This has a considerably drier climate than the west with average rainfalls of 400–500 mm. The prevailing summer winds are from the west, out to sea and away from the coast. There is little or no summer rainfall, and day after day of sunshine follow each other monotonously. The winters are correspondingly cooler than those of the west coast. The mean January temperature of Lamia is 7·6°C compared with 10·1°C at Patras on the west coast (5°C in London). Enclosed basins inland may become much colder in winter, and build up heat in the summer to temperatures of 32–8°C, as, for example, in the plain of Sparta. Local winds resulting from these temperature gradients occur periodically in winter and summer.

The Aegean Islands and Crete have the same over-all pattern of climate but are generally more equable, being tempered by the surrounding seas. Islands lying closer to the land masses come periodically under the influence of the mainland climates; for example, Crete receives warmer air from the North African coast. Likewise the eastern Aegean Islands are influenced by the adjacent Anatolian climates, and receive rainfall from the east, with the result that the eastern shores are much more heavily wooded and cultivated than the drier barren-looking western slopes.

4. *The north Aegean region and islands*

This region includes the coastal region of Thessaly and Macedonia and the northern Aegean islands. The climate is generally cooler owing not only to its more northerly position but also to its proximity to the continental land masses. In winter cold air comes down from the north through the valleys from the interior; the cold *vardarac* wind which blows down the Vardar basin in Yugoslavia is a notable example. Frosts may occur at night, but they are not sufficient to prevent the extensive cultivation of olives and tobacco. The summers have intermittent rainfall unlike the south, largely as a result of thunderstorms. At other times in summer the hot *livas* winds blow from the interior of Thessaly to the coast.

5. *The Black Sea coast and Thrace*

A climate of intermediate character occurs in the Thracian plain of Bulgaria south of the Stara Planina, and in much of Turkey-in-

Europe. Here the continental climate of the steppes has a strong influence, while at other times northern Mediterranean air masses penetrate some distance inland. In the winter, northerly winds prevail, bringing very cold air, much rain, and some snow. January means at Plovdiv are 0°C, and at Varna 1°C, while in Istanbul the January mean is 4·5°C and snow may fall on average on 18 days in the year. Periodic depressions in winter bring in warmer air from the south which removes the protective snow cover. In summer, northerly winds continue to prevail, bringing hot dry winds and a late-summer drought, with little rain between June and September. Thus in Turkey-in-Europe, for instance, the winters are too cold for many Mediterranean species and crops to survive, while the summers are too dry for a true central European forest flora to develop. The adjacent Istranca and Rhodope mountains have more rain and are forest-covered.

6. *The central European region*

This includes most of the interior of the Balkan Peninsula, from Macedonia northwards, as well as Bulgaria north of the Stara Planina. Extensions of this climatic region occur along the main mountain ranges of the Pindhos, Rhodopes, and even as far south, in a modified form, as the high mountains of the Peloponnisos. Rain falls throughout the year with maxima in May, June, and October. On higher ground much of it falls as snow in winter and continues to lie till early summer. In Greece, for example, above 1000 m, snow may lie from mid-December to early March, while above 2000 m continuous snow may lie till mid-May. Above this, snow may remain till June on the highest summits in the south.

In the plains the summers are hot and dry, and 'dusty conditions' prevail, broken by intermittent thunderstorms. The annual rainfall is between 500 and 600 mm, and temperatures in July average 22–4°C.

THE FLORA AND VEGETATION

The Flora

The plant-life of the Balkans is richer than any comparable area in Europe. Turrill[6] has estimated that in the Balkan area, as defined in this volume, but excluding the eastern Aegean Islands, there are over 6530 species of native seed plants. The precise figure does of course depend on the contemporary botanists' assessment of what constitutes a species or a subspecies (some of Turrill's species are not accepted in *Flora Europaea*). Also in the fifty years since he wrote his monumental

Plant Life of the Balkan Peninsula some new species have been described. The estimation of the total number of native European species given in *Flora Europaea* is 10 500, about 3500 of which are endemic to Europe. In addition there are about 800 naturalized alien species.[7]

The richness of the Balkan flora is a result of a number of conditions, the most important being:

(a) An old flora containing many Tertiary species which have survived the Quaternary Ice Ages.

(b) Isolation of land masses, islands, mountain ranges, etc. as a result of the changes in level and extent of the Mediterranean Sea. This has resulted in the fragmentation, isolation, and migration of species. New habitats have also been formed.

(c) The proximity of other floras, notably the central European, Anatolian, and Pontic floras from which migration has taken place.

(d) The profound influence of Man in destroying and changing the natural plant cover, creating new habitats, and introducing species from outside the Balkan Peninsula.

Thus the Balkans is the most important 'refuge' area of Europe, an important site of the formation of new species, and a valuable immigration route to and from the area. See map on page 27.

At the present day the Balkan flora consists of the following components (the numbers are given in round figures): European species, including those with wider transcontinental distribution, 1300; central European species (including 280 alpine species), 1000; Mediterranean species, 2000; Balkan species, including endemic species, nearly 2000. In addition there are 200 Pontic species, 300 Oriental species (Asia Minor), and about 17 with their main distribution in North Africa.

The numbers of species occurring in each of the botanical regions described by Turrill are shown on the map on page 27. The figures in bold represent the number of endemic species which occur in each region and which are exclusive to that region, not being found anywhere else in the Balkans or the rest of the world; they total 901. Nearly twice as many species are exclusive to the Balkans, but are more widely distributed and are found in several of the botanical regions. Turrill estimated the grand total of Balkan endemic species to be 1754. This represents nearly a quarter of the whole flora, and is a measure of the uniqueness of the Balkan flora. Endemic species are of great interest to the student of plant-life for they help to elucidate both the past history of the flora and its continuing development. Endemics may also be distinctive and of considerable beauty, and enhanced by their uniqueness they are often sought after by naturalists, who in

Distribution of Important Plants

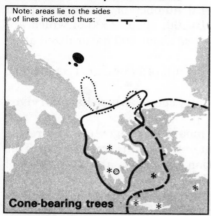

Note: areas lie to the sides of lines indicated thus:

Cone-bearing trees

— Abies cephalonica and hybrid — Greek Fir

⊚ Juniperus drupacea — Syrian Juniper

— — Pinus brutia — Calabrian Pine

⋯⋯ Pinus peuce — Macedonian Pine

●' Picea omorika — Serbian Spruce

∗ Cupressus sempervirens — Italian or Funeral Cypress

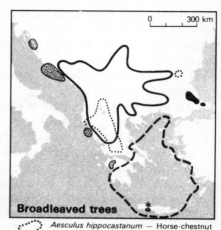

0 ____ 300 km

Broadleaved trees

⋯⋯ Aesculus hippocastanum — Horse-chestnut

● Fagus orientalis — Oriental Beech

∗ Zelkova abelicea

— — Acer sempervirens — Cretan Maple

⬭ Celtis tournefortii

— Corylus colurna — Turkish Hazel

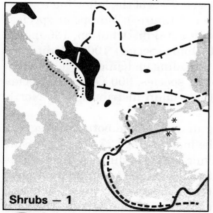

Shrubs — 1

● Daphne blagayana

— — Syringa vulgaris — Lilac

- - - - - Styrax officinalis — Storax

— Genista acanthoclada

⋯⋯ Petteria ramentacea

∗ Rhododendron luteum — Azalea

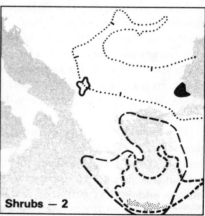

Shrubs — 2

⋯⋯⋯ Brukenthalia spiculifolia

● Rhododendron ponticum — Rhododendron

— Forsythia europaea — European Forsythia

⬭ Ebenus cretica

- - - - - Cistus parviflorus

— — Globularia alypum

24

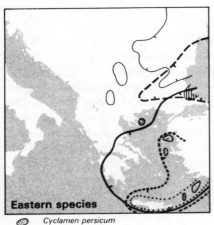

Eastern species

- 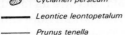 *Cyclamen persicum*
- ───── *Leontice leontopetalum*
- ───── *Prunus tenella*
- |||||||| *Daphne pontica*
- ▪▪▪▪▪ *Rosularia serrata*
- ── ── *Salvia nutans*   ········ *Ranunculus asiaticus*

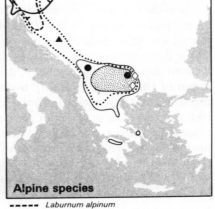

Alpine species

- ▪▪▪▪▪ *Laburnum alpinum*
- ───── *Helleborus niger*
- ◉ *Primula minima*
- ◇ *Soldanella pusilla / Gentiana pyrenaica*
- ◯ *Saxifraga exarata*   ········ *Gentiana punctata*
- ▲ *Primula glutinosa*   ● *Pulsatilla vernalis*

Balkan endemic species — 1

- ───── *Ramonda serbica*
- ▪▪▪▪▪ *Ramonda nataliae*
- ⦂⦂⦂ *Haberlea rhodopensis*
- ▲ *Jankaea heldreichii*
- ▒ *Petromarula pinnata*   ⬤ *Primula kitaibeliana*
- ＊ *Paeonia rhodia*   ◯ *Macrotomia densiflora*

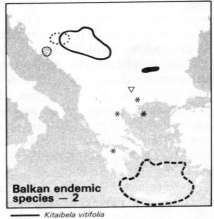

Balkan endemic species — 2

- ───── *Kitaibela vitifolia*
- ⦂⦂⦂ *Symphyandra hofmannii*
- ◎ *Edraianthus dalmaticus*
- ⬤ *Primula frondosa*
- ▽ *Primula deorum*
- ＊ *Viola delphinantha*   ▬▬▬ *Dianthus arboreus*

consequence have a special responsibility to ensure that they continue to survive.

There are many ancient endemics, often referred to as relict species or palaeoendemics, in the Balkans. These date from the Tertiary period some 20–50 million years ago. They probably once had a much wider distribution and lived in climates different from those of the present day. Such species have, it is thought, been exterminated over much of their range, largely owing to climatic changes, and at the present day they are only found in small isolated colonies. Conifers like the Omorica or Serbian Spruce, *Picea omorica*, and the Macedonian Pine, *Pinus peuce*, are examples. In the Gloxinia family, Gesneriaceae, which is now almost exclusively tropical and subtropical, the Tertiary relict generas *Haberlea*, *Jankaea*, and *Ramonda* survive in the cool mountains of the central Balkans (*Ramonda myconi* of the Pyrenees is the exception). There are many more examples of such relict species, but it is impossible to tell just how old they are as fossil records in the Balkans are very sparse. In general the Balkan flora can be said to have evolved from the Tertiary floras. Evolution has been continuous since that time and many new species and genera have been evolved. It is probable, however, that during the last 10 000 years there has been a further evolutionary 'spurt' as a result of man's impact on the vegetation. Such relatively 'recent' changes as the unprecedented destruction of the climax vegetation; the opening-up of new habitats; the introduction of the weed communities; the pressures on grazing-resistant species; the loss and change of soils; local climate changes resulting from the destruction of the natural plant cover—all these create conditions in which evolutionary change is speeded up and neoendemics are evolved. How important this contemporary evolutionary change is, in relation to the larger time-scale of the geological past, it is difficult to assess. An estimation of evolutionary change in the Cretan flora since its isolation is revealing (see page 60); it does not indicate rapid change on the island, but this may be an exceptional case.

Certain families seem, at the present day, to be very much on the evolutionary move, judging from the number of Balkan endemic species recorded for these families: Compositae 327, Caryophyllaceae 175, Labiatae 152, Scrophulariaceae 126, Leguminosae 107, Umbelliferae 107, Cruciferae 94, Liliaceae 93. While the genera with the largest number of endemic species are: *Centaurea* 114, *Dianthus* 64, *Verbascum* 55, *Thymus* 53, *Campanula* 50, *Silene* 46, *Viola* 33, *Astragalus* 31, *Stachys* 30, *Asperula* 27. Other genera with more than a few endemic species include: *Allium* 22, *Colchicum* 16, *Fritillaria* 15, *Crocus* 13, *Onosma* 11, *Tulipa* 10. These numbers do depend on the

Numbers and Distribution of Species

Botanical districts and the numbers of species in each district where known

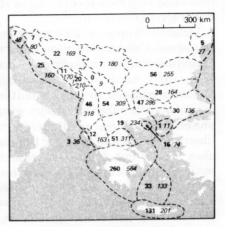

131 — **Numbers of species endemic to each botanical district**
201 — **Numbers of Balkan endemic species present in each botanical district**

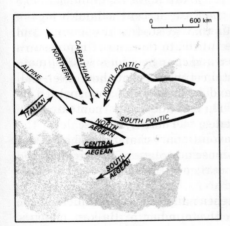

The chief routes of plant migration
Thickness of arrows indicates importance

Numbers of central European alpine species in each botanical district where known

(Based on W. B. Turrill, *The Plant-life of the Balkan Peninsula*, Oxford, 1929)

27

botanists' conception of the term species—these are Turrill's figures, not those of *Flora Europaea.*

Isolation of species is a conspicuous feature of the Balkan flora. This has several effects. Isolation often results in evolutionary stagnation: cross-breeding is unlikely and species change little over millions of years. As a result of interbreeding, isolated species may evolve local variants adapted to local conditions, or they may show 'drift' and evolve peculiar characteristics which seem to be of little evolutionary value, like change of colour, hairiness, size of flowers, etc. Consequently the plants of each isolated interbreeding community may show minor but distinctive differences. Examples are found in the many forms of certain alpine species, notably among the *Saxifraga.*

On the other hand, species with a wide distribution often have a richer genetic variability which, coupled with ready cross-breeding, results in a greater variety of forms, which can establish themselves in a greater variety of habitats. New habitats can be colonized; adaptation to old habitats can be more successful through the natural selection of the fittest forms.

The proximity of the ancient floras, notably those of Asia Minor and the Pontic regions surrounding the Black Sea, accounts for the un-European appearance of some of the Balkan plant communities. The gradual migration of species from the East, across the land bridge of Thrace and across the land masses that existed in the central and southern Aegean region before the ultimate flooding of the eastern Mediterranean Sea, was probably the means by which most eastern species have reached Europe. Turrill estimated that there are something like 680 of these species, and he lists 47 of them which he considered have arrived through Thrace, and more than 90 which migrated through Crete, the Cyclades, to Thessaly.

Man has not only influenced the evolutionary changes in the Balkan flora, but he may have been the unconscious cause of the survival and spread of many species—perhaps nearly one-fifth of the total—which flourish in waste ground, cultivated areas, vineyards, olive groves, etc. While at the same time he has intentionally introduced species from outside Europe, as crops, for ornament, etc., which are becoming established as an integral part of the contemporary flora.

The Vegetation

The plant communities of the Balkan region show as great a variety and complexity as any in Europe. They range from the high alpine vegetation of the mountain peaks, through dense central European forests of fir and beech, to the evergreen woods of the Mediterranean and the scorched dry phrygana of the Greek islands. With two such contrasting types of climate as the Mediterranean and the central European, with its intervening transitional zone, a great variety of local climates is developed. This, coupled with wide differences in bedrock formation, altitude, and aspect, as well as the varied origin of species, all makes for a great diversity in the plant communities. In consequence of these many factors, any over-all scheme describing the plant communities of the Balkans will inevitably be shot through with local variants and exceptions. In this account as simple a scheme as possible has been used, one which it is hoped will be comprehensible to the amateur naturalist, yet which does not run counter to the ideas of the central European plant sociologists, as recently outlined in *Vegetation Südosteuropas* by Horvat, Glavač, and Ellenberg (1974).

Main Types of Vegetation in the Balkan Peninsula

MEDITERRANEAN PLANT COMMUNITIES

Lowlands and hills (0–700 m approx.)
  1. Mediterranean evergreen forests
  2. Maquis (macchie)
  3. Phrygana (garrigue)
  4. Mediterranean mixed deciduous forests
  5. Herbaceous vegetation of stony ground—pseudo-steppe
  6. Vegetation of rock walls and gorges—chasmophytes
  7. Coastal terrestrial vegetation

Montane forests (700–1700 m approx.)
  8. Coniferous forests

Sub-alpine and alpine (1700–3000 m approx.)
  9. Scrub; alpine meadows; stony slopes, screes, rock vegetation

TRANSITIONAL PLANT COMMUNITIES

  10. Mediterranean transitional (sub-Mediterranean)
  11. East-central European transitional (Continental)
  12. Black Sea transitional (Pontic)
  13. Shiblyak (deciduous brushwood)

CENTRAL EUROPEAN PLANT COMMUNITIES

Lowlands and hills (0–700 m approx.)
  14. Mixed deciduous forests
  15. Vegetation of river valleys, marshes, lakes, etc.

Montane forests (700–1700 m approx.)
   16. Montane deciduous forests (beech)
   17. Montane coniferous forests

Sub-alpine and alpine (1700–3000 m approx.)
   18. Scrub; alpine meadows; stony slopes, screes, rock vegetation

Some interrelationships in Mediterranean plant communities

Altitude	Zones	Climax vegetation		Degraded vegetation
3000 m	ALPINE	*Vegetation of stony slopes, screes, rocks, and cliffs*		
	SUB-ALPINE	*Scrub zone* (Hedgehog zone)	⇌	Sub-alpine stony grasslands
1700 m	MONTANE	*Coniferous forests* Greek Fir Black Pine Stinking Juniper *Deciduous forests* Sweet Chestnut	⇌	Stony grasslands
300 m	HILLS LOWLANDS	*Mixed deciduous woods* Oak-Hornbeam Oriental Plane	⇌	Phrygana, stony grasslands
		Evergreen woods Holm Oak Kermes Oak Valonia Oak Aleppo Pine Stone Pine Cypress Carob-Olive zone	⇌	Maquis, phrygana, pseudo-steppe

MEDITERRANEAN PLANT COMMUNITIES

A climate of hot dry rainless summers and mild and wet winters and early springs constitutes the overriding influence affecting the vegetation surrounding the Mediterranean Sea and gives it its distinctive composition and appearance which is quickly appreciated by the traveller. The predominance of evergreen trees and shrubs; the abundance of small, often grey-leaved aromatic shrubs; the short-lived flush of brightly coloured annuals and bulbous plants in the spring, followed by another flush when the rains return in early winter—these features clearly distinguish the Mediterranean vegetation from all other types in Europe. But so much of the land has been deforested, or decimated by heavy grazing, while all potentially fertile ground is carefully cultivated, that it is often difficult to visualize the character and state of the natural vegetation. However, relict patches of more or less natural vegetation often do exist in some gorge, or steep hillside, mountain slope, or river valley, so that it is usually possible to ascertain one's whereabouts in the vegetation scheme outlined—and, incidently, to learn where to look for some of the more interesting plants.

Lowlands and hills (0–700 m approx.)

1. *Mediterranean evergreen forests*
Much of the Mediterranean zone, particularly in the coastal lowlands, is dominated by the Aleppo Pine, 'P. *halepensis*, which forms open patchy woods on most types of soils, up to altitudes of nearly 1000 m. These woods are well developed on the Adriatic coast and islands, in

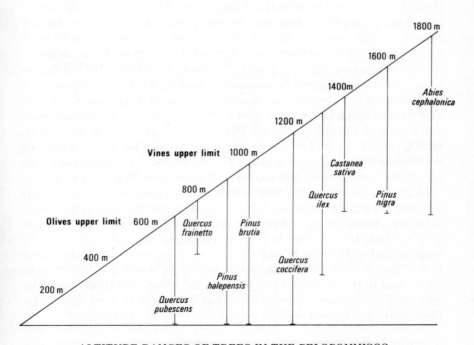

ALTITUDE RANGES OF TREES IN THE PELOPONNISOS
After A. Beuermann, 'Waldverhältnisse im Peloponnes unter besonderer Berücksichtigung der Entwaldung und Afforstung', *Erdkunde*, 10, 1956.

northern Greece, and many Aegean islands, but become rather scarce in the drier southern Aegean area. In Crete, the eastern Aegean Islands, and Thrace they are replaced by the eastern Mediterranean Calabrian Pine, *P. brutia*, which is very similar in its requirements.

The shrub layer which develops under these pine woods is largely of maquis and phrygana species, or in some cases the plants of open stony ground. Characteristic species include:

Juniperus oxycedrus	*Phillyrea latifolia*
J. phoenicea	*Olea europaea*
Quercus ilex	*Arbutus unedo*
Calicotome villosa	*Erica arborea*
Coronilla emerus	*E. manipuliflora*
Pistacia lentiscus	*Ruscus aculeatus*
Myrtus communis	

The evergreen Kermes Oak, *Q. coccifera*, is widespread in the hotter parts of the Mediterranean zone. It rarely forms pure stands but is often seen in cut and heavily grazed thickets with a few small scattered trees. It is often mixed with the Aleppo Pine. The Holm Oak, *Q. ilex*, is also widely dispersed and favours more humid climates as on the Adriatic coast, and here it may locally form small stands, with such under-shrubs as *Viburnum tinus*, *Rhamnus alaternus*, *Rosa sempervirens*, and *Lonicera implexa*. Another variant more typical of the Adriatic coast is characterized by the deciduous Manna Ash, *F. ornus*, with such shrubs as *Carpinus orientalis*, *Laurus nobilis*, and *Coronilla emerus*. On some Dalmatian Islands, for example on Hvar and Brac, are woods of Dalmatian Black Pine, *Pinus nigra* subsp. *dalmatica*, with *Quercus ilex* and *Fraxinus ornus*, in which *Juniperus oxycedrus* and *Salvia officinalis* are typical. The Stone Pine, *P. pinea*, occasionally forms stands near the sea as in the western Peloponnisos.

In the Peloponnisos, the southern Aegean region, and Crete there is a hotter drier zone which is characterized by the Olive and the Carob. It follows the coastline and penetrates inland no more than 12–20 km, and it is here largely that both these trees are carefully cultivated by man in orchards or scattered stands, often on terraces in the lower hills. Though these two trees never now form natural woods they do occur wild in the present-day semi-natural woods and thickets and by and large constitute a very distinctive feature of this climatic zone. Commonly associated species include:

Juniperus phoenicea	*Pistacia lentiscus*
Quercus coccifera	*P. terebinthus*
Calicotome villosa	*Myrtus communis*

Phillyrea latifolia　　　　　*Smilax aspera*
Salvia triloba　　　　　　　*Asparagus aphyllus*
Phagnalon rupestre

Other woodland communities of very limited distribution are the Cypress woods of *Cupressus sempervirens* in Crete, which are still found in the mountains at altitudes of 300–750 m, but which were once probably much more extensive. The Juniper woods of the Syrian Juniper, *J. drupacea*, limited to the northern slopes of Mount Parnon, is a unique European extension of this species which otherwise occurs in southern Turkey and Syria.

The evergreen woods of the Aegean coasts differ somewhat from those of the Adriatic, and are typified by the Eastern Strawberry Tree, *Arbutus andrachne*, associated with:

Juniperus oxycedrus　　　　*Arbutus unedo*
Pyrus amygdaliformis　　　　*Erica arborea*
Sarcopoterium spinosum　　　*E. manipuliflora*
Cercis siliquastrum　　　　　*Thymus capitatus*
Cistus salvifolius　　　　　　*Asparagus acutifolius*
C. incanus

in addition to the characteristic species of the Carob–Olive zone listed above.

The Mediterranean woodlands, such as they are, take a long time to reach maturity and relative stability, in what is known as the 'climatic climax'. They are in delicate balance with the environment, and any local change in the latter, and particularly man's activities, may quickly destroy these woods. They take a long time to rejuvenate, and in many cases, because of soil erosion or local climatic changes as a result of their destruction, they may never re-form. In the place of woods, a semi-natural cover of maquis, phrygana, or steppe-like vegetation may now be found, where cultivation is not possible.

2. *Maquis (macchie)*

This is a tall dense scrub often 2–3 m high, largely composed of hard-leaved evergreen shrubs, with a very characteristic collection of associated species, thus making even the smallest patch of true maquis readily identifiable. It occurs largely near the coast, in damper places, on the rather more acid soils; only in southern Greece and in Crete does it occur some distance inland. The main growing period of plants in the maquis is during late winter and spring when the majority of species flower; the summer is a time for hibernation with the minimum of growth activity. However, maquis cannot tolerate winters that are too

33

BALKAN MAQUIS, Trees and shrubs
1. *Quercus coccifera* 47   2. *Smilax aspera* 1659   3. *Juniperus oxycedrus* 13
4. *Laurus nobilis* 263   5. *Cistus incanus* 787   6. *Arbutus unedo* 924   7. *Spartium junceum* 515   8. *Paliurus spina-christi* 726

9. *Erica arborea* 927   **10.** *Quercus ilex* 48   **11.** Pinus halepensis 6   **12.** *Globularia alypum* 1262   **13.** *Phillyrea latifolia* (981)   **14.** *Cotinus coggygria* 706
15. *Juniperus phoenicea* 14   **16.** *Phlomis fruticosa* 1124

cold, and it is noticeably absent along those parts of the coast which are exposed to such cold north-easterly winds as the bora of Dalmatia and the varda of Thrace. It is most extensive on the wetter Adriatic sea-board, and those Aegean islands with more humid climates such as the northern Sporadhes, Thasos, Samos, etc.

Characteristic woody species of the maquis are:

Pinus halepensis	*Myrtus communis*
Juniperus phoenicea	*Arbutus unedo*
Quercus coccifera	*A. andrachne*
Q. ilex	*Erica arborea*
Laurus nobilis	*E. manipuliflora*
Cercis siliquastrum	*Phillyrea latifolia*
Calicotome villosa	*Olea europaea*
Spartium junceum	*Rosmarinus officinalis*
Cotinus coggygria	*Phlomis fruticosa*
Pistacia lentiscus	*Globularia alypum*
Paliurus spina-christi	*Asparagus acutifolius*
Cistus incanus	*Ruscus aculeatus*
C. salvifolius	*Smilax aspera*

Maquis may result from the clearing of the dominant trees, or may represent a local climax vegetation in its own right. Alternatively it may be a stage in the degradation of woods to phrygana, or in the natural succession of the establishment of mature evergreen woods thus:

Evergreen woods ⇌ Maquis ⇌ Phrygana ⇌ Pseudo-steppe
   (climax)

               Maquis ⤢
               (climax)

3. *Phrygana (garrigue)*

This is the most widespread dwarf scrub vegetation of dry slopes, hills, and islands in the Mediterranean climatic zone. It is the eastern form of the garrigues of France, and the tomillares of the Iberian Peninsula. The three variants have many circum-Mediterranean species in common. Phrygana tends to occur in drier climates than its western equivalents, and has a greater preponderance of small shrubs and shrublets with small leathery leaves, often spiny branches, commonly densely grey-hairy, and frequently aromatic. Phrygana is maintained in a relatively stable state by grazing, fires, and selective cutting by man, but it can be seen in all stages of degradation to a stony pseudo-steppe. In its most characteristic form it is composed of low thickets of dense rounded shrublets, about half a metre high, scattered over the hillsides. These give protection to many herbaceous plants, and something like 200 species have been recorded in the Greek phrygana. In early spring it can be very colourful with many annual species and a number of attractive bulbous and rhizomatous perennials in flower. This burst of colour is, however, very short-lived and by early summer these plants have died down and are seeding. Some of the larger herbaceous plants like thistles and mulleins flower in late spring, but by mid-summer their stems and leaves have shrivelled and some of the woody plants have lost their leaves, while the leaves of others curl up and become grey and dusty and the phrygana becomes monotonous and unattractive.

In its most degraded form, phrygana is composed of a few shrubs, scattered over large areas of soil-less stony ground which at most times of the year appears to be almost devoid of plants.

Phrygana is widely distributed in Crete, on the Aegean and Ionian islands, and in much of southern Greece, Thessaly, Macedonia, and Thrace. Some characteristic woody plants are:

Quercus coccifera	*Hypericum empetrifolium*
Sarcopoterium spinosum	*Cistus incanus*
Genista acanthoclada	*C. salvifolius*
Anthyllis hermanniae	*Satureja thymbra*
Euphorbia acanthothamnos	*Thymus capitatus*
Thymelaea tartonraira	*Globularia alypum*
T. hirsuta	

Other common plants of more southerly phrygana include:

Anthyllis tetraphylla	*Micromeria juliana*
Fumana thymifolia	*Origanum onites*
Thapsia garganica	*Helichrysum stoechas*
Teucrium polium	*Phagnalon rupestre*
T. brevifolium	*Pallenis spinosa*
Lavandula stoechas	*Asphodelus aestivus*
Phlomis fruticosa	*Urginea maritima*
Ballota acetabulosa	*Muscari comosum*

Spiny plants of the phrygana include the shrubs *Quercus coccifera*, *Sarcopoterium spinosum*, *Genista acanthoclada*, *Calicotome villosa*, *Euphorbia acanthothamnos*, *Verbascum spinosum*, and thistle-like plants such as *Eryngium campestre* and *Carlina corymbosa*, etc. Heather-like plants with tiny needle leaves are: *Hypericum empetrifolium*, *Satureja thymbra*, *Thymus capitatus*, *Micromeria* species, *Anthyllis hermanniae*, etc.; while aromatic plants include: *Satureja*

BALKAN PHRYGANA, Shrubs

1. *Cistus salvifolius* 790 2. *Anthyllis hermanniae* 619 3. *Salvia triloba* (1143)
4. *Micromeria juliana* 1153 5. *Lavandula stoechas* 1110 6. *Erica manipuliflora*
(934) 7. *Calicotome villosa* 498 8. *Helichrysum italicum* (1385) 9. *Rhamnus*
alaternus 720

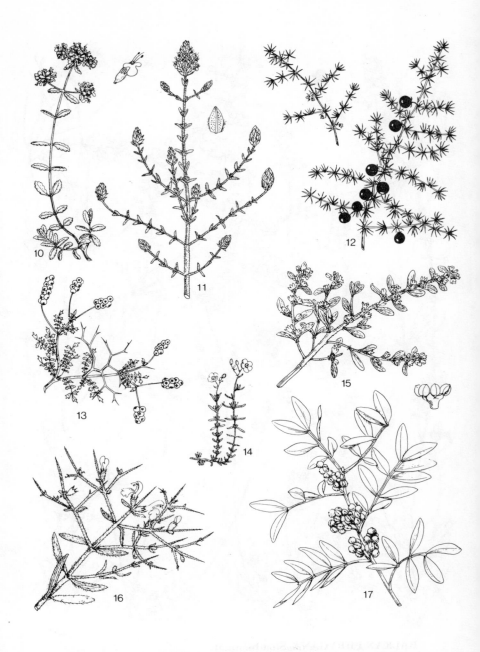

10. *Teucrium polium* 1097 **11.** *Thymus capitatus* 1162 **12.** *Asparagus acutifolius* 1650 **13.** *Sarcopoterium spinosum* 440 **14.** *Fumana thymifolia* 804 **15.** *Thymelaea tartonraira* 754 **16.** *Stachys spinosa* 1140b **17.** *Pistacia lentiscus* 703

PHRYGANA, Annuals and biennials

1. *Biscutella didyma* (349)   2. *Vicia lutea* (547)   3. *Trigonella monspeliaca* 578
4. *Coronilla cretica* 628b   5. *Melilotus sulcata* (575)   6. *Anchusa variegata* 1059b
7. *Trifolium angustifolium* 596   8. *Onobrychis aequidentata* 637a   9. *Valerianella echinata* 1310

10. *Campanula drabifolia* 1337a   **11.** *Ornithopus compressus* 623e   **12.** *Legousia pentagonia* (1347)   **13.** *Anthemis chia* (1410)   **14.** *Crupina crupinastrum* (1496)
15. *Medicago disciformis* 588b   **16.** *Medicago scutellata* 584a   **17.** *Hippocrepis unisiliquosa* 632   **18.** *Alyssoides sinuata* 323d

thymbra, Origanum onites, and many other members of the Mint family. Plants with grey- or white-woolly leaves are: *Phlomis fruticosa, Thymelaea tartonraira, Cistus incanus* subsp. *creticus, Verbascum* species, *Ballota acetabulosa, Teucrium polium,* and many others.

The dwarf shrub communities of Dalmatia differ somewhat from those of the south and are more closely related to the garrigues of Italy and France. Characteristic species include:

Juniperus phoenicea	*Myrtus communis*
Calicotome villosa	*Erica manipuliflora*
Spartium junceum	*E. multiflora*
Dorycnium hirsutum	*Rosmarinus officinalis*
Cistus spp.	*Thymus capitatus*

Some of these species may dominate the community, as for example *Rosmarinus officinalis,* which is often associated with *Erica multiflora* and *Cistus monspeliensis;* or *Calicotome villosa,* or *Cistus incanus* subsp. *creticus* may dominate the dwarf scrub. Other types result from the degradation of the maquis and contain depauperate examples of woody species of the maquis. There are many variants of these dwarf shrub communities.

4. Mediterranean mixed deciduous woods
These replace the evergreen oak and pine woods in cooler conditions inland and in the hills. They are referred to as the Sub-Mediterranean Winter Deciduous Forest Zone by plant sociologists. These mixed deciduous woods grade imperceptibly into the woods of the transitional zone.

The most widely dispersed and characteristic oak is the White Oak, *Quercus pubescens.* This forms low patchy woods which are rich in woody species, the most characteristic being:

Carpinus orientalis	*Colutea arborescens*
Ostrya carpinifolia	*Rhus coriaria*
Celtis australis	*Cotinus coggygria*
Laurus nobilis	*Pistacia terebinthus*
Pyrus amygdaliformis	*Acer monspessulanum*
Sorbus domestica	*Cornus mas*
Cercis siliquastrum	*Fraxinus ornus*

Another oak, the leathery-leaved deciduous Valonia Oak, *Q. macrolepis,* forms open woods in the lowlands, where it has been preserved mainly for its crop of acorn-cups which are used in dyeing and tanning. It is found in Greece and the Aegean region, usually as open woods with scattered trees and with herbaceous undergrowth which is used as grazing land.

In river valleys and on flood plains from the coast to the hills to nearly 1000 m are scattered stands of Oriental Plane, *P. orientalis*; it never forms woods, but is a very distinctive feature of the landscape in Greece and the Aegean region. Often accompanying the Planes are *Alnus glutinosa*, *Populus nigra*, and *P. alba*, and such willows as *Salix purpurea* and *S. elaeagnos*. *Dracunculus vulgaris*, *Arum italicum*, *Ranunculus ficaria*, and *Symphytum bulbosum* are rhizomatous perennials often found growing in the richer soils beneath the trees.

Coastal river valleys with wide flood beds filled with pebbles and boulders, dry much of the year but subjected to periodic flooding, have a distinctive brushwood community with the handsome pink-flowered Oleander, *Nerium oleander*, and the Chaste Tree, *Vitex agnus-castus*, often with Tamarisk species and the aromatic shrub *Dittrichia viscosa*.

5. *Herbaceous vegetation of stony ground—pseudo-steppe*

This occurs in the hill and lower mountain regions where the soil has been washed away, usually as a result of intensive over-grazing and clearing, thus exposing the bedrock. Large areas of this pseudo-steppe occur in the Balkan Peninsula particularly in southern Macedonia, Bulgaria, Dalmatia, Crete, and the Aegean region. All stages in the degradation of phrygana to pseudo-steppe occur, but these bare-looking areas are often rich in interesting species, particularly of bulbous and rhizomatous plants such as *Colchicum*, *Asphodelus*, *Gagea*, *Allium*, *Scilla*, *Ornithogalum*, *Muscari*, *Crocus*, *Romulea*, *Iris*, *Ophrys*, *Orchis*, *Serapias*, etc.

In Crete, for example, attractive species in the pseudo-steppe include:

Ranunculus asiaticus	*Chrysanthemum segetum*
Papaver rhoeas	*Coleostephus myconis*
Matthiola tricuspidata	*Calendula arvensis*
Anchusa azurea	*Asphodeline lutea*
Echium angustifolium	*Gagea graeca*
	Gynandriris sisyrinchium

There are large tracts of stony terrain in the karst country of Dalmatia, Istra, and Montenegro.

The so-called 'Felsentriften' of Dalmatia constitute a large part of the vegetation. The climate is Mediterranean and the plants are mostly geophytes, flowering in spring and early summer, or have a xeromorphic structure of the aerial parts . . . typically developed near Dubrovnik. In the spring species of

Romulea, Colchicum, Ornithogalum, Anemone, Crocus, Muscari, Iris, and *Helianthemum* flower, followed by species of *Bromus, Stipa, Briza, Polygala, Aethionema, Thymus, Asphodelus, Dianthus, Euphorbia, Edraianthus, Salvia, Campanula, Allium* and *Convolvulus.* In the summer the xeromorphic structure of the vegetation is very striking. Densely hairy species of *Inula, Stachys, Phlomis, Marrubium, Helichrysum, Verbascum* and *Origanum,* tall Umbelliferae (*Opopanax chironium, Cachrys ferulacea,* and *Ferulago campestris*), prickly species of *Eryngium* and *Echinops,* aromatic Labiatae (as *Micromeria juliana*) and xerophytic grasses . . . The beautiful white-leaved *Inula verbascifolia* is especially characteristic.[8]

Many examples of similar pseudo-steppes could be given in both the Mediterranean and the transitional zones; the lists of species would be long. The plant-hunter will find these some of the richest and most interesting localities in the Balkan Peninsula. It is always worth while stopping in spring or early summer to explore these dry stony barren-looking hillside or mountain slopes, particularly if one is moving from region to region.

6. *Vegetation of rock walls and gorges—chasmophytes*
Rock walls and gorges are a characteristic feature of most of the limestone mountains of the Balkans. Like the plant-life of the pseudo-steppes, the vegetation of rock walls and gorges is particularly rich in unusual species, but unlike the pseudo-steppes, they are relatively undisturbed by man and his grazing animals, and they act as refuges for many plants which cannot otherwise survive the competition of the more vigorous and aggressive plants of the surrounding habitats. These chasmophytes, as they are called, contain among their numbers some ancient Tertiary relict species which have only been able to survive in these very restricted habitats.

The gorge flora of Crete and the rock wall plants of the Rhodope Mountains are described elsewhere. Genera which particularly favour these habitats include *Cerastium, Silene, Dianthus, Onosma, Sempervivum, Sedum, Saxifraga, Potentilla, Linum, Hypericum, Asperula, Stachys, Verbascum, Ramonda, Campanula, Trachelium, Staehelina,* and other members particularly of the Cruciferae, Labiatae, and Compositae. Each gorge or cliff will have its own collection of species and many rarities are only to be found in such habitats; they will usually repay careful exploration.

7. *Coastal terrestrial vegetation*
Despite the enormous extent of the Balkan coastline and its hundreds of islands, the coastal vegetation is surprisingly uniform and devoid of interesting local species, by comparison with that of the cliffs and

gorges inland. This is largely due no doubt to the uniformity of climate and the relatively wide dispersal of many species. Weak Mediterranean tides, the absence of large rivers, and the relatively recent nature of the present-day coastline have tended to limit the variety of habitats. Much of the coastline has Aleppo or Stone Pine woods, maquis or phrygana extending right down to the shore-line, with hardly a single littoral species appearing between the land vegetation and high-water mark.

The main coastal habitats with their most characteristic species are as follows:

Stony or pebble beaches:

Arthrocnemum fruticosum	*Eryngium maritimum*
Salsola kali	*E. creticum*
Matthiola tricuspidata	*Crithmum maritimum*
Cakile maritima	*Inula crithmoides*

Sandy shores and sand-dunes of the south:

Pinus pinea	*Eryngium maritimum*
Polygonum maritimum	*E. creticum*
Glaucium flavum	*Echinophora spinosa*
Malcolmia flexuosa	*Cionura erecta*
Cakile maritima	*Calystegia soldanella*
Medicago marina	*Xanthium strumarium*
Euphorbia peplis	*Pancratium maritimum*
Tamarix spp.	

Salt-marshes likewise have many widely dispersed European species. They occur to a very limited extent on the Balkan coastline and are often flooded only in winter and dry out in summer. They grade almost imperceptibly into freshwater marshes. Characteristic salt-marsh plants of the Mediterranean region include:

Halimione portulacoides	*Echinophora tenuifolia*
Arthrocnemum fruticosum	*Linum maritimum*
A. glaucum	*Frankenia hirsuta*
Salicornia europea	*Plumbago europaea*
Suaeda vera	*Limonium oleifolium*
Salsola kali	*L. vulgare*
Spergularia marina	*Goniolimon collinum*

An example of a salt-marsh in Attica (Greece) showed a considerably richer flora than most, with in addition:

Mesembryanthemum nodiflorum	*Echium angustifolium*
Alhagi graecorum	*Lycium europaeum*
Lavatera cretica	*Cephalaria transylvanica*
Tamarix hampeana	*Scabiosa atropurpurea*
Bupleurum gracile	*Cardopatum corymbosum*
Limonium sinuatum	*Asphodelus fistulosus*
Heliotropium supinum	*A. ramosus*
	Urginea maritima

Freshwater marshes in the coastal region often lie inland or behind sand-dunes, but many inland marshes have been drained and are intensively cultivated. The marsh vegetation is in general largely composed of widespread European species such as:

Lythrum salicaria	*Juncus* spp.
Teucrium scordium	*Phragmites communis*
Lycopus europaeus	*Cyperus longus*
Mentha aquatica	*Scirpus maritimus*
M. longifolius	*S. holoschoenus*
Veronica anagallis-aquatica	*Carex* spp.
Alisma plantago-aquatica	*Typha angustifolia*
Butomus umbellatus	*Orchis laxiflora*
Iris pseudacorus	

The rocky coastal flora is likewise not well developed in the Balkan Peninsula; there is relatively little coastal erosion exposing bare cliffs which are subjected to periodic spray. The following are often seen:

Ephedra fragilis	*Crithmum maritimum*
Capparis spinosa	*Limonium* spp.
C. ovata	*Inula crithmoides*
Euphorbia paralias	*Dittrichia viscosa*

with such additional species in the Aegean region including:

Silene sedoides	*Trigonella balansae*
Malcolmia flexuosa	*Frankenia hirsuta*
Sedum litoreum	*Cichorium spinosum*

However, certain *Limonium* species are rather an exception to the rule for they do have a limited coastal distribution and are not widely distributed like most other species. For example, such species as *Limonium cancellatum* and *L. anfractum* are restricted to the Adriatic coasts, and *L. melium* to the Cyclades.

Montane forests (700–1700 m approx.)

8. Coniferous forests

Most of the higher mountains of Greece, with their summits above 2000 m, have on their moister exposures belts of Silver Fir forest ranging from approx. 900 to 1800 m. In the Peloponnisos and southern Greece the Greek Silver Fir, *Abies cephalonica*, is the dominant tree and forms pure stands, casting heavy shade. Few species are able to grow in the depth of the forest, but in open places and on forest verges in the Peloponnisos the following southern shrubs may be found: *Prunus cocomilia, Crataegus heldreichii, C. pycnoloba,* and *Lonicera nummulariifolia*, and with them in more open stony slopes:

Helleborus cyclophyllus	*Lamium garganicum*
Anemone blanda	*Digitalis viridiflora*
Corydalis solida	*D. lanata*
Cardamine graeca	*Campanula spathulata*
Lathyrus venetus	*C. persicifolia*
Cyclamen hederifolium	*Fritillaria graeca*
Symphytum ottomanum	*Scilla bifolia*

In most of central and northern Greece and Macedonia the hybrid *Abies borisii-regis* dominates, while further north, in the central European climatic zone, the Silver Fir, *Abies alba*, forms the forests.

Black Pine forests occur in the mountains above 800 m in the Peloponnisos and on Taiyetos and Parnon and become increasingly frequent further northwards in Yugoslavia, where they occur at considerably lower altitudes and are a component of the central European floristic region. Some of the plants of the Black Pine forests of Taiyetos and Olympus have been listed in the plant-hunting accounts.

Stinking Juniper, *J. foetidissima*, is a tree which occurs scattered in the upper montane zone, usually above the Fir and Pine woods on some of the higher mountains of Greece. It rarely forms stands of any extent owing in all probability to heavy cutting by man—the best example the author has seen is above the village of Papingon on Mount Timfi in the Pindhos.

Sub-alpine and alpine (1700–3000 m approx.)

9. Scrub; alpine meadows; stony slopes, screes, rock vegetation

Above the tree-line, which is often not clearly defined in the mountains of the Mediterranean zone, scattered forest trees may occur far up the mountainside. On more inaccessible rocky slopes and cliffs there is often the remains of a brushwood vegetation of taller shrubs and small trees of *Sorbus*, *Cotoneaster*, *Prunus*, *Acer*, etc.; these may occur scattered in the alpine zone though they are in all probability all that remains of an upper deciduous tree zone. However, grazing is so intensive on all these higher mountains that the natural plant communities cannot develop to maturity. Almost without exception, every mountain range that the author has explored in the Mediterranean zone has had its flocks of sheep grazing to the highest summits. Since it is on these often isolated mountains that quite a high proportion of Balkan endemic species are located, it is of paramount importance that legislators and biologists initiate effective conservation measures to ensure the survival of these mountain species.

The characteristic alpine scrub vegetation may be called colloquially the 'hedgehog' zone—comparable to that of the high Mediterranean mountains of the Iberian Peninsula—but with a different assemblage of plants.

In the Balkan Peninsula the spiny shrubs include: *Berberis cretica*, *Astragalus angustifolius*, *A. sempervirens*, *A. parnassi*, *A. creticus*, *Rhamnus prunifolius*, *Acantholimon androsaceum*, *Euphorbia acanthothamnos*, *Verbascum spinosum*, and, as a result of heavy grazing, *Acer sempervirens*. Other low shrubs commonly associated with the hedgehog zone include *Prunus prostrata*, *Juniperus communis* subsp. *hemisphaerica*, and *Daphne oleoides*.

The vegetation of open stony mountainsides, stony pastures, screes, and cliffs of the Mediterranean alpine region has been described in many of the plant-hunting accounts. Their uniqueness makes these perhaps the most rewarding of all Balkan plant communities to explore. The effort required to reach these remote places can be amply rewarded, and there still remain some almost unknown summits and ranges to be explored botanically.

TRANSITIONAL PLANT COMMUNITIES

Where the air masses of the Mediterranean, of eastern central Europe (Continental), and of the Black Sea (Pontic) intermingle with each other, a broad or narrow zone of transitional climates and transitional plant communities occur. These communities contain not only mixtures of species from more than one climatic zone, but in addition species which are restricted to this transitional zone. Some of the latter are of considerable importance and may form their own distinctive climax communities. Examples are the Bosnian and Macedonian Pine forests of the mountains of the central Balkans, the mixed Hungarian Oak forests, and the Oriental Beech and Rhododendron forests of Thrace, etc. Often associated with these communities are Balkan endemic species or other species which have 'penetrated' from outside their normal range into this transitional zone.

10. *Mediterranean transitional (sub-Mediterranean)*

The influence of the Mediterranean air masses may in places penetrate for a considerable distance inland, depending on the proximity of the coastal hills and mountains. In the Varda and Struma valleys Mediterranean air may penetrate as much as 60 km inland, and into Bulgaria, whereas in the Velebit Mountains of Dalmatia and in the Gulf of Corinth, where high mountains lie close to the coast, it may lose its influence in a matter of 10 km or so.

Characteristic of this transitional zone, though not exclusive to it, are the mixed deciduous woods of Manna Ash and Oaks which cover much of the middle hill region, and in particular much of the limestone karst country of the Balkan Peninsula—they have been called Karst-woods. They are light open deciduous woods showing all gradations into brushwood or shiblyak. The Manna Ash, *Fraxinus ornus*, and the White Oak, *Quercus pubescens*, may be locally dominant, but other characteristic trees and shrubs include:

Carpinus orientalis	*Pyrus communis*
C. betulus	*Sorbus torminalis*
Ostrya carpinifolia	*S. aucuparia*
Corylus colurna	*Prunus mahaleb*
Quercus cerris	*Acer monspessulanum*
Q. petraea	*A. campestre*
Celtis australis	

A number of Mediterranean shrubs, both evergreen and deciduous, are frequent components of these woods, and the ground flora is rich in herbaceous and bulbous species, many of them of Balkan origin.

Other transitional woodland communities, which occur very locally, are the Oriental Plane–Walnut woods of Macedonia, the Sweet Chestnut woods, and very limited Horse Chestnut stands of the montane region of Greece and Macedonia, as well as more extensive Macedonian Oak woods of *Q. trojana* which can be found in northern Greece and Macedonia.

11. *East-central European transitional (Continental)*

Much of the lowland and hill areas of central and western Bulgaria, eastern Yugoslavia, and north-eastern Greece are strongly influenced by continental air masses penetrating from the north-east. The forests of this transitional climatic zone are largely composed of mixed deciduous oaks. The most widespread species is the Hungarian Oak, *Quercus frainetto*, which occurs throughout the Balkan mainland. It forms forests which are characterized by the presence of *Pyrus communis* and *Tilia tomentosa* as well as many other small trees, of both Mediterranean and central European distribution.

Another type of oak forest of the lowlands and hills is distinguished by the presence of the Turkey Oak, *Q. cerris*, with such eastern species as *Q. pedunculiflora*, *Q. virgiliana*, and *Acer tataricum*, while in more montane areas the Durmast Oak, *Q. petraea*, forms forests in which *Acer obtusatum* and *A. hyrcanum* are characteristic.

In the true montane zone two very distinctive types of pine forest occur which are unique to the Balkan Peninsula and are only found within its transitional zone. They are the forests of Macedonian Pine, *Pinus peuce*, which occur in the mountains of south-west Bulgaria and Macedonia, and the forests of the white-barked Bosnian and Balkan pines, *P. leucodermis* and *P. heldreichii*, which have a wider distribution in the central Balkans. They form limited forests on many mountains up to altitudes of 1800 m or more, often mixed with Silver Fir and sometimes Spruce. The associated flora is largely central European in character.

12. *Black Sea transitional (Pontic)*

The influence of the climate of the Black Sea is marked in Thrace and south-east Bulgaria and the vegetation contains a considerable number of eastern species. The oak–ash forests of the river valleys have been noted elsewhere. The coastal hills have mixed deciduous oak forests of *Quercus polycarpa*, *Q. hartwissiana*, *Q. dalechampii*, and *Q. cerris*. In the Istranca Mountains on the borders of Bulgaria and Turkey-in-Europe are forests of Oriental Beech, *F. orientalis*, with the evergreen *Rhododendron ponticum*—a western outlier of a widely dispersed Anatolian community found nowhere else in Europe.

13. *Shiblyak (deciduous brushwood)*

Much of the land surface which was once covered by the mixed deciduous oak–hornbeam forests has been cleared by the settled communities and is now under permanent cultivation. What forests remain are usually heavily exploited by man, and the destruction of mature trees has left behind a brushwood which is often termed 'Shiblyak'. It is a useful term covering the extensive deciduous brushwoods that are encountered so frequently on hill-slopes and valley sides, where crop cultivation is not possible, in the central and northern Balkan region, in both the central European and transitional climatic zones.

With the removal of the dominant trees, more sunlight enters the shrub and ground layer, and this usually results in the establishment of more light-loving species and in a richer and more varied ground flora.

The following woody species are characteristic of various types of shiblyak:

Corylus avellana	Rhus coriaria
Quercus pubescens	Cotinus coggygria
Berberis vulgaris	Acer campestre
Crataegus monogyna	Paliurus spina-christi
Cercis siliquastrum	Cornus mas
Colutea arborescens	Syringa vulgaris
Coronilla emerus	Viburnum lantana

The central European climatic region extends southwards from the eastern Alps through most of central Yugoslavia to approximately the Albanian border and Macedonia and into Bulgaria, where it merges with the transitional continental climatic region.

Lowlands and hills (0–700 m approx.)

14. *Mixed deciduous forests*
The potential climatic climax vegetation over much of the lowlands and hills of this area is mixed deciduous forest. These forests are dominated by either one or more of the following oaks: *Q. cerris*, *Q. frainetto*, *Q. robur*, *Q. petraea*, and widely associated with them is the Hornbeam, *Carpinus betulus*. Collectively they can be called the oak–hornbeam forests of south-east Europe and they often include such typical trees as:

Fagus sylvatica	Acer campestre
Ulmus minor	A. tataricum
Pyrus pyraster	A. pseudoplatanus
Prunus avium	Tilia platyphyllos
	Fraxinus excelsior

Commonly associated shrubs are:

Corylus avellana	Cornus sanguinea
Crataegus monogyna	C. mas
Prunus spinosa	Ligustrum vulgare
Euonymus europaeus	Sambucus nigra
Frangula alnus	Viburnum lantana

There is a rich predominantly central European ground flora, with relatively few Balkan species. Some of the more interesting herbaceous species include:

Helleborus odorus	*Cyclamen purpurascens*
H. dumetorum	*Lamium orvala*
Eranthis hyemalis	*Lonicera caprifolium*
Epimedium alpinum	*Knautia drymeia*
Vicia oroboides	*Erythronium dens-canis*
Hacquetia epipactis	*Convallaria majalis*
Staphylea pinnata	*Galanthus nivalis*

Sweet Chestnut woods may occur locally in this climatic region, usually mixed with oaks. They are found particularly on the higher hills and lower montane zones.

15. *Vegetation of river valleys, marshes, lakes, etc.*

The vegetation of lakes and open waters is probably the most uniform of all Balkan communities, showing little change in species from the Mediterranean to the central European zone. Most species are ubiquitous, few if any are exclusively Balkan. These aquatic communities of Greece and the Aegean are very local, and sparse in species; many areas have been drained and are now under cultivation. In Macedonia permanent waters are more frequent and the following species can be found:

Nymphaea alba	*Stratiotes aloides*
Nuphar lutea	*Hydrocharis morsus-ranae*
Ranunculus aquatilis	*Vallisneria spiralis*
Trapa natans	*Potamogeton* spp.
Nymphoides peltata	

More frequent in the north are *Hottonia palustris* and *Sagittaria sagittifolia*.

Marshes are usually restricted to river valleys, often in elongated polja as in the karst region of Yugoslavia. In the river valley marshes of Greece and Crete, for example, in addition to sedges, rushes, etc. the following species occur:

Ranunculus sardous	*Hypericum perfoliatum*
R. muricatus	*Lythrum junceum*
Nasturtium officinale	*Samolus valerandi*
Trifolium resupinatum	*Blackstonia perfoliata*
T. nigrescens	*Veronica anagallis-aquatica*
	Alisma plantago-aquatica

While further north additional species include:

Tamarix tetrandra	*Gratiola officinalis*
Lythrum salicaria	*Petasites albus*
Lycopus europaeus	*Iris pseudacorus*
L. exaltatus	*Orchis laxiflora*
	Dactylorhiza majalis

In the polja marshes of the karst, which are flooded in winter but may dry out in the summer, are found:

Caltha palustris	*Cynara cardunculus*
Euphorbia palustris	*Butomus umbellatus*
Epilobium hirsutum	*Leucojum aestivum*
Vitex agnus-castus	*Iris pseudacorus*

as well as *Phragmites*, *Typha*, and sedges and rushes.

The very extensive marshes of the Danube and its tributaries may have a luxuriant marsh vegetation with many typically central European species dominating like *Phragmites communis*, with *Typha*, *Scirpus*, etc., and thickets of Alders, Willows, and Alder Buckthorn.

Montane forests (700–1700 m approx.)

16. Montane deciduous forests (beech)
Beech forests are very widespread in the central Balkan region, including the Stara Planina and Rhodope mountains of Bulgaria, and they spread southwards into Greece in the Pindhos Mountains, and down the 'stepping-stones' of Olympus, Ossa, and Pilion. Though cleared in the more accessible and fertile hill regions, they occupy many of the steeper slopes of the upper hills and mountains. The European Beech, *Fagus sylvatica*, occupies much of the northern and central Balkans while to the south and east the forests become progressively dominated by the hybrid Beech, *F.* × *moesiaca*, intermediate in character between the two parents, *F. sylvatica* and *F. orientalis*. The latter forms natural forests with *Rhododendron ponticum* in the Istranca Mountains of Thrace.

The montane beech forests are often almost pure. They are usually found on the dampest aspect of the mountain, at altitudes ranging from 700 to 1500 m or more. They regenerate relatively quickly after felling, unlike the Mediterranean forests, and in some instances have extended their range by replacing conifers after the latter have been felled.

Other trees often associated with these beech forests, but usually only as scattered specimens, are:

Abies alba	Corylus colurna
Picea abies	Ulmus glabra
Pinus sylvestris	Sorbus aucuparia
P. peuce	S. torminalis
Populus tremula	S. aria
Carpinus betulus	Acer platanoides
	A. pseudoplatanus

Shrubs include:

Juniperus communis	Daphne laureola
Salix caprea	D. mezereum
Alnus viridis	Vaccinium myrtillus
Ribes spp.	Lonicera spp.

The ground flora is often poor but in the Rila Mountains, for example, the following widely distributed species occur:

Asarum europaeum	Monotropa hypopitys
Cardamine bulbifera	Galium odoratum
Oxalis acetosella	Pulmonaria rubra
Geranium macrorrhizum	Paris quadrifolia
Euphorbia amygdaloides	Epipogium aphyllum
	Neottia nidus-avis

17. *Montane coniferous forests*

These are composed largely of the Silver Fir, *Abies alba*, the Spruce, *Picea abies*, Black Pine, *Pinus nigra*, and Scots Pine, *P. sylvestris*. They occur either mixed or in stands of the individual species. These forests cover wide areas of the higher mountain slopes, often forming a zone above the beech; or the conifers may be mixed with beech, or again the conifer or pure beech may occur as separate stands, depending on the slope and aspect of the mountain.

Pure Silver Fir forests are only found locally in the Balkans and they are more often mixed with other species, as for example in the fine coniferous forests of the Rhodope Mountains where *Picea abies*, *Pinus nigra*, and *P. sylvestris* are the commonest associates, with such others as:

Ostrya carpinifolia	Acer opalus
Fagus sylvatica	A. platanoides
Sorbus aucuparia	Tilia platyphyllos
S. torminalis	Fraxinus excelsior

Montane spruce forests occur only as far south as Macedonia, while Black Pine, *Pinus nigra*, forms forests on dolomitic limestone and serpentine in the mountains further south.

The Serbian Spruce, *Picea omorica*, is a very distinctive tall columnar tree which is a Tertiary relict and found only in the Balkans. It forms rather mixed forests in the Drin river basin of Montenegro and Bosnia-Hercegovina, with under-shrubs of *Corylus avellana*, *Spiraea cana*, *Cotinus coggygria*, *Rhamnus alpinus* subsp. *fallax*, and *Lonicera alpigena*.

Sub-alpine and alpine (1700–3000 m approx.)

18. *Scrub, alpine meadows, and vegetation of stony slopes, screes, rocks*
The tree-line of these Balkan mountains usually lies somewhere between 1700 and 2500 m. Above this there is often a layer of scrub, as for example in Bulgaria, Serbia, and northern Macedonia, consisting of *Pinus mugo*, *Juniperus communis* subsp. *nana*, *Alnus viridis*, and smaller shrubs such as *Rosa* species, *Genista tinctoria*, and *Daphne oleoides*.

On cooler damper slopes on acid soils, heath-like thickets of ericaceous shrublets occur comprising some of the following: *Arctostaphylos uva-ursi*, *Bruckenthalia spiculifolia*, *Vaccinium myrtillus*, *V. vitis-idaea*, *V. uliginosum*, often with *Juniperus communis* subsp. *nana*. Or *Chamaecytisus hirsutus* and *C. heuffelii* may form thickets on sunny stony slopes with *Rhamnus alpinus*. However, these scrub communities may often be very restricted because of grazing and clearing.

Many sub-alpine and alpine meadows have developed as a result of man's interference over long periods, such as the grazing of flocks, cutting, firing, and clearing of the upper forests and shrub communities. Consequently alpine meadows may descend as low as 1600–1700 m in the mountains, where some forest and shrub species may still persist. In general two types of meadow can be distinguished: those developed on predominantly alkaline soils over limestone and dolomite, and those on acid and siliceous soils. The former is dominated by the grass *Sesleria nitida*. It forms open grassy steppe-like mountain swards which in Macedonia are characteristically associated with:

Cerastium decalvans	*Sideritis scardica*
Onobrychis arenaria	*Teucrium montanum*
Helianthemum nummularium	*Thymus parnassicus*
H. oelandicum	*Achillea holosericea*

On acid soils *Poa*, *Festuca* species, and *Nardus stricta* are locally dominant in different aspects and soils. In Macedonia some characteristic species include:

Ranunculus oreophilus   *Thymus longicaulis*
Geum montanum   *Knautia dinarica*
Potentilla aurea   *Centaurea uniflora*
Genista tinctoria   *Scorzonera purpurea*
Gentiana verna

The alpine meadows of higher altitudes on both acid and alkaline soils have often been briefly recorded elsewhere. They are rich in alpine species and merge imperceptibly with the vegetation of rocky slopes and screes, many examples of which have been described in Chapter 2, in the mountains of central Yugoslavia, Macedonia, Pindhos, Rhodope, Rila, Pirin, and Stara Planina.

REFERENCES TO CHAPTER 1

[1] Geographical Handbook Series, Naval Intelligence Division, *Greece*, i, 1944, p. 3.

[2] Ibid.

[3] Ibid., p. 12.

[4] Geographical Handbook Series, Naval Intelligence Division, *Yugoslavia*, i, 1944, p. 51.

[5] Ibid., p. 219.

[6] W. B. Turrill, *The Plant Life of the Balkan Peninsula*, Oxford, 1929, p. 434.

[7] D. A. Webb, *'Flora Europaea—A Retrospect'*, *Taxon*, 27, 1978.

[8] W. B. Turrill, *The Plant Life of the Balkan Peninsula*, Oxford, 1929, p. 161.

2. The Plant-hunting Regions

1. CRETE

Crete, the largest of the many islands of the Aegean, is 256 km long, between 11 and 56 km in width, with an area of about 8700 square km. The main axis of the island lies east–west and it is part of a mountainous area which formed an arc of land stretching from the Peloponnisos to south-western Turkey in mid-Tertiary times. Today Crete and the islands of Kithira and Andikithira to the west, and Kasos, Karpathos, and Rhodes to the east, are all that remains above sea-level.

The dismemberment of the former land area was the consequence of complex tectonic processes. Large portions of the south Aegean continent were lowered by thousands of metres, and Crete soon became an island, or rather a group of islands. In all probability it was once more attached to the continents, for the last time, when towards the end of the Miocene period the Mediterranean Sea dried out. It was at the beginning of the Pliocene that the Mediterranean basin was again flooded from the west, 'changing it from a desert 10,000 feet below sea-level to its present form; . . . this occurred five and a half million years ago'.[1] Further changes of sea-level, renewed tectonic processes, and erosion leading to the formation of plains and valleys have produced the dramatic and varied landscapes of present-day Crete.

The mountains of the island bear evidence of a complex geological structure. In general they are of limestone or dolomite and rest on older metamorphic rocks. The hard limestones are mostly of Jurassic, Cretaceous, and Eocene origin. They weather very slowly, particularly in the low rainfall areas, and show many of the typical karstic features of the Balkans, with gorges, caves, polja, and vast underground drainage systems. The mountain summits are generally of grey eroded crags standing above stony deserts of fragmented limestone flakes. The highest mountains are Idhi (Psiloritis; 2456 m) in the centre, Pakhnes (2452 m) in the White Mountains or Levka Ori in the west, and Afendi Khristos (2148 m) in the Dhikti Ori to the east.

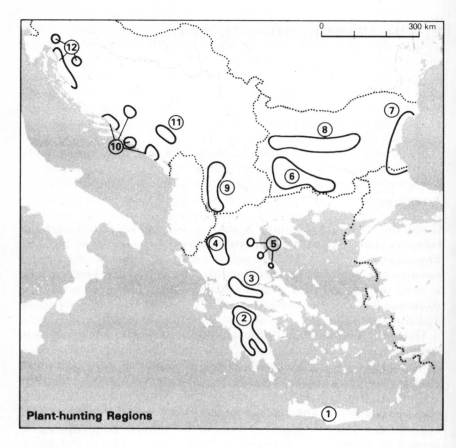

Plant-hunting Regions

1 Crete

2 Mountains of the Peloponnisos: Taiyetos, Parnon, Aroania (Chelmos), Killini, Panakhaikon

3 Southern Pindhos Mountains: Parnassos, Giona, Vardhousia, Timfristos

4 Northern Pindhos Mountains: Smolikas, Gamila, Astraka, and the Vicos Gorge

5 Olympus, Ossa, and Pilion

6 Rhodope Massif: Rila, Pirin

7 The Black Sea Coast and Hills: Istranca Mountains

8 Stara Planina (Balkan Range) and Western Bulgarian Mountains

9 Macedonian Mountains of Yugoslavia: Perister, Galicica, Mavrovo, Sar Planina

10 Southern Dalmatia: Orjen, Lovćen, Mljet, Neretva Gorge, Paklenica

11 Mountains of Central Yugoslavia: Durmitor, Sutjeska

12 Northern Dalmatia: Velebit, Plitvice Lakes, Risnjak

Chalks, sandstones, and marls of Tertiary and Quaternary origin occur in the lowlands at Khania, Rethimnon, Ierapetra, and around the Mesara plain. Metamorphic rocks forming the base rock of the island are sometimes exposed as outcrops on the lower slopes, or they may be exposed over wide areas as in the west, where the upper limestone has been eroded away. Quaternary conglomerates, composed of pebbles of varying sizes compressed together like coarse concrete, can often be seen in the coastal areas.

The Cretan climate is typically Mediterranean. The summers are hot and dry with very little if any rain from May to October, and the wettest months are December and January. However, the proximity of relatively high mountains close to the sea causes wide local variations in climate. Though snow is almost unknown at low altitudes, there are heavy falls in the mountains and snow may continue to lie on the summits till the end of May, or even later. Strong winds are common. Summer mists also make the northern slopes damper than those facing south, while intermittent rainstorms may occur locally in the mountains, but the surface water disappears almost at once through the porous rocks into underground drainage systems. In consequence many plants show distinctive xeromorphic features associated with the reduction of water loss, like small leathery leaves, hairiness, waxy surfaces, aromatic oils, spines, while many more have dormant summer storage organs such as bulbs, corms, and rhizomes. Spring and winter annuals evade the most difficult summer season as dormant seeds.

The flora of Crete consists of approximately 1600 species of wild seed plants. Its special interest lies in the very considerable number of species which are endemic or have an eastern Mediterranean or Anatolian distribution. They are predominantly plants of the mountains and of the dry phrygana of the hills and lowlands. Species with a distinctly eastern type of distribution include: *Rosularia serrata*, *Mattiastrum lithospermifolium*†, and *Lithodora hispidula*. Such endemics as *Zelkova abelicea*†, *Ricotia cretica*, *Ebenus cretica*, *Chionodoxa cretica* are representatives of genera whose centre of distribution lies much further east. The majority of other species have a circum-Mediterranean or Balkan distribution and these are mostly plants of man-made habitats, forests, and maquis. Few species are of North African affinity and they are mostly to be found on the drier southern coast, or on the adjacent islands. They include:

Astragalus peregrinus†	*Periploca laevigata* subsp.
Erodium hirtum†	*angustifolia*
Viola scorpiuroides†	*Stachys tournefortii*†
Helianthemum stipulatum†	*Androcymbium rechingeri*†

The endemic species of Crete are of the greatest interest. Not only do they comprise some of the most distinctive and beautiful plants, but their study has helped to unravel the geological past of the eastern Mediterranean.

Approximately 10 per cent of the Cretan vascular plant species are endemic—they are found nowhere else in the world except possibly on small neighbouring islands (Cyprus has a similar percentage). Most are ancient species which once had a much wider distribution and, having become isolated by the flooding of the Mediterranean, have managed to survive where their relatives perished elsewhere. *Petromarula pinnata*, for example, has no surviving close relatives. Surprisingly, despite Crete's long isolation (estimated at five and a half million years) relatively few new endemic species have evolved *in situ*, and it seems that the Cretan flora is at an evolutionary standstill. Rigorous competition between species, coupled with the absence of new habitats to which to adapt, has probably been the cause of this stagnation. Perhaps the most important new habitat has been that of the higher mountains which lowland species have been able to colonize successfully in competition with the existing flora. Something like 25 new forms—the new endemics (neo-endemics)—have evolved from lowland types since isolation, to become distinctive species, subspecies, or varieties.

Examples of new micro-species include:

Linum caespitosum †	*Linum arboreum*
(mountains)	(lowlands)
Colchicum cretense	*Colchicum pusillum*
(mountains)	(lowlands)

Other species have invaded the mountains to heights of over 2000 m without significant change, such as:

Euphorbia acanthothamnos	*Centaurea raphanina*
Onosma erecta	*Centaurea idaea*
Verbascum spinosum	

The two habitats which have above all else helped to preserve the ancient endemic species are the gorges and cliffs of the limestone mountains, and the mountain summits. Here, in all probability, they have for millions of years held their own against the more aggressive and vigorous species of the surrounding countryside, as well as against the changes in climate, and against animals and men. Intensive searching during the last few decades has revealed more endemic species in these habitats, and more may still be discovered.

Like most of the eastern Mediterranean coastlands, Crete has been inhabited by settled peoples for more than 5000 years, and they have played havoc with the primeval vegetation. The hills and valleys were once almost certainly covered with forests of cypress, pine, and oak, both deciduous and evergreen, where today there are great expanses of depauperate phrygana. On the other hand, the stony wastes of the high mountains—pale lunar landscapes of great beauty, but almost bereft of plant-life—appear to represent a natural state of the vegetation.

About half the total area of Crete is intensively grazed by sheep and goats, and 20 per cent of this is waste land. A few remains of cypress forest can still be found in the White Mountains. Evergreen oak forest relicts can be seen on Mount Psiloritis and elsewhere. Native Oriental Planes are to be found along some of the stream beds, and deciduous oak groves and scattered trees occur in the maquis and phrygana. There are also open stands of Calabrian Pine, *P. brutia*, on the south-facing slopes of the mountains.

Since earliest Minoan times the olive has been planted and grafted on to wild stock; it is the tree which now dominates the lowland valleys. In favourable soils vines, oranges, lemons, figs, quinces, apples, pears, almonds, carobs, and even bananas, as well as vege-tables and corn are grown intensively, often under irrigation from underground water sources; in places the land can have a wonderfully fertile appearance. During man's settlement it has been estimated that he has introduced something like 30 per cent of new species to the Cretan flora—species which are completely naturalized and are a now regular constituent of the wild flora.

The semi-natural scrub vegetation which covers so much of Crete at the present day consists of only a limited amount of true maquis, occurring in areas with schistose rock and relatively high rainfall, particularly in the west. It differs little from the mainland maquis, having *Arbutus unedo*, *Erica arborea*, *Cistus salvifolius*, *Lavan-dula stoechas*, and *Calicotome villosa* forming dense thickets, with Myrtle and Laurel occurring by the streamlets.

It is, however, the Cretan phrygana which dominates so extensively the non-cultivated poorer lowland and hill soils. It consists of a rich assemblage of drought-resistant shrubs, frequently aromatic, spiny, or with small leathery leaves, which in the summer heat give the land-scape a burnt-up lifeless appearance. There are 30 or more dwarf shrubs, many widely distributed in the Mediterranean area, but including such Cretan or south Aegean endemics as:

Ebenus cretica	*Origanum microphyllum*
Teucrium alpestre	*Verbascum spinosum*
Phlomis lanata	*Senecio gnaphalodes*†

There are also many interesting herbaceous or bulbous species to be searched for, notably the Turban Buttercup, *Ranunculus asiaticus*, in its common Cretan white form, or, less commonly, in soft shades of pink or a beautiful bright yellow. Orchid species are widespread in early spring; many are common, and the following are of particular interest:

Ophrys fusca subsp. *omegaifera*
*O. cretica
O. tenthredinifera
O. sphegodes subsp. *mammosa*
*O. sphegodes subsp. *aesculapii*
*O. spruneri
O. scolopax subsp. *cornuta*
*O. scolopax subsp. *heldreichii*
Orchis quadripunctata
*O. anatolica

*Orchis coriophora subsp.
fragrans
O. italica
*O.saccata
*O. boryi†
*O. lactea
*O. spitzelii
subsp. *nitidifolia*
Neotinea maculata
Limodorum abortivum

The open hillsides lying between the Minoan sites of Phaistos and Ayia Triadha on the plain of Mesara are rich in orchids in April.

Many herbaceous plants of the phrygana are widely distributed in the Mediterranean region; the following are of interest and are often common in Crete, including some endemic species:

*Silene cretica
*Anemone hortensis
*Anemone coronaria
Tetragonolobus purpureus
Euphorbia characias
*Polygala venulosa
*Cyclamen creticum
Alkanna tinctoria
*Salvia pomifera
Verbascum macrurum
Bellardia trixago
Acanthus spinosus
Pallenis spinosa

*Echinops spinosissimus
Centaurea idaea
*Centaurea raphanina
Tragopogon porrifolius
subsp. *australis*
Asphodelus aestivus
Asphodeline lutea
*Urginea maritima
Muscari comosum
*Romulea bulbocodium
Gynandriris sisyrinchium
*Arum creticum
Dracunculus vulgaris

Among the less frequent, largely endemic species of the phrygana are:

*Aristolochia cretica
Nigella doerfleri (E. Crete)
Nigella 'stricta' (W. Crete)†
Ranunculus cupreus†
*Paeonia clusii

*Onosma erecta
*Senecio gnaphalodes (E. Crete)†
*Fritillaria messanensis
*Tulipa cretica
Hermodactylus tuberosus

Autumn-flowering phrygana species include:

Euphorbia dimorphocaulon	**Sternbergia lutea* subsp. *sicula*
*Cyclamen graecum	**Biarum davisii*†
*Colchicum pusillum	

The disturbed ground of roadsides, fallow fields, and ditches has a distinctive collection of species, which are unable to survive in the surrounding phrygana; they are often associated with cultivation. Common are:

Papaver rhoeas	*Verbascum sinuatum*
Rubus ulmifolius	*Scabiosa atropurpurea*
Lavatera bryoniifolia	**Dittrichia viscosa*
*Alcea pallida subsp. *cretica*	*Dittrichia graveolens*
*Hypericum triquetrifolium	*Chrysanthemum coronarium*
*Eryngium creticum	**Picnomon acarna*
Anchusa azurea	*Carduncellus caeruleus*
*Echium italicum	*Scolymus hispanicus*
Echium plantagineum	

and in damp ditches *Lythrum junceum*.

Chasmophytes are plants that grow in gorges, on cliffs, or on rock-faces. They are of great interest in the Balkans, and particularly so in Crete, where gorges are numerous, very old, and one, the Samaria gorge, is said to be the deepest in Europe. Gorges act as refuges for many rare species. Each rock-face, ledge, or crevice, whether in sun or shade, creates its own special micro-habitat. Many species have distinctive habitat preferences, while each gorge has its own selection of species. Many are 'old' endemic species which have managed to survive precariously through all the fluctuations of ice ages and interglacials.

If you pass in June any sweltering gorge in Greece, you will find the hills brown as wild animals, thirsty and panting in the heat; but look at the cliffs, cool and shaded from the mid-day sun, you will see in the crevices plants which are not only freshly green and full of flower, but are often exceedingly rare and beautiful. They are confined, these ashen evangelical Inulas and Gothic-bellied Campanulas, to their selected crevices. Nowhere else are they found. They possess an indefinable charm, a grace and deportment achieved only by training for thousands of years in a hard school ... An air of greatness, whether hereditary or acquired, rests upon the chasmophytes.[2]

Exciting, unique, and often very beautiful chasmophytes are found in many Cretan gorges, notably:

*Ephedra fragilis subsp.	*Sanguisorba cretica*
campylopoda	**Ebenus cretica*

*Linum arboreum	*Staehelina arborea
Galium fruticosum	*Staehelina fruticosa†
*Origanum dictamnus	*Ptilostemon chamaepeuce
*Verbascum arcturus	*Centaurea argentea
*Petromarula pinnata	Scorzonera cretica
Achillea cretica	Lactuca acanthifolia

Other chasmophytes are very localized. For example, *Campanula tubulosa*, C. saxatilis, *Symphyandra cretica†, Helichrysum heldreichii†, and *Inula candida subsp. candida are found only in the west, while *Hypericum amblycalyx†, Campanula pelviformis, and Inula candida subsp. decalvans occur in the east.

Some interesting rare or recently discovered chasmophytic species include:

Hypericum 'jovis'†	Convolvulus argyrothamnos†
Hypericum aciferum†	Origanum tournefortii†
Eryngium amorginum	Campanula laciniata†

The higher mountain regions of Crete appear at first sight to be as inhospitable as any met with in Europe. Yet the flora, though relatively poor in species, consists of a high proportion of interesting endemics. A 'vegetable hedgehog' zone comparable to that on the Sierra Nevada of Spain and other high mountains bordering the Mediterranean is developed—and yet with almost totally distinct species.

Low, round, intensely spiny bushes, grazed 'to the quick' into domes or cushions, occur scattered over the stony slopes; common are:

*Berberis cretica	*Euphorbia acanthothamnos
*Astragalus angustifolius	*Acantholimon androsaceum
Astragalus creticus	

---

1 **Crete—Coast** The dominant land vegetation of Evergreen Oak, Juniper, *Cistus*, *Calicotome*, and *Pistacia* grows almost to the water's edge. There are few distinctive maritime cliff species. The tides are weak and the splash-zone very limited.

2 **Crete—Imbros Gorge** The many gorges of Crete act as a refuge for many rare and endemic chasmophytic plants seen nowhere else in Europe. They have survived the competition of more vigorous plants and the ravages of grazing animals. Each rockface, ledge, or crevice, whether in sun or shade, creates its own special micro-habitat where ancient relict species can survive.

3 **Crete—Lassithi Plain** A fertile polje in the mountains. Fruit trees and summer crops are irrigated with water pumped to the surface by numerous windmills. Field verges and the surrounding hills are rich in Greek phrygana plants; the Asphodel flourishes where grazing is most severe; it is unpalatable to animals.

Labiates such as *Sideritis syriaca* and *Satureja thymbra* and the endemics *Teucrium alpestre* and *Satureja spinosa* are often found among these bushes. Other attractive mountain plants often occurring in this zone are:

Prunus prostrata
Anchusa cespitosa†
Hypericum trichocaulon
Helichrysum italicum subsp.
  microphyllum

Lactuca viminea subsp.
  alpestris
Tulipa cretica

Common species which come into flower soon after the snow-patches melt, at different times and altitudes depending on the situation and orientation, are: *Crocus sieberi* in the typical form often known as 'heterochromus' with a deep golden-yellow throat and violet-purple stripes on white petals; the blue-flowered *Chionodoxa cretica*, the more delicate whitish-flowered *C. nana*, and the delicate pink-flowered *Corydalis rutifolia*.

On the highest slopes, above about 1800 m, where shattered flakes of limestone lie over hard rock, some plants have adapted themselves to the extreme conditions. Long woody roots penetrate the rock fissures, and slender, fragile stems, protected by the rock debris, produce shoots which just reach the surface—in consequence the mountain summits appear at first sight to be almost barren. On Mount Pakhnes, the highest in the White Mountains, the following are found:

Arenaria cretica
Minuartia verna subsp. *attica*
Paronychia macrosepala
Telephium imperati
Silene variegata
Gypsophila nana
Ranunculus brevifolius
  subsp. *pindicus*
Erysimum mutabile
Arabis alpina
Aubrieta deltoidea
Alyssum sphacioticum †

Alyssum fragillimum †
Draba cretica
Sedum tristriatum
Sedum laconicum
Cicer incisum
Andrachne telephioides
Euphorbia herniariifolia
Euphorbia myrsinites
  (rechingeri)
Viola fragrans
Peucedanum alpinum
Valantia aprica

---

4   **Delos**   Thousands of years of man's activity and intensive grazing have reduced the native vegetation to a phrygana of often spiny, aromatic shrubs with leathery leaves. Many colourful annuals, perennials, and bulbous plants flourish here, and there is a short-lived burst of colour each spring and early summer.

CRETE, CHASMOPHYTES

1. *Symphyandra cretica* 1344a 2. *Hypericum amblycalyx* 763c 3. *Ptilostemon chamaepeuce* 1488d 4. *Verbascum arcturus* 1196d 5. *Rosularia serrata* 397a 6. *Petromarula pinnata* 1348c 7. *Inula candida* 1388c 8. *Staehelina fruticosa* 1474b

9. *Ebenus cretica* 637b **10.** *Staehelina arborea* 1474c **11.** *Scutellaria sieberi* 1109a **12.** *Helichrysum orientale* 1384a **13.** *Scabiosa minoana* 1328g **14.** *Centaurea argentea* 1504d **15.** *Campanula tubulosa* 1331f **16.** *Linum arboreum* 658c **17.** *Teucrium divaricatum* 1099a

Cynoglossum sphacioticum †   *Scabiosa sphaciotica
*Mattiastrum   Senecio squalidus
    lithospermifolium †   (fruticulosus)
Thymus leucotrichus var.   Phagnalon 'pygmaeum'
    creticus   Muscari spreitzenhoferi
Veronica thymifolia

Western Crete

The White Mountains (Levka Ori) are without question the richest botanically, and although the highest summits are difficult to reach most of the endemic species are relatively accessible.

The most rewarding single excursion in Crete is to the Omalos plain in early summer. It is a flat, roughly triangular polje, surrounded by mountains, and at its southern end gives access to that great rift in the earth's surface—the gorge of Samaria. A good road from Khania, through the hill villages of Fournes and Lakki, climbs upwards from the orange groves of the plain through the olive groves to the open hillsides covered with phrygana. At about 1000 m the road winds up a steep open hillside before it crosses the col leading to the plain of Omalos (1160 m).

This hillside is a site for *Paeonia clusii*, with its large pure-white, fragrant flowers—one of the most splendid south Aegean endemics. Here too among the boulders are the pink-flowered, very fragrant *Daphne sericea*, the white-flowered *Cyclamen creticum*, the golden-yellow *Onosma erecta*, the chequered *Fritillaria messanensis*, and the orchids *Orchis quadripunctata* and *O. lactea*.

On the plain of Omalos itself, on the field verges and among the young corn, will be found the deep-pink flowers of *Tulipa saxatilis* (in the form often called *bakeri*) in April, with the blue Tassel Hyacinth, *Muscari comosum*. While in a cave mouth to the west of the point where the road drops down from the col on to the plain are *Arum creticum* and *Saxifraga chrysosplenifolia*.

The road crosses the plain and terminates abruptly on the threshold of the Samaria gorge. Great cliffs, to which gnarled and weather-beaten native cypress trees cling precariously, fall away at one's feet into what appears to be a bottomless chasm. Majestic mountains, whose tops at least are still snow-covered in spring, rise above tremendous precipices on all sides. On the steep stony hillsides above the tourist pavilion there is much of interest: ancient stands of *Cupressus sempervirens* are interspersed with small trees of *Acer sempervirens*, a small-leaved maple with pink-flushed fruits, which, when grazed to within an ace of its life, becomes a spiny domed shrub with tiny oval leaves, in this state amazingly similar to *Phillyrea latifolia. Unique is *Zelkova abelicea, a relative of the elms, which in one place not far from the road-head makes a fine grove of stately trees, under which the flocks gather for the night. Further up the mountainside the spring-flowering Cerastium scaposum, *Chionodoxa cretica, and *Crocus sieberi will be found by snow-patches, and on sunny loamy flats the low mats of *Anchusa cespitosa† will be studded with brilliant-blue stem-less chalices. Low clumps of *Berberis cretica, *Phlomis cretica, *Euphorbia acanthothamnos, and *Prunus prostrata spreading flat against the rocks are common. On the mountain slopes several endem-ics can be found such as Calamintha cretica, *Verbascum spinosum, Satureja spinosa, and *Origanum microphyllum, as well as the more widespread *Sideritis syriaca, *Satureja thymbra, and *Acantholimon androsaceum. In June and July it is well worth following the track further up the hill past the freshwater spring called Linoselli to the col above. On a cliff near the source are the endemics Scabiosa albocincta† and the extremely rare Bupleurum 'kakiskalae'† with large rosettes of spear-shaped leaves, which were both recently discovered and are known only from this one site, and other endemics such as Onobrychis sphaciotica†, Asperula incana, and A. rigida. Other local and interest-ing species here are: Dianthus juniperinus†, *Odontites linkii, Cephalaria squamiflora, *Trachelium jacquinii†, and Crepis auriculifolia†.

And then there is the gorge itself. Walk only a short way down the path and in spring you will see the yellow hanging bells of *Onosma erecta and the fragrant white flowers of *Cyclamen creticum under the pine trees. *Polygala venulosa scrambles up through the cistus shrubs, and the orchids *Ophrys fuciflora and *Orchis provincialis subsp. pauciflora, O. quadripunctata, *O. anatolica, and *O. lactea may also be found.

On the cliffs above the path are hanging mats of *Ebenus cretica, covered in spikes of beautiful pink flowers, and the shrubby *Staehelina arborea with laurel-shaped leaves, silvery-white under-neath, and with pink flower clusters. Yellow-flowered *Linum arboreum and *Hypericum empetrifolium, and the blue spikes of *Pet-romarula pinnata nestle in crevices in the rocks. On the stony slopes are the rosettes of the stemless *Centaurea raphanina and prostrate *Aubrieta deltoidea.

Further down, near the deserted village of Samaria at the junction of two branches of the gorge, are *Paeonia clusii and the Giant Orchid, Barlia robertiana.

In early July a different flora greets the walker; fine pink clumps of

CRETE, Mountain plants
1. *Acer sempervirens* 712c 2. *Berberis cretica* 261b 3. *Mattiastrum
lithospermifolium* 1083i 4. *Veronica thymifolia* 1229d 5. *Viola fragrans* 786a
6. *Astragalus angustifolius* 528i 7. *Chionodoxa cretica* 1643b 8. *Acantholimon
androsaceum* 971a 9. *Corydalis rutifolia* 278b 10. *Prunus prostrata* 476b

11. *Ranunculus subhomophyllus* 239g **12.** *Crocus sieberi* 1677d **13.** *Verbascum spinosum* 1192b **14.** *Anchusa cespitosa* 1057a **15.** *Centaurea raphanina* 1500c **16.** *Scabiosa sphaciotica* 1329b **17.** *Aubrieta deltoidea* 320 **18.** *Arum creticum* 1818c **19.** *Tulipa cretica* 1627b **20.** *Gypsophila nana* 178e
21. *Colchicum pusillum* 1588d

Putoria calabrica hang from the rocks, while in clearings among the trees are:

Ononis spinosa	**Verbascum spinosum*
*Salvia pomifera	**Lamyropsis cynaroides*
*Satureja thymbra	*Centaurea idaea*
*Origanum microphyllum	*Asphodeline lutea*

and *Hypericum hircinum* in moist places lower down by the side of streamlets.

In the lower parts of the gorge, 'On the shady cliffs dropped down, like German bedding hung out in the air, the great white bells of **Symphyandra cretica*†. Each bell is nearly 2″ long.'[3] Between the villages of Samaria and Ayia Roumeli near the sea the following chasmophytes are found on the cliffs:

*Silene gigantea	**Campanula tubulosa*
Brassica cretica	*Campanula laciniata*†
Sanguisorba cretica†	**Petromarula pinnata*
Eryngium ternatum†	**Helichrysum orientale*
*Teucrium flavum subsp.	*Helichrysum heldreichii*†
gymnocalyx	**Inula candida* subsp. *candida*
*Scutellaria sieberi	*Achillea cretica*
*Origanum dictamnus	*Scorzonera cretica*
*Verbascum arcturus	

The road south from Khania to Khora Sfakion on the south coast passes through the village of Imbros. Near the village is the beginning of the Imbros gorge, which cuts through the Sfakia Mountains on its way to the sea. It is an ancient mule track and is relatively easy going for the walker. The modern road, built in the early thirties, traverses the western flank of the gorge and then drops in wide sweeps to the southern sea, past acres of pink Oleander.

For us botanists, these cliffs are marvellous not only because of their beauty; they are in a way, the paradise of Cretan floristics. Here most of the non-alpine endemic and relict species grow side by side, in greater concentration than anywhere else outside the gorges of the Sphakia district, except in the summital mountain areas.[4]

The gorge has many treasures, including:

Celtis tournefortii	*Petrorhagia dianthoides*†
Cerastium scaposum	*Dianthus juniperinus*†
Silene pinetorum†	**Ranunculus creticus*†
Petrorhagia candica	*Ranunculus cupreus*†

Erysimum raulinii
*Ricotia cretica
Umbilicus parviflorus
*Rosularia serrata
*Saxifraga chrysosplenifolia
Sanguisorba cretica
*Coronilla globosa
*Ebenus cretica
*Linum arboreum
Eryngium ternatum †
Ferulago thyrsiflora †
Galium fruticosum
Procopiania cretica
*Scutellaria sieberi

Calamintha cretica
*Origanum dictamnus
*Verbascum arcturus
*Verbascum spinosum
*Campanula tubulosa
*Petromarula pinnata
*Inula candida
*Staehelina arborea
*Ptilostemon chamaepeuce
*Centaurea argentea
Centaurea redempta
*Cichorium spinosum
*Allium callimischon subsp.
haemosticum

With Khora Sfakion as a base there are two worthwhile excursions from the small village of Anopolis, which lies up in the hills to the north-west at about 600 m. The first is to the Aradhena gorge. It is but a short walk westwards, over the rocky phrygana, when quite suddenly one finds oneself at the top of a cliff so steep that one cannot see the bottom. Down it zigzags a beautifully paved mule-track—possibly an ancient Minoan way. At the top of the cliff, in April, *Ophrys lutea* subsp. *murbeckii*, with small flowers and narrow hairless margin to the brown lip, and *Orchis quadripunctata* can be found and over on the far side of the gorge, high up on a vertical cliff, the shiny leaves and pink blooms of *Tulipa saxatilis*. Walk down into and along the gorge southwards and in the crevices of the rock-faces you will see:

*Ranunculus creticus †
Erysimum candicum
*Rosularia serrata
*Linum arboreum
Teucrium cuneifolium

Valeriana asarifolia
Helichrysum heldreichii †
Ptilostemon gnaphaloides
subsp. *pseudofruticosus*

while on the floor of the gorge are *Scrophularia peregrina*, *Anchusa variegata*, *Smyrnium apiifolium*, and, tucked away at the base of the cliffs in cool shady places, the ubiquitous sweet-scented *Cyclamen creticum*.

The second excursion from Anopolis is more exacting as it involves sleeping out, and is only to be attempted by good walkers. From the village a rough stony track leads northwards, up through fragrant forests of *Pinus brutia* to a spring at about 1400 m. Here there are very fine bushes of *Berberis cretica* in full flower in late May. Between this and a cistern, at Ammoutzara, at about 1850 m, are compact hummocks of the superb blue-flowered Cretan endemic *Anchusa cespitosa* †. One can camp at this cistern, and from here it is possible to walk to the top of Pakhnes, the highest peak in the White Mountains, and back to Anopolis in a day (but it is a long day).

In late May near this camping-place *Viola fragrans*, white with a yellow centre and streaked with dark purple, can be found among the rocks, and, often protected by the spiny bushes of *Berberis* and *Astragalus*, the blue-flowered *Chionodoxa cretica*. *Centranthus nevadensis* subsp. *sieberi* † is an endemic only known from this particular spot, and in the craggy craters of the karst desert to the north can be found a few bushes of the recently discovered *Clematis 'elizabethae-carolae'* † covered with flowers scented like orange-blossom. Several rare and interesting small ferns are also here. On the summit screes, between the patches of melting snow, the following will be seen in flower: *Bufonia stricta, Dianthus sphacioticus* †, *Corydalis rutifolia, *Arabis alpina, *Aubrieta deltoidea, Draba cretica, *Helianthemum hymettium, Cynoglossum sphacioticum* †, *Scutellaria hirta, Thymus leucotrichus* var. *creticus,* and *Veronica thymifolia.*

Central Crete

The most interesting short excursion from Iraklion is to an isolated mountain—that of Zeus lying on his back, so the Cretans say—which is a conspicuous feature of the southern horizon as seen from the city. Fifteen kilometres south of the city lies a fertile grape-growing region centred round the prosperous village of Arkhanai, and above this is the mountain of Youktas (810 m), which has on its summit the small Church of the Transfiguration. 'Doubtless this was an ancient islet of the ancient Pliocene sea, and the numerous relict species growing there are remnants of the Miocene era'.[5] The limestone cliffs, facing west, and the surrounding phrygana-clothed slopes, are where the most interesting species are to be found, including the following:

*Silene gigantea	*Origanum dictamnus
Dianthus juniperinus †	*Petromarula pinnata
*Alyssoides cretica	Helichrysum stoechas
*Rosularia serrata	*Phagnalon graecum
*Coronilla globosa	*Echinops spinosissimus
*Ebenus cretica	subsp. bithynicus
Asperula incana	*Staehelina arborea
Asperula tournefortii †	*Ptilostemon chamaepeuce
Ballota pseudodictamnus	Scorzonera cretica
*Teucrium divaricatum	Asparagus aphyllus
*Phlomis lanata	*Iris unguicularis

and the recently described species *Erysimum 'candicum'*† and *Hypericum 'jovis'*†.

Autumn-flowering species include: *Euphorbia dimorphocaulon*, **Atractylis gummifera*, *Lactuca acanthifolia*, *Allium tardans*, *A. chamaespathum*, **Sternbergia lutea* subsp. *sicula*, **Biarum davisii*†.

Other areas of interest lie to the south-west of Iraklion towards the heights of Mount Ida (Idhi Oros). Beyond the village of Anoyia a very rough road southwards leads to the Nidha plain at 1317 m—a polje surrounded by mountains like Omalos. At its head, under Mount Psiloritis (2456 m), the highest mountain in Crete, is the cave known as Idheon Andron, famous in antiquity as the birthplace of Zeus. Grazing by flocks of sheep and goats is very heavy throughout the region; the summit areas are desert-like stony wildernesses with a poorly grown vegetation with few species. As the snows melt in spring the following are to be found: **Corydalis rutifolia*†, *Draba cretica*, *Gagea bohemica*, *G. amblyopetala*, **Chionodoxa cretica*, **Crocus sieberi*, and the endemic *Alyssum idaeum*†.

By early July other species include:

Minuartia verna subsp. *attica*	**Astragalus angustifolius*
Bufonia stricta	*Astragalus creticus*
Silene variegata	*Ononis spinosa*
*Silene saxifraga	*Hypericum trichocaulon*
Telephium imperati subsp. *orientale*	*Asperula idaea*
	*Scabiosa sphaciotica
*Ranunculus brevifolius	*Anthemis abrotanifolia*
Arabis serpillifolia subsp. *cretica*	**Cichorium spinosum*

Another area worth exploring is to the west of Mount Ida in the region of the monastery of Asomaton and the village of Fourfouras. Here **Tulipa orphanidea*, **Cyclamen creticum*, **Iris unguicularis*, **Romulea bulbocodium*, and several interesting orchid species can be seen in the spring. Later in the year can be found on the cliffs of Mount Kedhros the recently described *Dianthus 'pulviniformis'*†, which is known only on this one mountain, and in the phrygana the peculiar annual thistle *Ptilostemon stellatus*.

Eastern Crete

South-east of Mallia lies the large and richly cultivated plain of Lasithi, famous both for its water-pumping windmills and for the Dictaean cave, the mythical hiding-place of Zeus. The Dhikti Ori surround the plain; Afendi Khristos (2148 m) is the highest peak. Most of the species found on Mount Ida occur here; the following found on the higher screes and rocks are of particular interest:

Polygonum idaeum	Vincetoxicum 'creticum'
Silene dictaea†	Galium incanum subsp.
Alyssum lassiticum†	creticum
Sedum tristriatum	Teucrium alpestre
Astragalus idaeus†	Scutellaria hirta
Euphorbia deflexa	Phagnalon 'pygmaeum'
*Lysimachia serpyllifolia	Scorzonera idaea†

The easternmost extension of the island is dominated by the backbone of the Sitia Mountains which rise steeply to summits of over 1400 m. An interesting and spectacular road running parallel to the coast traverses the northern flanks of these mountains and leads to Sitia, and from this road several interesting branch roads lead off to villages in the mountains. The road to Rousa-Ekklisia is a good choice. Above the village can be seen fine plants of the shrubby endemic *Senecio gnaphalodes*†, and by the cliffs facing the road such interesting plants as *Linum arboreum*, *Hypericum amblycalyx*†, *Seseli gummiferum*, *Origanum tournefortii*†, *and Asperula tournefortii*†. On the road south from Sitia, near the village of Dafni, are found *Scabiosa minoana*† and *Aster creticus*.

On the far eastern coast at Vai is the now famous Date Palm grove of endemic *Phoenix theophrasti*†, which is well worth a visit. It will also give one an opportunity to find, near the monastery of Toplou, the shrubby *Viola scorpiuroides*†. Further south lies the Minoan site at Kato Zakros on the coast and just before it, where the road crosses a stream bed, is a short gorge. The following can be found flowering on the steep cliffs:

Ephedra fragilis subsp.	Nepeta melissifolia
campylopoda	Origanum onites
*Hypericum amblycalyx†	Campanula pelviformis
Asperula tournefortii†	*Staehelina fruticosa
*Scutellaria sieberi	

In summer *Dianthus juniperinus*† can be seen flowering on a cliff just behind a banana plantation, while the boulder-strewn stream bed is bright with the blue-violet inflorescences of *Vitex agnus-castus* and the pink-flowered clusters of *Nerium oleander*.

2. MOUNTAINS OF THE PELOPONNISOS

The Peloponnisos was described by Strabo as like a plane-leaf with a five-lobed leaf blade, each lobe with a central mid-rib of mountains, and attached to the mainland by its 'stalk' at the narrows of Rion-Antirion across the Gulf of Corinth. It is largely 'a region of wild and rugged highlands of limestone, fringed on the north and west by a zone of sandstone foothills and by a narrow coastal plain'.[6]

The central 'lobe' has the Taiyetos Mountains as its mid-rib running southwards for 100 km and ending in the dramatic and wild peninsula of the deep Mani and the southernmost tip of mainland Greece. To the west and east of the Taiyetos are the low-lying plains of Messinia and Sparta while to the north lies the remarkable basin of Tripolis which is 'one of the largest areas without superficial drainage in Greece'. The eastern 'lobe' has a rib of lower, less rugged mountains which reach their highest point towards the north in Parnon (1935 m). At the 'base' of the leaf blade, towering above the Gulf of Corinth, are the mountains of Aroania (Chelmos) and Killini, each over 2300 m and of great botanical interest. The relative isolation of the Peloponnisos from the mainland, its long isolation from Crete and Asia Minor as a result of the flooding of the eastern Mediterranean, as well as its southern position, have all contributed to its unique flora. Many endemic species are found here which occur nowhere else in the world. Turrill[7] estimates that no less than 260 endemics have been described in the Peloponnisos and southern Greece south of the 39° parallel (roughly south of Lamia).

Taiyetos

The Taiyetos Mountains, seen from the plain of Sparta, are a magnificent range of steep, jagged limestone and dolomitic peaks, covered in snow until late spring, rising in the south to the conical summit of Prophitis Ilias (2404 m). Lower forested foothills of semi-metamorphic rock surround the main massif and rise steeply out of the plain. These are pierced on their eastern flanks by deep gorges through which the snow-melt water floods down into the Evrotas river in the plain of Sparta.

There are at least six major gorges which are of considerable botanical interest and are well worth exploring. The largest is the gorge at Tripi, which is called Langarda, and it is up this that the main Sparta–Kalamata road winds its way over the spine of the Taiyetos Mountains and gives ready access to this part of the range.

Between Sparta and the village of Tripi at the entrance to the gorge lies the medieval Byzantine site of Mistras—the last stronghold of Byzantium in the south—one of the most impressive ruins in Greece. It lies on the steep flank of the Taiyetos foothills, overlooking the plain of Sparta, and during its heyday was itself protected by its frowning Frankish castle which still perches on a crest high above the ruins, with the snow peaks of the Taiyetos beyond. The terraces, old walls, rubble of houses, churches, and ruins of palaces of Mistras form a natural rock-garden in a superb setting.

In spring the pink flowers of Judas Trees, borne on the bare branches, are set against the silvery-grey of the olives in the plain below. Clumps of golden-yellow Jerusalem Sage, mixed with the stately candelabras of the pale-pink Asphodel, are resplendent against the rich red of Byzantine walls. Scarlet anemones with dark centres, *A. pavonina*, and the blue and khaki Tassel Hyacinth, *Muscari comosum*, occur in grassy places among the old walls. Sometimes there are pink swards of *Silene colorata*, or *Crepis rubra*, and later the little Star Clover, *T. stellatum*, often turns patches bright brick-red as its fruits mature. Pink-flowered *Convolvulus althaeoides* subsp. *tenuissimus*, with its neatly cut silvery-haired leaves, spreads over the rubble. In grassy patches there are some interesting orchids to be searched for. The yellow Bee Orchid, *Ophrys lutea*, is common, but more unusual is *O. spruneri* with a hairy brown bumble-bee lip with two parallel metallic-blue reflective patches, and the rather similar *O. argolica*† which has bright-pink petals and a knob at the end of the lip. The Giant Orchid, *Barlia robertiana*, can also be seen here; it often flowers earlier than the Bee Orchids.

Mistras still remains a very flowery site, by contrast with the majority of classical sites where modern sprays are used to clear the ruins of 'weeds', yet which a decade or so ago were equally flowery. Long may Mistras remain inviolate from such killers.

On the sunny walls grow such beautiful plants as *Campanula andrewsii*—pressed flat against the stones, with rich blue trumpets and silvery-grey foliage—and *Onosma frutescens* with its pendent reddish and yellow bells set amid clusters of bristly grey leaves hanging from cracks and crevices, together with the familiar yellow *Alyssum saxatile*. While in sheltered places are found the delicate white-flowered *Saxifraga chrysosplenifolia*, violet-flowered *Cymbalaria microcalyx*, and the strange striped brown Friar's Cowl, *Arisarum vulgare*. There is much else besides which adds to the colour and interest of the old walls and terraces.

The Langarda gorge near Tripi is a magnificent spot. It is easily explored from the main road, and is rich in interesting plants. The gorge begins above the village of Tripi and at first runs under vertical cliffs with gushing springs and rocks dripping with water and covered with Maidenhair Fern. Oriental Planes line the narrow valley floor, and steep bush- and tree-covered slopes and towering cliffs rise up on each side. In spring the sunny side is brilliant with the yellows of Jerusalem Sage and Spanish Broom, while the sheltered slopes are covered with heavy forests of Black Pine and Greek Fir.

On the more shaded cliffs above the village can be found the snowy-leaved *Stachys candida*†, with pale-purple, spotted flowers, often growing with the endemic Catchfly, *Silene goulimyi*†, which has small pink flowers borne on slender stems above lax cushions of leaves. Two southern woody plants grow here, the Cretan Maple, *Acer sempervirens*, with evergreen leaves, already showing bright reddish fruits as early as June, and the shrubby *Lonicera nummulariifolia* with paired, creamy flowers. Down on the floor of the valley under the plane trees are swards of *Cyclamen repandum* in spring, while about Christmas time the flowers of the rare *Galanthus nivalis* subsp. *reginae-olgae*† appear leafless· from the leaf-mould. *Saxifraga chrysosplenifolia* grows on shaded cliffs.

On the sunny slopes and cliffs there is much to be found, including a silvery-leaved, mat-forming scabious with solitary pink flower heads, *Scabiosa crenata* subsp. *breviscapa*, endemic to this mountain. Other species of note are:

Minuartia pichleri†	*Hypericum empetrifolium
*Silene integripetala†	Malabaila aurea
S. echinosperma†	Bupleurum fruticosum
*S. gigantea	Heptaptera colladonioides
Bolanthus laconicus†	*Onosma erecta
*Petrorhagia illyrica subsp.	*O. frutescens
taygetea	O. montana
P. glumacea	*Scutellaria rubicunda
*Thalictrum orientale†	*Cymbalaria microcalyx
Malcolmia flexuosa	Campanula andrewsii
Arabis muralis	*C. versicolor
*Aubrieta deltoidea	Colchicum boissieri
*Sedum sediforme	Gagea graeca
*Lathyrus grandiflorus	*Scilla messeniaca†
*Trifolium aurantiacum†	

The main road over the Taiyetos range to Kalamata crosses the watershed at about 1300 m where there is a hotel which is open only in the summer. The summital ridge here is covered in dense Black Pine forests and from this pass forest roads branch off north and south. In

79

May the clearings in the forests are ablaze with brightly coloured flowers. *Campanula spathulata* has large blue-mauve bells on slender stems; it grows in abundance as does the white-flowered *Anthemis cretica* with silvery-greyish leaves. In places *Genista halacsyi*† forms bright-yellow prostrate mats. Other notable plants are:

Cerastium illyricum
Lychnis viscaria
Dianthus viscidus
Ranunculus rumelicus
*Saxifraga chrysosplenifolia
*S. graeca
*Vicia melanops
*Lathyrus digitatus
*L. laxiflorus
*Trifolium physodes
Geranium asphodeloides
Polygala anatolica
Helianthemum
   nummularium subsp.
   obscurum

H. umbellatum
Galium rotundifolium
Symphytum ottomanum
Myosotis sylvatica subsp.
   cyanea
*Veronica austriaca subsp.
   teucrium
Doronicum orientale
*Centaurea triumfetti
Tragopogon crocifolius
*Orchis provincialis
O. tridentata
Neotinea maculata
Platanthera bifolia

It is less easy to reach the highest parts of the range and to climb to the bold summit of Prophitis Ilias. About 10 km south of Sparta on the main road to Yithion there is a signpost to the H.A.C. refuge, or katafigion, and to Prophitis Ilias. The branch road takes one through the village of Paliopanayia, and from here a dirt road climbs steeply up the hillside to the hamlet of Poliana—a collection of summer houses set among terraces of corn and hidden by walnut trees. The rough road continues upwards through forests of pine and fir to a spot where it crosses a clear stream running through a glade of old and gnarled Oriental Planes. A short distance before this stream is reached there is a rusty signpost to the katafigion which is about 1½–2 hours from this

---

5 **Mani** The extreme south of Greece is wind-blown and scorched by the summer sun. The rocks and stony slopes are dominated by tufts of *Ballota, Salvia, Cistus, Phlomis*, and Shrubby Spurge. The cliffs harbour some interesting plants which can survive in such extreme conditions.

6 **Mycenae** Set in a typical Peloponnisos landscape of rounded hills; the lower slopes are carefully cultivated with Olives and summer corn. Tall Cypresses stand round the farms; the gullies and steep slopes are covered with bushes of yellow Jerusalem Sage, and of *Calicotome*, the eastern equivalent of gorse.

point, while the summit is a further 2–2½ hours' climb. Once this path to the refuge has been found, it is a steady but easy ascent through woods of the Black Pine, *P. nigra* subsp. *pallasiana*, and open grassy glades to the refuge which stands at 1600 m among magnificent weather-beaten Black Pines, below the towering summit of Prophitis Ilias. Along this path, and on the sunny south-facing grassy slopes on the opposite side of the valley, the following are to be found:

Cerastium candidissimum
Petrorhagia prolifera
Dianthus cruentus
D. corymbosus
Hypericum olympicum
Cyclamen repandum
Armeria canescens
Alkanna graeca
Melittis melissophyllum

Verbascum macrurum
Digitalis laevigata
Morina persica
Campanula spatulata
Hymenonema laconicum
Lilium candidum
Anacamptis pyramidalis
Cephalanthera rubra

In the vicinity of the refuge, at the tree-line, and above to about 1800 m, is the plant 'association' of *Scabiosa taygetea† and *Onosma leptantha* described by Quézel,[8] and unique on Taiyetos. It is rich in species of which perhaps the most interesting and unusual are:

Anemone blanda
Berberis cretica
Erysimum pusillum
Malcolmia bicolor
Aubrieta deltoidea
Aethionema saxatile
Ribes uva-crispa
Prunus prostrata

Onobrychis alba subsp.
laconica
Helianthemum hymettium
Lysimachia serpyllifolia
*Scutellaria rubicunda†
subsp. *rupestris*
Pterocephalus perennis
Centaurea raphanina
Lactuca graeca

According to Quézel the main plant communities of the higher summits of the Taiyetos range are as follows: on calcareous cliffs the association of *Saxifraga marginata* and *Potentilla speciosa*, ranging from 1800 m to 2200 m; on high screes that of *Valantia aprica* and *Minuartia juniperina*, from 2300 m to 2400 m; and on rocks and

---

7 **Taiyetos** The highest and most isolated mountain range in the Peloponnisos, seen in spring from the olive groves in the Sparta valley. Dense forests of Black Pine, its southernmost site, cover the lower slopes. Deep gorges which cut through the outer range are rich in endemic species flowering in spring. In summer the summit flora is rich and varied.

screes above about 2300 m the association of *Acantholimon androsaceum* and *Rindera graeca*†. On deeper less rocky soils, from 1850 to 2200 m, another association characterized by *Sideritis clandestina*† with *Astragalus* tussocks is widely distributed.

Many species are common to some or all of these communities, but other species are confined to restricted habitats like screes, cliffs, rocks, and deeper soils respectively.

The track above the refuge is quite well marked with signs leading to the chapel on the summit, but there is no distinct path. From the refuge one climbs diagonally in a north-easterly direction up the flank of the south-east-facing slope—not directly up the summit cone from the refuge—past a fine rocky corrie with late-lying snow and vertical cliffs above.

Here can be seen the *Saxifraga marginata*—*Potentilla speciosa* association, with *Gypsophila nana*, *Arenaria cretica*, *Silene pusilla*, and *Lysimachia serpyllifolia*, and in damper places under cliffs *Ranunculus millefoliatus*, *Saxifraga sibthorpii*, *S. chrysosplenifolia*, *Primula vulgaris* subsp. *sibthorpii*, and *Cyclamen repandum*. Past this cirque the track turns back on itself, southwards, along a limestone ledge, to a small gash in the summit ridge at its lowest point. Above this there is a well-worn, partly made-up, zigzag track climbing up the north flank of the great scree cone of the summit of Prophitis Ilias to 2404 m. On the top is a small church, some monks' cells, and a well of drinking water. On all sides magnificent views spread before one, particularly southwards to the lower mountains stretching away in the distance towards the Mani peninsula. The summit screes have a sparse vegetation with dwarf weather-beaten plants sheltering amongst the stones which include *Minuartia juniperina*, *Cerastium candidissimum*, *Aethionema saxatile*, *Euphorbia herniariifolia*, *Galium incanum*†, *Valantia aprica*, *Verbascum acaule*, *Scrophularia heterophylla*, and *Achillea aegyptiaca*. In more consolidated rocky places the association of *Acantholimon androsaceum*, with rounded spiny cushions and pink flowers, and the silvery-leaved tufts of *Rindera graeca*†, with purplish flowers, occurs together characteristically, with *Alyssum taygeteum*†, *Thlaspi graecum*, *Asperula lutea*†, *Veronica thymifolia*†, *Campanula papillosa*, and *Jurinea taygetea*†; all small but interesting species.

On more consolidated earthy slopes on the west side—where cows and sheep graze under the summit ridge—are the characteristic 'hedgehog' heaths of high Mediterranean mountains consisting of spiny bushes of the taller *Astragalus creticus* subsp. *rumelicus* with pinkish flowers, and the cushion-like *A. angustifolius* with white flowers. In grassy patches by the last melting snows are *Crocus* *sieberi* and *Scilla bifolia*. Lower down on similar soils clumps of the silvery-leaved *Sideritis clandestina*† are common, associated with such species as *Cerastium candidissimum, *Erysimum pusillum, *Malcolmia bicolor, *Sedum sartorianum, *Prunus prostrata, *Daphne oleoides, *Viola parvula, *Lamium garganicum, *Verbascum epixanthinum, Veronica verna, *Senecio thapsoides,* and *Muscari botryoides*.

Parnon

Mount Parnon (1935 m) is the highest point of the considerably lower, less rugged range of limestone mountains which form the central backbone of the easternmost of the three southern peninsulas of the Peloponnisos. It is renowned for its many rare wild flowers, most of which grow between the highest peak and the village of Kastanitsa.[9] They include the white-flowered *Astragalus lacteus*†, the blue-flushed yellowish-flowered dwarf shrubby plant *A. agraniotii*†, the rare yellow-flowered *Centaurea macedonica*† subsp. *parnonia*, and the pink-flowered *Matricaria rosella*.

The highest point of the range is reasonably accessible from the town of Ay. Petros to the north. Continue eastward along the main road towards Ay. Ioannis, cross a river bridge, and turn into a dirt road leading southwards—at this point the H.A.C. refuge is signposted 15 km. Should one continue along the main road towards Ay. Ioannis one passes through fine thickets of trees of *Juniperus drupacea*† stretching up the mountainside. This is its only station in Europe; otherwise it occurs in Turkey and Syria. It is quite unmistakable in having the largest cones of any juniper species—often as much as 2·5 cm across, with seeds united into a stone—and the longest needles. *Acer sempervirens* and *Pyrus amygdaliformis* are common in these thickets.

The refuge is set amid forests of Black Pine, some distance from the highest point of Parnon which rises no more than 500 m or so above the last trees. Rough forestry tracks through stands of Greek Fir and some Stinking Juniper, *Juniperus foetidissima*, take one in half an hour to the bare-looking summit. The stony slopes, though quite heavily grazed, have a number of interesting species such as *Sideritis clandestina*†—one of the several related species used to make a local tea—with silvery-white foliage and small yellow flowers; *Stachys chrysantha*† with lemon-yellow flowers and silvery foliage; the handsome *Centaurea laconica*† with pink flowers; and *Verbascum mallophorum* with a branched inflorescence of yellow flowers with violet stamen hairs.

The delicate white-flowered, pink-veined *Gypsophila nana*, *Achillea holosericea* with yellow flower heads, *A. umbellata* with white flower heads and white foliage, and the pink, quite prostrate *Asperula boissieri* were found in the rocky ridges and crevices of the summit.

Other distinctive plants include:

Paronychia kapela subsp. *chionaea*	*Astragalus angustifolius*
	Onobrychis alba
Cerastium candidissimum	Polygala nicaeensis
Malcolmia ramosissima	*Helianthemum cinereum*
Draba lasiocarpa	*Armeria canescens*
Sedum rubens	*Linaria peloponnesiaca*
S. sartorianum	Anthemis spruneri†

Aroania (Chelmos)

The base of the 'plane-leaf' in Strabo's description of the Peloponnisos is taken up by three large mountain groups which tower majestically over the Gulf of Corinth, which lies to their north. From west to east are Panakhaikon (1926 m), Aroania (2341 m), better known as Chelmos to the outside world, and Killini (2376 m).

The isolation of the two latter mountains is such that not only do they harbour a number of ancient endemic species, but Chelmos is unique in having two boraginaceous species which have their only station in Europe on this mountain—occurring otherwise in Asia Minor, at least 500 km away. They are *Solenanthus stamineus†* and *Macrotomia densiflora†*, and presumably, before the flooding of the eastern Mediterranean, they had a much wider natural range stretching from southern Greece to southern Turkey across the land bridge.

The valley of the Styx—renowned in classical mythology—cuts into the north face of Aroania. It is famous also for its very rich flora containing among others a number of plants which otherwise range no further south in Europe than Mount Olympus in northern Greece; as, for example, *Viola delphinantha†*, *Gentiana verna*, and *Asphodeline taurica*. The best time to visit this wild and remote valley is in mid-June, and to reach it it is necessary to climb up to the H.A.C. refuge, over the summit ridge, and down a steep path in a north-westerly direction to the valley 1000 m or so below.

The guides led us on to what seemed the edge of a sheer cliff, beyond which one saw vacancy—and then the valley floor miles further on. But at the left corner a zig-zag path worked giddily down beneath a huge 800 foot precipice of dark reddish brown rock. Over this fell a little stream in a curtain of spray—the Styx! In old Greek legend the Styx was the river over which Charon ferried the dead to their abode in Hades . . . at the foot of the cliff lies a deeply overhanging cave, from the back of which spurts an icy spring the *mavroneri* or black waters.—But thoughts of gloom would soon be banished, for in the wet, stony soil at the edge of the cave grew a glorious drift of *Aquilegia ottonis* with big blue and white flowers and fresh green leaves.[10]

This beautiful Columbine with curved spurs is found only in this spot and in Sicily!

In the Styx valley, on the crags and cliffs, is to be found a most unusual assemblage of plants, including locally endemic species, both northern and southern Balkan plants, as well as others from the mountains of central Europe. These include:

*Gypsophila achaia†	*Globularia stygia†
*Thalictrum orientale†	*Pinguicula hirtiflora
*Aquilegia amaliae†	Lonicera hellenica
*Prunus prostrata	*Valeriana crinii
*Linum aroanium†	*Centranthus longiflorus
Polygala subuniflora†	Scabiosa graminifolia
*Viola chelmea	*S. crenata
*V. delphinantha†	*Trachelium asperuloides
Primula vulgaris	*Aster alpinus
*Omphalodes luciliae†	*Staehelina uniflosculosa
*Rindera graeca†	*Asphodeline taurica
*Macrotomia densiflora†	*Allium callimischon
*Solenanthus stamineus†	*Scilla messeniaca†
*Teucrium aroanium†	

The summit of Chelmos is most easily reached from Kalavrita. A rough forest road climbs up the hillside above the town, past a large white memorial cross in a grove of trees, ascending in a south-easterly direction towards the fir forests. A left-hand turn, signposted to Helmon, continues upwards and ends in a little clearing among stately Greek Firs, at about 1450 m. From this point a short track upwards brings one on to the open mountainside among flocks of sheep with their shepherds. From here there is a fine view of the summit ridge of Chelmos. A steady climb upwards through spiny patches of *Astragalus parnassi* and *Berberis cretica*, past gnarled and grazed bushes of *Prunus cocomilia* and *Crataegus pycnoloba*, and past the last stand of Greek Firs, brings one up a dry stream bed to a clearly defined semicircular moraine at about 1850 m. Above this the H.A.C.

refuge stands out boldly on the saddle at about 2100 m. The direct path up the valley to the refuge is not so rewarding as the ridge on the left, which leads step by step up to the summit plateau. On the north side of this ridge, overlooking the Styx valley 1000 m below, are magnificent limestone cliffs, with extensive views across the lower mountains to the Gulf of Corinth beyond. In mid-June a few patches of snow still lay in the hollows along the ridge, but on the exposed places were fine mats of the pink-flowered *Asperula boissieri*, silvery mats of *Paronychia kapela* subsp. *chionaea*, the delicate blue spikes of *Asyneuma limonifolium*, and the tiny prostrate *Euphorbia hernariifolia*. But the most exciting plant found was surely the smallest of all mulleins, *Verbascum acaule*†, with a tiny rosette of leaves flat on the ground, with large red flower buds and several unmistakable short-stalked deep-yellow mullein flowers.

On the north-facing cliffs were:

Silene auriculata	*Saxifraga taygetea*
Arabis bryoides	*S. glabella*
Sempervivum marmoreum	*Potentilla speciosa*
Saxifraga scardica	*Viola delphinantha*†
S. sempervivum	*Helianthemum cinereum*

On open slopes where there was a greater accumulation of soil among the stones and rocks, from about 1800 m to the summit, the following among others were seen:

Minuartia juniperina	*Daphne oleoides*
Cerastium candidissimum	*Viola chelmea*
Ranunculus sartorianus	Gentiana verna
R. pseudomontanus	Scrophularia canina
Aubrieta intermedia	*Globularia stygia*†
Aethionema saxatile	*Fritillaria graeca*
Astragalus angustifolius	Tulipa sylvestris subsp.
A. depressus	australis
Linum elegans	*Crocus sieberi*
L. punctatum	Dactylorhiza sambucina

Quézel and Katrabassa[11] describe the vegetation of mountain cliffs on Chelmos above 2000 m as characterized by *Viola chelmea* and *Valeriana crinii*, and another community by *Saxifraga scardica* and *Saxifraga exarata*. Another association has *Aquilegia ottonis* and *Saxifraga spruneri* as distinctive species, while a fourth association of *Aquilegia amaliae*† with *Pinguicula hirtiflora* occurs on damp water-flushed cliffs and ledges, together with the delicate yellow-flowered *Saxifraga sibthorpii*.

On screes above 1800 m is the association of *Valantia aprica* and *Minuartia juniperina*, with *Drypis spinosa*, *Corydalis bulbosa* subsp. *blanda*, *Ranunculus brevifolius*, *Astragalus depressus*, *Trifolium parnassi*, and the dwarf *Verbascum acaule†.

On *pelouses écorchées* they describe the associations of *Galium lucidum* and *Cirsium hypopsilum*, with *Astragalus parnassi* subsp. *cylleneus* and other species including:

Juniperus communis subsp.	*Marrubium cylleneum*
hemisphaerica	*Morina persica*
Dianthus corymbosus	*Pterocephalus perennis*
Ribes uva-crispa	*Carduus tmoleus* subsp.
Prunus prostrata	*armatus*
Daphne oleoides	*Allium roseum*
Lysimachia serpyllifolia	*Arum italicum*
Scutellaria orientalis	

On the rocks are communities of *Asperula nitida* and *Euphrasia salisburgensis*; and of *Globularia stygia†* and *Aster alpinus*, associated with *Silene radicosa*, *Saponaria bellidifolia*, *Linum elegans†*, *Macrotomia densiflora†*, and *Scabiosa graminifolia*.

Killini

Mount Killini is reached from Trikkala, a holiday resort situated at 1100 m on its northern slopes. The H.A.C. refuge A lies at an altitude of 1650 m, about 2½ hours from the village of Ano Trikkala; there is also a smaller refuge at 1750 m, about half an hour further on.

On calcareous cliffs above 1700 m, Quézel[12] describes a community characterized by the tight cushion-forming *Minuartia stellata* with *Valeriana olenaea†*—a creeping woody Valerian with dense, usually unbranched flower clusters. Associated with these are:

Silene auriculata	*Asperula arcadiensis*
Arabis caucasica	*Myosotis suaveolens*
Sedum magellense	*Scrophularia heterophylla*
Potentilla speciosa	*Achillea umbellata*
Viola heterophylla subsp.	*A. holosericea*
graeca	*Hieracium pannosum*

On rocks and screes from about 2100 m and up to the summit is the community of *Aster alpinus* and *Globularia stygia†*. The former occurs in a form described by some botanists as a distinct species, *A. cylleneus*, and the latter is a creeping shrublet with almost stalkless blue flower heads, which is restricted to the mountains of the north

SOUTH PELOPONNISOS, Mountains

1. *Arenaria cretica* 128c   2. *Scabiosa taygetea* 1325d   3. *Silene goulimyi* see 166b   4. *Silene integripetala* 165c   5. *Campanula spathulata* 1334a   6. *Genista halacsyi* see 508h   7. *Scabiosa crenata* 1329a   8. *Thalictrum orientale* 255c 9. *Helianthemum hymettium* 801h

10. *Anthemis cretica* 1409b   **11.** *Dianthus cruentus* 187c   **12.** *Lysimachia serpyllifolia* 962a   **13.** *Onosma leptantha* 1077g   **14.** *Sideritis clandestina* 1115e   **15.** *Valantia aprica* 1030b   **16.** *Scilla messeniaca* 1635b   **17.** *Cyclamen repandum* 960   **18.** *Malcolmia bicolor* 298g   **19.** *Minuartia juniperina* 132g

Peloponnisos. Other species associated with these two are *Draba lacaitae*, **Sempervivum marmoreum*, **Helianthemum hymettium*, **Acantholimon androsaceum*, and *Veronica thymifolia*. On similar terrain at lower altitudes the following can also be seen: **Minuartia juniperina*, **Ptilotrichum cyclocarpum*, *Draba athoa*, **Iberis sempervirens*, **Erodium chrysanthum*†, **Asperula boissieri*, *A. nitida*, **Galium incanum*, **Valantia aprica*, **Veronica austriaca*. On screes the characteristic species are **Drypis spinosa* and **Ranunculus brevifolius*, often with *Campanula albanica*.

On deeper soils between 1700 and 2100 m is a community of spiny plants dominated by **Astragalus parnassi* subsp. *cylleneus* (with leaflets hairy on both sides), associated with the whitish-flowered thistle *Cirsium hypopsilum*, the blue spiny **Eryngium amethystinum*, and the pinkish-purple, cluster-flowered thistle **Carduus tmoleus* subsp. *armatus*. Here also grew the endemic *Verbascum cylleneum*† which is now probably extinct, and other species including: **Crataegus pycnoloba*, **Geranium cinereum* subsp. *subcaulescens*, **Marrubium cylleneum*, **Sideritis clandestina*†, **Morina persica*, **Campanula spathulata*, **Centaurea triumfetti*, etc.

Panakhaikon

Mount Panakhaikon (1926 m) can be easily reached from Patras. It is well worth a visit as an introduction to the mountain flora of the Peloponnisos. Take a good road southwards, past the new University of Patras, which climbs up to the cherry-growing village of Kastritsi. Above this, a rough road zigzags up the mountainside, revealing magnificent views over the Gulf of Corinth and the mountains of Aitolia to the north. The road continues past scattered stands of the Greek Fir to terminate above the tree-line at the O.T.E. station. A steep scramble above this, through the hedgehog zone of spiny *Astragalus*, thistles, and *Eryngium*, brings one in about an hour to the northern summit. There is also an H.A.C. refuge at 1500 m which is reached from the village of Romanos in 3 hours, and another at 1800 m 45 minutes further on.

Among the more distinctive species collected on the mountain were:

Minuarta verna	**Astragalus angustifolius*
*Cerastium candidissimum	**Eryngium amethystinum*
Silene caesia	*Myosotis sylvatica*
*Dianthus pinifolius	*Verbascum macrurum*
*Ranunculus psilostachys	**Carlina corymbosa* subsp.
*Malcolmia bicolor	*graeca*
*Aubrieta intermedia	*Onopordum illyricum*

Centaurea triumfetti   *Crocus cancellatus*
Colchicum turcicum

A visit to the Botanical Institute of the University of Patras can be very rewarding; the localities of many of the more interesting Greek species can be traced here, and helpful advice given to bona fide students of Greek vegetation.

3. SOUTHERN PINDHOS MOUNTAINS

Three massive mountain groups, Parnassos, Giona, and Vardhousia, all over 2400 m, form the culminating peaks of the southern part of the Pindhos range, north of the Gulf of Corinth. The fourth, Timfristos, which is slightly lower, lies further north. They are composed largely of dolomites and limestone, with their summits rising well above the tree-line; snow covers the heights until April and early May, and snow-patches remain well into July. In consequence of this relative isolation and height an interesting and often unique Balkan mountain flora is developed.

Parnassos

Parnassos, the sacred mountain of Apollo and the Muses, lies nearest the sea. Seen from the south it is a massive dome-shaped group of peaks, about 25 km long, with Liakoura the highest summit (2457 m) and Gerontovrakhos and Kotrona over 2400 m. From the north-east the massif appears more craggy with glacial cirques under each summit.

Parnassos has recently become a relatively easy mountain to reach. The main Athens–Levadhia–Delphi road passes through the village of Arakhova, from where a new road climbs over the shoulder of Parnassos across the Livadi plain northwards to Gravia and Lamia. Eleven kilometres from Arakhova there is a well-signposted road to the Athens Ski Club and H.A.C. refuges which climbs up through the fir forest to these two refuges at about 1900 m, just above the tree-line. From here it is about 1–1½ hours to the skyline ridge, and a further 1½ hours to the northernmost summit of Gerontovrakhos.

Parnassos was decreed a National Reserve area as long ago as 1938, but up to the time of writing no practical steps have been taken to conserve its rich flora. Careful selective conservation would surely have a startling effect on some of the choice and rare species, which at the present time manage to hold on precariously under intensive grazing.

NORTH PELOPONNISOS, Mountains
1. *Marrubium cylleneum* 1113e   2. *Adonis cyllenea* 231b   3. *Teucrium aroanium* 1101b   4. *Ranunculus sartorianus* 241b   5. *Globularia stygia* 1263c   6. *Linum aroanium* 661b   7. *Scrophularia myriophylla* 1216c

8. *Astragalus depressus* 530b   **9.** *Allium callimischon* 1611d   **10.** *Aster alpinus* 1363   **11.** *Valeriana olenaea* 1315e   **12.** *Solenanthus stamineus* 10831.   **13.** *Viola chelmea* 774c   **14.** *Euphorbia herniariifolia* 684a   **15.** *Aquilegia ottonis* 253c
16. *Valeriana crinii* 1315d   **17.** *Gypsophila achaia* 178f

Before climbing the slopes of Parnassos there is much to be seen in spring in the vicinity of the ancient site of Delphi. The rock walls above the site are a veritable rock-garden. The best way to explore them is to take a steep and ancient zigzag track which climbs up the cliffs; it commences to the west of the stadium, outside the confines of the site, and eventually takes one up to the Livadi plateau and among the Greek Firs. On these cliffs can be seen a number of yellow-flowered shrubs including *Anagyris foetida, *Genista acanthoclada, *Medicago arborea, *Coronilla emerus, *Euphorbia acanthothamnos, and the rare and beautiful pale-purple-flowered *Daphne jasminea†. Other perennials and small shrubs growing in rock crevices and stone gullies include:

*Osyris alba	*Cerinthe retorta
Isatis lusitanica	*Teucrium divaricatum
Cardamine graeca	*Scrophularia heterophylla
*Alyssum saxatile	*Campanula topaliana subsp.
*Hypericum rumeliacum	delphica
*Ferulago nodosa	*Inula verbascifolia
*Alkanna orientalis	Asphodeline lutea
*Onosma frutescens	

On the cliffs by the Kastalian spring, east of the ancient site, can be seen: *Silene congesta, *S. gigantea, *Smyrnium orphanidis, Centranthus ruber, *Alkanna graeca, *Stachys swainsonii†, *Campanula topaliana subsp. delphica, *C. versicolor, *Ptilostemon chamaepeuce, and *Centaurea pelia. It is well worth exploring the terraces and olive groves round and below the site for their profusion of annuals and perennials associated with disturbed ground, while the rocks and terrace walls support many interesting shrubs and perennials and add colour to the spring landscape of this, one of the most beautiful of all classical sites.

On Parnassos the Greek Fir zone commences at approximately 800 m; heavy, almost pure forests clothe the middle slopes to the west and north, while the southern and eastern slopes are almost devoid of forest. In open places quite a rich flora can be seen, but as soon as the snow has cleared it becomes very heavily grazed, and careful search may be necessary to find some of the following more interesting plants:

Aristolochia pallida	*Aubrieta deltoidea
*Helleborus cyclophyllus	Thlaspi bulbosum†
*Anemone blanda	*Crataegus heldreichii
Ranunculus spruneranus	*Prunus cocomilia
Malcolmia flexuosa	*Genista parnassica†

Vicia onobrychioides
*V. dalmatica
*Lathyrus digitatus
*Geranium macrostylum
*Erodium chrysanthum†
Euphorbia myrsinites
Vinca herbacea

*Lonicera nummulariifolia
*Centaurea triumfetti
*Scorzonera mollis
Ornithogalum montanum
Muscari comosum
Iris pumila subsp. *attica*

Where the new road has recently been constructed, later in the year in high summer, magnificent stands of tall thistles and other plants line the embankment and cuttings such as **Onopordum tauricum*, **Cirsium candelabrum*, *C. eriophorum*, **Ptilostemon afer*, *Echinops ritro*, also the very striking candelabra Mullein, **Verbascum delphicum*, and an abundance of the brown-flowered Foxglove **Digitalis laevigata* subsp. *graeca*.

The Greek Fir forest itself is often dark and forbidding but interesting species like **Helleborus cyclophyllus*, **Lilium chalcedonicum*, and **Fritillaria graeca* may be found under its canopy or on the edge of clearings.

The two refuges, with a small ski-lift, are situated at about 1900 m, where the last battered Greek Firs find a foothold in the almost lunar landscape, formed by glaring-white limestone rocks. The effects of heavy grazing are everywhere to be seen. Most of the predominating plants owe their present success to being in some way self-protective against the intense grazing of sheep and goats. Some contain a poisonous or obnoxious substance and are obviously unpalatable, others are intensely spiny, while some seem to camouflage themselves by evolving silvery-grey foliage which matches the colour of the weathered limestone. The following seemed to be particularly successful in resisting grazing:

*Cerastium candidissimum
*Astragalus angustifolius
*Daphne oleoides
*Marrubium velutinum

Nepeta nuda
*Echinops spinosissimus
*Senecio thapsoides

and the shrub or small tree **Juniperus foetidissima*, Stinking Juniper.

The significance of this selective grazing and the very heavy pressure it brings to bear on the palatable species can hardly be overstressed. Most of the mountain tops of the southern Balkans that I visited suffer this same persistent and severe grazing. Unfortunately, it is often the rare, sometimes endemic plants which may suffer the most, and many of them only manage to hold out precariously on inaccessible cliffs, or in particularly favourable spots. It is imperative

PARNASSOS and DELPHI, Rock plants

1. *Euphorbia acanthothamnos* (670)   2. *Jurinea mollis* 1476   3. *Alkanna orientalis* 1061b   4. *Ferulago nodosa* 898a   5. *Anagyris foetida* 495   6. *Stachys swainsonii* 1136g   7. *Onosma frutescens* 1076e   8. *Alyssum saxatile* 324   9. *Centaurea pelia* 1503g

10. *Osyris alba* 70   **11.** *Medicago arborea* 585   **12.** *Coronilla emerus* 642
13. *Smyrnium orphanidis* 866a   **14.** *Ephedra fragilis ssp. campylopoda* 18
15. *Silene gigantea* 167b   **16.** *Inula verbascifolia* 1388b   **17.** *Phagnalon graecum* 1386a
18. *Campanula topaliana* 1331i

and urgent that a proper conservation policy be brought into being to ensure that these unique species are not lost for ever. However, our present knowledge of the delicate balances set up between these mountain plant communities and the environment may not yet be sufficient to prepare practical plans for conservation. For example, we do not yet know what would be the effect of prohibiting grazing in limited areas, or for limited periods: would aggressive species take over? Nor are the daily and seasonal habits or the selective feeding of sheep and goats on different species known to us—scientists and shepherds, as well as conservationists, need to get together!

In the vicinity of the refuges the mountainside is covered with swallow-holes or dolina—small steep-sided depressions with flat bottoms—which are characteristic of weathered limestone. These dolina present a variety of micro-climates and show characteristic 'profiles'. On the cliff-sides, often out of reach of grazing animals, are found the most varied and interesting species. *Geranium macrorrhizum* fills the clefts with pale foliage and beautiful pink flowers with long purple stamens, while the delicate *Campanula rupicola*, with dark-blue bells, often lines vertical crevices. Towards the base of the cliffs *Senecio thapsoides* is usually present with striking silvery-grey foliage and branched heads of small yellow flowers. Other cliff species include:

Juniperus communis subsp.	*Prunus prostrata*
hemisphaerica	*Rhamnus saxatilis*
Minuartia verna	*Frangula rupestris*
Silene saxifraga	*Daphne oleoides*
Sedum magellense	*Lysimachia serpyllifolia*
Saxifraga sempervivum	*Galium scabrifolium*
S. sibthorpii	*Campanula versicolor*
Rosa orientalis	*Doronicum columnae*
Cotoneaster nebrodensis	*Hieracium pannosum*

On the almost flat floor of the doline, where a rich terra rossa accumulates from the weathering slopes above, quite a different assemblage of species is found including:

Dianthus viscidus	*Campanula radicosa*
Ranunculus demissus	*Carduus tmoleus*
Trifolium parnassi	*Scilla bifolia*
Stachys germanica	*Crocus sieberi*
Acinos alpinus	*C. veluchensis*
Plantago atrata	*Arum maculatum*

Later in the year it is the grazing-resistant species like *Cerastium candidissimum*, *Marrubium velutinum*, and *Nepeta nuda* which are most conspicuous.

On the exposed tops of the doline is a very xerophytic flora of spiny bushes of *Astragalus baldaccii*, with *Teucrium montanum*, *Asperula lutea†*, *Thymus parnassicus*, and the rare and beautiful, quite prostrate, *Convolvulus boissieri*, subsp. *compactus*, with silvery leaves and pink flowers, which is much sought after by the sheep and goats. The most memorable sight was perhaps that of the tough woody branches of *Prunus prostrata* smothered with deep-pink flowers, lying flat over every rise and fall in the underlying limestone rocks, as if its very survival depended on hugging the surface, while up through this carpet pushed clusters of dark rosy-purple flowers of the handsome Dead-nettle, *Lamium garganicum*.

On the steep slopes and cliffs above the refuges, *en route* for the summit ridges, the following can also be found:

Cerastium decalvans	*Nepeta parnassica*
Dianthus haematocalyx	*Satureja parnassica*
Arabis caucasica	*Verbascum epixanthinum*
Potentilla speciosa	*Pterocephalus perennis*
Astragalus creticus subsp.	*Morina persica*
rumelicus	*Campanula spatulata*
Hypericum rumeliacum	*Achillea umbellata*
Viola brachyphylla	*A. holosericea*
Asperula boissieri	*Colchicum variegatum*
A. pulvinaris†	*Sternbergia lutea* subsp.
Sideritis syriaca	sicula

Quézel, who has studied the plant communities of most of the higher mountains in Greece, describes the richness of south-facing cliffs on Parnassos and Giona, where he has recorded over 70 species, as 'un des joyaux les plus remarquables de la végétation des hautes montagnes du sud de la Grèce tant par sa richesse en éléments endémiques, qui n'est sans doute égalée nulle part ailleurs dans cette région, que par la beauté de quelques uns des elements qui la constituent'.[13] He characterizes it by the presence of Campanula aizoon† and Campanula rupicola†, with *Edraianthus graminifolius*, Linum elegans, etc., and many of the species listed above. On Parnassos this community is well developed on the cliffs above Arakhova, near Gourna, and particularly above the plateau of Ay. Nicolaos, at altitudes ranging from 1500 to 2000 m.

On the screes below the summits of Parnassos, *Ranunculus brevifolius*, with fleshy glaucous leaves matching the stones through which it grows, carries its neat solitary yellow flowers. By contrast, the tufts of *Drypis spinosa* are bright green and stand out conspicuously in the screes, its prickly pointed leaves perhaps protecting it from grazing. Stabilized screes support large, hard, pale-green mats of *Minuartia stellata*, composed of tiny saxifrage-like rosettes each bearing tiny white flowers. These mats give protection and support to other species such as *Iberis sempervirens*, *Draba parnassica*, *Asperula boissieri*, *Thymus teucrioides*, *Veronica thymifolia*, *Erigeron alpinus*, and *Colchicum variegatum*.

By the end of May most of the arêtes and cliffs are clear of snow but there are still extensive patches on the higher sheltered slopes. As the snow recedes, masses of *Crocus veluchensis* in all shades of purple and violet appear, always with the deep-blue *Scilla bifolia*, and less frequently with *Corydalis bulbosa* subsp. *blanda* with pink flowers above glaucous leaves. Other plants include: *Corydalis solida*, *Erysimum pusillum*, *Thlaspi microphyllum*, *Aethionema saxatile*, *Iberis pruitii*, *Euphorbia rigida*, *Ornithogalum oligophyllum*, *Muscari neglectum*.

On the summit cliffs and glacial corries is a most interesting community of plants characterized by the rock-loving *Viola poetica* and *Saxifraga spruneri*, with *Edraianthus parnassicus*, found mostly on the damper east and north faces. On the summit of Gerontovrakhos the following were found in this community:

Silene acaulis	*Saxifraga sibthorpii*
S. auriculata	*S. scardica*
S. saxifraga	*Potentilla speciosa*
Erysimum pusillum	*Astragalus depressus*
Arabis bryoides	Viola brachyphylla
Draba parnassica	Erigeron alpinus
Saxifraga sempervivum	*Doronicum columnae*

Giona

Giona (2510 m) lies to the west of Parnassos and is much less accessible. It can be reached from the village of Kaloskopi, which is signposted off the main Amfissa–Lamia road. North of Kaloskopi lies Karknoz and from there a new forest road enables one to reach the high grazing of Skasmada Diacelo at 2100 m, in the east of the range, but some distance away from the summit of Giona. The H.A.C. refuge lies in the heart of the circle of mountains, with Giona its highest summit, at the head of the Reka valley, which can be reached from the village of Prosilion.

As one would expect, there are many similarities between the plant communities of Giona and Parnassos, and Quézel[14] has distinguished the following associations, each named after their two most characteristic species:

On limestone cliffs

Parnassos 700–1100 m	*Sideritis syriaca* and *Alkanna graeca*
Giona 1000–1350 m	*Asperula chlorantha* and *Daphne jasminea*†
Parnassos 1700–1850 m	*Satureja parnassica* and *Sedum magellense*, particularly on north-facing slopes
Parnassos and Giona 1500–2300 m	*Campanula aizoon*† and *Campanula rupicola*†, widely scattered
Parnassos and Giona 2200–2400 m	*Viola poetica* and *Saxifraga spruneri*

On screes

Parnassos and Giona 1700–1950 m	*Geranium macrorrhizum* and *Senecio thapsoides*
Parnassos and Giona 1500–2100 m	*Sclerochorton junceum* and *Euphorbia deflexa* (with *Ranunculus brevifolius*)
Parnossos and Giona 2200–2400 m	*Corydalis bulbosa* subsp. *blanda* and *Astragalus hellenicus*

On dry grassland

Parnassos and Giona 1700–1800 m	*Convolvulus boissieri* and *Astragalus* spp.
Parnassos and Giona 1700–2050 m	*Astragalus creticus* and *Marrubium velutinum*
Parnassos 2300–2400 m	*Minuartia stellata* and *Erysimum pusillum*

Quézel describes an interesting community characterized by *Globularia meridionalis* and *Anthyllis montana*, with *Thymus teucrioides*, *Arabis bryoides*, *Asperula boissieri*, and *Satureja parnassica* on Giona, on cliffs above Skasmada, and elsewhere. Exploration of the cliffs below the pass leading to the high grazing of Skasmada Diacelo from the head of the road revealed an interesting collection of plants in mid-June including:

*Cerastium decalvans
*Silene saxifraga
*Helleborus cyclophyllus
*Ranunculus psilostachys
 Malcolmia angulifolia
*Arabis bryoides
*A. caucasica
*Aubrieta deltoidea
*Iberis sempervirens
*Saxifraga sibthorpii
*S. sempervivum
*S. adscendens subsp.
   parnassica

Linum elegans
Polygala nicaeensis
Rhamnus lycioides
*Hypericum rumeliacum
*Lysimachia serpyllifolia
 Asperula nitida
*Lamium garganicum
*Globularia meridionalis
 Valeriana tuberosa
*Achillea umbellata
 A. fraasii
*Allium roseum
*Muscari armeniacum

On the grassy alpine meadow of Skasmada the following were seen:

*Silene roemeri
 Dianthus viscidus
 Ranunculus demissus
 Corydalis solida
 Astragalus creticus subsp.
   rumelicus
*Geranium cinereum subsp.
   subcaulescens

*Eryngium amethystinum
 Gentianella ciliata
 Campanula radicosa
 C. tymphaea
*Scorzonera purpurea
 Ornithogalum oligophyllum
*Crocus sieberi
*C. veluchensis

Vardhousia

Vardhousia (2406 m) is not an easy mountain to reach but it is well worth exploring, being rich in mountain species and relatively less heavily grazed than Parnassos. A rough, and at times rather difficult, road runs northward from Lidokhorikion up the Mornos valley between Giona and Vardhousia to the village of Mousounitsa—in the summer there is a daily bus service to this village. The village lies in the Greek Fir zone at 1000 m, and above it a rough road continues to within 20 minutes' walk of the H.A.C. refuge at 1750 m, but characteristically heavy rain prevented our reaching it in the third week in June.

In the dense Fir forests above the village the following were seen:

 Minuartia eurytanica
*Umbilicus rupestris
*Sedum laconicum
*S. magellense

*Cyclamen repandum
 Galium rotundifolium
 Calamintha grandiflora
 Limodorum abortivum

In the pastures above the forest were:

*Silene roemeri
Dianthus biflorus
*Helleborus cyclophyllus
*Potentilla recta
*Hypericum rumeliacum
H. barbatum
*Ferulago nodosa
*Primula veris subsp.
    columnae

*Armeria canescens
Vincetoxicum hirundinaria
    subsp. nivale
*Digitalis lanata
Rhinanthus pubescens
Campanula tymphaea
*C. spatulata
Lilium martagon
*Fritillaria graeca

On the cliffs at about 1700 m the most interesting species were the small trees *Acer heldreichii* and *Sorbus graeca*, and many rock-loving species such as:

*Arenaria filicaulis subsp.
    graeca
*Silene auriculata
Thalictrum minus
Malcolmia bicolor
Draba lacaitae
*Saxifraga sempervivum
*S. adscendens
*S. chrysoplenifolia
*Potentilla speciosa
Euphorbia deflexa

Malabaila involucrata
Galium thymifolium
Anchusa serpentinicola †
Salvia argentea
*Veronica austriaca subsp.
    dentata
*Campanula rupicola †
*Edraianthus graminifolius
*Achillea holosericea
A. fraasii

Timfristos

This is a relatively easy mountain to reach, once the snow has left the lower slopes. Karpension is the centre for this mountain and a short distance outside the town, by a petrol station on the Lamia road, a rough road is signposted to the H.A.C. refuge. A good time for a visit is mid-June before the arrival of the flocks, when one can drive almost up to the H.A.C. refuge at 1850 m. From here it is about an hour's climb to the summit at 2315 m.

The road at first zigzags steeply up the mountainside, through a sparse forest of Greek Fir, to a high col under the summit. On the lower slopes of the mountain there is much to be seen, including such shrubs as *Juniperus oxycedrus*, *Crataegus heldreichii*, *Sorbus umbellata*, *Spartium junceum*, and *Daphne oleoides*, while in open places between the thickets are herbaceous species such as *Cerastium candidissimum*, *C. decalvans*, *Saponaria calabrica*, *Aubrieta gracilis*,

Asperula boissieri, *Ajuga chamaepitys*, *Scutellaria orientalis*, *Salvia argentea*, and *Asyneuma limonifolium*. We spent the night below the refuge at the head of the pass where the snow-patches were still melting and the last of the *Crocus veluchensis* was still flowering, together with *Ranunculus sartorianus*, *Viola aetolica*, *Plantago atrata*, and *Scilla bifolia*.

On the screes and upper slopes other distinctive plants in flower were:

Minuartia verna	*Geranium cinereum subsp.
*Dianthus corymbosus	subcaulescens
*Ranunculus brevifolius	*G. macrostylum
*Erysimum pusillum	*Myosotis suaveolens
*Malcolmia angulifolia	Calamintha nepeta subsp.
*Arabis bryoides	glandulosa
*Ptilotrichum cyclocarpum	Valeriana tuberosa
	*Centaurea triumfetti

4. NORTHERN PINDHOS MOUNTAINS

The mountains of the Pindhos range occupy much of the centre of northern Greece. It is a wild country of remote valleys, usually densely forested, with small widely dispersed villages joined only by rough tracks which for many months in the year are impassable to wheeled vehicles. Craggy mountains and grassy hills rise in fold after fold above the forested valleys, each summit with its own shepherds and flocks of sheep, which every summer come up from their winter grazing below—each flock belongs to one of the many isolated villages. The higher mountains, of which there are many, rise to well over 2100 m, but there are many more which reach 1800 m. The majority are predominantly of limestone but there are a few of serpentine, like Smolikas. Access to many of these summits is not easy, but there are now two good, though seasonally vulnerable, roads crossing the northern Pindhos—a southern road from Ioannina through Metsovon to Trikkala, and a more recently constructed northern road from Konitsa, near the Albanian border, through Eptahorion to Kozani. Both roads traverse magnificent unspoiled and sparsely populated country, but to reach the higher peaks, where some of the more unusual species are to be found, usually involves hiring a guide and requires a considerable expenditure of time, and can only be undertaken in the summer months from about mid-June to early September. There are some areas still to be explored botanically.

Vicos gorge, Astraka

Probably the easiest high mountain range to reach is Timfi, lying south of Konitsa in the north-west corner of Greece. It has a rich montane and alpine flora characteristic of the Pindhos range, much of it Balkan in origin, but with a number of more northern species with their main distribution in the Alps. About 19 km south of Konitsa a good road branches off the main Ioannina–Konitsa road eastwards up into the hills to Aristi and Papingon, and from the latter village the summit of Timfi can quite easily be reached. *En route* for Papingon the road drops down steeply into a deep wooded valley to cross the crystal-clear Voldomatis river. Above and below this crossing the river cuts a deep and magnificent gorge, the Vicos gorge, through the southern flank of the mountain. It has been compared to the Samaria gorge in Crete; it is 20–5 km long with a depth varying from 300 to 1100 m and is composed mainly of dolomites and limestone. It is richly wooded with a mixed deciduous forest of *Carpinus betulus*, *Ostrya carpinifolia*, *Corylus avellana*, *Acer pseudoplatanus*, *Tilia tomentosa*, *Cornus mas*, and *Ilex aquifolium*, with occasional trees of native *Aesculus hippocastaneum*. In the valley bottom are fine stands of *Platanus orientalis* with *Alnus glutinosa*, and the willows *S. elaeagnos*, *S. purpurea*, and *S. alba*. Herbaceous species include:

*Helleborus cyclophyllus	Salvia glutinosa
Hesperis matronalis	Melissa officinalis
Cardamine bulbifera	*Digitalis lanata
Lathyrus venetus	*Acanthus balcanicus
*Geranium macrorrhizum	Lonicera etrusca
Convolvulus cantabrica	Campanula persicifolia
Symphytum ottomanum	*Asphodeline liburnica
Myosotis sylvatica	*Lilium candidum
Melittis melissophyllum	

The cliffs and ledges have a rich flora including:

*Saponaria calabrica	*Salvia candidissima
Hesperis laciniata	*Digitalis lanata
*Sedum cepaea	Ramonda serbica
Saxifraga paniculata	Campanula tymphaea
*Frangula rupestris	*Pterocephalus perennis
*Hypericum rumeliacum	*Inula verbascifolia
Athamanta macedonica	*Achillea holosericea
*Moltkia petraea	*Staehelina uniflosculosa
Onosma sp.	*Centaurea graeca
Nepeta spruneri	

On the drier slopes above the gorge up to about 1600 m are the oaks *Q. coccifera*, *Q. pubescens*, and *Q. trojana*, with *Juglans regia*, *Carpinus orientalis*, *Ostrya carpinifolia*, *Ulmus* sp., *Celtis australis*, *Clematis flammula*, *Pyrus cordata*, *Pyrus amygdaliformis*, *Malus sylvestris*, *Prunus cocomilia*, *Prunus avium*, *Cercis siliquastrum*, *Coronilla emerus*, *Pistacia lentiscus*, *Acer obtusatum*, *Acer monspessulanum*, *Cornus sanguinea*, *Fraxinus angustifolia*, and *Sambucus nigra*: as fine a mixture of central European and Balkan woody species as one is likely to encounter anywhere in the Balkans. While there are no pines or firs above the village of Papingon, the upper tree-line consists of fine, almost pure open stands of *Juniperus foetidissima*, with some magnificent old trees up to 14 m in height.

The village of Papingon is a cluster of well-built stone houses, with narrow cobbled lanes, vine-covered terraces, and stout doorways opening on to sheltered courtyards. It lies under the formidable cliffs of Astraka (2486 m), perched high above the Vicos gorge. From here a rough track, sometimes signposted, leads through mixed woods of the species listed above, to the H.A.C. refuge at 1950 m, on a saddle to the west of the summit of Astraka—it can be seen from the village. It is about 2½ hours climb to the refuge. Beyond the saddle the ground falls away to a plateau of glacial lakes and moraines at about 1700 m, to rise upwards again to the rounded summit of Gamila (2483 m) and, further to the east, the summits of Goura (2466 m) and Tsoumako (Tsouka) (2151 m).

From the refuge I climbed directly up the steep cliffs to the summit ridge of Astraka where I came upon a sight common on these remote heights—a lusty shepherd with his flock which was grazing to the detriment of the mountain-top flora. About the summit, on cliffs, fractured rock ridges, crevices, and screes there was a rich alpine flora including the following:

Erysimum pusillum
Cardamine carnosa
Arabis alpina
Aubrieta intermedia
Alyssum montanum
Ptilotrichum cyclocarpum
Saxifraga taygetea
S. adscendens subsp.
    parnassica
S. marginata
Cotoneaster integerrimus
Astragalus sirinicus
Trifolium pallescens

Armeria canescens
Gentiana verna
Myosotis suaveolens
Thymus cherlerioides
Veronica austriaca subsp.
    dentata
Veronica serpyllifolia subsp.
    humifusa
Pedicularis brachyodonta
Edraianthus graminifolius
Centaurea triumfetti subsp.
    cana

Gamila

The village of Skamneli, at 1250 m, south of the Gamila–Tsoumako ridge, is an alternative approach to these summits. Above the village, to heights of 1600 m, are wooded areas of Black Pine with a potentially rich vegetation in open areas but 'destroyed by grazing animals during the summer months'.

Between 1700 and 2000 m the vegetation consists mostly of *Malcolmia angulifolia, Rosa heckeliana, *R. glutinosa, *Geranium cinereum* subsp. *subcaulescens, *Viola magellensis, *Thymus cherlerioides, Rhinanthus pubescens, *Pedicularis graeca, *Achillea holosericea*, and the tea-giving plant *Sideritis scardica*. Another attractive plant is *Linaria peloponnesiaca* with large yellow flowers in elongated inflorescences, forming small lovely groups.

Above 2000 m the vegetation is alpine, with abundant grass species mixed with shrubby vegetation of *Astragalus angustifolius* and *Globularia cordifolia*, which also grows at lower elevations. *Centaurea deustiformis* subsp. *ptarmicifolia* is also found here, with solitary purple flower heads and greyish-woolly leaves, as well as *Crocus veluchensis* earlier in the year. On the extremely steep and massive cliffs at this elevation a very interesting chasmophytic flora consisting of *Silene saxifraga, *Aubrieta gracilis, *Saxifraga marginata, *Potentilla speciosa, *Valeriana crinii, Campanula albanica, *Achillea abrotanoides*, etc. can be found.

Goulimis describes the plants on the Goura–Tsoumako end of the ridge as follows:

This ridge, about the middle of July, is covered with many alpines including *Gentiana verna* ssp. *pontica*; *Linum punctatum*, a delicate procumbent azure-flowered flax; pale blue and brown forms of the lovely *Viola magellensis*, several alpine saxifrages including *S. marginata, *S. taygetea, *S. sempervivum* and *S. oppositifolia*; *Sempervivum marmoreum* an alpine Houseleek with purple-striped, pink petals; several species of *Sedum*; *Artemisia eriantha* an alpine species of *Absinthium* . . . Another ridge lies on the northern slopes of Mount Gamila not very far from the Goura–Tsouka ridge. The upper part of the slope is composed of huge cliffs more than 500 metres high, which drop abruptly from the ridge. The cliffs are indented by precipitous ravines where snow, many metres thick, persists until the middle of July. On the shelves and terraces which protrude from these cliffs grows a luxuriant vegetation, among which fine specimens of Horse Chestnut, *Aesculus hippocastanum*, are particularly prominent. Both cliffs and ravines are inaccessible to man and serve as refuges for the wild goat and brown bear . . . The whole scenery is of wild beauty and one of the finest in Greece.

On my visits to these slopes in early July I found a great many species of wild flowers in full bloom. Their list is very long and I will confine myself to the following only: *Soldanella hungarica*; two lilies *L. carniolicum* and L. mar-tagon; *Rosa villosa* a mountain wild rose with large rose-coloured flowers; two cream-coloured species of *Pedicularis*, *P. hoermanniana* and P. comosa; several *Saxifraga* including *S. marginata* and *S. stribrnyi*; a number of cranes-bills, notably *G. macrorrhizum*, G. sylvaticum, *G. cinereum* ssp. subcaules-cens and G. aristatum, the last with reflexed petals; and also several orchids including *Epipactis atropurpurea* a helleborine with black-red flowers; and *Cephalanthera rubra* the Red Helleborine; *Jurinea glycacantha* with large, purple flower heads; *Viola orphanidis*; *Hesperis dinarica*, a white-flowered Dame's Violet; a fine rock scabious *Pterocephalus papposus*; several campanulas . . .; the magnificent *Lathyrus grandiflorus*; the rare *Ranunculus platanifolius*, a fine buttercup with white flowers . . .; *Lembotropis nigricans*, a shrub somewhat like a Laburnum with smaller flowers; several composites such as the showy *Doronicum austriacum* and the dainty *Achillea abrotanoides*; and *Aubrieta gracilis*, the most graceful of all aubrietias. On the streams of scree which fan out of these slopes there occur also the long-stemmed *Centranthus longiflorus* related to the Red Valerian, a weird *Vincetoxicum fuscatum* (?) of the Asclepias family, *Dianthus sylvestris* with rose or purple flowers; *Campanula hawkinsiana*; large cushions of *Drypis spinosa* . . . and many others.[15]

Smolikas

This is one of the largest massifs in the Pindhos range; its summit of 2637 m is second only in height to Mount Olympus in Greece. It lies to the north of Gamila and is separated from it by the deep valley of the Aoos river. Another deep valley, up which the main road from Konitsa runs northwards, separates Smolikas from Mount Grammos (2523 m) on the Albanian border, another interesting and little-explored mountain. Smolikas is of particular interest being largely composed of serpentine, a rock rarely encountered in Greece, and which in consequence has a distinctive flora. This rather impenetrable bedrock, in combination with a high precipitation, accounts for an unusual abundance of water; travellers on the mountain need not bring any water but may supply themselves from the numerous springs and brooks flowing down the slopes. The mountain is perhaps most easily accessible from Samarina or Pades, where mule paths lead to the top. In both villages it is possible to hire mountain guides and mules. The summit area may also be reached by foot from the winding road running around the mountain, usually above the 1000 m contour, and ascending to nearly 1700 m north-west in the vicinity of Samarina.

Forests of *Pinus nigra* or mixed forests occur on serpentine from about 1100 m. A fine example of mixed forest can be seen at about 1300 m in the great south-eastern ravine which can be reached by the timber road from Armata to Samarina. On these steep slopes there is a luxuriant forest of *Pinus nigra, Taxus baccata, Ostrya carpinifolia, Fagus sylvatica, Sorbus aucuparia, Acer obtusatum, Euonymus latifolius*, and *Fraxinus excelsior*, and a field layer which includes *Digitalis viridiflora, Campanula trachelium, Solidago virgaurea, Prenanthes purpurea, Tanacetum corymbosum*, and *Convallaria majalis*. On slopes or rocks moistened by seeping water the following species are typical: *Parnassia palustris, Gentiana asclepiadea, *Rhynchocorys elephas*, and *Pinguicula hirtiflora*.

From 1500 to 1700 m *Pinus nigra* is gradually replaced by *Pinus heldreichii*, which becomes the main forest-former to the timber-line. It is a rather open woodland, heavily grazed and often reduced to open stony ground or to a scrub of *Buxus sempervirens*, and *Juniperus communis*. In this type of vegetation the typical serpentine cruciferous plant *Peltaria emarginata* is abundant. Other species in more or less degraded forest of *Pinus heldreichii* are:

Silene schwarzenbergeri	Gentiana lutea
*Helleborus cyclophyllus	*Acinos alpinus
Euphorbia myrsinites	Scabiosa webbiana
*Daphne oleoides	Echinops ritro
Moneses uniflora	Carlina acanthifolia

On schists and limestones the pine forest is usually replaced by forest of beech and *Abies borisii-regis*. On the eastern slopes of Smolikas, 5 km south of the small village of Paraskevi, is probably the finest forest of this type to be seen on the mountain, covering the ravines surrounding a conspicuous limestone peak to altitudes of 2000 m. The field layer consists mainly of northern and central European species such as:

*Ranunculus platanifolius	Lapsana communis
Cardamine bulbifera	Lilium martagon
Geranium sylvaticum	*Polygonatum verticillatum
Sanicula europaea	Corallorhiza trifida
Galium odoratum	Neottia nidus-avis
Mycelis muralis	

Balkan species occurring are:

Geranium reflexum	Euphorbia heldreichii
G. versicolor	Campanula trichocalycina

In the clearings on moist ground sub-alpine meadows are developed, and here stout perennials such as *Cirsium appendiculatum,

Doronicum austriacum, and *Veratrum album* are dominant. Along the small rivulets stands of *Tozzia alpina* are found; a very rare plant in Greece.

A few isolated individuals of *Pinus heldreichii* occur as high as 2200 m. Here begin the alpine meadows and screes, which are heavily grazed by the sheep which are everywhere, even on the highest peaks. Many species restricted to northern Greece and Albania are found in this alpine zone. Species typical for north- and east-facing rocks in the summit area are: *Silene pindicola†, *Aubrieta gracilis var. degeniana, Sedum athoum, *Saxifraga taygetea, *S. exarata, S. paniculata, *Potentilla speciosa, Artemisia eriantha, *Doronicum columnae, and the strictly serpentine species *Cardamine plumieri*. On drier rocks are found the small scrubby *Euphorbia glabriflora* and *Asperula aristata*.

One of the finest species of the screes is *Viola magellensis*. In July and August the beautiful pink flowers of this violet are in abundance on the east-facing slope of the summit cone, along the path from Samarina, at an altitude of about 2300 m. Of the numerous other species found in scree and on stony ground the following may be mentioned:

Cerastium 'smolikanum'	Linum punctatum subsp.
Rumex scutatus	pycnophyllum
Arenaria conferta	*Euphorbia herniariifolia
Minuarta verna	*Valantia aprica
Silene vulgaris	Thymus teucrioides
*Dianthus haematocalyx	*Campanula hawkinsiana
*Cardamine glauca	*Asyneuma limonifolium
Alyssum smolikanum†	*Fritillaria epirotica
*Bornmuellera baldaccii	Muscari botryoides
T. epirotum†	Lilium albanicum
*Aethionema saxatile	Tulipa sylvestris subsp.
*Iberis sempervirens	australis

Along streams and brooks, and round the margins of pools and depressions, where ground moisture is often held throughout the growing season, are closed communities dominated by sedges and grasses, and usually grazed by sheep to resemble well-trimmed lawns. Balkan species such as *Soldanella pindicola†, *Pinguicula hirtiflora, *P. balcanica*, and *Veronica serpyllifolia* subsp. *humifusa* can be found as well as *Parnassia palustris* and *Alchemilla acutiloba*.

Northern Pindhos

The two good roads which now cross the northern Pindhos Mountains traverse wild and sparsely populated country that is otherwise difficult of access. Much of the country is heavily forested with Black Pine and Fir, and less commonly Beech. Rough forest roads, passable only during the summer months, when the snow has melted and the ground has dried out, trail off up remote intermittently cultivated valleys, or across wooded slopes where earth-slides, rock-falls, and unstable slopes make permanent roads so difficult to maintain. It is not a beautiful country; there are few bold features, but in the summer from mid-June onwards there is a wealth of colour from the many attractive herbaceous species which grow in the clearings, by the roadsides, and on rough field verges.

The recently constructed road running northwards from Ioannina through Konitsa and Eptahorion (Eptachori) and over the main ranges to Neapolis and Kozani should be visited in early July for the height of the flowering period. By comparison with the burnt and dried-up Mediterranean coast and hills at this time of the year, it presents a welcome contrast of flowery meadows, banks, and forest-clearings. By the roadsides are stands of the majestic, pale-flowered thistle *Cirsium candelabrum* mixed with mulleins and with foxgloves such as *Digitalis grandiflora*, *D. ferruginea*, *D. lanata*, and several white-flowered Compositae including *Achillea nobilis*, *A. pannonica*, *Tanacetum corymbosum*, and the yellow-flowered *Achillea coarctata* and *Anthemis tinctoria*. Other upstanding and distinctive perennials include:

Lychnis coronaria	*Haplophyllum patavinum*
Dianthus cruentus	*Onosma heterophylla*
D. viscidus	Anchusa officinalis
Potentilla detommasii	*Salvia sclarea*
P. argentea	*S. nemorosa*
Onobrychis ebenoides	*Knautia orientalis*
*O. alba	Scabiosa webbiana
Linum hirsutum	*Tremastelma palaestinum*
L. nodiflorum	*Asyneuma limonifolium*
L. aroanium	Scorzonera parviflora

On earthy unstable slopes are notably: *Salvia candidissima*, *Scabiosa crenata*, and *Dianthus sylvestris*. While on more rocky ground the following stonecrops occur: *Sedum sediforme*, *S. ochroleucum*, *S. album*. In the forest itself there is less of interest but in glades the following should be mentioned:

*Potentilla recta
*Lathyrus laxiflorus
*L. grandiflorus
L. venetus
*Trifolium alpestre
T. patulum
Trifolium medium subsp.
   balcanicum
Geranium versicolor
Hypericum spruneri

Stachys scardica
Prunella laciniata
*Linaria genistifolia subsp.
   dalmatica
*L. peloponnesiaca
Knautia drymeia
Campanula persicifolia
C. patula
Anthericum liliago

Orchids include: Anacamptis pyramidalis, *Himantoglossum hircinum, Cephalanthera rubra, etc.

The better-known southern road, which crosses the Pindhos range at the Katara pass (1705 m), a few kilometres east of the village of Metsovon, is also a good hunting-ground. On banks in clearings, two unusual white-flowered crucifers can be seen: *Peltaria emarginata, *Bornmuellera tymphaea, with the white *Minuartia baldaccii and yellow Cytisus procumbens and *Chamaecytisus supinus. In clearings among the beech trees are *Helleborus odorus, *Cardamine glauca, Thlaspi praecox, Iberis pruitii, and *Doronicum columnae, and the tiny Muscari botryoides, also *Fritillaria pontica and Tulipa sylvestris subsp. australis. In more sheltered places the more northern species Cardamine bulbifera, Moneses uniflora, Pyrola chlorantha, and Paris quadrifolia may be found. Perhaps the choicest plant is *Daphne blagayana growing under box bushes along the sides of streams. Here also in wet flushes are: *Pinguicula balcanica and Soldanella pindicola†. On more open ground under pines grow *Dianthus haematocalyx subsp. pindicola, *Verbascum mallophorum, Scabiosa webbiana, *Carduus tmoleus subsp. armatus, C. macrocephalus, and Allium flavum.

Grassy meadows in early June are filled with masses of sweet-

---

8 **Olympus** The giant of the Balkans, standing in isolation, has an endemic flora of rare distinction. July and August are the months to see the Olympian Potentilla, Campanula, Omphalodes, Aquilegia, Viola, and saxifrages, while its most famous endemic, Jankaea heldreichii, flowers on rocks in the forests earlier in the year.

9 **Parnassos** The tree-line is at about 1800 m where scattered trees of Greek Fir and Stinking Juniper can still survive. Above stretches a lunar landscape of glaring white limestone with clumps of Astragalus, Daphne, Nepeta, Marrubium, and the prostrate Cherry, Prunus prostrata. Summer grazing is very intensive on all Greek mountains though some plants have become unpalatable to animals.

scented *Narcissus poeticus*, with clumps of *Cardamine pratensis*, *Chamaecytisus supinus*, *Geranium tuberosum*, *Linum hologynum*, *Viola tricolor* subsp. *macedonica*, *Primula veris*, *Symphytum bulbosum*, *Ajuga reptans*, *Campanula tymphaea*, *Veratrum album*, *Asphodelus albus*, *Dactylorhiza sambucina*, but heavy grazing and drying summers make the riot of colour short-lived.

In this difficult country there are other mountains to explore, notably Peristeri (2295 m) and Tzoumerka (2469 m) south of Metsovon; also Grammos (2523 m) on the Albanian border; Vourinos (1866 m) south of Siatista; Pieria (2194 m) west of Katerini; and north of Drama in eastern Macedonia towards the Greek–Bulgarian border, Boz-Dagh (Falakron; 2232 m), but the traveller should be warned that permission may not be granted for visiting some of these more remote mountains situated near the frontiers.

5. OLYMPUS, OSSA, AND PILION

Olympus

Mount Olympus is a giant of the Balkans. Standing in isolation, situated on the borders of Thessaly and Macedonia in northern Greece, it is undoubtedly one of the most interesting botanical localities of the whole of our area.

The combination of long geographical isolation, an unusual range of climatic conditions—where a high mountain is situated close to the sea—and the meeting of Mediterranean and central European floras, has resulted in a unique vegetation.

Over 1500 species of seed plants have been recorded on its flanks and summits; but more significant is its role as a major refuge for some of the most exciting ancient endemic species of Europe. About 18 to 22 unique species—depending on different botanists' interpretation of the concept of a species—are found in this relatively small area, and all are exclusively Balkan species: a high proportion of very unusual plants.

The summit of Olympus, Mytikas (2917 m), is the second highest in the Balkans, and there are two other summits reaching 2900 m. They

---

10   **Lake Prespa**   The frontiers of Greece, Albania, and Yugoslavia meet somewhere in the waters of this lake. The mountains to the right are part of the Galicica National Park where *Thlaspi bellidifolium, Viola allchariensis, Crocus cvijicii,* and *Colchicum hungaricum* are to be seen. The Macedonian wayside flora of disturbed ground is of considerable interest and variety in July.

NORTH PINDHOS, Mountains

1. *Achillea clavennae* 1420b   2. *Erigeron glabratus* 1370c   3. *Silene pindicola*
170f   4. *Campanula hawkinsiana* 1338c   5. *Dianthus petraeus* 189a   6. *Potentilla
speciosa* 445c   7. *Satureja parnassica* 1152f   8. *Astragalus vesicarius* 531

9. *Hypericum rumeliacum* 772f **10.** *Bornmuellera baldaccii* 329b **11.** *Senecio squalidus* 1452a **12.** *Viola magellensis* 786f **13.** *Sempervivum marmoreum* 384k **14.** *Saxifraga taygetea* 403b **15.** *Cardamine glauca* 312a **16.** *Fritillaria graeca ssp. thessala* 1622c **17.** *Alkanna pindicola* 1061g

lie only 20 km from the sea to the east and rise abruptly from a plateau of about 300 m. The high summits of Olympus form a cone which at the 1000 m contour has a diameter of about 20 km. This cone is dissected by deep valleys, the most important of which are the wide Enipefs valley in the east and the narrow Papa Rema and Xerolaki ravines in the north. Because of its proximity to the sea, the eastern and northern slopes receive most moisture and have a more luxuriant, often forest, vegetation than that of the drier southern and western slopes. Species belonging to the alpine and upper montane zones, however, often descend far below their normal range in the protected valleys.

Olympus is almost exclusively limestone, dramatically folded in the summit area, and weathered into extensive screes in much of the alpine zone. The limestone is very porous, and although there is a fair amount of precipitation—the rainfall and snow vary within wide limits from year to year—the alpine area becomes very dry in summer. The maximum precipitation recorded at the Meteorological Station on Ay. Antonios (2815 m) between 1966 and 1973 for the months July to September was 542 mm, and the minimum 86 mm. Melting snow also accounts for some moisture, but melt-water quickly disappears into fissures and the ground becomes dry only a short distance below a melting snowfield. Some snow may remain in sheltered hollows throughout the summer; there are no true glaciers.

Access to Olympus has improved greatly in recent years. A good road from the village of Litokhoron in the east takes one to the refuge at Stavros (944 m), and then across the flank of the main valley, through forests of Black Pine, to the spring at Prionia, where there is a parking-place and a taverna in the summer. This lies in the heart of the Enipefs valley, and from here a steep climb of about $2\frac{1}{2}$ hours brings one to the H.A.C. Refuge A at 2100 m, perched on a spur of the mountain among magnificent Bosnian Pines (open during the summer; sleeps 60 persons; supplies food and basic equipment). A further steep climb up a rough track for $2\frac{1}{2}$–$3\frac{1}{2}$ hours enables one to climb to the summits of Mytikas (2917 m), Skolio (2911 m), and Prof. Ilias (2786 m). High walking over barren and very rough limestone screes, for a further hour or so, is necessary to reach the outlying peaks such as Ay. Antonios, Kalogeros, and Kakavrakos. Good weather, stout walking-boots, and plenty of time are required to visit all these peaks (see H.A.C. pamphlet on Olympus).

Refuge C is situated at the foot of the summit of Prof. Ilias at 2650 m. It can be reached in about 6 hours by following a signposted path which branches off from the road about an hour's walking above Stavros. No water is to be found on this route. Refuge C can also be reached from Refuge A in about 2 hours.

To the south of the mountain a road leads westwards from Leptokaria to Karia, past the villages of Sikaminea and Kriavrissi. Seven kilometres beyond Kriovrissi a road branches off to the right, northwards, and climbs to Refuge B at 1850 m, on the southern slopes of Ay. Antonios. This road can also be reached from the south-west, from the town of Elasson via the villages of Kallithea and Olimbias. The area around Refuge B is deforested and dominated by box scrub, but there is an interesting flora in some of the ravines.

The north-western side of the mountain can be reached by following the main road from Katerini towards Ay. Dimitrios, and after 21 km turning left to the village of Petra. From here a forest road leads to Kokinoplos at 1100 m. Eleven kilometres from Petra, at an altitude of 710 m, another forest road branches off to the left and follows the eastern side of Xerolaki Rema until it crosses the upper part of the ravine and returns on the western side to the Petra road. This road passes through fine stands of pure Black Pine forest.

The mouth of the Papa Rema ravine can be reached by road from the village of Vrondou. Above it, a path follows the eastern side of the ravine. It is one of the finest plant-hunting sites on the mountain with many of the rare plants growing at altitudes of only 500–800 m.

About forty square kilometres of the summit area of Olympus and its eastern slopes down to the spring of Prionia were proclaimed a National Reserve area as long ago as 1938, but so far the laws of the Royal Decree have never been fully implemented, though restrictions on hunting have saved the Chamois. They may still occasionally be seen in the upper zones. There is a rich bird life which includes such large birds of prey as the Imperial Eagle and the Bearded Vulture or Lammergeier.

Few if any of the rare plants on Olympus are at present seriously endangered, as most of them occur in adequate quantities in inaccessible places. But the alpine habitat is very vulnerable and the rapidly increasing numbers of visitors are a matter for concern. At present there is apparently little serious effort to conserve this unique natural area, but steps must be taken to do so before it is too late. There is every reason to show the utmost care with camp-fires, while collecting of specimens should generally be prohibited or be strictly limited to scientific studies.

In general terms, the flora of Olympus consists of three basic components: Mediterranean species, central Europe species, and Balkan mountain species; and these may be found in close proximity—a rare occurrence.

The main vegetational zones are often clearly distinguishable, but variations in local climate and topography may cause considerable overlapping and intermingling. The zones may be outlined as follows:

Mediterranean
        0–300 m   Phrygana of the coastal plateau
     300–700 m   Maquis of the lower slopes

Central European (montane)
   700–1400 m   Forests of Black Pine, Greek Fir, and Beech

Balkan mountain (montane, alpine)
  1300–2100 m   Forests of Beech and Bosnian Pine
  2100–2300 m   50–80% cover of low shrubs and herbaceous
                     plants
  2300–2900 m   10–40% cover of low prostrate or cushion plants

Many of the 1500 species of Olympus are confined to the maquis and phrygana, or to agricultural land at low altitudes. As can be seen from the graph below, about 600 species have been recorded above an altitude of 1200 m. From here on there is a regular and fairly slow decline in the number of species to 1800 m, followed by a more rapid fall in numbers to 2200 m when many of the forest species disappear.

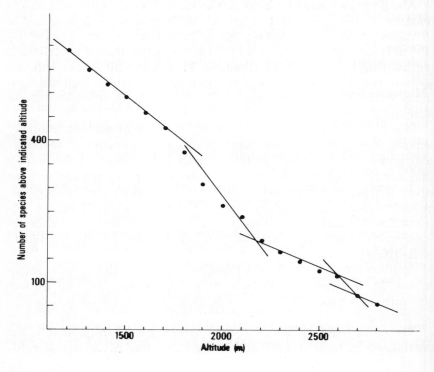

There is another rapid decline between 2600 and 2700 m, when alpine meadows disappear. Above 2700 m the terrain is almost exclusively rocks and scree and only 78 species have been recorded here, though careful search might reveal a few more.

Mediterranean maquis. The foothills between about 300 and 700 m are covered with a well-developed maquis, generally 2–3 m high, dominated by the Kermes Oak, *Quercus coccifera*, and including other evergreen shrubs like *Juniperus oxycedrus* (occasionally with the parasitic *Arceuthobium oxycedri*), *Quercus ilex*, *Erica arborea*, *Arbutus unedo*, *A. andrachne*, and *Phillyrea latifolia*. Common deciduous shrubs are: *Ostrya carpinifolia*, *Cotinus coggygria*, *Pistacia terebinthus*, *Acer monspessulanum*, and *Fraxinus ornus*. The spring flora includes many small annuals like: *Cardamine graeca*, *Arabis verna*, *Clypeola jonthlaspi*, *Sherardia arvensis*, *Valantia muralis*, *Parentucellia latifolia*, and *Crepis sancta*, and bulbous and tuberous species like *Tulipa sylvestris* subsp. *australis*, *Iris reichenbachii*, *Ophrys scolopax* and *O. sphegodes*, *Orchis simia*, *O. quadripunctata*, *O. provincialis*.

Deciduous oak scrub, predominantly of White Oak, *Quercus pubescens*, replaces the maquis in damper localities with small deciduous trees of *Carpinus orientalis*, *Celtis australis*, *Cercis siliquastrum*, *Cornus mas*, and *Pyrus amygdaliformis*, often mixed with evergreen species. *Platanus orientalis* is common along streams, and *Paliurus spina-christi* forms thickets in openings and disturbed ground.

Montane forests of Black Pine, Macedonian Fir, and Beech. The first trees of the Black Pine, *P. nigra*, appear at 300–400 m, and from 600–700 m up to 1400 m they form forests in many places, especially on the dry ridges. The Macedonian Fir, *Abies borisii-regis*, is less abundant and tends to avoid dry ridges. It reaches altitudes of up to 2000 m on the northern slopes.

Between 700 and 1200 m a mixed forest of Black Pine and Macedonian Fir is found with scattered trees of *Taxus baccata*, *Carpinus betulus*, Beech, Sweet Chestnut, *Ulmus glabra*, and Holly. The lower layer consists of more central European deciduous trees and shrubs such as *Corylus avellana*, *Acer campestre*, *Cornus mas*, *Hedera helix*, and *Sambucus nigra*, with occasional Mediterranean evergreens such as *Juniperus oxycedrus* and *Arbutus andrachne*. The Mistletoe, *Viscum album*, is a common parasite on both pines and firs, but is not found on deciduous trees. In central Europe, by contrast, its hosts are usually deciduous trees. The related *Loranthus europaeus* is an occasional parasite on deciduous oaks and Sweet Chestnuts. Fine examples

119

of such forest may be seen around the ruins of the monastery of Ay. Dionyssios in the Enipefs valley. A common, often dominant, under-shrub is *Staehelina uniflosculosa*, with silvery-grey leaves and clus-ters of pink flowers, together with *Genista radiata*, *Cistus incanus*, and *Daphne laureola*.

The following perennials occur largely in clearings:

Thalictrum minus	*Phlomis samia*
Astragalus monspessulanus	*Salvia ringens*
Geranium sanguineum	*Verbascum graecum*
Ferulago sylvatica	*Campanula persicifolia*
Primula veris	*Asphodeline liburnica*
P. vulgaris	*Lilium calcedonicum*
Teucrium chamaedrys	*Fritillaria messanensis*

One could well start one's exploration of the mountain proper with a visit to the waterfall at Prionia. Water in considerable volume gushes from under vertical cliffs and pours down limestone gullies. On mossy ledges, dripping with water, are clusters of *Pinguicula hirtiflora*, with pale-violet flowers and glistening rosettes spreading over the moss. *Trachelium jacquinii*† has pale-blue globular flower clusters, and trails from the dry rocks above. Handsome clumps of *Campanula versicolor*, with pale-blue salver-shaped flowers with darker blue centres also grow here, with the blue Willow Gentian, *G. asclepiadea*, a central European alpine plant. But the choicest plant, well out of reach on the cliffs, is none other than Olympus's most famous, *Jankaea heldreichii*†. Its silvery rosettes fit snugly into mossy crevices in the vertical rock-face under the shade of the beech trees, and in early summer it must be a most wonderful sight with its large violet chalices hanging from every ledge and crevice. It is a 'living fossil'—a Tertiary relict—one of five surviving European members of an otherwise tropi-cal family, the Gloxinia family. It has somehow managed to hold out, probably for millions of years, on the cliffs of Olympus. Fortunately, *Jankaea* has a wide altitudinal range, occurring from about 400 to 2400 m on damp shady rocks in many places on the mountain. In suitable localities, for instance in the Papa Rema and Xerolaki ravines, it may occur in great quantity. It flowers from early May to August, according to its altitude and exposure.

Exposed limestone rocks in this zone have an interesting assemb-lage of species including:

Sedum ochroleucum	*Onosma echioides*
S. sartorianum	*Teucrium chamaedrys*
Asperula purpurea	*T. polium*

Micromeria juliana	*Achillea ageratifolia*
Campanula versicolor	*Inula verbascifolia*
Trachelium jacquinii†	Allium flavum

Beech woods do not form a continuous belt around the mountain, but they are found in patches in the valleys where the soil is deep and there is a fair amount of moisture. The path from Prionia to Refuge A passes through a fine stretch of pure beech wood between 1200 and 1600 m.

Shrubs in these forests include: *Cotoneaster nebrodensis*, *Euonymus latifolius*, and *Ruscus hypoglossum*, in addition to other more widespread northern species.

Herbaceous species are mostly central European and include:

Cardamine bulbifera	Monotropa hypopitys
Arabis turrita	Galium odoratum
Saxifraga rotundifolia	G. rotundifolium
Potentilla micrantha	*Asperula muscosa†*
Mercurialis ovata	Convallaria majalis
Viola reichenbachiana	Polygonatum odoratum
V. riviniana	Cephalanthera rubra
Sanicula europaea	Limodorum abortivum
Pyrola rotundifolia	Neottia nidus-avis
Orthilia secunda	

A Balkan mountain forest zone occurs above the Beech and Black Pine zone, from about 1300 to 2400 m, and it is dominated by the Bosnian Pine, *P. heldreichii*; but weather-beaten bushes of this species may grow as high as 2600 m. This pine, with characteristically pale-grey bark and massive conical shape, forms rather open forests, possibly as a consequence of repeated fires. It is well developed around Refuge A.

The flora of this zone, comprising as it does many Balkan species, is of great interest and beauty. For example, in the rocky gullies which fall away on each side of the pine-covered ridge on which Refuge A stands there are many of the best species that Olympus has to offer. On the cliffs of these gullies grows in some abundance one of the most unusual and beautiful of all violets, *V. delphinantha†*, with its long-spurred pink flowers which line the cracks and crevices. Often growing near at hand is the beautiful blue-trumpeted *Campanula oreadum*, one of the mountain's choicest endemics. Below the cliffs on the scree verges, where snow often lies late, is the delicate blue-and-white flowered *Aquilegia amaliae†*, while in the scree itself may be found the white-flowered *Cardamine carnosa* and *Saxifraga glabella*.

Perhaps with good fortune you might catch sight of another plant, famous though rather uncommon on Olympus, its pale-blue flowers with glaucous leaves peeping from shaded crevices—it is the rare *Omphalodes luciliae†.

Common gully plants include: *Silene chromodonta*, *Geranium macrorrhizum*, *Smyrnium rotundifolium*, *Achillea grandifolia*, and *Doronicum columnae*. On more consolidated scree where some soil has accumulated is a very rich flora including:

*Cerastium banaticum
*Saponaria bellidifolia
*Dianthus minutiflorus
*D. haematocalyx
*Ranunculus sartorianus
*Alyssoides utriculata
 Alyssum corymbosum
*Iberis sempervirens
*Sedum magellense
*S. ochroleucum
*Jovibarba heuffelii
 Rosa heckeliana
 Chamaecytisus polytrichus
*Anthyllis aurea
 Hypericum olympicum
*Viola gracilis
 Gentiana verna subsp.
  pontica
*Gentianella crispata
*Asperula aristata subsp.
  thessala

*Teucrium montanum
*Marrubium thessalum
*Sideritis scardica
*Thymus cherlerioides
 T. sibthorpii
*Linaria peloponnesiaca
*Globularia meridionalis
 C. albanica
*Edraianthus graminifolius
*Erigeron epiroticus
*Achillea holosericea
*Centaurea pindicola
 Lactuca graeca†
 Hieracium pannosum
*Linum spathulatum, L.
  elegans
*Cotoneaster integerrimus
*Daphne oleoides
*Arctostaphylos uva-ursi
*S. scardica

The Balkan alpine vegetation occurs above the tree-line; there are several distinctive plant communities developed respectively on rock crevices, screes, solifluction soils, and by snowbeds.

The following species are primarily found on rocks, though they may occur elsewhere:

*Arenaria cretica
*Arabis bryoides
*Aubrieta gracilis
 Draba athoa
 Saxifraga moschata

*S. scardica
*S. spruneri
*S. sempervivum
*Omphalodes luciliae†
*Campanula oreadum

Two unique associations of rocks and cliffs have been described by the plant sociologists on Olympus. At altitudes of between 1200 and 2200 m a characteristic species is *Saxifraga sempervivum*. At higher altitudes from 2300 m to the summit cliffs another association is dominated by the Olympus endemic *Potentilla deorum* with *Saxifraga scardica*. Over large areas of moderately sloping ground where the rock has been weathered and fragmented into extensive screes the landscape has a desolate almost lunar appearance. Few species can tolerate these extreme conditions, and the plant coverage is often as low as 5–10 per cent.

The most interesting and representative species here are:

Arenaria conferta
Cerastium 'theophrastii'
*Ranunculus brevifolius
*Alyssum handelii
Rhynchosinapis nivalis†
*Sedum magellense
*Anthyllis vulneraria subsp.
   pulchella

*Euphorbia capitulata
*E. herniariifolia
Viola 'striis-notata'
Galium anisophyllon
*Thymus cherlerioides
Linaria alpina
Veronica thessalica
Achillea ambrosiaca†

Terraced solifluction soils, where there is a slow downhill movement as a result of alternate freezing and thawing, may develop in places where the ground slopes some 5–10°. Such areas are often dominated by the grass *Sesleria korabensis*. Other species include:

*Cerastium banaticum
*Iberis sempervirens
Viola heterophylla subsp.
   graeca

*Acinos alpinus
*Senecio squalidus
*Centaurea pindicola
*Carduus tmoleus

A well-defined plant community is found on flat ground and in shallow depressions, where the snow does not melt until July and there is a good layer of fine soil. Such patches can be recognized from a long distance by their fresh green colour owing to the continuous cover of grasses and low herbs which in late summer is reduced to close-cut swards by grazing sheep. The dominant grass is usually *Alopecurus gerardii*, and other species are:

Beta nana†
Herniaria parnassica
Trifolium parnassi
T. pallescens
Gentiana verna

Myosotis suaveolens
*Erigeron epiroticus
Omalotheca supina
Scilla nivalis
*Crocus veluchensis

The best time of year to see most of these high alpine species is

123

during the last weeks of July and the first weeks of August, when to traverse the high slopes of Olympus is veritably to walk 'amongst the Gods' with the far distant sea beyond—an experience long to be remembered.

Ossa and Pilion

Both mountains are easily accessible by road and are well worth visiting, particularly between June and August. Like Olympus they lie close to the sea, receiving moisture-laden winds from the north and east, and supporting heavy forests on the seaward flanks. In the heat of summer, when the surrounding plains become parched and devoid of flowers, the mountain glades and valleys are full of colour.

The summit of Ossa (Kissavos) is at a height of 1978 m. The forest vegetation ceases at about 1600 m and above this a Mediterranean sub-alpine zone of limited extent occurs which contains some interesting Balkan species. Pilion is lower (1651 m) and the summit does not rise above the tree-line. Both have forests of Beech, Macedonian Fir, and Sweet Chestnut.

Ossa can be reached from the village of Sikourion in the plain. A steep exposed road leads through heavily grazed slopes covered with depauperate maquis to the village of Spilia at about 900 m. From here a rougher road traverses round the north of the mountain towards the sea, overlooking the valley of Tempi a long way below. North of Spilia is a fine stand of scattered Kermes Oaks, *Quercus coccifera*, with dark crowns up to 15 m high and trunks nearly 1 m across—unusually well-grown trees. As one traverses further east towards the sea, stands of Macedonian Fir and Beech appear, and on the seaward slope these form dense forests. A continuation of this road leads to the H.A.C. refuge at about 1600 m, above the tree-line. It is about 1–1½ hours' climb to the conical summit of Ossa, across heavily grazed grasslands with low spiny bushes of *Astragalus angustifolius*, and *Berberis cretica*, with the silvery-leaved herbaceous *Marrubium thessalum* and the blue *Eryngium amethystinum*, still flowering at the end of August. At this time of year the first flowers of *Colchicum bivonae*, with pink chequered tepals, were appearing in the dried grass.

A forest road from the village of Karitsa on the eastern side of Ossa climbs up to the O.T.E. station not far below the summit. It rises through heavily forested slopes, where the Laurel, *Laurus nobilis*, is frequent in ravines at low altitude, while at 300–400 m there are extensive, almost pure stands of Sweet Chestnut, *Castanea sativa*, which are occasionally parasitized by *Loranthus europaeus*. Above about 400 m Beech becomes more and more frequent. Other trees like

Taxus baccata, Acer pseudoplatanus, Aesculus hippocastanum, and *Ilex aquifolium* occur. The Horse Chestnut, *Aesculus hippocastanum,* is generally thought to be native only in a small area on the borders of Greece, Albania, and Yugoslavia, but here it is found far from human habitation, and does not appear to have been introduced. At higher altitudes the Macedonian Fir becomes more frequent, often mixed with **Juniperus oxycedrus, Ostrya carpinifolia, Sorbus torminalis, *Crataegus orientalis, *Prunus cocomilia, Acer monspessulanum,* and **A. hyrcanum,* while in August *Cyclamen hederifolium* flowers in abundance among the rocks which cover the floor of the forest.

Below the O.T.E. station are mountain meadows with **Ranunculus sartorianus, *R. millefoliatus,* and *Tulipa sylvestris* subsp. *australis* flowering in May. The summit has rocky outcrops and some screes, and from it are magnificent views, with the 'steps of the giant', Olympus, Ossa, and Pilion ranging along the eastern coast of Greece.

*Saxifraga scardica is common in rock crevices; it shows a pronounced variation, the rosettes becoming smaller and denser with increasing altitude and exposure. One of the most conspicuous plants of rocky places is *Erodium absinthoides* with large red flowers. In depressions and on sheltered slopes there are patches of mountain meadow with species like *Ranunculus ficaria, Corydalis solida, Scilla nivalis, *Crocus veluchensis, Gagea minima,* and *Ornithogalum oligophyllum*—all flowering at the end of May, often very close to melting snowfields. By the end of August there were still some species in flower in the summit area including: **Dianthus haematocalyx, *Jovibarba heuffelii, *Sedum sartorianum, *Eryngium amethystinum,* and *Centaurea attica* subsp. *ossaea†.*

Pilion, like Ossa, with its heavy afforestation and well-developed Mediterranean maquis, makes a worthwhile visit during the summer months. There are fine forests of Sweet Chestnut on the eastern flank and Beech woods up to the summits. A good road from Volos to Hania climbs over the summit ridge to join a narrow contouring road which traverses the seaward slopes through heavily wooded valleys high above the sea. From the village of Hania a road climbs steeply northwards towards the highest point where there is a naval installation. Below this on the open hillsides can be found what I consider the finest of all Balkan campanulas, **C. incurva†*—high praise indeed in a country rich in beautiful campanulas with up to 90 species, over half of which are endemic. Domes of glorious pale blue-violet flowers, sometimes a metre across, with several hundred large bells opening to the sun at the same time, grow on the rocky hillside. Each erect, flask-shaped flower is as much as 5 cm across and 7 cm long. Like the Canterbury Bell which it resembles it is a biennial but infinitely more

OLYMPUS, OSSA, PILION

1. *Sideritis syriaca* 1115d 2. *Jankaea heldreichii* 1268c 3. *Centaurea pindicola* 1501e 4. *Campanula oreadum* 1330b 5. *Viola delphinantha* 781c 6. *Achillea ageratifolia* 1419a 7. *Saxifraga sempervivum* 410g 8. *Potentilla deorum* 445e 9. *Cardamine carnosa* 312b 10. *Aquilegia amaliae* 253d 11. *Saxifraga scardica* 398e

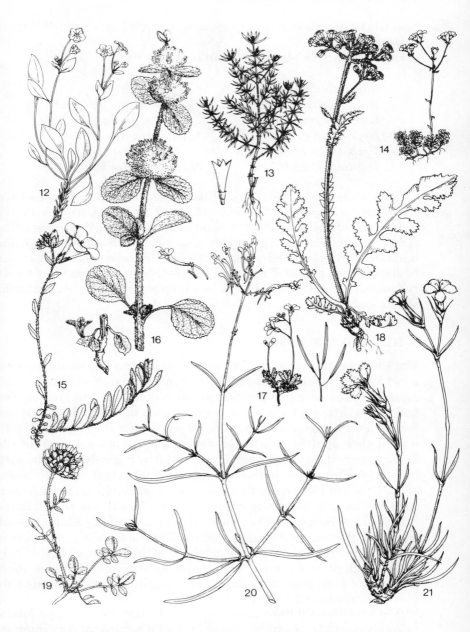

12. *Omphalodes luciliae* 1047b   **13.** *Asperula muscosa* 1017f   **14.** *Saxifraga glabella* 409a   **15.** *Linum spathulatum* 662b   **16.** *Marrubium thessalum* 1113d   **17.** *Arabis bryoides* 318a   **18.** *Achillea holosericea* 1421b   **19.** *Anthyllis vulneraria* ssp. *pulchella* 622   **20.** *Centranthus longiflorus* 1316b   **21.** *Dianthus haematocalyx* 193e

beautiful in form. It is a rare species restricted to a limited area of eastern Greece.

Other attractive plants noted in mid-July were:

*Silene compacta	*D. laevigata
Dianthus viscidus	*Siphonostegia syriaca†
*Lathyrus grandiflorus	Achillea grandiflora
*Hypericum olympicum	*Centaurea alba subsp.
Verbascum eriophorum	princeps†
Digitalis grandiflora	*Lilium chalcedonicum

Rare or endemic species on Pilion requiring protection include: *Siphonostegia syriaca†, Salvia eichlerana†, and Verbascum pelium†.

The hills above the village of Makrinitsa are worth exploring in the spring, when there are fine swards of scarlet *Anemone pavonina, and handsome chocolate-and-yellow-flowered Hemodactylis tuberosa can be seen in open places. Thickets of yellow *Spartium junceum vie with the massive clumps of the largest of our spurges, Euphorbia characias subsp. wulfenii.

6. RHODOPE MASSIF

The Rhodope Massif forms the core of the Balkan Peninsula, being a great wedge of ancient igneous and metamorphic rock including much crystalline limestone. It is situated north of the present Aegean Sea and shows little evidence of having ever been submerged. Thus it has formed a very ancient land mass. In all probability the Rhodope Massif was further uplifted and folded during the Alpine upheaval of the Tertiary period. At the present day it is relatively low-lying in the east, in Turkish Thrace, but builds up to nearly 3000 m to the highest peaks of the Rila and Pirin mountains in south-west Bulgaria. From here it runs northwards, becoming gradually lower in altitude, to end in the hills south of Belgrade. The new-fold mountains of the Alpine period, the Stara Planina to the north and the Dinaric folds of Macedonia to the west, have risen against the hard core of the Rhodope Massif. Only in the highest mountain groups, such as Rila and Pirin, is there evidence of local glaciation during the recent ice ages, and this only found above about 2000 m.

In consequence of their past geological history the Rhodope Mountains have acted as an important refuge for a number of very interesting and ancient endemic species, while at the same time the relative isolation of the area has resulted in the recent evolution of new species. Turrill[16] has estimated that of the 1850 species or so in this area, something like 286 species are Balkan endemics, of which 47 are known only for the Rhodope.

In addition, the Rhodope have acted as a natural bridge allowing north Aegean and Pontic species from northern Turkey to spread into Bulgaria and Macedonia. Turrill gives a list of over 50 species which may have come into Europe this way and suggests that there are probably others: migration can of course occur in both directions.

Rila and Pirin mountains

These mountains are situated in the south-west corner of Bulgaria. At lower altitudes their southern flanks have a sub-Mediterranean vegetation of forests of *Quercus pubescens*, *Carpinus orientalis*, *Fraxinus ornus*, etc., while the northern, western, and eastern slopes, which are subjected to a wetter, colder climate, have typically central European forests of beech, hornbeam, pine, and spruce. Above 1800–2000 m a distinctive alpine vegetation is developed which is snow-covered for several months each year. It consists of a surprising admixture of northern European, central European alpine, and Balkan mountain species, and includes even a few more southerly Mediterranean mountain species. Plant-hunting in these mountains can lead to unexpected finds, and one can see here unusual assortments of species growing together in close proximity, which elsewhere can only be seen scattered over diverse regions many hundreds of kilometres apart.

Rila Planina. This is a cluster of mountains situated about 115 kilometres south of Sofia. Here lies the highest of all the Balkan summits, that of Musala, (2925 m)—only a few metres higher than Olympus. There are several other peaks of over 2700 m, and in all more than 100 summits of over 2000 m. There are also something like 150 small lakes situated above 2000 m. In the Rila truly alpine conditions prevail above about 2000 m; there are heavy snowfalls in winter and considerable amounts of snow continue to lie as late as July. These mountains are well supplied with alpine huts and marked tracks, and they have become important centres for recreation: for skiing, mountaineering, and walking. They are easily reached via Samokov, and sound roads take the traveller up through the forests of pine and beech to the three main hotel and refuge complexes, one at Vada in the west, Malyovitsa centrally, and in the east the village of Borovec for the main valley leading to Musala. They can also be reached from the south, via Rila monastery or Yakoruda.

The valley of Malyovitsa is accessible by a road passing through the village of Govedarci, after which it climbs up through the pine forests and ends at a complex of hotels and restaurants at about 2000 m. This lies at the entrance to a deep glacial valley which leads up to the cleft peak of Malyovitsa (2729 m) at its head. These buildings lie in forests of Scots Pine and the long-coned Macedonian Pine, *Pinus peuce*, a distinctive endemic of the Balkans, together with some Spruce and Silver Fir.

In these forests can be found:

Aconitum variegatum	Geranium sylvaticum
A. lamarckii	*Pulmonaria rubra
*Ranunculus platanifolius	Senecio nemorensis
*Saxifraga rotundifolia	Doronicum austriacum
Chrysosplenium alternifolium	*Centaurea triumfetti

By June there is much of interest in the more open forest clearings and bushy places on the way to the alpine pastures. In particular the beautiful creamy-yellow *Aquilegia aurea*—the only yellow-flowered European species—which grows in damp places by streams. Flushes of *Caltha palustris* grow with scarlet *Geum coccineum* and pink *Polygonum bistorta*, together with a tall thistle, *Cirsium appendiculatum*, with rosy-purple flower heads and dark-purple buds, another Balkan endemic. *Soldanella hungarica*, distinguished by the stalked glands on the leaf stalk, and its deeply frilled, dark-purple corolla, grows on wet mossy banks by the side of streamlets.

In July the following can be seen in flower:

Rumex alpinus	Daphne mezereum
Lychnis viscaria	Viola biflora
*Thalictrum aquilegifolium	V. orbelica
Saxifraga paniculata	*Bruckenthalia spiculifolia
Parnassia palustris	Gentiana punctata
Rosa pendulina	*Globularia meridionalis
Geum montanum	*Pinguicula balcanica
*Geranium macrorrhizum	Cicerbita alpina
*Polygala major	

Higher up the valley the taller conifers are replaced by thickets of Dwarf Mountain Pine, *P. mugo*, where in June patches of snow still lie in the clearings. Here *Crocus veluchensis*, in all shades of mauve, pale violet to almost white, is abundant. The choicest plant of all to be seen in these mountains is undoubtedly *Primula deorum*, named from the Turkish for the mountain, Mus-Allah or Allah's Mount (hence Musala), and thus known as God's Primrose. It is an ancient endemic, or palaeoendemic plant quite unlike any other European species, and more like some Himalayan nivalid primula with its sheaf of tightly packed narrow erect leaves arising from a stout root-stock, and with a long-stemmed cluster of beautiful mauve flowers. It is only found in the Rila Mountains. Yet surprisingly, growing not far off in this valley is another primula, *P. farinosa*, which is a native of the Alps and of the far north of Scotland! There are a number of alpine species which spread no further south or east than to these mountains of south-west Bulgaria; they include among others: *Salix lapponum*, *Pulsatilla vernalis*, *Rhodiola rosea*, *Dryas octopetala*, *Primula minima*, *Soldanella pusilla*, *Gentiana punctata*, *Gentiana pyrenaica*, *Swertia perennis*, *Homogyne alpina*, and *Listera cordata*.

At higher altitudes in the peaty meadows, above the last of the Dwarf Mountain Pines, are patches of brilliant-blue *Gentiana pyrenaica*, which also has a unique distribution including the eastern Pyrenees, the Carpathians, and the central Balkans, but missing out the Alps altogether. Often close at hand grows *Gentiana verna* with a much wider and less fragmented distribution. The attractive dwarf cushion-forming *Dianthus microlepis*, with carmine, pink, or white flowers, is common. *Thlaspi praecox*, *Plantago gentianoides*, *Soldanella hungarica*, *Ranunculus crenatus*, a neat dwarf white-flowered buttercup with rounded fleshy leaves, and the small yellow-flowered *Pedicularis oederi* are conspicuous. Other choice plants of these higher regions include the Balkan endemic *Geum bulgaricum*, with handsome rather leathery leaves and drooping clusters of yellow flowers, like a Water Avens. On mossy banks and ledges are mats of *Primula minima* bearing stalkless rosy-purple flowers, and in rocky crevices are cushions of *Saxifraga juniperifolia* with yellow flowers, and clusters of the cream-coloured *Arabis alpina*.

Other alpines worthy of note include the following:

Salix herbacea	*Chamaecytisus eriocarpus*
S. reticulata	Astragalus australis
Rheum rhaponticum†	Empetrum nigrum
*Silene acaulis	*Gentianella bulgarica*
Pulsatilla vernalis	Armeria maritima subsp.
Draba carinthiaca	alpina
Rhodiola rosea	*Senecio abrotanifolius*
Saxifraga carpathica	S. doronicum
Geum reptans	Artemisia eriantha
Sibbaldia procumbens	Lloydia serotina

The alpine meadows commence above 2000 m, and as late as September there are still many plants in flower including: the dwarf dark-blue, solitary-flowered form of *Campanula alpina* subsp. *orbelica* which is common, with *Minuartia recurva*, *Dianthus* *microlepsis, *Primula halleri, *Androsace hedraeantha, *Pedicularis orthantha, *Veronica bellidioides, Plantago gentianoides,* and **Jasione laevis* subsp. *orbiculata.*

The highest peak of the Balkans, Musala (2925 m), is most easily reached from the village of Borovec. Here a chair-lift will take one up under the crest of Mount Deno (2790 m), from which a high-level track leads to the Musala hut at 2376 m. It is another 2 hours approximately to the summit. Alternatively, a path up the valley through the forests from Borovec takes one to the hut in 4 hours. Cliffs in the vicinity of the hut support some interesting chasmophytes including:

Oxyria digyna	**Symphyandra wanneri†*
*Silene lerchenfeldiana	**Senecio abrotanifolius* subsp.
S. waldsteinii	*carpathicus*
Saxifraga oppositifolia	

On the north face of Musala the following were recorded by Halda:[17]

Cerastium alpinum subsp.	**Geum bulgaricum*
lanatum	*G. reptans*
*Silene lerchenfeldiana	**Soldanella pusilla*
*Ranunculus crenatus	*Gentiana frigida*
Draba carinthiaca	**Pedicularis verticillata*
Sedum alpestre	*Omalotheca supina*
Rhodiola rosea	*Homogyne alpina*
*Saxifraga pedemontana	**Doronicum columnae*
subsp. *cymosa*	*Senecio doronicum*
S. retusa	

Other species present include *Minuartia recurva, Cerastium rectum,* and **Silene acaulis.*

Pirin Planina. These mountains lie in the south-west corner of Bulgaria, roughly 50 km south of Rila and about 150 km north of the Aegean Sea. The highest summit is Vikhren (2915 m). Like Rila, Pirin has a mixture of central European, alpine, Balkan mountain, and sub-Mediterranean species, but in addition there are some endemic species not found elsewhere. The mountains are mostly composed of granite and slates, but local areas of limestone lie between the summits of Vikhren and Kaminitsa as well as in the central Pirin. The presence of these limestone rocks and the more southerly position of the range, and closer proximity to the Aegean coupled with its relative isolation have made the Pirin Planina a most important refuge, like the Rila.

Pirin is less easily reached from the north than Rila, but can best be approached via Bansko, from which a rough but sound road rises through fine forests of pine including the Balkan Pine, *P. leucodermis*, beech, and spruce, up into the zone of Dwarf Mountain Pine, *P. mugo*, where it ends at the Banderitsa chalet. Here can be seen the oldest specimen of the Balkan Pine which is estimated to be 1200 years old. The Vikhren chalet is 500 m higher at 1950 m. As late as July snow-patches were lying above the chalet and *Crocus veluchensis*, *Scilla bifolia*, *Primula elatior* subsp. *intricata*, and *P. farinosa* were flowering.

The alpine meadows above the highest trees are rich in interesting species including:

Polygonum alpinum	Anchusa officinalis
Arenaria pirinica	*Sideritis scardica
Silene waldsteinii	*Verbascum longifolium
Dianthus tristis	*Digitalis viridiflora
Herniaria nigrimontium	Veronica saturejoides
*Aquilegia aurea	Bartsia alpina
Arabis ferdinandi-coburgi†	*Campanula alpina subsp.
*Saxifraga ferdinandi-coburgi	orbelica
*Chamaecytisus eriocarpus	*Jasione bulgarica
*Genista carinalis	*Centaurea napulifera
G. tinctoria	C. achtarovii
Oxytropis urumovii	Allium schoenoprasum
Daphne cneorum	

In higher meadows and wet banks can be found *Arenaria biflora*, *Dianthus microlepis*, *Ranunculus montanus*, *Linum capitatum*, *Ligusticum mutellina*, *Primula minima*, *Soldanella pusilla*, *Gentiana pyrenaica*, *Ajuga pyramidalis*, *Veronica alpina*, *Pinguicula balcanica*, *Plantago gentianoides*, *Homogyne alpina*, and *Leucorchis frivaldi*.

The limestones of Vikhren are exposed from about 2000 m upwards. Here the following saxifrage species are to be found: *S. ferdinandi-coburgi* with large hard mats of rosettes of narrow silvery leaves and compact clusters of bright-yellow flowers; *S. luteoviridis* with greenish-yellow flowers and glaucous crowded rosettes; *S. pedemontana* subsp. *cymosa* with palmately lobed leaves and white flowers; *S. marginata* with dense compact rosettes of tiny recurved leaves, with white or pale-pink flowers; and *S. moschata* with greenish flowers; as well as *S. oppositifolia* and *S. paniculata*. Other species include:

Silene waldsteinii	*S. acaulis
*S. roemeri	

Papaver pyrenaicum subsp.
   degenii
Aubrieta intermedia
Dryas octopetala
Potentilla apennina
**Viola grisebachiana*
V. perinensis
**Androsace villosa*
Gentiana verna
Galium stojanovii†

**Scutellaria alpina*
**Thymus cherlerioides*
**Pedicularis orthantha*
**Globularia cordifolia*
*Aster alpinus
Erigeron uniflorus
Leontopodium alpinum
   subsp. *nivale*
**Senecio abrotanifolius*

Artemisia eriantha (*petrosa*), a silky-haired dwarf wormwood, is typical of this mountain, often growing with **Anthyllis montana*. Local endemic species are *Brassica jordanoffii*, *Verbascum davidoffii*†, and *Centaurea achtarovii*. Other plants found not exclusively on limestone summits include:

Alyssum cuneifolium
**Thlaspi bellidifolium*
**Iberis sempervirens*
**Saxifraga exarata*

Armeria maritima subsp.
   alpina
Galium anisophyllon

Rhodope Mountains

These extensive mountains lie south of the plain of Thrace and form a southern bastion along which the Bulgarian-Greek border runs. They are very different in character from the truly 'alpine' Rila and Pirin mountains. They are ancient mountains with rounded summits which rise in wave upon wave, but rarely to above 2000 m. Most of them are heavily forested to their summits. Deep river valleys cut into these mountains and form gorges and shady ravines. Autumn and winter rains, cool spring weather, and abundant water from the melting snows all help to give a high humidity for much of the year—though the short summers may be hot and dry. These gorges and ravines are covered with dense forests of deciduous trees, with here and there an exposed cliff or rock wall towering above the steep forested slopes.

A good example of this type of country can be seen near Asenovgrad, which lies 19 km south of Plovdiv. Ten kilometres south of Asenovgrad is the eleventh-century monastery of Backovo, situated deep in the valley of the Asenitsa river. Above the monastery and the tree-covered slopes are the rocks of the Cervenata Stena or Red Wall, formed of Palaeozoic marble, rising to a height of 1500 m. This area has as recently as 1962 been declared a Botanical Reserve and is a protected site under the Ministry of Forestry. It is readily accessible to the public, and there are rough paths through the forests, but its many rare plants and animals are protected.

The reserve commences a short way above the monastery. Dense forests of beech, hornbeam, *Ostrya carpinifolia*, and *Quercus dalechampii* cover the cooler more north-facing slopes and deep ravines, and at higher altitudes they are replaced by the conifers *Pinus nigra* and *Abies alba*. Warmer slopes have a more sub-Mediterranean vegetation, with more open forests of *Quercus pubescens*, *Carpinus orientalis*, *Fraxinus ornus* with some *Pinus nigra*, *Juniperus oxycedrus*, and *Cotinus coggygria*. Much of the forest is young, consisting of brushwood or shiblyak, resulting largely from coppicing in recent times. As one climbs upwards through meadows, clearings, and forest verges a rich flora of both central European and Balkan species can be seen, including:

Cerastium moesiacum	*Lamium garganicum
Thalictrum flavum	Scutellaria altissima
*Corydalis bulbosa	Salvia glutinosa
Lunaria rediviva	*S. verticillata
*Alyssum murale	*Digitalis viridiflora
*Sedum sartorianum	*D. lanata
*S. hispanicum	*Pedicularis brachyodonta
S. annum	subsp. grisebachii
*Frangula rupestris	*Campanula lingulata
*Genista januensis	C. persicifolia
*G. carinalis	Aster amellus
Dictamnus albus	Tanacetum corymbosum
*Hypericum rumeliacum	Chondrilla pauciflora
H. annulatum	Polygonatum odoratum
Lysimachia punctata	Ophrys scolopax subsp.
Symphytum ottomanum	cornuta

In the sombre beech forests one encounters enormous moss-covered bolders and cliff-faces thickly covered with flat hairy rosettes of *Haberlea rhodopensis*, a very special plant of the Rhodope Mountains, with violet trumpet-shaped flowers. Like *Ramonda* and *Jankaea* it is a 'living fossil' which has survived for millions of years, probably in isolation in these same mountains, long after its nearest relatives have become extinct. Perhaps at the present day the closest relatives of these plants are to be found no nearer than among the foothills of the Himalaya! Other forest plants include such saprophytic plants as *Monotropa hypopitys*, *Neottia nidus-avis*, and *Corallorhiza trifida*, as

BULGARIA, RHODOPE RANGE, VITOSA, RILA, PIRIN
1. *Silene lerchenfeldiana* 171a   2. *Arenaria biflora* 128   3. *Centaurea napulifera*
1501g   4. *Saxifraga pedemontana ssp. cymosa* 409b   5. *Viola grisebachiana*
785f   6. *Symphytum tuberosum* 1053   7. *Saxifraga rotundifolia* 403   8. *Pedicularis*
orthantha 1253a   9. *Genista carinalis* 508g

10. *Gentianella bulgarica* 999c   **11.** *Primula farinosa* 943   **12.** *Lilium carniolicum* "jankae" (1619)   **13.** *Viola dacica* 785d   **14.** *Soldanella pusilla* (956) **15.** *Cardamine rivularis* 310b   **16.** *Lathyrus alpestris* 565h   **17.** *Jasione bulgarica* 1355g   **18.** *Senecio pancicii* 1452b

well as the green-leaved *Goodyera repens* and the very local *Cypripedium calceolus*†.

Other cliff plants, usually found in more open sunny aspects, include: *Sedum ochroleucum*, *Jovibarba hueffelii*, *Micromeria cristata*, *Trachelium jacquinii*, and *Inula verbascifolia*.

Higher up, open scree slopes formed from the crumbling cliffs appear to slice down through the forests. On the more stabilized screes are to be found swards of *Geranium macrorrhizum*, *Saxifraga rotundifolia*, and *Aquilegia vulgaris*, while on the exposed rocks on their flanks are the following: *Saxifraga stribrnyi* with nodding clusters of red-hairy flowers, *Saxifraga sempervivum*, and *Scabiosa rhodopensis*.

On the sunnier slopes in spring can be found the dark-violet *Pulsatilla halleri* subsp. *rhodopaea*, the yellow *Draba lasiocarpa*, and *Primula veris*, *Muscari botryoides*, and *Crocus flavus*. While in early summer *Lathyrus laxiflorus*, the blue-flowered *Linum hirsutum*, and *L. tenuifolium* with pink or almost white flowers, and a yellow-flowered flax of the *flavum* group, together with *Echium russicum*, *Teucrium chamaedrys*, and *T. polium*, can be found in flower.

Shrubs in the forest include: *Cotoneaster integerrimus*, *Euonymus verrucosus*, *E. latifolius*, *Rhamnus saxatilis*, *Syringa vulgaris*, *Lonicera xylosteum*, and *Amelanchier ovalis*, and in rocky places *Frangula rupestris*.

Many central European herbaceous species are found often in shady and bushy places, such as:

Aconitum variegatum	Ajuga genevensis
Anemone sylvestris	Echinops ritro
Cardamine bulbifera	Anthericum liliago
Lunaria rediviva	Lilium martagon
Daphne mezereum	Erythronium dens-canis
*D. oleoides	Scilla bifolia
Moneses uniflora	Paris quadrifolia
Pyrola chlorantha	Platanthera bifolia
Gentiana asclepiadea	

Stanev[18] has listed what he calls the interesting 'relict' species of this reserve as:

Taxus baccata	Prenanthes purpurea
*Arceuthobium oxycedri	Corallorhiza trifida
Ostrya carpinifolia	*Epipogium aphyllum
Syringa vulgaris	Epipactis atrorubens
*Haberlea rhodopensis	Neottia nidus-avis

The following species are found very sparingly elsewhere in Bulgaria:

*Ranunculus spruneranus, Daphne laureola, Lathraea rhodopea, Scabiosa rhodopensis, *Trachelium jacquinii* subsp. *rumelianum*, *Morina persica*, and *Goodyera repens*, while *Cypripedium calceolus*† is only found here.

The summit of the Red Wall is covered with open woods of Black Pine and on the steep shady cliffs below the summit are mats of rosettes with clusters of the beautiful violet trumpets of *Haberlea rhodopensis* in full bloom in mid-June.

Other rock-loving plants on these cliffs, mostly Balkan endemic species, include:

Silene flavescens
*Dianthus petraeus
*Saxifraga ferdinandi-coburgi
*Astragalus angustifolius
Hippocrepis comosa
*Linum thracicum
*Viola grisebachiana
*Teucrium montanum
*Marrubium velutinum
*Globularia cordifolia
*Morina persica

Scabiosa rhodopensis
Campanula rotundifolia
*Trachelium jacquinii subsp.
  rumelianum
*Inula verbascifolia subsp.
  aschersoniana
*Achillea clypeolata
*A. ageratifolia
*Centaurea napulifera subsp.
  pseudaxillaris

Other plants of interest recorded in this area are: *Silene compacta*, *S. bupleuroides*, *Dianthus pinifolius*, *D. petraeus*, *Polygala supina* subsp. *rhodopea*, *Daphne oleoides*, *Huetia cynapioides*, *Verbascum decorum*, *V. humile*, *Scabiosa triniifolia*, *Campanula moesiaca*, *Carduus adpresseus*, *Merendera attica*, *Fritillaria pontica*, *F. orientalis*, *Tulipa urumoffii*, *Sternbergia colchiciflora*, and *Iris reichenbachii*.

The Rhodope foothills bordering the southern flank of the plain of Thrace are of considerable botanical interest, being composed largely of calcareous rocks and acting as a refuge where a number of more southerly sub-Mediterranean species can be found. In the mountains near Asenovgrad, and in particular on Dobrustan mountain, are such species as:

*Anemone blanda
Astragalus pugionifer
Vicia bithynica
Lathyrus pannonicus
Medicago saxatilis
Euphorbia myrsinites
E. taurinensis
Ruta graveolens

*Polygala supina subsp.
  rhodopea
Helianthemum ledifolium
Peucedanum aegopodioides
Cyclamen hederifolium
Jasminum fruticans
Galium rhodopeum
Symphytum ottomanum

Alkanna primuliflora
A. stribrnyi
*Onosma thracica
Salvia aethiopis

Verbascum nigrum subsp.
abietinum
*Linaria pelisseriana
Tragopogon elatior
*Iris reichenbachii

Forest reserve Beglica

This lies in the western Rhodope, south-west of Plovdiv, near Batak. It is situated up in the hills at about 1700 m at the head of the Suisuza river valley. It is largely composed of extensive and little-exploited dense forests of Spruce and Scots Pine, interspersed with open montane meadows. It is not a countryside of great botanical richness, but it is easily reached, and in the reserve can be seen a number of Rhodopian endemic species of considerable interest. In the open meadows, coming into flower after the early flushes of the lilac-blue flowers of **Crocus veluchensis*, which is common in this area, are such Balkan endemic species as the yellow-flowered **Geum rhodopeum* (considered by some to be a hybrid with the red-flowered *G. coccineum*) and the large yellow-flowered *Hypericum cerastoides* with black glands on the petals and on the unequal sepals.

Bulgarian endemic species include the attractive pansy **Viola rhodopeia* with narrow leaves and yellow flowers and slender pale-violet spurs; *Veronica rhodopaea†*, a creeping plant with deep-blue flowers and elliptic to rounded leaves; *Sedum kostovii†* with pale-yellow flowers and dark-red fruits spreading in a star; and **Jasione bulgarica* with bluish-lilac flower heads and hairless stems, leaves, and involucral bracts.

Not far from the forest hut in the reserve is an interesting relict colony of *Potentilla fruticosa*, forming small low thickets in a meadow, where there is also an abundance of one of the most attractive of the flaxes, **Linum capitatum*, with dense clusters of large yellow flowers. Other interesting plants of the area include:

*Arenaria biflora
*Saponaria bellidifolia
*Saxifraga adscendens
*Aremonia agrimonoides
Potentilla rupestris
*Onobrychis alba
Gentiana utriculosa
Gentianella ciliata
*Pulmonaria rubra
Ajuga pyramidalis

*Teucrium montanum
Campanula patula subsp.
epigaea
Antennaria dioica
Allium flavum
*Fritillaria graeca
Muscari botryoides
*Crocus veluchensis
*C. flavus

Relict species which are outside their normal range, or on their periphery here are:

Clematis alpina	Astragalus centralpinus
Geum rivale	Viola palustris
Potentilla palustris	Potamogeton lucens
Genista germanica	*Ornithogalum collinum

In the central Rhodope in the district of Smoljan in the south is a calcareous region between the villages of Trigrad and Tesel. Here there is a deep and beautiful gorge with a fine cave which has been selected by the Balkan Tourist Committee as of particular interest to the tourist. It is at the same time a refuge for many interesting plants such as:

Arenaria 'rhodopaea'	*Haberlea rhodopensis
*Saxifraga luteoviridis	*Morina persica
*S. stribrnyi	*Campanula moesiaca
S. aizoides	*Symphyandra wanneri
*Polygala supina subsp.	*Trachelium jacquinii
rhodopea	*Cirsium appendiculatum
Hypericum linarioides	Centaurea affinis
H. montbretii	*C. napulifera subsp.
*Marrubium velutinum	pseudaxillaris
*Sideritis scardica	Crepis viscidula
Verbascum nobile	*Tragopogon balcanicus

Near here in the reserves of Kastranlii and Sabavica are some of the finest and wildest forests of Spruce and Black Pine in the whole of Bulgaria.

7. BLACK SEA COAST AND HILLS

The Black Sea and the surrounding hills and plains are of particular interest to western European naturalists because many species will in all probability be new to them, as these plants have their main centres of distribution further east. Ukrainian, Moldavian and south-west Russian, Crimean, and even Caucasian species may be found here, while other species which have migrated across the isthmus of Thrace from Turkey and Asia Minor also occur. Others again have their centres of distribution in the Black Sea basin, others are Mediterranean in origin.

It is a varied, but not a spectacular, coastline, with many wide sandy bays, low cliffs, marshes and lagoons, and small fertile river valleys falling gradually from the low wooded hills of the interior. North of the

Stara Planina one passes into the flatter country of the Danube basin where low cliffs and small sandy bays form the coastline. The climate of the Black Sea coast is predominantly continental, but with a strong Pontic influence, the latter giving milder winters but cold and windy springs. The prevailing winds are east and north-east. In consequence the spring is too cold for many more southerly species and crops, while the summers are hot and dry, with little reliable rain between June and September, and this is unsuitable for many central European plants and crops.

On the Balkan part of the Black Sea coast, the influence of the warmer Mediterranean air masses is felt across the low hills of Thrace and the Bosporus, and into the Thracian plain and northwards as far as Burgas and the coastal hills of the Stara Planina.

The vegetation to the south of the Stara Planina is largely central European in character, but with strong Mediterranean as well as Pontic influences. Here are mixed forests of oaks and Oriental Beech, with *Carpinus orientalis* and *Fraxinus ornus* and many Pontic plant communities. North of these hills the vegetation becomes more characteristic of east-central Europe, with mixed oak forests of *Quercus cerris*, **Q. frainetto*, *Q. pubescens*, **Q. virgiliana*, *Q. brachyphylla*, *Carpinus betulus*, *Sorbus domestica*, **Pyrus elaeagrifolia*, *Crataegus* spp. etc. Much of the natural forest has, however, been destroyed and shiblyak takes its place.

As one travels northwards along the 380 km of coastline the flora changes. It loses many of its Anatolian and southern species and becomes more easterly in character with species of Dacian (Romania to Crimea), Pannonian (Hungary and east-central Europe), Pontic (Black Sea), and Caucasian origins.

Towards the Romanian border, north of Cape Kaliakra one passes into quite a different floristic region, that of the stony steppes, grass-steppes, and steppe-woodlands of the Danube basin, known as the Dobrogea—the home of the Great Bustard—vegetation which is now rapidly disappearing as much of the steppe has been turned over to crop cultivation.

Ropotamo

The National Park of Ropotamo lies on the Black Sea coast about 55 km south of Burgas. The small Ropotamo river flows down from the heavily forested Istranca Mountains in a wide, densely forested valley, and runs into the sea between sand-dunes and marshy tracts. The guide-book paints the picture: 'Powerful oaks, elms and ash-trees, dense willows, winding ivies, grasses and bushes lend the Ropotamo landscape the appearance of a real jungle.' To the naturalist the Black Sea coast seems to be imminently threatened by the massive recreational developments of the Balkan Tourist Committee, as, for example, at Golden Sands (Zlatni Pyassatsi), Sunny Beach (Slunchev Bryag), and Druzba, and many smaller developments. As the flora and fauna are unique in Europe, and have been diminishing as a result of climatic changes since the Tertiary period, they are consequently of great interest to naturalists from all over the world. It is hoped fervently that there will be no further destruction of these natural habitats.

The flood-plain forests, or *longos*, of the Ropotamo river are largely composed of Narrow-leaved Ash, *Fraxinus angustifolia* subsp. *oxycarpa*, *Quercus robur*, *Q. pedunculiflora*, and *Ulmus minor*, with some *Populus alba* and *Carpinus betulus*. Smaller trees include: *Acer tataricum*, *A. campestre*, *Pyrus communis*, and *Alnus glutinosa*. The shrub layer comprises: *Crataegus monogyna*, *C. pentagyna*, *Euonymus latifolius*, *Corylus avellana*, *Sambucus nigra*, and *Ligustrum vulgare*. The jungle-like character of these forests is undoubtedly given by the number of climbing plants, both woody and herbaceous, which make the forests very dense and intertwined, with little light reaching the forest floor. Woody climbers of relict origin include the deep-purple-flowered *Clematis viticella* and the purplish-brown-flowered *Periploca graeca*, while more widespread climbers are the vine *Vitis vinifera* subsp. *sylvestris*, *Smilax excelsa*, *Clematis vitalba*, *Cynanchum acutum*, *Hedera helix*, and the herbaceous climbers *Humulus lupulus*, *Calystegia sepium*, *Cucubalus baccifer*, and *Tamus communis*.

The heavy shade limits the ground flora, but the following may be seen:

Ranunculus serbicus	*Lilium martagon*
Buglossoides purpurocaerulea	*Ruscus aculeatus*
Symphytum orientale	*Polygonatum latifolium*

At the mouth of the river lie sand-dunes which are unfortunately threatened by a large rest camp and a resort. The dunes are low and fairly stabilized, with damp marshy ground in the hollows. Plants recorded in the dunes which are predominantly Black Sea or south-west Russian species include:

Silene exaltata	*Scabiosa argentea*
Lepidotrichum	*Jurinea albicaulis*
uechtritzianum†	*Centaurea arenaria*
Goniolimon collinum	*Lactuca tatarica*
Stachys angustifolia	

143

Other sand-dune species, largely Balkan in origin, are:

*Silene thymifolia	*Trachomitum venetum
S. dichotoma	*Cionura erecta
*Dianthus cruentus	Stachys maritima
*Glaucium leiocarpum	S. milanii
*Maresia nana	*Verbascum purpureum†
Medicago marina	V. humile
*Linum tauricum	*Linaria genistifolia

More widely distributed southern European sand-dune plants include:

Polygonum maritimum	*Teucrium polium
Beta vulgaris	Plantago arenaria
Halimione pedunculata	Aster tripolium
Silene vulgaris subsp.	Otanthus maritimus
maritima	Artemisia maritima
Euphorbia paralias	Centaurea arenaria
E. peplis	Pancratium maritimum
Eryngium maritimum	Ammophila arenaria
Anchusa officinalis	

In the damp and swampy areas the majority of plants are widely distributed European species, but the following more eastern species were recorded: *Rorippa prolifera*, *Dipsacus laciniatus*, and *Leucojum aestivum*. In the dunes were clumps of the Lime, *Tilia tomentosa*, whose leaves showed strikingly white undersides when turned by the wind.

Drier bushy places in the area often had clumps of *Carpinus orientalis*, *Paliurus spina-christi*, with some *Acer tataricum*, the Maple with almost entire or very shallow-lobed leaves. Herbaceous species included:

Minuartia glomerata	*Trifolium purpureum
*Astragalus ponticus	T. lappaceum

---

11 **Astraka** One of the summits of the Timfi range in the northern Pindhos Mountains, with the Hellenic Alpine Clube refuge below at 1950 m. The cliffs and limestone summits have a rich alpine flora of *Aubrieta, Alyssum, Saxifraga, Edraianthus, Globularia*, etc.

12 **VicosGorge** The Voldomatis river cuts deeply into the Timfi Mountains to a depth of 1100 m, to join the Aoos river on its course to the Adriatic through southern Albania. The flanks of the gorge are covered with woods of Hornbeam, Hop-Hornbeam, Maple, Lime, Horse Chestnut, Macedonian and other oaks; Oriental Planes, Alders, and Willows grow on the valley floor.

Linum trigynum	**Knautia orientalis*
Prunella laciniata	*Campanula macrostachya*
Veronica spicata subsp.	*C. sparsa*
orchidea	*Ornithogalum pyrenaicum*

In the lake of Arkution, not far from Ropotamo, can be seen the Lotus, *Nelumbo nucifera*, with large pink scented flowers 15–20 cm across and large circular leaf blades up to one metre across, held well above the surface of the water on flexible stalks. It is an introduced plant, one with edible seeds and rhizomes, brought in from further east. Plants which have in all probability spread across from Turkey into this corner of Bulgaria include **Silene thymifolia, Lepidotrichum uechtritzianum*†, and *Symphytum orientale*. Endemic to this part of Bulgaria and Turkey-in-Europe is the yellow-flowered umbellifer *Heptaptera triquetra*, and also the small yellow-flowered *Hypecoum ponticum*†.

A similar vegetation and flora to that found in the Ropotamo reserve can be seen in the Kamchiya river valley, though with the loss of some southern and Turkish species. It is the main river draining the eastern end of the Stara Planina, and flows, unlike the larger Maritsa and Tundzha rivers, directly into the Black Sea. It lies about 30 km south of Varna, and is the largest *longos* in Bulgaria, having an area of about 500 hectares.

Istranca (Strandza) Mountains

These relatively low-lying mountains bordering the Black Sea and running westwards from the Bosporus through Turkey-in-Europe to the south-east of Bulgaria have a unique type of vegetation found nowhere else in Europe. In Turrill's words, it is 'characteristic of the northern parts of Asia Minor and the western Caucasus. It is very probable that we have here, and in parts of the Rodopes, the nearest

---

13 **Black Sea Coast** North of Cape Kaliakra lie the wide grassy steppes of the Dobrogea, must of it now under cultivation. The influence of the Mediterranean is no longer felt and the flora is east-central European in character with plants like *Paeonia tenuifolia, Matthiola odoratissima, Salvia nutans*, and *Iris pumila*.

14 **Stara Planina** Walking along the 'moors' of the mountains running through the heart of Bulgaria. Beech forests clothe the slopes, while the summits are covered with *Vaccinium, Juniperus*, and *Bruckenthalia spiculifolia*, the eastern equivalent of heather. The rock-outcrops have an interesting assortment of Balkan and central European species.

existing approach to the vegetation which covered at least much of the Balkan Peninsula in Tertiary times.'[19]

The forests of the Istranca, which lie on the northern slopes bordering the Black Sea, are dominated by oak woods of *Quercus polycarpa*, with some *Q. dalechampii*, *Q. hartwissiana*, *Q. cerris*, and *Sorbus domestica*. In damper situations the Oriental Beech, *Fagus orientalis*, forms forests, in sometimes more or less pure stands, as on the highest summit Mahya Dagi (1030 m), but elsewhere with an unusual admixture of both European and Asiatic species. The former include *Carpinus betulus*, *Ulmus glabra*, *Acer pseudoplatanus*, *A. platanoides*, *A. campestre*, *Tilia cordata*, and locally *Taxus baccata*, with several *Quercus* species. But of particular interest are the Pontic or euxine species which spread from the Caucasus in the east, along the Black Sea coast of Anatolia to as far west as the Istranca Mountains. They include the following trees and shrubs:

Mespilus germanica	*Rhododendron ponticum*
Prunus laurocerasus	*Vaccinium arctostaphylos†*
Daphne pontica	*Smilax excelsa*
Hypericum calycinum	*Ruscus hypoglossum*

as well as *Acer trautvetteri* and the low shrubby holly, *Ilex colchica*, which occur only in Turkey-in-Europe in our area. Interesting herbaceous plants, mostly of euxine distribution, include:

Epimedium pubigerum	*Cyclamen coum*
Helleborus orientalis†	*Symphytum tauricum*
Hypericum bithynicum	*Trachystemon orientalis*
Cicer montbretii	*Salvia forskaohlei*
Chaerophyllum byzantinum	*Verbascum bugulifolium*
Primula vulgaris subsp.	
sibthorpii	

The interior drier parts of the mountains are dominated by forests of deciduous oaks such as *Q. cerris*, *Q. frainetto*, and *Q. pubescens*, with *Celtis australis* and *Carpinus orientalis*, and such Mediterranean shrubs as *Pyracantha coccinea*, *Phillyrea latifolia*, *Cistus incanus*, and *Asparagus acutifolius*, and the herbaceous *Anemone pavonina*. In damper places are *Quercus hartwissiana* and *Fraxinus angustifolia* subsp. *oxycarpa*.

Where the forests have been partially cleared, a rather surprising shrubby community occurs which is dominated by Ling, *Calluna vulgaris*, but with such Mediterranean species as:

Osyris alba	*Dianthus pinifolius*

*Hypericum rumeliacum
  H. montbretii
*Cistus incanus
*C. salvifolius
  Opopanax bulgaricus
*Erica arborea

*Ajuga laxmannii
  Stachys angustifolia
  Verbascum lagurus†
  Centaurea thracica
*Asparagus acutifolius
*Romulea bulbocodium

On rocks the following endemics are found: *Saponaria sicula* subsp. *stranjensis* and *Anthemis jordanovii* which is possibly a local variant of **A. tenuiloba*.

Along the Black Sea coast of Turkey-in-Europe are the following euxine species:

Isatis arenaria
Lepidotrichum
  euchtritzianum†
Peucedanum obtusifolium
Convolvulus persicus

Argusia sibirica
Centaurea kilaea
Jurinea albicaulis subsp.
  kilaea

Several *Colchicum* and *Tulipa* species are known from this region including **Colchicum triphyllum*, **C. bivonae*, **C. turcicum*, *C. micranthum* and **Tulipa boeotica*†, **T. orphanidea*, *T. praecox*.

Balchik and Cape Kaliakra

North of Burgas the influence of the Mediterranean becomes very much less prevalent and more plants from central Europe and southwest Russia are found. South of the town of Balchik the main road from Varna passes close to the coast, under stark white cliffs of chalk reminiscent of parts of central Anatolia. These cliffs are very friable and are continually slipping seawards, to end in vertical coastal chalk cliffs. It is perhaps because of their lack of stability that they have not been reafforested like many of the wilder areas in the lowlands of Bulgaria. Consequently they remain in a semi-natural state with bare cliffs and open grassy slopes with clumps of scrub of **Paliurus spinachristi* and some *Colutea arborescens* in the hollows. The willow-leaved Oleaster, *Elaeagnus angustifolia*, is found on the cliffs near the village of Bjala.

In the spring such colourful species as *Paeonia peregrina* and **P. tenuifolia*, *Adonis volgensis*, **Hyacinthella leucophaea*, and *Iris pumila* in a wide range of colours, are to be found in this area and on the steppe-like plains of the Dobrogea to the north of Balchik.

In June many plants can be seen on these chalky slopes near Balchik. The list is long but of considerable interest to those who visit the

BULGARIA, Black Sea coast and hills
1. *Symphytum tauricum* 1052a 2. *Artemisia pedemontana* 1434a 3. *Satureja coerulea* 1152g 4. *Lactuca tatarica* 1541c 5. *Dianthus pseudarmeria* 186d
6. *Asyneuma anthericoides* 1349b 7. *Maresia nana* 298i 8. *Thymus pannonicus* 1164b 9. *Goniolimon besseranum* 974i 10. *Verbascum purpureum* 1192c

11. *Trigonella procumbens* 577a **12.** *Trachomitum venetum* 1007a **13.** *Salvia grandiflora* 1143d **14.** *Glaucium leiocarpum* 272a **15.** *Jurinea albicaulis* 1476e **16.** *Genista sessilifolia* 510a **17.** *Silene thymifolia* 170g **18.** *Onosma thracica* 1077f **19.** *Stachys leucoglossa* 1136j

Black Sea for the first time. The following are mostly eastern European species:

Dianthus pseudarmeria
D. *moesiacus*
Erysimum comatum
Alyssum tortuosum
Syrenia cana
Matthiola fruticulosa
M. *odoratissima*
Crambe tataria†
Lembotropis nigricans
Genista sessilifolia
Astragalus spruneri
A. vesicarius
A. *exscapus*
A. onobrychis
A. *varius*
Onobrychis gracilis
Linum tenuifolium
Hypericum elegans
Nonea pulla
Onosma visianii
O. taurica

O. thracica
Ajuga laxmannii
Marrubium peregrinum
Stachys leucoglossa
Salvia grandiflora
S. ringens
S. nemorosa
S. nutans
S. verticillata
Thymus pannonicus
Verbascum purpureum†
Campanula sibirica
Achillea clypeolata
Tanacetum millefolium
Helichrysum arenarium
Artemisia pedemontana
Jurinea stoechadifolia
Allium scorodoprasum subsp.
rotundum
Ornithogalum collinum

The cape of Kaliakra has unfortunately been landscape-gardened into a park with an 'interesting restaurant in the caves of the rocks, and archeological museum, a camp site etc.'. One of Bulgaria's two colonies of seals live in the caves below the red cliffs of the cape! On the flat steppe-like country between Kavarna and the cape are wide open steppes with grassland and stony valleys where the following were recorded among others:

Ephedra distachya
Paronychia cephalotes
Silene 'caliacrae'
Alyssum sibiricum
A. *caliacrae*
Crambe maritima
Chamaecytisus austriacus
Astragalus onobrychis
Euphorbia nicaeensis
E. *myrsinites*
Ruta graveolens

Teucrium polium
Scutellaria orientalis
Sideritis montana
Asyneuma anthericoides
Inula oculus-christi
Achillea clypeolata
Artemisia pedemontana
Centaurea marschalliana
Asphodeline lutea
Iris pumila

On the cliffs there is poor flora by comparison with the Mediterranean, and characteristic species include: *Crambe maritima, Crithmum maritimum, Ephedra distachya*, and the reddish-flowered *Limonium gmelinii*.

Inland from Varna, 15 km along the main road to Novi Pazar and Shumen, is an interesting natural phenomenon, the Standing Stones, or Pobitite Kamini. Enormous columns of grey rock up to 5 m high and 50 cm to 3 m thick stand erect above the sandy soil looking from a distance like the ruins of a Greek temple. The soil has been eroded by wind from around these rock columns leaving them exposed; there are several hundred of them and they stretch for nearly a kilometre to the north and south of the main road.

The vegetation is interesting and contains many more easterly European species, and some Balkan endemics. The most unusual are:

Moehringia grisebachii†	*Astragalus pugionifer*
Silene frivaldszkyana	*A. glaucus*
Dianthus nardiformis†	*A. varius*
D. pontederae	*Verbascum purpureum*†
Syrenia cana	*Anthemis regis-borisii*

8. STARA PLANINA (BALKAN RANGE) AND WESTERN BULGARIAN MOUNTAINS

The Stara Planina is the main mountain backbone which runs in an east–west direction through the heart of Bulgaria. It has a profound influence on the country, for not only does it separate physically the north of Bulgaria and the Danubian platform from the south, but it also acts as an important climatic barrier. North of the range the climate is continental in character, while to the south, in the plain of Thrace, a milder climate prevails under the influence of the Mediterranean air masses. To the north, crops like wheat, barley, maize, sunflower, and sugar beet are grown, while to the south, cotton, tobacco, rice, fruit, and vegetables are the main crops.

The Stara Planina are Tertiary fold mountains, being a southern extension of the Carpathians. They are largely composed of limestone, with some igneous and crystalline rocks exposed in the west and central areas. Their general height is below 2000 m, and consequently there is no truly alpine region and no evidence of glaciation. Their crests are rounded and plateau-like; the highest summit is Botev (2376 m), with Vezhen (2197 m) in the centre and Midzor (2169 m) on the Yugoslav border. To the east, beyond the town of Sliven, the mountains fall away into a number of minor ranges, one of which ends in Cape Emine, north of Burgas. The Balkan foothill region, at average altitudes of 400–700 m, and largely calcareous, lies to the north of the main Stara Planina.

To the immediate south of the Stara Planina is a deep valley, a Tertiary fault depression, in which the towns of Sofia, Karlovo, and Kazanlŭk are situated. South of this fault, running parallel to the Stara Planina, is a lower more dissected range, the Sredna Gora, or anti-Balkans. This is an outlier of the ancient Rhodope Massif and is composed of igneous and crystalline rocks; Bogdan (1604 m) is its highest summit. In the west it breaks into smaller irregular mountain masses which can best be called the western Bulgarian mountains. South of the Sofia basin is the mountain massif of Vitoša with Cherni Vrŭkh (2290 m) the highest summit. South of the Sredna Gora is the wide and fertile plain of Thrace—such an important feature in Bulgarian topography. It is drained by the Maritsa river which eventually breaks through the Rhodope range to enter the Aegean Sea as the Evros river of the Greeks, or the Meric of the Turks. The plain of Thrace is a Tertiary depression which is covered by more recent deposits; these produce the fertile black *chernozem* soils: it can truly be called the 'Garden of Bulgaria'.

The Stara Planina has characteristically a central European vegetation. Extensive forests of beech are found in the valleys and on hill and mountain slopes; at lower altitudes is a zone of mixed oaks and hornbeam, often reduced to deciduous scrub or shiblyak. There are few evergreen coniferous forests, but some do occur in the central and western regions, dominated by Silver Fir, Spruce, and the Scots and Macedonian Pines. In the lower hills of the eastern end of the range are deciduous oak forests of *Quercus cerris*, **Q. frainetto*, with *Acer monspessulanum* particularly on south-facing aspects in the foothills. On more northerly and less sunny slopes are forests of **Quercus polycarpa*, *Carpinus betulus*, **Fagus orientalis*, with *Acer pseudoplatanus*, *A. platanoides*, *A. campestre*, *Tilia tomentosa*, as well as *Ulmus glabra* and *Corylus colurna*.

The central region of the Stara Planina is the most interesting; it can be quite easily reached by road by the two main passes which cross the main range, the Troyanski prohod south of Lovech in the west, and the Shipka pass (1330 m) south of Gabrovo in the east. In this area below 1000 m are relict forests of *Fagus × moesiaca*—probably a hybrid between the western Beech and the Oriental Beech—together with the Cherry Laurel, *Prunus laurocerasus*, a Pontic shrub. Here are found such interesting relict species as **Rhynchocorys elephas*, a strange lousewort-like plant with bright-yellow flowers and a central projecting beak, sometimes with two darker spots—the eyes—at the base of the beak, like a miniature elephant at bay. Here also is *Haberlea rhodopensis* with handsome tubular violet flowers which grows on cliffs. Endemic species include the delicate pink-flowered primrose *Primula frondosa†, which grows on damp sheltered rock ledges, and the small thyme-like *Micromeria frivalszkyana†. In these mountains the white-flowered *Daphne blagayana* as well as *D. oleoides* are found, and the primula-like *Cortusa matthioli* with drooping flowers, which has a distribution as far east as the Himalaya. *Rhododendron myrtifolium* with clear pink flowers, a small shrub of the Carpathians, is also found in the central Stara Planina and in the eastern Rila Mountains.

An interesting area to explore is a fine tract of country around the highest summit, Botev. From the village of Kalofer a sound walking track climbs up through dense beech forests to bring one in about 1½–2 hours to the chalet 'Gendema'. The chalet is well situated on a small grassy plateau, surrounded by beech forests, and above it rise the magnificent cliffs of the southern face of Botev, down which a lace-like waterfall pours in a drop of something like 200 m. A climb up to the waterfall revealed the fine Turk's-Cap Lily, *Lilium carniolicum*, in the form which is called locally *L. jankae*. It is quite typical of this part of the Balkans with its rich golden-yellow flowers. On the vertical cliffs grow an abundance of *Geranium macrorrhizum* with fresh green foliage and large bright-pink flowers, and the white-flowered *Potentilla rupestris*. Other plants of interest recorded here were:

Lychnis viscaria	*Primula frondosa†
Silene lerchenfeldiana	*Haberlea rhodopensis*
Dianthus corymbosus	*Verbascum pannosum*
Thalictrum aquilegifolium	*Pedicularis brachyodonta*
Saxifraga juniperifolia	*Anthemis carpatica*
S. pedemontana	*Iris reichenbachii*
Chamaecytisus supinus	*Allium victorialis*

Other species on the summit slopes of Botev included:

Juniperus sabina	*Genista tinctoria (depressa)*
Minuartia saxifraga†	*Bupleurum longifolium*
Cerastium moesiacum	*Viola dacica*
Silene ciliata	*Rhododendron myrtifolium*
S. waldsteinii	*Androsace villosa*
Dianthus moesiacus	*Gentiana acaulis*
Saxifraga paniculata	*Verbascum lanatum*
Potentilla haynaldiana	*Veronica bellidioides*

Phyteuma confusum   *Nigritella nigra*
Allium flavum   **Crocus veluchensis*

And such endemic species as *Alchemilla achtarowii*, *A. jumrukczalica*, *Centaurea kernerana*†, and *Campanula velebitica* (*bulgarica*).

The Troyanski prohod is reached from the north from the town of Troyan. The road climbs up through the beech forests and out on to the grassy slopes of the summit ridge. From here stretch fine views, both eastwards and westwards, along the Stara Planina. Rounded grassy summits rise in fold upon fold above deep valleys, up the sides of which dense forests climb almost to the top. There are no spectacular mountain peaks, only rounded heights which in places are scarred by small cliffs and ridges where the limestone bedrock breaks through to the surface. It is this broken ground which is of most interest as many of the less common plants can be seen there. The grassy slopes and summits have patches of *Juniperus communis* subsp. *nana*, and the dwarf ericaceous shrublets *Vaccinium myrtillus*, *V. vitis-idaea*, and **Bruckenthalia spiculifolia* taking the place of *Calluna*; otherwise the similarity of the vegetation with that of western Atlantic moorlands is striking.

Limestone crags and cliffs lie about half an hour's walk westwards from the Troyanski prohod, along the summit ridge. Here is the Kozjata Stena—the Goat Wall—where the following interesting plants were seen:

Salix caprea	Cortusa matthioli
Minuartia verna	Gentiana verna
*Cerastium banaticum	*Myosotis alpestris
Lychnis viscaria	*Teucrium montanum
Anemone narcissiflora	Satureja alpina
Ranunculus nemorosus	*Thymus striatus
*Arabis alpina	Verbascum lanatum
*Saxifraga adscendens	Antennaria dioica
*S. marginata	Leontopodium alpinum
Dryas octopetala	'slavianum'
*Potentilla aurea	*Achillea ageratifolia
*Cotoneaster integerrimus	Homogyne alpina
*Linum capitatum	*Centaurea napulifera subsp.
*Daphne blagayana	nyssana
*Viola dacica	Allium moschatum
Helianthemum canum	*Ornithogalum collinum

Of special interest are the beautiful Balkan endemics *Silene 'trojanensis'*, *S. balcanica*, and *Seseli degenii*†.

At lower altitudes on the road between Karnare and Troyan and in the region of the Troyanski monastery we saw the following:

Minuartia saxifraga
*Silene saxifraga
*Dianthus giganteus
*D. petraeus subsp.
    simonkaianus
*Helleborus odorus
Rorippa sylvestris
Saxifraga paniculata
Spiraea chamaedryfolia
Chamaespartium sagittale

*Chamaecytisus banaticus
Trifolium velenovskyi
Geranium pyrenaicum
Lysimachia punctata
Symphytum ottomanum
Digitalis grandiflora
Campanula sparsa
*Centaurea napulifera
Muscari tenuiflorum

Further west in the Stara Planina towards the Yugoslav border the following may be found:

Aconitum firmum
Vicia truncatula
Daphne laureola
*D. oleoides
*Primula halleri

Androsace obtusifolia
*A. hedraeantha
*Symphyandra wanneri†
*Scutellaria alpina
Ramonda serbica

At the last point on the border in the north-west, on the hills of Vrâška Čuke, are some interesting species including *Eranthis hyemalis* var. *bulgaricus* and *Centaurea atropurpurea*.

In the Balkan foothill region north of the main Stara Planina between Lovech and Tŭrnovo are fine limestone areas, the best example of which is the labyrinthine region known as 'Korudere' near Gabrovo. Here the following can be found:

Lathyrus pannonicus
*Linum tauricum
Hypericum rochelii
H. umbellatum

Cyclamen hederifolium
Micromeria frivaldszkyana†
Parentucellia latifolia
*Acanthus balcanicus

At the eastern end of the Stara Planina where the lower hills finally drop down to the sea at Cape Emine there is heavily forested country. The main road running northwards from Burgas to Varna crosses these hills and passes through mixed forests of *Quercus polycarpa*, an eastern European oak distinguished by the conspicuously swollen brownish scales of its acorn cups.

Other trees and shrubs in these forests include:

Carpinus orientalis
Corylus colurna

*Fagus orientalis
Sorbus domestica

155

Sorbus 'bulgarica'
S. torminalis
Colutea arborescens
*Coronilla emerus
*Cotinus coggygria

*Acer hyrcanum
Tilia tomentosa
Cornus mas
Fraxinus ornus
Jasminum fruticans

Herbaceous species include the Anatolian Sage, Salvia forskaohlei, with conspicuous violet-blue flowers with white or yellow markings, and the yellow-flowered umbellifer, Heptaptera triquetra, endemic of the Thracian peninsula growing to 1½ metres in height, as well as a handsome garlic, *Nectaroscordum bulgaricum, which has umbels of large drooping pinkish bells borne on a stout stem and grows in bushy places. Other herbaceous species seen were:

*Lychnis coronaria
Lathyrus vernus
L. aureus
*Primula vulgaris subsp.
    sibthorpii
Trachystemon orientalis

Melittis melissophyllum
Physalis alkekengi
*Asphodeline liburnica
A. lutea
Polygonatum latifolium

Vitoša

This is an undistinguished, ancient, rounded mountain lying south of Sofia and filling its southern horizon. It is largely composed of syenite—predominantly alkaline feldspars. From Sofia one of several hotel and restaurant complexes on the mountain can easily be reached by local bus. Alternatively above the village of Dragalevci, to the south of the city, a good road winds up at first through hornbeam and beech forests, then higher through mixed forests of pines, spruces, and firs to a cluster of hotels and huts at Aleko, at 1860 m. This is a popular walking and skiing centre, and from here a small chair-lift takes one higher to the grassy slopes above the forests. But from Aleko it is an easy climb into the alpine zone and to the summit of Cherni Vrŭkh at 2286 m, where there is a meteorological station.

In the forests and clearings on the way up to Aleko the following are conspicuous in the summer:

*Silene roemeri
Lychnis viscaria subsp.
    atropurpurea
Dianthus superbus
*Saxifraga rotundifolia
*Lathyrus alpestris
*Polygala major

Atropa bella-donna
*Linaria genistifolia subsp.
    dalmatica
*Campanula moesiaca
Antennaria dioica
Senecio nemorensis
Orchis purpurea

Gymnadenia conopsea   *Limodorum abortivum*
Leucorchis albida

In wet flushes grows the beautiful and striking scarlet *Geum coccineum*, with purple *Cardamine rivularis*, and golden-yellow *Trollius europaeus*, often growing among the large leaves of *Rumex alpinus*. While in the spring before the beech and hornbeams break into leaf the following can be found: *Anemone ranunculoides*, *Corydalis bulbosa*, *Pulmonaria officinalis*, *Erythronium dens-canis*, and *Scilla bifolia*.

Above the forests in sheltered places the snow still lies as late as June, and round the melting patches are sheets of the large purple crocus, *C. veluchensis*—an unforgettable sight. Later a dark form of *Lilium martagon* is found mainly in the forests and brushwood, while the beautiful golden-yellow *Lilium carniolicum* grows mainly in meadows and pastures. On warmer slopes in June, where the snow has long since melted, *Campanula alpina*, with a conical spike of dark-blue nodding flowers, and the attractive cushion-forming *Dianthus microlepis*, in varying shades of pink, are both frequently encountered. In wet flushes *Primula farinosa*, a northern plant, *Tozzia alpina* of the Alps, and *Pinguicula balcanica* of the Balkans, all grow in close proximity with *Geranium pyrenaicum* and occasionally with *Gentiana punctata* which has large yellow-and-purple-spotted flowers. Balkan species like *Pedicularis orthantha*, which is pink with a dark-purple 'hood', *Viola dacica* with large violet or yellow flowers, and the white *Thlaspi avalanum* and yellow *Senecio pancicii* occur with such central European plants as:

Arenaria biflora   *Myosotis sylvatica*
Caltha palustris   *Veronica bellidioides*
Filipendula ulmaria   *Campanula patula*
Geum montanum   *Antennaria dioica*
Pyrola minor   *Homogyne alpina*

A steady climb up grassy meadows between stands of pine and spruces brings one to the rounded summit, which is scattered with huge rounded granitic bolders, reminiscent of the tors of Dartmoor. Peaty patches and drier ridges are covered with low bushes of *Juniperus communis* subsp. *nana*, *Vaccinium myrtillus*, *V. uliginosum*, *V. vitis-idaea*, and *Arctostaphylos uva-ursi* which in the swirling mists only heightens the illusion of being in the western Atlantic world. But instead of the heathers of Dartmoor are found the small pink-flowered Balkan heath, *Bruckenthalia spiculifolia*—which quickly dispels the illusion.

The *Flora of Vitosa* by Kitanov and Penev (in Bulgarian) illustrates, among others, the following species which are predominantly Balkan and which grow on the mountain:

*Comandra elegans	J. heldreichii
*Silene lerchenfeldiana	Echinops bannaticus
*Helleborus odorus	Carduus kerneri
*Aquilegia aurea	*Cirsium appendiculatum
*Gentianella bulgarica	Centaurea kotschyana
*Onosma heterophylla	*C. napulifera
*Verbascum longifolium	*Hyacinthella leucophaea
*Jasione bulgarica	*Iris reichenbachii

Golobardo (Bare Mountain)

This range lies south of Pernik, about 35 km south-west of Sofia. It forms rolling country, covered with scrub and grassland on the summits and with woodlands in the valleys. The highest point is Vetrushka (1158 m). It is largely composed of limestone, and though it has predominantly a continental climate, it is influenced to a small extent by Mediterranean air masses which penetrate deep into Bulgaria up the Struma valley. In consequence an interesting collection of species can be seen here: central European, Mediterranean, and in addition some eastern European species which find an outlying refuge in these relatively dry limestone areas. Part of Golobardo—Ostriza—has been declared a botanical reserve, and grazing has been very much restricted locally.

The grasslands are particularly rich in species and include such plants as: *Edraianthus serbicus which has a dense terminal cluster of deep-bluish-purple bells, surrounded by leafy bracts, and long narrow leaves; *Onosma visianii, another Balkan species, with long tubular, pendulous, yellow bells, so characteristic of the genus, as well as O. taurina, a tufted plant with even longer pale-yellow trumpets. The silvery-white-leaved *Cerastium banaticum, with starry white flowers, is another plant which stands out strikingly in the stony grassland.

Very characteristic of the Ostriza reserve are the slopes which are covered with dwarf scrub, either of the spiny *Astragalus angustifolius, or with a mixture of the white-woolly aromatic Artemisia alba and the nearly hairless, narrow-leaved under-shrub *Satureja montana subsp. kitaibelii growing in close association with each other.

There are also small clumps of residual oak forests with *Q. frainetto, Q. dalechampii, Q. cerris, and Q. pubescens. Some beech,

Oriental Hornbeam, and Hornbeam (on northern slopes) also occur.

Thickets of taller shrubs include almost pure stands of the Lilac, *Syringa vulgaris*, and the Wig Tree, *Cotinus coggygria*, while on earthy slopes are low patches of a dwarf almond, *Prunus tenella*, with bright-pink flowers and tiny woolly almond fruits. Elsewhere are mixed thickets of Hawthorn, *Crataegus monogyna*, and the largely Mediterranean shrubs *Paliurus spina-christi*, *Colutea arborescens*, *Coronilla emerus* subsp. *emeroides*.

On the margins of some of these thickets is *Helleborus odorus* with large green flowers which surprisingly are scented, as its name implies.

The best time to visit this area is in May and the first weeks of June when most plants are flowering and Ostriza is 'like a flowering rock garden'. Notable are:

*Saponaria bellidifolia	*Ajuga laxmannii
Dianthus stribrnyi†	Phlomis tuberosa
Adonis vernalis	Salvia virgata
*Cotoneaster nebrodensis	*Satureja montana subsp.
*Genista januensis	kitaibelii
G. subcapitata	*Micromeria cristata
Astragalus monspessulanus	Hyssopus officinalis
A. wilmottianus†	*Veronica austriaca
*Anthyllis aurea	Pedicularis petiolaris
*Onobrychis alba subsp.	Globularia punctata
calcarea	*Achillea ageratifolia
*Linum tauricum	*A. clypeolata
L. austriacum	Artemisia alba
*Haplophyllum suaveolens	A. pontica
Dictamnus albus	*Jurinea mollis subsp.
*Polygala major	anatolica
Daphne cneorum	Serratula radiata
Eryngium palmatum	*Centaurea orientalis
Cachrys alpina	

Monocotyledonous bulbs and rhizomatous species include:

*Asphodeline taurica	Fritillaria orientalis
*Asphodelus albus	Tulipa urumoffii
Gagea arvensis	Muscari comosum
G. lutea	M. neglectum
Allium moschatum	*Hyacinthella leucophaea
A. cupani	Sternbergia colchiciflora

> Crocus biflorus   *I. reichenbachii*
> Iris pumila

In shaded places are:

> Isopyrum thalictroides   Geranium sylvaticum
> Anemone sylvestris   Primula vulgaris
> A. ranunculoides   *P. veris
> Hepatica nobilis   Asperula taurina
> *Clematis recta   Erythronium dens-canis

About 30 km south-west of Golobardo is the highest summit of the western Bulgarian mountains, Konjavsca (1487 m). Many of the plants listed above are found on it as well as the endemic Mullein, *V. anisophyllum†*, which has densely greyish-yellow woolly leaves and a branched inflorescence of yellow flowers each 3 cm across, with violet filament hairs. From the viewpoint of the Bulgarian flora, this mountain is noteworthy because it comes under considerably stronger Mediterranean influence and supports a number of Mediterranean species which do not occur elsewhere in the country, such as:

> *Juniperus oxycedrus   Oxytropis pilosa
> Ostrya carpinifolia   Jasminum fruticans
> *Aubrieta intermedia   Symphytum ottomanum

9. MACEDONIAN MOUNTAINS OF YUGOSLAVIA

These mountains lying within the boundaries of Macedonia (Makedonija) of southern Yugoslavia are situated in the transitional vegetation zone and consequently have an interesting mixture of central European alpine species and Balkan endemic species.

National Park of Perister

This is situated about 15 km west of Bitola in southern Macedonia in the Baba Mountains not far from the Greek-Yugoslav border. The highest summit is Perister (2601 m) which can quite easily be reached from Magarevo by a rough but passable road climbing through the forests, to end at the refuge of Begova Cesma. From here it is 2–3 hour's climb to the summit.

The lower mountain slopes are heavily forested with the Balkan endemic long-coned Macedonian Pine, *P. peuce. It forms fine, nearly pure stands on the northern slopes of the mountain from 1100 to 2000 m. The Bracken, *Pteridium aquilinum*, is characteristic of the ground flora in the lower forests to about 1500 m, while at higher altitudes the Wortleberry, *Vaccinium myrtillus*, is dominant. These forests contain some interesting shrubs including: *Rosa pendulina*, **Cotoneaster integerrimus*, **C. nebrodensis*, **Crataegus laciniata*, **Acer heldreichii*, *A. obtusatum*, **A. hyrcanum*, *Euonymus latifolius*, *E. verrucosus*, *Tilia tomentosa*, *Daphne mezereum*.

Herbaceous species of interest include the following:

*Helleborus cyclophyllus
*Ranunculus platanifolius
*Thalictrum aquilegifolium
Corydalis solida
*C. bulbosa
*Aremonia agrimonoides
Genista lydia
*Lathyrus laxiflorus
*L. grandiflorus
Coronilla elegans
Geranium asphodeloides
*G. macrorrhizum
Pulmonaria mollis
Calamintha grandiflora
*Digitalis viridiflora

*D. ferruginea
D. grandiflora
Knautia midzorensis
K. macedonica
Campanula foliosa
C. bononiensis
C. persicifolia
*Doronicum columnae
D. orientale
Centaurea deustiformis
Scilla bifolia
Galanthus nivalis
*Crocus chrysanthus
*C. veluchensis
*Iris sintenesii

On the western slopes of the mountain round the village of Maloviste are extensive beech forests.

In damp places and by streamlets in forest clearings are interesting herbaceous plants like:

*Silene asterias
Saxifraga stellaris
*S. rotundifolia
*Geum coccineum
Geranium sylvaticum

Adenostyles alliariae
Doronicum austriacum
subsp. *'giganteum'*
*Cirsium appendiculatum

A rough track runs westwards from Begova Cesma through the forest for some distance and ends in a path which proceeds upwards through the forest to the alpine refuge, which is situated above the tree-line. At this altitude a dense low scrub covers the mountain-side with *Juniperus communis* subsp. *nana*, the ericaceous plants **Bruckenthalia spiculifolia*, *Vaccinium myrtillus*, and *V. uliginosus; Rubus fruticosa*, and *Genista tinctoria* which are less widespread. In grassy places the rosy-purple-flowered creeping **Dianthus myrtinervius*†, looking like *Silene acaulis*, and the tall Martagon Lily are found.

Above the refuge the upper slopes of the mountain are covered with heavy grassland and much *Vaccinium myrtillus*, with little evidence of grazing, and here there seemed to be little of interest. But in the broken ground on the ridges and round the summit are many interesting plants. Near late snow-patches under the northern slopes of the ridges in mid-July were *Crocus veluchensis*, the white-flowered *Ranunculus crenatus*, *Arenaria biflora*, and *Gentiana punctata* with large yellow flowers spotted with reddish-purple, also *Potentilla aurea*, *Armeria canescens*, and *Rosa pendulina*.

On stony ground in the upper grasslands the following were recorded:

Minuartia recurva	*V. austriaca subsp. vahlii
*Dianthus minutiflorus	Pedicularis heterodonta†
Ranunculus croaticus	*Jasione laevis subsp.
Geum montanum	orbiculata
*Geranium cinereum subsp.	*Senecio abrotanifolia subsp.
subcaulescens	carpathicus
*Linum capitatum	*Centaurea triumfetti
Viola beckiana†	*Scorzonera purpurea
Thymus praecox	*Ornithogalum collinum
*Veronica bellidioides	

On the summit cliffs grows the white-flowered *Saxifraga pedemontana* subsp. *cymosa*.

Autumn-flowering plants on the mountain include: *Cyclamen hederifolium*, *Plumbago europaea*, *Odontites lutea*, *O. glutinosa*, *Colchicum autumnale*, *Scilla autumnalis*, *Crocus pulchellus*, *Spiranthes spiralis*.

Other species of predominantly Balkan or south-east European origin which were recorded on the mountain by Todorovski[20] include:

Silene waldsteinii	*Jovibarba heuffelii
S. trinervia	*Genista carinalis
*S. roemeri	Hypericum barbatum
*Dianthus pinifolius	*H. rumeliacum
*D. petraeus	*H. olympicum
*Ranunculus psilostachys	Viola orphanidis
R. serbicus	V. eximia
Erysimum cuspidatum	Eryngium palmatum
E. comatum	*Soldanella hungarica
*Sempervivum kindlingeri†	Stachys plumosa
*S. marmoreum	S. angustifolia

*Verbascum longifolium
*Linaria peloponnesiaca
Pedicularis limnogena†
P. leucodon
*P. brachyodonta
*Pinguicula balcanica

Scabiosa webbiana
*Campanula lingulata
*Achillea coarctata
A. crithmifolia
Senecio macedonicus
Fritillaria gussichiae

West of Bitola, on a pass of 1169 m which crosses the shoulder of the Baba range, the following were recorded in meadows and clearings:

Minuartia bosniaca
Petrorhagia saxifraga
*Dianthus corymbosus
Rosa gallica
Vicia grandiflora
Lathyrus latifolius
Geranium asphodeloides
*Armeria canescens
Gentiana utriculosa

Gratiola officinalis
Orobanche caryophyllacea
Knautia macedonica
Matricaria trichophylla
Centaurea nigrescens?
*Allium carinatum subsp.
    pulchellum
*Orchis laxiflora
*O. coriophora

In late spring Narcissus poeticus and Leucojum aestivum are abundant in the meadows below the pass, with Doronicum cordatum and Colchicum hungaricum.

Further down the road towards Ohrid the rocky outcrops by the roadside are festooned with brilliant golden *Alyssum saxatilis and yellow *Coronilla emerus var. emeroides making an upright bush 3–4 ft tall, while along the rocky ledges here and there nestled the common Sedum album. Colchicum leaves dotted one slope and will probably prove to be the C. hungaricum collected before in S. Yugoslavia. The Yugoslav-Albanian border area is especially interesting botanically. There are quite a few plants which occur practically wholly in Albania but 'sneak' across the border in odd places. Plants like the rare *Crocus scardicus, *Astragalus baldaccii and *Petteria ramentacea (a rather cytisus-like shrub) come readily to mind. Others like *Crocus cvijicii and Fritilaria macedonica occur only along the border, but have not been collected for many years, and remain to be rediscovered and brought into cultivation. The inaccessibility of Albania to botanical exploration at the present time makes the Yugoslav side of the border all the more interesting.[21]

National Park of Galicica

This is a large mountainous tract about 30 km long lying between Lake Prespa and Lake Ohrid, stretching southwards from the town of Ohrid to the Yugoslav-Albanian border. The general height of the

MACEDONIA, Roadside plants
1. *Alcea pallida* 748 **2.** *Coronilla varia* 627 **3.** *Silybum marianum* 1492
4. *Psoralea bituminosa* 537 **5.** *Xeranthemum annuum* 1464 **6.** *Centaurea salonitana* 1503 **7.** *Centaurea solstitialis* 1499 **8.** *Salvia nemorosa* (1147)

9. *Echium italicum* 1081   **10.** *Picnomon acarna* 1488   **11.** *Achillea coarctata* 1421e   **12.** *Onopordum tauricum* (1494)   **13.** *Marrubium peregrinum* 1113 **14.** *Cirsium candelabrum* (1484)   **15.** *Salvia sclarea* 1144   **16.** *Scolymus hispanicus* 1510

range is about 1500 m but it builds up towards the south to the summit of Galicica (2255 m). The most interesting part of the park is in the south, in the summit region and particularly on the steep rock slopes of Zli Dol, which fall away to the west to Lake Ohrid, above the monastery of Sveti Zaum. This part is easily reached by a good road which crosses the park from Prespa to Ohrid at about 1600 m, and from this pass in fine weather there are magnificent views across the lakes of Prespa to the east and Ohrid to the west, while southwards lie the Mali Thate Mountains of Albania.

If one takes the road from the east along the shore of Lake Prespa passing the village of Otesevo, one climbs up through some fine wooded country. From the road one can see the beautiful pink spikes of *Morina persica with its rosettes of thistle-like leaves growing on the banks. It is also worth stopping and searching for the mat-forming *Thymus cherlerioides and the blue bells of *Edraianthus serbicus and *Onosma echioides with its yellow bells.

The lower mountain slopes have mixed deciduous wood of Carpinus orientalis with Acer monspessulanum, Pistacia terebinthus, Fraxinus ornus, and *Phillyrea latifolia, a distinctly transitional Mediterranean association, reaching an altitude of 1200 m in places. The upper slopes of the park are often well wooded, predominantly by oaks, such as the Macedonian Oak, Quercus trojana, which may form almost pure stands, and there are also mixed oak woods with the Hungarian Oak, *Q. frainetto, and the Turkey Oak, Q. cerris. Beech woods, dominated by the hybrid Fagus × moesiaca, are found above the oak forests to 1900–2000 m. Other trees such as *Pinus leucodermis, *Juniperus excelsa, *J. foetidissima, and Aesculus hippocastanum are also to be found in the park. Above the tree-line are extensive sub-alpine moors and grasslands, often heavily grazed by sheep and goats.

From the pass it is an easy climb southwards, at first through dense beech forests to the open grassy summit of Galicica in about 2½–3 hours. The dense shade and heavy leaf litter of the beech allow few species to flourish but *Helleborus cyclophyllus, *Anemone apennina, Viola orphanidis, and Cardamine bulbifera were seen. In the heavily grazed meadows in the forest zone are fine clumps of the spear-headed *Asphodelus albus, as well as:

Lychnis viscaria	*Primula veris
*Arabis caucasica	Stachys germanica
*Astragalus sericophyllus	*Veronica austriaca subsp.
Genista radiata	vahlii
Helianthemum canum	*Lonicera alpigena
Euphorbia myrsinites	*Senecio ovirensis
*Daphne oleoides	

In the stony grasslands are mats of juniper harbouring many interesting plants, such as the beautiful little yellow-flowered pansy, *V. allchariensis*, which is at its best by melting snow-patches. The violet-flowered *Thlaspi bellidifolium* forms handsome clumps among the rocks, and the creamy-white crocus, *C. cvijicii*, with yellow styles can also be seen near snow-patches; it is an endemic species of the mountains of Macedonia and eastern Albania. Other interesting plants included:

Minuartia verna
Cerastium decalvans
Dianthus minutiflorus
Erysimum pusillum
Arabis alpina
A. bryoides
Alyssoides utriculata
Iberis saxatilis
Saxifraga adscendens
S. scardica
Potentilla recta
P. speciosa
Cotoneaster niger

Geranium cinereum subsp.
  subcaulescens
Daphne alpina
Gentiana verna
Acantholimon androsaceum
Myosotis alpestris
Pedicularis friderici-augusti
Edraianthus tenuifolius
Anthemis tenuiloba
Achillea chrysocoma
A. ageratifolia
Colchicum hungaricum
Scilla bifolia
Crocus chrysanthus

The following additional species were recorded by Chater[22] in a walk south-east of Ohrid in the national park at between 1300 and 1650 m:

Dianthus cruentus
Anemone apennina
Ranunculus millefoliatus
R. psilostachys
Erysimum diffusum
E. pectinatum
Sempervivum kindingeri†
Cytisus decumbens
Astragalus baldaccii

Oxytropis purpurea
Viola elegantula†
Anchusa serpentinicola†
Ajuga orientalis
Salvia argentea
Asphodeline taurica
Ornithogalum refractum
Iris graminea

From Struga on the northern shore of Lake Ohrid the main road northwards runs close to the Albanian border and follows the valley of the Crni Drim river. This takes the outflow from Ohrid through north Albania to drain into the Adriatic south of Lake Shkodër. It passes through forested country, and by the roadside north of Debar are some interesting Balkan plants including the unusual *Kitaibela vitifolia*, a tall mallow with large white flowers and vine-like leaves, growing 2–3 m high. It is the only species of this genus and is endemic to

167

MACEDONIA, PELISTER and GALICICA

1. *Morina persica* 1319a　**2.** *Primula veris ssp. columnae* 939　**3.** *Pedicularis frederici-augusti* 1249f　**4.** *Centaurea triumfetti* 1501d　**5.** *Alyssoides utriculata* 323　**6.** *Viola allchariensis* 786e　**7.** *Dianthus minutiflorus* 194d　**8.** *Ranunculus crenatus* 246b　**9.** *Onobrychis alba* 636j

10. *Veronica austriaca ssp. valhii* 1229   **11.** *Anthyllis aurea* 621c   **12.** *Erysimum pusillum* 294a   **13.** *Astragalus sericophyllus* 526d   **14.** *Cerastium decalvans* 140d   **15.** *Senecio abrotanifolius* 1452d   **16.** *Helleborus cyclophyllus* (200) **17.** *Juniperus foetidissima* 14a   **18.** *Geranium cinereum ssp. subcaulescens* 641a **19.** *Crataegus laciniata* 462a   **20.** *Pinus peuce* 10a   **21.** *Crocus chrysanthus* 1679

Yugoslavia, though it has become naturalized in adjacent countries to the east. Other plants seen in the vicinity of Debar and on the road north to Lake Mavrovo and on towards Gostivar include:

Corydalis ochroleuca
*Peltaria alliacea
Draba aizoides
Lembotropis nigricans
*Chamaecytisus heuffelii
*Haplophyllum coronatum
Acer tataricum

*Lavatera thuringiaca
Scutellaria altissima
*Digitalis laevigata
*D. ferruginea
*Ramonda serbica†
Telekia speciosa
Allium flavum

National Park of Mavrovo

This is the largest national park of Yugoslavia, comprising 7600 hectares and including the three mountain ranges of Bistra, Šar, and Korab. The main road from Ohrid to Gostivar runs through the centre of the park, past the large artificial lake of Mavrovo. The lake is roughly central and is surrounded by mountains. To the north lie the rounded summits of the southern part of the Šar range. To the west, along the Albanian border, are the Korab Mountains with a broken alpine relief rising to 2764 m in Golem Korab. In the south are the lower grassy summits of the Bistra Planina with the highest point Curkov Dol (2111 m).

The national park has fine forests, and is of particular interest in having the southernmost example of spruce forest, as well as some Horse Chestnut woods which are in all probability native here. The main spruce forests lie south of Volkovija, east of the main road on the slopes of the Bistra Planina. There are also woods of the Turkey Oak, *Q. cerris*, with *Acer obtusatum* and the Hop Hornbeam, *Ostrya carpinifolia*, while the Oriental Hornbeam, *Carpinus orientalis*, is common at lower altitudes. Beech forests are extensive in the park but there is very little fir forest.

To reach Bistra Planina take the new tarmac road which branches off southwards, before reaching the half-submerged church of the old village of Mavrovo, and it will climb up through dense beech forests and take you on to the wide grassy slopes of the mountain. In damp flushes in the clearings in the forest are such attractive plants as: *Silene asterias*, *Geum coccineum*, *Primula veris* subsp. *columnae*, *Veratrum album*, and *Dactylorhiza majalis*, while *Lathyrus laxiflorus* and *Platanthera chlorantha* grow in the forest. In meadows are: *Dianthus giganteus*, *Vicia narbonensis*, *Linaria angustissima*, and *Campanula patula*.

Above the grassland on rocky outcrops and ridges the following were recorded:

Minuartia graminifolia
   subsp. *clandestina*
M. verna
Cerastium decalvans
Saponaria bellidifolia
Ranunculus sartorianus
Thalictrum minus
Erysimum pectinatum
Malcolmia illyrica
Iberis sempervirens
Sedum flexuosum
Saxifraga sempervivum
S. marginata
S. adscendens
S. paniculata
Alchemilla flabellata
Potentilla apennina
Astragalus vesicarius
Onobrychis montana subsp.
   scardica

*Oxytropis prenja†
Geranium cinereum subsp.
   subcaulescens
Linum perenne subsp.
   extraaxillare
Daphne oleoides
Viola gracilis
Armeria rumelica
Asperula doerfleri
Thymus cherlerioides
Scrophularia canina
Veronica austriaca subsp.
   vahlii
Pedicularis brachyodonta
Edraianthus graminifolius
Senecio ovirensis
Lilium carniolicum
Coeloglossum viride

While on more exposed cliffs were *Arabis alpina*, *A. bryoides*, *Anthyllis vulneraria*, and *Achillea ageratifolia*; and by late snow-patches:

Ptilotrichum cyclocarpum
Draba aizoides
Viola frondosa†

Primula veris subsp.
   columnae
Gentiana verna
Plantago atrata
Scilla bifolia

Šar Planina

This range of mountains runs roughly in an arc in a north-easterly direction from the north of the Mavrovo National Park and forms part of a natural boundary between Macedonia and Serbia to the north. It is a long range of high mountains of about 60 km with many summits rising to 2400–2600 m. The town of Tetovo lies approximately at the centre of the range and is probably the best point from which to explore these mountains. A cable-car from Tetovo climbs the southern flanks taking the skier and tourist to the mountain resort of Popova Sapka.

From here it is not difficult to reach the highest summits of Titov Vrh (2747 m) and Turcin (2762 m) on the subsidiary range of Rudoka Planina.

The ski-lift from Tetovo takes one high above the beech woods on to the open grassy slopes of the mountain to 1720 m at Popova Sapka. From here rounded slopes stretch upwards towards Titov Vrh with but one small open stand of Silver Fir. A typical alpine scrub association of *Juniperus communis* subsp. *nana*, *Bruckenthalia spiculifolia*, *Vaccinium myrtillus*, *V. uliginosum* with *Sorbus chamaemespilus* and *Daphne mezereum* is widespread. In more open grassy places the following are conspicuous:

Cerastium alpinum	*Androsace hedraeantha*
Saponaria bellidifolia	*Soldanella dimoniei*
Anemone narcissiflora	*Myosotis alpestris*
Dryas octopetala	*Pinguicula balcanica*
Geum coccineum	*Verbascum scardicola*
Polygala supina	*Senecio nemorensis*
	Crocus veluchensis

On more stony slopes above about 2000 m are:

Minuartia verna	*Viola elegantula*†
Silene saxifraga	*Thymus cherlerioides*
Dianthus minutiflorus	*Scutellaria alpina*
Draba tomentosa	*Pedicularis petiolaris*
Aethionema saxatile	*Edraianthus graminifolius*
Iberis saxatilis	*Achillea ageratifolia*
Jovibarba heuffelii	*Homogyne alpina*
Saxifraga sempervivum	*Centaurea epirotica*
S. marginata	*Muscari botryoides*
Linum viscosum	

At about 2500 m additional species on screes and rocky places are:

Salix reticulata	*Potentilla aurea* subsp.
Silene acaulis	*chrysocraspeda*
Cardamine glauca	*Helianthemum oelandicum*
Thlaspi bellidifolium	subsp. *alpestris*
Saxifraga scardica	*Pedicularis oederi*
S. oppositifolia	*Plantago atrata*
	Antennaria dioica

Of particular interest, found growing near melting snow-patches, is a golden-yellow crocus with purple in the throat, *C. scardicus*, which is an endemic of Yugoslavia. It grows with the much commoner and more widespread purple-flowered mountain crocus *C. veluchensis. The former has leaves 1 mm wide or less without a median white stripe and a corm with finely netted fibrous scales, and with a fibrous neck. Another Macedonian endemic is *Alkanna noneiformis*† with purplish-violet tubular flowers, over ½ cm across; it is closely related to *A. pindicola but differs in having a corolla tube just a little longer than the calyx instead of twice as long.

Another approach to the Šar Planina is from the smaller unmetalled road which runs across its northern slopes to the town of Prizren. It passes Brezovica which is a good centre for exploring this whole region, with a ski-lift to the south up the slopes of Livadice and towards Ljuboten, two summits of nearly 2500 m. West of the village is the Prevalac pass of 1540 m, from where it is an easy climb of 2½–3 hours to the summit ridge of the Šar Planina. From here are magnificent views to the south and west into the heart of Macedonia.

A rather different assortment of species occurred on this summit ridge, such as:

Minuartia recurva
Silene chromodonta
*Dianthus scardicus
*Arabis caucasica
*Sempervivum macedonicum†
Saxifraga bryoides
S. moschata
*S. glabella
S. aizoides
Hypericum richeri
*Armeria canescens

*Gentianella bulgarica
Veronica alpina
V. aphylla
*Pedicularis verticillata
*Campanula alpina
*Phyteuma confusum
*Jasione laevis subsp.
    orbiculata
Anthemis carpatica
*Senecio abrotanifolius subsp.
    carpathicus

Flowering by melting snow-patches were *Ranunculus crenatus, *Androsace hedraeantha, *Cardamine glauca, and Rumex nivalis.

The road westwards from Prizren after leaving Peć, climbs upwards into the magnificent wild country of Crna Gora (Montenegro) over the high Čakor pass at 1849 m on the way to the Dalmatian coast. The road at first passes through a fearsome gorge, the Rugovo gorge, and then climbs steeply up through Silver Fir forests, becoming rougher and narrower as it ascends to the pass. Here there is a small rest house and from the head of the pass walks both to the north and south can be taken, but there are no high mountains in the vicinity. In May, lower down on the rock walls of the gorge will be found the creamy-flowered Corydalis ochroleuca which is restricted to the western Balkans and to Italy. And in meadows and by Juniper thickets:

Helleborus cyclophyllus   *Lilium martagon*
 Hepatica nobilis   *Galanthus nivalis*
 Corydalis solida   *Crocus vernus*
 C. bulbosa subsp.
   marschalliana

Later in the year up on the pass, gentians such as *G. acaulis*, *G. verna*, *Gentianella crispata* are to be seen with the handsome yellow *Linum capitatum*, and tall spikes of *Aconitum 'pentheri'* (*divergens*) growing up through the Juniper scrub.

To the south lie the high mountains of the Prokletije on the borders of Albania. From Murino a road runs south to the holiday resort of Plav on the lake of that name; it is a good base for exploring these wild mountains.

10. SOUTHERN DALMATIA

The most distinctive feature of the southern Dalmatian coastline is the great fjord-like inlet of the Gulf of Kotor, a flooded valley running into the heart of the coastal mountains. Narrow straights between high mountains lead into wide expanses of open waters which have been used as natural harbours for millenniums.

The inner bays of Risan and Kotor are surrounded by steep mountains rising to 1200 m, with massive limestone cliffs which seem almost to overhang the narrow coastal cultivated belt. To the north-west lies the rounded massif of Orjen (1895 m), while to the south-east are the five Lovćen peaks, the highest being 1749 m. From the town Kotor, situated at the innermost protected part of the gulf, a well-engineered road, with seventeen hairpin bends, climbs to about 1000 m, over the watershed into the interior to Cetinje and Titograd. Views from the top of this pass are magnificent.

These coastal mountains receive the heaviest rainfall of any part of the Mediterranean region. Over 3000 mm falls yearly, much of it in the winter with the passage of vigorous depressions. There may be up to a hundred days of rain, but these are often quickly followed by bright periods; July and August are the driest months.

Orjen

This mountain is most easily reached by a road, at first metalled, which climbs northwards above Herceg Novi at the entrance to the Bay of Kotor. Take the branch road, which is a rough, stony, but sound road to the village of Vrbanje. This crosses some typical karst country with bare limestone rocks interspersed with patches of wood of White

)ak, Hop-hornbeam, and Montpellier Maple, or the characteristic crub of the maquis and garique. There are stony meadows in the unken dolina of the karst, or the dolina may be walled and cultivated vith corn or potatoes, and contrast vividly with the stony slopes and ocky scrubland all round.

Above the village of Krusevica on the way to Vrbanje, which is ituated on a small fertile polje, the attractive blue-flowered oraginaceous plant, *Moltkia petraea*, grows by the roadside. It is a hrublet with narrow bristly leaves and pendulous clusters of narrow-ubular, deep blue-violet flowers with projecting stamens. It is estricted to the mountains of the central Balkan region. Also in this area grows the endemic laburnum-like shrub, *Petteria ramentacea*, vith erect clusters of yellow flowers and trifoliate leaves. The creeping *Globularia cordifolia* with blue spherical heads is abundant. *Campanula ramosissima*, a low-growing annual with trumpet-shaped blue lowers with white centres and *Thymus striatus*, both species estricted to the Balkan Peninsula and Italy, grow here with the more videspread purple-flowered Common Sage, *Salvia officinalis*.

Above Vrbanje the road climbs upwards to the pass over the shoulder of Orjen. At first it passes through heavy beech forests with occasional open grassy dolina where *Thalictrum aquilegifolium*, *Ajuga enevensis*, *Rorippa lippizensis*, and *Scilla litardierei†* were seen; he last two restricted to the Balkans, the Squill being endemic to Yugoslavia.

Above the Beech forests are scattered open stands of Bosnian Pine, *P. leucodermis*, with deep-purple young cones set among stiff light-green needles. The striking bark is silvery-white and old trunks become fissured into regular angular block-like segments, like the scales of the skin of a lizard; old trees had their tops flattened and twisted by exposure to the elements.

The open stony screes and rocks were rich in species and the following were flowering at the end of June:

Cerastium grandiflorum	*Salvia officinalis*
Cardamine glauca	*Scrophularia heterophylla*
Aethionema saxatile	subsp. *laciniata*
Rosa pendulina	*Valeriana montana*
Anthyllis vulneraria	*Edraianthus graminifolius*
Hippocrepis comosa	*E. tenuifolius*
Bupleurum ranunculoides	*Senecio thapsoides* subsp.
subsp. *gramineum*	*visianianus*
Vincetoxicum hirundinaria	*Ornithogalum collinum*
Stachys alopecuros	

Earlier in the year there are several species of crocus in flower including *C. vernus* subsp. *albiflorus*, an eastern European form with white flowers and the style usually much shorter than the stamens and *C. malyi* which also has white flowers but a deep-yellow throat. This is the classical locality of this latter species; it was discovered here, and is restricted to western Yugoslavia.

Other spring-flowering species include *Euphorbia characias* subsp. *wulfenii*, a robust species growing to nearly 2 m, and the spiny domed bushes of *Euphorbia spinosa*; also *Scilla bifolia*, *Muscari neglectum*, *Galanthus nivalis*, *Orchis provincialis*, *Orchis simia*, and *O. morio*.

From the top of the pass it is an easy scramble up steep grassy stony slopes and through clumps of pines to the summit. Magnificent views spread out before one of bleached white limestone crags of the surrounding Dinaric ranges, up which scattered clumps of pine seem to cling precariously to any sheltered fold or gully in the stark cliffs. Far below an occasional patch of bright green catches the eye where half a hectare or so of doline bottom is cultivated, contrasting vividly with the otherwise wild inhospitable landscape. A huge fallen limestone block prevented our progress south over the pass, but no doubt it has been cleared and one can continue along this wild road to Risan at the head of the Gulf of Kotor.

About the summit of Orjen in June the following plants were the most distinctive:

*Alyssoides utriculata	*Plantago argentea
*Iberis sempervirens	*Achillea abrotanoides
*Euphorbia capitulata	*Fritillaria messanensis

Lovćen

The National Park of Lovćen is in many ways set in similar country to that of Orjen, and the vegetation is also very similar. But on Lovćen grazing has been heavier and much of the natural vegetation has been destroyed.

The park lies to the south of the Kotor–Cetinje road, which runs round the northern periphery of the park as it crosses the rough karst

15 **Rila Planina** Like the Pirin Planina it has an alpine climate and an unusual assortment of species. In clearings between the Black and Macedonian Pines, and the Dwarf Pine at higher altitudes, can be found such plants as *Aquilegia aurea*, *Geum coccineum*, *Soldanella hungarica*, and the unique 'primula of the Gods', *Primula deorum*.

on its way east to Cetinje. The park itself can be reached from the village of Krstac by a road turning southwards to the hamlet of Ivanova Korita. From here a fairly steep road takes one up to within ten minutes' walk of the summit of Jezerski Vrh (1660 m) where there is a mausoleum to Peter II, Njegoš. Štirovnik (1749 m) is the highest summit of Lovćen; it lies about 5 km further north-west. From these summits there are grand views over the Gulf of Kotor and the Adriatic coast, and inland as far as Durmitor, while to the south-east lies the lake of Shkodër and the mountains of Albania, but fine weather is needed in this heavy rainfall zone where over 5300 mm has been recorded in a single year.

Unusual species that should be searched for on Lovćen, Orjen, and the surrounding hill country include *Lonicera glutinosa*, a shrub with glandular-hairy elliptic leaves and small yellowish two-lipped flowers tipped with red. It is endemic to the mountains of western Yugoslavia. *Amphoricarpos neumayeri*, a thistle-like plant with pink flower heads and neat green involucral bracts with papery margins, and narrow strap-shaped but spineless basal leaves. *Iris pallida* with large lilac or violet flowers and silvery-white papery spathes; it is a native of western Yugoslavia but is often planted as an ornamental. Another species which is endemic to this part of Yugoslavia and Albania is Pyrethrum, *Tanacetum cinerariifolium*, which is now widely culti- vated throughout the world as an insecticide and known as 'Dalmatian powder'. It has silvery-grey dissected foliage and solitary white flower heads with white strap-shaped ray-florets. *Leucanthemum chloroticum*†, an endemic Ox-eye Daisy, and the thyme-like plants *Micromeria dalmatica*, *M. thymifolia*, and *Satureja montana* subsp. *illyrica* are others worth hunting.

On the rocks of the pass over which the old Kotor–Cetinje road crosses the main watershed, on the perimeter of the national park, the following can be seen:

Dianthus sylvestris	*Asperula scutellaris*
Eryngium amethystinum	*Teucrium arduini*†

---

16 **Pirin Planina** This mountain range lies in the extreme south-west corner of Bulgaria and supports a very interesting mixture of central European, alpine, Balkan mountain, and some Mediterranean species. The climate has many similarities with the Alps, snow lies late, yet it is situated only about 150 km north of the Aegean Sea.

17 **National Park of Plitvice** Natural dams of turfa are formed by the clear water, creating numerous cascades as it flows from the upper to the lower lakes. Dense mixed forests of Beech and Silver Fir, with Manna Ash, Hop-Hornbeam, and *Sorbus*, flourish in the humid atmosphere.

YUGOSLAVIA, ORJEN, LOVCEN
1. *Lonicera glutinosa* 1302b 2. *Achillea abrotanoides* 1420d 3. *Satureja montana* 1152 4. *Teucrium montanum* 1103 5. *Euphorbia capitulata* 670b 6. *Valeriana montana* (1315) 7. *Thalictrum aquilegifolium* 255 8. *Stachys alopecuros* 1142 9. *Micromeria thymifolia* 1153c 10. *Edraianthus tenuifolius* 1354a

11. *Aethionema saxatile* 346 **12.** *Thymus striatus* 1163g **13.** *Anemone apennina* (214) **14.** *Bupleurum ranunculoides* 883b **15.** *Plantago argentea* 1290a **16.** *Cerastium grandiflorum* 140c **17.** *Salvia officinalis* 1143 **18.** *Frangula rupestris* 724a **19.** *Tanacetum cinerariifolium* 1429a

Stachys annua   *Calamintha nepeta*
Micromeria parviflora†   **Campanula versicolor*

Mljet

The National Park of Mljet is the finest example existing of a richly
wooded island off the Dalmatian coast. It is a long island running
parallel to the mainland and is all that remains above water of a low
range of hills, one of the Dinaric folds thrown up during the Tertiary
Alpine upheavals, with the intervening valleys flooded during the late
Quaternary times. It is a limestone island showing many of the typical
karstic features of the mainland; notable are the almost land-locked
gulfs opening by narrow channels to the sea. On the south coast is the
deep inlet of Veliko Jezero, while on the north coast the Gulf of Polace
is protected by low outlying islands. The national park covers the
whole of the north-west of the island with its wooded hills, inlets, and
islands. The wooded areas are dominated by the evergreen Holm Oak,
*Quercus ilex, which with the Manna Ash, *Fraxinus ornus*, forms the
natural vegetational climax under these conditions, but which elsewhere has been devastated by man and his animals. This has a very
characteristic Mediterranean flora of small trees and shrubs such as:
*Juniperus oxycedrus, *Laurus nobilis*, **Pistacia lentiscus*, **Phillyrea
latifolia*, **Arbutus unedo*, *Viburnum tinus*, with **Coronilla emerus*,
Myrtus communis, **Erica manipuliflora*, and *Teucrium fruticans*.
Climbing and scrambling up through these thickets are the typical
Mediterranean climbers: *Clematis flammula*, *Rubia peregrina*,
*Lonicera implexa, **Smilax aspera*, *Tamus communis*, and
*Asparagus acutifolius.

   Herbaceous plants include: **Cyclamen repandum*, and **Cephalaria
leucantha*, while *Limonium cancellatum*, *Pancratium maritimum*, and
Calystegia soldanella are on the coastal rocks and sands.

   The Aleppo Pine, **P. halepensis*, forms the finest stands of
mature trees in the Adriatic, with typical maquis beneath, in
which the Myrtle, *M. communis*, is particularly well developed in
these woods.

The Neretva gorge

The Neretva river flowing down from the heartland of Bosnia and
Hercegovina is the largest river on the Yugoslav coast draining into
the Adriatic. It cuts a deep gorge through the successive ranges of the
Dinaric Mountains passing the towns of Jablanica, Mostar, and Met-
ković near the coast, to debouch into kilometres of low coastal water-

logged flats opposite the islands of Mljet and Korčula. Down the Neretva valley the main road and rail communications run between the Adriatic and the interior to Sarajevo. Around the upper regions of the Neretva gorge, south of Jablanica, are the limestone mountains of Prenj (2155 m), Čvrsnica (2228 m), and Cabulja (1780 m), which are well worth exploring. Some of the distinctive species of this whole limestone area are:

Taxus baccata	*Daphne alpina
Corylus colurna	*Androsace villosa
Saxifraga sedoides subsp.	*Moltkia petraea
prenja	*Satureja montana
Dryas octopetala	Linaria alpina
Potentilla apennina	Scabiosa columbaria subsp.
*P. speciosa	cinerea
*Petteria ramentacea	*Senecio thapsoides subsp.
*Oxytropis prenja†	visianianus
Rhamnus pumila	*Amphoricarpos neumayeri

Also *Edraianthus serpyllifolius, an attractive small campanula-like plant of limestone crevices restricted to the Balkans, with solitary dark-violet bell-shaped flowers and spathulate ciliate leaves.

The Neretva valley allows the penetration of many Mediterranean species into the interior and some as far inland as Mostar and beyond. Such species as the Pomegranate, Punica granatum, Ephedra major, *Arbutus unedo, *Salvia officinalis, and *Inula verbascifolia can be seen some way up the slopes of Mount Prenj, for example. But where the mountains run parallel to the coastline the penetration of the Mediterranean flora inland may be restricted to a narrow coastal belt only.

Further north up the Dalmatian coast lies the small mountain range of Biokova. It runs parallel to the coast, and within 5 km or so rises steeply to summits of 1600–1700 m, opposite the islands of Hvar and Brač. The bora winds, bringing cold air down from these coastal mountains, are intense in this region, particularly behind Makarska. They have a profound effect on the vegetation, and some interesting and unique plants are found here, such as Edraianthus pumilio†, a dwarf tufted plant with solitary stalkless blue-violet campanula-like flowers, and the similar E. dinaricus which is a taller plant with flowering stems 2–6 cm, also Centaurea biokovensis† with pink flower heads.

Other interesting, largely Balkan species include:

Pinus nigra subsp. dalmatica	Ephedra major
Juniperus sabina	*Cerastium grandiflorum

Paronychia kapela
Dianthus sylvestris subsp.
　tergestinus
Ranunculus illyricus
Corydalis ochroleuca
Peltaria alliacea
Genista radiata
Astragalus purpureus
Geranium macrorrhizum

Euphorbia characias subsp.
　wulfenii
Rhamnus intermedius †
Portenschlagiella
　ramosissima
Ferulago campestris
Peucedanum longifolium
Onosma visianii
Thymus striatus

11. MOUNTAINS OF CENTRAL YUGOSLAVIA

National Park of Durmitor

The high limestone plateau of the Dinaric range that runs through the heart of Yugoslavia rises to its loftiest point in the Durmitor Massif of Montenegro. A cluster of high, round-shouldered, limestone summits rises boldly out of the plateau which itself is about 1500 m. The highest summit is the craggy Bobotov Kuk (2522 m) which lies at the centre with an outer ring of only slightly lesser peaks surrounding it, including Veliki Štulac (2196 m), Crvena Greda (2100 m), Suva Rtina (2294 m) to the north, Prutas (2400 m) to the west, Šljeme (2477 m), Savin Kuk (2312 m), and Medjed (2280 m) to the east, and Stozina to the south. There are also mountain lakes of glacial origin such as the Crno and Zmijinje-Skrcko, often surrounded by dense fir forests. 'Outstanding features are the rich glacial relief in the form of numerous kettles and cirques, valleys and lakes, as well as a number of interesting karst phenomena (funnels, uvulas, clefts and caves).'[23]

To the north of this massif is the great gorge of the Tara river which has eroded a cleft in the plateau to over 1000 m in depth through the limestone bedrock on its way northwards to join the Drina and the Danube at Belgrade. To the west is another deep canyon, the Sušica, which has cut through the plateau to nearly 700 m above sea-level.

It is a magnificent, unspoiled country which should be visited between June and August for its flora, and an energetic week or more is certainly required for its exploration. The best centre from which to reach the heart of this country is the small summer village of Žabljak, where there are hotels and a camping site. The village is situated on the meadowland plateau below the forested slopes, with the towering peaks of Medjed and Savin Kuk as a background.

The meadows in the vicinity of Žabljak and those on lower slopes of the mountains between the fir forests are rich in attractive herbaceous plants. The finest of the milkworts, *Polygala major*, is common, its clusters of blue-violet or rich-pink flowers showing up conspicuously in the grass; sometimes the flowers are variegated and have pink petals and blue 'wings'. One of the finest of the flaxes, *Linum capitatum*, also grows here in abundance with dense rounded heads of bright golden-yellow flowers, borne on erect stems, with small lance-shaped leaves. The small annual gentian, *G. utriculosa* is also found here. It has intense blue flowers and a calyx which is conspicuously winged, thus distinguishing it from the similar-looking more central European *G. nivalis.*

Other meadow species include:

Polygonum bistorta	*Veronica austriaca
Cerastium moesiacum	*Pedicularis brachyodonta
*Silene roemeri	Campanula patula
Dianthus moesiacus	Hypochoeris maculata
Ranunculus oreophilus	*Scorzonera purpurea subsp.
Biscutella laevigata	rosea
Sanguisorba minor	Nigritella nigra
*Genista januensis	Traunsteinera globosa
Trifolium montanum	

The forest vegetation is well developed on the lower slopes of the massif, particularly on the northern-facing slopes. Most interesting are the beech forests of distinctive Montenegrin character which appear to be in their natural state particularly in the area of the Mlinski stream which runs into the Crno Jezero or Black lake, which is about 3 km from Žabljak. In the vicinity of the lake and the path that encircles its shores can be seen:

Lychnis viscaria	Digitalis grandiflora
Geranium sylvaticum	Veronica urticifolia
Euphorbia myrsinites	Melampyrum nemorosum
Moneses uniflora	Lilium martagon
Gentiana asclepiadea	Paris quadrifolia

Coniferous forests of *Picea abies* and *Abies alba* occupy the lower mountain slopes and sub-alpine beech forests here form the uppermost zone and tree-line below the open mountain grassland.

The upper forests are unusual in having *Acer heldreichii*, which is easily distinguished from all other European maples by its very deeply lobed leaves with the middle lobe cut almost to the base and parallel-sided. Other shrubs and small trees include: *Cotoneaster nebrodensis*, *Sorbus aucuparia*, *S. torminalis* and *S. chamaemespilus*, *Rhamnus alpinus* subsp. *fallax*, and *Lonicera alpigena*.

There are a number of rough paths starting from the vicinity of the

YUGOSLAVIA, DURMITOR, MAGLIC, VOLUJAK, ZELENGORA
1. *Minuartia hirsuta* 132a   2. *Anthyllis montana* 621   3. *Gentiana cruciata* 989
4. *Silene acaulis* 170   5. *Primula halleri* (943)   6. *Helianthemum oelandicum* 801g
7. *Arabis alpina* 318   8. *Aster bellidiastrum* 1363a   9. *Androsace villosa* (954)
10. *Potentilla chrysantha* 453f

11. *Gentianella crispata* 1001a **12.** *Orthilia secunda* 914 **13.** *Polygonatum verticillatum* 1656 **14.** *Sedum atratum* 392 **15.** *Globularia cordifolia* 1263
16. *Viola calcarata ssp. zoyzii* 786 **17.** *Soldanella alpina* 956 **18.** *Arctostaphylos uva-ursi* 925 **19.** *Scabiosa silenifolia* 1325e **20.** *Acinos alpinus* (1157)
21. *Myosotis alpestris* 1067

Black lake, leading to the south to Medjed and Savin Kuk, to the east to Bobotov Kuk, and northwards to Crvena Greda. The climbs to the summits are steep, but the final rise is rarely more than 1000 m so that they can be easily reached within the day from Žabljak.

Above the tree-line are open grassy slopes with clumps of *Juniperus communis* subsp. *nana* and some of the shrubs mentioned above.

The alpine flora, though not particularly rich in Balkan species, has many good central European mountain species. The Balkan species include:

*Minuartia hirsuta	*Potentilla chrysantha
*Cerastium decalvans	Euphorbia myrsinites
*Ranunculus sartorianus	*Viola calcarata subsp. zoysii
Thlaspi praecox	*Pedicularis brachyodonta
*Iberis sempervirens	*Edraianthus graminifolius

These were all seen on Savin Kuk, above 2000 m.

Other interesting plants of limited distribution found in this region are the woolly-haired, yellowish-leaved *Verbascum durmitoreum*† which is restricted to the mountains of Montenegro and Bosnia; it usually has a simple inflorescence and yellow flowers with violet stamen hairs and kidney-shaped anthers. Also found are *Centaurea kotschyana* with dark-purple flower heads and dark-brown or black involucres, occurring only in Macedonia and the Carpathians; the white-flowered umbellifer *Pimpinella serbica*; *Gentinaella crispata*, distinguished by the black crisped margins of the calyx lobes and its violet or whitish flowers; the lilac-blue *Scabiosa silenifolia* with flower heads with long marginal florets, and calyx bristles 2–3 times as long as the 20–4-veined crown; and the tiny *Omalotheca (Gnaphalium) pichleri*.

Plants with a more central European distribution seen on Savin Kuk were:

*Silene acaulis	*Androsace villosa
*Dianthus sylvestris	A. chamaejasme
*Saxifraga adscendens	*Soldanella alpina
S. paniculata	*Myosotis alpestris
Dryas octopetala	*Veronia austriaca subsp.
*Astragalus vesicarius	vahlii
A. australis	*Pedicularis verticillata
*Anthyllis vulneraria	*Globularia cordifolia
*A. montana	*Aster bellidiastrum
*Helianthemum oelandicum	Muscari botryoides
Polygala alpestris	Nigritella nigra
Gentiana verna	

While on screes near the summit were:

Salix retusa
Moehringia ciliata
Papaver kerneri
*Cardamine glauca

*Arabis alpina
Linaria alpina
*Valeriana montana

National Park of Sutjeska

This lies north of the Durmitor Massif, over the provincial boundary of Bosnia-Hercegovina. The main Dubrovnik–Sarajevo road runs through the Sutjeska valley and the village of Tjentište which lies within the boundary of the national park. It is a fine area of heavily forested country rising to a rim of mountains ranging in general height from 1800 to 2200 m, with mountain grassland on their summits. To the south are the Zelan Gora and Volujak mountains while Maglić, the highest summit (2387 m), lies to the south-east on the Bosnia-Hercegovina border.

On the north-western slopes stretching downwards from Maglić to the main road in the Sutjeska valley is the Reserve of Peručica where some of the finest, almost virgin, forests of Yugoslavia can be seen.

The most extensive forest community in this reserve is that of mixed Silver Fir – Beech, while above this on the upper slopes are extensive sub-alpine Beech forest with patches of Spruce, Picea excelsa, on the steeper slopes. Above this again are thickets of the Dwarf Mountain Pine, Pinus mugo, climbing nearly to the summits. Mountain grassland covers much of the highest ground.

At lower altitudes in the park are more limited forest communities such as Oak – Hornbeam; Hop-Hornbeam–Manna Ash–Black Pine; Maple – Ash; and Alder thickets in damp ground of Alnus glutinosa and A. incana.

The park is rich in small trees and shrubs including the following largely central European species:

Rosa pendulina
Sorbus chamaemespilus
Amelanchier ovalis
*Cotoneaster nebrodensis
*C. integerrimus
*Cotinus coggygria
Euonymus latifolius
Rhamnus alpinus subsp.
    fallax

*Acer heldreichii
Daphne laureola
Cornus mas
Fraxinus ornus
Sambucus nigra
*Lonicera alpigena
L. xylosteum

Herbaceous species include:

Asarum europaeum
Silene sendtneri
Anemone narcissiflora
*Ranunculus platanifolius
Papaver kerneri
Cardamine bulbifera
C. enneaphyllos
Lunaria rediviva
*Saxifraga rotundifolia
Aruncus dioicus
*Aremonia agrimonoides
*Potentilla clusiana
P. micrantha
Chamaespartium sagittale
Vicia oroboides
Geranium phaeum
*Linum capitatum

*Viola calcarata subsp. zoysii
*Orthilia secunda
*Soldanella alpina
Salvia glutinosa
Veronica urticifolia
Digitalis grandiflora
Scabiosa columbaria subsp.
  cinerea
Telekia speciosa
Homogyne alpina
Adenostyles alliariae
*Asphodelus albus
Allium victorialis
A. ursinum
*Polygonatum verticillatum
Nigritella nigra

12. NORTHERN DALMATIA

The Velebit Mountains run close to the coast for over 100 km in northern Dalmatia. Their summits of 1000–1700 m lie within 5 km of the coast and they rise steep and stark from narrow coastal terraces, which at one time formed the coastline when the waters of the Adriatic were much higher than they are today. It is a wild forbidding country, with an almost deserted rocky coastline with occasional villages tucked under the steep crags of the Velebit. Here the cold winds of the bora are particularly penetrating and insistent, and trees are only able to grow in sheltered places. To seaward stretches the low inhospitable-looking island of Pag with its bleached limestone ridges.

There are only three good mountain roads from the Adriatic highway to the interior over the Velebit range. The northern one climbs above Senj to the Vratnik pass of 698 m, while the second ascends the centre of the range above Karlobag, rising steeply to the Stara Vrata pass of 928 m before entering the karst plateau. The southernmost is above Obrovac and crosses the Velebit by the Mali Halan pass. Inland of the Velebit lie the large Ličko and Grako Polja. They are drained by surface rivers which flow under the Velebit Mountains; the Lika river disappears for 30 km or more underground before it drains into the Adriatic.

National Park of Paklenica

This lies in the heart of the southern Velebit Mountains, inland from Starigrad. It is easily reached from the Adriatic highway, but when first seen from the highway the terrain looks bleak and forbidding and gives no hint of what lies behind. The entrance to the park is above the village of Marasovići, and above it a narrow gorge cuts through the outer ramparts of the Velebit to lead into a richly forested area which is well worth exploring. A couple of hours' steady but easy climb up a well-made hill track takes one to the mountaineers' refuge which lies in the heart of the forest area. From here tracks lead upwards to the rim of mountains which encircle the whole park. These are the stark limestone summits of Bulima (1559 m), Babin Kuk (1435 m), Crni Vrh (1115 m), and Golić (1266 m); Sveto Brdo (1753 m) and Vaganski Vrh (1758 m) which lie just outside the park are the highest of the Velebit summits. The latter is about 5 hours' walk from the refuge.

The entrance to the park up the Velika Paklenica is dramatic. Sheer bare cliffs of over 400 m rise on each side of the gorge, while the floor of the gorge is blocked by huge boulders round which the track twists and turns. At first there is no stream, but after passing through the gorge one reaches more open ground with slopes covered with shiblyak brushwood of Hop-Hornbeam, Oriental Hornbeam, White Oak, and Manna Ash, and soon a small stream appears. Further up, one is walking through shady beech woods, with a clear stream in the valley below. The contrast between the barren weather-beaten world of the coast and the green and shady forest in the park is breath-taking in its suddenness.

Before entering the beech woods two interesting endemic campanulas can be found growing in the rocks and cliffs. *Campanula fenestrellata*† has heart-shaped, toothed basal leaves and numerous erect blue goblet-shaped flowers with spreading petals. There are two subspecies, *fenestrellata* which is a hairless plant with smaller flowers, and *istriaca* which is woolly-haired, with larger flowers 2 cm long. Both grow in the park. Similar in general form is another endemic campanula, **C. waldsteiniana*, which has the upper stem leaves lanceolate and stalkless (not heart-shaped and short-stalked as in the former species); it also grows in rock crevices. **Teucrium arduini*†, a woolly-haired plant with a very dense spike of small whitish flowers, grows here; it is endemic to western Yugoslavia and northern Albania. **Cymbalaria muralis* subsp. *visianii*, a delicate woolly-haired creeping Toadflax with lilac flowers, was found here, as well as the white-flowered **Peltaria alliacea* with its pendent flattened fruits.

The forests of beech are dense; there is little undergrowth, and huge

YUGOSLAVIA, VELEBIT RANGE, PAKLENICA, RISHJAK
1. *Scutellaria alpina* 1107   2. *Arenaria gracilis* 128a   3. *Peltaria alliacea* 323b
4. *Omphalodes verna* 1047   5. *Dianthus sylvestris* 194   6. *Aubrieta columnae* 320c
7. *Degenia velebitica* 323e   8. *Veronica spicata* 1219   9. *Cymbalaria muralis* 1210

10. *Primula kitaibeliana* 946d   **11.** *Rosa pimpinellifolia* 434   **12.** *Aquilegia kitaibelii* 253f   **13.** *Lonicera alpigena* (1303)   **14.** *Stachys recta* 1136   **15.** *Silene saxifraga* 170c   **16.** *Inula hirta* 1389a   **17.** *Phyteuma orbiculare* 1352   **18.** *Gentiana clusii* (992)   **19.** *Campanula cochleariifolia* 1338

limestone boulders often make passage difficult. The following are often encountered in the forest:

Cardamine bulbifera
C. enneaphyllos
*Aremonia agrimonoides
*Symphytum tuberosum
Salvia glutinosa

Phyteuma spicatum
Prenanthes purpurea
Polygonatum latifolium
Allium victorialis

On the path above the tourist house towards the summit of Bulima there is more open ground with an open woodland of small trees of White Oak and Manna Ash, mixed with other trees such as *Acer obtusatum* and *Sorbus aria*.

On the drier more exposed rocky slopes are stands of Black Pine, **Pinus nigra*, with *Erica herbacea*, *Teucrium chamaedrys*, **Satureja montana* subsp. *illyrica*, and **Globularia meridionalis*.

Forests of beech, oak, and pine cover much of the park and spread up the sides of the surrounding mountains, wherever they can get a footing, leaving only the bare craggy crests glistening like snow in the sun. The park is a beautiful unspoiled area which is easily accessible and would repay several days of exploration. The following plants, among others, should be searched for:

*Drypis spinosa subsp.
 jacquiniana
Ranunculus carinthiacus
*Aquilegia kitaibelii†
Arabis scopoliana
*Aubrieta columnae subsp.
 croatica

*Sempervivum marmoreum
*Sibiraea altaiensis var.
 croatica†
Genista holopetala†
*Euphorbia capitulata
*Arctostaphyllos uva-ursi
*Primula kitaibeliana

---

18 **Dalmatian Coast** The limestone summits of Orjen lie behind the fiord-like gulf of Kotor. Beech forests and Bosnian Pine clothe the steep slopes. In the rocks and screes several species of *Crocus* can be found early in the year, and later such interesting species as *Petteria ramentacea*, *Moltkia petraea*, *Thymus striatus*, *Edraianthus tenuifolius*, and *Scilla litardierei*.

19 **Lake Shkodër** The lake of Shkodër is considered to be a huge polje in the karst which has remained permanently flooded. The surrounding hills are covered with woods of White Oak, Hornbeam, Manna Ash, etc. The lake has a rich vegetation of widely distributed aquatic species such as the Water Soldier, Frogbit, Flowering Rush, and the Water Chestnut.

20 **Adriatic Coast** The Adriatic has, in recent geological times, flooded the limestone valleys of the outer Dinaric ranges, leaving exposed ridges as islands and promontories. A garrigue and maquis of shrubs like *Cistus*, *Erica*, *Juniperus*, Strawberry Tree, and Evergreen Oaks spread down to the sea.

Gentiana clusii
Scutellaria alpina
Edraianthus graminifolius

Leontopodium alpinum
Tulipa sylvestris subsp.
    australis

Another easily accessible area is the country inland above Karlobag over the Stara Vrata pass on the road to Gospić. A rough road branches northwards to the hills of Rusovo (1333 m), crossing open country with karstic grasslands; woodlands and brushwood cover the moister northern and eastern slopes.

On a short visit to this area a number of predominantly Balkan species were found growing in the limestone crevices of the eroded craggy summit ridges of these hills. *Potentilla clusiana* is one of several white-flowered potentillas with silvery leaves which inhabit such cliffs. This is one of the most attractive and has relatively large flowers with the white petals much longer than the sepals, borne on short stems with leaves with 5 leaflets. It is a plant of the eastern Alps as well as of Yugoslavia and Albania. *Genista sericea*, another largely Balkan species, forms small rounded bushes, growing from crevices, with simple leaves and slender-branched clusters of small yellow flowers with silvery-haired petals.

The beautiful pink- or lilac-flowered primula, *P. kitaibeliana†*, endemic to west-central Yugoslavia, was another inhabitant of shady rock crevices; it is distinguished largely by its glandular-sticky leaves. The handsome tufted *Cerastium grandiflorum*, with white-felted leaves and dense clusters of 7–15 white flowers—one of the finest of this genus—was also found in the limestone screes and crevices. It is restricted to the western part of the Balkan Peninsula.

Other interesting plants in the vicinity were:

Arenaria gracilis
Silene saxifraga
Dianthus sylvestris
Alyssoides sinuata
Sedum ochroleucum

Rosa glauca
R. pimpinellifolia
    Amelanchier ovalis
Anthyllis montana
    Rhamnus pumilus

---

21 **The Karst of Yugoslavia** A dolin in the limestone mountains. The fertile terra rossa resulting from the weathering of limestone accumulates in the dolin floor; only here is it possible to grow crops of vegetables or grain. Heavy grazing, and the erosion of soil cover, restrict the rejuvenation of woodlands.

22 **Velebit** The eroded limestone summits of Dalmatia look forbidding and infertile but they contain many interesting and unique Balkan species. Here you can find *Cerastium grandiflorum, Degenia velebitica, Potentilla clusiana, Primula kitaibeliana*, and endemic hellebores, crocuses, and colchicums.

Daphne oleoides
  Athamanta turbith
Stachys recta
Veronica spicata
  Orobanche gracilis
Lonicera alpigena
Phyteuma orbiculare
Edraianthus tenuifolius

Inula hirta
  I. britannica
Achillea clavennae
Centaurea alba
  Lilium bulbiferum
  Fritillaria orientalis
  Convallaria majalis

The Velebit range

This has been described as:

a rich country for bulbous plants which flower in the autumn or very early spring. In April and May they are mostly long over, and one can only locate them by searching for the drying leaves and making a guess at the name until they flower in cultivation. Among the more interesting species to be found are *Sternbergia colchiciflora*, the smallest *Sternbergia* with tiny yellow flowers before the leaves. *Colchicum visianii*, an autumn flowering species with chequered flowers, and *C. hungaricum* with both pink and white flowered forms in the early spring. Crocus is well represented here by *C. dalmaticus*, a pale lavender one, and *C. reticulatus* which has very slender flowers of a deep purple inside and buff-coloured outside, often striped with purple. These both occur low down on the coastal slopes and are spring flowering species. The third spring-flowering species is a dark purple *C. vernus* which prefers the top of the Velebit Mountains in the deeper soils around trees. The autumn flowering species are represented by *C. sativus* var. *pallasii* which appears to be rather uncommon. It has lavender flowers and a bright red stigma and the flowers are produced together with the leaves. *Anemone pavonina* grows in the scrub by the sea, and is rather small-flowered compared with the Greek forms, but has some attractive colour variants. Also preferring the turf just above the shore is *Romulea bulbocodium*, but to see this and the other bulbous plants mentioned one would have to visit the area as early as February.[24]

Other largely Balkan species found in these north-western mountains are *Helleborus multifidus* subsp. *istriacus* and *H. odorus*; the endemic crucifer *Degenia velebitica*†, *Cerastium dinaricum*, *Scabiosa silenifolia*, and *Micromeria croatica*.

National Park of Plitvice

This is a natural area of great beauty which is unique in the Balkans. It is a series of large and small lakes of brilliant emerald green and startling blue set amongst a richly forested mountain area. A small stream in a karstic valley in the Mala Kapela Mountains has, during its fall from 639 to 483 m, built up by natural accretion a series of dams and curtains of tufa or travertine from the dissolved calcium and magnesium salts in the water, which hold back the water at successive levels. Through or over these natural dams the crystal-clear water has carved runnels and spouts, falling in cascades from the lake above to the lake below. The lakes and waterfalls are surrounded by dense forests and the lower branches of overhanging Beech, Hornbeam, and Maple sparkle with spray. The roar of waterfalls, the flash of swift-flowing lustrous waters, the dappled sunlight of the forests contrive to make it a magical world which is deservedly popular with visitors in spring and summer.

The park is about 20 000 hectares in extent, 2000 of which are lakes and 14 000 forests. There are 16 larger lakes and many waterfalls and cascades: the highest, of 30 m, is at Galovac. It is easily reached from the coast by the mountain road which climbs above Senj over the Velebit on its way to Bihać. The village of Plitvice-Leskovac lies on this road at the head of the lakes which stretch northwards for 5 km. The park can also be reached from Karlovac by the main road running south to Zadar. The park is well laid out with leafy walks which sometimes skirt the shores of the lakes or climb up through the dense forest high above the sparkling water; elsewhere there are walkways across the cascades or through the reed beds to the base of waterfalls. A road runs relatively unobtrusively the whole length of the lakes and there is a hotel complex above the largest lake, Kozjak. Though much frequented, the park is well maintained and the natural vegetation seems to be little disturbed.

The forests are largely of beech, with or without Silver Fir. Other common trees include *Ostrya carpinifolia*, *Sorbus aucuparia*, *Acer obtusatum*, and *Fraxinus ornus*. Most of the plants in the park are widespread and can be seen in most other Croatian Beech–Fir forests. There are also open grassy areas and heath-like communities. However, the following should be listed:

Helleborus niger	*Campanula persicifolia*
Hepatica nobilis	*C. latifolia*
*Clematis recta	*Cirsium erisithales*
Daphne mezereum	*Anthericum ramosum
Erica herbacea	*Convallaria majalis*
*Primula vulgaris	*Cephalanthera rubra*
Cyclamen purpurascens	*Neottia nidus-avis*
Digitalis grandiflora	

National Park of Risnjak

This is situated about 30 km inland from Rijeka. It is a fine area of forested mountain country of about 3000 hectares, showing many limestone karstic features such as funnels, springs, and streams which disappear underground. It lies in a heavy rainfall area with an annual average of 3600 mm, with rain occurring at most seasons of the year. The highest summit is Risnjak (1528 m) which just raises its head into the alpine zone. There are a number of limestone peaks in the park ranging from 1200 to 1400 m, many of them forested to their summits.

Access to the park is either from Crni Lug, with a forest track of 10 km to Markov Brlog; from here it is 1½ hours by mountain path, across Medvedja Vrata to the refuge at 1418 m, below the summit of Risnjak. The refuge can also be reached from the Gornje Jelenje-Segine forest road. Several tracks lead upwards to the refuge which can be reached in 1½ hours. The refuge is open from May to October and has 24 beds.

The main forests are of beech mixed with fir, and they cover large areas of the park. At lower altitudes there are transitional woods of Oriental Hornbeam and Hop-Hornbeam, while on the highest summits, particularly on Risnjak, are thickets of Dwarf Mountain Pine, *P. mugo*, interspersed with mountain grassland. There are also patches of spruce forests round many of the summits. Sometimes the upper zones are inverted, as for example in the largest karst funnel on Risnjak. Here the beech forests reach the tops of the ridges to over 1300 m, while the Dwarf Mountain Pine occurs on the lower slopes at 1200 m.

In the Beech–Fir forests the following characteristic species occur:

Anemone ranunculoides	**Omphalodes verna*
Cardamine bulbifera	*Salvia glutinosa*
C. enneaphyllos	*Telekia speciosa*
C. waldsteinii	*Lilium martagon*
*C. trifolia	**L. carniolicum*
Lunaria rediviva	*Paris quadrifolia*
Cyclamen purpurascens	*Iris variegata*
Gentiana asclepiadea	

In the areas of spruce forest are:

Asarum europaeum	*Galium rotundifolium*
Clematis alpina	*Lonicera caerulea*
Rosa pendulina	*Valeriana tripteris*
Moneses uniflora	*Cirsium erisithales*
Erica herbacea	*Listera cordata*

In the mountain grassland and among the clearings in the thickets of Dwarf Mountain Pine the following mostly central European alpine species are present:

Silene pusilla
Trollius europaeus
Pulsatilla alpina
Ranunculus thora
Saxifraga paniculata
Dryas octopetala
*Trifolium alpestre
T. montanum
Viola biflora
Eryngium alpinum
Rhododendron hirsutum
*Arctostaphyllos uva-ursi
*Androsace villosa
*Soldanella alpina
*Gentiana clusii

G. verna subsp. *tergestina*
G. lutea
Thymus praecox
Bartsia alpina
Globularia punctata
Pinguicula alpina
*Campanula cochlearifolia
C. scheuchzeri
*Aster alpinus
Antennaria dioica
Lentopodium alpinum
*Achillea clavennae
Homogyne sylvestris
Nigritella nigra

Other more restricted species are: *Arabis scopoliana*, *Genista holopetala*†, and the local *Valeriana elongata*.

NATIONAL PARKS AND NATURE RESERVES IN THE BALKANS

A national park by definition should have the following characteristics. It should be a relatively large area where one or several ecosystems are not materially altered by man, where plant and animal species, geomorphological sites and habitats are of special scientific, educative, and recreative interest, and which contains a natural landscape of great beauty. To qualify it must be maintained in this condition and visitors allowed to enter under special conditions, for inspirational, educative, cultural, and recreative purposes.[25]

Certainly not all the national parks listed below satisfy these conditions and qualify for this title; many are in reality 'nature parks' where recreational activities take precedence over conservation; others again are reserves where a certain ecosystem, species, or geological formation is preserved either in a 'strict reserve' with only limited access, or which is freely open to visitors.

The national parks and nature reserves of Bulgaria and Yugoslavia are among the finest in Europe, and many contain tracts of almost primeval country where man has had a minimal influence on the natural vegetation. The building of roads, hotel complexes,

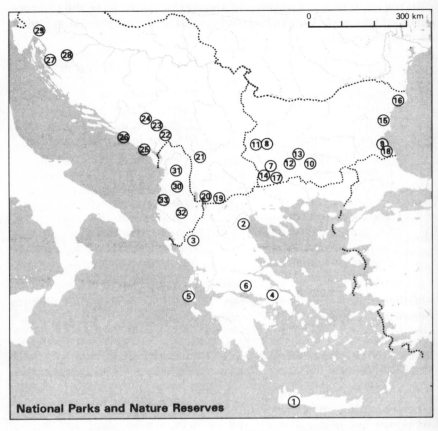

National Parks and Nature Reserves

Greece and Crete
1 National Park of Samaria Gorge
2 National Park of Olympus
3 National Park of Pindhos
4 National Park of Mount Parnis
5 National Park of Mount Ainos
6 National Park of Parnassos

Bulgaria
7 National Park of Pirin (Vikhren Reserve)
8 National Park of Vitoša
9 National Park of Ropotamo
10 Reserve Cervenata Stena (Red Wall)
11 Reserve Ostriza
12 Forest Reserve Beglica
13 Reserve Izgorjaloto Gjune
14 Reserve Tisova Barchina
15 Reserve Kamchiya
16 Reserve Kaliakra
17 Reserve Ali Botus

18 Reserve Silkossija and Reserve Uzun Budzak

Yugoslavia
19 National Park of Perister
20 National Park of Galicica
21 National Park of Mavrovo
22 National Park of Biogradska Gora
23 National Park of Durmitor
24 National Park of Sutjeska
25 National Park of Lovćen
26 National Park of Mljet
27 National Park of Paklenica
28 National Park of Plitvice Lakes
29 National Park of Risnjak

Albania
30 National Park of Dajti
31 National Park of Lura
32 National Park of Tomori
33 National Park of Divjaka

rest-houses, refuges, and camping areas in many parks is undoubtedly rapidly increasing the pressure on these parks, but careful management and control of tree-felling, grazing, hunting, and man's other destructive activities have helped to redress the balance, and they still remain some of the most awe-inspiring and dramatic places where man can relax from the pressures of the modern world and study nature in the wild.

Further south—particularly in southern Greece—the situation is very different. Under Mediterranean climates, where the density of population has for milleniums brought heavy pressure to bear even on the more remote natural areas, and where rejuvenation of vegetation under much harsher climatic conditions is very slow or irreversible, there remains no relatively primeval terrain with its natural vegetational cover. The national parks that have been designated, but hardly qualify for this title, show all the marks of man's ceaseless activity in the destruction of forests and the grazing of his flocks to an intensity known perhaps nowhere else in Europe. Here, as in the Iberian Peninsula, which are the two floristically richest areas of Europe, it is imperative that national parks and nature reserves should be created and maintained to conserve this unique flora and fauna. The problems are many, both ecological and social, and Broussalis[26] has suggested that the three most immediate protective measures for the preservation of the flora in Greece should be: the suppression or reduction of the flocks of goats, and possibly sheep, that graze freely over the countryside; the gradual eradication of the habit of gathering indiscriminately the wild flowers both by professional and amateurs—an educational problem; and the protection of rare and endangered plants by law.

GREECE and CRETE

National Park of Samaria Gorge, Crete, 4850 hectares (see p. 68).
National Park of Olympus, 3998 hectares (see p. 113).
National Park of Pindhos (Pindus), 45 000 hectares (see p. 104)
National Park of Mount Parnis (Parnes), 3700 hectares [Attic mountain and forest species]
National Park of Mount Ainos, Cephalonia, 2841 hectares [Greek Fir Forests and mountain species]
National Park of Parnassos, 3512 hectares (see p. 91)

BULGARIA

National Park of Pirin (Vikhren Reserve), 6736 hectares (see p. 132)
National Park of Vitoša, 22 800 hectares (see p. 156)
National Park of Ropotamo, 847 hectares (see p. 142)

Nature Reserves of botanical interest include:

Reserve Cervenata Stena (Red Wall), Rhodope Mountains, 230 hectares (see p. 134)

Reserve Ostriza, Golobado Mountain (see p. 158)

Forest Reserve Beglica, Western Rhodope Mountains (see p. 140)

Reserve Izgorjaloto Gjune, Central Rhodope Mountains, near Krichim [*Juniperus excelsa*, Balkan species, and sub-Mediterranean species]

Reserve Tisova Barchina, Struma valley, 19 hectares [*Juniperus excelsa* and many Mediterranean and sub-Mediterranean species]

Reserve Kamchiya, outfall of Kamchiya river, south of Varna, 524 hectares [river valley forest]

Reserve Kaliakra, Black Sea, 53 hectares (see p. 147)

Reserve Ali Botus, Slavijanka Mountain, near Paril, Blagoevgrad District, 542 hectares [Balkan endemic species; Mediterranean and sub-Mediterranean species]

Reserve Silkossija and Reserve Uzun Budzak, Istranca Mountains, near Kosti, Burgas District (see p. 145)

YUGOSLAVIA

National Park of Perister (Macedonia), 10 400 hectares (see p. 160)

National Park of Galicica (Macedonia), 23 000 hectares (see p. 163)

National Park of Mavrovo (Macedonia), 79 070 hectares (see p. 170)

National Park of Biogradska Gora (Montenegro), 3600 hectares [Beech-Fir forests, Spruce, mountain scrub and grassland.]

National Park of Durmitor (Montenegro), 32 000 hectares (see p. 182)

National Park of Sutjeska (Bosnia-Hercegovina), 17 250 hectares (see p. 187)

National Park of Lovćen (Montenegro), 2000 hectares (see p. 176)

National Park of Mljet (Croatia), 3100 hectares (see p. 180)

National Park of Paklenica (Croatia), 3616 hectares (see p. 189)

National Park of Plitvice Lakes (Croatia), 19 172 hectares (see p. 194)

National Park of Risnjak (Croatia), 3014 hectares (see p. 196)

ALBANIA

National Park of Dajti, 3000 hectares [Beech and Bosnian Pine Forests with Mediterranean maquis]

National Park of Lura, 3000 hectares [forests of Bosnian, Macedonian, Scots and Black Pine, Beech, and Spruce; alpine meadows]

National Park of Tomori, 3000 hectares [Beech forests and Bosnian Pine; alpine meadows]

National Park of Divjaka, 1000 hectares [Aleppo and Stone Pine forest on the Adriatic coast]

REFERENCES TO CHAPTER 2

[1] *Britannica Yearbook of Science and the Future*, 1974, p. 36.

[2] P. H. Davis, 'On the Rocks', *Bull. Alp. Gard. Soc.* 15, 1947, p. 37.

[3] P. H. Davis, 'A Collector in Crete', *Bull. Alp. Gard. Soc.* 5, 1939, p. 399.

[4] W. Greuter, *Guide to the Excursions*, 1st Optima Meeting, Geneva, 1975, p. 17.

[5] Ibid., p. 15.

[6] Geographical Handbook Series, Naval Intelligence Division, *Greece*, i, 1944, p. 27.

[7] W. B. Turrill, *The Plant Life of the Balkan Peninsula*, Oxford, 1929.

[8] P. Quézel, 'Végétation des hautes montagnes de la Grèce méridionale', *Vegetatio*, xii, 1964.

[9] N. A. Goulandris and C. N. Goulimis, *Wild Flowers of Greece*, Kiffisia, 1968, p. xxvi.

[10] H. P. Thomson, 'Plant-hunting in South Greece', *Bull. Alp. Gard. Soc.* 8, 1940, p. 58.

[11] P. Quézel and M. Katrabassa, 'Premier aperçu sur la végétation du Chelmos Peloponèse', *Rev. Biol. Écol. Médit.* i, 1974.

[12] P. Quézel, 'Vegetation des hautes montagnes de la Grèce méridionale', *Vegetatio*, xii, 1964.

[13] Ibid. 300.

[14] P. Quézel, 'Vegetation des hautes montagnes de la Grèce méridionale', *Vegetatio*, xii, 1964.

[15] N. A. Goulandris and C. N. Goulimis, *Wild Flowers of Greece*, Kiffisia, 1968, p. xxiii.

[16] *Plant Life of the Balkan Peninsula*.

[17] J. Halda, 'Plant-hunting in the Bulgarian Mountains', *American Rock Gdn. Soc. Bull.* 31, 1973, p. 57.

[18] S. Stanev, *Our Reserves and Natural Sites*, Sofia, 1970, p. 86.

[19] *Plant Life of the Balkan Peninsula*.

[20] A. G. Todorovski, *National Park Pelister*, Bitola, 1972.

[21] B. Mathew and C. Grey-Wilson, 'Some Flowers of Yugoslavia', *Bull. Alp. Gard. Soc.* 39, 1971, p. 116.

[22] A. O. Chater, Lists of Species collected in Macedonia (personal communication).

[23] T. Wojterski, 'National Parks of Yugoslavia', *Zaklad Ochrony Przyrody Polskiej Akademii Nauk*, 1971, p. 126.

[24] B. Mathew and C. Grey-Wilson, 'Some Flowers of Yugoslavia', *Bull. Alp. Gard. Soc.* p. 198.

[25] *1975 United Nations List of National Parks and Equivalent Reserves*, International Union for Conservation of Nature and Natural Resources, Switzerland, 1975.

[26] P. Broussalis, 'The Protection of the Greek Flora and its Problems', *Nature Bull. Hellenic Soc. for the Protection of Nature*, 12, 1977, p. 31.

3. The Identification of Species

GYMNOSPERMAE
PINACEAE | Pine Family

ABIES 1. *A. alba* Miller SILVER FIR. Mountain forests. YU.AL.GR(north).BG. ****(1).**A. *cephalonica* Loudon GREEK FIR. A forest-forming tree to 30 m, which replaces 1 in the mountains of Greece. Distinguished by its stiff, spiny-pointed leaves arranged all round the hairless twigs, and by its resinous buds. Cones erect, 12–16 cm; scales falling when ripe, leaving a persistent axis. Mountains between 800 and 1700 m. GR+Euboea. **1b.** *A. borisii-regis* Mattf. Probably a hybrid with intermediate characters between *alba* and *cephalonica* which occurs in the S. Balkans. Young twigs densely hairy; buds and leaves like (1). **1c.** *Pseudotsuga menziesii* (Mirbel) Franco DOUGLAS FIR. A native of N. America; planted in YU.GR.BG.

PICEA **2. *P. abies* (L.) Karsten NORWAY SPRUCE. A forest-forming, tall pyramidal tree of the mountains. Leaves spirally arranged, without white lines, diamond-shaped in section, spiny-tipped. Cones pendulous. Mountains. YU.AL.BG. **2a.** *P. omorika* (Pančič) Purkyně OMORICA or BOSNIAN SPRUCE. A slender, very local tree of central Yugoslavia, in the Drina river basin, the Viogora Mountains, and Radomislje; distinguished by its flattened leaves with 2 white bands above and dark green beneath. Cones small 3–6 by 1·5–2·5 cm. YU(central). Pl. 1.

PINUS | Pine

1. Lvs. in pairs
  2. Seeds wingless or nearly so; tree
     umbrella-shaped.   *pinea*
  2. Seeds winged; tree usually not
     umbrella shaped.
    3. Twigs grey in first year.
      4. Lvs. less than 1 mm wide.   *halepensis*
      4. Lvs. more than 1 mm wide.   *heldreichii, leucodermis*
    3. Twigs not grey.
      5. Buds not resinous.
        6. Resin canals of lvs. median,
           cone 8–20 cm (introd. YU.
           GR.TR.).   ****4.** *P. pinaster* Aiton
        6. Resin canals of lvs. marginal;
           cone 5–11 cm (E. Aegean Is.
           Thrace. CR.TR.),   **6a.***P. brutia* Ten. Pl.1.

5. Buds resinous.
   7. Lvs. usually more than 8 cm
      long.   *nigra*
   7. Lvs. usually less than 8 cm
      long.
      8. Tree; branches and upper
         trunk pale red-ochre (YU.
         AL.GR.BG.).   **8.** *P. sylvestris* L.
      8. Shrub to 3·5 m; branches
         dull brown (YU.AL.BG.).   **9.** *P. mugo* Turra
1. Lvs. in clusters of 5.   *peuce*

5. *P. pinea* L. STONE or UMBRELLA PINE. A dense bright-green, umbrella-shaped tree, bearing heavy, nearly globular, mahogany-brown cones 8–15 cm. Leaves rigid, 10–20 cm. Sandy and gravelly soils, on the littoral and coastal hills. Widespread in the Med. region. YU.AL.GR.CR.TR. ****6.** *P. halepensis* Miller ALEPPO PINE. Usually a rather slender tree with silvery-grey branches and delicate pale-green foliage. Leaves 6–15 cm by less than 1 mm broad. Cones conical, 5–12 cm. Often forming open forests in the Med. region. YU.GR+Is. CR? Page 35.

**7. *P. nigra* Arnold BLACK PINE. A very variable tree, distinguished by its dark blackish bark, its rigid, dark-green, and usually very long leaves, and its shining, oval-conical, stalkless, solitary or clustered cones. Subsp. *nigra* has spiny-pointed, straight or incurved leaves 7–18 cm. Cones 5–8 cm. YU.AL.GR(north). Subsp. *pallasiana* (Lamb.) Holmboe has rigid, twisted, or curved leaves 12–18 cm. Cones 5–12 cm. YU(east). GR.BG.TR. Page 205. Subsp. *dalmatica* (Vis.) Franco is a small tree with rigid leaves 4–7 cm. Cones 3·5–4·5 cm YU.

7a. *P. leucodermis* Antoine BOSNIAN PINE. A massive pyramidal tree with spreading branches, grey bark, and oval-conical stalkless, horizontal cones. Leaves 7–9 cm, rigid, spiny-pointed. Cones 7–8 cm; cone scales with recurved tips. Forming forests in high limestone mountains in central and west Balkans. YU.AL.GR.BG. Page 205. **7b.** *P. heldreichii* Christ WHITE-BARK PINE. Distinguished from 7a by its rounded pyramidal crown, its young twigs which are grey only in the first year, then becoming brown, and its cone scales which have very short, straight tips. Limestone mountains of central Balkans. YU.AL.GR(north). Page 205. **10a.** *P. peuce* Griseb. MACEDONIAN PINE. A slender, densely pyramidal tree with grey bark, slender leaves in clusters of 5, and long cylindrical, pendulous cones 8–15 cm by 2–4 cm. Leaves rigid, 6–12 cm, deep green, densely crowded. Forming forests on siliceous mountains of central Balkans. YU(South.AL.GR?.BG. Page 169.

CUPRESSACEAE | Cypress Family

CUPRESSUS ****11.** *C. sempervirens* L. FUNERAL CYPRESS. A tree with two distinctive forms, either conical with massive spreading branches, or slender and columnar with erect branches. Twigs covered with small triangular overlapping scale-like leaves about 1 mm, dark green (recalling 14). Female cones globular, 2·5–4 cm, at length separating into flat-topped, angular scales. Rocky places in mountains. Widely cultivated (particularly the columnar form) and naturalized in the Med. region. Native in CR.

JUNIPERUS | Juniper Cones globular, berry-like, not separating into scales.

1. Lvs. all needle-like; cones
    axillary.
  2. Lvs. with 2 white bands above.
    3. Cones 8–15 mm.   *oxycedrus*
    3. Cones 2–2·5 cm.   *drupacea*
  2. Lvs. with 1 white band above
      (Widespread on mainland).   ****12.** *J. communis* L.
1. Lvs. of adult plants scale-like;
    cones terminal.
  4. Scale-like lvs. blunt, with a
      narrow papery border; ripe cones
      dark red.   *phoenicea*
  4. Scale-like lvs. without a narrow
      papery border; ripe cones dark
      purplish or blackish.
    5. Ripe cones 7–12 mm.
      6. Twigs 1 mm wide, quadrangular;
          scale-like lvs. 1·5–2 mm; cones
          almost black when ripe; seeds 1–2.   *foetidissima*
      6. Twigs 0·6–0·8 mm wide, rounded;
          scale-like lvs. 1–1·5 mm; cones
          dark purplish-brown when ripe;
          seeds 4–6.   *excelsa*
    5. Ripe cones 4–6 mm
        (YU.AL.GR.BG.).   **15.** *J. sabina* L.

13. *J. oxycedrus* L. PRICKLY JUNIPER. A shrub or small tree with silvery-green, sharp-pointed, spreading leaves. Cones 6–15 mm, yellow, becoming reddish when ripe. Rocky places mostly within the influence of the sea. Widespread. Page 34. **13a. *J. drupacea* Labill. (*Arceuthos d.*) SYRIAN JUNIPER. A pyramidal tree like 13, but distinguished by its much larger cones which are brown to blackish-purple, and bloomed. Leaves 1–2·5 cm by 2–4 mm, pointed. Forest-forming in mountains. GR(Parnon). Page 205.

14. *J. phoenicea* L. PHOENICIAN JUNIPER. A dense conical, dark-green shrub or small tree. Adult leaves about 1 mm, scale-like, lying flat. Cones 8–14 mm, dark red when ripe. Rocky places, and evergreen scrub in the Med. region. YU.AL.GR+Is.CR. Page 35. **14. *J. foetidissima* Willd. A tree with a narrow conical crown, with strongly smelling leaves when crushed, and distinctly quadrangular twigs. Apex of scale-like leaves free. Ripe cones black. Rocky places in mountains, usually as scattered trees. YU(Macedonia).AL.GR+Euboea, Lesbos. Page 169. Pl. 1. **14b.** *J. excelsa* Bieb. GRECIAN JUNIPER. Like 14a but differs in having more slender, rounded twigs, and dark purplish-brown cones. Rocky places. YU(Macedonia).GR(north)+E.Aegean Is.CR.BG. Page 205.

TAXACEAE | Yew Family

TAXUS ****17.** *T. baccata* L. YEW. Woods on calcareous soils in mountains. Scattered throughout the mainland.

1. *Juniperus excelsa* 14b **2.** *Pinus nigra* subsp. *pallasiana* 7 **3.** *Juniperus drupacea*
3a **4.** *Pinus heldreichii* 7b **5.** *Pinus leucodermis* 7a

EPHEDRACEAE | Joint-Pine Family

EPHEDRA
1. Twigs flexible; fruit 8–9 mm.   *fragilis*
1. Twigs rigid; fruit 5–7 mm.
   2. Scale-like lvs. green on back;
      shrub to 50 cm (YU.AL.GR.BG.).   **(18).** *E. distachya* L.
   2. Scale-like lvs. almost
      completely papery; shrub to 2 m
      (YU.AL.GR.TR.).   **18a.** *E. major* Host

**18. *E. fragilis* Desf. A scrambling or spreading shrub to 5 m, with greenish, flexible brittle stems and conspicuous globular red fruits. Subsp. *campylopoda* (C. A. Meyer) Ascherson & Graebner is a climbing plant with pendulous branchlets. Bushy places. Widespread in the Med. region. YU.AL.GR+Is.CR.BG.TR. Page 97.

ANGIOSPERMAE
DICOTYLEDONES
SALICACEAE | Willow Family

SALIX | Willow   About 25 species are found in our area; the majority are widespread European species.

POPULUS | Poplar   The following are widespread on the mainland.   **30.** *P. alba* L WHITE POPLAR;   ****32.** *P. tremula* L. ASPEN;   **33.** *P. nigra* L. BLACK POPLAR. Other species are sometimes cultivated, including **31.** *P. canescens* (Aiton) Sm. GREY POPLAR.

JUGLANDACEAE | Walnut Family

JUGLANS   ****35.** *J. regia* L. WALNUT. A large deciduous tree with pale-grey trunk compound leaves, and large green ovoid fruits. Leaflets 7–9, large 6–15 cm, oval-elliptic aromatic when crushed. Native in much of the Balkans; cultivated elsewhere AL.GR.CR.BG. Introd. YU.

BETULACEAE | Birch Family

BETULA   ****36.** *B. pendula* Roth COMMON SILVER BIRCH. Distinguished by its pendulous, hairless twigs, and its hairless nutlets. Woods. YU.AL.BG.   **37.** *B. pubescens* Ehrh. BIRCH. Woods, moors. YU.

ALNUS   ****40.** *A. glutinosa* (L.) Gaertner ALDER. Distinguished by its blunt or notched leaves which are green and hairless beneath except in the axils of the veins. Catkins appearing before the leaves. Trunk dark brown, fissured. Riversides, marshes. Widespread on mainland and larger islands. **41. *A. incana* (L.) Moench GREY ALDER. Like 40 but leaves acute, pointed, greyish-green, and finely hairy beneath. Trunk pale grey or yellowish, smooth. Riversides. YU.AL.BG. **39. *A. viridis* (Chaix) DC. GREEN ALDER. Catkins appearing with leaves. A shrub. Mountains. YU.BG.

CORYLACEAE | Hazel Family

CARPINUS **42. *C. betulus* L. HORNBEAM. A tree with grey fluted bark, distinguished by its double-toothed, ovate leaves, 4–10 cm. Fruit with a three-lobed bract, the middle lobe longest. Woods in mountains. Widespread on mainland.

**(42). *C. orientalis* Miller ORIENTAL HORNBEAM. Like 42 but usually a shrub, leaves smaller 2·5–6 cm, and bract of fruit ovate, toothed, not three-lobed. Bushy places in hills. Widespread on mainland+Euboea, Thasos, Samothrace.

OSTRYA **43. *O. carpinifolia* Scop. HOP-HORNBEAM. Distinguished by its ovoid, hoplike fruits composed of papery overlapping bracts which completely enclose the nuts. A small tree with leaves and male catkins like 42. Bushy places in limestone hills and mountains. Fairly widespread on mainland.

CORYLUS 44. *C. avellana* L. HAZEL. Distinguished by the bracts investing the nut being about as long as the nut. Bushy places and woods. Widespread on mainland. (44). *C. maxima* Miller Like 44 but investing bracts longer than the nut, constricted at the apex, and cut into lanceolate, entire lobes. Stipules blunt. Mainland woods; cultivated elsewhere for its fruits. YU.GR.TR? 44a. *C. colurna* L. CONSTANTINOPLE HAZEL. Like 44 but bracts much longer than the nut, not constricted at apex, and divided almost to the base into long-pointed, toothed lobes. Stipules long-pointed. Usually a small tree. Woods. Scattered mainly on mainland (except S.Greece).

FAGACEAE | Beech Family

FAGUS 45. *F. sylvatica* L. BEECH. Leaves ovate, with 5–8 pairs of lateral veins; cup of fruit with awl-shaped scales. Forming forests in hills and mountains of mainland. YU.AL.GR.BG. (45). *F. orientalis* Lipsky EASTERN BEECH. Very like 45 but leaves obovate, with 8–12 pairs of lateral veins; scales of lower part of cup of fruit spathulate. Replacing 45 in the east and forming forests in hills and mountains of mainland. GR.BG.TR. Hybrids between 45 and (45) occur in the central and western regions of the Balkans; they are named *moesiaca*. Page 214.

CASTANEA **46. *C. sativa* Miller SWEET CHESTNUT. Forming forests locally in mountains. Scattered throughout mainland, also Euboea, Thasos, Samothrace, Lesbos, Crete, and Rhodes.

QUERCUS 17 species native to our area. A difficult genus with frequent hybridization.

1. Lvs. evergreen, tough and leathery.
  2. Mature lvs. hairless beneath.   *coccifera*
  2. Mature lvs. hairy beneath.
    3. Bark very thick, corky (YU?).   **49.** *Q. suber* L.
    3. Bark not thick and corky.   *ilex*
1. Lvs. deciduous or semi-evergreen.
  4. Lvs. semi-evergreen, very rigid,
     persisting through the winter; a
     shrub (GR(north).TR.).   **55d.** *Q. infectoria* Olivier
  4. Lvs. deciduous.
    5. Frs. ripening in second year
      (borne on leafless part of twig).
      6. Mature lvs. hairless, shining;
        lv. stalk 2–5 mm.   *trojana*
      6. Mature lvs. finely hairy or
        rough, dull; lv. stalk usually
        more than 10 mm.
        7. Teeth on leaf lobes with fine
          bristle-like tips; upper lv.
          surface smooth.   *macrolepis*
        7. Teeth on leaf lobes blunt, or with
          short points; upper lv. surface
          rough.   *cerris*
    5. Frs. ripening in first year (borne
      among the lvs. at the base of the
      twig).
      8. Scales of cup fused together except
        at apex; frs. usually not crowded,
        borne on a fairly long stalk.
        9. Lvs. with 8–14 pairs of straight,
          parallel lateral veins; inter-
          calary veins absent (BG(east).
          TR.).   **52a.** *Q. hartwissiana* Steven
        9. Lvs. with 5–9 pairs of often
          rather irregular lateral veins;
          intercalary veins usually present.
          10. Lvs. hairless; lateral veins
            mostly straight (YU.AL.BG.).   ****52.** *Q. robur* L.
          10. Lvs. greyish, minutely downy
            beneath; lateral veins arising
            at an acute angle with the mid-
            rib, but then curving outwards
            towards margin (YU.GR.
            BG.TR.).   **52b.** *Q. pedunculiflora* C. Koch
      8. Scales of cup distinct, not fused; frs.
        usually crowded, borne on a short
        stalk, or stalkless.
        11. Young twigs tomentose; lv.
          stalk not grooved.
          12. Lvs. with usually more than 8
            pairs of parallel lateral veins,

intercalary veins absent;
hairs brownish. *frainetto*

12. Lvs. usually with less than 8
pairs of lateral veins which
are not strictly parallel, and
often mixed with intercalary
veins; hairs white or grey.

13. Scales of cup unequal, the
lower broader and more
or less hairless, the upper
narrower, hairy; lateral
veins of lvs. diverging
from mid-vein almost at a
right angle (GR+Is.CR.). **54b.** *Q. brachyphylla* Kotschy

13. Scales of cup more or less
equal; lateral veins of lvs.
diverging from mid-vein
at an acute angle.

14. Lv. stalk 15–25 mm;
scales of cup loosely
adpressed. *virgiliana*

14. Lv. stalk 5–12 mm;
scales of cup closely
adpressed. *pubescens*

11. Young twigs hairless or silky-
haired; lv. stalk grooved.

15. Scales of cup minutely hairy;
mature lvs. hairy beneath or
at least in vein axils (YU.AL.
GR.BG.TR.). **53.** *Q. petraea* (Mattuschka) Liebl.

15. Scales of cup almost hairless,
with distinct swellings;
mature lvs. hairless beneath.

16. Lvs. somewhat leathery,
regularly wavy-margined;
acorn cup brownish. *polycarpa*

16. Lvs. thin, deeply, or
irregularly lobed; acorn
cup greyish (YU.AL.GR.
BG.). **53c.** *Q. dalechampii* Ten.

**47. *Q. coccifera* L. KERMES or HOLLY OAK. A holly-like tree or small shrub often forming
dense thickets. All leaves with undulate spiny margins, quite hairless when mature.
Acorn cup with spreading or recurved spiny scales. Evergreen thickets. Widespread in
the Med. region. YU.AL.GR+Is.BG.TR. Page 34. **48.** *Q. ilex* L. HOLM OAK, ILEX. Like 47
but leaves grey-felted beneath, usually with smooth margins (sucker shoots often with
spiny leaves). Acorn cup with adpressed, non-spiny scales. Evergreen thickets, sometimes forming forests. Widespread in the Med. region. YU.AL.GR+Is.CR.TR. Page
35.

**(50). *Q. trojana* Webb MACEDONIAN OAK. A small half-evergreen tree readily distinguished by its rather stiff, shining, hairless, bristly-toothed, narrowly oblong leaves, and its rather large acorn cups 2–2·5 cm across, with recurved scales. Nuts 3–4·5 cm. Leaves small 3–7 cm, with 8–14 pairs of teeth. Bushy places. YU.AL.GR.BG.TR. **50. *Q. macrolepis* Kotschy (*Q. aegilops*) VALONIA OAK. A small tree distinguished from other oaks by its extremely large acorn cups to 4 cm across, with tough spreading or recurved woody scales. Leaves grey-felted beneath, dull dark green above, with 3–7 pairs of deep triangular bristly-tipped lobes, often further divided; leaf stalk about 1 cm; twigs woolly-haired. Forming open woods in the south Balkans. YU.AL.GR+Is.CR.TR. Page 211. **51. *Q. cerris* L. TURKEY OAK. A large tree distinguished by its slender persistent stipules, its rough, deeply lobed leaves, and small acorn cups with recurved pointed scales, borne on last year's twigs. Leaves variable but usually with 4–8 pairs of deep narrowly triangular teeth or lobes; grey-haired beneath. Woods and bushy places. Widely scattered on mainland.

54a. *Q. frainetto* Ten. HUNGARIAN OAK. A medium-sized deciduous tree to 30 m, with hairy brown branchlets. Buds large, hairy. Leaves very variable, usually very large to 20 cm, deeply cut into 7–9 pairs of oblong, often lobed segments, auricled at base and softly hairy beneath. Acorn cups hairy, scales loosely adpressed. Widespread on mainland. Page 211. **54. *Q. pubescens*. Willd. WHITE OAK. A small deciduous tree or shrub, with densely woolly-haired twigs and young leaves which later become finely hairy beneath. Leaves usually small 4–12 cm, with rounded lobes. Acorn cups grey-haired, scales lanceolate, closely adpressed. Sunny hills, thickets. Widespread on mainland +Aegean Is.

54c. *Q. virgiliana* (Ten.) Ten. A deciduous tree with rather large leaves to 16 cm with rounded lobes and long leaf stalks. Leaf lobes 5–7 pairs, often further lobed. Acorn cups rather large, scales of cup ovate-lanceolate with erect apices. Forming woods. YU.GR.BG.TR. Page 211. **53b.** *Q. polycarpa* Schur A deciduous tree distinguished by its hairless twigs, buds, and mature leaves. Leaves somewhat leathery with 7–10 pairs of equal, shallow, rounded lobes; leaf stalks 1·5–3·5 cm. Scales of acorn cup ovate-acute, with numerous small knob-like swellings, brownish. Forming woods. YU(east). GR.BG.TR. Page 211.

ULMACEAE | Elm Family

ULMUS **56. *U. glabra* Hudson WYCH ELM. Woods in mountains. Scattered throughout mainland. **57.** *U. procera* Salisb. ENGLISH ELM. Bushy places. YU.GR.BG. **58.** *U. minor* Miller SMOOTH-LEAVED ELM. Woods, particularly by riversides. Widespread on mainland+CR, Rhodes. **59. *U. laevis* Pallas FLUTTERING ELM. Distinguished by its long-stalked, pendulous fruits, which are ciliate and have a centrally placed nut. Woods particularly by riversides. YU.AL.GR.BG.

ZELKOVA Like *Ulmus* but fruit a dry drupe, not winged; perianth segments fused.

59a. *Z. abelicea* (Lam.) Boiss. (*Z. cretica*) A small elm-like tree or shrub with small oval, coarsely toothed, nearly stalkless leaves 1–2·5 cm, which are densely hairy beneath. Flowers white, sweet-scented, densely clustered or solitary. Fruit hairy. Rocky places in mountains. CR. Page 214.

1. *Quercus frainetto* 54a   **2.** *Quercus polycarpa* 53b   **3.** *Quercus virgiliana* 54c
4. *Quercus macrolepis* 50

CELTIS
1. Lvs. with acute teeth; nut with a
   network of ridges.
  2. Lvs. rounded or heart-shaped at
   base.   *australis*
  2. Lvs. wedge-shaped at base (YU. BG
   (east).).   **60b.** *C. caucasica* Willd.
1. Lvs. with rounded or broad blunt
   teeth; nut with 4 ridges.   *tournefortii*

60. *C. australis* L. NETTLE TREE. A slender graceful tree with a smooth greyish trunk and oval-lanceloate, long-pointed, acutely toothed leaves. Flowers green. Leaves 4–15 cm, softly hairy beneath, recalling those of elm but distinguished by the marginal vein which connects up the lateral veins. Fruit globular 9–12 mm, solitary, long-stalked, blackish, edible. Bushy places in the Med. region. YU.AL.GR.CR.BG. **60a. *C. tournefortii* Lam. Distinguished from 60 by the blunt, rounded teeth of the leaves and its yellowish-brown ripe fruits. Bushy places, uncommon. YU.GR.CR.

MORACEAE | Mulberry Family

MORUS   **61.** *M. nigra* L. COMMON MULBERRY, and ****(61).** *M. alba* L. WHITE MULBERRY are both often cultivated for their fruit and leaves, and are sometimes naturalized.

FICUS   **62.** *F. carica* L. FIG. Distinguished by its large three- to seven-lobed leaves, its smooth grey stems and branches, and its green or purplish pear-shaped fruits. Widely cultivated for its fruit in the Med. region; often naturalized in rocky places; possibly native in GR.CR.

CANNABACEAE | Hemp Family

HUMULUS   ****63.** *H. lupulus* L. HOP. Hedges, bushy places, on river banks. Widespread on mainland.

CANNABIS   ****64.** *C. sativa* L. HEMP. A slender, erect, leafy, annual to 2·5 m, with leaves deeply cut into 3–9 narrow, acute, toothed segments, and with branched clusters of green 'dock-like' flowers. Widely cultivated and naturalized on mainland.

URTICACEAE | Nettle Family

URTICA   Widespread are: **65.** U. *dioica* L. NETTLE; **66.** *U. urens* L. SMALL NETTLE.
**67. *U. pilulifera* L. ROMAN NETTLE. Distinguished by its globular hanging green female

flower clusters. Leaves painfully stinging. Waste places, waysides. Widespread in the Med. region. **(66). *U. dubia* Forskål Distinguished by the flowers of the male spike which are borne on the upper side of a swollen axis. Stipules 2 at each node. Waste places, cultivated ground in the Med. region. YU.AL.GR.CR+Rhodes.TR.

PARIETARIA Widespread are: **68. *P. officinalis* L. ERECT PELLITORY-OF-THE-WALL; (68). *P. diffusa* Mert. & Koch PELLITORY-OF-THE-WALL; 69. *P. lusitanica* L.

69c. *P. cretica* L. Distinguished by its small leaves, usually less than 1 cm, and its bracts which become brown, hard, and fused to form a five-lobed investment round the fruit. Rocks and walls. GR(south)+Is.CR.

SANTALACEAE | Sandalwood Family

COMANDRA Flowers bell-shaped with 4–5 lobes (in *Osyris* lobes 3–4); stamens 4–5; flowers bisexual. Fruit globular, fleshy, with persistent dry perianth.

70b. *C. elegans* (Reichenb.) Reichenb.fil. An erect, shrubby plant with slender usually unbranched stems, with small flat-topped clusters of cream-coloured or greenish bell-shaped flowers. Flowers about 8 mm across. Leaves numerous, oblong-acute, about 2 cm long. Bushy places, sandy ground. YU.AL.GR.BG. Page 214.

OSYRIS **70. *O. alba* L. A small switch-like shrub to 1 m, with slender green branches and narrow leathery leaves 15 mm by 2–5 mm. Flowers yellowish, with 3 spreading petal-like lobes. Fruit fleshy, bright red, 5–7 mm. Stony and bushy places in the Med. region. Widespread. Page 97.

THESIUM | **Bastard Toadflax** Inconspicuous semi-parasitic plants of low, slender habit, with pale yellowish-green foliage, narrow leaves, and tiny whitish or greenish flowers. About 11 species in our area.

LORANTHACEAE | Mistletoe Family

LORANTHUS 72. *L. europaeus* Jacq. A parasitic shrub up to 1 m with dark-green deciduous leaves, growing on the branches of oak, Sweet Chestnut, and beech. Flowers greenish-yellow, in terminal clusters 2·5–4 cm long. Leaves oval, 1–5 cm, leathery, short-stalked. Fruit yellow, fleshy and sticky within, globular, about 1 cm. Widespread on mainland. Page 214.

VISCUM **73. *V. album* L. MISTLETOE. Parasitic shrub on many species of trees including conifers. Fruit white. Widespread on mainland+CR.

ARCEUTHOBIUM 73a. *A. oxycedri* (DC.) Bieb. A tiny, densely branched, yellowish-green shrublet 5–20 cm, growing parasitically on juniper. Leaves triangular, scale-like, sheathing the stem; branches jointed. Flowers greenish, unisexed, the males with 3 petals. Fruit green, exploding when ripe to eject the sticky seeds. Frequently parasitizing juniper in the Med. region. Widespread on mainland. Pl. 1.

1. *Zelkova abelicea* 59a **2.** *Comandra elegans* 70b **3.** *Fagus orientalis* (45)
4. *Loranthus europaeus* 72

ARISTOLOCHIACEAE | Birthwort Family

ASARUM ****74.** *A. europaeum* L. ASARABACCA. Woods. YU.AL.BG.

ARISTOLOCHIA | Birthwort 13 species, including 4 species with limited distribution, which are little known.

Flowers 2 or more in the leaf axils
**76. *A. clematitis* L. BIRTHWORT. A robust leafy perennial to 1 m, with whorls of yellow flowers in the upper leaf axils. Flowers 2–3 cm, short-stalked. Stony and bushy places, vineyards. Widespread on mainland.

Flowers solitary in the leaf axils
(a) Tube of perianth strongly curved
75. *A. sempervirens* L. (incl. *A. altissima*) CLIMBING BIRTHWORT. A slender, twining, woody perennial with glossy evergreen leaves, and yellowish-brown to dull-purple, strongly curved, tubular flowers. Flowers 2–5 cm. Leaves triangular-heart-shaped, 2–10 cm long. Hedges and shady thickets in the Med. region. GR.CR. **75a.** *A. cretica* Lam. CRETAN BIRTHWORT. An erect or spreading, distinctly hairy perennial with strongly curved, large, dull-purple flowers 5–12 cm long. Perianth with long white hairs within, hairy outside, base of tube conspicuously inflated into an oblong swelling. Leaves 2–5·5 cm, oval-heart-shaped, blunt, stalked. Sunny rocks. CR+Rhodes, Karpathos. Pl. 1. **75b.** *A. hirta* L. Distinguished from 75a by the ovate-triangular leaves with at least the upper leaves acute, and by the ovate lip of the perianth. E. Aegean Is.

(b) Tube of perianth more or less straight
  (i) Leaves stalkless, clasping stem
**77. *A. rotunda* L. ROUND-LEAVED BIRTHWORT. Distinguished by its almost stalkless leaves which have rounded basal lobes encircling the stem. Flowers with a yellowish corolla tube and dark-brown strap-shaped lip. Tuber globular. Among herbaceous plants, bushy places. Widespread.

  (ii) Leaves stalked, not clasping stem
(77). *A. pallida* Willd. Differs from 77 in having green, yellow, or pale-brown flowers with brownish or purplish stripes and often darker lip. Flowers 3–6 cm. Tuber globular. Stony and bushy places. Widespread on mainland. **(77).** *A. longa* L. Distinguished by its cylindrical tuber. Flowers 3–5 cm, brownish or yellowish-green with a brownish-purple lip. Stony places in the Med. region. AL.GR.CR.BG. Pl. 1. **77c.** *A. microstoma* Boiss. & Spruner Quite distinctive with its brownish club-shaped flowers 1·5–3 cm with an apical pore and without a lip like other species. Leaves heart-shaped. Sunny stony places. GR(south)+Cyclades.

RAFFLESIACEAE | Rafflesia Family

CYTINUS Quite unmistakable parasitic plants growing on the roots of *Cistus* species and usually found near these bushes.

78. *C. hypocistis* (L.) L. Flowers yellow, in dense rounded heads and contrasting with the orange or red bracts. YU.GR+Is.CR.TR. **(78). *C. ruber* (Fourr.) Komarov Grows on pink-flowered *Cistus* species. Like 78, but flowers ivory-white or pale pink with contrasting crimson or carmine bracts. YU.GR.

POLYGONACEAE | Dock Family

POLYGONUM About 23 species occur in our area. The following are widespread on mainland: ****79.** *P. maritimum* L. SEA KNOTGRASS; **80.** *P. aviculare* L. KNOTGRASS; **80a.** *P. patulum* Bieb.; **80b.** *P. mite* Schrank; ****81.** *P. hydropiper* L. WATER-PEPPER; ****(82).** *P. lapathifolium* L. PALE PERSICARIA.
In the central European climatic region the following appear to be widely distributed: **82.** *P. persicaria* L. PERSICARIA, REDLEG; ****83.** *P. amphibium* L. AMPHIBIOUS BISTORT; ****84.** *P. bistorta* L. BISTORT, SNAKE-ROOT; **86.** *P. alpinum* All. ALPINE KNOTWEED.

79a. *P. equisetiforme* Sibth. & Sm. A slender, usually procumbent plant with long wiry, branched stems arising from a woody rhizome, with numerous small white or pink flowers. Leaves narrow, soon falling. Roadsides and rocks in the Med. region. GR.CR.BG.TR.

BILDERDYKIA Widespread on mainland are: ****87.** *B. convolvulus* (L.) Dumort. BLACK BINDWEED; ****(87).** *B. dumetorum* (L.) Dumort. COPSE BINDWEED.

FAGOPYRUM ****89.** *F. esculentum* Moench BUCKWHEAT. Sometimes grown as a crop, and naturalized in waste ground and roadsides in the central Balkans.

OXYRIA **90.** *O. digyna* (L.) Hill MOUNTAIN SORREL. Stony pastures, by melting snow. YU.BG.

RUMEX | **Dock, Sorrel** 25 mostly widespread European species occur in our area.

CHENOPODIACEAE | Goosefoot Family

A family well adapted to soils which are rich in salt, usually by the sea in our region. They are often shrubby, and their monotonous grey-green foliage may cover extensive areas around lagoons, estuaries, mud-flats, salt steppes, and any ground periodically flooded by salt water. The GLASSWORTS, the shrubby *Arthrocnemum* species, and annual *Salicornia* species have fleshy cylindrical branched stems which 'crackle underfoot'. The SEABLITES: **115.** *Suaeda vera* J. F. Gmelin, and ****(115).** *S. maritima* (L.) Dumort. have fleshy, narrow, pointed leaves, and grow with the Glassworts nearest the open saline water. The SALTWORTS: ****116.** *Salsola kali* L.; ****(116).** *S. soda* L. are common on sands and salt-flushed mud; they are distinguished by their stiff, half-cylindrical, usually spiny-pointed leaves. Two shrubs having silvery-white, oval to oblong flattened leaves are: **107.** *Atriplex halimus* L. SHRUBBY ORACHE, and ****112.** *Halimione portulacoides* (L.) Aellen SEA PURSLANE, which grow on mud-flats. Common weeds of cultivation, waste places, and roadsides are: GOOSEFOOTS, *Chenopodium* (14 spp.); ORACHES, *Atriplex* (7 spp.); and *Polycnemum* (3 spp.).

BETA ****101.** *B. vulgaris* L. BEET. Widespread on the littoral. Also cultivated in our region quite extensively as Sugar Beet.

101a. *B. trigyna* Waldst. & Kit. A distinctive, erect dock-like plant to 1 m, with branched stems bearing dense clusters of small yellowish-white flowers, the clusters being crowded towards the apex and separated lower down. Stem leaves oval to lanceolate, to 20 cm long. Waysides, waste ground. YU(east).BG. Pl. 2. **101b.** *B. nana* Boiss. & Heldr.

A small rosette-forming perennial of the highest mountains of Greece, with lax, spreading, leafless flowering stem 5–10 cm bearing greenish flowers. Flowers solitary, subtended by bracts. Leaves ovate-oblong 15–20 mm. By snow-patches. GR(south).

AMARANTHACEAE | Cockscomb Family

AMARANTHUS About 9 species, mostly alien, are found in waste places and cultivated ground.

PHYTOLACCACEAE | Pokeweed Family

PHYTOLACCA **119. *P. americana* L. VIRGINIAN POKE. A tall, rather fleshy herbaceous perennial with wide-spreading branches, large oval leaves 12–15 cm, and dense cylindrical spikes of greenish to pinkish flowers arising from the angles of the branches, or opposite the leaves. Fruit globular 1 cm, fleshy, at first reddish then purplish-black. Waysides, bushy places, ditches. Native of N. America; naturalized throughout.

AIZOACEAE | Mesembryanthemum Family

CARPOBROTUS **(120). *C. acinaciformis* (L.) L. Bolus Forming dense mats of bright-green, fleshy, triangular-sectioned leaves, and bearing very large brilliant red-carmine flowers up to 12 cm across. Petals very numerous; stamens purple. Native of S. Africa; naturalized on rocks by the sea. GR. + Ionian Is.

MESEMBRYANTHEMUM 121. *M. nodiflorum* L. A procumbent annual to 20 cm, with fleshy cylindrical leaves and small white or yellowish flowers 1·5 cm across, with numerous narrow petals shorter than the sepals. Waste places, on the littoral. YU.GR+Is.CR. (121). *M. crystallinum* L. ICE PLANT. Differs from 121 in having flat, oval leaves which are densely covered with shining crystalline swellings, like hoar-frost. Flowers 2–3 cm across; petals longer than sepals. On the littoral. YU.GR.CR.

MOLLUGINACEAE | Mollugo Family

GLINUS 121e. *G. lotoides* L. A woolly-haired, spreading leafy annual 10–40 cm, with opposite or whorled, obovate, stalked leaves and yellowish flowers in dense axillary clusters. Sepals yellow; 'petals' white, numerous; stamens 12; stigmas 5. Fruit opening by 5 valves. Waste places in the Med. region. YU.AL.GR.CR.

MOLLUGO 121f. *M. cerviana* (L.) Ser. A hairless, branched, slender-stemmed annual

217

to 40 cm, with whorls of linear leaves and stalked flat-topped clusters of greenish flowers. Perianth five-lobed. Leaves with papery margins. Fruit three-angled with 3 valves. Sandy places in the Med. region. GR.BG.

PORTULACACEAE | Purslane Family

PORTULACA ****122.** *P. oleracea* L. PURSLANE. A creeping, fleshy, shiny-leaved annual with usually solitary yellow, stalkless flowers 1–2 cm across. Leaves 1–2 cm, oval-oblong. Cultivated ground, sandy places, waysides. Widespread, local on Islands.

CARYOPHYLLACEAE | Pink Family

Quite an important family with 31 genera and about 380 species in our area. The following genera, largely of small inconspicuous species, are not included: *Moehringia* 12 spp.; *Bufonia* 3 spp.; *Pseudostellaria* 1 sp.; *Holosteum* 1 sp.; *Moenchia* 3 spp.; *Myosoton* 1 sp.; *Sagina* 7 spp.; *Scleranthus* 3 spp.; *Corrigiola* 1 sp.; *Herniaria* 5 spp.; *Illecebrum* 1 sp.; *Polycarpon* 3 spp.; *Spergula* 3 spp.

ARENARIA | Sandwort Distinguished from *Minuartia* by the fruit capsule which splits into twice as many teeth—usually 6—as there are styles. Leaves rounded to lanceolate, rarely narrower or rarely with a bristle-like apex. 18 species in our area.

Sepals hairless on surface
128. *A. biflora* L. A slender creeping perennial, rooting at the nodes, with tiny rounded leaves 3–4 mm, and solitary or paired white flowers on short slender stalks. Flowers about 1 cm across; petals slightly longer than the usually hairless sepals. Damp places, often by snow-patches in mountains. YU.AL.GR.BG. Page 136. **128a.** *A. gracilis* Waldst. & Kit. A usually loosely tufted perennial with lanceolate-acute leaves and solitary or paired white flowers on stalks 2–5 times as long as the sepals. Flowers 1–1·5 cm across; petals about twice as long as sepals. Leaves 2–8 mm, usually hairless, margins finely toothed. Mountain rocks. YU(south and west). Page 190.

Sepals hairy on surface
128b. *A. filicaulis* Fenzl A densely tufted, glandular-hairy perennial of the mountains, with 3–10 white flowers 1–1·5 cm across, borne on slender stems 5–15 cm. Sepals acute. Leaves 3–10 mm, narrowly oval, densely glandular-hairy, closely clustered on stems below. Rocks. YU(Macedonia).GR.BG(south). Pl. 2. **128c.** *A. cretica* Sprengel A densely tufted perennial with slender fragile stems, blunt oblong, usually hairless leaves 3–10 mm, and white flowers with petals twice as long as the sepals or more. Flowers 5–10 mm across, in clusters of 1–5; sepals obtuse. Mountain rocks. AL(South).GR.CR. Page 88.

MINUARTIA | Sandwort Distinguished from *Arenaria* by its fruit capsule which splits into as many teeth as there are styles—usually 3. Leaves usually linear, bristle-tipped. About 30 species in our area.

Cushion-forming mountain and alpine plants
132a. *M. hirsuta* (Bieb.) Hand.-Mazz. A lax cushion-forming, glandular-hairy perennial

218

with linear-pointed leaves 0·5–3 cm. Flowers white, in lax clusters of 3 to many, on stems, to 20 cm; petals narrow-elliptic. Sepals ovate-lanceolate, five- to seven-veined, spreading during flowering, margins papery. Mountains. YU.GR.BG. Page 184. **132b.** *M. recurva* (All.) Schinz & Thell. A dense cushion-forming, woody-based perennial distinguished by its curved, sickle-shaped, linear-pointed leaves 4–10 mm. Flowering stem erect, to 12 cm; flowers white; petals blunt; flower stalks up to three times as long as sepals. Sepals 3–6 mm, ovate-lanceolate, five- to seven-veined, little shorter than petals. Alpine regions. YU.AL.GR.BG.

132c. *M. stellata* (E. D. Clarke) Maire & Petitmengin A very dense, firm, pale-green, domed cushion plant. Distinguished by its numerous rosettes of rigid, triangular-lanceolate, bristly-tipped leaves to 1 cm. Flowers nearly stalkless, solitary, white, with narrow elliptic petals about one and half times as long as the calyx. Sepals 6 mm, lanceolate, five-veined. Rocks in high mountains. AL(south).GR. Pl. 2.

Erect or spreading, often tufted plants, in lowlands or mountains
(a) Sepals green, with papery margins
132d. *M. baldaccii* (Halácsy) Mattf. A low, creeping perennial with dense clusters of curved, pointed leaves and relatively large white flowers with petals 2–2½ times as long as the sepals, which are 4–7 mm long, blunt, and oblong. Leaves 0·5–1·5 cm, linear, ciliate at base, often arranged to one side, not rigid. Rocks in mountains. YU.AL.GR(north). Pl. 2. **132g.** *M. juniperina* (L.) Maire & Petitmengin A densely tufted perennial with numerous stiff, almost spiny, spreading linear leaves 1·5–2·5 cm on erect rigid stems with thickened nodes. Flowers in flat-topped clusters, white; petals about 1½ times as long as the lanceolate-acute sepals, which are 4–5 mm. Alpine and sub-alpine rocks. GR(south and west). Page 89.

(b) Sepals predominantly white, with a green central strip
132e. *M. setacea* (Thuill.) Hayek A small tufted plant with slender bristly leaves and numerous erect stems to 20 cm bearing several white flowers with conspicuous green-striped sepals. Flowers about 1 cm; petals as long as the sepals, which are lanceolate-acute, with a white mid-vein separating 2 narrow green bands. Leaves about 1 cm. Rocks, sandy places. YU.AL.GR.BG. **132f.** *M. glomerata* (Bieb.) Degen Distinguished from 132e by its glandular-hairy stems and leaves, and its bunched flat-topped clusters of white flowers. Petals shorter than sepals, which are 4–6 mm; flower stalks to 2 mm. Sandy places. YU.GR.BG.TR.

STELLARIA | Stichwort 8 mostly widespread central European species in our area including: **134.** *S. nemorum* L. WOOD STITCHWORT; **135.** *S. media* (L.) Vill. CHICKWEED; ****136.** *S. holostea* L. GREATER STITCHWORT; **(138).** *S. alsine* Grimm BOG STITCHWORT; ****138.** *S. graminea* L. LESSER STITCHWORT.

CERASTIUM | Chickweed 29 species in our area, including 7 endemic to it. Many are small annuals.

Plants with long, soft white or greyish hairs
(a) Leaves more than three times as long as wide
140b. *C. candidissimum* Correns Readily distinguished by its snowy-white, densely felted stems, leaves, and sepals, and its numerous conspicuous white flowers in a dense flat-topped cluster. Petals about 1 cm, twice as long as woolly sepals, which are blunt; individual flower stalks not more than twice as long as sepals. Leaves narrow-lanceolate, 2–3·5 cm by 3 mm wide, straight or curved, with inrolled margins; stems 14–20 cm. Rocks and stony places in high mountains. GR(west and south). Pl. 2. **140c.**

C. grandiflorum Waldst. & Kit. Like 140b but leaves narrower 0·5–1·5 mm, margins inrolled; ovary hairy (hairless in 140b). Flowers 7–15 in a many branched, flat-topped cluster; petals to 18 mm, twice as long as sepals or more. Rocks in mountains. YU(west).AL. Page 179.

140d. *C. decalvans* Schlosser & Vuk. Recalling 140b with greyish-white, woolly-haired leaves, but flower clusters lax, with longer undivided flower stalks, which are recurved after flowering, then erect. Sepals and bracts acute. Leaves linear, 3–5 mm wide, at least three times as long as wide. Rocks in hills and mountains. YU.AL.GR.BG. Page 169.

(b) Leaves less than three times as long as wide
140e. *C. moesiacum* Friv. Like 140d with densely woolly white or greyish leaves, but leaves broader, up to 1 cm, less than three times as long as wide. Flowering stems erect to 40 cm, many-flowered, with white petals up to three times as long as calyx. Sepals and bracts blunt, with papery margins. Wet places in hills. YU.AL.GR(north).BG. ****142.** *C. alpinum* L. ALPINE MOUSE-EARED CHICKWEED. Alpine meadows. YU.AL.GR.BG.

Plants with fine short hairs, or hairless
140f. *C. banaticum* (Rochel) Heuffel A perennial with numerous leafy branches and erect stems to 40 cm, with a branched inflorescence of rather large white flowers about 2 cm across. Bracts conspicuous, silvery, with wide papery margins. Sepals 5–9 mm, with conspicuous papery margins. Leaves linear to lanceolate, with bristly hairs at the base. Fruit stalks and capsule straight. Among rocks. YU.AL.GR(north).BG. Page 221.

PARONYCHIA Low creeping plants, usually distinguished by their shining silvery bracts which hide the tiny flowers. 7 species in our area.

Calyx with papery margin, awned
**149. *P. argentea* Lam. Rocky places. GR+Is.CR.

Calyx with green margin, not awned
150. *P. kapela* (Hacq.) Kerner Flower clusters 7–15 mm across, conspicuous, owing to the silvery bracts which encircle the flowers. Calyx less than 1½ times as long as the ripe fruit. Bracts 3–5 mm. A spreading, mat-forming perennial. Rocky places. Widespread. P1.2. **150a.** *P. cephalotes* (Bieb.) Besser Very like 150 but leaves broader, oblong to linear-lanceolate; bracts larger 5–7 mm; calyx 2·5–4 mm, twice as long as the ripe fruit. Rocky places. Widespread on mainland. Page 221.

SPERGULARIA | Spurrey 7 species in our area. Widespread coastal species are: **155.** *S. media* (L.) C. Presl GREATER SEA SPURREY; **(156).** *S. marina* (L.) Griseb. LESSER SEA SPURREY. ****157.** *S. rubra* (L.) J. & C. Presl SAND SPURREY is a common sand-loving species found largely inland.

TELEPHIUM Distinguished by the 3 separate styles and the three-angled fruits, which split into 3 valves.

156c. *T. imperati* L. A hairless, glaucous, prostrate shrublet 15–40 cm, with herbaceous stems bearing numerous fleshy, oval leaves 0·5–1·5 cm long, and dense terminal clusters of small white flowers. Petals little longer than sepals, which have conspicuous white papery margins. Rocky places. GR(central and south).CR.

LYCHNIS | Catchfly
158. *L. coronaria* (L.) Desr. ROSE CAMPION. A shaggy, white-haired perennial with erect, little-branched stems 30–100 cm bearing a few large, usually bright-red flowers about

1. *Paronychia cephalotes* 150a   **2.** *Saponaria calabrica* 180b   **3.** *Petrorhagia illyrica* 184c   **4.** *Bolanthus graecus* 178d   **5.** *Saponaria sicula* 181b   **6.** *Cerastium banaticum* 140f   **7.** *Gypsophila paniculata* 178c

3 cm across. Petals spreading, with lanceolate scales; calyx shaggy-haired with acute, twisted teeth. Sunny, stony, and bushy places. Widespread on mainland. Pl. 3. **160. *L. flos-cuculi* L. RAGGED ROBIN. Damp meadows, marshes. YU.AL.GR.BG.

**161. *L. viscaria* L. RED GERMAN CATCHFLY. An attractive campion-like perennial 15–90 cm, with a lax interrupted cluster of reddish-purple flowers, and with very characteristic brown sticky patches on the stem below. Flowers about 2 cm across; petals notched, each with 2 conspicuous scales. Grassy places, shady banks. Widespread on mainland.

AGROSTEMMA **163. *A. githago* L. CORN COCKLE. Cornfield weed. Widespread.

SILENE | Campion, Catchfly
A large and quite important genus, with about 104 species in our area, 10 of which are restricted to it. Small differences and combinations of characters of seed, fruit, calyx, and inflorescence often distinguish otherwise similar-looking plants. Petals with a narrow basal claw and a wider usually spreading limb, often with flap-like scales at the junction of claw and limb.

Mountain perennials, cushion- or mat-forming
169b. *S. auriculata* Sibth. & Sm. A dense tufted perennial of the mountains, with a branched woody stock and stout unbranched stems bearing one or few rather large creamy flowers with conspicuous calices. Petals with a two-lobed limb and claw longer than calyx. Calyx inflated, 12–15 mm long, glandular-hairy, with red veins. Basal leaves thick, lanceolate to 12 cm, stem leaves 1–3 pairs and much smaller. Rocks in mountains. GR. Pl. 3. **170.** *S. acaulis* (L.) Jacq. MOSS CAMPION. Damp rocks, stony pastures in mountains. YU.BG. Page 184.

170c. *S. saxifraga* L. A loose, glossy-green, mat-forming perennial with numerous slender erect, usually sticky stems to 20 cm, each bearing 1–2 dull whitish-green flowers, which are greenish or reddish on the underside of the petals. Calyx club-shaped, 8–13 mm long, hairless, usually green, claw of petal usually longer. Limestone grasslands and screes in mountains. YU.AL.GR.CR.BG. Page 191. **170e.** *S. waldsteinii* Griseb. A densely tufted dwarf campion with numerous erect stems to 25 cm and small solitary or few whitish flowers, like 170c, but with a narrow club-shaped, hairless calyx which is 2–2·5 cm long. Calyx teeth spreading, acute, claw of petals longer. Capsule longer than calyx. Rocks in mountains. YU.AL.GR.BG. **170f.** *S. pindicola* Hausskn. A dwarf tufted perennial of serpentine rocks in the mountains with slender sticky stems to 10 cm bearing solitary brownish flowers. Calyx 18–24 mm, wine-coloured, claw of petals much longer. Leaves linear, to 1·5 cm. GR(north). Page 114.

Perennials of lowlands or mountains, not cushion- or mat-forming
(a) Plants hairy, at least at base
Widespread on mainland are: **166.** *S. italica* (L.) Pers. ITALIAN CATCHFLY; **167.** *S. nutans* L. NOTTINGHAM CATCHFLY; **168.** *S. otites* (L.) Wibel SPANISH CATCHFLY.

166b. *S. paradoxa* L. Like 166 in general habit but inflorescence lax, flowers larger; petals yellowish beneath; calyx cylindrical 2·5–3 cm long, glandular. Rocky places. YU.AL.GR. **166c.** *S. goulimyi* Turrill A small woody-based perennial related to 166, of rock gorges in the Taiyetos Mountains of the Peleponnisos. Stems slender bearing one or few pinkish flowers; calyx 11–12 mm. GR(south). Page 88. **167a.** *S. viridiflora* L. Distinguished by its lax inflorescence of drooping greenish-white flowers 2 cm across, with 3 violet styles. Claw of petal longer than calyx, petal-lobes linear; calyx 1·5–2 cm, hairy. A softly hairy, robust perennial 40–90 cm with stems sticky above;

lower leaves oblong-spathulate, long-stalked. Shady bushy places in hills. YU.AL.GR. BG.TR.

168a. *S. roemeri* Friv. Distinctive with its dense globular terminal cluster about 1·5 cm across of tiny cream-coloured flowers, or with several rounded clusters, one above the other. Calyx 3–4 mm long, almost hairless. A shortly hairy perennial to 50 cm; basal leaves oblong-spathulate, with long, ciliate leaf stalks. Mountain meadows. YU.AL.GR.BG. Pl. 3. **168b.** *S. densiflora* D'Urv. Recalling 168. *S. otites* in habit with numerous tiny creamy flowers, but whole plant larger, up to 2 m. Flowers almost stalkless, in dense clusters ranged along the branched inflorescence. Calyx 4–6 mm. Open stony places YU.GR.BG. Page 224.

169d. *S. congesta* Sibth. & Sm. A woody-stemmed, hairy perennial to 30 cm, with a branched inflorescence with clusters of 3–5 rather small yellowish flowers. Calyx 6 mm long, densely glandular-hairy. Rocky places in mountains. GR. **169e.** *S. cephallenia* Heldr. Plant woody at the base, with linear or linear-lanceolate, shortly hairy basal leaves. Stem erect, usually with rather crowded three-flowered clusters (or a single cluster in small specimens). Flowers white, petals two-lobed; calyx 1–1·8 cm long, glandular-hairy. Rocky places. AL.GR(west). **170g.** *S. thymifolia* Sibth. & Sm. A spreading, tufted, woody-based perennial of maritime sands of the Black Sea, with fleshy, hairy ovate leaves and few white flowers. Petals bilobed, the claw longer than the calyx. Calyx 12–15 mm, glandular-hairy. BG.TR. Page 149.

(b) Plants hairless, or with a few hairs below

169. *S. vulgaris* (Moench) Garcke BLADDER CAMPION. Very variable, usually hairless. Meadows, grassy places, sands, and rocks on the littoral. Widespread. **169c. *S. fabarioides* Hausskn. A robust, hairless, and glaucous perennial to 1 m, with upper part of stem often leafless. Flowers in a lax branched inflorescence, white or pinkish-purple; calyx conspicuous, inflated, ten-veined, 8–10 mm long. Capsule longer than calyx. Rocky places. YU?AL.GR.BG. **171a.** *S. lerchenfeldiana* Baumg. A slender, hairless perennial with lanceolate rosette leaves, and leafy stems to 25 cm bearing a branched inflorescence of small reddish or purplish flowers, with petals deeply cut into linear lobes. Petal limb oblong; calyx 9–12 mm long, narrow club-shaped. Rocky places in mountains. YU.GR.BG. Page 136.

173a. *S. frivaldszkyana* Hampe A hairless perennial to 80 cm, with lanceolate-spathulate basal leaves. Inflorescence elongated, with pairs of opposite, solitary, whitish flowers borne along its length. Calyx 12–16 mm long. Mountain meadows. YU.AL.GR.BG. **176a.** *S. asterias* Griseb. A beautiful perennial of wet flushes with very dense hemispherical heads of bright-crimson flowers. Flower heads encircled by numerous papery bracts (recalling **176b.** *S. compacta* but bracts of latter leafy). Basal leaves large, spathulate; stem leaves few, lanceolate, all pale green; stems to 1 m. Mountains. YU.AL.GR.BG. Page 224. Pl. 3.

Annuals or biennials of lowlands and lower mountains
(a) Plants hairy

165b. *S. pinetorum* Boiss. & Heldr. A wiry branched annual up to 20 cm, downy below, hairless and sticky above, with pink flowers in lax branched clusters. Calyx 4–8 mm long, hairless, teeth rounded. Maritime sands, lower mountain slopes. CR. **165c.** *S. integripetala* Bory & Chaub. A branched, rather densely hairy annual 10–40 cm, with a lax inflorescence of few or many pink flowers. Flowers 1–1·5 cm across; petals not notched; calyx 13–17 mm long, cylindrical, reddish, glandular-hairy. Lower leaves ovate, upper narrower. Sunny hill slopes. GR(south). Page 88.

1. *Dianthus corymbosus* 193g **2.** *Dianthus myrtinervius* 193b **3.** *Silene conica* subsp. *sartorii* 175 **4.** *Silene cretica* 165e **5.** *Silene asterias* 176a **6.** *Dianthus arboreus* 186a **7.** *Silene densiflora* 168b

165d. *S. sedoides* Poiret A small, much-branched, densely hairy annual to 30 cm, with small reddish or whitish flowers. Calyx 6–8 mm long, greenish or rarely reddish, with short stiff hairs. Coasts. YU.AL.GR+Is.CR.   **167b.** *S. gigantea* L. A robust, finely hairy biennial to 1 m, with a basal rosette of leaves, and a dense or lax branched inflorescence of small white, pink, or greenish flowers. Calyx tubular 8–12 mm long, teeth blunt. Rosette leaves thick, spathulate. Bushy places. YU.AL.GR+Is.CR.BG. Page 97.

174. *S. gallica* L. SMALL-FLOWERED CATCHFLY. Stony places, cultivations. Widespread.   **(174).** *S. colorata* Poiret A finely hairy annual 10–50 cm, with pink or white flowers 1–2 cm across, in lax clusters. Limb of petal 5–9 mm, deeply two-lobed; calyx 11–17 mm long, cylindrical, becoming club-shaped in fruit, calyx teeth ovate, blunt, ciliate. Seeds with 2 undulate wings. Often abundant in sandy waste places, rocks. Widespread in the Med. region. AL.GR+Is.CR.BG.TR.   **175.** *S. conica* L. STRIATED CATCHFLY. Readily distinguished by its distinctive ovoid calyx which is strongly ribbed, particularly in fruit. Flowers pink, to 1 cm across; petals notched; calyx 8–18 mm long, with 30 parallel veins, calyx teeth slender, long-pointed. Lower leaves acute. Sandy places, cultivation. Widespread on mainland. Subsp. *sartorii* (Boiss. & Heldr.) Chater & Walters is glandular-hairy, with blunt lower leaves, and is found in maritime sands. Page 224.

(b) Plants hairless, or with few hairs below

176. *S. armeria* L. SWEET-WILLIAM CATCHFLY. An attractive annual or biennial to 40 cm, with broad hairless, glaucuous leaves and lax, rather flat-topped clusters of bright-pink flowers, each 1·5 cm across. Petals notched, scales 2, acute; calyx 12–15 mm long, hairless, cylindrical-club-shaped, teeth blunt. Sands and rocks. Widely scattered on mainland.   **176b. *S. compacta* Fischer A robust biennial similar to 176 but with stouter stem and wider leaves, the upper stem leaves closely surrounding the base of the many-flowered, densely clustered inflorescence. Bushy places. GR.BG.TR. Pl. 3.

165e. *S. cretica* L. A slender, little-branched annual to 60 cm, with few conspicuous pink flowers with notched or two-lobed petals. Calyx 9–16 mm long, hairless, with triangular-pointed teeth. Plant hairless except for some deflexed hairs below. Fruit ovoid. Weedy fallow land, waste ground. Widespread in the Med. region; introd. YU. Page 224.   **165f.** *S. graeca* Boiss. & Spruner A hairless, glaucous annual 20–40 cm, dichotomously branched, with white or flesh-coloured flowers in one-sided clusters. Petals large, bilobed, with the blade 5–10 mm; calyx 9–15 mm, greenish, teeth blunt. Stem leaves dense but not overlapping. Cultivated ground in the Med. region. YU.AL.GR.

165g. *S. muscipula* L. A hairless annual 15–40 cm with stem leaves always overlapping and pink flowers with deeply cleft petals. Inflorescence lax; calyx 13–17 mm long, greenish or reddish, with prominent greenish or reddish branched veins. Cultivated fields. GR.   **165h.** *S. behen* L. A hairless, leafy annual 15–20 cm, with pink flowers, distinguished in fruit by its conspicuous ovoid calyx 11–17 mm long, with branched whitish or reddish veins. Stem leaves glaucous, usually overlapping. Cultivated hillsides. GR+Is.CR.TR.

CUCUBALUS   ****177.** *C. baccifer* L. BERRY CATCHFLY. Damp shady places. Widespread on mainland.

DRYPIS   Distinguished by its one-seeded fruit capsule which splits transversely. Stamens 5; stigmas 3.

177a. *D. spinosa* L. A much-branched, pale-green, spiny perennial 8–30 cm, with brittle quadrangular stems and terminal clusters of small white or pink flowers with apparently 10 narrow petals, set among spine-tipped bracts. Petals in reality 5 but deeply lobed. Leaves awl-shaped, spiny-tipped. Stony places in mountains, by the sea. YU.AL.GR. Pl. 3.

GYPSOPHILA Distinguished by the five-veined calyx with pale papery seams between the veins. Capsule with 4 teeth. 13 species, including 4 endemic species in our area.

Flowers many in dense globular clusters or in much-branched lax clusters
178a. *G. petraea* (Baumg.) Reichenb. A low densely tufted plant to 20 cm, with narrow basal leaves and nearly leafless stems terminating in a globular head, 1–2 cm, of tiny white or pale purplish flowers. Calyx 2–4 mm, petals 1½ times as long. Leaves 2–5 cm by 1 mm. Limestone mountains. BG(Rhodope). **178b.** *G. glomerata* Bieb. A tall erect perennial to 1 m, branched above, with long-stalked, tiny white globular flower heads about 8 mm across. Bracts surrounding flower heads papery. Leaves glaucous, 2–10 cm by 1–4 mm. Dry grassy places. BG.

178c. *G. paniculata* L. BABY'S BREATH. A tall glaucous perennial 50–90 cm, much branched above with numerous slender branches bearing myriads of minute white or pink flowers, giving a cloud-like appearance. Flowers 2–3 mm across; calyx conspicuously green and white. Leaves lanceolate, pointed. Dry sandy and stony places. YU.BG. Page 221.

Flowers few in a short, lax inflorescence
178e. *G. nana* Bory & Chaub. A small somewhat tufted, sticky-haired dwarf perennial of mountain rock crevices. Flowers about 1 cm across, pale purplish with darker-veined petals; one or several on short stems usually only 2–4 cm. Leaves narrow-oblong, 5–10 mm. GR(south).CR. Page 71. **178f.** *G. achaia* Bornm. Like 178e but flowers white and stems and leaves hairless. Petals 2–3½ times as long as calyx (about twice as long in 178e). Mountain rocks. GR(South). Page 93.

BOLANTHUS Often included in *Gypsophila* but distinguished by its tubular calyx with 5 projecting ribs, and petals with a small spreading limb abruptly contracted into a long narrow claw. 4 species in our area.

178d. *B. graecus* (Schreber) Barkoudah (*Gypsophila polygonoides*) A small prostrate, glandular-hairy, woody-stemmed perennial with dense hemispherical clusters of usually purple-striped, or rarely white, flowers on stems 5–25 cm. Flowers about 5 mm across. Calyx conspicuous, white with green ribs, hairy, 5–6 mm. Leaves 5–10 mm, linear-elliptic. Rocky places. GR+Is.BG(south). Page 221.

SAPONARIA | Soapwort Distinguished by its conspicuous oblong or cylindrical five-toothed calyx with 15–25 parallel veins, without papery seams. Styles 2, rarely 3.

1. Fls. yellow, in a dense cluster. *bellidifolia*
1. Fls. reddish, purplish, rarely white, never yellow; inflorescence various.
  2. Calyx less than 1·5 cm long.
    3. Perennials, woody below, with conspicuous non-flowering leafy shoots; calyx with soft glandular hairs (YU.). ****180.** *S. ocymoides* L.

3. Annuals, without leafy non-flowering
    shoots; calyx with rough glandular
    hairs.   *calabrica*
2. Calyx more than 1·8 cm long.
    4. Low tufted plants; lvs. one-veined;
        fls. often few.   *sicula*
    4. Erect branched plants; lvs. three-
        veined; fls. usually many.
        5. Plant more or less hairless; petal
            limb about 1 cm (Widespread on
            mainland).   **181.** *S. officinalis* L.
        5. Plant very glandular-sticky;
            petal limb about 5 mm.   *glutinosa*

(180). *S. bellidifolia* Sm. A tufted woody-based perennial distinguished by its pale-yellow flowers in a dense rounded head, surrounded by a pair of bracts. Stem 20–40 cm, unbranched, usually with one pair of linear stem leaves; basal leaves spathulate. Alpine meadows and rocks. YU.AL.GR.BG. Pl. 4. **180b.** *S. calabrica* Guss. A glandular-hairy annual with wide-spreading branches bearing rather lax clusters of pink flowers with conspicuous cylindrical, often purple-flushed, calyx. Flowers about 1·5 cm across, claw of petals longer than calyx; calyx 6–10 mm, teeth very blunt. Lower leaves spathulate, upper oblong-ovate. Rocks in the Med. region. AL.GR+Is.CR.TR. Page 221.

181a. *S. glutinosa* Bieb. A robust annual or biennial to 50 cm, readily distinguished by its very sticky, hairy stems and leaves. Flowers purple, in branched pyramidal clusters; petal limb 5 mm, bilobed, reflexed; calyx narrow-tubular, 2–2·5 cm. Meadows on limestone. Widely scattered on mainland. Pl. 4. **181b.** *S. sicula* Rafin. Distinguished by its conspicuous somewhat inflated red-flushed calyx and pink flowers with petals with small limbs and long claws longer than the calyx. Calyx 2–3 cm, hairy, teeth with hairy, pointed apices. A densely tufted plant with narrow one-veined leaves, and erect branched stems 10–50 cm, with few leaves. Grassy places in mountains. AL.GR(north).BG(south-east). Page 221.

VACCARIA Distinguished by its inflated, angular green-winged calyx; petals without scales.

182. *V. pyramidata* Medicus COW BASIL An erect, branched leafy annual 30–60 cm, with pink flowers and distinctive white and green inflated calyx. Leaves glaucous, ovate to lanceolate. Cornfields. Widespread.

PETRORHAGIA Distinguished by its calyx which has papery seams. Flowers with or without bracts (or epicalyx). 15 species, including 8 endemic in our area.

Flowers in branched clusters, solitary, without bracts
184c. *P. illyrica* (L.) P. W. Ball & Heywood (*Tunica i.*) A slender branched perennial to 40 cm, usually with a lax inflorescence of rather numerous white or rarely pale-yellow flowers, spotted with purple at the base. Flowers about 8 mm across; petals entire; calyx 4–8 mm, green-veined with white seams. Leaves linear-pointed, three-veined. Dry stony places in the Med. region. YU.AL.GR.CR? BG. Page 221.

Flowers in dense terminal clusters surrounded by bracts
184a. *P. glumacea* (Chaub. & Bory) P. W. Ball & Heywood (*Kohlrauschia g.*) A quite hairless, pink-like annual to 50 cm, with tight terminal heads of several pink flowers surrounded by broad, blunt, brown papery bracts. Flowers 1–1·5 cm across; petals

usually toothed. Lower leaves spathulate, upper linear. Seeds warty or smooth. Rocky places in the Med. region. YU.AL.GR.CR.BG? **184.** *P. prolifera* (L.) P. W. Ball & Heywood (*Tunica p., Dianthus p.*) PROLIFEROUS PINK. Distinguished from 184a by its obcordate or blunt petals; its outermost bracts with a fine point; and its netted seeds. Sandy and rocky places. Widely scattered on mainland.

(184). *P. velutina* (Guss.) P. W. Ball & Heywood Distinguished from 184a by the acute or fine-pointed (*muronate*) outer bracts; middle part of stem densely glandular-hairy; leaf sheaths at least twice as long as wide. Petals pink, two-lobed. Rocky places in the Med. region. Widespread. **184b.** *P. thessala* (Boiss.) P. W. Ball & Heywood A slender wiry perennial 10–35 cm, with few linear leaves and little-branched erect stems bearing tiny heads of white or pinkish flowers. Heads less than 1 cm long, up to ten-flowered; bracts brown with white papery margin, or completely white. Petals toothed, notched, or entire. Rocks. GR(east).

DIANTHUS | Pink About 69 species, including 16 species which are restricted to our area. Many are very similar in general appearance; the dimensions and details of the calyx, and epicalyx scales at the base of each flower, are important in distinguishing species; the petals may have hairs on their upper surface, in which case they are said to be *bearded*.

Flowers in densely clustered heads surrounded by bracts
185. *D. barbatus* L. SWEET WILLIAM. Meadows and woods. YU.BG.TR.   ****186. *D. armeria* L. DEPTFORD PINK. Roadsides, shady places. Widespread on mainland.   ****187.** *D. carthusianorum* L. CARTHUSIAN PINK. YU.AL.TR.

186a. *D. arboreus* L. Small shrub up to 50 cm, with tortuous branches and narrow fleshy leaves. Flowers scented, about 2 cm across; petals pink, toothed, and bearded. Epicalyx scales 10–20, obovate and tapering to a rigid point. Calyx 18–22 mm, narrowed from about the middle. Rocky places. GR(south-west).CR + Karpathos. Page 224.   **186b.** *D. fruticosus* L. Like 186a but flowers not scented; epicalyx scales 8–10; calyx 20–30 mm. Rocky places. GR(Sikinos, Serifos, Folegandros).

186c. *D. juniperinus* Sm. A small shrub with erect woody branches, herbaceous flowering stems up to 15 cm, and spiny-pointed leaves. Petals pale pink, toothed, and bearded. Epicalyx scales 4–8, obovate, $\frac{1}{3}$–$\frac{1}{2}$ as long as the more or less cylindrical calyx, which is 13–20 mm. Rocky places. CR.   **186d.** *D. pseudarmeria* Bieb. Like 186 with a dense cluster of flowers surrounded by leafy bracts, but flowers pink. Epicalyx scales ovate abruptly contracted with a bristle-like point (lanceolate in 186); calyx 10 by 1·5 mm (15–20 by 2–3 mm in 186). Sandy places. YU.BG.TR. Page 148.

187a. *D. pinifolius* Sibth. & Sm. A densely tufted perennial up to 40 cm, with numerous fine basal leaves, recalling pine needles. Flowers in dense heads; petals purple or violet, toothed. Epicalyx scales usually 6, obovate, with a long bristle-like apex; calyx 10–20 mm, tapering somewhat above the middle. Rocky places. Widespread on mainland. Pl. 4.   **187b.** *D. giganteus* D'Urv. A tall slender perennial to 1 m, with dense terminal clusters of numerous bright-purple flowers and contrasting brown or whitish epicalyx. Petals toothed, limb 5–8 mm. Epicalyx variable, papery or leathery, about half as long as calyx; calyx 15–22 mm, often purple-flushed. Meadows. Widespread on mainland. Pl. 4.

187c. *D. cruentus* Griseb. A bluish-green perennial up to 1 m. Bracts and epicalyx scales oblong to ovate-oblong, with long awn-like apex, up to as long as the calyx, which is 18–20 mm. Petals deep purple or pink-purple, toothed, and more or less bearded. Steep grassland, bushy places. Widespread on mainland. Page 89.

Flowers solitary, few, or in lax clusters
(a) *Petals deeply cut into many narrow strap-shaped lobes*
188. *D. superbus* L. SUPERB PINK. Woody hills, dry meadows. YU.BG. **189a. *D. petraeus* Waldst. & Kit. A green or bluish-green, loosely tufted, variable perennial to 30 cm. Petals very variable, bearded or not, deeply cut, toothed, or almost entire, white or less frequently pinkish. Epicalyx scales 2–8, elliptic to ovate, pointed, one-fifth to one-quarter as long as the calyx; calyx characteristically long and narrow, 12–32 mm, tapering upwards from below the middle. Rock fissures. YU.AL.GR.BG. Page 114.

(b) *Petals with shallow teeth, or entire*
  (i) *Petals hairy above (bearded) (see also 189a.)*
193a. *D. degenii* Bald. Like **193.** *D. deltoides* MAIDEN PINK but calyx smaller 8–10 mm (14–18 mm in 193), and with small solitary, stalkless, pink flowers about 1 cm across. Mountains. YU.AL.GR. **193b.** *D myrtinervius* Griseb. A dwarf, sprawling, densely leafy perennial with solitary, short-stemmed purple flowers. Epicalyx scales 2–4, the outer usually leaf-like, about half as long as the calyx; calyx 5–8 mm, bell-shaped. Leaves elliptic, 2–5 mm long. Alpine pastures in Macedonia. YU.GR. Page 224. **193c.** *D. sphacioticus* Boiss. & Heldr. A perennial to 10 cm, with a stout woody rootstock and solitary pale-pink flowers. Epicalyx scales usually 6, outer ovate and leaf-like, inner obovate, with a rigid point, about one-third as long as the calyx; calyx 13–16 mm, tapering from below the middle. Mountains. CR.

193d. *D. gracilis* Sibth. & Sm. A hairless perennial to 40 cm, with solitary or grouped flowers which are deep pink above and yellow or purplish beneath. Epicalyx scales 4–6, obovate, with rigid points, from a half to one-third as long as the calyx; calyx 10–20 mm by 3–5 mm wide, more or less cylindrical. Rocky places in mountains. YU.AL.GR.BG. **193e.** *D. haematocalyx* Boiss. & Heldr. An attractive, low tufted perennial with rather large conspicuous bright pinkish-purple flowers which are yellow beneath. Epicalyx scales 4–6, tapering gradually to a point, or abruptly contracted into a fine point; calyx 16–25 mm, narrowed upwards. Subsp. *pindicola* (Vierh.) Hayek Flowers solitary, short-stalked. Subsp. *haematocalyx* has stems to 30 cm which are one- to five-flowered. Mountains, rocky places. YU(Macedonia).AL.GR. Page 127. Pl. 4.

193f. *D. biflorus* Sibth. & Sm. A loosely tufted perennial to 40 cm with one or several, but often paired flowers with brick-red petals. Petals with numerous short hairs and glands all over the surface. Epicalyx scales 6, obovate, with a hair-like tip, leathery, about one-third to half as long as the calyx; calyx 20–5 mm, tapering from about the middle. Mountains. GR(central and south). **193g.** *D. corymbosus* Sibth. & Sm. An annual or perennial to 40 cm, more or less downy and often glandular, with many erect stems ending in a more or less dense cluster of 2 to several purple flowers which are often yellow beneath. Epicalyx scales ovate-lanceolate, with a green apex, from half to three-quarters as long as the calyx; calyx 10–20 mm, almost cylindrical, and usually downy. Meadows. Widespread on mainland. Page 224.

  (ii) *Petals without hairs (not bearded) (see also 189a)*
194. *D. sylvestris* Wulfen WOOD PINK. A tufted variable perennial with pale-pink flowers, often with entire, hairless petals, and glaucous calyx, but flowers very variable. Epicalyx scales 2–8, one-quarter as long as the calyx, which is 12–22 mm. Rocky places. YU.AL.GR. Page 190. **(194). *D. caryophyllus* L. CLOVE PINK, CARNATION. Widely cultivated; naturalized in GR. **194d.** *D. minutiflorus* (Borbás) Halácsy. A loosely tufted, slender-stemmed perennial with small usually solitary white flowers about 1 cm across. Epicalyx scales 4, broadly oval with an awl-shaped apex, about one-third as long as the calyx, which is 7–12 mm by 3–4 mm wide and somewhat inflated on one side. Leaves

2–3 cm long, rough. Rocks and grassy places in mountains. YU(south).AL.GR. Page 168.

194e. *D. serratifolius* Sibth. & Sm. A loosely tufted perennial to 30 cm, with long, trailing woody stems recalling, 194d but with pink flowers which are often darker purple beneath. Epicalyx scales 4–6, ovate, pointed, usually brown, about one-third as long as the calyx; calyx 12–15 mm, tapering from near the base. Leaves 1·5–3 cm, conspicuously three-veined beneath. Rocks in mountains. GR(south). Pl. 4. **194f.** *D. scardicus* Wettst. A low cushion-like tufted plant with long non-flowering stems and many short flowering stems bearing usually solitary, pink, scented flowers about 1·5 cm across. Epicalyx scales 2, oval with a green three-veined point, about half as long as the calyx, which is 10–12 mm and widens upwards. Mountain pastures. YU(Sar Planina). Pl. 4.

194g. *D. microlepis* Boiss. Like 194f but the short flowering stems leafless or with 1 or 2 pairs of scale-like leaves only. Basal leaves narrower, 1–1·5 mm wide (2–3 mm wide in 194f). Flowers purple. Alpine meadows. YU? BG. Pl. 4. **194h.** *D. xylorrizus* Boiss. & Heldr. A lax perennial to 15 cm, with a thick woody stock, and dirty white flowers about 1·5 cm across. Epicalyx scales usually 4, ovate-pointed, leathery, about one-quarter as long as the calyx; calyx 20–5 mm, widest below the middle. Basal leaves rather thick and soft, to 4 mm wide. Rocky places. CR+Kasos.

VELEZIA Like *Dianthus* but without epicalyx; calyx long cylindrical.

195. *V. rigida* L. A stiff much-branched wiry annual to 15 cm with tiny stalkless pink flowers with conspicuous long cylindrical calyx tubes. Petals with 2 narrow lobes. Sunny places in the Med. region. Widespread. **195a. *V. quadridentata* Sibth. & Sm. Like 195 but petals four-toothed; calyx narrowly elliptic, slightly inflated near the middle. Sunny places in the Med. region. GR+Is.TR.

NYMPHAEACEAE | Water-Lily Family

NYMPHAEA ****196.** *N. alba* L. WHITE WATER-LILY. Stagnant waters throughout our area. **196a.** *N. candida* C. Presl Stagnant or slow-running waters. YU.

NUPHAR ****197.** *N. lutea* (L.) Sibth. & Sm. YELLOW WATER-LILY. Stagnant waters. Throughout mainland. **(197).** *N. pumila* (Timm) DC. LEAST YELLOW WATER-LILY. Stagnant waters. YU.

RANUNCULACEAE | Buttercup Family

HELLEBORUS
1. Fls. pure white or pinkish; bracts entire (YU(north)).   ****201.** *H. niger* L.
1. Fls. greenish, reddish, or purplish; bracts divided and leaf-like.
  2. Carpels free to the base, shortly stalked.
    3. Fls. green; lvs. not over-

wintering. *cyclophyllus*
3. Fls. cream, becoming greenish-
yellow-brown; lvs. over-
wintering. *orientalis*
2. Carpels fused at their base for
about one quarter of their length.
4. Fls. greenish with no reddish or
purplish tinge.
5. Fls. at most 3·5–4 cm across,
nodding. *multifidus, dumetorum*
5. Fls. at least 4·5–7 cm across,
not or slightly nodding at
flowering. *odorus*
4. Fls. reddish or purplish at least
outside.
6. Lvs. entirely hairless. *dumetorum*
6. Lvs. finely hairy, at least on the
veins beneath. *multifidus, pupurascens*

(200). *H. cyclophyllus* Boiss. GREEK HELLEBORE. A robust perennial to 60 cm, with clusters of 3–4 pale glaucuous-green flowers, each about 6 cm across, and large divided leaves produced anew each spring. Leaves with 5–9, usually undivided, toothed segments, the uppermost leaves similar. Carpels free, narrowed to a short stalk at base. Woods, bushy places. YU(south).AL.GR.BG. Page 169. **200b. *H. orientalis* Lam. Like (200) but flowers pale greenish-cream and becoming brownish-yellow-green. Leaves over-wintering. Leaf segments 5–11, double-toothed. Woods, thickets. GR(Thrace).TR. Pl. 5.

200c. *H. odorus* Waldst. & Kit. Like (200) but carpels distinctly fused to each other at their bases. Flowers clear green, scented, 5–7 cm across. Basal leaf usually 1, over-wintering, leathery, finely hairy beneath, with 7–11 mostly undivided segments. Grassy, bushy places. YU.AL.BG. Pl. 5. **200d.** *H. multifidus* Vis. Like 200c with leaves finely hairy beneath but leaves with 9–15 segments which are further deeply divided almost to the base into linear lobes 3–10 mm wide. Flowers smaller 3–4 cm across, green, with petals scarcely overlapping. Grassy, bushy places. YU.AL. Subsp. *serbicus* (Adamović) Merxm. & Podl. has purplish-violet flowers, and is found in Serbia. Pl. 5.

200e. *H. dumetorum* Waldst. & Kit. Distinguished by its partly fused carpels; hairless or nearly hairless leaves which do not over-winter; and its green or violet flowers. Flowers 2–3, 3·5–5·5 cm across. Leaves thin, with 7–11 undivided segments which are distinctly stalked. Grassy places. YU. Pl. 5. **200f.** *H. purpurascens* Waldst. & Kit. Distinguished by its large flowers 5–7 cm across, with spreading purplish-violet petals and its non-over-wintering leaves with usually 5 segments further divided into 2–5 lobes, which are finely hairy at least on the prominent veins beneath. Woods, bushy places. YU.

ERANTHIS ****202.***E. hyemalis* (L.) Salisb. WINTER ACONITE. Woods. YU(north).BG.

NIGELLA | Love-in-a-Mist 9 species in our area, including 4 restricted to Crete and the Islands.

203. *N. damascena* L. LOVE-IN-A-MIST. Distinguished from most other species by its involucre of dissected leaves which closely surrounds each flower. Flowers bluish; petals shortly stalked. Carpels fused together their whole length, and much inflated in fruit. Stony, sunny places in the Med. region. Widespread. **(203).** *N. arvensis* L. Distinguished from 203 by the pale-blue flowers often veined with green which are usually long-stalked and without an involucre (subsp. *aristata* (Sibth. & Sm.) Nyman has an involucre closely investing each flower. GR+Cyclades). Petals with a stalk (claw) at least as long as the blade. Anthers with a fine point. Carpels three-veined, fused for half their length. Fields, sunny hills. Widespread on mainland. **(203).** *N. sativa* L. Flowers whitish, stalked, and without an involucre. Carpels fused their whole length and readily distinguished by the numerous small swellings covering the surface, not inflated. Cultivated; naturalized and scattered throughout our area.

TROLLIUS **204.** *T. europaeus* L. GLOBE FLOWER. Damp mountain meadows. YU.AL.BG.

ISOPYRUM **205.** *I. thalictroides* L. RUE-LEAVED ISOPYRUM. Woods. YU.BG.

ACTAEA **206.** *A. spicata* L. BANEBERRY. Woods. YU.AL.GR.BG.

CIMICIFUGA Like *Actaea* but differing in its fruit which has 2–8 dry, splitting carpels.

206a. *C. europaea* Schipcz. (*C. foetida*) A tall erect, foetid perennial to 1 m or more, with very large, two- or three-times-pinnate leaves and long slender terminal spike-like clusters of greenish flowers. Leaflets oval-lanceolate, double-toothed. Fruit hairy. Woods. BG.

CALTHA 207. *C. palustris* L. MARSH MARIGOLD. Marshes, riversides. Widespread on mainland, except TR.

ACONITUM | Monkshood
Flowers yellowish
208. *A. vulparia* Reichenb. WOLFSBANE. Flowers yellowish with a conical-cylindrical helmet about three times as long as wide. Terminal inflorescence small and few-flowered. Leaves deeply divided into segments which are three-lobed to the middle. Woods, bushy places. YU.BG? **208a.** *A. lamarckii* Reichenb. A stout erect perennial to 1 m, like 208, but with large many-flowered terminal inflorescence of yellowish, hairy flowers 1·5–2 cm long. Helmet conical-cylindrical. Leaves light green, with 7–8 segments divided to below the middle into several linear lobes. Woods, bushy places in mountains. YU.AL.GR? BG.

209. *A. anthora* L. YELLOW MONKSHOOD. Flowers usually yellowish, rarely blue, and helmet more or less hemispherical, about as high as wide. Perianth persisting. Leaves with narrow strap-shaped segments, not more than 3 mm broad. Carpels 5, usually densely hairy. Rocks in mountains. YU.BG.

Flowers blue or purple, sometimes white
(a) Helmet of flowers usually higher than wide
(210). *A. variegatum* L. BRANCHED MONKSHOOD. An erect hairless perennial to 1 m, with blue-violet flowers, in usually branched inflorescences which are hairless or at least without glandular hairs. Helmet about twice as high as wide. Fruit hairless. Seeds winged on one angle. Bushy places in mountains. YU.BG. **210g.** *A. toxicum* Reichenb. Like (210) with blue-violet flowers but inflorescence glandular-hairy; bracts ovate. Bushy places. YU(west and central).

(*b*) *Helmet usually wider than high* (*A. nepellus* group)
210d. *A firmum* Reichenb. Inflorescence usually simple and few-flowered with violet or blue flowers and with leaves not crowded below the inflorescence. Leaves with segments divided to about halfway into oblong lobes. Hills, mountain grasslands. YU.BG? **210e.** *A. tauricum* Wulfen Like 210d but leaves crowded below inflorescence, which is simple. Leaf segments with rather wide lobes. Mountains. YU. **210f.** *A. 'pentheri'* Hayek=(*divergens*) Very like 210d but differing in being glandular-hairy and in having up-curved flower stalks. Hills, mountains. YU.BG. Pl. 5.

DELPHINIUM

1. Perennials; inner lateral perianth segments with hairs on upper surface.
  2. Inner perianth segments contrasting and darker than outer; tubers absent; seeds not covered with scales, winged at angles (YU.).   ****211.** *D. elatum* L.
  2. Inner perianth segments similar-coloured to outer; tubers present; seeds covered with papery scales, not winged.
    3. Fls. 24–7 mm (including spur), blue or lilac; bracts linear.   *fissum*
    3. Fls. 19–23 mm (including spur), pale blue to whitish-yellow; bracts oval-lanceolate (GR(north). BG.).   **211j.** *D. albiflorum* DC.
1. Annuals or biennials; inner lateral perianth segments hairless.
  4. Spur much shorter than perianth segments (YU.AL.GR+Is.CR.).   **212.** *D. staphisagria* L. Pl.6.
  4. Spur as long as perianth segments, or longer.
    5. Blade of inner lateral perianth segments wedge-shaped at base gradually narrowed into a stalk (claw).
      6. Spur up to twice as long as perianth segments.   *peregrinum*
      6. Spur about equalling perianth segments (Aegean Is.).   **211g.** *D. hirschfeldianum* Heldr. & Holzm.
    5. Blade of inner lateral perianth segments abruptly contracted into a stalk (claw).
      7. Fr. hairless; blade of inner lateral perianth segments heart-shaped at base (GR.).   **211h.** *D. hellenicum* Pawl.
      7. Fr. with long spreading hairs; blade of inner lateral perianth segments rounded.   *balcanicum*

211k. *D. fissum* Waldst. & Kit. A slender erect, usually unbranched perennial to 1·5 m, with dark blue-violet or lilac flowers in a slender spike. Flowers rather large 24–7 mm, including the more or less horizontal spur; bracts linear. Leaves deeply cut into many linear lobes, 1–3 mm wide. Fruit hairy or hairless. Grassy places in mountains. YU.AL.BG.TR. ****(211).** *D. peregrinum* L. A slender annual with blue-violet flowers and usually upward-pointing spurs, in a compact cylindrical spike. Uppermost leaves linear entire, lower usually once-cut into linear segments. Carpels 3, usually finely hairy. Stony ground, cultivation. Widespread. Pl. 6

211i. *D. balcanicum* Pawl. BALKAN LARKSPUR. Like (211) with usually a very compact cylindrical spike with numerous flowers with long, up-turned spurs 2–3 times as long as the perianth segments. Blade of inner lateral perianth segments like a ping-pong bat, the outer segments hairy outside. Uppermost leaves linear-pointed. Fruit 5–9 mm. Open ground, cultivated places. YU.GR.BG. Page 238.

CONSOLIDA Like *Delphinium* but always annual, and inner perianth segments only 2 and fused at the base into a single long spur which is included in the outer spur. Carpel solitary. 10 similar-looking species, including 4 endemic to Greece and the Balkans, and 2 restricted to Turkey-in-Europe.

213. *C. ambigua* (L.) P. W. Ball & Heywood LARKSPUR. An erect little-branched annual to 1 m, with spike-like clusters of usually deep-blue flowers with a more or less straight spur 13–18 mm. Flowers sometimes white or pink. Leaves deeply cut into linear segments, the upper stalkless. Carpel 1·5–2 cm, finely hairy; seeds black. Sunny, stony places. YU.AL.GR+Is.CR.TR. **(213). *C. orientalis* (Gay) Schrödinger Like 213 but spur shorter, not more than 12 mm; carpel very abruptly contracted at apex; seeds reddish-brown. Cultivated ground. Throughout mainland. Pl. 5.

(213). *C. regalis* S. F. Gray Like 213 but bracts subtending flowers linear, entire, not cut; carpel usually hairless. Flowers dark or light blue; spur 12–25 mm; inflorescence either dense and little-branched, or lax and much-branched. Fields, stony ground. Widespread on mainland. **213a. *C. tenuissima* (Sibth. & Sm.) Soó SLENDER LARKSPUR. A slender, widely branched annual to 50 cm, with blue-violet flowers which are the smallest of the genus. Perianth limb 5–7 mm, spur straight, 7–10 mm; upper lobe of inner perianth segments broad, two-lobed. Upper leaves with linear segments. Carpel 5–8 mm long, rounded, less than twice as long as wide, hairless. Stony places in mountains. GR(south-east). Page 238.

ANEMONE The coloured petal-like perianth segments are referred to as petals in the following descriptions.

Stem leaves shortly stalked, more or less similar to basal leaves
(a) Flowers yellow
(214). *A. ranunculoides* L. YELLOW WOOD ANEMONE. Distinguished by its solitary, or 2–3, yellow flowers borne above three-lobed, deeply divided, and toothed stem leaves. Flowers 1·5–2 cm across. Woods, bushy places. YU.BG.

(b) Flowers blue
(214). *A. apennina* L. BLUE WOOD ANEMONE. Like 214 in habit, but flowers blue, with 8–14 petals which are finely hairy beneath. Leaves finely hairy beneath. Fruit head erect. Bushy, grassy places. YU.AL.GR(north).BG. Page 179. **(214).** *A. blanda* Schott & Kotschy Very like *A. apennina* but petals hairless beneath and more numerous. Leaves hairless beneath. Fruit head nodding. Rocks, bushy places. AL.GR+Is.BG.TR. Pl. 6.

(c) Flowers white or pinkish

214. *A. nemorosa* L. WOOD ANEMONE. Flowers white or pink-flushed, 2–4 cm across, solitary, borne above 3 palmately lobed leaves with toothed or lobed segments, similar to basal leaves. Anthers yellow. Fruit head nodding. Woods, shady places. Widespread on mainland. **214a. *A. trifolia* L. Differs from 214 in having usually only stem leaves which are divided into 3, toothed (not lobed) segments and buds in the axils of the leaves. Flowers white; anthers blue. Woods. YU. Pl. 6.

215. *A. sylvestris* L. SNOWDROP WINDFLOWER. Distinguished by its large, usually solitary, white flower 4–7 cm across, and its deeply lobed and cut, short-stalked stem leaves. Petals 5, broadly ovate, silky beneath. Fruit head densely woolly-haired. Dry meadows, bushy places. YU(north).BG. **215a.** *A. baldensis* L. Like 215 but a smaller plant to 12 cm, with smaller white flowers 2·5–4 cm across, with more than 5 petals. Leaves more divided, three times cut. Stony alpine pastures. YU.

Stem leaves stalkless and dissimilar to basal leaves
(a) Stem leaves bract-like, entire or lobed

(216). *A. hortensis* L. Very like 216 *A. coronaria* in habit but distinguished by its stem leaves which are bract-like, linear-lanceolate, and usually undivided. Petals usually 15 but ranging from 12 to 19, narrowly elliptic and usually not overlapping, pink, purplish, scarlet, or white. Grassy, stony places in the Med. region. YU.AL. **(216).** *A. pavonina* Lam. Like *A. hortensis* with bract-like linear-lanceolate stem leaves which are entire or often three-lobed, but petals usually 8–9, broader and overlapping, scarlet, pink, or purple, often with a yellowish base. Stony places in the Med. region. Widespread, except AL. Pl. 6.

(b) Stem leaves deeply divided into narrow segments

**216. *A. coronaria* L. CROWN ANEMONE. Flowers large, solitary, red, pink, blue, or white, 3·5–6·5 cm across, and distinguished from similar species by the stem leaves which are deeply cut into numerous narrow segments. Petals 5–8, elliptic, overlapping; anthers blue. Stony and bushy places in the Med. region. YU.GR+Is.CR.TR. Pl. 6.

**218. *A. narcissiflora* L. NARCISSUS-FLOWERED ANEMONE. Readily distinguished by its umbel of 3–8 white or pinkish flowers, each 2–3 cm across, and its stem leaves which are deeply cut into narrow linear segments. A rather robust, hairy perennial to 50 cm, with several deeply palmately cut basal leaves. Meadows in the mountains. YU.AL.BG.

HEPATICA ****219.** *H. nobilis* Miller HEPATICA. A low-growing perennial with usually bluish-violet or purple solitary flowers 1·5–2·5 cm across, arising on slender stems direct from the stock. Petals usually 5–7; calyx-like involucre of 3 green oval bracts lying close to the petals. Leaves distinctive, heart-shaped with 3 broad shallow lobes, evergreen. Montane woods and coppices. YU.AL.BG.

PULSATILLA Like *Anemone* but with long feathery styles. Differentiation of species often not clear-cut; all are very similar in habit. 7 species in our area, mostly restricted to the north in Yugoslavia and Bulgaria.

1. Stem lvs. shortly stalked, resembling
   the basal lvs. but smaller.
   2. Terminal segments of mature lvs.
      not divided to mid-rib; lv. blade
      distinctly hairy (YU.).   ****220.** *P. alpina* (L.) Delarbre
   2. Terminal segments of mature lvs.
      divided to the mid-rib; lv. blade

almost or quite hairless (YU.).

1. Stem lvs. stalkless, divided into linear segments and not closely resembling the basal lvs.

    220a. *P. alba* Reichenb.

    3. Basal lvs. palmately divided (BG.).

    223b. *P. patens* (L.) Miller

    3. Basal lvs. pinnately divided.

        4. Basal lvs. evergreen, once-pinnate; segments lobed; fls. usually white within (YU.BG.).

        **221. *P. vernalis* (L.) Miller

        4. Basal lvs. withering in autumn; two–four-times pinnate; fls. usually purple.

            5. Fls. erect (YU.GR.BG.).

            221a. *P. halleri* (All.) Willd.

            5. Fls. nodding.

                6. Petals less than 1½ times as long as the stamens, recurved at apex (YU.BG?).

                222. *P. pratensis* (L.) Miller Pl.6.

                6. Petals at least twice as long as the stamens, not recurved at apex (YU.BG.).

                222a. *P. montana* (Hoppe) Reichenb.

CLEMATIS

Plants climbing or scrambling, usually woody

(a) Flowers white or yellowish

224. *C. vitalba* L. TRAVELLER'S JOY, OLD MAN'S BEARD. Flowers greenish-white, with petals hairy on both sides. Leaves once-pinnate with 3–9 oval leaflets. Woods, bushy places. Widespread on mainland. **224a. *C. orientalis* L. Differs from 224 in its yellowish flowers with pointed petals, and its ciliate stamens. Leaflets glaucous, narrower, oblong to linear, three-lobed. Bushy rocky places. GR.(Tinos,Kos,Rhodes).

225. *C. flammula* L. FRAGRANT CLEMATIS. Flowers pure white, in terminal- and axillary-branched clusters, very fragrant. Leaves twice-cut into numerous small, oval, or oblong, entire or three-lobed leaflets. Bushy places. Widespread on mainland +CR. **226. *C. cirrhosa* L. VIRGIN'S BOWER. A climbing or scrambling plant distinguished by its large, solitary, hairy, cream-coloured nodding flowers with conspicuous green cup-shaped involucres. Leaves clustered, shiny, variable with entire, three-lobed, or divided leaves often occurring on the same plant. Bushy places. GR+Is.CR.

(b) Flowers bluish, purple, or violet

226a. *C. viticella* L. A deciduous, partly woody climber with solitary fragrant, bluish or purplish, long-stalked, nodding flowers with spreading petals. Petals 1·5–3·5 cm, obovate, margins undulate. Leaves pinnate with three-lobed leaflets. Bushy places; often naturalized from cultivation. Scattered throughout mainland+CR. Pl. 8. ****227.** *C. alpina* (L.) Miller ALPINE CLEMATIS. Distinguished from all other species by its large solitary, nodding, violet flowers, with pointed petals and with numerous nectaries in place of the outer stamens. Leaves twice-ternate; leaflets coarsely toothed. Woods and bushy places in mountains. YU.BG.

Plants erect, herbaceous, woody only at base, not climbing

228. *C. recta* L. An erect, hollow-stemmed perennial to 1·5 m, with white fragrant flowers in a terminal-branched cluster. Petals hairless except on margins. Leaves with 5–7 large oval leaflets. Bushy places, sunny slopes. YU.BG.Pl.8. ****229.** *C. integrifolia* L.

An unmistakable erect herb with oval-acute, entire, stalkless leaves and solitary terminal, purple, open-bell-shaped flowers. Petals acute, hairless except near margins. Pastures. YU.BG.

ADONIS
Perennials; petals more than 10

231. *A. vernalis* L. YELLOW ADONIS. Flowers large and solitary with many shining, narrow yellow, spreading petals, and hairy sepals half as long. Basal leaves reduced to scales; stem leaves cut into linear segments. Beak of fruit short, curved. Grassy places. YU.BG. **231b. *A. cyllenea* Boiss., Heldr. & Orph. Like 231 but differing in absence of stem scales, and in having a freely branched stem to 60 cm and long-stalked basal leaves. Fruit with a long hooked beak. Flowers yellow. Mountains. GR(south). Page 92.

Annuals; petals 8 or less

230. *A. annua* L. PHEASANT'S EYE. Flowers scarlet with a black basal patch, rather small, 1·5–2·5 cm across; sepals hairless. Carpels 3–5 mm, hairless, without a projection, with or without a transverse ridge, beak straight. Grassy places, cultivation. Widespread. **(230). *A. flammea* Jacq. Flowers deep scarlet, rarely yellow, 2–3 cm across; sepals hairy. Carpels with a rounded projection on the inner margin just below the ascending beak. Grassy places, fields. Widespread on mainland.

(230). *A. aestivalis* L. Like 230 but carpels larger 5–6 mm, with a transverse ridge, inner margin with 2 projections, the lower acute and the upper obtuse, one-third the distance from the base to apex, broad and rounded. Fields, uncultivated ground. Widespread on mainland. **(230).** *A. microcarpa* DC. Flowers usually yellow. Carpels 3–4 mm with or without a transverse ridge, and 2 projections on inner margin, the upper projection blunt and close to the style. Grassy, sunny uncultivated ground. YU.GR.CR.BG.TR.

RANUNCULUS | Buttercup, Crowfoot About 80 species occur in our area. They can be conveniently divided into: (a) shining yellow-flowered species, ranging from lowland marshes and meadows to the highest mountains (about 65 species); (b) white-flowered species of ponds, ditches, lakes, and damp mud, often with both submerged dissected leaves and rounded floating leaves (8 species); (c) white-flowered terrestial plants mostly of the mountains, or with flowers occasionally red, pink, or purplish (7 species).

Yellow-flowered species
(a) *Leaf blades about as long as broad, mostly shallowly lobed to less than half their width*
  (i) *Carpels with spines or conspicuous swellings on sides*
232. *R. muricatus* L. SPINY-FRUITED BUTTERCUP. Distinguished by its head of rather large carpels each 7–8 mm, with spiny sides, broad sharp-edged margin, and nearly straight style 2–3 mm. Flowers pale yellow, 3–6 mm across. A pale, rather shiny annual to 50 cm with rounded or kidney-shaped leaves with shallow lobes. Damp grassy places. Widespread. **232a. *R. chius* DC. A pale-greenish, softly hairy annual with tiny pale-yellow flowers 6–8 mm across, and kidney-shaped three-lobed leaves. Fruit distinctive, having carpels with conical swellings on each face, a conspicuous border, and a short curved beak; fruit stalk swollen. Rather damp grassy places in the Med. region. Widespread, except AL. Page 238.

  (ii) *Carpels with smooth sides*
**233. *R. ficaria* L. LESSER CELANDINE. Widespread on mainland. Subsp. *ficariiformis* Rouy & Fouc. (*Ficaria grandiflora*). Much larger in all its parts, with flowers 3–5 cm across, and leaves 3–5 cm broad. Sepals yellowish-white. Bushy places, woods in the

1. *Ranunculus psilostachys* 239i **2.** *Consolida tenuissima* 213a **3.** *Ranunculus millefoliatus* 239e **4.** *Ranunculus chius* 232a **5.** *Delphinium balcanicum* 211i

Med. region. YU(Macedonia).GR+Is.CR. **233a.** *R. ficarioides* Bory & Chaub. Like 233 but carpels hairless and leaves bluntly lobed with 11–15 lobes, or coarsely toothed. Mountains, rocky places. GR. **234a.** *R. bullatus* L. Flowers yellow, scented, rather large 2·5 cm across, 1–2 borne on leafless stems. Leaves distinctive, all basal, broadly ovate, margin with rounded teeth, bristly-haired beneath and more or less embossed (bullate). Dry scrub. GR.CR+Rhodes.

234b. *R. brevifolius* Ten. A tiny plant of mountain screes with glaucous, fleshy, kidney-shaped, toothed, three-lobed leaves, and 1 or 2 yellow flowers 1·5–2·5 cm across. Carpels swollen, hairless, with a long curved beak. YU.AL.GR.CR. Pl. 7. **239d.** *R. creticus* L. A distinctive and attractive species with rather few large yellow flowers 2·5–3 cm across, and rather thick, softly hairy rounded basal leaves which are divided from a quarter to a half into 5 or more broad, toothed lobes. Stem leaves few, deeply cut into 3 narrow lobes. Carpels flattened, winged, with a short curved beak. Shady limestone rocks. CR.+Karpathos, Rhodes. Pl. 7.

(*b*) *Leaf blades about as long as broad, deeply lobed to more than half their width, or to the base*

(Key to the more distinctive broad, dissected-leaved, yellow-flowered buttercups)

1. Roots of 2 kinds, some slender fibrous,
   some fleshy and forming spindle-shaped to
   ovoid tubers.
   2. Sepals not reflexed at flowering.
      3. Basal lvs. lobed almost to base.   *millefoliatus*
      3. Basal lvs. more or less lobed
         to about three-quarters.   *spruneranus, subhomophyllus*
   2. Sepals reflexed at flowering.
      4. Lvs. without adpressed hairs.   *gracilis, neapolitanus*
      4. Lvs. with adpressed or silky hairs.
         5. Tubers spindle-shaped or cylindrical.   *psilostachys*
         5. Tubers ovoid or shortly oblong
            (YU.GR.BG.TR.).   **239j.** *R. rumelicus* Griseb.
1. Roots all fibrous, sometimes thickened
   but not tuberous.
   6. Carpels not or slightly compressed;
      annuals.
      7. Head of frs. cylindrical
         (YU.GR.BG.TR.).   **236.** *R. sceleratus* L.
      7. Head of frs. globular (Widespread).   **242.** *R. auricomus* L.
   6. Carpels strongly compressed.
      8. Carpels spiny; hairy annuals; sepals
         erect (Widespread).   **235.** *R. arvensis* L.
      8. Carpels smooth.
         9. Receptacle hairless; perennials.
            10. Sepals reflexed at flowering.
               11. Lv. segments wedge-shaped.   *velutinus*
               11. Lv. segments broadly ovate.   *constantinopolitanus*
            10. Sepals not reflexed at flowering.
               12. Lvs. more or less silky beneath;
                  carpels twice as long as beak.   *serbicus*
               12. Lvs. hairy to nearly hairless
                  beneath; carpels as long as beak.   *brutius*

9. Receptacle hairy.
 13. Sepals reflexed at flowering.
  14. Annuals (YU.AL.GR.BG.). **(237)** *R. sardous* Crantz
  14. Perennials.
   15. Several of the lower stem lvs.
    similar to the basal lvs. but
    smaller (YU.AL.GR? BG.). **237.** *R. bulbosus* L.
   15. Stem lvs., except the lowest,
    small and bract-like. *neapolitanus*
 13. Sepals not reflexed at flowering.
  16. Fl. stalk furrowed.
   17. Creeping stolons present
    (YU.AL.GR.BG.). **238** *R. repens* L.
   17. Stolons absent. *nemorosus*
  16. Fl. stalk cylindrical, not furrowed;
   plant with rhizome. *oreophilus, sartorianus*

237a. *R. neapolitanus* Ten. A perennial 10–50 cm, covered with soft spreading hairs Flowers 2–3 cm across; sepals strongly reflexed. Basal leaves tripartite with rounded to wedge-shaped, toothed lobes. Carpels rounded, smooth, bordered, with a very short beak. Damp places. Widespread. **238a.** *R. velutinus* Ten. Very like 237a in general appearance, but covered with adpressed silky hairs above and spreading hairs below and distinguished by its hairless receptacle and rounded flower stalk. Flowers 1·5–2·5 cm across; sepals reflexed. Fruit head depressed globular; carpels compressed strongly bordered, hairless, with very short straight beak. Damp places. Widespread.

238b. *R. constantinopolitanus* (DC.) D'Urv. A robust, coarse perennial to 75 cm, with dense spreading hairs below and adpressed hairs above, and basal leaves triangular-ovate, cut to three-quarters into 3 broadly ovate segments which are further cut into toothed lobes. Flowers 1·5–2·5 cm across. Fruit head globular; carpels with beak curved. Thorn scrub, rocky places. GR.BG.TR. **238c.** *R. nemorosus* DC. A much-branched perennial 20–80 cm, with basal leaves three-lobed, with broadly ovate, lobed and toothed segments. Petals 1·5–2 cm, golden-yellow; flower stalk grooved; sepals spreading. Carpels with a strongly curved beak 1·5 mm long; receptacle hairy. Mountain woods. YU.AL.GR.BG.

239e. *R. millefoliatus* Vahl A delicate somewhat hairy perennial to about 15 cm, with ovoid or shortly oblong tubers. Basal leaves parsley-like, much divided, with numerous narrow lobes less than 1 cm long. Flowers small about 1·5 cm across; sepals hairless Carpels compressed with hooked beak half as long; receptacle hairless. Stony places YU.AL.GR.BG. Page 238. **239f.** *R. spruneranus* Boiss. A stout, much-branched almost hairless to hairy, erect perennial 20–45 cm, with cylindrical tubers. Flowers numerous petals 1–1·5 cm. Basal leaves with spreading hairs, circular-heart-shaped, with 3–5 toothed lobes. Fruiting head oblong; carpels with bulbous-based hairs, and with a hooked beak nearly equalling the carpel. Grassy hill slopes. YU.GR+Is.CR.BG.

239g. *R. subhomophyllus* (Halácsy) Vierh. A slender little-branched nearly hairless perennial with 1–2 yellow flowers; petals 5–10 mm. Basal leaves hairless rounded or wider than long, divided into 3 segments which are each further bluntly lobed. Fruit head cylindrical; carpels with slender curved beak. Tubers oblong. Grassy, stony places GR+Is.CR. Page 71. **239h.** *R. gracilis* E. D. Clarke. A rather delicate perennial 6–18 cm, with almost circular, shallowly three-lobed, outer basal leaves, and inner more deeply cut leaves with wedge-shaped toothed segments. Stems one- three-flowered petals 6–10 mm; sepals strongly reflexed. Fruit head cylindrical; carpels flattened, smooth, beak short, upturned. Tubers ovoid. Stony grassland, hills, fallow fields. YU.GR.CR.BG.TR.

239i. *R. psilostachys* Griseb. Like 239h but with leaves silky beneath and tubers spindle-shaped. Flowers few, large to 3 cm across, with strongly reflexed sepals. Leaves mostly longer than wide, deeply cut into narrow wedge-shaped lobes; stems 15–35 cm. Carpels rounded, warty, with a long more or less curved beak. Mountain meadows, stony grassland. YU.AL.GR.BG.TR. Page 238.

240a. *R. serbicus* Vis. An erect robust perennial 1–1·5 m, with a long stout rhizome, and with silky leaves, at least when young. Basal leaves divided into 3 broad wedge-shaped, sometimes stalked, variously cut segments; stem leaves similar but smaller. Flowers few, in a little-branched cluster; filaments of stamens hairy. Carpel with a nearly straight beak shorter than itself. Mountain woods. YU.AL.GR.BG. **240b.** *R. brutius* Ten. Distinguished by its large roughly pentagonal basal leaves which are cut into 3–5 often short-stalked segments, which are characteristically divided into large ovate or lanceolate, saw-toothed, lobes. Stem leaves smaller, similar. Flowers few, about 2 cm across; filaments of stamens hairless. Carpels compressed, keeled, with a slender curved beak about as long as itself. Mountain woods. GR.TR.

241a. *R. oreophilus* Bieb. One of several similar small alpine grassland species closely related to **241 *R. montanus*. They can generally be distinguished by their rounded (in outline), broad-lobed basal leaves, contrasting with the deeply cut stem leaves with narrow lobes. Flowers usually 1–3, relatively large 2–4 cm across. Carpels flattened, with a hooked beak. **241a** has hairy leaves and carpels with a very short adpressed beak. YU.AL. **241b.** *R. sartorianus* Boiss. & Heldr. Very similar to 241a but carpels with a stout rigid beak about one-third as long as the carpel. YU.AL.GR.BG. Page 92.

(c) Lea blades several times longer than broad, usually entire
Fairly videspread on mainland, in damp places, marshes, riversides are: 243. *R. flammula* L. LESSER SPEARWORT; **(243). *R. ophioglossifolius* Vill. SNAKESTONGUE CROWFOOT; **244. *R. lingua* L. GREAT SPEARWORT.

White-flowered aquatic species
Widespread submerged and floating species are: **250. *R. peltatus* Schrank POND CROWFOOT; (250). *R. aquatilis* L. COMMON WATER CROWFOOT; 251. *R. trichophyllus* Chaix

White- to red-flowered terrestrial species
(246). *R. platanifolius* L. LARGE WHITE BUTTERCUP. A tall plant to 1 m or more with large five- seven-lobed leaves and a much branched, nearly leafless inflorescence of many white flowers. Flowers 1–2 cm across, long-stalked; sepals reddish or purple. Middle lobe of leaf not free to the base. Carpels 5 mm, beak slender, hooked. Mountain woods. YU.AL.GR.BG. Pl. 7. **246. *R. aconitifolius* L. WHITE BUTTERCUP. Like (246) but the middle segment of the leaf is free to the base. Flowers white, 1–2 cm across, the flower stalks hairy above, 1–3 times as long as the subtending leaf (4–5 times as long and hairless in (246)). By springs. YU(central).

246b. *R. crenatus* Waldst. & Kit. A tiny hairless perennial 4–10 cm, with solitary or paired white flowers and more or less circular, toothed, leaves, found growing in high mountains by melting snow. Flowers comparatively large 2–2·5 cm across. Basal leaves with rounded blades 1–1·2 cm, stem leaves bract-like, linear. YU.AL.BG. Page 168. **246a.** *R. asiaticus* L. TURBAN BUTTERCUP. SCARLET CROWFOOT. A very beautiful large-flowered plant with purple, scarlet, yellow, or white flowers very like 216 *Anemone coronaria* but distinguished from it by the 5 sepals lying beneath the petals.

Flowers 3–6 cm across, cup-shaped, with a central boss of dark stamens. Outer basal leaves with 3 broad, toothed lobes, inner basal and stem leaves deeply cut into narrow lobes. Fruiting head elongated; carpels with a hooked beak. Dry stony places. GR(Peloponnisos).CR+Rhodes. Pl. 7.

CERATOCEPHALUS Like *Ranunculus* but carpels with an empty cell on either side of the seed, and with a long-pointed, up-curved beak.

251a. *C. falcatus* (L.) Pers. A small downy annual 2–10 cm, with solitary yellow flowers 10–15 mm across, and with distinctive cylindrical fruiting heads of numerous carpels with long sickle-shaped beaks. Leaves all basal, deeply cut into many narrow segments. Fields in lowlands. YU.GR.BG.TR. **251b.** *C. testiculatus* (Crantz) Roth Very like 251a but beak of fruit nearly straight; flowers smaller 5–10 mm across. Cultivated ground, waste places. GR(Thrace).BG.

MYOSURUS 2 species in our area. ****252.** *M. minimus* L. MOUSE-TAIL. Damp fields. Scattered on the mainland.

AQUILEGIA | Columbine

1.	Fls. yellow.		*aurea*
1.	Fls. purple, blue, white, or bicoloured.		
2.	Spurs shorter than limb of inner petals (YU.).		**253f.** *A. kitaibelii* Schott Page 191
2.	Spurs as long as or longer than limb of inner petals.		
3.	Fls. bicoloured.		
4.	Limb of inner petals 9 mm or more.		
5.	Basal lvs. once-ternate.		*dinarica*
5.	Basal lvs. twice-ternate.		
6.	Limb of inner petals 13–14 mm, rounded at apex; stamens not projecting.		*amaliae*
6.	Limb of inner petals 16–20 mm, cut off at apex; stamens projecting (GR.).		**253c.** *A. ottonis* Boiss. Page 93
4.	Limb of inner petals less than 8 mm (YU(Serbia)).		**253h.** *A. pancicii* Degen
3.	Fls. one-coloured.		
7.	Spurs strongly hooked.		
8.	Stems glandular-hairy.		*nigricans*
8.	Stems without glandular hairs (YU.).		****253.** *A. vulgaris* L.
7.	Spurs more or less straight (YU.).		**253e.** *A. grata* Zimmeter

253i. *A. aurea* Janka The only yellow-flowered species in Europe. Outer petals 2–3 cm; spurs 13–15 mm, hooked. Basal leaves twice-ternate, hairless above, finely hairy beneath, uppermost leaflets linear. Fruit finely hairy. Mountain woods. YU(Macedonia).BG. Pl. 8.

253g. *A. dinarica* G. Beck Flowers nodding, with outer petals intense blue, and inner white or bluish; spurs 10–15 mm, blue, hooked. Leaves greyish, covered with soft spreading hairs. Rocks in mountains. YU.AL.

253d. *A. amaliae* Boiss. Flowers nodding, with outer petals pale blue-violet, and inner white; spurs 13–14 mm, pale violet, strongly hooked. Leaves hairless above, sparsely hairy beneath. Rocks in mountains. YU.AL.GR. Page 126. Pl. 8. **253j.** *A. nigricans* Baumg. Distinguished by its very large, nodding, dark-purple flowers with outer petals 25–35 mm; spurs stout, hooked, 13–15 mm. Stems glandular hairy above. Woods in hills. YU.BG.

THALICTRUM | **Meadow Rue** 7 species in our area.

255. *T. aquilegifolium* L. GREAT MEADOW RUE. Flowers lilac or whitish. Woods in the mountains. YU.AL.GR.BG. Page 178. **255c. *T. orientale* Boiss. A delicate hairless perennial 10–30 cm, with few conspicuous white flowers, restricted to the foothills of the Taiyetos Mountains of the Peloponnisos. Petals 8–10 mm, longer than the stamens. Leaves 2–3 pinnate, leaflets rounded, lobed, or toothed. Carpels 2–6, grooved. Rare, shady rocks. GR(south). Page 88. ****256.** *T. minus* L. LESSER MEADOW RUE. Rocks, bushy places. Widespread on mainland, except TR. ****257.** *T. flavum* L. COMMON MEADOW RUE. Damp meadows, bushy places. YU.AL.BG.

PAEONIACEAE | Peony Family

PAEONIA

1. Fls. white. (Rhodes).
   (Crete, Karpathos).

    258a. *P. rhodia* W. T. Stearn Pl. 9.
    258b. *P. clusii* F. C. Stern Pl. 9.

1. Fls. red or deep pink.
   2. Segments of lower lvs. numerous, more than 40, each less than 5 mm wide

    tenuifolia

   2. Segments of lower lvs. less than 40, broader than 5 mm.
      3. Most leaflets undivided, the lower lvs. with 9–16 elliptic, ovate, or rounded segments.
         4. Leaflets rounded, margins undulate. (YU.)

    259c. *P. daurica* Andrews

         4. Leaflets broadly to narrowly elliptic, not undulate. (YU.AL.GR.BG.)

    **259. *P. mascula* (L.) Miller

      3. Most leaflets divided, the lower lvs. with 17–30 narrowly elliptic to lanceolate segments.
         5. Lvs. with minute bristles along the veins on the upper surface.

    peregrina

         5. Lvs. without minute bristles. (YU.AL.).

    **258. *P. officinalis* L.

(259). *P. tenuifolia* L. NARROW-LEAVED PEONY. Readily distinguished by its feathery leaves which have numerous narrow linear segments less than 5 mm wide. Flowers red, 6–8 cm across, appearing to rest on the leaves; stamen filaments yellow. Fruit of 2–3 woolly-haired carpels. Dry grassy places. YU(Serbia).BG. Pl. 9.

258c. *P. peregrina* Miller (*P. decora*) Flowers red, 7–13 cm across, stamen filaments red. Lower leaves divided into 17–30 narrowly elliptic segments, further cut into broadly triangular ultimate lobes and giving a toothed appearance. Fruits of 2–3 white-woolly carpels. Fields in the mountains. YU.AL.GR(Thrace).BG.

BERBERIDACEAE | Barberry Family

LEONTICE Herbaceous perennials with deeply rooted tubers; leaves two to three times ternate; flowers in terminal unbranched clusters; fruit dry and inflated.

260b. *L. leontopetalum* L. A weed of stony cultivated ground with elongated clusters of numerous long-stalked, yellow flowers subtended by usually conspicuous leafy bracts on stems 30–50 cm, and fleshy two- to three-times-ternate leaves. Flowers about 1·5 cm across; petals 6; sepals 6. Leaflets oval, entire, often glaucous. Fruit ovoid, 2·5–4 cm. GR+Is.CR.BG.TR. Pl. 8.

BONGARDIA Differs from *Leontice* in having all leaves basal and pinnate; petals shorter than sepals.

260c. *B. chrysogonum* (L.) Griseb. (*Leontice c.*) Flowers yellow, in much-branched lax clusters, with long ascending branches, and long flower stalks borne on stems to 60 cm. Sepals 3–6, reddish and soon falling. Leaves with whorled or opposite, oblong-wedge-shaped often lobed, stalkless, reddish-blotched leaflets. Fruit ovoid, 15 mm. Fallow fields. Chios. Rhodes. Pl. 8.

EPIMEDIUM
1. Inflorescence shorter than stem leaf;
   inner petals dark red (YU(north).AL.).   ****260.** *E. alpinum* L.
1. Inflorescence longer than stem leaf;
   inner petals pink (BG.TR.).   **260a.** *E. pubigerum* (DC.) Morren & Decne.

BERBERIS ***261.** *B. vulgaris* L. BARBERRY. A densely branched erect shrub 1·5–3 m, with branches with three-pronged spines, and drooping clusters of bright-yellow flowers. Leaves oblong, spiny-margined, 2·5–6 cm, longer than the stem spines. Fruit an orange-red berry 10 mm long. Bushy, rocky places. YU.AL.GR.BG.TR. **261b.** *B. cretica* L. Like 261 but a low bush less than 1 m with usually spineless leaves 1–2 cm long, shorter than the stem spines. Flower clusters 7–15 mm long, with 3–8 flowers, scarcely longer than the leaves. Fruit at length black, 6–7 mm long. Stony ground in mountains. GR+Is.CR. Page 70.

LAURACEAE | Laurel Family

LAURUS **263.** *L. nobilis* L. LAUREL. An evergreen shrub or small tree with dark-green leathery leaves, which are very aromatic when crushed. Flowers greenish-yellow, in short clusters, shorter than the oblong-lanceolate, entire leaves 5–10 cm. Fruit black, ovoid 10–15 mm. Woods and thickets in the Med. region; often cultivated. YU.AL.GR+Is.CR.TR. Page 34.

PAPAVERACEAE | Poppy Family

PAPAVER | Poppy 12 species in our area. The following are widespread, usually as weeds of cultivation: ****264.** *P. somniferum* L. OPIUM POPPY; ****265.** *P. rhoeas* L. CORN POPPY; **(265).** *P. dubium* L. LONG-HEADED POPPY; **266.** *P. argemone* L. PRICKLY LONG-HEADED POPPY; **(266).** *P. hybridum* L. PRICKLY ROUND-HEADED POPPY.

Fruit bristly-haired

266b. *P. nigrotinctum* Fedde Like 266. *P. argemone* but plant smaller to 20 cm, many-stemmed, with a rosette of leaves, and pink flowers with a large dark basal blotch to each petal. Bud nearly globular, hairy. Fruit 1–1·5 cm, oblong-elliptic, not club-shaped as in 266. Stony ground. GR(Corinth)+Is. Pl. 9. **266c.** *P. apulum* Ten. Distinguished by its bright scarlet, black-blotched flowers; broadly ovoid and nearly hairless buds, and elliptic fruit to 1 cm. Grassy and stony places. Widespread on mainland. **267b.** *P. erneri* Hayek Distinguished from ****267.** *P. rhaeticum* with similar golden-yellow flowers, by its glaucous leaves with narrow leaflets usually 1 mm wide. Fruit 1 cm, broadly club-shaped. Mountain screes. YU(central).

Fruit hairless

(265). *P. pinnatifidum* Moris Like (265) *P. dubium* but anthers yellow (violet in (265)), and leaf segments triangular-oval, entire or toothed. Flowers brick red; petals 2–2·5 cm; filaments dark violet. Calcareous slopes. GR(Thrace).BG.

ARGEMONE ****270.** *A. mexicana* L. PRICKLY POPPY. A glaucous annual to 90 cm, with prickly stems and leaves, and pale-yellow or orange poppy-like flowers 5–6 cm across. Fruit spiny. Native of America; naturalized in GR(Thrace).BG.

ROEMERIA ****271.** *R. hybrida* (L.) DC. VIOLET HORNED-POPPY. Readily distinguished by its deep-violet poppy-like flowers with large crumpled petals which soon fall, and by its long slender fruits. Cornfields, waste ground in the Med. region. YU.GR+Is.CR.BG.TR.

GLAUCIUM | Horned-Poppy

Flowers yellow; fruit hairless

*272. *G. flavum* Crantz YELLOW HORNED-POPPY. Distinguished by its glaucous, bristly leaves, large yellow flowers 6–9 cm across, and its slender, often curved, fruits 15–30 cm. Fruit covered with small swellings. Sands and gravels on the littoral. Widespread. **272a.** *G. leiocarpum* Boiss. Like 272 but flowers deeper yellow and fruit not more than 10 cm, with small swellings only at the apex. Dry stony places, seashores. YU.GR.CR.BG. Page 149.

Flowers red or orange; fruit hairy

*273. *G. corniculatum* (L.) J. H. Rudolph RED HORNED-POPPY. Flowers small 2·5–5 cm, often with a blackish basal blotch. Fruit 10–20 cm, usually straight, bristly-haired. Waste places, cultivated ground. Scattered throughout.

CHELIDONIUM 274. *C. majus* L. GREATER CELANDINE. Bushy and waste places. Widespread on mainland.

HYPECOUM

. Stem finely ribbed; outer petals
   almost as long as wide,
   three-lobed.

2. Outer petals with middle lobe
    largest.   *procumbens*
2. Outer petals with 3 equal lobes
    or lateral lobes largest (Widespread).   **(276).*H. imberbe* Sibth. & Sm.
1. Stem smooth; outer petals almost
    twice as long as wide, unlobed.
    3. Fruit pendulous; cultivated
        ground (YU?GR.BG.TR.).   **276a.** *H. pendulum* L.
    3. Fruit erect; maritime sands (BG.TR.).   **276b.** *H. ponticum* Velen.

276. *H. procumbens* L. A rather delicate, glaucous spreading annual 5–30 cm, of cultivated ground and sandy places, with lax branched clusters of small yellow flowers.
Flowers 0·5–1 cm across, with characteristic three-lobed petals. Leaves thrice cut into
narrow lobes. Fruit 4–6 cm, curved, jointed. YU.GR+Is.CR.BG.TR.

CORYDALIS   8 species in our area.

Aerial stems with many leaves; roots not tuberous
(278). *C. ochroleuca* Koch Distinguished by its dense clusters of cream-coloured
flowers which are yellowish towards the tips of the petals. Flowers 1·5 cm long. Leaves
three-pinnate, fern-like, glaucuous, with narrowly winged leaf stalks. Fruits erect.
YU.AL.GR(Macedonia).   **278.** *C. lutea* (L.) DC. YELLOW CORYDALIS. A perennial with
golden-yellow flowers, unwinged leaf stalks, and pendulous fruits. Shady rocks and
screes. YU(Croatia). Pl. 9.   **278a.** *C. acaulis* (Wulfen) Pers. Like 278 but an annual with
very glaucous leaves on both sides, and white flowers with yellowish-green tips. Walls
and rocks. YU(Croatia).

Aerial stems with only 1 or 2 leaves; roots tuberous
(a) Stem with a conspicuous scale below lowest leaf
279. *C. solida* (L.) Swartz A small perennial 10–20 cm, with a dense cluster of pink, or
whitish, spurred flowers borne above a pair of twice-ternate leaves. Flowers 1·5–2·5 cm
long; bracts subtending lowest flowers deeply lobed. Woods, bushy places. Widespread
on mainland.   **(279).** *C. intermedia* (L.) Mérat Like 279 but all bracts undivided, and
flowers smaller, 1–1·5 cm long, purple or rarely white. Woods, bushy places.
YU(Croatia, Bosnia).

(b) Stem without a conspicuous scale below lowest leaf
(279). *C. bulbosa* (L.) DC. Like 279 with 10–20 flowers but flowers rather large 2–3 cm
long, white, cream, yellow, or purplish, with curved spurs. All bracts entire; leaves
alternate, twice-ternate. Woods, thickets. YU.AL.GR.BG. Pl. 9.   **278b.** *C. rutifolia*
(Sibth. & Sm.) DC. (*C. uniflora*) Flowers 1–3, whitish marked with purple and spur
8–12 mm, borne on stems 5–15 cm. Bracts entire; leaves opposite. Mountains.
CR+Rhodes. Page 70.

FUMARIA | Fumitory   About 19 similar-looking species in our area. Widespread are
281. *F. officinalis* L. COMMON FUMITORY; **(281).** *F. parviflora* Lam. SMALL-FLOWERED
FUMITORY.

280. *F. capreolata* L. RAMPING FUMITORY A climbing and scrambling annual with
long-stalked axillary clusters of up to 20 creamy or pinkish flowers, with petals tipped
with dark reddish-purple. Flowers 10–14 mm long. Leaves compound. Fruit smooth
when dry. Bushy places in the Med. region. YU.AL.GR+Is.CR.TR.   **280a.** *F. flabellata*
Gaspar. Like 280 with creamy flowers, rarely purple-tipped, in clusters of 10–30. Fruit
densely warty when dry. Bushy places in the Med. region. YU.GR.CR.

CAPPARIDACEAE | Caper Family

CAPPARIS **283.** *C. spinosa* L. CAPER. Easily recognized by its large white flowers, 5–7 cm across, with numerous, projecting violet stamens, its rounded and sometimes fleshy leaves, and its usually spiny trailing stems. Fruit a large fleshy berry with pink flesh and dark-purple seeds. Rocks, waste places, on the littoral in the Med. region. YU.AL.GR+Is.CR. **(283).** *C. ovata* Desf. Very like 283 but leaves oblong to elliptic or ovate, with a fine projecting point—the continuation of the mid-rib. Flowers 4–5 cm across, zygomorphic (with petals distinctly unequal). Walls, sunny places in the Med. region. YU.AL.GR+Is.CR.TR.

CLEOME Distinguished from *Capparis* by the clustered flowers, 6 stamens, and dry splitting fruit.

283b. *C. ornithopodioides* L. A glandular-hairy sticky annual 15–50 cm, with trifoliate lower leaves and simple upper leaves, and small axillary whitish or yellow flowers. Petals 4, 3 mm; stamens longer. Fruit slender, 1·5–2·5 cm. Amongst herbaceous vegetation. GR(north).BG.TR.

CRUCIFERAE | Mustard Family

An important family in the Balkans with 79 genera and about 340 species. Only conspicuous-flowered or distinctive species have been included and many yellow- and white-flowered species have been omitted, including the following genera: *Sisymbrium* 11 spp.; *Descurainia* 1 sp.; *Alliaria* 1 sp.; *Arabidopsis* 1 sp.; *Myagrum* 1 sp.; *Isatis* 5 spp.; *Bunias* 1 sp.; *Syrenia* 1 sp.; *Chorispora* 1 sp.; *Euclidium* 1 sp.; *Barbarea* 11 spp.; *Rorippa* 9 spp.; *Armoracia* 1 sp.; *Nasturtium* 1 sp.; *Cardaminopsis* 3 spp.; *Berteroa* 5 spp.; *Lepidotrichum* 1 sp.; *Lobularia* 2 spp.; *Schivereckia* 1 sp.; *Erophila* 1 sp.; *Petrocallis* 1 sp.; *Andrzeiowskia* 1 sp.; *Kernera* 1 sp.; *Camelina* 4 spp.; *Neslia* 1 sp.; *Capsella* 4 spp.; *Hutchinsia* 1 sp.; *Hymenolobus* 1 sp.; *Hornungia* 1 sp.; *Teesdalia* 2 spp.; *Coronopus* 2 spp.; *Subularia* 1 sp.; *Moricandia* 1 sp.; *Diplotaxis* 3 spp.; *Brassica* 11 spp.; *Sinapis* 2 spp.; *Erucastrum* 1 sp.; *Rhynchosinapis* 1 sp.; *Hirschfeldia* 1 sp.; *Carrichtera* 1 sp.; *Rapistrum* 2 spp.; *Didesmus* 1 sp.; *Calepina* 1 sp.; *Enarthrocarpus* 2 spp.; *Raphanus* 1 sp. Ripe, often dry fruits are essential for the identification of many genera and species.

ERYSIMUM | **Treacle Mustard** A very difficult genus with about 23 species in our area.

Fruit flattened in section.

294a. *E. pusillum* Bory & Chaub. A tufted perennial with green to grey-green narrow leaves and many leafy flowering stems to 40 cm with yellow flowers. Petals 10–15 mm, finely hairy on the backs; sepals 5–8 mm, pouched at base. Fruit 2–6 cm, on fruit stalks up to 6 cm. A very variable species. Cliffs, rocks. YU.AL.GR.BG. Page 169. **294b.** *E. olympicum* Boiss. A biennial 15–35 cm, with yellow flowers with petals 10–16 mm. Leaves linear-lanceolate, coarsely toothed or lobed, the upper entire. Fruits long, 6–10 cm, grey with greenish angles, long-stalked, erect; style 1–3 mm. Mountain woods. GR(Olympus).

Fruit quadrangular in section.

294c. *E. raulinii* Boiss. Distinguished by its rather large golden-yellow flowers with petals 11–15 mm, and its grey fruits 1·5–4 cm, which lie closely pressed to the stem. A biennial 15–50 cm, with green, lanceolate, toothed leaves, and lower part of stem distinctly angled. Rocks in mountains. CR. **294d.** *E. graecum* Boiss. & Heldr. Distinguished at maturity by its spreading or deflexed fruits 3·5–7·5 cm. Petals 8–12 mm bright yellow; sepals 5–8 mm. Cultivation, waste places. AL.GR.TR. **294e.** *E. pectinatum* Bory & Chaub. A perennial 10–50 cm, with yellow flowers with petals 11–13 mm, finely hairy on back. Sepals pouched at base. Leaves deeply lobed. Fruit 2–4·5 cm, with star-shaped hairs. Mountain cliffs. GR. Pl. 10.

HESPERIS 7 species in our area, including 3 endemic species.

Flowers yellow or sometimes purple-flushed
(296). *H. laciniata* All. CUT-LEAVED DAME'S VIOLET. A tall biennial or perennial to 80 cm with yellow and purple-flushed, or sometimes reddish-purple flowers, and conspicuously pinnately lobed leaves, the upper stalkless. Petals 1·5–3 cm by 3–9 mm. Fruit 5–15 cm lower fruit stalks up to 1·5 cm. Rocks in the Med. region. YU.AL.GR+E.Aegean Is.BG. **296a.** *H. tristis* L. Like (269) but lower fruit stalks much longer 4–9 cm, and petals narrower 2–4 mm wide, linear-lanceolate, yellow or rarely purplish. Leaves not lobed. Dry sunny scrub. YU.BG.

Flowers purple, pink, or white
296b. *H. sylvestris* Crantz An erect, glandular-hairy perennial to 1 m with pink or purple petals 13–18 mm. Middle and upper leaves stalkless and more or less clasping, toothed, the lower shallowly lobed. Fruit soon hairless. Woods, bushy places. YU.AL? BG. **296c.** *H. dinarica* G. Beck Like 296b but flowers white and petals 17–20 mm. Leaves toothed, not lobed. Woods. YU.AL.GR.BG. ****296.** *H. matronalis* L. DAME'S VIOLET. Distinguished from 296b by its middle and upper stem leaves which are shortly stalked—never clasping. Plant glandular or not, variably hairy, rarely hairless. Flowers 2–2·5 cm across, purple or white. Wood verges; sometimes grown for ornament. YU.AL.

MALCOLMIA Individual species largely distinguished by their ripe fruits.
1. Sepals not pouched at the base.
   2. Petals 4–8 mm; fr. constricted between seeds, with adpressed hairs; fr. stalks 2–7 mm. (GR.).   **297a.** *M. ramosissima* (Desf.) Thell.
   2. Petals usually 8–10 mm; fr. not constricted between seeds, with spreading rigid hairs; fr. stalks 1–2 mm (GR.CR.TR.).   **297b.** *M. africana* (L.) R. Br.
1. At least 2 sepals strongly pouched at the base.
   3. Calyx 1–1·5 cm; style 5–10 mm in fr. (GR.).   **298a.** *M. macrocalyx* (Halácsy) Rech. fil. Pl.10.
   3. Calyx not more than 1 cm; style not more than 5 mm in fr.
      4. Inflorescence with bracts, at least below.
         5. Lower leaves heart-shaped at base; petals 11–25 mm.   *angulifolia*
         5. Lower leaves wedge-shaped at base; petals 10–15 mm (YU.AL.).   **298c.** *M. illyrica* Hayek

4. Inflorescence without bracts.
  6. Hairs all with 3–4 branches (GR(south)). **298d.** *M. graeca* Boiss. & Spruner
  6. Hairs mixed, with many two-branched hairs attached at the middle.
    7. Style more than 1·5 mm in fr.; petals 10–25 mm; sepals 5–16 mm.
      8. Fr. stalks distinctly narrower than the fr. *maritima*
      8. Fr. stalks about as wide as the fr. *flexuosa*
    7. Style not more than 1·5 mm in fr.; petals 5–12 mm; sepals 3–6 mm.
      9. Petals pink or violet; fr. 3–7 cm (GR+Is.CR.TR.). **298b.** *M. chia* (L.) DC.
      9. Petals pink with a yellow base; fr. 1–3·5 cm.
        10. Fr. straight, rigid, spreading (AL.GR.). **298g.** *M. bicolor* Boiss. & Heldr. Page 89
        10. Fr. curved, flexuous (GR(Idhra, Euboea)). **298h.** *M. hydraea* (Halácsy) Heldr. & Halácsy

298. *M. maritima* (L.) R.Br. VIRGINIA STOCK. A slender branched annual 10–35 cm with violet, pink, or rarely white flowers about 1 cm across. Petals 12–25 mm; sepals 6–10 mm. Fruit 1–2 mm wide, stalk 4–15 mm long, narrower than fruit. Sandy places by the sea. Widely cultivated and often self-seeding. AL.GR (south and west). **298e. *M. flexuosa* (Sibth. & Sm.) Differs from 298 in its shorter thicker fruit stalks and broader fruit, up to 3 mm wide. Seashores, maritime cliffs. AL.GR+Is.CR.

298f. *M. angulifolia* Boiss. & Orph. Distinguished by its broad rounded to ovate, toothed or cut leaves, the lowest of which have heart-shaped bases, but these soon die. Petals pink, 11–25 mm; inflorescence with leafy bracts below. Fruit stalks 5–10 mm; fruit curved outwards. Rocks in mountains. YU.AL.GR.BG. Pl. 10.

MARESIA Very like *Malcomia* but differing in having star-shaped hairs, and a distinct style with a club-shaped or notched stigma (stigma deeply two-lobed, but the lobes pressed together in *Malcolmia*).

298i. *M. nana* (DC.) Batt. A small annual of maritime sands with violet or pink flowers. Petals 4–4·5 mm. Leaves oblong, entire or toothed. Fruit 10–18 mm, constricted between seeds; style 0·5–1 mm. (Very like 297a but distinguished by its styles and stigma.) YU.AL.GR.BG.TR. Page 148.

CHEIRANTHUS ****299.** *C. cheiri* L. WALLFLOWER. A yellow-flowered native of the Aegean region growing on rocks and walls. Cultivated for ornament and naturalized elsewhere, with flowers in a variety of colour forms, ranging from yellow, orange, purple, to brown. GR+Is.CR.

MATTHIOLA | Stock 6 species in our area.

Fruit with 3 equal horns at apex, each at least 2 mm
(301). *M. tricuspidata* (L.) B.Br. THREE-HORNED STOCK. Distinguished by the 3

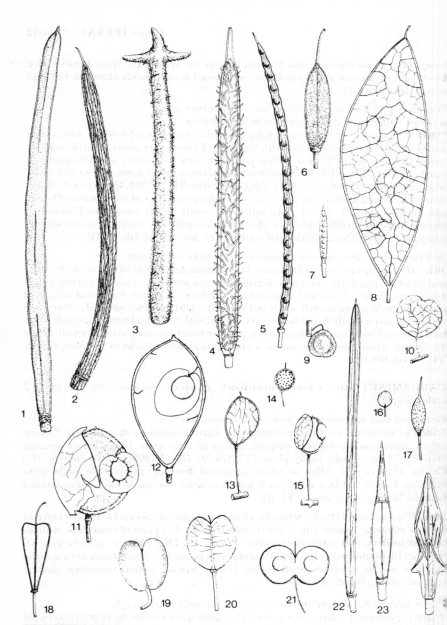

CRUCIFERAE fruits

1. *Hesperis laciniata* (296) 2. *Malcolmia flexuosa* 298e 3. *Matthiola tricuspidata*
(301) 4. *Cardamine graeca* 313a 5. *Arabis caucasica* (318) 6. *Aubrieta deltoidea*
320 7. *Ricotia cretica* 321a 8. *Lunaria rediviva* 321 9. *Clypeola jonthlaspi*
330 10. *Peltaria emarginata* 323c 11. *Alyssoides utriculata* 323 12. *Fibigia
lunarioides* 327a 13. *Ptilotrichum cyclocarpum* 327b 14. *Alyssum murale*
324b 15. *Bornmuellera baldaccii* 329b 16. *Lepidium latifolium* 352 17. *Draba
lasiocarpa* 331c 18. *Thlaspi praecox* 343a 19. *Aethionema saxatile* 346 20. *Iberis
sempervirens* 347 21. *Biscutella didyma* (349) 22. *Conringia orientalis* 355
23. *Eruca vesicaria* 363 24. *Cakile maritima* 366

conspicuous equal conical horns 2–6 mm long at the apex of the fruit. Petals violet, 15–22 mm. A greyish annual 7–40 cm, with lobed leaves. Sandy shores in the Med. region. GR+Is.CR.TR.

Fruit without horns, or with only 2 horns longer than 2 mm
(a) Fruit flattened in section; fruit stalks up to 2·5 cm
300. *M. incana* (L.) R.Br. STOCK. A robust biennial or perennial 10–80 cm, with a stout woody-based stem covered below with leaf scars. Flowers very sweet-scented, reddish-purple to white; petals 2–3 cm. Leaves lanceolate, entire or nearly so, white-woolly to nearly hairless. Fruit 4·5–16 cm, without conspicuous apical horns. Rocks and walls; often cultivated and naturalized. YU.GR.CR+Rhodes.TR.   ****301. *M. sinuata* (L.) R.Br. SEA STOCK. Distinguished from 300 by the wavy-margined or lobed lower leaves. Flowers pale purple; petals 17–25 mm. A densely white-woolly, rarely woody-based, spreading biennial 8–60 cm. Fruits 5–15 cm, with conspicuous yellow or black glands and without conspicuous apical horns. Rocks and walls by the sea. YU.AL.GR+Is.TR.

(b) Fruit more or less rounded in section; fruit stalks up to 3 mm
301c. *M. longipetala* (Vent.) DC. (incl. *M. bicornis*) An annual to 50 cm, with wavy-margined or pinnately cut lower leaves. Flowers stalkless, pink or purple; petals 1·5–2·5 cm. Fruit 7–15 cm with 2 upcurved apical horns 2–10 mm. Fields and cultivated places in the Med. region. GR(south and central).   **(301).** *M. fruticulosa* (L.) Maire SAD STOCK. Usually a woody-based perennial to 60 cm, with rosettes of narrow greyish-white basal leaves. Flowers variously coloured, greenish to reddish-purple, or brownish; petals 12–28 mm. Fruit 2·5–12 cm, with or without apical horns. Rocks in the Med. region. YU.AL.GR.BG.TR.

CARDAMINE | **Bitter Cress, Coral-Wort**   27 species in our area, including 2 endemic species.

Perennials with subterranean, fleshy scale-covered rhizomes
308. *C. bulbifera* (L.) Crantz CORAL-WORT. Distinguished by its pale-lilac flowers, entire uppermost leaves, with purplish bulbils in their axils. Lower leaves pinnate; stem 35–70 cm. Woods, bushy places. YU.GR.BG.TR.   **(309).** *C. enneaphyllos* (L.) Crantz. Flowers pale yellow or white, somewhat drooping; petals little longer than stamens. Leaves 2–4 in a whorl, each with 3 or more oval-lanceolate, double-toothed leaflets. Woods, bushy places. YU.AL.

309a. *C. quinquefolia* (Bieb.) Schmalh. Flowers deep purple; petals 14–18 mm, twice as long as stamens. Leaves 3 in a whorl, pinnate with 2–3 pairs of lanceolate, double-toothed leaflets, and a terminal leaflet. Woods. BG.TR.   **309b.** *C. glanduligera* O. Schwarz Distinguished by its purple flowers, with petals 12–22 mm, twice as long as the stamens, and its whorl of 3 stem leaves. Leaves ternate; leaflets lanceolate, double-toothed, ciliate. Woods. YU.BG.

Plants without fleshy subterranean rhizomes; annuals to perennials
310. *C. pratensis* L. LADY'S SMOCK. A very variable species particularly in its leaves and the branching of its inflorescence. Flowers white, tinged pink, or violet. This and closely related species are widely scattered in damp meadows throughout the mainland.   **310b. *C. rivularis* Schur Very like 310 but distinguished by its purple petals and anthers, and its more numerous stem leaves (4 or more) with 4–6 pairs of rounded equal leaflets. Basal leaves with terminal leaflet about equalling the rest (terminal larger in 310). Mountains. BG. Page 137.

312a. *C. glauca* Sprengel A delicate glaucous, hairless, flexuous-branched annual, or

perennial, 5–25 cm, with small white flowers which have deeply notched petals 5–8 mm
Lowest leaves entire or three-lobed, sometimes rosetted, middle leaves with 1–3 pairs o
blunt leaflets, the upper with 2–5 pairs of narrow acute leaflets. Fruit 2–3·5 cm, hairless
beak 1–2 mm. Stony places, screes. YU.AL.GR.BG. Page 115. **312b.** *C. carnos*
Waldst. & Kit. A hairy perennial 20–30 cm, with a few thick pinnate leaves with 3–'
pairs of obovate leaflets. Flowers white; petals 6–9 mm. Fruit 2–3 cm, with stiff adpres
sed hairs; beak 1–3 mm. Mountain rocks. YU.AL.GR. Page 126. **313a.** *C. graeca* L. A
rather delicate annual or biennial 10–30 cm, with small white flowers and pinnat
leaves. Petals 4–6 mm; sepals blunt. Lower leaves with 4–5 pairs of broadly wedge
shaped three-lobed leaflets. Fruit 3·5–5 cm, in a somewhat one-sided cluster. Rocks
YU.AL.GR+Is.CR.BG.

ARABIS | Rock-Cress   About 25 species in our area, including 4 endemic species.

**317. *A. turrita* L. TOWER ROCK-CRESS. Distinguished by its very long slender wand-like
inflorescence which is at length curved over to one side. Flowers pale yellow; petals
6–8 mm. Basal leaves in a rosette, long-stalked, upper clasping, all hairy; stem
20–80 cm. Fruit arched, 10–14 cm. Woods, rocky places. YU.AL.GR.BG. ****(319).** *A.*
verna (L.) R.Br. A small rosette-forming annual 5–40 cm, with often several slender
nearly leafless stems bearing tiny violet flowers with yellowish centres. Petals 5–8 mm.
Rosette leaves oblong-oval, toothed. Rocks and uncultivated places in the Med. region.
Widespread.

**318. *A. alpina* L. ALPINE ROCK-CRESS. A loosely tufted mountain perennial with few lax
rosettes of coarsely toothed leaves, and erect flowering stems to 40 cm bearing open
clusters of white flowers. Petals 6–10 mm, twice as long as sepals. Rosette leaves with
4–7 coarse teeth, stem leaves ovate and clasping with heart-shaped base, all densely
covered with star-shaped hairs. Fruit 2–3·5 cm. Rocks and screes in mountains.
YU.AL.GR?BG. Page 184.

(318). *A caucasica* Schlecht. Like 318 but with larger, more conspicuous petals 9–18 mm,
and having stem leaves with arrow-shaped bases clasping the stem. Fruit 4–7 cm. Rocks
in mountains. YU.AL.GR.CR.BG. Page 255. **318a.** *A. bryoides* Boiss. A low cushion
plant of mountain rocks with tiny hairy rosettes and slender leafless stems to 5 cm
bearing a lax cluster of 3–6 white flowers. Petals 6–7 mm, broadly oval. Rosettes about
5 mm across; leaves ovate-spathulate, densely hairy, ciliate at apex. Fruit 1–2 cm.
YU.AL.GR. Page 127. Pl. 10.

AUBRIETA   6 very similar-looking species including 2 very local; distinguished
largely by their fruits.

1. Fr. with long unbranched hairs as
    well as branched hairs   *deltoidea*
1. Fr. with star-shaped hairs only.
    2. Fr. 2–3·5 cm, 6–15 times as long
        as wide.   *gracilis*
    2. Fr. less than 2 cm, 2–6 times as
        long as wide.
        3. Fr. usually not more than 12 mm,
            2–4 times as long as wide, usually
            not strongly compressed (YU.AL.).   **320c.** *A. columnae* Guss. Page 190
        3. Fr. 13–18 mm, usually 4–6 times
            as long as wide, strongly compressed.   *intermedia*

320. *A. deltoidea* (L.) DC. AUBRETIA A tufted or low straggling perennial 5–10 cm, having reddish-purple to violet flowers with petals 12–28 mm. Distinguished by the long unbranched hairs on the fruits mixed with star-shaped hairs. Rocks mostly in hills. YU? GR+Is.CR. Page 71. **320a** *A gracilis* Boiss. Flowers violet; petals 12–18 mm; sepals 5–7·5 mm. Leaves entire, or with 2 teeth. Rocks in hills and mountains. YU.AL.GR.BG. Pl. 11. **320b.** *A. intermedia* Boiss. Flowers violet; petals 15–22 mm; sepals 6–12 mm. Leaves variable, usually obovate or rhombic, with 1–3 pairs of teeth. Rocks in mountains. YU.AL.GR.BG. Pl. 10.

RICOTIA Fruit flattened, usually borne on curved fruit stalks. Flowers pink or pale violet. 2 species in our area.

321a. *R. cretica* Boiss. & Heldr. A hairless, branched annual 10–25 cm, with twice-cut leaves, and pink flowers with petals 10–12 mm. Leaf segments ovate or elliptic. Fruit 3–5 cm by 8–9 mm, not winged. Stony places in mountains. CR. Page 255.

LUNARIA | Honesty 3 species in our area.

321. *L. rediviva* L. PERENNIAL HONESTY. A robust perennial to 140 cm, with pale-purple or violet flowers and distinctive large, wafer-thin, usually elliptic fruits. Petals 12–20 mm. Leaves ovate, all stalked, finely toothed. Fruit 3–9 cm. Woods. YU.AL.BG. ****322.** *L. annua* L. HONESTY. Like 321 but a biennial with the upper leaves stalkless or nearly so, and coarsely toothed, and flowers usually a bright reddish-purple. Fruit usually almost circular in outline, 2–7 cm. Bushy places, rocks; often naturalized from gardens. YU.AL.GR.CR.BG.

PELTARIA Distinguished by its small flat, wafer-thin, rounded pendulous fruits each with a single central seed.

323b. *P. alliacea* Jacq. A yellow-flowered perennial 20–60 cm, easily distinguished in fruit by its disk-like rounded fruits hanging on very slender stalks from a branched inflorescence. Petals 3–5 mm. Upper leaves ovate to lanceolate, clasping with rounded or arrow-shaped bases, hairless. Fruit 6–10 mm across. Hills, woods. YU.AL. Page 190. **323c.** *P. emarginata* (Boiss.) Hausskn. Like 323b, but upper stem leaves with a wedge-shaped base, not clasping; petals 3 mm; fruit deeply notched at apex. GR(central) and east). Page 255.

ALYSSOIDES Distinguished by its globular or ovoid inflated fruits.

323. *A. utriculata* (L.) Medicus A yellow-flowered, woody-based, rosette-forming perennial to 40 cm, with conspicuous globular fruits with long styles. Petals 2 cm, blade rounded with a long stalk (claw); sepals 8–12 mm. Basal leaves oblong-spathulate, green, with star-shaped hairs; stem leaves lanceolate, hairless. Fruit 10–12 mm. Rocks, cliffs. YU.AL.GR.BG. Page 168. **323a.** *A. cretica*(L.) Medicus Like 323 but leaves grey- or white-woolly, with dense star-shaped hairs. Petals 12–20 mm, pale yellow, blade elliptic; sepals 7–11 mm. Fruit hairy, with fruit stalks as long as the globular capsules. Rocks, cliffs. CR+Karpathos. Pl. 10. **323d.** *A. sinuata* (L.) Medicus A little-branched grey-hairy perennial 15–50 cm, with yellow flowers, of rocks, cliffs, and walls of the north-western Balkans. Distinguished from 323 by its smaller sepals 3–4 mm, and notched petals 5–8 mm. Fruit hairless, 7–12 mm; style 2–4 mm. YU.AL. Page 41.

DEGENIA Flowers yellow, petals long-clawed. Fruit inflated, ellipsoid; style long; seeds winged, 2 in each chamber.

323e. *D. velebitica* (Degen) Hayek A rare and restricted endemic perennial of mountain screes (1200–1300 m) in the Velebit Mountains of Yugoslavia. A silvery-grey, tufted plant with narrow lanceolate leaves in rosettes. Flowers yellow in a flat-topped cluster on stems to 10 cm; sepals 7–8 mm; petals 15 mm. YU. Page 190.

ALYSSUM   A difficult genus with about 48 species in our area.

Seeds 2–4 in each chamber
324. *A. saxatile* L. GOLDEN ALYSSUM. A woody-based shrubby perennial 10–40 cm, with flat-topped clusters of numerous small yellow flowers on branched stems. Petals 3–6 mm, notched or bilobed. Lower leaves in rosettes, obovate to spathulate, grey-haired, upper leaves lanceolate. Fruit flattened, hairless. Seeds 2 in each chamber. Rocks. YU.AL.GR+Is.CR.BG.TR? Page 96.   **324a.** *A. corymbosum* (Griseb.) Boiss. An erect perennial 30–50 cm, with a flat-topped branched inflorescence of small yellow flowers, and conspicuous in fruit with numerous small globular inflated capsules 4–6 mm across. Petals about 4 mm; sepals 2–2·5 mm. Leaves obovate-lanceolate, toothed or entire. Seeds 4 in each chamber, winged. Rocks in the Med. region. YU.AL.GR.

Seeds 1–2 in each chamber
324b. *A. murale* Waldst. & Kit. A tufted but very variable perennial 25–70 cm, with long non-flowering stems or dense rosettes of leaves. Flowers yellow, numerous in simple or branched clusters; petals 2–4 mm. Leaves grey-green above, white or grey beneath, from obovate to lanceolate. Fruit 2–6 mm by 2–4 mm, hairy, valves flat, often undulate. Rocks. YU.AL.GR.CR.BG. Page 255.   **324c.** *A. scardicum* Wettst. A grey-green or green spreading perennial 5–20 cm, of stony places and rocks, in hills and mountains, with long or short clusters of yellow flowers. Petals 4–6 mm; sepals 2–4 mm; flower stalks 5–8 mm, spreading. Lower leaves elliptic, with five- to nine-rayed hairs. Fruit 3–6 mm by 2–4 mm, densely hairy; style 3–4 mm; seeds 2 in each chamber. YU.AL.BG.   **324d.** *A. handelii* Hayek A tiny tufted silvery-grey or greenish perennial with bright-yellow flowers endemic to Mount Olympus, above 2000 m. Petals 7–8 mm. Leaves obovate, long-stalked. Fruiting clusters dense, short, fruit 8–9 mm ovate, finely hairy. GR(Olympus). Pl. 11.

FIBIGIA   Distinguished by its flattened bat-shaped fruits, with conspicuous styles. 3 species in our area.

327. *F. clypeata* (L.) Medicus A densely grey-felted biennial 30–75 cm, with small yellow flowers and conspicuous, flattened elliptic fruits in an elongated cluster 10–20 cm. Petals 8–13 mm. Leaves oblong, green or grey-green. Fruit 14–28 mm, grey-felted; style long. Among rocks. YU.AL.GR(north-east).BG.   **327a. *F. lunarioides* (Willd.) Sibth. & Sm. A woody-based, somewhat tufted shrubby plant 5–30 cm, with ash-white lower leaves, and flowering stems less than 30 cm. Petals yellow, 12–16 mm. Fruit in a short cluster not more than 5 cm long. Rocks, cliffs. GR(Cyclades).

PTILOTRICHUM   Hairs star-shaped or scale-like. Fruit rounded with flat or convex sides and 1 or 2 seeds in each chamber; style short.

327b. *P. cyclocarpum* Boiss. (*Alyssum rupestre*) A densely tufted woody-based plant to 20 cm, with silvery-haired linear to lanceolate leaves, and a dense rounded cluster of white flowers which elongates in fruit. Petals 5 mm. Fruits obovate-obicular, 5 mm long, covered with small scales, or scaleless. Cliffs in mountains. YU.AL.GR. Page 255.

1. *Thlaspi bellidifolium* 345b 2. *Arabis caucasica* (318) 3. *Alyssum murale*
324b 4. *Peltaria emarginata* 323c 5. *Ptilotrichum cyclocarpum* 327b 6. *Ricotia
cretica* 321a

BORNMUELLERA Distinguished from *Ptilotrichum* by the hairs which are attached in the middle, and the filaments of the stamens with a tooth-like appendage at the base. 3 species in our area.

329b. *B. baldaccii* (Degen) Heywood A low creeping perennial, recalling a candytuft, with numerous crowded, linear-lanceolate leaves, and short erect leafless stems to 10 cm, with a dense terminal cluster of small but conspicuous white flowers. Petals 5 mm. Fruit ovate, somewhat inflated, with a short style, on spreading or reflexed fruit stalks. Serpentine rocks in mountains. Al.GR(north-west). Page 115. **329c. *B. tymphaea*** (Hausskn.) Hausskn. Like 329b, but with broader oblong-spathulate leaves 2–4 cm by 3–8 mm, which are densely silvery-hàired, and inflorescence branched. Petals white. Fruit globular 7–8 mm, inflated. Serpentine or schistose rocks. GR(north). Pl. 11.

CLYPEOLA 330. *C. jonthlaspi* L. DISK CRESS. A tiny erect annual to 20 cm, with minute yellow flowers, only distinctive because of its rounded disk-like, pendulous fruits borne in elongated clusters. Petals 1–2 mm. Fruit 2–5 mm. Stony ground in the Med. region. Widespread.

DRABA 15 species, including 6 endemic to our area. A difficult genus.

Flowering stems hairless
****331. *D. aizoides*** L. YELLOW WHITLOW-GRASS. A tiny tufted alpine perennial with dense rosettes of stiff bristly leaves, and leafless stems 5–10 cm, bearing rather dense clusters of bright-yellow flowers. Petals 4–6 mm. Leaves linear, hairless, except for the conspicuously ciliate margin. Fruit 6–12 mm, ellipsoid, in an elongated cluster. Rocks in mountains. YU.AL.GR? BG. **331b. *D. athoa*** (Griseb.) Boiss. A relatively robust perennial to 12 cm, with broadly linear rosette leaves which have bristly hairs all over and strongly ciliate margins. Flowers comparatively large, yellow; petals and stamens 6–7 mm. Fruit flat, 6–10 mm long, usually bristly; style more than 3 mm. Rocks, and by melting snow in alpine regions. YU.AL.GR.

331c. *D. lasiocarpa* Rochel Distinguished from 331b by its fruits with styles rarely more than 1·5 mm long, and its deep-yellow petals, 4–5 mm, which are longer than the stamens. Leaves up to 3 mm wide. A relatively robust, usually densely tufted plant to 20 cm. Mountain rocks. YU.AL.GR.BG. **331d. *D. scardica*** (Griseb.) Degen & Dörfler A dwarf slender perennial of mountains with small yellow flowers on hairless stems 5–10 cm. Petals about 4 mm. Leaves 1–2 mm wide, ciliate. Fruit 3–4 mm, with rounded sides, usually hairless; style 3–7 mm. Alpine rocks. YU.AL.GR.BG.

Flowering stems hairy
331e. *D. parnassica* Boiss. & Heldr. A dwarf tufted plant with many small bristly-haired rosettes and woolly flowering stems bearing a compact cluster of yellow flowers. Petals 5–9 mm; sepals woolly-haired. Leaves with bristly margin. Fruit densely bristly-haired; style less than 1·5 mm. Alpine rocks. YU.AL.GR. Pl. 11. **331f. *D. cretica*** Boiss. & Heldr. Like 331e but yellow petals smaller 3·5–4·5 mm, and fruits covered with dense star-shaped hairs; style less than 1 mm. Mountains. CR.

THLASPI | Pennycress Distinguished by its flattened fruit which is notched and usually winged, and its stem leaves which are more or less clasping. 17 species in our area, including 4 endemic species.

Flowers white
(343). *T. perfoliatum* L. PERFOLIATE PENNYCRESS. Fields, undisturbed ground. Wide-

spread. **343a.** *T. praecox* Wulfen A usually tufted perennial 10–20 cm, with crowded rosettes of ovate, stalked, more or less glaucous leaves, and several erect stems with terminal, usually rounded, clusters of small white flowers. Petals 5–7 mm, twice as long as sepals, or more. Stem leaves few, clasping. Fruit broadly winged above, with a wide notch and style slightly longer or considerably so. Stony or grassy places, especially on limestone. YU.AL.GR.BG?

Flowers pinkish, violet, or purple

345b. *T. bellidifolium* Griseb. A dwarf, densely tufted plant 1–3 cm, of alpine pastures and by melting snow, with a rosette of oblong, toothed leaves and a dense cluster of intense purple flowers with yellow anthers. Basal leaves stalked, stem leaves clasping. Fruiting stem not elongating; fruits not winged. YU.AL.GR? BG? Page 255. **345c.** *T. bulbosum* Boiss. An attractive glaucous tufted perennial 5–10 cm, with several ascending stems terminating in compact elongated clusters of dark-violet or dark-lilac flowers with bright-violet anthers. Petals 6–8 mm, twice as long as stamens. Leaves of two kinds, rosette leaves ovate long-stalked, stem leaves clasping oblong-ovate. Fruiting stems elongating; fruits obcordate, broadly winged; style longer than notch. Roots swollen, carrot-like. Mountain woods. GR. **345d.** *T. microphyllum* Boiss. & Orph. A tiny tufted plant 1–3 cm, of high mountains, with a compact almost stemless cluster of pale-violet to pinkish flowers. Sepals violet; petals 3–5 mm, whitish. Rosette leaves obovate, 3–6 mm. Fruiting stem not elongating. By melting snow. AL.GR.CR. Pl. 11.

AETHIONEMA Filaments of 4 inner stamens flattened and winged, and sometimes toothed above. Fruit flattened, with a broad wing, which is notched at the apex.

1. Petals 7 mm, yellow; style 3 mm or more (GR(central)).   **346a.** *A. cordatum* (Desf.) Boiss.
1. Petals less than 7 mm, not yellow; styles less than 3 mm.
   2. Fr. entire at apex (GR(Taiyetos, Athos)).   **346b.** *A. orbiculatum* (Boiss.) Hayek
          (GR(Poros, Euboea).).   **346c.** *A. polygaloides* DC.
   2. Fr. notched at apex.
      3. Fr. 10 mm or more (BG.TR.).   **346d.** *A. arabicum* (L.) O.E. Schulz
      3. Fr. less than 10 mm.
         4. Lvs. of lower part of stem acute, 3–5 mm (GR(Euboea)).   **346e.** *A. iberideum* (Boiss.) Boiss.
         4. Lvs. of lower part of stem obtuse, more than 5 mm.   *saxatile*

**346. *A. saxatile* (L.) R.Br. BURNT CANDYTUFT. A lowly hairless annual or perennial to 30 cm, with glaucous leathery ovate to oblong leaves and a dense terminal cluster of pink, pale-violet, or white flowers. Petals 2–5 mm, longer than sepals. Fruit flattened and broadly winged, notched. Stony, rocky places, mountains. YU.AL.GR+Is.CR.BG. Page 179.

IBERIS | Candytuft Petals unequal, the two outer much larger than the inner. Fruit compressed, winged, and often notched; seed solitary in each cell. 7 species occur in our area; they are distinguished by small botanical differences.

Fruit in elongated clusters

347. *I. sempervirens* L. EVERGREEN CANDYTUFT. An evergreen shrublet 10–25 cm, with thick, blunt leaves and flat-topped clusters of white flowers which are borne on lateral stems. Leaves oblong-spathulate, 2–5 mm wide. Fruit 6–7 mm, broadly winged. Rocks in mountains. YU.AL.GR.CR. Pl. 11. **(347).** *I. saxatilis* L. Like 347 but with narrower linear acute leaves 1–2 mm wide, semi-cylindrical on non-flowering stems, flat on flowering stems. Flower clusters terminal. Calcareous rocks. YU.

Fruit in flat-topped clusters
(a) Leaves entire
347d. *I. pruitii* Tineo A very variable tufted perennial or annual 3–15 cm, usually with small lilac flowers, but flowers sometimes white, in a flat-topped cluster. Leaves rather fleshy, blunt, entire or with a few teeth, the lower obovate-spathulate, the upper narrower. Fruit 6–8 mm, rectangular-elliptical, with wide, erect, acute lobes. Rock crevices in mountains in the Med. region. YU.AL.GR.BG. **(348).** *I. umbellata* L. An erect hairless annual 20–70 cm, branched above, with a dense flat-topped cluster of pink to purplish flowers. Flowers 7–12 mm across. Leaves linear-lanceolate, long-pointed, mostly entire. Fruit head remaining compact, fruits to 1 cm, lobes acute. Bushy and rocky places; also grown for ornament and sometimes naturalized. YU.AL.GR. Pl. 12.

(b) Leaves pinnately lobed
348b. *I. pinnata* L. An annual cornfield weed of the Med. region with leaves deeply cut into 1–3 pairs of linear segments. Flowers white or lilac, fragrant, in a short dense cluster on erect stems 10–30 cm. Fruit 5–6 mm, almost square, lobes blunt. Stony places. YU.GR. **348f.** *I. odorata* L. A branched bristly-haired annual 15–30 cm, with a dense flat cluster of white flowers surrounded by the upper leaves. Outer petals scarcely larger than inner. Leaves linear-spathulate, cut into 1–2 pairs of segments near the apex. Fruit with acute lobes. Limestone mountains. GR.CR.TR.

BISCUTELLA Readily distinguished by its paired, disk-shaped fruits. Flowers yellow. 3 species in our area.

349. *B. laevigata* L. BUCKLER MUSTARD. Rocks and grassy places in hills and mountains. YU.CR.BG. **(349).** *B. didyma* L. A slender annual to 40 cm, with a dense sparsely branched inflorescence of small yellow flowers, which becomes more distinctive in its fruiting state. Petals 4 mm. Basal leaves often in a rosette, obovate-wedge-shaped, toothed, hairy. Fruit of 2 flattened disks, with thickened margins, each 5–7 mm across, placed edge to edge. Stony ground, dry hills in the Med. region. YU.AL.GR+Is.CR. Page 40.

LEPIDIUM | Pepperwort About 9 species in our area; distinguished largely by their fruits.

352. *L. latifolium* L. DITTANDER, BROAD-LEAVED PEPPERWORT. Salt-rich ground. Throughout the mainland. **(352).** *L. graminifolium* L. GRASS-LEAVED PEPPERWORT. Stony ground, waste places. Widespread on mainland. **352a.** *L. perfoliatum* L. A pale yellow-flowered annual or perennial 20–40 cm, distinctive because of the 2 contrasting forms of leaf on the same plant. Basal leaves 2–3 times cut into narrow segments, long-stalked, the upper stem leaves ovate-entire and completely encircling the stem. Petals 1–1·5 cm. Fruit globular, 4 mm. Salt-rich soils, waste places. YU.AL.GR.BG.TR.

CARDARIA ****353.** *C. draba* (L.) Desv. HOARY PEPPERWORT, HOARY CRESS. Waste places, salty ground. Widespread.

CONRINGIA 2 species in our area. ****355.** *C. orientalis* (L.) Dumort. HARE'S EAR

CABBAGE. Distinguished by its glaucous, elliptic, clasping stem leaves, its yellowish or greenish-white flowers, and its long slender fruits. Petals 8–13 mm long. Plant hairless, 10–15 cm. Fields, uncultivated ground. Widespread on mainland.

ERUCA ****363.** *E. vesicaria* (L.) Cav. A rough-haired, mustard-like annual 20–100 cm, distinguished by its whitish or yellowish petals with violet veins. Petals 1·5–2 cm, wide-spreading. Leaves deeply lobed, with the terminal lobes largest. Fruit with a flattened sabre-like beak. Cultivated and disturbed ground. Widespread.

ERUCARIA Petals lilac; fruit jointed transversely, with both segments fertile.

366a. *E. hispanica* (L.) Druce (*E. myagroides*, etc.) An erect, nearly hairless annual or biennial to 60 cm, with deeply cut, somewhat fleshy leaves with narrow lobes, and lilac flowers 1–1·5 cm long. Fruit 1–2 cm, the lower part cylindrical, the upper part compressed and sword-shaped. Cultivated ground, waste places. GR(south)+Is.CR.

CAKILE ****366.** *C. maritima* Scop. SEA ROCKET. Coasts of Mediterranean and Black Sea.

CRAMBE 3 species in our area. **368c.** *C. tataria* Sebeók A stout coarse-growing, usually hairy perennial to 1·5 m, with a much-branched inflorescence of white flowers, and large irregularly pinnate leaves. Petals 3–6 mm. Fruit with lower segment short and upper segment globular 3–6 mm. Bushy, grassy places. YU.BG. Pl. 12.

RESEDACEAE | Mignonette Family

RESEDA 8 species in our area, including 3 that are very local.

1. Sepals 4; lvs. undivided
   (Widespread).   371. *R. luteola* L.
1. Sepals 5–7, rarely 4, then
   lvs. divided.
   2. Lower lvs. undivided; upper lvs.
      undivided, ternate, or with 1–2
      pairs of lobes.
      3. Sepals to 5 mm in fr.   *inodora*
      3. Sepals 5–13 mm in fr.
         (YU.AL.GR.).   ****374.** *R. phyteuma* L.
   2. All leaves pinnately lobed.
      4. Petals white; lvs. with
         5–15 pairs of lobes.   *alba*
      4. Petals yellow; lvs. usually
         with 1–2 pairs of lobes
         (Widespread).   ****372.** *lutea* L.

374c. *R. indora* Reichenb. A white-flowered, scentless, erect, hairless biennial or perennial 20–60 cm. Leaves mostly lanceolate, some upper leaves with 1–2 pairs of lobes. Petals 3 mm, with 11–17 narrow lobes. Carpels 3, often pendulous. Bushy and grassy places. YU.BG. ****373.** *R. alba* L. UPRIGHT MIGNONETTE. A robust often glaucous annual, biennial or perennial 30–80 cm, distinguished by its long slender spikes of numerous whitish flowers, and its deeply pinnately-cut leaves with narrow lobes. Sepals and petals usually 5, the latter with 3 lobes; stamens 10–12. Carpels 4, elliptic or obovate, erect. Uncultivated or rocky ground. YU.GR.+Is.CR. Pl. 12.

DROSERACEAE | Sundew Family

ALDROVANDA 378a. *A. vesiculosa* L. A submerged, rootless, animal-catching plant, with crowded whorls of 6–9 leaves each 1–1·5 cm. Leaf with a wedge-shaped base, a rounded blade, and 4–6 bristles, the blade being hinged along the mid-vein and capable of closing rapidly in a trap. Flowers rarely present. Shallow, still waters. YU.BG.

DROSERA 376. *D. rotundifolia* L. COMMON SUNDEW. Peat moors. YU.BG. ****377.** *D. anglica* Hudson GREAT SUNDEW. Peat moors. YU.GR? ****378.** *D. intermedia* Hayne LONG-LEAVED SUNDEW. Peat moors. YU.

CRASSULACEAE | Stonecrop Family

UMBILICUS | Pennywort

1. Petal lobes as long or longer
    than the corolla tube.
    2. Fls. 9–13 mm; petal lobes about
        as long as the corolla tube.   *erectus*
    2. Fls. 3–6 mm; petal lobes considerably
        longer than the corolla tube.
        3. Fls. 4–6 mm, obconical (GR(south).
            CR.).   **380a.** *U. parviflorus* (Desf.) DC.
        3. Fls. 3–4 mm, bell-shaped or almost
            spherical.   *chloranthus*
1. Petal lobes considerably shorter than
    the corolla tube.
    4. Fls. usually pendulous, occupying   ****379.** *U. rupestris* (Salisb.)
        more than half the length of the   Dandy Pl. 13.
        stem (Widespread).
    4. Fls. usually horizontal, occupying not
        more than half the length of the stem
        (Widespread except TR.)   **(379).** *U. horizontalis* (Guss.) DC.

380. *U. erectus* DC. A stout perennial 20–60 cm, of damp shady rocks, with rounded basal leaves, and long dense spikes of numerous small greenish-yellow flowers which become reddish-brown on drying. Corolla 9–13 mm, tubular, with narrow lobes about as long as the tube. Leaves rounded-heart-shaped, toothed, to 7 cm across, stalked, becoming progressively smaller up stem. YU(south).AL.GR.CR.BG(south). **380b.** *U. chloranthus* Boiss. Differs from 380 in having smaller bell-shaped or almost spherical yellowish flowers 3–4 mm long. Corolla with lobes 1½–2 times as long as the tube. Basal leaves 2–5 cm across. Rocks and walls. YU.GR.CR+Rhodes. Page 266.

SEMPERVIVUM | Houseleek 12 species, including 6 often local species endemic to Macedonia. Hybridization is common where both parents are present.

1. Upper surface of petals yellowish at
   least in the apical half.
   2. Bristly marginal hairs of rosette
   lvs. 2–4 mm, stiff, interwoven with
   those of neighbouring lvs.   *ciliosum*
   2. Bristly marginal hairs of rosette lvs.
   2 mm, not interwoven.
      3. Petals tinged with pink or purple
      towards base; filament of stamens
      more or less purple.
         4. Flowering stem up to 10 cm.
            5. Petals with a white margin,
            otherwise mainly pink but
            yellow towards the apex (YU.
            (Macedonia)).   **384b.** *S. thompsonianum* Wale
            5. Petals without white margin,
            mainly yellow but with a
            lilac spot near the base (YU
            (Macedonia)).   **384c.** *S. octopodes* Turrill
         4. Flowering stem 12–25 cm.
            6. Petals pale yellow or cream-
            coloured; filaments pale yellow,   **384d.** *S. kindingeri* Adamović
            striped with red (YU.GR.).   Pl. 13.
            6. Petals bright greenish-
            yellow; filaments purple
            (YU? BG.).   **384e.** *S. zeleborii* Schott Pl. 13.
      3. Petals not tinged with purple or
      pink; filaments pale (BG.).   **384f.** *S. leucanthum* Pančić
1. Upper surface of petals predominantly
   red, pink, or purple.
   7. Mature lvs. finely downy on both surfaces.
      8. Rosettes well spaced; stolons 4–12 cm.
         9. Rosettes 6–8 cm. across; stolons
         stout (YU(south-west)).   **384g.** *S. kosaninii* Praeger
         9. Rosettes 3–5 cm; stolons slender.   *macedonicum*
      8. Rosettes crowded; stolons about 2 cm
      (BG.).   **384i.** *S. erythraeum* Velen.
   7. Mature lvs. hairless except for the
   marginal bristles.
      10. Young lvs. minutely downy on both
      surfaces.
         11. Mature lvs. without marginal bristly
         hairs in the apical third (GR
         (north-west)).   **384j.** *S. ballsii* Wale
         11. Mature lvs. with marginal bristly
         hairs to the apex.   *marmoreum*
      10. Young lvs. hairless except for marginal
      bristles (introd. YU.).   **383.** *S. tectorum* L.

384a. *S. ciliosum* Craib Rosettes grey-green, 2–3·5 across, with incurved leaves and interwoven hairs from leaf tip to leaf tip, and with lemon-yellow flowers. Petals 12–14, 10–12 mm; filaments whitish. Stolons slender. Rocks in hills and mountains in Macedonia. YU.GR.BG. **384h.** *S. macedonicum* Praeger Flowers dull pinkish-lilac, with 11–12 petals which are 8–10 mm long, in a compact cluster on stems 7–10 cm. Filaments lilac. Leaves reddish towards tip, with a short point, margins ciliate; rosettes 3–5 cm. Rocks in mountains. YU(Macedonia). Page 266. **384k.** *S. marmoreum* Griseb. Flowers with 12 red petals with white margins, in a cluster up to 7 cm across, borne on stems to 20 cm. Rosette leaves olive-green, often red-tinged, with reflexed marginal bristles, stem leaves narrower, long-pointed, to 3 cm. Rosettes large, about 6 cm, with stout stolons. Rocks in mountains. YU.AL.GR.BG. Page 115.

JOVIBARBA Differs from *Sempervivum* in having sepals and petals 6, and pale-yellow bell-shaped flowers with the petals fringed with glandular hairs.

385. *J. hirta* (L.) Opiz Rosettes globular, 3–7 cm across, with hairless leaves except for the marginal bristles, and with stolons. Flowers pale yellow, in a dense rounded cluster; petals 15–17 mm, fringed. Rosette leaves dark green, not red-tipped; stem leaves glandular-hairy, clasping. Calcareous rocks. YU.AL. **385a.** *J. heuffelii* (Schott) Á. & D. Löve Summer rosettes with leaves spreading like a star, 5–7 cm across, without stolons. Flowers yellow; petals 10–12 mm, usually with a three-tipped apex; sepals glandular-hairy. Leaves dark green or glaucous, with white margins with stiff bristles. Calcareous mountains. YU.AL.GR.BG. Pl. 13.

AEONIUM ****386.** *A. arboreum* (L.) Webb & Berth. A branched succulent plant, with thick brown, rather woody stems and terminal flattened rosettes of overlapping, broadly strap-shaped, bristly-margined leaves. Flowers yellow, numerous. Naturalized; GR.CR.

SEDUM | Stonecrop 35 species in our area. Most species grow in dry stony and rocky places, ranging from the lowlands to high mountains. The following 25 species are the most widespread and most likely to be encountered:

Leaves fleshy, cylindrical (not flattened)
(a) *Flowers yellow, cream, or greenish-white*
   (i) *Carpels erect; leaves usually fine-pointed; petals usually 6–9.*

1. Fls. few, in lax one-sided, flat-topped clusters (YU.GR.CR.BG.TR.).   (388). *S. tenuifolium* (Sibth. & Sm.) Strobl Page 266
1. Fls. numerous, in crowded flat-topped clusters.
   2. Sepals 5–7 mm, glandular-hairy; inflorescence flat-topped in bud and fr. (YU.AL.GR.CR.BG.).   387. *S. ochroleucum* Chaix Pl.14.
   2. Sepals 2–5 mm, hairless; inflorescence rounded in bud, concave in fr.
     3. Inflorescence erect in bud; sepals ovate; lvs. about 4 mm wide (YU.AL. GR.CR.TR.).   (387). *S. sediforme* (Jacq.) Pau Pl. 14.
     3. Inflorescence drooping in bud; sepals lanceolate; lvs. 2 mm wide (YU.AL.GR.).   ****388.** *S. reflexum* L.

(ii) Carpels spreading; leaves blunt; petals 5.
389. *S. acre* L. STONECROP. Widespread on mainland. **(389). *S. alpestre* Vill. ALPINE STONECROP. YU.BG. **390.** *S. annuum* L. ANNUAL STONECROP. Widespread on mainland +CR. **390a.** *S. litoreum* Guss. A small hairless annual 4–15 cm, of the littoral, with pale-yellow flowers in long lax clusters. Petals 5, small 2–4 mm, acute, as long as or little longer than the sepals. Leaves 1–2 cm, alternate, flattish, blunt. Rocks. YU.GR+Is. CR. **389a.** *S. sartorianum* Boiss. Like 389 *S. acre* with yellow flowers but lower part of stems clothed with dead leaves which are white only at the base, and grey or black towards the apex and rather leathery (entirely white, soft, and papery in 389). Rocky mountain slopes. YU.AL.GR.BG.TR. Pl. 14. **389b.** *S. laconicum* Boiss. Like 389 *S. acre* but leaves cylindrical and more or less parallel-sided instead of ovoid, also differing in having bright-yellow petals with red mid-veins, 3–4 mm long (yellow and 6–8 mm long in 389). Mountains. GR. Pl. 13.

(b) Flowers white, pink, red or violet
  (i) Perennials with non-flowering shoots present at flowering
391. *S. album* L. WHITE STONECROP. Flowers pink or white. Widespread. **(391). *S. dasyphyllum* L. THICK-LEAVED STONECROP. Flowers white streaked with pink. Rocks, walls. Widespread, except TR. **391c.** *S. stefco* Stefanov A small hairless perennial with alternate, bright-pink leaves. Flowering stem about 7 cm, with a rather dense cluster of short-stalked, four-petalled, pale-pink flowers. Petals 4–5 mm, pointed, more or less erect; stamens 8. Mountains. BG(Rhodope).

(ii) Annuals or biennials without non-flowering shoots at flowering
  (x) Stamens twice as many as petals
392. *S. atratum* L. DARK STONECROP. Flowers pale yellow and red-tinged, or greenish-yellow and scarcely red-tinged. Mountains. YU.AL.GR.BG. Page 185. **393.** *S. hispanicum* L. Flowers white with a pink mid-vein. Rocky places. Widespread. Pl. 14. **393a.** *S. pallidum* Bieb. Like 393 but usually hairless, with five-petalled pink flowers (6–9 petals in 393), and nearly erect, dark-red fruits, (spreading, pale-pink or whitish fruits as in 393). Rocky places. YU(south).AL.GR+Is.CR.BG.TR.

  (xx) Stamens usually equal in number to petals
392a. *S. rubens* L. RED STONECROP. A glandular-hairy, usually reddish, somewhat glaucous annual 5–12 cm, with flat-topped clusters of white or pink flowers. Petals 5 mm, sharp-pointed; stamens usually 5, rarely 10. Leaves 1–2 cm, cylindrical-linear, spreading. Carpels glandular-hairy, long-pointed, spreading from the base. Rocky places, walls, sandy fields. Scattered throughout, except TR. Pl. 14. **392b.** *S. caespitosum* (Cav.) DC. (*S. rubrum*) A tiny, hairless, usually reddish annual to 5 cm, with stalkless white flowers tinged with pink. Petals 3 mm, pointed; stamens 4 or 5. Leaves ovoid 3–6 mm, overlapping. Carpels hairless, spreading. Sandy places. YU.GR+Is.CR. BG. **392c.** *S. aetnense* Tineo A glaucous annual 2–6 cm, with erect-adpressed, conical-oblong leaves with ciliate teeth, and a thin, colourless spur at the base. Flowers stalkless, axillary, with 4–5 ciliate sepals and white or pink petals, 2 mm. Sandy places. YU.AL? BG.

Leaves flattened, fleshy
395. *S. telephium* L. ORPINE, LIVELONG. A robust erect perennial 15–80 cm, with numerous rounded to narrowly oblong, flat but fleshy leaves. Flowers numerous, in dense flattened heads, reddish-purple, lilac, or whitish. Subsp. *maximum* (L.) Krocker is possibly the most widespread in our area. It has greenish- or yellowish-white flowers, and ovate, often clasping leaves. Rocks, dry woods. YU.AL.GR? BG. **396. *S. cepaea* L. Flowers pale pink with a red, finely hairy, mid-vein, in a long lax cluster. Leaves

obovate, flat, entire. Damp woods, rocks in mountains. Widespread, except TR. Pl. 13. **(396).** *S. stellatum* L. STARRY STONECROP. Flowers pink. Leaves 1–1·5 cm, rounded, toothed, flat. Rocks, walls. YU.AL.GR.CR.

396c. *S. magellense* Ten. A hairless perennial with either alternate or opposite, obovate, flat leaves 6–10 mm long. Flowering stems 6–15 cm, bearing stalked flowers with 5 whitish, pointed petals, and 10 stamens. Mountains. YU.AL.GR.CR.BG. Pl. 14. **396d.** *S. creticum* Boiss. A glandular-hairy biennial, or sometimes an annual 5–8 cm, with pink, short-stalked flowers ranged along the length of the branched stem. Petals 5, pointed; stamens 10. Leaves oblong-spathulate, flat, alternate, mostly forming a basal rosette. Rocky places. CR+Karpathos. Pl. 14. **389c.** *S. idaeum* D. A. Webb A tiny hairless, often reddish-tinged, perennial 1–4 cm, somewhat resembling 389 *S. acre* but with the leaves distinctly flattened on the upper side and with a dense cluster of cream or whitish flowers irregularly striped with red. Leaves 3–7 mm. Fruit spreading in a star. Mountains. CR.

RHODIOLA **397.** *R. rosea* L. Alpine rocks. YU.BG.

ROSULARIA Distinguished by its tubular or bell-shaped flowers like *Umbilicus*, but differing in having a basal rosette of stalkless leaves, and with several flowering stems arising from their axils.

397a. *R. serrata* (L.). A. Berger (*Cotyledon s.*) Distinguished by its rosette of stalkless, blunt, leathery-margined, oblong leaves from which arise several narrow clusters of small purplish-red flowers. Flowers about 6 mm; petal-lobes pointed, erect, somewhat longer than the tube. Rocks, cliffs. CR+Karpathos, E. Aegean Is. Page 66.

SAXIFRAGACEAE | Saxifrage Family

SAXIFRAGA | **Saxifraga** 34 species in our area of which 7 are endemic. The following 26 species are the most distinctive in the Balkans.

Sepals longer or about equalling the petals and partially concealing them
(a) Petals pale yellow
410f. *S. luteoviridis* Schott & Kotschy A tufted perennial with crowded glaucous leaf rosettes, distinguished by its rather numerous pale yellowish-green flowers on stems 4–13 cm. Petals about 3 mm; sepals densely glandular-hairy. Limestone rocks above 1500 m. BG. Page 266.

(b) Petals pink to reddish
410g. *S. sempervivum* C. Koch A very attractive alpine plant of rocks and ledges, with compact silvery-green rosettes, from the centre of each arises a velvety-red, nodding flowering stem with several glandular, velvety-red flowers. Calyx bell-shaped, densely red-glandular; petals as long, pinkish-purple; flowering stem 8–14 cm, with 7–20 flowers. Rosette leaves lime-encrusted, linear-oblong, 8–15 mm by 1–3 mm, not, or barely, widening toward the acute apex; the upper stem leaves oval, densely red-glandular. YU.AL.GR.BG. Page 126. Pl. 15. **410h.** *S. grisebachii* Degen & Dörfler Differs from 410g in having flatter, regular, silvery rosettes of blunt spathulate leaves 10–30 by 4–7 mm which are wider towards the tip. Inflorescence spike-like, taller 10–18 cm, with 15–25 red flowers. Limestone rocks. YU(south).AL.GR(north).BG. Page 266.

410i. *S. stribrnyi* (Velen.) Podp. Densely tufted plant of limestone cliffs with rather flat rosettes of leathery, oblong-spathulate, entire, bluish-green, lime-encrusted leaves 1–2 cm. Inflorescence branched 5–10 cm, reddish above, with 10–30 stalked flowers with individual flower stalks 1–1·5 cm. Petals pinkish-purple, as long as the bell-shaped calyx, which is densely covered with dark-red glandular hairs. Limestone rocks. GR.BG. Pl. 15.

Sepals shorter than petals
(a) *Flowers bright yellow or orange*
402. *S. aizoides* L. Damp places, bogs in mountains. YU.AL. **411j. *S. sibthorpii* Boiss. A delicate annual 5–10 cm, with fleshy, shining kidney-shaped, shallowly lobed, stalked leaves and tiny orange-yellow flowers. Flowers to 1 cm across, axillary, stalked; sepals reflexed; fruit stalks often reflexed. Damp or shady rocks above 1500 m. GR. Pl. 15. **410d.** *S. juniperifolia* Adams A dense hard cushion-forming plant of shady mountain rocks, with short stems 2–4 cm, bearing a rather dense cluster of yellow flowers. Flowers 3–11, nearly stalkless; petals bright yellow, 5–6 mm, shorter than the stamens. Leaves in compact rosettes, leathery, with sharp prickly tips, margin of lower part of blade finely toothed. GR(north).BG. Pl. 15. **410e.** *S. ferdinandi-coburgi* J. Kellerer & Sünd. Like 410d in forming a dense hard cushion of rosettes, but with the flowering stem longer, 3–7 cm, with usually 7–13 stalked flowers; petals bright yellow, 5–7 mm, longer than the stamens. Leaves with an incurved point, not prickly-tipped. Mountains. GR(north).BG. Page 266.

(b) *Flowers purple, pink, white, or pale yellow*
   (i) *Leaves with lime glands on margin*
398. *S. oppositifolia* L. PURPLE SAXIFRAGE. Mountains. YU.AL.BG. **398b. *S. marginata* Sternb. Densely tufted with rosette-like or columnar leafy shoots, with blunt, leathery leaves 3–12 mm long, which have white margins and glandular hairs towards the base. Flowers 2–8, pale pink or white, in a compact cluster, on stems to 9 cm; petals 7–14 mm; sepals reddish-purple. Limestone mountains. YU.AL.GR.BG. Page 266. **398e.** *S. scardica* Griseb. Like 398b but leaves grey-green, stiff and pointed, keeled beneath, and rosettes often forming a hard compact cushion. Flowers 4–12; petals about 7 mm, white or pink. Mountains. YU.AL.GR. Page 126. Pl. 15.

398d. *S. spruneri* Boiss. Plant with tufted, columnar, leafy shoots forming large, hard cushions composed of numerous rosettes each about 1 cm across. Flowering stems erect 4–8 cm, densely glandular-hairy, bearing 6–12 white flowers. Petals about 5 mm. Leaves oblong-obovate, blunt, 4–6 mm, with a white marginal row of lime-secreting glands, ciliate, glandular-hairy beneath. Mountains. YU?AL.GR.BG. Pl. 15. ****400.** *S. paniculata* Miller LIVELONG SAXIFRAGE Distinguished by its medium-sized glaucous, lime-encrusted rosettes 1–3 cm across, and its branched cluster of white or creamy flowers often spotted with red, on flowering stems 12–40 cm. Mountains. YU.AL.GR.BG.

   (ii) *Leaves without lime glands on margin*
      (x) *Leaf blades as broad or broader than long, usually long-stalked*
403. *S. rotundifolia* L. ROUND-LEAVED SAXIFRAGE. Mountain streams, marshy places. YU.AL.GR.CR.BG. Page 136. **403a. *S. chrysosplenifolia* Boiss. Very like 403 with rounded coarsely toothed leaves more than 2·5 cm wide, but with a horny white border to the blade which in the basal leaves continues along the leaf stalk. Flowers white, usually spotted with yellow or red, on stems 15–40 cm. Mountain woods, grassland. YU.AL.GR.CR.BG. Pl. 15. **403b.** *S. taygetea* Boiss. & Heldr. Like 403 but smaller; stems about 10–15 cm, basal leaves only 1·5–2·5 cm wide, with 5–9 coarse teeth. Petals white, spotted with red towards the base. Exposed mountain rocks. AL.GR. Page 115.

1. *Saxifraga ferdinandi-coburgi* 410e **2.** *Saxifraga marginata* 398b **3.** *Saxifraga luteoviridis* 410f **4.** *Saxifraga adscendens* 411k **5.** *Umbilicus chloranthus* 380b **6.** *Sempervivum macedonicum* 384h **7.** *Saxifraga graeca* 404j **8.** *Saxifraga grisebachii* 410h **9.** *Sedum tenuifolium* (388)

403c. *S. hederacea* L. An erect or spreading, very slender plant 4–14 cm, with solitary white flowers borne on thread-like stalks, and ovate or kidney-shaped, toothed leaves. Petals 3 mm, scarcely longer than the erect sepals. Damp shady rocks. YU.GR+Is.CR. ****(404).** *S. bulbifera* L. BULBIFEROUS SAXIFRAGE. Flowers white; bulbils present in axils of all leaves. Grasslands. YU.AL.GR.BG.TR. **404j.** *S. graeca* Boiss. Like (404) in habit but with 7–12 stem leaves (stem leaves 10–20 in (404)), a less compact inflorescence of white flowers, never with bulbils in the upper part of the plant. Mountain meadows. YU(south).AL.GR.BG. Page 266.

(xx) Leaf blade longer than broad, stalkless or short-stalked

407. *S. stellaris* L. STARRY SAXIFRAGE. Mountains, damp places, streamsides, springs. YU.AL? GR.BG. ****409. *S. androsacea* L. Mountains, damp pasture, scree, near snowpatches. YU.AL.BG. **409a.** *S. glabella* Bertol. A slender, delicate, laxly tufted perennial with hairless, narrowly spathulate, entire leaves. Flowering stems 3–10 cm, with several flowers; petals white, rounded, 3 mm; sepals hairless. Screes in mountains. YU.AL.GR. Page 127. Pl. 15.

409b. *S. pedemontana* All. A small white-flowered, rosette-forming mountain saxifrage of shaded siliceous rocks. Subsp. *cymosa* Engler occurs in the Balkan Peninsula. It has short flowering stems to 8 cm bearing a fairly compact cluster of several flowers; petals 11–14 mm. Leaves wedge-shaped, thin, soft, with wide, blunt lobes. YU.GR?BG. Page 136. **411k.** *S. adscendens* L. A small erect biennial to 25 cm, with a branched cluster of small white or yellowish flowers, and with a compact rosette of basal leaves which persists during flowering. Petals 3–5 mm, usually notched. Basal leaves stalkless, wedge-shaped, with 2–5 short, apical teeth, stem leaves similar. Mountains. YU.AL.GR.BG. Page 266.

411. *S. moschata* Wulfen MUSKY SAXIFRAGE. Rocks, screes in mountains. YU.AL? BG? **(411). *S. exarata* Vill. FURROWED SAXIFRAGE. Mountains. YU.AL? GR.BG. Pl. 15. ****(411).** *S. tridactylites* L. RUE-LEAVED SAXIFRAGE. Rocky places. Widespread.

CHRYSOSPLENIUM **412.** *C. alternifolium* L. ALTERNATE-LEAVED GOLDEN SAXIFRAGE. Wet shady places. YU.BG.

PARNASSIACEAE | Grass of Parnassus Family

PARNASSIA ****413.** *P. palustris* L. GRASS OF PARNASSUS. Readily distinguished by its solitary, conspicuously veined, white flower borne on an erect unbranched stem, with a broad oval, stalkless leaf towards its base. Flower 1·5–3 cm across, with branched nectaries and glistening yellowish glands. Damp meadows. YU.AL.GR.BG.

GROSSULARIACEAE | Gooseberry Family

RIBES | **Currant, Gooseberry** 6 species occur in our area. The following are scattered throughout the central European climatic region of the Balkans: ****(414).** *R. alpinum* L. MOUNTAIN CURRANT; **415.** *R. nigrum* L. BLACK CURRANT; **(415).** *R. petraeum* Wulfen ROCK RED CURRANT; ****416.** *R. uva-crispa* L. GOOSEBERRY.

415a. *R. multiflorum* Roemer & Schultes A deciduous shrub of the Balkan Peninsula, growing to 2 m, with long very dense, drooping clusters to 12 cm, of numerous yellowish-green flowers. Sepals strap-shaped hairless, reflexed, much longer than the reflexed petals. Leaves about 10 cm, heart-shaped, shallow-lobed and toothed, hairless. Fruit red, hairless. Rocks and bushy places in mountains. YU.GR.BG.

PITTOSPORACEAE | Pittosporum Family

PITTOSPORUM **417.** *P. tobira* (Thunb.) Aiton fil. A robust evergreen shrub to about 3 m, with oval-oblong entire, blunt, leathery leaves. Flowers creamy-white, very sweet-scented, in flat-topped clusters 5–8 cm across. Petals blunt. Native of E. Asia; often grown for ornament in the Med. region.

PLATANACEAE | Plane Tree Family

PLATANUS ****418.** *P. orientalis* L. ORIENTAL PLANE. Often a massive tree with mottled bark, distinguished by its deeply cut five- to seven-lobed leaves with the middle lobe much longer than broad. But in the wild trees often found growing from stools with several trunks. Flowering and fruiting heads 3–6, borne on long hanging stalks. Woods, riversides. Native of the southern Balkans to about 42°N.; often planted as a village shade tree. Widely scattered on the mainland+Cr. **(418).** *P. hybrida* Brot. LONDON PLANE. Leaves shallowly cut to less than half the width of the blade. Fruiting heads usually 2. Often grown as an ornamental shade tree in villages and by roadsides.

ROSACEAE | Rose Family

PHYSOCARPUS Like *Spiraea* but carpels fused together at their bases, each carpel splitting along both sides.

419a. *P. opulifolius* (L.) Maxim. A deciduous shrub to 3 m, with long-stalked, three- to five-lobed, double-toothed leaves, and white flowers in dense flat-topped clusters. Flowers about 1 cm across. Carpels 4–5, hairless. Native of N. America; cultivated for ornament; naturalized in YU. Page 270.

SPIRAEA Carpels 5, free at base, splitting along one side. 6 species in our area.

Flowers pink
**419. *S. salicifolia* L. WILLOW SPIRAEA. An erect, suckering shrub to 2 m, readily distinguished by its bright-pink flowers in dense terminal, cylindrical clusters 4–8 cm long. Flowers about 8 mm across, with projecting stamens. Leaves oblong-elliptic 3–8 cm, toothed, hairless. Bushy places. BG.; naturalized in YU.

lowers white

1) Leaves with 3 longitudinal veins

19b. *S. crenata* L. An erect shrub to 1 m, with stalked, lateral, flat-topped clusters
bout 2 cm across, of 10–20 white flowers. Flowers about 8 mm across. Leaves glaucous,
anceolate to obovate-acute, entire or toothed, with 3 very conspicuous veins, up to 4 cm
ong. Bushy places, among rocks; sometimes cultivated. YU.GR.BG.

5) Leaves pinnately veined

19c. *S. chamaedryfolia* L. (*S. ulmifolia*) A densely branched shrub to 2 m, with angular,
rown branches, and hemispherical lateral flower clusters about 4 cm across. Petals
bout 6 mm, white; sepals reflexed. Leaves ovate to elliptic, irregularly or double-
oothed, hairless, up to 7 cm long. Fruit hairless, shining. Woods, rocky places. YU.BG.
age 270. **419d.** *S. media* Franz Schmidt Leaves sparsely hairy or grey-woolly beneath
vhen young, hairless when mature, broadly elliptic, entire or toothed, up to 5 cm long.
nflorescence stalked, almost spherical to 4 cm; petals rounded about 3 mm, white or
ale yellow. Fruit hairless or thinly hairy. Shrub to 1·5 m; branches rounded, not angled.
3ushy, rocky places. YU.BG. Pl. 16.

SIBIRAEA Like *Spiraea* but flowers both one-sexed and bisexual on different
lants; inflorescence a terminal-branched cluster. Carpels fused at the base; leaves
ntire.

19e. *S. altaiensis* (Laxm.) C. K. Schneider (incl. *S. croatica*) A densely branched,
eciduous, procumbent shrub to 1 m with grey-brown scaly bark, and clusters of grey-
reen, oblong-entire, blunt, hairless leaves to 8 cm. Flowers white, in terminal-
longated clusters about 3 cm long, elongating further in fruit. Petals 2–3 mm. Fruit
–5 mm. Calcareous cliffs, rocks in mountains. YU(Mostar, Velebit). Pl. 16.

ARUNCUS **420.** *A. dioicus* (Walter) Fernald GOAT'S-BEARD. A tall leafy perennial to
m, distinguished by its conspicuous large, branched, pyramidal inflorescence of very
umerous tiny white flowers each 5 mm across. Leaves very large to 1 m, fern-like, twice-
ut into large ovate toothed leaflets. Woods in mountains. YU.AL.

FILIPENDULA **421.** *F. ulmaria* (L.) Maxim. MEADOW-SWEET. Marshes, riversides.
Widespread on mainland. **422.** *F. vulgaris* Moench DROPWORT. Dry grassy places.
Widespread on mainland.

ALCHEMILLA **423.** *A. vulgaris* agg. LADY'S MANTLE. In this aggregate over 30 species
re listed in our area. Meadows. Widespread on mainland.

APHANES 3 species in our area. **425.** *A. arvensis* L. PARSLEY PIERT. Fields, culti-
ated ground. Widespread on mainland+Cr.

RUBUS | **Blackberry, Bramble,** etc. A very difficult group, mostly of the central
?uropean climatic region of the Balkans. About 24 species in our area.

28. *R. idaeus* L. RASPBERRY. Woods in mountains. YU.AL.GR.BG. **429.** *R. fruticosus*
gg. BLACKBERRY. Bushy places. Widespread. **(429).** *R. caesius* L. DEWBERRY. Damp
ushy places. YU.AL.GR.BG.

ROSA About 40 species are native or are naturalized in our area; including 12 in the
anina' group, 7 in the 'tomentosa' group, and 9 in the 'rubiginosa' group. Probably the
aost likely species to be encountered are the following:

1. *Physocarpus opulifolius* 419a **2.** *Spiraea chamaedryfolia* 419c **3.** *Potentilla detommasii* 453d **4.** *Prunus cocomilia* 477a

(430). *R. sempervirens* L. Flowers white, usually 3–7, in flat-topped clusters. Leaves evergreen, leathery, shining. Stems climbing; prickles sparse. Fruit broadly ovoid, red. Bushy places in the Med. region. YU.AL.GR.CR.TR. ****430.** *R. arvensis* Hudson FIELD ROSE. Bushy places. Widespread on mainland. **431.** *R. gallica* L. Distinguished by its solitary, sweet-scented, bright-red or purple flowers 6–9 cm across. A creeping shrub forming large patches, with stems less than 1 m high. Leaves leathery, paler, finely hairy and glandular beneath. Sunny hills, bushy places. YU.AL.GR.BG.TR.

434. *R. pimpinellifolia* L. (*R. spinosissima*) BURNET ROSE. Flowers usually white, solitary 2·5–5 cm across; sepals narrow lanceolate. A low shrub to 1 m, forming dense clumps, with stems with abundant long prickles; leaves with 5–11, rounded, toothed leaflets 5–15 mm. Fruit blackish-purple. Rocks and cliffs. YU.AL.GR.BG. Page 191. **432. *R. canina* L. DOG ROSE. Bushy places. YU.AL.GR.CR.BG.

433a. *R. glutinosa* Sibth. & Sm. A rose of the 'rubiginosa' or SWEET BRIAR group. Distinguished by the stalked glands on the flower stalks, hairy styles, and its stout stem prickles mixed with glandular bristles and needle-like prickles. Flowers pink; petals 1–1·5 cm. Leaflets densely glandular, usually finely hairy or woolly above, 7–15 mm; stems to 50–100 cm. Fruit red, globular, usually glandular-hairy. Rocks in the Med. region. Widespread, except TR. Pl. 16.

433b. *R. turcica* Rouy One of many sweet briars with glandular leaves smelling of apples when crushed. Flowers small white with narrow petals 1–1·5 cm long, solitary. A low-growing shrub often forming mats, with stems with many stout curved hooked prickles interspersed with glandular bristles. Leaflets 5–7, 8–20 mm, toothed. Fruit scarlet, globular to ovoid, smooth. Dry open places, rocky slopes in mountains. GR.BG.TR. Pl. 16.

*435. *R. pendulina* L. (*R. alpina*) ALPINE ROSE. Flowers bright carmine, 3·5–5 cm across, solitary. Leaflets 7–11, 2–6 cm long. Branches usually without prickles, sometimes reddish; stems erect to 2 m. Fruit pendulous, glandular-hairy. Woods, bushy places in mountains. YU.AL.GR.BG. **435a.** *R. glauca* Pourret Distinguished by its deep-pink flowers in clusters of 1–5, and its bluish-green or purplish leaflets and stems. Petals 8–22 mm, rather narrow; styles free, white-woolly-haired; sepals usually with a few linear lobes, hairless. Leaflets 5–9, elliptic, toothed, 2–4·5 cm. Fruit globular, reddish-brown, usually hairless. Mountains. YU.AL.BG. Pl. 16.

AGRIMONIA | Agrimony

436. *A. eupatoria* L. AGRIMONY. Woods, bushy places. Widespread on mainland. **(436).** *A. procera* Wallr. FRAGRANT AGRIMONY. Differs from 436 in having only long, non-glandular, hairs on the stem (instead of both long and short hairs), and mature fruits grooved for half their length (grooved for three-quarters in 436). Woods. YU.BG.

AREMONIA Like *Agrimonia* but flowers with epicalyx, and with an involucre.

437. *A. agrimonoides* (L.)DC. BASTARD AGRIMONY. Flowers yellow, 7–15 mm across, in a few-flowered, leafy cluster, each flower surrounded by a leafy involucre of 6–10 sepal-like lobes. Leaves with 2–4 pairs of oval, toothed lower leaflets and 3, 5, or 7 larger terminal leaflets. Woods in mountains. Widespread on mainland. Pl. 16.

SANGUISORBA | Burnet 4 species in our area, including 2 local species.

*438. *S. officinalis* L. GREAT BURNET. Damp fields. YU.AL.GR.BG. **439.** *S. minor* Scop. SALAD BURNET. Dry grassy or stony places. Widespread on mainland+CR.

SARCOPOTERIUM ****440.** *S. spinosum* (L.) Spach (*Poterium* s.) THORNY BURNET. A low domed bush to 60 cm, with numerous stiff greyish, interwoven, spiny branches. Flower heads globular or oblong, to 3 cm; female flowers above, with purple feathery styles, the males below. Leaves with 9–15 tiny lobed leaflets, soon falling. Fruit fleshy, bright red. Dry stony ground in the Med. region. YU.AL.GR+Is.CR.TR. Page 39.

DRYAS ****441.** *D. octopetala* L. MOUNTAIN AVENS. Screes and rocks in mountains. YU.AL.GR(north).BG.

GEUM | Avens   9 species in our area.

Fruit with hooked styles, the hooks remaining after the style-tip falls off
442. *G. urbanum* L. HERB BENNET, WOOD AVENS. Flowers yellow, petals 4–7 mm. Woods, bushy places. Widespread on mainland. ****443.** *G. rivale* L. WATER AVENS. Damp woods, marshy fields. YU.AL.GR(north).BG. **442c.** *G. coccineum* Sibth. & Sm. A striking plant 25–45 cm, readily distinguished by its erect inflorescence bearing 2–4 large scarlet flowers 2–4 cm across. Petals 10–18 mm. Basal leaves with 2–3 pairs of leaflets and a much larger terminal kidney-shaped leaflet to 8 cm. Meadows in mountains. YU.AL.GR.BG. Pl. 17.

442e. *G. rhodopeum* Stoj. & Stefanov An endemic of the Rhodope Mountains related to 442c but with yellow petals 7–14 mm, and leaves with 3–6 pairs of lateral leaflets. Meadows in mountains. BG. Pl. 17. **442d.** *G. molle* Vis. & Pančić Flowers pale yellow, 3–5, borne on erect stem 30–40 cm. Petals 10–12 mm long, elliptic, spreading. Basal leaves pinnate, stem leaves three-lobed; whole plant covered in dense soft hairs. Thickets and meadows in mountains. YU.AL.GR.BG.

Styles not hooked and style-tip persisting
444. *G. montanum* L. ALPINE AVENS. A rosette-forming plant with short flowering stems 3–10 cm, bearing bright-yellow flowers, 2–4 cm across. Basal leaves with a terminal heart-shaped leaflet, much larger than the lateral leaflets, which are progressively smaller towards the base. Stolons absent. Alpine meadows. YU.AL.GR.BG. **(444).** *G. reptans* L. CREEPING AVENS. Like 444 but stolons present, and lateral leaflets more or less equal. Alpine screes. YU.AL.BG. **443a.** *G. bulgaricum* Pančić Like ****443.** *G. rivale* with several nodding bell-shaped flowers, but flowers pale yellow or whitish, about 2·5 cm across. Sepals green (brownish-purple in 443). Basal leaves pinnate with a large terminal heart-shaped leaflet 10–15 cm. Damp places in mountains. YU.AL.BG. Pl. 16.

WALDSTEINIA   Like *Geum* but leaves lobed, not pinnate; epicalyx small or absent.

444c. *W. geoides* Willd. A small leafy herbaceous plant 12–25 cm, with a lax cluster of few small yellow flowers, and long-stalked leaves with rounded, lobed blades. Flowers 1–1·5 cm long. Leaves broadly heart-shaped with 5–7 coarsely toothed lobes. Styles soon falling. Woods. YU.BG.

POTENTILLA | Cinquefoil   A difficult genus with about 43 species in our area; hybridization is common. Distinguished from *Ranunculus*, with which it is sometimes confused, by the presence of an additional row of lobes, or epicalyx, which alternate with the true calyx. Also easily confused with *Fragaria* from which it is distinguished by the receptacle not being swollen in fruit.

Flowers red or purple
446. *P. palustris* (L.) Scop. MARSH CINQUEFOIL. Flowers reddish-purple; petals shorter than the sepals. Leaves pinnate, leaflets 3–6 cm. Marshes, bogs. YU.BG. **445b. *P.*

kionaea Halácsy A tufted perennial, with 1–3 purple flowers and petals longer than the sepals, on grey- or white-woolly stems to 10 cm. Leaves trifoliate, leaflets 4–8 mm, silvery-haired. Limestone cliffs and rocks above 1800 m. GR(Giona).

Flowers white, cream, or pink
(a) Leaves trifoliate
445c. *P. speciosa* Willd. A silvery white-haired mountain perennial to 30 cm, with numerous white flowers. The trifoliate leaves have obovate leaflets 1·5–3 cm, with scalloped margins for at least the apical two-thirds, and are white-haired beneath. Petals 10 mm, slightly longer than the sepals. Epicalyx lobes as long as or longer than the sepals. Mountains. YU.AL.GR.CR. Page 114. **445d.** *P. apennina* Ten. A white-haired tufted perennial to 20 cm, with few white or pale-pink flowers. The trifoliate leaves have leaflets 7–15 mm that are toothed at the apex and are silvery-silky above and below. Petals 8–12 mm, longer than sepals. Epicalyx as long as or slightly shorter than the sepals. Mountains. YU.AL.BG.

445e. *P. deorum* Boiss. & Heldr. A silvery white-haired tufted plant, with 3–6 white flowers congested at the end of stem to 25 cm. Petals 10–12 mm, longer than the sepals; epicalyx equalling the sepals. Leaves trifoliate, leaflets elliptic 8–20 mm, toothed at the apex, silvery-silky or sometimes grey-green above. Limestone cliffs, stony ground above 2000 m. GR(Olympus). Page 126. Pl. 17.

(b) Leaves with 5–7 leaflets
448. *P. rupestris* L. ROCK CINQUEFOIL. A robust erect hairy, glandular perennial to 60 cm, distinguished by its rather large white flowers 1–1·5 cm across, borne in a branched cluster. Petals 8–14 mm, longer than sepals; epicalyx shorter than sepals. Leaves pinnate with 5–7 ovate to rounded, toothed or double-toothed leaflets. Rock fissures in lowlands and mountains. YU.AL.GR.BG. ****447. *P. alba* L. WHITE CIN-QUEFOIL. A slender hairy perennial to 15 cm, with white flowers with petals 7–10 mm, longer than sepals. Lower leaves digitate with 5 oblong leaflets, green above, silvery-silky beneath. Meadows. YU.BG.

447d. *P. haynaldiana* Janka A tufted perennial 10–50 cm, with long spreading hairs and numerous glandular hairs, and digitate leaves with usually 7 leaflets which are green above, silvery-silky below. Flowers white, numerous in a dense head; petals 6–7 mm, shorter than sepals. Epicalyx distinctly longer than the sepals. Mountain rocks. BG.

447e. *P. doerfleri* Wettst. A densely softly hairy perennial somewhat similar to 447d, but with 5 leaflets which are scalloped in the apical half, green or grey-green. Flowers white, numerous; petals 5–7 mm, shorter than the sepals. Epicalyx about as long as sepals. Siliceous cliffs above 2000 m. YU(Sar Planina). **447f.** *P. clusiana* Jacq. A tufted perennial 5–10 cm, with leaves with 5 small leaflets, and white flowers with petals 9–10 mm much larger than the sepals. Petals notched, spreading; flowers 1–3. Leaflets hairy 7–12 mm, obovate, with 3–5 teeth towards the apex. Limestone rocks in mountains. YU.AL. Pl. 17.

Flowers yellow
(a) Leaves, at least some, pinnate
456. *P. anserina* L. SILVERWEED. Damp places. YU.BG. **448a.** *P. geoides* Bieb. A softly hairy perennial 15–50 cm, often confused with ****448.** *P. rupestris*, with pinnate leaves with 7–9 nearly rounded, double-toothed leaflets, but with pale-yellow flowers. Petals 10–12 mm, longer than the often three-lobed sepals. Damp places, roadsides. YU? GR(north).BG. ****457.** *P. fruticosa* L. SHRUBBY CINQUEFOIL. Rocky places in mountains. BG. **456b.** *P. visianii* Pančić A downy perennial with long spreading glandular hairs,

and flowering stems to 40 cm, with numerous yellow flowers in a lax terminal cluster. Petals 8–10 mm, much longer than the sepals. Epicalyx as long or longer than the sepals. Leaflets 11–17, obovate-wedge-shaped, with 2–7 coarse teeth. Cliffs, dry grassland, usually on serpentine. YU.AL.

(b) *Leaves trifoliate, rarely a few digitate*

449. *P. erecta* (L). Räuschel COMMON TORMENTIL. Woods, grassland. YU.AL.BG. **453e.** *P. montenegrina* Pant. A densely hairy perennial 30–80 cm, with 3 leaflets 3·5–8 cm long, broadly obovate to obovate-oblong, coarsely toothed and hairy on the veins beneath; the central leaflet distinctly stalked. Flowers yellow, numerous; petals 10–12 mm, about twice as long as the sepals. Mountain grassland. YU.AL.BG.

(c) *Leaves mostly digitate (with 5–7 leaflets arising from the apex of the leaf stalk)*

450. *P. reptans* L. CREEPING CINQUEFOIL. Grassland, waysides. Widespread. **455.** *P. aurea* L. GOLDEN CINQUEFOIL. Grassland, rocky places in mountains (1400–2600 m) on acid soils. YU.AL.GR.BG. Pl. 17. **452.** *P. argentea* group Distinguished by its digitate leaves with 5 toothed leaflets, which are green or grey above and densely white-woolly beneath. Flowers rather small, 1–1·5 cm across, in spreading clusters, on woolly-haired stems; petals 4–7 mm, as long as or longer than the woolly-haired sepals and epicalyx. Grassy, sandy, rocky places in mountains. Widespread on mainland.

452a. *P. inclinata* Vill. (*P. canescens*) A perennial 15–50 cm, very variable in the degree of hairiness, and in the division of the 5–7 leaflets; rather similar to 452, from which it may be distinguished by the spreading, long, simple hairs on the flowering stem (curled hairs in 452). Petals yellow, 5–7 mm. Rocky places, dry grassland. Widespread on mainland. **452b.** *P. collina* Wibel A variable perennial to 30 cm, somewhat similar to 452 from which it may be distinguished by the presence of club-shaped, but often somewhat distorted styles (styles conical tapering to apex in 452). Petals yellow, 4–7 mm. Plant hairless to white-woolly. Calcareous rocks in mountains. YU.BG.

453. *P. recta* L. SULPHUR CINQUEFOIL. A robust perennial to 70 cm, with sparse, short glandular or non-glandular hairs below, distinguished by its digitate leaves with 5–7 large oblong, deeply toothed or lobed leaflets, each 1·5–10 cm long. Flowers yellow, numerous, in a lax cluster; petals 6–12 mm, as long as or longer than sepals. A very variable species. Thorn scrub. Widespread on mainland. Pl. 17. **453d.** *P. detommasii* Ten. Distinguished from 453 by its obovate leaflets 2–5 cm long which are silky or shaggy-haired below with many crowded short hairs. Flowers yellow, usually numerous, crowded; petals 12–14 mm, as long as or longer than the sepals. Thickets. YU.AL.GR.BG.TR? Page 270. **453f.** *P. chrysantha* Trev. A densely hairy, laxly branched perennial 10–50 cm, with yellow flowers, and leaves usually with 5 toothed leaflets. It is closely related to 453 which may be one of its parents. Petals 7–10 mm, 1½–2 times as long as sepals; epicalyx about as long as sepals. Bushy places. YU.BG. Page 184.

SIBBALDIA 2 species in our area. ****458.** *S. procumbens* L. Alpine pastures. YU.BG.

FRAGARIA **459.** *F. vesca* L. WILD STRAWBERRY. Woods, bushy places. Widespread on mainland. **(459).** *F. moschata* Duchesne HAUTBOIS STRAWBERRY. Like 459 but creeping runners absent or few. Flowers white, about 2 cm across, on stems distinctly longer than the leaves. Deciduous woods, bushy places. YU.AL.BG.TR. **(459).** *F. viridis* Duchesne Differs from 459 in having short, very slender runners. Flowers creamy-white; sepals recurved after flowering. Fruit about 1 cm, fleshy receptacle without carpels near its base. Bushy places. YU.AL.GR.BG.

MESPILUS 460. *M. germanica* L. MEDLAR. A shrub or small deciduous tree to 6 m, with solitary white flowers 3–4 cm across, and dull-green oblong leaves 5–12 cm. Petals and sepals equal. Fruit brown, flattened-globular, open at top, 2–3 cm across. Hedges, woods, often introduced from cultivation. GR.BG; introd. YU. Pl. 18.

CRATAEGUS | Hawthorn A difficult genus, with considerable amount of hybridization. The leaves and stipules of the flowering shoot are referred to below; leaves on non-flowering shoots are often larger and more deeply cut.

Young twigs and leaves hairless
461. *C. monogyna* Jacq. HAWTHORN. Flowers white 8–15 mm across. Leaves with 3–7 lobes, which are entire or sparsely toothed near the apex of each lobe; stipules entire. Fruit dark or light red. A very variable species. Woods, bushy places, or rocks. Widespread. Subsp. *azarella* (Griseb.) Franco has twigs and young leaves densely hairy. Dry mountain thickets. Widely distributed. **461a.** *C. calycina* Peterm. Distinguished from 461 by the larger white flowers, 1·5–2 cm across. Leaves usually with 3–5 lobes, with lobes finely toothed almost to the base; stipules toothed. Woods, bushy places, calcareous rocks. YU.BG.

Young twigs and leaves with woolly or silky hairs (see also 461)
(a) Fruit black
461b. *C. pentagyna* Willd. A small shrub or tree with spiny, grey-brown branches with woolly hairs when young. Flowers white, 1·2–1·5 cm across, in a compound cluster. Leaves dark green, leathery, at first hairy but becoming hairless, deeply lobed to two-thirds, with 3–7 broad blunt, toothed lobes. Fruit dull, blackish-purple. Woods. YU.AL.BG.TR? **461c.** *C. nigra* Waldst. & Kit. Like 461b but leaves more shallowly lobed to half-way into 7–11 lobes, which are acute, sharply and irregularly toothed, woolly-haired. Flowers few, in a simple cluster. Fruit black, shining. Woods. YU.AL.

(b) Fruits red or yellow

1. Lvs. not more than 3 cm; fr. 7–10 mm.
  2. Lvs. and twigs woolly-haired; styles 1–3.   *heldreichii*
  2. Lvs. and twigs silky-haired; styles 5 (GR(south)).   **462c.** *C. pycnoloba* Boiss & Heldr. Pl.19.
1. Lvs. 3–5 cm; fr. 1·5–2·5 cm.
  3. Twigs and lvs. woolly-haired, but soon nearly hairless; fr. dark red (GR.).   **462d.** *C. schraderana* Ledeb.
  3. Twigs and lvs. persistently hairy; fr. orange-red or yellow.
    4. Lvs. with 3–7 acute lobes which are deeply toothed; styles 3–5.   *laciniata*
    4. Lvs. with 3 blunt entire lobes; styles 1–2 (CR.; naturalized elsewhere).   ****462.** *C. azarolus* L.

462b. *C. heldreichii* Boiss. A shrub with woolly twigs and small leathery leaves which are woolly-haired on both sides. Flowers 15–18 mm across, in clusters about 3 cm across; flower stalks and calyx woolly-haired. Leaves 1·5–3 cm, cut to two-thirds into 3–5 lobes, which are sparsely toothed towards the apex. Fruit red, 7 mm. Hills, mountains. AL.GR.CR. Pl. 19. **462a.** *C. laciniata* Ucria Distinguished from 462b by its larger woolly-haired leaves 3–5 cm which are deeply cut to seven-eighths into 3–7 long, narrow, toothed lobes. Fruit larger 1·5–2 cm, red to orange, woolly when young. Bushy places in mountains. YU.AL.GR.CR.BG. Page 169. Pl. 19.

CYDONIA **463. *C. oblonga* Miller QUINCE. A shrub or small tree to 6 m, with large solitary white or pale pink flowers 4–5 mm across, and large ovate entire leaves. Leaves 5–10 cm, hairless above, white-woolly beneath. Fruit globular or pear-like, yellow, woolly-haired. Woods, bushy places. Native of Asia; widely naturalized.

PYRUS Styles free at base; fruit with gritty 'stone-cells' in the flesh.

1. Fr. large 6–15 cm long, fleshy, sweet-tasting (cultivated throughout).	**464.** *P. communis* L.
1. Fr. small not more than 5·5 cm long, hard, not sweet-tasting.	
2. Lvs. not more than 1½ times as long as wide.	
3. Lvs. tapering gradually into a rigid point (cuspidate) (GR(Thrace).TR?).	**464e.** *P. caucasica* Fedorov
3. Lvs. acute or with a very short point.	*pyraster*
2. Lvs. more than 1½ times as long as wide.	
4. Mature lvs. hairless beneath.	*amygdaliformis*
4. Mature lvs. hairy beneath.	
5. Styles densely hairy at least at base.	
6. Styles hairy to the middle; fr. 2–3 cm long.	*elaeagrifolia*
6. Styles hairy only at base; fr. 3–5 cm long (YU.BG.).	**464g.** *P. nivalis* Jacq.
5. Styles more or less hairless (YU.GR.).	**464h.** *P. salvifolia* DC.

(464). *P. pyraster* Burgsd. WILD PEAR. A small- to medium-sized tree 8–20 m, with usually spiny branches, and small yellow, brown, or black, rounded or top-shaped fruits 1–3·5 cm long borne on slender stalks. Flowers white; petals 10–17 mm. Leaves 2·5–7 cm, elliptic to rounded, finely toothed, usually hairless at maturity. Woods, bushy places. Widespread on mainland. **464b.** *P. amygdaliformis* Vill. A shrub or small tree to 6 m, with grey twigs, distinguished by its narrow lance-shaped to obovate, usually entire, leaves 2·5–8 cm which are hairless beneath when mature. Flowers white; petals 7–8 mm, usually notched. Fruit globular 1·5–3 cm, yellowish-brown, borne on a stout stalk about as long. Dry bushy places. Widespread. Pl. 18.

464f. *P. elaeagrifolia* Pallas A shrub or small tree with stout spiny branches and grey-woolly twigs. Flowers numerous, on white-woolly stalks; petals about 1 cm. Leaves lanceolate entire, 3·5–8 cm, densely greyish- to white-woolly beneath. Fruit globular, or top-shaped, 2–3 cm, stalk stout. Dry bushy places. AL.GR.BG.TR. Pl. 18.

MALUS Styles fused at base; fruit with few or no gritty 'stone-cells'.

1. Lvs. lobed.	
2. Leaf stalk 2–7 cm; sepals 7–10 mm,	**465a.** *M. trilobata* (Labill.) C.K.

persisting (GR(Thrace)).

2. Leaf stalk 0·5–2 cm; sepals 3–4 mm,
   soon falling.
1. Lvs. not lobed.
   3. Mature lvs. hairless on both sides.
   3. Mature lvs. woolly-haired beneath.
      4. Fr. large, more than 5 cm (widely
         cultivated).
      4. Fr. small, less than 5 cm,
         yellowish.

Schneider

florentina

sylvestris

465. *M. domestica* Borkh.

dasyphylla

465b. *M. florentina* (Zuccagni) C. K. Schneider A small spineless tree to 4 m, with broadly ovate, toothed, deeply and irregularly lobed leaves 3–6 cm, which are white-woolly beneath. Flowers white, 1·5–2 cm across, in umbels. Fruit red, about 1 cm. YU(south).AL.GR(north). **(465).** *M. sylvestris* Miller CRAB APPLE. Distinguished by its small yellowish-green, sour fruits, and its leaves which soon become quite hairless. Flowers white or pink, 3–4 cm across. Leaves 3–11 cm, ovate to rounded, toothed. Fruit 2·5–3 cm. Woods. Widespread on mainland. **465c.** *M. dasyphylla* Borkh. A medium-sized, sparsely spiny tree with elliptic-pointed, toothed leaves 3·5–11 cm, which are woolly-haired beneath. Flowers pink or white. Fruit 4 cm, yellowish, sour. Woods. YU.AL.GR.BG.

SORBUS A very difficult genus, with much hybridization followed by the doubling of chromosomes and loss of sexual processes (apomixis) in the formation of the seeds.

Leaves pinnate; leaflets toothed

****466.** *S. aucuparia* L. ROWAN, MOUNTAIN ASH. Woods. YU.AL.GR.BG.BG. ****(466).** *S. domestica* L. SERVICE TREE. Distinguished from 466 by its shredding bark; styles 5 (3–4 in 466); and its larger, 2 cm, greenish or brownish fruit (fruit 1 cm, scarlet in 466). Flowers white, 16–18 mm across. Woods, bushy places; often cultivated for its fruits. YU.AL.GR.BG.TR.

Leaves entire, or conspicuously toothed, but not lobed

****467.** *S. aria* (L.) Crantz WHITEBEAM. Leaves white-woolly beneath. Flowers white, 1–1·5 cm across. Fruit scarlet with numerous small lenticels, 8–15 mm. A tree to 25 m. Dry woods, bushy places. YU.AL.BG. **468.** *S. chamaemespilus* (L.) Crantz FALSE MEDLAR. A shrub to 1·5 m distinguished from 467 by its pink flowers, leaves which are hairless beneath, and its orange-red fruit. Mountain woods. YU.AL.GR? BG. **467c.** *S. graeca* (Spach) Kotschy A shrub or small tree. Distinguished by its rather leathery leaves which are obovate or nearly rounded, broadest above the middle, with broadly wedge-shaped base, double-toothed, and usually a thick greenish-white mat of hairs beneath. Fruit crimson, lenticels few, less than 12 mm. A very variable species. Woods. YU.AL.GR.CR.BG.

Leaves deeply or shallowly lobed

469. *S. torminalis* (L.) Crantz WILD SERVICE TREE. A tree to 25 m with large ovate leaves 5–9 cm, with 3–4 pairs of broad-triangular, long-pointed, saw-toothed lobes, green above and beneath. Flowers white, 1–1·5 cm across. Fruit brown, with numerous lenticels, 12–18 mm. Woods. YU.AL.GR.BG.TR. **469a.** *S. umbellata* (Desf.) Fritsch Distinguished from 469 by its shallowly lobed leaves which are densely white-woolly beneath, and with 4–7 pairs of lateral veins, but leaves very variable in lobing and size. Flowers about 1·5 cm across. Fruit yellowish. A shrub or small tree. Woods and conifer forests in mountains. YU.AL.GR.BG. Pl. 19 **469b.** *S. austriaca* (G. Beck) Hedl. Like 469a with shallowly lobed leaves, but lobes somewhat overlapping, whitish-grey-haired beneath, with 8–11 pairs of lateral veins. Fruit red, with large and numerous lenticels. Woods and bushy places in hills. YU.GB.

ERIOBOTRYA **470. *E. japonica* (Thunb.) Lindley LOQUAT. A small, distinctive tree to 10 m, with its large dark-green, strongly veined, laurel-like leaves, which are rusty-haired beneath, and with orange, plum-like edible fruits. Flowers white, in reddish-brown, densely woolly-haired clusters. Native of China; widely cultivated in the Med. region for its fruits; naturalized in CR.

AMELANCHIER **471. *A. ovalis* Medicus SNOWY MESPILUS. A small shrub 1–3 m, easily recognized when in flower by its narrow, widely separated white petals and its 'snowy' young foliage which is covered with dense white hairs. Flowers in clusters; petals 1–1·5 cm. Leaves ovate, coarsely toothed, at length hairless. Fruit small, bluish-black, sweet. Rocky places, stony woods. Widespread on mainland+CR.

COTONEASTER

1. Fr. black (YU.BG.). 473a. *C. niger* (Thunb.) Fries Pl. 18.
1. Fr. red or purplish.
   2. Sepals hairless or finely hairy on
      margin. *intergerrimus*
   2. Sepals densely hairy or woolly-
      haired.
      3. Petals erect, scarcely longer
         than calyx; fr. hairy *nebrodensis*
      3. Petals spreading, twice as long
         as calyx; fr. hairless at maturity 473b. *C. nummularia* Fischer &
         (CR.). C. A. Meyer

473. *C. integerrimus* Medicus A branched shrub to 2 m, distinguished by its neat oval leaves 2–5 cm, which are densely grey-woolly beneath and green and hairless above. Flowers pinkish, 2–4, in short axillary clusters. Fruit 8 mm, shining red. Stony woods, rocky places. YU.AL.GR(north).BG. **474.** *C. nebrodensis* (Guss.) C. Koch (*C. tomentosus*) Like 473 but a shrub to 3 m with larger, rounded leaves 3–6 cm; calyx and inflorescence densely hairy. Flowers in nodding clusters of 3–12. Fruit red, hairy. Rocks. YU.AL.GR.BG. Pl. 18.

PYRACANTHA 475. *P. coccinea* M. J. Roemer FIRETHORN. A dense, very spiny shrub to 2 m or more, with elliptic, finely toothed, shining evergreen leaves 2–4 cm. Flowers white or pinkish-yellow, 7–8 mm across, numerous in dense flat-topped clusters. Fruit ovoid 5–7 mm, bright red. Hedges, bushy places. Widespread on mainland. Pl. 18.

PRUNUS The following fruit trees are widely cultivated in a variety of forms and often occur naturalized in our area. Asian species include: **479.** *P. dulcis* (Miller) D. A. Webb ALMOND; **478.** *P. persica* (L.) Batsch PEACH; **(478).** *P. armeniaca* L. APRICOT. Native species include: **477.** *P. domestica* L. PLUM, BULLACE, DAMSON, GREENGAGE; **480.** *P. avium* L. CHERRY.

The remaining native *Prunus* species in our area can be distinguished as follows:

1. Fr. and ovary hairy.
   2. Small shrub without spines. *tenella*
   2. Tree or spiny shrub. *webbii*

1. Fr. and ovary hairless.
   3. Fls. in elongated spike-like
      clusters.
      4. Lvs. evergreen.   *laurocerasus*
      4. Lvs. deciduous (YU.BG.).   **482. *P. padus* L.
   3. Fls. solitary, or in rounded clusters,
      or umbels.
      5. Fr. stalk at least twice as long as
         ripe fr.
         6. Fls. in a stalkless umbel, with
            papery bud scales at base.   *cerasus, fruticosa*
         6. Fls. in a short, more or less
            elongate cluster, without papery
            scales at base.   *mahaleb*
      5. Fr. stalk shorter, or only slightly
         longer than ripe fr.
         7. Fls. pink; lvs. 1–2·5 cm.   *prostrata*
         7. Fls. white; lvs. usually more than
            2·5 cm.
            8. Plant quite hairless, spineless.   *cocomilia*
            8. Plant somewhat hairy, usually with
               spines.
               9. Fr. 2–3 cm, red or yellow;
                  petals 8–10 mm.   *cerasifera*
               9. Fr. 1–1·5 cm, bluish-black,
                  bloomed; petals 5–8 mm
                  (Widespread on mainland.).   **476 *P. spinosa* L.

479a. *P. tenella* Batsch (*P. nana*) A small suckering shrub to 1·5 m, with grey, hairless branches bearing solitary bright-pink, or rarely white flowers, with petals 1–1·5 cm. Leaves hairless, lanceolate, toothed, to 5 cm. Fruit rounded 12–20 mm, yellowish-grey, densely shaggy-haired. Dry grasslands, bushy places. YU.BG. Pl. 19.   **479b.** *P. webbii* (Spach) Vierh. Like wild plants of **479.** *P. dulcis* ALMOND but leaves smaller and narrow, 3·5 cm by 6–9 mm wide, and smaller densely woolly fruit 2–2·5 cm. Flowers deep pink; petals about 1 cm. A shrub 2–4 m. Sunny rocks. YU(south). AL.GR.CR.BG.

481. *P. cerasus* L. SOUR CHERRY. A suckering shrub 1–5 m, with rather glossy dark-green, hairless leaves, and umbels of white or pinkish flowers. Petals 9–15 mm. Fruit bright red, rarely black or yellowish, sour. Bushy places, margins of vineyards. Native of Asia; cultivated and naturalized in YU.AL.GR.BG.   **481a.** *P. fruticosa* Pallas A low, quite hairless shrub 30–150 cm, distinguished from 481 by its smaller white flowers with petals 5–7 mm. Fruit 7–10 mm, dark red. Sunny places. YU.BG.   **484.** *P. mahaleb.* L. ST. LUCIE'S CHERRY. Usually a deciduous shrub, with short, erect, branched leafy clusters of 3–10 white, fragrant flowers. Petals 5–8 mm. Leaves 4–7 cm, broadly ovate, toothed, nearly hairless. Fruit small, 8–10 mm, red then black, bitter. Bushy places, rocks, dry hillsides. YU.AL.GR.BG.

476b. *P. prostrata* Labill. A low spreading, tough and twisted, woody shrub to 1 m, with bright-pink to pale-pink, solitary flowers appearing before the leaves. Petals about 7 mm. Leaves oblong to obovate, usually 9–12 mm, coarsely toothed, hairless above, grey-woolly beneath. Fruit red, about 8 mm. Rocks in mountains in the Med. region. YU.AL.GR.CR. Page 70.   **477a.** *P. cocomilia* Ten. A quite hairless small tree, or shrub, procumbent at high altitudes, with white flowers in clusters of 2–4, appearing with the leaves. Petals about 6 mm. Leaves elliptic-ovate, toothed, 2–4 cm. Fruit yellow, flushed red-purple, 1·2–4 cm; fruit stalk very short. In mountains. YU.AL.GR. Page 270. Pl. 19.

(477). *P. cerasifera* Ehrh. CHERRY PLUM. Distinguished from 477 *P. domestica* by its hairless and glossy young twigs, and its leaves which are hairless and glossy above and hairy on the veins beneath. Flowers white, mostly solitary, stalked; petals 8–10 mm. Fruit globular, yellow or red, 2–3 cm; nut smooth; fruit stalk about 1·5 cm. Mountain woods. YU.AL.GR.BG.TR. ****483. *P. laurocerasus* L. CHERRY-LAUREL. An evergreen shrub to 8 m with large glossy, laurel-like leaves 5–12 cm, and pale-green twigs. Flower spikes axillary, scarcely as long as the subtending leaves; flowers white, numerous, about 1 cm across. Fruit shining black, about 12 mm. Mountain woods. YU(Serbia).BG.TR.

LEGUMINOSAE | Pea Family

A very important family in the Balkans comprising about 61 genera and nearly 500 native species. In many of the larger genera, limitation of space has resulted in the selection of only the most widely encountered, or the most conspicuous and distinctive species. Trees often planted for ornament by roadsides, in hedges, parks, and gardens and occasionally naturalized include: **487.** *Gleditsia triacanthos* L. HONEY LOCUST, THREE-THORNED ACACIA; *Gymnocladus dioica* (L.) K. Koch KENTUCKY COFFEE TREE; *Poinciana regia* Bojer FLAMBOYANT, PEACOCK-FLOWER; *Caesalpinia gilliesii* Wall; **494.** *Sophora japonica* L. PAGODA-TREE; ****493.** *Albizia julibrissin* Durazz. PERSIAN ACACIA, PINK SIRIS; *Albizia lophantha* (Willd.) Benth.

CERCIS ****485.** *C. siliquastrum* L. JUDAS TREE. A small tree to 10 m with several main stems, and with distinctive rounded to kidney-shaped, hairless leaves 7–12 cm. Flowers very conspicuous, bright rose-purple, appearing in clusters on twigs and branches before the leaves. Pod reddish-brown, strap-shaped 6–10 cm. Rocky places, by rivers, in the Med. region. Widespread.

CERATONIA ****486.** *C. siliqua* L. CAROB. A robust but low-growing spreading evergreen tree to 10 m, with dark-green leaves with 4–10 oval, leathery leaflets 3–5 cm, and numerous broad, strap-shaped pods arising from the old branches. Flower clusters greenish, appearing in autumn. Pods 10–20 cm by 1·5–2 cm broad. Native in the Med. region; also widely cultivated for its sugary pods, and often naturalized. YU.AL.GR+Is.CR.

ACACIA Leaves either twice-pinnate, or reduced in the adult state in some species to simple 'phyllodes'. Mostly natives of Australia. The following are most commonly planted in the Med. region and may become locally naturalized:

Leaf-like phyllodes, lanceolate
489. *A. cyanophylla* Lindley BLUE-WATTLE. A small tree to 10 m with glaucous leaf-like phyllodes 10–20 cm, and pendulous twigs. Flower heads bright yellow, globular, 1–1·5 cm, in short axillary clusters of 2–6. Pod constricted between seeds. Often planted for ornament and soil-stabilization in the Med. region; naturalized in GR. Pl. 20. **489a.** *A. retinodes* Schlecht. Like 489 but phyllodes light green, 6–15 cm, and twigs not

pendulous. Flower heads pale yellow, smaller 4–6 mm, in axillary clusters of 5–10. Pod not or slightly constricted between seeds. Planted for ornament in the Med. region.

Leaves twice-pinnate

490. *A. dealbata* Link SILVER WATTLE, MIMOSA. A tree to 30 m with greyish-white twigs and young leaves, and twice-pinnate glaucous-green leaves. Pinnae 8–20 pairs, each pinna with 30–50 pairs of tiny leaflets, 3–4 mm long. Flower heads pale yellow, 5–6 mm, numerous in much-branched clusters longer than the leaves. Planted for ornament, timber, and soil-stabilization in the Med. region; naturalized in YU. **491.** *A. farnesiana* (L.) Willd. A shrub to 4 m, with twice-pinnate, bright-green leaves. Pinnae 3–8 pairs, each pinna with 10–25 pairs of leaflets 3–5 mm. Flower heads bright yellow, fragrant, 10–12 mm, 2–3 together in the axils of older leaves. Often cultivated in the Med. region.

ANAGYRIS   495. *A. foetida* L. BEAN TREFOIL. A fetid, poisonous deciduous shrub to 4 m, with trifoliate leaves, green twigs, and clusters of yellowish, laburnum-like flowers borne on last year's branches. Flowers 2–2·5 cm; standard dark-blotched at base, about half as long as the wings. Pod 10–20 cm; seeds poisonous, violet or yellow. Fields, rocky places in the Med. region. YU.AL.GR+Is.CR.TR. Page 96.

LABURNUM   **496. *L. anagyroides* Medicus LABURNUM. A small deciduous shrub or tree to 7 m, with smooth trunk and branches, trifoliate leaves, and long pendulous clusters of yellow flowers. Flowers about 2 cm, streaked with brown. Twigs with adpressed hairs, greyish-green. Pod 4–6 cm, with adpressed hairs, becoming hairless. Bushy places in mountains. YU(west).   **497.** *L. alpinum* (Miller) Berchtold & J. Presl Like 496 but twigs almost hairless, green, and pods hairless. Flower clusters longer 15–40 cm; flowers about 1·5 cm, yellow. Bushy places. YU(west).AL.

BROOMS and their allies with yellow, white, or rarely purple flowers. Key to the following broom-like genera: *Podocytisus, Calicotome, Lembotropis, Cytisus, Chamaecytisus, Teline, Genista, Chamaespartium, Gonocytisus, Spartium, Petteria, Adenocarpus, Lotononis, Argyrolobium, Caragana, Astragalus*:

Plants spiny

1. Calyx with 5 short teeth, the upper portion breaking away at flowering time to leave a cup-shaped remnant.   *Calicotome* Page 283
1. Calyx funnel-shaped, two-lipped, the upper part not breaking away.
  2. Calyx distinctly two-lipped, the teeth unequal.
    3. Upper lip of calyx deeply two-lobed.   *Genista* Page 285
    3. Upper lip of calyx with 2 short teeth; lvs. trifoliate.   *Chamaecytisus* Page 283
  2. Calyx tubular or bell-shaped with 5 more or less equal teeth; spines formed from stalks of old leaves.
    4. Pod long and narrow, splitting.   *Caragana* Page 289
    4. Pod usually not more than twice as long as broad, not splitting.   *Astragalus* Page 289

Plants without spines

1. Young stems broadly winged.   *Chamaespartium* Page 286
1. Young stems not broadly winged.

2. Fr. with prominent glandular
swellings. *Adenocarpus* Page 287
2. Fr. without glandular swellings
(sometimes with glandular hairs).
  3. Fr. broadly winged. *Podocytisus* Page 283
  3. Fr. not broadly winged.
    4. Lvs. simple or one-bladed,
sometimes very small.
      5. Calyx split to the base. *Spartium* Page 286
      5. Calyx not split to the base.
        6. Upper lip of calyx with short
teeth. *Cytisus* Page 283
        6. Upper lip of calyx two-cleft or
deeply toothed. *Genista* Page 285
    4. Lvs. at least some, with 3 – many
leaflets.
      7. Lvs. with more than 5 leaflets. *Astragalus* Page 289
      7. Lvs. at least some, with 3 leaflets.
        8. Calyx tube distinctly shorter
than lips. *Argyrolobium* Page 287
        8. Calyx tube as long as or longer
than lips.
          9. Fls. in long terminal leafless
spike-like clusters.
            10. Upper lip of calyx deeply
two-cleft.
              11. Lv. stalks 1·5–5 cm;
fr. 3·5–5 cm. *Petteria* Page 287
              11. Lv. stalks not more than
1·5 cm; fr. not more than
2·5 cm. *Gonocytisus* Page 286
            10. Upper lip of calyx with 2
short teeth. *Lembotropis* Page 283
          9. Fls. not in long leafless spike-
like clusters.
            12. Upper lip of calyx deeply four-
toothed, lower lip one-toothed. *Lotononis* Page 287
            12. Upper lip of calyx not four-
toothed.
              13. Upper lip of calyx deeply
two-cleft.
                14. Fl. stalks 0·5–1 cm; fr.
hairless. *Cytisus* Page 283
                14. Fl. stalks 1–3 mm; fr.
hairy.
                  15. Fls. in umbel-like
heads. *Genista* Page 285
                  15. Fls. axillary, or in
axillary clusters.
                    16. Standard dis-
tinctly shorter

than keel.	*Genista* Page 285
16. Standard longer	
than keel.	*Teline* Page 285
13. Upper lip of calyx with	
2 short teeth.	
17. Calyx tubular.	*Chamaecytisus* Page 283
17. Calyx bell-shaped.	*Cytisus* Page 283

PODOCYTISUS   Distinguished by its broadly winged pod which does not split open.

497a. *P. caramanicus* Boiss. & Heldr. A shrub to 2 m, with trifoliate leaves and terminal, often pyramidal, clusters 5–15 cm long of yellow flowers. Flowers 1·5–2 cm; standard, wings, and beaked keel equal in length. Leaves stalked; leaflets obovate 5–15 mm. Pod 5–7 cm, curved, winged. Bushy places in the Med. region. YU(Macedonia).AL.GR. Pl. 20.

CALICOTOME   **498. *C. villosa* (Poiret) Link SPINY BROOM. A very spiny, much-branched shrub to 3 m, with silvery-hairy twigs, and with short-stalked clusters of 2–15 yellow flowers. Flowers 12–18 mm; calyx densely hairy. Leaves trifoliate; leaflets 5–15 mm, softly hairy beneath. Pod about 3 cm, densely hairy. Thickets, sunny slopes, rocky places in the Med. region. YU.AL.GR+Is.CR.TR. Page 38.

LEMBOTROPIS   **499. *L. nigricans* (L.) Griseb. BLACK BROOM. A slender erect shrub to about 1 m, with trifoliate leaves and characteristic long terminal, leafless, spike-like clusters of numerous yellow flowers. Petals 7–10 mm, turning black when dry, keel beaked; calyx hairy. Leaflets 1–3 cm, obovate to linear, paler beneath. Pod 2–3·5 cm, adpressed-hairy. Bushy places. YU.AL.GR.BG.

CYTISUS | Broom   Not easy to distinguish in the field, particularly from some species of *Chamaecytisus* and *Genista*. Calyx two-lipped, the upper lip with 2 teeth, the lower lip with 3 teeth, the teeth usually much shorter than the lips. 6 species in our area.

Leaves simple

500. *C. decumbens* (Durande) Spach A spreading shrublet 10–30 cm, with slender, five-ribbed stems which have spreading hairs, simple leaves, and yellow flowers in the axils of the upper leaves. Flowers in clusters of 1–3; standard 10–14 mm. Pod 2–3 cm, with spreading hairs, becoming black. Margins of woods, grassy places, banks. YU.AL. **500a.** *C. procumbens* (Willd.) Sprengel A spreading shrublet 20–40 cm, rarely up to 80 cm. Similar to 500 but the young twigs with adpressed hairs, strongly ribbed, with 8–10 wings that are T-shaped in cross-section. Flowers solitary or 2–3, golden-yellow; standard 12–15 mm. Pod 3 cm, with adpressed hairs. Dry grasslands, open woods. YU.AL?GR.BG.

Leaves mostly trifoliate

501. *C. villosus* Pourret An erect, very leafy, hairy shrub to 2 m, with yellow flowers arising singly, or in groups of 2–3, and forming a terminal leafy spike-like cluster. Flowers rather large; standard 15–18 mm, veined with red at the base; flower stalks hairy. Leaflets elliptic 1·5–3 cm, the central leaflet longer, all with long shaggy hairs beneath. Woods, scrub, preferring acid soils. YU.AL.GR. **505.** *C. scoparius* (L.) Link BROOM. Woods, scrub. YU.

CHAMAECYTISUS   A difficult genus not easily distinguished from *Cytisus* and *Genista*. Calyx tubular, two-lipped, the upper lip with 2 teeth, the lower with 3 teeth, the teeth much smaller than the lips (upper lip deeply divided in *Genista*). About 24 species in our area, including 9 which are restricted to it. The following are the most distinctive or widespread species:

Flowers in leafy elongated clusters, sometimes solitary
1. Older branches becoming spiny.
    2. Young branches hairless; fr.
        white-haired (CR.).   **506a.** *C. creticus* (Boiss. & Heldr.) Rothm.
    2. Young branches with adpressed
        hairs; fr. with adpressed grey
        hairs (CR.).   **506b.** *C.subidaeus* (Gand.)Rothm. Pl. 20.
1. Older branches not becoming spiny.
    3. Petals purple (YU.AL.).   **(506).** *C. purpureus* (Scop.) Link
    3. Petals yellow.
        4. Branches erect; fr. sparsely to
            densely hairy.
            5. Leaflets usually 6–20 by 4–10 mm.   *hirsutus*
            5. Leaflets usually 20–30 by 10–15 mm.   *ciliatus*
        4. Branches spreading; fr. with long,
            weak hairs.   *polytrichus*

506. *C. hirsutus* (L.) Link HAIRY BROOM. A usually erect hairy plant 20–100 cm, with few-flowered clusters of large yellow, usually dark-blotched flowers, ranged along the previous year's branches. Flowers 2–2·5 cm; calyx hairy. Leaves trifoliate; leaflets obovate to elliptic 6–20 mm, usually hairy above and densely so below. A very variable plant. Woods, bushy places, in mountains. Widespread on mainland. **506c. *C. ciliatus* (Wahlenb.) Rothm. Distinguished from 506 by its larger leaflets 2–3 cm, and its yellow flowers without brown spots. Bushy places. YU.AL.GR.BG. **506d.** *C. polytrichus* (Bieb.) Rothm. Distinguished by its prostrate stems and yellow flowers with brown spots on the standard. Calyx with spreading hairs. Leaflets 1–1·5 cm, with spreading hairs above and below. Hills, mountains. YU.GR.BG.

Flowers in dense heads surrounded by an involucre of leaves
(a) Leaflets hairless or sparsely hairy above
507. *C. supinus* (L.) Link CLUSTERED BROOM. A spreading hairy shrub 50–100 cm, with trifoliate leaves, and a terminal head of 2–8 yellow flowers, usually spotted with brown. Petals 2–2·5 cm; calyx with long dense spreading hairs. Upper leaves surrounding flower head; leaflets large 1·5–3·5 cm. Shrubby hillsides. Widespread on mainland. Pl. 20. **507h. *C. heuffelii* (Wierzb.) Rothm. An erect shrub to 50 cm, with adpressed-hairy branches, and yellow flowers in heads of 2–8. Flowers 1·5–2·5 cm; standard sparsely hairy outside; calyx adpressed-hairy. Leaflets 1·3–3 cm, oblong to obovate, or lanceolate, hairless or sparsely adpressed-hairy above, usually densely so beneath. Pod silvery-silky. Sandy, grassy places. YU.AL.GR.BG. Pl. 20.

(b) Leaflets with adpressed hairs or silky-haired above and beneath
507e. *C. eriocarpus* (Boiss.) Rothm. Distinguished from 507 *C. supinus* by its smaller elliptic to lanceolate leaflets 1–1·7 cm, which are densely adpressed-silky or woolly on both sides. Flowers yellow, about 2 cm; calyx with spreading hairs. Bushy places. YU.GR.BG. Page 288. **507f.** *C. austriacus* (L.) Link An erect or prostrate shrub 15–70 cm, with densely adpressed-hairy leaves, and usually numerous deep-yellow flowers in dense heads. Standard 1·5–2·2 cm, silky-haired; calyx with spreading hairs. Leaflets 1–2·5 cm, variable, oblong, obovate, or lanceolate, more or less densely adpressed-hairy above, whitish-silky beneath. Pod 2–3 cm, silky-haired. Bushy places,

among herbaceous vegetation. YU.AL.GR.BG.TR? **507d.** *C. albus* (Hacq.) Rothm. Distinguished by its white flowers in heads of 5–8, with standard 1·6–2 cm. Leaflets 2–3 cm, oblong-obovate, with adpressed hairs on both sides, or only beneath. Branches erect or ascending, with both spreading and adpressed hairs. Pod 2–3 cm, with adpressed silky hairs. Sunny hills. YU.AL.GR.BG.TR.

507g. *C. banaticus* (Griseb. & Schenk) Rothm. Like 507d but flowers pale yellow; calyx adpressed-hairy (hairs semi-spreading in 507d). Leaflets 2–3 cm, obovate to lanceolate, always adpressed-hairy on both sides. Pod 3–4 cm. Sunny hills. YU.BG. Pl. 21. **507i.** *C. danubialis* (Velen.) Rothm. Distinguished by its linear, silvery-hairy leaflets, 2 cm by 2–3 mm. Flowers pale yellow, about 2 cm, minutely hairy outside; calyx with spreading hairs. Plant 40–70 cm, much branched, branches silvery-silky-haired. Pod 2 cm, densely woolly. Grassy hills. BG.

TELINE 507b. *T. monspessulana* (L.) C. Koch A much-branched, erect leafy shrub 1–3 m, with numerous, stalked trifoliate leaves and dense axillary clusters of rather small yellow flowers. Standard 10–12 mm, longer than keel and wings, hairless like the wings; keel sparsely silky-haired; calyx densely silky-haired. Leaves 8–20 mm, stalked; leaflets obovate, hairy like the young branches. Pod about 2 cm, densely silky-haired. Woods and scrub in the Med. region. YU.GR.TR?

GENISTA Distinguished by the calyx which is two-lipped with the upper lip deeply two-lobed and the lower lip with 3 teeth. Flowers yellow. Plants of the genus are often not easy to identify in the field because they show considerable variability. About 21 species occur in our area, 5 of which are restricted to it.

Shrubs or shrublets without spines
(a) *Leaves simple*
  (i) *Petals hairy*
508. *G. pilosa* L. HAIRY GREENWEED. A small, much-branched shrub to 150 cm, with slender, ribbed branches, simple narrow-oval leaves, and lax leafy clusters of small yellow flowers. Flowers 1–2 in the axils of each bract; standard 8–10 mm; calyx, standard, and keel with dense silvery hairs. Pod 1·5–2·5 cm, covered with dense adpressed hairs. Heaths, forests, rocky places, often in mountains. YU.AL.BG. **508g.** *G. carinalis* Griseb. A slender spreading shrub to 20 cm, with ribbed branches, simple linear-oblong leaves, and yellow flowers in terminal clusters. Standard 5–7 mm, hairless; keel longer, sparsely hairy. Pod ovoid-pointed, with 1–2 seeds. Mountain woods. YU.GR.BG.TR. Page 136.

508h. *G. albida* Willd. A spreading shrub with young branches and leaves with short hairs, and flowers borne singly, or in pairs, in axils of each bract, forming dense clusters. Standard 9–12 mm, equalling keel, silky-haired. Leaves 3–10 mm, elliptic to obovate, sometimes hairless above. Rocky ground. YU?GR.BG. Page 88. The plant illustrated is *G. halacsyi* Heldr. from S. Greece, and it is probably a form of 508h. **508i.** *G. sericea* Wulfen A much-branched tufted shrub with narrow leaves with inrolled margins, silky-haired beneath and with small bright-yellow flowers with silky hairs. Petals 10–14 mm; flowers solitary in leaf axils but with 2–5 at the ends of the branches; calyx silvery-hairy. Leaves 5–25 mm, narrow-elliptic to obovate, hairless above. Rocks in mountains. YU.AL.GR? Pl. 20.

  (ii) *Petals hairless*
509. *G. tinctoria* L. DYER'S GREENWEED. An erect or spreading variable shrub to about 1 m or more, with straight, green, ribbed branches, and often numerous yellow flowers in

long, simple or branched, somewhat leafy terminal clusters. Leaves simple, variable in shape, oblong to ovate, 1–5 cm, hairless or densely silky. Petals equal 8–15 mm, hairless; calyx hairless or silky. Pod 2·5–3 cm, hairless or silky. Woods, grassy places. Widespread on mainland. **509a.** *G. januensis* Viv. A spreading or erect shrub to 50 cm, usually with a three-winged stem and branches. Flowers in short lateral clusters; petals 9–10 mm, hairless; calyx almost hairless. Leaves elliptic to obovate, 5–40 mm, all hairless, with a narrow, colourless, obscurely toothed margin. Sunny slopes, bushy places on limestone. YU.AL.GR.BG. Pl. 20. **509b.** *G. lydia* Boiss. A spreading or erect shrub to 1 m, somewhat similar to 509 G. *tinctoria* but leaves on main stem smaller, narrower, linear-oblanceolate, 3–10 mm, nearly hairless. Scree slopes. YU.GR.BG.TR.

(b) Leaves trifoliate

510. *G. radiata* (L.) Scop. RAYED BROOM. A densely branched, tufted, apparently leafless, green, switch-like shrub to 50 cm, with a dense terminal cluster of 4–12 yellow flowers. Standard 8–14 mm, hairless or with a central row of silky hairs. Leaflets 5–20 mm, oblanceolate, almost hairless above, silky beneath, soon falling and leaving tough, thickened leaf stalks; branches strongly furrowed. Pod silky-haired, beaked. Rocks in limestone hills. YU.AL.GR. **510a.** *G. sessilifolia* DC. An erect silvery-green shrub 20–50 cm, with a few whippy branches arising from near the base, and trifoliate leaves with linear leaflets, with little or no leaf stalk. Leaflets 5–25 mm, often inrolled, silky-haired. Flowers yellow, in long lax silky terminal clusters; calyx, standard, and keel densely silky-haired. Standard 7–10 mm, silky, about half as long as keel. Limestone hills. YU(south),BG. Page 149.

Spiny shrubs

510b. *G. acanthoclada* DC. An erect spiny shrub to about 50 cm, with dense opposite branches ending in a spine, the older branches with prominent swollen bases to the leaf stalks. Standard 6–10 mm, silky; calyx silky. Leaves trifoliate; leaflets 0·5–1 cm, narrowly oblanceolate. Pod with 1–2 seeds. Stony hillsides. GR+Is.CR. Page 288. **510c.** *G. anatolica* Boiss. A spreading or erect shrub 10–15 cm, readily distinguished by the short terminal cluster of flowers ending in a spine. Standard about 8 mm, triangular, notched, hairless, shorter than keel; calyx hairy. Leaves simple, 5–10 mm, narrowly elliptic, silky or hairy beneath. Pod ovoid pointed, one-seeded. Dry hill slopes. BG.TR. **511g.** *G. parnassica* Halácsy A low densely spiny bushy shrub, distinguished by its silvery-silky simple leaves, and its silky-haired standard 9–13 mm. Branches ribbed, ending in stout spines, young branches silky. Rocks, stony places GR(Parnassos). Pl. 21.

CHAMAESPARTIUM **513.** *C. sagittale* (L.) P. Gibbs WINGED BROOM. A low matted, woody-based, dwarf shrub 10–50 cm, which is unmistakable with its broadly winged and flattened green stems and branches, and its dense terminal cluster of yellow flowers. Flowers about 1 cm; calyx silvery-haired. Leaves simple, elliptic, 5–20 mm. Meadows, rocky places in mountains. YU.AL.GR+Is.BG.

GONOCYTISUS Calyx two-lipped with upper lip divided to the base, with asymmetrical teeth. **513e.** *G. angulatus* (L.) Spach A shrub to 5 m, with trifoliate leaves and short terminal leafless clusters of yellow flowers each 4–6 mm. Leaflets 4–22 mm, oblong-ovate, adpressed-hairy. Pod 1–1·5 cm, hairy. Hillsides. TR.

SPARTIUM **515.** *S. junceum* L. SPANISH BROOM. Unmistakable with its erect, glaucous, rush-like, flexible, leafless stems 1–3 m, and its showy, sweet-scented, rich-yellow flowers. Flowers 2–2·5 cm, in lax terminal spikes. Leaves narrow, soon falling. Pod flattened, 5–8 cm by 7 mm, silky-haired, becoming hairless. Rocky, bushy places in the Med. region. Widespread. Page 34.

PETTERIA Calyx two-lipped, with upper lip divided to about two-thirds, the lower lip with 3 teeth.

515a. *P. ramentacea* (Sieber) C. Presl An erect shrub to 2 m, with large stalked trifoliate leaves, and upright, dense, terminal clusters of yellow flowers. Flowers 1·6–2 cm; standard notched; calyx with adpressed hairs. Leaflets unequal, elliptic to rounded, 2–7 cm, hairless above; twigs light brown. Pod 3·5–5 cm, light brown, hairless. Bushy places in the mountains. YU(Dalmatia, Montenegro).AL. Page 288.

ADENOCARPUS **518.** *A. complicatus* (L.) Gay An erect, straggly, spineless shrub to 4 m, with small trifoliate leaves and long terminal clusters of yellow flowers, and characteristic pods covered with raised glandular swellings. Standard 1–1·5 cm, silvery-haired; calyx with or without glandular swellings. Leaflets 0·5–2·5 cm, oblanceolate. Bushy places in mountains. GR.

LOTONONIS Calyx weakly two-lipped, the upper lip deeply four-toothed, the lower one-toothed.

518d. *L. genistoides* (Frenzl) Bentham A grey or silvery, tufted dwarf shrub to 30 cm, with trifoliate leaves and pale-yellow flowers. Flowers 8–9 mm, in clusters of 1–3, sometimes slightly reddish. Leaflets oblong-lanceolate, 5–8 mm. Pod 6–10 mm, hairy, little longer than the persisting calyx. Dry slopes in pine forests. GR(Thrace).BG (south-east).

LUPINUS | Lupin

Flowers yellow
519. *L. luteus* L. YELLOW LUPIN. Cultivated for fodder; naturalized in YU.BG.

Flowers white to blue
(a) *Upper lip of calyx shallowly two-toothed; seeds 8–14 mm*
521. *L. albus* L. WHITE LUPIN. A hairy annual to 120 cm, with a few-flowered spike of white or deep-blue flowers. Flowers alternate; petals 15–16 mm. Pod yellow, 6–10 cm. Disturbed ground; cultivated for its edible seeds and for fodder. Widespread.

(b) *Upper lip of calyx deeply two-cleft; seeds 4–10 mm*
520. *L. angustifolius* L. NARROW-LEAVED LUPIN. Distinguished by its narrow leaflets which are hairless above, and its spike of rather small dark-blue flowers each 11–13 mm, which are arranged alternately. An annual 20–80 cm; leaflets 2–5 mm wide. Cultivated ground, rocks in the Med. region. YU?GR+Is.CR.BG.TR. **(520).** *L. micranthus* Guss. HAIRY LUPIN. Distinguished from 520 by its obovate leaflets and its blue flowers with the standard white at the centre, and the tip of the keel blackish-violet. Flowers 10–14 mm. Plant covered in brown spreading hairs. Cultivated ground, stony places on acid soils in the Med. region. Widespread, except BG. **520a.** *L. varius* L. Like (520) with obovate leaflets, but blue flowers larger 15–17 mm, irregularly whorled, and standard blue with a white and yellow or pale-purple blotch. Stems and leaves silvery, or white with shaggy hairs. Rocky places in the Med. region. YU.AL.GR+Is.CR. Pl. 21.

ARGYROLOBIUM **522.** *A. zanonii* (Turra) P. W. Ball A silvery-grey shrublet to 25 cm, with solitary or clusters of 2–3 terminal golden-yellow flowers, and trifoliate

1. *Astragalus parnassi* 528g   **2.** *Astragalus baldaccii* 526b   **3.** *Petteria ramentacea* 515a   **4.** *Genista acanthoclada* 510b   **5.** *Chamaecytisus eriocarpus* 507e

leaves with narrow elliptical leaflets which are densely silvery-haired beneath. Flowers 9–12 mm. Pod 1·5–3·5 cm, silvery-haired. Rocky places in the Med. region. YU.AL.

ROBINIA **523. *R. pseudacacia* L. FALSE ACACIA. A well-known tree to 25 m, with pendulous clusters of white 'pea-flowers', and large pinnate leaves with 3–10 pairs of oval leaflets. Trunk greyish, deeply furrowed; sucker shoots spiny. Pod 5–10 cm by 1 cm, hairless. Native of N. America; widely grown for ornament and soil-stabilization on the mainland, and often naturalized by roadsides and waste places.

GALEGA **524. *G. officinalis* L. GOAT'S RUE, FRENCH LILAC. A robust, nearly hairless perennial to 1·5 m, with many-flowered axillary clusters of white, lilac, or purple flowers. Flowers 1–1·5 cm; calyx mostly hairless. Leaves pinnate with 4–10 pairs of elliptic to lanceolate leaflets 1·5–5 cm. Pod 2–5 cm, beaked. Damp meadows, by rivers. Widespread on mainland.

COLUTEA **525. *C. arborescens* L. BLADDER SENNA. A hairless much-branched shrub to 6 m, distinguished by its inflated bladder-like pods 5–7 cm, which become brittle and parchment-like. Flowers yellow, 1·5–2 cm, in pendulous clusters of 3–8. Leaves pinnate with 4–5 pairs of broadly elliptical leaflets to 3 cm. Bushy places. YU.AL.GR.BG.TR? **525b.** *C. cilicica* Boiss. & Balansa Like 525 but flowers larger to 22 mm, and wings distinctly longer than keel, and with a distinct spur on their lower edges. Ovary quite hairless. In mountains, open woods. GR(Thrace).TR. (Intermediates between 525 and 525b occur in central Greece.) **525c.** *C. orientalis* Miller Distinguished by its orange-red flowers, with beaked keels, and pods which are curved upwards at the apex. A native of the Caucasus; grown for ornament and sometimes naturalized in YU.

CARAGANA Calyx tubular or bell-shaped with 5 more or less equal teeth. Pod linear.

525d. *C. frutex* (L.) C. Koch A dwarf shrub to 1 m, with pinnate leaves with 2 pairs of obovate spine-tipped leaflets. Flowers yellow, 1·5–2·5 cm, solitary or two- to three-clustered in leaf axils. Pod linear 2·5–4·5 cm. Bushy places. BG(east).

ASTRAGALUS | Milk-Vetch An important but difficult genus with 63 species in our area, including 29 which are restricted to it.

Spiny plants; leaves ending in a spine
(a) Calyx with dense spreading woolly hairs
528a. *A. sempervirens* Lam. MOUNTAIN TRAGACANTHA. A lax, tufted woody shrublet 5–40 cm, with leaves ending in a spine and with 4–10 pairs of linear-oblanceolate leaflets, and clusters of white to purple flowers. Flowers 3–8, in short-stalked clusters; standard 1–2 cm; calyx woolly 7–15 mm, slightly inflated in fruit, with teeth as long as the tube. Stipules joined to the leaf stalk for about half their length. Alpine rocks. GR. **528f.** *A. creticus* Lam. A perennial, woody at the base, forming large hemispherical tussocks, with spine-tipped leaves with 5–7 pairs of narrow leaflets. Flowers paired, white or pinkish, in the axils of the leaves, forming a dense cluster; calyx 6–10 mm, densely white-woolly with teeth nearly as long as the tube. Bracteoles absent. Mountains. YU? AL.GR.CR.

528g. *A. parnassi* Boiss. A woody-based plant forming large, hemispherical tussocks, with spine-tipped leaves with 5–15 pairs of linear-lanceolate to elliptical leaflets, and purplish or pink, or rarely yellow, flowers. Flowers stalkless, 3–4 in the leaf axils; calyx 10–12 mm, densely woolly-haired. Bracteoles about as long as calyx. Mountain rocks.

YU.AL.GR.TR. Page 288. **528h.** *A thracicus* Griseb. Like 528g but a more robust erect plant, not tussock-forming, and calyx large 15 mm or more. Flowers white or purplish, up to 7 in each axillary cluster. Leaves to 9 cm, spine-tipped; stipules conspicuously woolly-haired. Rocky places, fields. YU.GR.BG.TR, Pl. 21.

(b) Calyx with hairs adpressed
528i. *A. angustifolius* Lam. A dense, woody, very spiny, cushion plant having leaves with 6–10 pairs of tiny narrow-elliptical to linear leaflets, the terminal leaflet present on young leaves but soon falling and leaving a spine. Flowering stalks shorter or slightly longer than the leaves and bearing ovoid clusters of 3–12 white flowers. Standard 1·3–2·3 cm, keel 8–15 mm, sometimes purplish; calyx 5–9 mm, with adpressed hairs, and teeth one-quarter to half as long as the tube. Leaflets 2–7 mm, with dense adpressed hairs. Pod 1–1·5 cm. Mountain rocks. YU.AL.GR.CR.BG. Page 70.

Plants not spiny; leaves with an odd terminal leaflet
(a) Annuals or biennials
(527). *A. hamosus* L. A green- or greyish-hairy, sprawling annual up to 60 cm, readily distinguished in fruit by its stalked clusters of 5–14, wide-spreading but sharply upcurved, sickle-shaped pods 2–5 cm long. Flowers white or pale yellow, in short-stalked globular clusters, on stems half as long as the leaves; standard 7–8 mm; calyx 5–6 mm. Leaflets oblong-obovate, notched, 9–11 pairs. Dry grassland. Widespread. **527e.** *A. haarbachii* Boiss. A robust, spreading or ascending annual, biennial or perennial 5–50 cm, with leaves with 8–12 pairs of oblong to ovate, notched leaflets. Flowers yellowish, in dense clusters of 7–18, borne on stalks as long or slightly longer than the leaves. Standard 2–2·5 cm; calyx 10–13 mm, with teeth two-thirds as long as the tube. Pod oblong, woolly, 2·5–3 cm, beak straight. Grassland, stony places. GR.CR.BG.

527j. *A. boeticus* L. An erect annual to 60 cm, with pinnate leaves with 10–15 pairs of narrow-oblong to oblong-obovate leaflets, blunt or notched. Flowers yellow, in rather dense clusters of 5–15, on stalks half as long as the leaves. Standard 1·2–1·4 cm; calyx 5–7 mm, with teeth as long as the tube. Pod oblong, adpressed-hairy, 2–4 cm, beak hooked. Arable fields. YU.GR.CR+Karpathos,Rhodes.

(b) Perennials (see also 527e)
  (i) Plants with well-developed stems
    (x) Flowers yellow to cream-coloured (see also (532))
527. *A. glycyphyllos* L. MILK VETCH. Bushy places. Widespread on mainland. **527f. *A. ponticus* Pallas A robust, erect perennial 50–100 cm, with several large ovoid clusters of handsome yellow flowers in the axils of the upper leaves. Flower clusters about 5 cm long. Standard 1·5–2·2 cm; calyx pale, 1–1·5 cm; finely hairy, with teeth less than half as long as the tube. Leaves 10–30 cm, with 15–25 pairs of oblong-elliptic leaflets, hairless except on the mid-vein below. Pod included in the slightly inflated calyx. Dry meadows. BG. Pl. 21. **527g.** *A. centralpinus* Br.-Bl. Like 527f but leaflets 20–30 pairs, elliptic to ovate-lanceolate, hairless above, sparsely hairy below. Flowers yellow, in stalkless clusters (stalk to 1 cm in 527f). Standard 1·5–2 cm; calyx 1·4–2 cm, with long dense hairs and teeth two-thirds as long as the tube. GR.BG (Rhodope).

527i. *A. graecus* Boiss. & Spruner A robust, erect perennial 30–40 cm, with clusters of 5–10 large yellow flowers, on stalks to 2 cm. Standard 3–4 cm; calyx 13–20 mm, woolly, with teeth half as long as the tube. Leaflets small, 20–35 pairs, ovate or cordate-orbicular, hairless above, woolly below. Pod 2·5–3 cm, oblong-ovate, hairy. Fields, fallow lands. GR(south and central)+Rhodes, Samos. ****528.** *A. cicer* L. WILD LENTIL A softly hairy, spreading perennial up to 1 m, with pale-yellow flowers in dense clusters on long stalks shorter than the leaves. Standard 1·4–1·6 cm; calyx 7–10 mm, with adpressed black hairs. Leaves with 8–15 pairs of lanceolate to ovate-lanceolate, hairy leaflets. Pod ovoid, inflated, 1–1·5 cm, densely hairy. Grassy places. YU.BG.

(xx) Flowers pink, purple, blue

(532). *A. onobrychis* L. SAINFOIN MILK-VETCH. A robust leafy perennial 10–60 cm, recalling Sainfoin, with dense, oval, long-stalked, terminal clusters of usually bright-violet or pink flowers, but less commonly flowers white or yellowish. Distinguished by its narrow standard 1·5–3 cm, which is much longer than the notched wings; calyx 6–8 mm; flower stalks much longer than the leaves. Leaflets 8–15 pairs, elliptic-lanceolate, with adpressed hairs. A very variable species. Dry grassland. Widespread on mainland. Pl. 21. **532a.** *A. austriacus* Jacq. A slender perennial to 60 cm, with long lax clusters of small blue or violet flowers, on flower stalks about as long as the leaves. Standard 5–8 mm; calyx 2–3 mm, with teeth about one-quarter as long as the tube. Leaflets 5–10 pairs, usually linear, almost hairless. Pod linear-oblong 0·5–1·2 cm. Dry grassland. YU.BG.

(ii) Plants with stems poorly developed or absent

526. *A. monspessulanus* L. MONTPELLIER MILK-VETCH. Readily distinguished by its rosy-purple to violet-purple flowers, in dense ovoid clusters, borne on stems which arise directly from the rootstock. Inflorescence longer than the leaves; standard 2–3 cm entire or lobed, upturned, longer than the wings; calyx 9–16 mm. Leaves with 7–20 pairs of oblong leaflets. Pod 2·5–4·5 cm, linear, slightly curved, sparsely hairy. Arid hills, rocky ground in mountains. YU.AL.GR.BG.TR? **526c. *A. spruneri* Boiss. Like 526 but leaflets usually 5–8 pairs; calyx 12–18 mm. Flowers white, pale purple, or violet. Pod 1·2–2 cm, obovoid, curved, wrinkled. Open stony places. YU(Macedonia). AL.GR+Is.,Rhodes.BG.TR.

526b. *A. baldaccii* Degen A small stemless plant with stalkless clusters of white flowers nestling amongst grey-hairy leaves. Flowers sometimes lilac-flushed; standard 2·3–2·5 cm; calyx 1–1·4 cm, pale, densely adpressed-hairy, with teeth a quarter as long as tube. Leaflets 4–9 pairs, oval to lanceolate 6–9 mm, with adpressed hairs. Mountains. Al.GR(central and north). Page 288. **526d.** *A. sericophyllus* Griseb. A dwarf tufted, woody-based perennial with small pinnate leaves and much longer flower stalks bearing a lax cluster of reddish-purple or whitish flowers. Standard 1·8–2 cm; calyx about 1 cm, with adpressed black, and spreading white, hairs. Leaves 2–3 cm; leaflets 4–6 pairs, 5–8 mm, silvery-hairy. Pod 2·5 cm, densely hairy. In mountains. YU.AL.GR. Page 169.

527h. *A. hellenicus* Boiss. A stemless, silky-leaved perennial with very short-stalked clusters of 3–10 yellow flowers. Standard 2–2·7 cm, hairless, wings shorter; calyx 1–1·5 cm, silky-haired, with teeth from one-third to half as long as the tube. Leaflets 7–20 pairs, ovate to ovate-elliptic, densely silky on both surfaces. Pod 1·5–2·2 cm, ovate, grooved, densely hairy. Mountains. GR(south). **530b.** *A. depressus* L. A very variable perennial, stemless or with short prostrate stem, and short-stalked cylindrical clusters of 6–14 whitish or bluish-purple flowers. Standard 1–1·2 cm; calyx 3–6 mm, the teeth two-thirds as long as the tube. Leaflets 6–14 pairs, obovate to obcordate, hairless above, adpressed-hairy below. Pod linear-lanceolate, 1·5–2·2 cm. Stony places in hills and mountains. YU.AL.GR.CR.BG. Page 93.

531. *A. vesicarius* L. Recognized by its strongly inflated calyx in fruit. A short-stemmed perennial, woody at the base, with often silvery-grey foliage, and terminal clusters of bicoloured flowers with the standard purplish and the wings paler or whitish. Flower stalks 2–3 times as long as the leaves. Standard 1·7–2·3 cm; calyx with dense spreading black and white hairs. Leaflets 4–10 pairs. Pod 8–15 mm, as long as calyx. Dry grassland, stony places. YU.AL.GR.BG. Page 114.

OXYTROPIS Distinguished from *Astragalus* by the keel of the flowers having a tooth at the apex on the underside. Pod oblong to ovoid. 8 similar-looking species in our area, including 3 endemic species.

Flowers white or yellow
533. *O. campestris* (L.) DC. MEADOW BEAKED MILK-VETCH. Flowers yellow, whitish, or pale violet. Alpine meadows. YU.BG. **533c. *O. urumovii* Jáv. A dense cushion-forming, silky-haired stemless perennial, with dull-yellow flowers which become reddish at the apex, in oval clusters borne on stalks about as long as the leaves. Flowers 4–15; standard 1–1·4 cm; calyx with white hairs. Leaflets 8–15 pairs, densely silky-haired. Pod 1–1·5 cm, twice as long as calyx, densely white-hairy. Rocks in hills and mountains. YU(west).AL.BG(south-west). **533d.** *O. pilosa* (L.) DC. An erect or ascending perennial 20–50 cm, with many pale-yellow flowers in an ovoid, long-stalked cluster. Standard 1·2–1·4 cm; calyx teeth as long or longer than tube. Leaflets 7–15 pairs, oblong to linear-oblong, with adpressed hairs. Pod 1·5–2 cm, white-hairy. Among herbaceous vegetation. YU.BG.

Flowers purple (see also 533)
534a. *O. prenja* (G. Beck) G. Beck A stemless, hairy perennial with dense few-flowered clusters of purple flowers on stalks about as long as the leaves. Standard 1·5–2 cm, notched; calyx with dense black and white hairs. Leaflets 6–8 pairs, oblong-ovate, with adpressed hairs. Pod about 1·5 cm, ovoid. Mountains. YU(south-west).AL. Pl. 21.

BISERRULA **533b.** *B. pelecinus* L. A small annual 10–40 cm, with clusters of small bluish or pale-yellow flowers 4–6 mm, with blue tips. Unmistakable in fruit on account of its unique pod, which looks like a two-edged saw. Leaves with 7–15 pairs of oblong, notched leaflets. Stony and grassy places in the Med. region. AL.GR + Is.CR.BG(south).

GLYCYRRHIZA ****536.** *G. glabra* L. LIQUORICE. A robust perennial 50–100 cm, with pinnate leaves and long lax stalked spikes of numerous bluish or violet flowers. Corolla 8–12 mm; calyx glandular-hairy. Leaflets 9–17, elliptic to oblong, 2–5 cm, often glandular-sticky. Pod to 3 cm, glandular or not, sometimes with bristles. Bushy places, damp ditches. Widespread on mainland + CR,Rhodes. **(536).** *G. echinata* L. SPINY-FRUITED LIQUORICE. Like 536 but distinguished by the smaller flowers 4–6 mm, which are borne stalkless or short-stalked in dense rounded heads. Fruiting heads distinctive, having globular clusters of elliptic pods, usually with dense glandular, reddish-brown spines. Muddy places by rivers. YU.GR.CR.BG.TR. Pl. 22.

AMORPHA **536b.** *A. fruticosa* L. A leafy shrub to 6 m, with pinnate leaves, and long dense terminal spikes of numerous small bluish or purplish flowers. Standard 6 mm; wings and keel absent; calyx 3 mm. Leaflets 5–12 pairs, oval to elliptic, 1·5–4 cm. Pod glandular. Native of N. America; naturalized in YU.AL.BG.

PSORALEA ****537.** *P. bituminosa* L. PITCH TREFOIL. Readily distinguished by its compact clover-like heads of blue-violet flowers, its trifoliate leaves, and, when crushed, its strong penetrating smell of tar. Flowers 1·5–2 cm. Leaflets of upper leaves lanceolate, those of lower leaves nearly rounded. Sandy and stony ground. Widespread. Page 164.

PHASEOLUS Keel of flower with a spirally coiled beak. Climbing plants.

538. *P. vulgaris* L. KIDNEY or HARICOT BEAN. Flowers white, bluish, or yellow. Pod smooth, 10–50 cm. Native of S. America; widely cultivated as a vegetable crop. **539.** *P.*

coccineus L. SCARLET RUNNER. Flowers all scarlet or with some white on wings and keel. Pod rough, 10–30 cm. Native of S. America; widely cultivated as a vegetable crop.

VIGNA Like *Phaseolus* but keel with a recurved, not spirally coiled beak.

539a. *V. unguiculata* (L.) Walpers A nearly hairless scrambling annual to 2 m, with large trifoliate leaves and clusters of white, pale-yellow, pink, or red flowers. Corolla 2–2·5 cm; calyx 7–8 mm. Pod pendulous, 1·5–3 cm, rounded in section. Native of tropical Africa; cultivated for its seed and for fodder in GR.

GLYCINE **540.** *G. max* (L). Merr. SOJA or SOYA BEAN. A rough-haired, non-climbing annual 30–200 cm, with large trifoliate leaves and inconspicuous, almost stalkless clusters of whitish or violet flowers. Corolla 6–7 mm. Pod oblong, 2·5–8 cm, constricted between seeds, pendulous, densely hairy. Origin unknown; sometimes cultivated in the Med. region for its oil and edible seeds.

CICER | **Chick-Pea** 4 species in our area, including 1 endemic species.

541. *C. arietinum* L. CHICK-PEA. A hairy annual 20–50 cm, with solitary bluish or white flowers 10–12 mm, on jointed stalks. Leaves with 13–17 oval, toothed, glandular-hairy leaflets 1–1·5 cm. Pod 2–3 cm, very inflated, glandular-hairy. Native of Asia; cultivated for its edible seed; sometimes naturalized. **541b.** *C. incisum* (Willd.) K. Malý A procumbent alpine perennial 10–30 cm, with pinnate leaves and solitary or paired violet flowers 7–10 mm. Leaflets 1–3 pairs, three- to five-lobed, 2–10 mm. Pod ellipsoid, 1 cm, glandular-hairy. Screes. GR(south).CR. **541a.** *C. montbretii* Jaub. & Spach. An erect hairy perennial 25–40 cm, distinguished by its larger white flowers 25–8 mm, with a violet spot on the standard. Calyx teeth unequal. Pod 2–5 cm. Bushy places. AL.BG(south-east).TR.

VICIA | **Vetch** Distinguished from *Lathyrus* by the usually unwinged stems. Leaves with usually more than 3 pairs of leaflets. Style hairless, or equally hairy all round, or hairy on the lower side (hairy on upper side only in *Lathyrus*). About 42 species in our area, including 5 restricted to it. The following species are most likely to be encountered:

Flower stalks absent or shorter than flowers
(a) Standard hairy on back
547a. *V. pannonica* Crantz Flowers purple or yellow, in an almost stalkless cluster of 1–4. Petals 1·4–2·2 cm; calyx teeth more or less equal, shorter than the tube. A hairy annual 10–60 cm; leaves with 4–10 pairs of oblong or linear-oblong leaflets 0·8–3 cm; stipules entire, spotted. Pod 2–3·5 cm, yellow, hairy. Cultivated fields. Widespread on mainland. **547.** *V. hybrida* L. HAIRY YELLOW VETCHLING. Flowers solitary, pale yellow or purplish, 1·8–3 cm; calyx teeth unequal. Leaves with 3–8 pairs of elliptic-oblong, notched or blunt leaflets. Pod 2·5–4 cm, brown, hairy. Grassy places, disturbed ground. Widespread. Pl. 22.

(b) Standard hairless on back
    (i) Calyx teeth unequal
1. Petals yellow, sometimes tinged
    with purple.
  2. Pod hairless, except for the margin
      (Widespread on mainland). **(547).** *V. melanops* Sibth. & Sm. Pl. 22.
  2. Pod densely hairy (Widespread). **(547).** *V. Lutea* L. Page 40
1. Petals purple.

3. Leaflets not more than 3 mm wide.   *peregrina*
3. Leaflets more than 4 mm wide.
   4. Perennials; leaflets 3–9 pairs
     (YU.GR.BG.).   **548.** *V. sepium* L.
   4. Annuals; leaflets 1–3 pairs (Wide-
     spread on mainland + Aegean Is.).   **550.** *V. narbonensis* L. Pl. 22.

549b. *V. peregrina* L. Flowers purple, solitary or paired; petals 1–1·6 cm; calyx teeth unequal, the lowest equalling the tube. An annual to 1 m, with few hairs, and leaves with 3–7 pairs of linear or oblong leaflets 0·8–3 cm, appearing three-lobed at their apices. Pod 3–4 cm, brown, hairy. Stony places, fields, cultivation. Widespread.

  (ii) Calyx teeth equal, or nearly so
1. Standard and keel yellow, sometimes
   tinged with purple.
  2. Petals 2·3–3·5 cm; wings yellow,
    sometimes with black tips.   *grandiflora*
  2. Petals 1·8–2·2 cm; wings blue.   *barbazitae*
1. Petals purple, sometimes with whitish
   wings and keel.
  3. Leaflets 1–3 pairs; wings and keel
    whitish (Widespread).   **551.** *V. bithynica* (L.) L.
  3. Leaflets of upper lvs. 3 or more
    pairs; wings and keel purplish
    (Widespread).   ****549.** *V. sativa* L.

(547). *V. grandiflora* Scop. LARGE YELLOW VETCH. Flowers large, yellow, sometimes with a purple tinge and wings sometimes black at tip, 1–4 on a short stalk. Calyx teeth equal, shorter than the tube. A hairy annual 30–60 cm; leaves with 3–7 pairs of linear to rounded leaflets 1–2 cm. Pod 3–5 cm, black, hairy. Meadows, fields, stony places. Widespread on mainland. **547b.** *V. barbazitae* Ten. & Guss. Like (547) *V. grandiflora* but flowers yellow with blue wings; calyx teeth about as long as the tube; pod brown, glandular, hairless at the ends. Mountain woods. YU.GR.BG.

Flower stalks longer than flowers
(a) Calyx teeth equal, all equalling or longer than calyx tube
542. *V. hirsuta* (L.). S. F. Gray HAIRY TARE Flowers tiny 4–5 mm, pale violet or whitish. Leaves with tendrils. Pod not constricted between seeds. Grassy places. Widespread on mainland. **(542).** *V. ervilia* (L.) Willd. Differs from 542 having larger flowers 6–9 mm, white tinged with red or purple. Leaves without tendrils. Pod constricted between seeds. Disturbed ground. Widespread.

(b) Calyx teeth unequal, at least the upper shorter than the calyx tube
  (i) Stipules toothed
1. Leaflets 2–3 pairs; fls. in
   clusters of 1–3 (Widespread).   **551.** *V. bithynica* (L.) L.
1. Leaflets 3 or more pairs; fls. in
   clusters of 4–20.
  2. Calyx tube pouched at the
    base (GR.).   ****(545).** *V. benghalensis* L.
  2. Calyx tube not pouched.
  3. Stipules crescent-shaped.   *dumetorum*
  3. Stipules half arrow-shaped
    (Widespread on mainland).   ****(544).** *V. onobrychioides* L.

544a. *V. dumetorum* L. Flowers blue or purple, in clusters of 2–14; petals 1·2–2 cm. Calyx teeth unequal, shorter than the tube. Leaves with 3–5 pairs of ovate leaflets, 1·2–4 cm. Pod 2·5–6 cm, brown, hairless. Woods, scrub. YU.GR.BG.

(ii) *Stipules entire*

1.   Calyx strongly pouched at the base.
  2.   Petals reddish-purple (GR.).   **(545). *V. benghalensis* L.
  2.   Petals violet or purple
       (Widespread).   **(545). *V. villosa* Roth
1.   Calyx not or slightly pouched
    at the base.
  3.   Petals 1·8 cm or more
       (Widespread on mainland).   **(544). *V. onobrychioides* L.
  3.   Petals less than 1·8 cm.
    4.   Plants with long shaggy
       hairs.   *sibthorpii*
    4.   Plants hairless, or shortly
       hairy.
      5.   Pod 3–6 mm wide.
        6.   Petals 0·4–1·3 cm.
          7.   Stem hairless or with
            adpressed hairs (Widespread
            on mainland).   ****545.** *V. cracca* L.
            (Widespread)   **(542).** *V. terasperma* (L.) Schreber
         7.   Stems with dense,
            spreading hairs.   *incana*
        6.   Petals 1·2–1·8 cm.
          8.   Leaflets 2–7 mm wide
            (YU.AL.GR? CR.BG.TR.).   **(545).** *V. tenuifolia* Roth
                                  pinetorum
        8.   Leaflets 1–2 mm wide.   *dalmatica*
      5.   Pods 6–8 mm wide.   *cassubica*

545a. *V. incana* Gouan Like 545 *V. cracca* but stems with dense spreading hairs, and leaves with 10–22 pairs of leaflets. Flowers bluish-violet, 8–12 mm, in clusters of 20–40. Bushy places in mountains. YU.AL.GR.BG.   **545b.** *V. dalmatica* A. Kerner Distinguished from 545 *V. cracca* by the larger bluish-violet flowers 1·2–1·8 cm in clusters of 8–20, and by the limb of the standard longer than the stalk (claw). Leaflets 5–13 pairs, linear or bristle-like, 1–2 mm wide. Grassy places. Widespread, except TR. Page 302.

545c. *V. sibthorpii* Boiss. Flowers bluish-purple, sometimes with white wings, in clusters of 8–25. Petals 1·2–1·5 cm, limb of standard equal or longer than the stalk (claw). A densely hairy perennial to 1 m, with leaves with 6–12 pairs of elliptical leaflets 5–20 mm long. Pod 1·2–2 cm, brown, hairy. Cultivated fields, grassland. YU? GR+Is.CR.   **545d.** *V. cassubica* L. Flowers purple or blue with whitish wings and keel, in clusters of 4–15. Petals 1–1·3 cm; calyx teeth unequal, shorter than the tube. A more or less hairless perennial to 1 m, with 5–16 pairs of linear-lanceolate to elliptical leaflets 7–30 mm. Pod 1·5–3 cm, yellow, hairless. Woods. Widespread on mainland.   **545e.** *V. pinetorum* Boiss. & Spruner Flowers yellow in long-stalked clusters of 8–20; petals 1·2–1·9 cm. Leaves without tendrils, with 10–16 pairs of elliptic-lanceolate leaflets, 1–2 cm, each with a fine point. Pod 1·5–1·8 cm, dark brown, sparsely hairy. Mountain woods. GR.CR. Pl. 22.

LENS | Lentil Like *Vicia* but seeds flat; calyx teeth at least twice as long as tube; plants not climbing.

1. Stipules half spear-shaped or toothed
   2. Pod hairless; fl. stalk ending in a fine point (YU.AL? GR.CR.).   **(552).** *L. nigricans* (Bieb.)Godron
   2. Pod finely hairy; fl. stalk not ending in fine point (Widespread).   **552a.** *L. ervoides* (Brign.)Grande
1. Stipules oblong-lanceolate, entire.
   3. Pod 1·2–1·6 by 0·6–1·2 cm; fl. clusters about equalling the lvs.   *culinaris*
   3. Pod 0·7–1·1 by 0·4–0·6 cm; fl. clusters longer than the lvs.(GR(south)).   **552b.** *L. orientalis* (Boiss.)M.Popov

**552. *L. culinaris* Medicus LENTIL. Distinguished by its few, tiny whitish flowers about 5 mm, borne in the axils of the leaves, and its flattened almost rectangular pods. Leaflets 3–8 pairs, entire, tendril usually present. Origin not known; widely cultivated in the south for its seeds; sometimes naturalized.

LATHYRUS | Pea, Vetch Distinguished from *Vicia* by its often winged stem, its 1–2 pairs or few pairs of parallel-veined leaflets, and by its style which is usually hairy on the upper side only, but the distinction between these two genera is not always clear. About 45 species, mostly widespread, including 6 restricted to our area. The following 32 species are most frequently seen in our area:

Flowers yellow, cream, brown, or orange
(a) Stems winged
554. *L. ochrus* (L.) DC. WINGED VETCHLING. Cornfields, dry places. Widespread, except BG. **(555). *L. annuus* L. ANNUAL YELLOW VETCHLING. Fields, waysides, waste ground in the Med. region. Widespread.

(b) Stems not winged
553. *L. aphaca* L. YELLOW VETCHLING. Fields, dry places. Widespread. **555. *L. pratensis* L. MEADOW VETCHLING. Grassy places, waysides. Widespread on mainland. **565a.** *L. laevigatus* (Waldst. & Kit.) Gren. A robust perennial 20–60 cm, with long-stalked clusters of rather striking pale-yellow flowers which turn to an ochre colour as they mature. Flowers 1·5–2·5 cm, 2–20 in a cluster, pendulous and out-turned. Leaves without tendrils; leaflets large, 3–10 cm, 2–6 pairs, pinnately veined. Pod hairless. Mountain pastures and woods. YU. Pl. 23. **565e.** *L. aureus* (Steven) Brandza A sparsely hairy perennial, readily distinguished by its large ovate leaflets 2·5–5 cm wide with brownish glands on the underside. Petals brown- or orange-yellow, 1·7–2·2 cm. Leaflets 3–6 pairs. Young pods glandular. Meadows. Black Sea region. GR? BG. **565g.** *L. pallescens* (Bieb.) C. Koch Flowers pale cream with reddish or purplish tinge. Petals 1·7–2·2 cm; calyx hairy. Leaflets 1–3 pairs. Pod 3–6·5 cm, hairless. Meadows. YU.BG.

Flowers red, purple, blue, or two-coloured (rarely yellowish)
(a) Stems not winged
  (i) Some leaves with tendrils
1. Petals 2·5–3 cm; pod 6–9 cm.   *grandiflorus*
1. Petals less than 2·5 cm; pod less than 6 cm.
   2. Fl. clusters two- to many-flowered; calyx teeth more or less distinctly unequal.

3. Leaflets 1 pair (YU.AL.GR.BG.).   **559. *L. tuberosus* L.
3. Leaflets 2–5 pairs (YU.AL.BG.TR.).   (562). *L. palustris* L.
2. Fls. solitary (rarely 2); calyx
    teeth usually equal.
   4. Fl. stalks 2–5 mm (Widespread on
      mainland).   **557d.** *L. inconspicuus* L.
   4. Fl. stalks 5–70 mm.
    5. Pod 7–11 mm wide, downy when
      young (Widespread, except TR.).   (557). *L. setifolius* L.
    5. Pod 3–7 mm wide, hairless.
     6. Fl. stalks 0·5–2 cm; pod 4–7 mm
       wide, with prominent longitudinal
       veins.   *sphaericus*
     6. Fl. stalks 2–7 cm; pod 3–4 mm wide,
       with indistinct network of veins.   *angulatus*

559a. *L. grandiflorus* Sibth. & Sm. A robust climbing perennial with very large and striking bright rosy-purple flowers with a darker purple keel, in clusters of 1–4; petals 2·5–3 cm. Leaflets usually 1 pair, ovate 2·5–5 cm. Pod 6–9 cm, brown, hairless. Shady places in mountains. YU(Macedonia).AL.GR.CR.BG. Pl. 23.   **557a.** *L. sphaericus* Retz. A slender annual 10–50 cm, with very narrow, pointed paired leaflets, and small solitary, short-stalked, orange-red flowers, usually 6–13 mm. Flower stalks 0·5–2 cm, ending in an awn; calyx teeth equal. Leaflets 2–10 cm; stipules linear. Pod brown, with prominent longitudinal veins. Vineyards, fields, among crops. Widespread.   **557b.** *L. angulatus* L. Like 557a but flowers purple or pale blue, on longer stalks, 2–7 cm. Cornfields, fields, uncultivated places in the Med. region. YU.GR.CR.

(ii) All leaves without tendrils
1. Leaflets absent (Widespread on
    mainland).   **556. *L. nissolia* L.
1. Lvs. with 2 or more leaflets
   2. All lvs. with 1 pair of leaflets   *laxiflorus*
     (Widespread on mainland).   **557d.** *L. inconspicuus* L.
   2. At least some lvs. with more than
     one pair of leaflets.
    3. Fls. solitary, with stalks 2–10 mm;
      calyx teeth more or less
      equal.   *saxatilis*
    3. Fls. in clusters, rarely solitary
      and then with stalks more than 1
      cm and calyx teeth unequal.
     4. Leaflets pinnate-veined or very
       feebly parallel-veined, the
       lateral veins much weaker than
       the midvein.
      5. Pod covered with brown glands;
       fl. clusters with 6–30 fls.   *venetus*
      5. Pod without glands; fl. clusters
       with 1–10 fls.
       6. Leaflets fine-pointed; stipules
        1–2·5 cm (YU.AL.GR.BG.).   **564. *L. vernus* (L.) Bernh.
       6. Leaflets blunt or almost acute;

stipules 4–10 mm (Widespread
on mainland).   **563.** *L. niger* (L.) Bernh.

4. Leaflets parallel-veined, the
lateral veins reaching nearly
to the apex of the leaflets.

   7. Leaflets 1–2 pairs; roots spindle-
shaped.   *digitatus*

   7. Leaflets 2–4 pairs; roots slender.

     8. Leaflets 5–11 mm wide,
usually less than ten
times as long as wide.   *alpestris*

     8. Leaflets 2–6 mm wide,
always more than ten
times as long as wide
(YU.AL.).   **565d.** *L. bauhinii* Genty

557c. *L. saxatilis* (Vent.) Vis. A hairy annual 10–30 cm, with solitary pale-blue or yellowish flowers, 6–9 mm. Leaflets 1–3 pairs, those of the lower leaves obcordate, those of the upper leaves linear. Hills, waysides in the Med. region. YU.GR+Is.CR. **557e.** *L. laxiflorus* (Desf.) O. Kuntze Readily distinguished perennial 20–50 cm, with only one pair of lanceolate to rounded leaflets, 2–4 cm. Clusters two- to six-flowered; petals 1·5–2 cm, blue-violet. Pod 3–4 cm, hairy. Mountain woods. Widespread. Page 302.

564a. *L. venetus* (Miller) Wohlf. Like 564 *L. vernus* but leaflets ovate-orbicular, pointed; stipules ovate-orbicular. Flowers reddish-purple, 1–1·5 cm, in clusters of 6–30 flowers. In fruit readily distinguished by the brown glands covering the pod. Woods. YU.AL.GR.BG. **565f.** *L. digitatus* (Bieb.) Fiori A hairless perennial 10–40 cm with fleshy, spindle-shaped roots, and distinctive almost digitate leaves, with 1–2 pairs of linear-pointed leaflets. Flowers bright reddish-purple 1·5–3 cm, in long-stalked clusters of 4–10. Pod 4–7 cm, brown, hairless. Widespread on mainland. Pl. 23. **565h.** *L. alpestris* (Waldst. & Kit.) Čelak. A hairy or hairless perennial 15–60 cm, with stem sometimes narrowly winged at the apex, and reddish-purple flowers in clusters of 3–6. Petals 12–16 mm. Leaves with 2–3 pairs of linear to elliptic leaflets. Pod 3–4 cm, brown, hairless. Mountain woods. YU(Macedonia).AL.GR(Macedonia).BG. Page 137.

(b) Stems winged
   (i) Lowest leaves without leaflets
561. *L. clymenum* L. Flowers crimson with violet or lilac wings. Upper leaves with 2–4 pairs of leaflets usually 6–11 mm wide. Grassland, stony ground in the Med. region. Widespread, except BG. **561a.** *L. articulatus* L. Like 561 but flowers crimson with white or pink wings. Leaflets usually 1–5 mm wide. Grassland in the Med. region. YU.AL.GR+Is.CR.TR. Pl. 22.

   (ii) All leaves with 2 or more leaflets
1. At least some lvs, with 2 or more
pairs of leaflets; fl. clusters two-
to many-flowered.

   2. Lvs. without a tendril.

     3. Calyx hairy.   *alpestris* (see **565h.**)

     3. Calyx hairless.   *montanus*

   2. Lvs. with a tendril (YU.AL.BG.TR.).   **(562).** *L. palustris* L.

1. All lvs. with only 1 pair of leaflets;

or rarely some with 2 pairs and then
fls. solitary.

4. Fl. clusters five- to many-flowered
(rarely 3).
   5. Stipules less than half as wide
as stem (YU.AL.GR.BG.).   **560.** *L. sylvestris* L.
   5. Stipules at least half as wide
as stem.   *latifolius*
4. Fl. clusters one- to three-flowered
(rarely 4).
   6. Fls. 2 cm or more (Widespread).   **(557).** *L. sativus* L.
   6. Fls. less than 2 cm.
      7. Calyx teeth not or only slightly
longer than tube.
         8. Sparsely downy; fls. crimson
with blue wings (Widespread
on mainland).   **558.** *L. hirsutus* L.
         8. Hairless; fls. orange-red
(Widespread, except TR.).   (557). *L. setifolius* L.
      7. Calyx teeth 1½–3 times as
long as tube. Fls. white,
pink, or purple.
         9. Pod with 2 keels on upper
margin; fl. stalks 1–3 cm
long (Widespread).   **557.** *L. cicera* L.
         9. Pod with 2 wings on upper
margin; fl. stalks 3–6 cm
(Widespread).   **(557).** *L. sativus* L.

(560). *L. latifolius* L. EVERLASTING PEA. Easily recognized by its large, bright reddish-carmine flowers 2–3 cm, its large, oval to linear, paired leaflets 4–15 cm, and its broadly winged leaf stalks and stem. Flowers in clusters of 5–15. Stipules leaf-like, 3–6 cm long. Climbing perennial 60–300 cm. Hedges, vineyards, fields, and uncultivated places. YU.AL.GR.BG. **562.** *L. montanus* Bernh. BITTER VETCH. A low-growing, non-climbing perennial 15–50 cm, recognized by its rather small flowers about 1·5 cm, in clusters of 2–6, which are at first bright crimson, then becoming greenish-blue with age. Leaves usually with 2–4 pairs of linear to elliptic leaflets, without tendrils. Pod red-brown. Mountain pastures, bushy places. YU.AL.

PISUM **566.** *P. sativum* L. PEA. A hairless climbing annual to 1 m, with large pink, violet, or dark-purple flowers with black-purple or whitish wings. Flowers solitary or 2–3, 1·5–3·5 cm. Leaflets 1–3 pairs, oval, smaller than the very distinctive oblong-oval, often black-blotched stipules. Bushy places in the Med. region; widely cultivated for its seeds and for fodder. Widespread on mainland.

ONONIS | Restharrow 17 species occur in our area, with only 1 restricted to it. Mostly annuals with inconspicuous flowers.

Perennials
(a) Flowers pink
568. *O. spinosa* L. SPINY RESTHARROW. A dwarf shrub 10–80 cm, usually spiny, with trifoliate leaves. Flowers 6–20 mm, pink or purple, usually borne singly at each node

and forming lax clusters. Dry, grassy places, waysides. Widespread. **568b.** *O. arvensis* L. A shrubby, erect spineless perennial to 1 m, with pink flowers in stalked axillary pairs in each node, forming dense terminal clusters. Petals 1–2 cm. Leaves mostly trifoliate, with elliptic to ovate leaflets 1–2·5 cm. Pod 6–9 mm. A variable species. Dry grassy places. YU.AL.GR.BG.

(b) Flowers yellow

571. *O. natrix* L. LARGE YELLOW RESTHARROW. A dwarf shrub 20–60 cm with yellow flowers, frequently red- or purple-veined, and with usually trifoliate leaves. Flowers 12–20 mm. Dry slopes, scrub. YU.CR. **571g. *O. adenotricha* Boiss. A glandular-hairy, multi-stemmed perennial to 30 cm, with long-stalked yellow flowers 7–9 mm. Lower leaves pinnate, but uppermost leaves with 3 or 1 leaflet; leaflets ovate, 6–7 mm. Pod 1–1·2 cm. Dry stony places, limestone mountains. BG(south). ****572.** *O. pusilla* L. Flowers small 5–12 mm, almost stalkless, in a dense terminal leafy cluster; petals shorter or scarcely longer than narrow-pointed calyx teeth. Leaves trifoliate, long-stalked; leaflets 5–13 mm. Pod 6–8 mm. Dry banks, gravelly places. Widespread on mainland.

Annuals

(a) Flowers pink or purple

567. *O. reclinata* L. SMALL RESTHARROW. A glandular-hairy procumbent annual 2–15 cm, with trifoliate leaves and solitary stalked pink or purple flowers in the upper leaf axils. Petals 5–10 mm, equalling calyx. Pod 8–14 mm. Dry, grassy places. Widespread. **567c.** *O. verae* Širj. Differs from 567 in having all leaves consisting of a single leaflet only, those of the lower leaves rounded, and of the uppermost linear. Flowers pink, 1–1·2 cm, little longer than the calyx. Pod 6–8 mm. Coastal hills. CR. **567e.** *O. mitissima* L. An erect or procumbent annual 15–60 cm, with dense terminal spike-like clusters of short-stalked pink flowers. Petals 1–1·2 cm, longer than tubular calyx. Leaves trifoliate; leaflets obovate to elliptical, 1–2 cm; upper bracts papery, without leaflets. Pod 5–6 mm. Dry hill slopes in the Med. region. YU.GR.CR+Rhodes.TR.

(b) Flowers yellow

567a. *O. viscosa* L. A very variable, densely and softly hairy glandular annual 10–80 cm, with solitary, long-stalked, axillary yellow flowers forming a long leafy terminal cluster. Petals to 12 mm, often red-veined; longer or shorter than calyx; flower stalk 1–2 cm, ending in a short awn. Leaves mostly with one elliptic to obovate leaflet 1–2 cm. Dry banks and fields in the Med. region. Widespread, except BG. **567d.** *O pubescens* L. A viscid and bristly hairy annual 15–35 cm, the upper and lower leaves with one leaflet, the middle trifoliate; leaflets elliptic 1–2·5 cm. Flowers yellow, 1·5 cm, in a dense cluster; flower stalks one-flowered, awnless. Pod 8–10 mm. Fallow land, stony places in the Med. region. GR+Is.CR.TR. Pl. 23.

567f. *O. ornithopodioides* L. An erect, glandular-hairy annual to 30 cm, with trifoliate leaves with leaflets 1–1·5 cm, obovate to oblanceolate, the terminal leaflet long-stalked. Petals 6–8 mm, shorter than the calyx; flower stalks one- to two-flowered, awned. Pod 1·2–2 cm, curved, knobby. Stony ground in the Med. region. YU.AL.GR. **567g.** *O. biflora* Desf. An erect, glandular-hairy annual 10–50 cm, with trifoliate leaves and fleshy, elliptical leaflets. Petals 12–16 mm, yellow, sometimes with pink veins; flower stalks one- to two-flowered, awned. Pod 2 cm. Coastal meadows. GR.

MELILOTUS | Melilot 12 similar-looking species occur in our area.

Flowers white

**573. *M. alba* Medicus WHITE MELILOT. An erect, branched annual or biennial

0–150 cm, with lax many-flowered clusters of tiny white flowers. Petals 4–5 mm. Pod –5 mm, net-veined, greyish-brown. Open disturbed ground. Widespread on mainland.

lowers yellow
idespread are: **575.** *M. officinalis* (L.) Pallas COMMON MELILOT; **574.** *M. indica* (L.) All.; 575). *M. sulcata* Desf. FURROWED MELILOT Page 40.

74a. *M. italica* (L.) Lam. An annual 20–60 cm, with relatively large yellow flowers –9 mm, in a many-flowered lax cluster. Stipules of middle leaves toothed. Pod 5–6 mm, rongly net-veined. Dry open places in the Med. region. YU.AL.GR+Is.CR. **574b.** *M. eapolitana* Ten. Distinguished from 574a by its stipules which are lanceolate, entire. lowers bright yellow 4–6 mm, in lax clusters of about 1 cm long. Pod 3–4 mm, net-eined, with a conical beak. Dry open places. Widespread.

RIGONELLA | Fenugreek 15 species occur in our area, 3 are restricted to it.

alyx 5–12 mm, tubular.
) Flowers blue
30a. *T. coerulescens* (Bieb.) Halácsy A hairy, clover-like plant to 40 cm, with rounded eads of small blue flowers, and trifoliate leaves. Recognized by its long, hairy, tubular alyx 5–8 mm, with teeth as long as the tube. Flowers 11–16 mm. Leaflets obovate, aggy-haired, 8–20 mm. Pod lanceolate, shaggy-haired, 1–1·5 cm, beak 3–5 mm. Dry ll slopes. GR.

) Flowers creamy, yellowish, often tinged purple
580. *T. foenum-graecum* L. FENUGREEK. Flowers creamy, purple-tinged, 12–18 mm. eaflets 2–5 cm. Pod 6–11 cm, with a beak 2–3 cm. Cultivated for fodder; native of Asia; aturalized throughout. **580b.** *T. gladiata* Bieb. A densely hairy annual 5–25 cm, with eamy purple-tinged flowers and sickle-shaped beaked fruits, distinguished from 580 the smaller flowers, leaflets, and fruits. Flowers 8–10 mm, solitary. Leaflets –12 mm. Pod 1·5–4 cm, with a beak 1–2 cm, finely hairy. Hills, dry grassland. Wide-read, except AL.

lyx 2–5 mm, usually cup-shaped.
) Flowers blue, rarely white
577. *T. caerulea* (L.) Ser. BLUE FENUGREEK. Flowers blue or white, about 6 mm, many in obular clusters on stalks as long as the leaves. Leaflets 2–5 cm, ovate, finely toothed; em 20–60 cm. Pod rounded, swollen, beaked. Origin not known; cultivated for fodder; idely naturalized in our area. **577a.** *T. procumbens* (Besser) Reichenb. Like 577 with usters of blue flowers but stems solid (hollow in 577), and the leaflets smaller, 2–2·8 cm, linear-lanceolate. Inflorescence more or less globular but becoming oblong d lax. Damp places, thickets. YU.GR.BG.TR. Page 149.

) Flowers yellow, sometimes tinged purple
(i) Flower clusters stalkless, or with a stalk less than 5 mm
8. *T. monspeliaca* L. STAR-FRUITED FENUGREEK A small yellow-flowered annual to cm, readily distinguished by its clusters of pendulous up-curved, sickle-shaped fruits, th thick oblique veins. Flowers about 4 mm, in clusters of 4–14. Leaves trifoliate, aflets 4–10 mm, obovate, toothed. Pod 7–17 mm. Dry stony places, rocks. Widespread. age 40.

(ii) Flower clusters stalked
9. *T. corniculata* (L.) L. SICKLE-FRUITED FENUGREEK. Distinguished by cylindrical heads small yellow flowers borne on stalks much longer than the leaves, and by its clusters of ooping, sickle-shaped pods. Flowers about 6 mm; calyx teeth very unequal. Leaflets

1. *Lathyrus laxiflorus* 557e **2.** *Trifolium tomentosum* 599 **3.** *Vicia dalmatica* 545b **4.** *Trigonella graeca* 579b **5.** *Trigonella balansae* 579a **6.** *Trifolium hirtum* 607a

ɔovate, the upper narrower. Pod 1–1·6 cm by 2–3 mm, long-pointed, hairless, with thin ransverse veins. Grassy places, cultivation in the Med. region. YU.AL.GR.CR? ·G. **579a.** *T. balansae* Boiss. & Reuter Like 579 but flower cluster globular during ɔwering. Wings of flowers equalling the keel (shorter in 579). Pod wider 2–4 mm, ore or less acute. Grassland, cultivation in the Med. region. GR+Is.CR Page 302. ₮9b. *T. graeca* (Boiss. & Spruner) Boiss. Quite unmistakable with its clusters ˈ flattened, disk-like papery pods 1–2 cm across, with a broad wing and conspicuous ansverse veins. Flowers yellow, 7–10 mm, in a dense long-stalked cluster. ınnual 10–30 cm. Rocky hills in the Med. region. GR(south and west). Page ɔ2.

EDICAGO | Medick About 25 very similar-looking species are found in our area. ı fruit the species are very distinctive and show small but constant differences in shape, ·gree of coiling, and spininess of the pod.

ɔds *without spines* (*see also* 589b, 588)
ˈ**584.** *M. orbicularis* (L.) Bartal. LARGE DISK MEDICK. Distinguished by its large flat-ned, disk-shaped pods 1–2 cm across, formed of 4–6 smooth turns, closed at the centre, ıd with conspicuous radiating veins when dry. An almost hairless procumbent annual ɔ–90 cm, with axillary clusters of yellow flowers, each 3 mm long. Grassy places, ıltivation in the Med. region. Widespread, except TR. **584a.** *M. scutellata* (L.) Miller ɔd cup-shaped, 10–18 mm across, formed of a spiral of 5–7 turns, each coil enclosing the ≥xt, glandular-hairy. Flowers in clusters of 1–3; petals 6–7 mm, yellow. Cultivated nd. YU.GR.CR. Page 41.
ˈ5. *M. arborea* L. TREE MEDICK. A dense silky-leaved shrub 1–4 m, with rather dense ·ect clusters of golden-yellow flowers. Petals 1·2–1·5 cm. Leaves with oval leaflets, ‖ky-haired beneath. Pod flattened, sickle-shaped, in an open spiral. Rocks, cliffs in the .ed. region; sometimes planted for ornament. AL.GR+Is.CR. Page 97. ****583.** *M. tiva* L. LUCERNE, ALFALFA. Distinguished by its dense long-stalked cylindrical clusters purple or blue flowers and its open-spiralled pod of 1–3½ turns. Petals 5–11 mm. ˈidely cultivated as a forage crop in our area. ****582.** subsp. *falcata* (L.) Arcangeli ϽKLE MEDICK has yellow flowers and sickle-shaped or nearly straight pods. Hybrids ·tween 583 and 582 may have greenish or almost blackish petals. Grassy places, aysides. Widespread.
ɔds *with prickly spines or projections*
ˈ7. *M. minima* (L.) Bartal. Pod 3–5 mm, globular or disk-like, in 3–7 lax spiral turns, ·ually with a double row of hooked spines. Flowers 4 mm. A woolly-haired annual. Dry assy places, waysides. Widespread. ****588.** *M. polymorpha* L. ҢAIRY MEDICK. Pod ·9 mm across, flattened, somewhat disk-like, with 1½–6 close spiral turns, with a ırrow outer margin having 2 rows of spines (or spines absent), and a strongly netted ·rface. Flowers solitary, or in clusters of 2–8, on stalks as long as the leaves. Cultivated ɔund, waysides. Widespread.
ˈ8a. *M. coronata* (L.) Bartal. Pod tiny 2–4 mm across, recalling a two-edged band-saw ·iled in 2 lax spiral turns, with a flat outer margin bearing 2 rows of spines, one ·-pointing and the other down-pointing. Flowers yellow, 3–8 in a dense cluster; petals ·3 mm. Hills, grassy and sandy places in the Med. region. Widespread. **588b.** *M. sciformis* DC. Distinguished by its disk-like, smooth pod about 6 mm across, in a spiral 5 lax turns, with a single row of slender, somewhat curved long marginal spines on the ·wer coils, the upper coils spineless. Flowers yellow, 4–5 mm. Grassy places, cultivation the Med. region. GR+Is.CR.BG. Page 41.

589. *M. marina* L. SEA MEDICK. An easily recognized littoral plant, usually growing prostrate over the sand, with silvery, almost white woolly foliage and stems. Flowers pale yellow, 6–8 mm, in short-stalked rounded clusters. Pod woolly-haired, coiled, with a hole in the middle, with 2 rows of short conical spines. Widespread. **589a. *M. rigidula* (L.) All. Pod disk-like to cylindrical, 5–8 mm across, in a tight spiral of 4–7 turns, nearly always densely glandular-hairy with spines on the margins, but variable. Flower yellow, 6–7 mm. Meadows, wasteland. Widespread. **589b.** *M. littoralis* Loisel. Distinguished from 589a by the hairless fruit 4–6 mm across in a disk-like or cylindrical spiral of 3–6 turns, spiny or not. Flowers yellow, 5–6 mm. Rocky places in the Med. region Widespread, except TR.

TRIFOLIUM | **Clover, Trefoil**   About 85 species are found in our area, 23 of which are restricted to it. A large and difficult genus.

Flowers pale yellow, bright yellow, or orange-yellow
590. *T. dubium* Sibth. SUCKLING CLOVER, LESSER YELLOW TREFOIL. Dry grassland waysides. YU.AL.GR.BG. **591.** *T. campestre* Schreber HOP TREFOIL. Grassy places waysides. Widespread **592.** *T. aureum* Pollich LARGE HOP TREFOIL. Thickets, grassy places. YU.AL.GR.BG. **592a.** *T. velenovskyi* Vandas Like 592 but flowerheads laxer and fewer-flowered, on longer flower stalks, and flowers becoming darker after flowering. Stipules dilated and rounded below (stipules of upper leaves not dilated at the base in 592). Alpine meadows in N. Balkans. YU.AL.BG.

592b. *T. aurantiacum* Boiss. & Spruner A hairy annual distinguished by the upper leaves having the terminal leaflets with a stalk 1–2 mm; leaflets obovate-wedge-shaped apex blunt or indented; stipules oblong-lanceolate. Flowers orange, 8–9 mm, forming relatively large ovoid heads, which in fruit become 2–2·5 cm long. Alpine meadows GR.CR. Pl. 24   ****593.** *T. badium* Schreber BROWN TREFOIL. Meadows and rocky places in mountains. YU.AL.BG. **593b.** *T. boissieri* Boiss. Like 593 but with spreading hair (adpressed or hairless in 593), and stalks of flower clusters shorter than or equalling the leaves. Petals pale yellow (golden-yellow turning to brown in 593), 8–10 mm. Dry hillsides. GR+Is.CR.

Flowers purple, pink, white, cream, or yellowish-white
(a) Flower heads elongate, at least twice as long as broad
594. *T. arvense* L. HARE'S FOOT CLOVER. Sandy places. Widespread. **606a. *T. pan nonicum* Jacq. Flowers yellowish-white, large 2–2·5 cm, in cylindrical heads up to 5 cm or more long, on stems 4–8 cm. Lowest tooth of calyx twice as long as other teeth Leaflets elliptic or oblong-lanceolate; perennial 20–50 cm. Meadows, dry grassland open scrub. YU.AL.GR.BG.TR? Pl. 23.

595. *T. incarnatum* L. CRIMSON CLOVER. Flowers 1–1·2 cm in a cylindrical cluster 1–4 cm long, blood-red, pink, to white. Leaflets obovate to rounded. Grassy, bushy places, cultivation; also grown as a fodder crop and naturalized. Widespread on main land. **596. *T. angustifolium* L. NARROW-LEAVED CRIMSON CLOVER. Flowers pale reddish purple 1–1·2 cm, in a cylindrical cluster 2–8 cm long. Distinguished from 595 by its linear-lanceolate, acute leaflets. Dry hills, banks, rocky places. Widespread. Page 40. **(596).** *T. purpureum* Loisel. PURPLE CLOVER. Flowers bright reddish-purple 1·6–2·5 cm, much longer than calyx (petals shorter or equalling calyx in 596). Leaflet linear to oblong-elliptic acute. Meadows. Widespread, except CR. Pl. 24.

596a. *T. rubens* L. Flower heads purple, rarely white, cylindrical to 8 cm long by 2·5 cm wide, solitary or paired on stems 30–60 cm. Leaflets to 7 cm by 1 cm, oblong-lanceolate with bristly teeth; stipules lanceolate green, often toothed. Flowers about 1·5 cm; calyx usually hairless, teeth hairy. A rhizomatous perennial. Scrub, open woods. YU.AL. Pl. 24. **596b.** *T. alpestre* L. Flower heads purple, rarely pink or white, globular to ovoid 1·5–2·5 cm, solitary or paired on stems 14–40 cm. Leaflets lanceolate to narrowly elliptic 2–5 cm; stipules linear papery, apex finely hairy. Dry pastures, shrub, open woods. YU.AL.GR.BG. Pl. 24.

(b) Flower heads globular (see also 596b)
   (i) Flower heads stalkless or stalked for less than 1 cm above the nearest leaf
      (x) Calyx not inflated
The following common European species are widely scattered in our area: **598.** *T. striatum* L. SOFT TREFOIL; ****(598).** *T. scabrum* L. ROUGH TREFOIL; **(598).** *T. subterraneum* L. SUBTERRANEAN CLOVER; **605.** *T. pratense* L. RED CLOVER; **606.** *T. ochroleucon* Hudson SULPHUR CLOVER; ****607.** *T. medium* L. ZIG-ZAG CLOVER.

598a. *T. uniflorum* L. A low cushion-like perennial 1–3 cm, with conspicuous relatively large, white, cream, purple, or multi-coloured, almost stalkless flowers 1·2–2·7 cm, nestling amongst the leaves. Leaflets rounded, toothed, 4–10 mm; stipules broadly triangular, overlapping. Dry grasslands, stony places. GR+Is.CR.TR. Pl. 23. **607a.** *T. hirtum* All. A hairy, erect or ascending annual to 35 cm, with solitary, globular, pale-purple, densely hairy heads 1·5–2·5 cm, with an involucre of enlarged stipules and 1 or 2 trifoliate leaves directly beneath. Petals 12–15 mm, longer than the densely hairy calyx; calyx teeth bristly-haired. Leaflets obovate-wedge-shaped 8–20 mm. Dry grassy places in the Med. region. Widespread. Page 302. **608a.** *T. pignantii* Fauché & Chaub. A handsome perennial with creeping stems, and erect stems 15–45 cm bearing large solitary globular heads of yellowish-white flowers 2–3 cm, with stipules and trifoliate leaves directly beneath. Petals 1·5–1·8 cm; calyx tube hairless, teeth bristle-like with long spreading hairs. Leaflets 1–3 cm, elliptic or obovate. Mountain woods and thickets. YU.AL.GR.BG.

   (xx) Calyx inflated in fruit
599. *T. tomentosum* L. WOOLLY TREFOIL. A small spreading, tufted annual to 15 cm, distinguished by its globular, woolly-white, stalkless fruiting heads 0·7–1·4 cm across, in the axils of the leaves. Flowers pink, 2–8 mm; calyx becoming much inflated and spherical in fruit. Rocky places, cultivation, roadsides. Widespread except BG. Page 302.

   (ii) Flower heads distinctly stalked for at least 1 cm above the attachment of the nearest leaf
      (x) Calyx not inflated
593a. *T. speciosum* Willd. A small erect annual 10–30 cm, recognized by the violet flowers in lax ovoid flower heads, which enlarge up to 3 cm in fruit. Petals 8–10 mm, standard strongly veined; calyx teeth very unequal. Stalks of flower heads 2–3 times as long as the leaves. Leaflets oblong-elliptical, the middle leaflet short-stalked. Dry hill slopes. Widespread.

601. *T. hybridum* L. ALSIKE CLOVER. Grassy places; often grown as a fodder crop. YU.GR.CR.BG.TR. **(601).** *T. montanum* L. MOUNTAIN CLOVER. Dry grassy places. YU.BG. ****602.** *T. stellatum* L. STAR CLOVER. A Mediterranean annual 8–20 cm, readily distinguished by its conspicuous globular fruiting heads with brightly coloured, often reddish calyx teeth spreading in a star. Flowers pink, inconspicuous; calyx densely silky-haired. Dry places, sands, tracksides. Widespread, except BG. **606b.** *T. echinatum* Bieb. A rather variable erect or spreading annual 10–60 cm, readily recognized in fruit by its globular heads with stiff spiny, erect, unequal calyx teeth. Flower heads cream or pink, globular 8–15 mm, borne on stalks 2–8 cm long; petals 8–12 mm, at least twice as long as calyx. Grassy often damp places. Widespread on mainland.

598b. *T. pilulare* Boiss. An ascending, a prostrate much-branched annual, 10–60 cm, distinguished by its rounded densely grey-hairy, ball-like fruiting heads 8–10 mm. Flowers white 6–7 mm; calyx with long bristle-like teeth with cottony-white spreading hairs. Grassy and rocky slopes, fields. E. Aegean Is. Pl. 24. **606c.** *T. clypeatum* L. A rather robust leafy annual 10–30 cm, with large ovoid heads 1–2·5 cm of pink or whitish flowers. Very distinctive in fruit owing to the much-enlarged lanceolate teeth which spread in a star, the lowest tooth twice as long as the remainder, and throat of calyx tube closed. Flowers 2–3 cm; calyx tube strongly ten-ribbed, the teeth spine-tipped. Fields, roadsides. Karpathos, E. Aegean Is.

(xx) Calyx inflated in fruit

604. *T. fragiferum* L. STRAWBERRY CLOVER. Grassy places, fields, waysides. Widespread. **604a.** *T. resupinatum* L. A hairless annual 10–30 cm, readily distinguished in fruit by its globular heads with inflated calices spreading in a star each with 2 bristly teeth. Petals pink, rarely reddish-purple, 2–8 mm; calyx strongly net-veined, becoming hairless. Very variable. Grassy places, disturbed damp ground; doubtfully native. Widespread. Pl. 24.

604b. *T. physodes* Bieb. Like 604 with swollen calyx, forming a strawberry-like, often pinkish, globular fruiting head, but bracts only 1 mm, free (4–5 mm, fused into an irregular involucre in 604). Flowers pink, 8–14 mm. A spreading or erect perennial 5–25 cm. Grassland in the Med. region. Widespread, except BG. Pl. 24. **604c.** *T. spumosum* L. A sprawling or ascending annual 10–30 cm, with globular or ovate heads of small pink flowers, and with very distinctive, pale, much inflated pear-shaped calyx with longitudinal and transverse ridges. Petals little longer than the calyx; flower head stalks usually 1–4 cm. Leaflets 1–2 cm, broadly obovate, finely toothed. Dry grassy places, disturbed ground. GR + Is.CR.TR.

DORYCNIUM Leaves with 5 leaflets, the upper 3 separated from the lower 2 by a rachis, or all 5 leaflets arising from the same position.

Flowers 1–2 cm

**609. *D. hirsutum* (L.) Ser. A perennial or shrublet 20–50 cm, with globular heads of white or pink flowers. Leaves densely hairy, with 5 leaflets, rachis 1–3 mm. Bushy and grassy places in the Med. region. Widespread, except BG.

Flowers 3–8 mm

(a) Calyx equally five-toothed; pod with 2–8 seeds

610. *D. rectum* (L.) Ser. A perennial or shrublet similar to 609 with white or pink flowers, but flowers 4–7 mm, and leaves with a distinct rachis 5–11 mm. Pod 8–14 mm, with valves contorting when ripe. Damp bushy places, marshes in the Med. region. AL.GR.CR.TR. **610a. *D. graecum* (L.) Ser. Like 610 but leaves with a much shorter rachis 1–3 mm; pod 5–9 mm, and valves not contorting. Flowers white, 5–8 mm. Bushy places in the Med. region. GR.BG.TR.

(b) Calyx with unequal teeth; pod with one seed

611. *D. pentaphyllum* Scop. Like 610 but leaves without a rachis. Flowers 3–6 mm, white, pink, or purple at base. Pod globular 3–5 mm. Woods, bushy places. Widespread on mainland.

LOTUS | Birdsfoot-Trefoil About 22 rather similar-looking species in our area.

Calyx tubular-bell-shaped, with 5 more or less equal teeth

Widespread mainly on the mainland are: **(612).** *L. angustissimus* L. SLENDER BIRDSFOOT-

TREFOIL; **613.** *L. corniculatus* L. BIRDSFOOT-TREFOIL; **613b.** *L. tenuis* Willd.; ****614.** *L. uliginosus* Schkuhr LARGE BIRDSFOOT-TREFOIL.

614a. *L. edulis* L. Readily distinguished by its inflated curved pod, 2–4 cm by 4–8 mm wide, which is grooved on the back. A spreading, sparsely hairy annual 10–50 cm, with 1–2 axillary yellow flowers, borne on a stalk longer than the subtending leaf. Petals 1–1·6 cm, curved. Leaflets obovate 5–16 mm. Sandy, stony, and rocky places. YU.AL.GR+Is.CR.TR. **614b.** *L. aegaeus* (Griseb.) Boiss. A robust, often rather hairy perennial 20–60 cm, with long-stalked axillary heads of 1–5 pale-yellow, rather large flowers 1·8–2·5 cm, with standard much longer than the keel. Calyx teeth longer than the calyx tube. Pod 3–5 cm. Grasslands. YU.GR.BG. **612a.** *L. strictus* Fischer & C. A. Meyer Recognized by its axillary heads of 2–10 flowers on stalks longer than the subtending leaves, with white or pale-yellow petals 1·5–2 cm. Calyx teeth shorter than the calyx tube. A hairless or almost hairless annual up to 1 m. Pod 2·5–3 cm, straight or curved at apex. Salt-rich ground. GR(north-east).BG(east).

Calyx two-lipped, teeth unequal

612. *L. ornithopodioides* L. A hairy annual 10–50 cm, with long-stalked axillary clusters of yellow flowers; petals 7–10 mm. Distinctive in fruit, having 2–5 long curved pods 2–5 cm, hanging below 3 conspicuous rounded leafy bracts. Leaflets 8–30 mm, broadly oval-rhombic. Grassy, sandy, and stony places. YU.AL.GR+Is.CR.TR. **612b.** *L. peregrinus* L. Like 612 but with smaller leaflets 5–15 mm, smaller yellow flowers 6–9 mm, and straight pods not constricted between seeds (constricted in 612). Grassland, maritime sands. GR+Is.CR.

615. *L. creticus* L. SOUTHERN BIRDSFOOT-TREFOIL. A densely silvery-haired, spreading perennial to 50 cm, with heads of 2–6 yellow flowers. Calyx distinctive: two-lipped, the upper 2 teeth curved upwards, the lateral teeth acute, about as long as the upper but shorter than the lower tooth. Flowers 12–18 mm; the wings much longer than the keel, which has a long, straight, purple beak. Maritime sands. YU.GR. **615a. *L. cytisoides* L. Like 615 but 2 lateral teeth of calyx blunt, much shorter than the upper 2 teeth. Flowers 8–14 mm; standard notched, wings slightly longer than the keel, which has a short, curved, purple-tipped beak. Rocky, stony, and sandy places. YU.GR+Is.CR. Pl. 25.

TETRAGONOLOBUS Distinguished from *Lotus* by its angled or winged pod. Leaves trifoliate. 4 species in our area.

616. *T. maritimus* (L.) Roth WINGED PEA. A low creeping perennial 10–40 cm, easily distinguished by its large solitary, long-stalked, pale-yellow flowers 2·5–3 cm, and its large four-winged pods which are 3–6 cm by 3–5 mm. Fields, damp grassy places, sands on the littoral, brackish soils. YU.BG. **616a. *T. biflorus* (Desr.) Ser. Like 616 but an annual with bright deep-orange flowers 1·7–2·5 cm, in a cluster of 1–4. Pod 2–4 cm, finely hairy. Grassy places. GR(north-west). ****617.** *T. purpureus* Moench ASPARAGUS PEA. Unmistakable with its large, short-stalked, solitary or paired crimson flowers 15–22 mm, with a blackish keel, and its large four-winged pods. Calyx teeth as long as or up to twice as long as calyx tube. Cultivated places, vineyards in the Med. region. GR+Is.CR.

HYMENOCARPUS **618.** *H. circinnatus* (L.) Savi DISK TREFOIL. Distinguished by its circular, disk-like pods 1–1·5 cm, often with short scattered spines on the undulate margin. Flowers yellow, 5–7 mm, in a stalked cluster. Lower leaves simple, the upper pinnate, with 3–5 pairs of oval leaflets and a much larger terminal leaflet. A hairy annual, to 30 cm. Grassy places in the Med. region. Widespread.

SECURIGERA Distinguished by its linear pods with thickened margins and with a long beak.

618a. *S. securidaca* (L.) Degen & Dörfler An erect hairless annual 10–50 cm, distinctive in fruit with long fruit stalks bearing a cluster of 4–8 long slender pods each long-beaked and with curved or hooked apices. Flowers yellow, 8–12 mm. Leaflets 4–7 pairs obovate-oblong. Pod 5–10 cm including beak. Fields, waysides, grassy places in the Med region. Widespread. Page 310.

ANTHYLLIS | Kidney-Vetch
Shrubs or under-shrubs with woody branches
619. *A. hermanniae* L. A dense shrub to 0·5 m, with interwoven spiny-tipped branches and small yellow flowers in leafy clusters. Flowers about 5 mm; calyx hairy. Leaves trifoliate or with one leaflet; leaflets 1–2 cm, linear-oblong, silky-haired. Rocky places in the Med. region. YU.AL.GR+Is.CR.TR. Page 38. **620. *A. barba-jovis* L. JUPITER'S BEARD. A dense, silvery-white, leafy shrub to 1 m, with pinnate leaves and with bright yellow flowers in terminal globular heads. Flowers numerous, about 1 cm; calyx 4–6 mm, pale, woolly-haired. Leaflets 9–19, narrowly elliptic to obovate. Maritime rocks in the Med. region. YU.AL? GR.CR. Pl. 25. **620b.** *A. aegaea* Turrill Differs from 620 in having very narrowly elliptic to linear leaflets, long calyx 6–9 mm, and flower clusters with 5–9 flowers. GR(Cyclades).CR.

Herbaceous plants, sometimes woody at the base, branches herbaceous
(a) Calyx inflated, constricted at apex after flowering
**622. *A. vulneraria* L. KIDNEY-VETCH. A very polymorphic species with up to 24 distinctive subspecies in Europe, many of which have been recognized as species by some botanists. Subsp. *praepropera* (A. Kerner) Bornm. (*A. spruneri*) usually has red or purplish flowers, and calyx purple at apex with shining silky hairs. Grassy places rocks in the Med. region. Widespread, except BG. Pl. 25. Subsp. *pulchella* (Vis. Bornm. A small, delicate, often procumbent, adpressed-silky plant 6–20 cm of mountain pastures, with red or yellow flowers. Calyx 3–9 mm, reddish above, with spreading shaggy, not shining hairs. YU.AL.GR.BG. Page 127. Subsp. *bulgarica* (Sagorski) Culler Flowers yellow to almost whitish; calyx 10–11 mm, greenish. A robust plant to 40 cm YU.AL.GR.BG.

**623. *A. tetraphylla* L. BLADDER VETCH. A low spreading annual, readily distinguished by its very swollen calyx, which enlarges and becomes almost globular in fruit. Flowers pale yellow, tipped with red, 1–7 in dense stalkless clusters. Leaflets 3–5, the terminal leaflet much larger. Grassy places and olive orchards in the Med. region YU.AL.GR+Is.CR.

(b) Calyx not swollen, or constricted at apex at flowering
621. *A. montana* L. MOUNTAIN KIDNEY-VETCH. A mat-forming, woody-based, perennial with globular heads of pink flowers on short nearly leafless stems. Leaflets numerous 17–61, narrow-elliptic. Mountain rocks. YU.AL.GR.BG. Page 184. **621c. *A. aurea* Welden A montane or alpine slightly woody perennial distinguished by its dense globular golden-yellow flower heads. Calyx 6–8 mm, ridged, silky-haired. Leaves mostly basal or on the lower third of the flowering stem; leaflets 13–19, ovate to elliptic, silky-haired. Mountain rocks. YU.AL.GR.BG. Page 169. Pl. 25.

ORNITHOPUS Pod elongate, with a network of veins, usually constricted between the seeds and splitting into one-seeded portions. 2 species in our area.

623e. *O. compressus* L. A hairy annual 10–50 cm, with heads of 3–5 small yellow flowers, pinnate leaves, and distinctive pods which are slender, flattened, curved, jointed, and end in a long sickle-shaped beak. Flower heads subtended by pinnate bracts; flowers 5–8 mm. Leaflets elliptic, 7–18 pairs, hairy. Rocks, waysides. Widespread. Page 41.

CORONILLA Distinguished by its long slender jointed pods which are not constricted between the seeds, and its compact 'umbel' of flowers. 11 species in our area, including one endemic to it.

Shrubs usually over 50 cm

624. *C. emerus* L. SCORPION SENNA. A robust leafy shrub to 1 m or more, with pinnate leaves and with long-stalked, globular clusters of pale-yellow flowers and long slender, pendulous, jointed pods 5–11 cm. Corolla 14–20 mm, petal claw 2–3 times as long as calyx. Leaflets 5–9, obovate, 1–2 cm. Thickets, rocky places in the hills. Widespread. Page 97.   **625. *C. valentina* L. SHRUBBY SCORPION-VETCH. Like 624 but leaflets shallowly notched at apex, and claw of petals little longer than calyx. Pod 1–5 cm, angled. Scrub, cliffs in the Med. region. YU.AL.GR.CR.   **630.** *C. vaginalis* Lam. SMALL SCORPION-VETCH. A small shrub to 50 cm, with yellow flowers 6–10 mm, in heads of 4–10, with claw of petals little longer than calyx. Leaflets 2–6 pairs, 4–10 mm, shortly stalked, with narrow papery margins. Pod 1·5–3·5 cm, winged. Dry grassland, scrub, open woods in limestone mountains. YU.AL.

Herbaceous perennials

627. *C. varia* L. CROWN VETCH. A robust, spreading, leafy herbaceous perennial 20–120 cm, distinguished by its globular, long-stalked clusters of 10–20 white, pink, or often multi-coloured flowers each 1–1·5 cm. Leaflets 7–12 pairs, oblong, 6–20 mm. Pod four-angled, 2–6 cm. Grassy and bushy places; sometimes cultivated. Widespread. Page 164.   **627a. *C. globosa* Lam. Like 627 but flowers usually white, in globular heads of 15–40. Leaflets larger, 1·5–3 cm. Cliffs. CR. Page 310.   ****629.** *C. coronata* L. SCORPION-VETCH. Flowers yellow, 12–20 in rounded heads; petals 7–11 mm. Leaves with 3–6 pairs of elliptic leaflets. Dry woods, scrub, and grassland. YU.AL.GR.

Annuals

628. *C. scorpioides* (L.) Koch ANNUAL SCORPION-VETCH. A common annual weed to 40 cm of cultivation with rounded glaucous leaflets, long-stalked clusters of tiny yellow flowers, and long curved pendulous jointed pods. Petals 4–8 mm. Pod 2–6 cm. Widespread.   **628c.** *C. rostrata* Boiss. & Spruner An erect or ascending annual with pink, white, or pale-yellow flowers in heads of 3–9. Petals 7–11 mm; flower stalks elongating in fruit. Leaves with 5–8 obovate-oblong notched leaflets. Pod 3–8 cm, strongly curved, with a straight beak. Among herbaceous vegetation. AL.GR+Is.CR.TR. Page 310.   **628b.** *C. cretica* L. Differs from 628c in having smaller flowers 4–7 mm which are white or pink. Pod straight, with a curved beak, and segments four-angled. Grassy places, cultivation. Widespread. Page 40.

HIPPOCREPIS | Horseshoe Vetch 5 species in our area.

631. *H. comosa* L. HORSESHOE VETCH. Grassy and rocky places in mountains. YU.AL.GR.BG.   **632.** *H. unisiliquosa* L. A slender yellow-flowered annual to 40 cm, easily distinguished by the unusual pods which look like a string of horseshoes placed side by side. Flowers usually solitary, 4–7 mm. Leaflets 3–7 pairs, 2–12 mm. Pod 1·5–4 cm, with 7–10 horseshoe-shaped constrictions. Grassy, bushy places in the Med. region. Widespread. Page 41.   **632a.** *H. ciliata* Willd. Distinguished by its pods which are curved, with 4–9 horseshoe-shaped constrictions opening on the concave edge, and with long rounded swellings and hairs. Flowers yellow, 2–6, on a stalk

1. *Securigera securidaca* 618a   **2.** *Ebenus sibthorpii* 637c   **3.** *Coronilla globosa* 627a   **4.** *Coronilla rostrata* 628c   **5.** *Onobrychis arenaria* (636)

about equalling the leaves; petals 3–5 mm. Sunny, grassy places in the Med. region. Widespread.

HAMMATOLOBIUM Distinguished by its jointed pods with 2–6 segments, the segments rounded and finely net-veined.

632b. *H. lotoides* Fenzl A much-branched, woody-based, densely hairy procumbent perennial 10–15 cm of mountain cliffs, with 1–5 purple or yellow flowers in a head. Petals 1–1·4 cm. Leaflets 3–5, broadly ovate. Pod linear 1–2·5 cm, shortly beaked. GR(south).

SCORPIURUS Readily distinguished by its pods which look very like swollen, coiled caterpillars.

**(633). *S. muricatus* L. An annual to 80 cm, with entire, parallel-veined lanceolate leaves, and long-stemmed axillary clusters of 2–5 small yellow flowers, and distinctive coiled fruits. Flowers 5–10 mm. Pod smooth, or with spines or protuberances on the outer ridges. Fields, cultivated ground. Widespread.

HEDYSARUM Distinguished by its flattened pods which are strongly constricted between the seeds into segments which look like a series of disks placed edge to edge. 8 species in our area.

Annuals

(634). *H. spinosissimum* L. A low spreading annual 15–35 cm, with few-flowered clusters of pinkish-purple to whitish flowers. Corolla 8–11 mm; calyx 4–6 mm. Leaflets 2–8 pairs, often oblong. Pod with 2–4 spiny and finely hairy segments. Limestone hills, dry stony places in the Med. region. GR+Is.CR. ****(634).** *H. glomeratum* F. G. Dietrich Like (634) *H. spinosissimum* but flowers more showy, pinkish-purple, 14–20 mm, 2½–5 times as long as calyx. Leaflets broader, sometimes obovate. Pod distinctive, with netted segments covered with fine grey hairs and pale hooked spines. Dry places in the Med. region. YU.GR.

Perennials

634. *H. coronarium* L. ITALIAN SAINFOIN, FRENCH HONEYSUCKLE. A robust perennial 30–100 cm, with handsome cylindrical clusters of 10–15 large bright carmine flowers, and leaves with 3–5 pairs of large elliptical or rounded leaflets. Petals 12–15 mm. Pod with 2–4 spine-covered segments. Cultivated ground. Native of W.Europe; grown for fodder and naturalized in YU.GR. **635a.** *H. tauricum* Willd. A sparsely hairy perennial to 20 cm, with 6–8 pairs of hairy oblong, elliptic, or obovate leaflets, and stalked clusters of 8–15 purple flowers. Petals 11–18 mm. Pod with 2–5 hairy segments. Grassy hills. BG(north-east). **635b.** *H. grandiflorum* Pallas A stemless perennial with clusters of large yellow flowers, arising on stalks direct from the woody rootstock. Petals 1·8–2·5 cm. Leaflets 2–5 pairs, elliptic 1·5–4 cm, silky-haired beneath. Pod with bristles. Dry grassy places. BG(north-east).

ONOBRYCHIS Distinguished by its rounded, non-splitting, somewhat flattened pods, which have very conspicuous toothed margins and are netted and spiny on the flanks. 14 similar-looking species in our area.

The following are most frequently encountered:
1. Annuals; fls. small, petals little longer
    than the calyx; fl. clusters usually with
    1–6 fls.

2. Fls. 7–8 mm (Widespread in the Med. region).

2. Fls. 10–14 mm (Widespread in the Med. region).

1. Perennials; fls. conspicuous; fl. clusters with at least 10 fls.

   3. Standard at least 1½ times as long as keel.

      4. Calyx teeth 2–3 times as long as tube.   *ebenoides*

      4. Calyx teeth 1–2 times as long as tube.   *gracilis, pindicola*

   3. Standard shorter or slightly longer than keel.   *alba, arenaria*

637. *O. caput-galli* (L.) Lam.

637a. *O. aequidentata* (Sibth. & Sm.) D'Urv. Page 40

636g. *O. ebenoides* Boiss. & Spruner A greyish adpressed-haired, erect or spreading perennial to 45 cm, with leaves with 4–7 pairs of linear-elliptic leaflets, and long-stalked clusters of pale-pink veined flowers. Petals 7–10 mm, rarely white; calyx densely hairy, with long teeth. Pod 5–7 mm, densely woolly-haired, and with 4–5 spines on the margin. Rocky hillsides, waysides. GR(central and south). Pl. 25. **636h.** *O. gracilis* Besser A hairless to sparsely hairy perennial 25–75 cm, with leaves with 3–8 pairs of linear leaflets 1·2–2·5 cm. Petals pink, 5–7 mm; calyx hairless except on margin. Sunny hill slopes. YU.GR.BG.TR.

636i. *O. pindicola* Hausskn. Similar to 636h but leaflets oblong or elliptical, usually 7–11 mm. Petals pink, 6–9 mm; calyx shortly hairy. Mountains. YU.GR.BG? **636j.** *O. alba* (Waldst. & Kit.) Desv. A variable hairy perennial to 60 cm, with lower leaves with 6–10 pairs of linear to elliptic leaflets up to 3·5 cm. Petals white or pink, often veined, 8–12 mm; calyx teeth 2–4 times as long as calyx tube. Pod with spreading hairs, toothed sides, and 2–6 marginal teeth. Sunny hill slopes, meadows YU.AL.GR.BG. Page 168. **(636).** *O. arenaria* (Kit.) DC. Flowers pink with purple veins, in long spikes. Petals 8–12 mm; calyx hairy or hairless, teeth hairy. Leaflets 3–12 pairs, linear-oblong to elliptic. Pod finely hairy, with toothed sides, and usually with 3–8 teeth on margin. Rocky places sandy ground. Widespread on mainland. Page 310.

EBENUS Pods one- to two-seeded, not splitting, and included in the calyx. Flowers pink or purple, in heads, or in spike-like clusters, subtended by papery bracts.

637b. *E. cretica* L. A very beautiful, small, grey-leaved shrub to 50 cm, with dense ovoid clusters of bright-pink flowers and long silvery-haired calyx teeth which are longer than the buds and so give the young flowering cluster a fluffy greyish apex. Flowers 1–1·5 cm, occasionally white. Leaves with 3–5, oblong-elliptic leaflets 1·5–3 cm; stipules two-toothed at apex. Cliffs, gorges, rocky ground. CR. Page 67. Pl. 25. **637c.** *E. sibthorpii* DC. Distinguished from 637b by its long-stalked, globular head of reddish-purple flowers with papery bracts at its base. Flowers 1–1·5 cm. Leaves with 2–4 pairs of leaflets; stipules with 3–4 teeth at apex. A herbaceous plant 15–30 cm with a woody base. Cliffs, rocks, GR(south-east)+Rhodes. Page 310.

ARACHIS Distinguished by its pods which after fertilization are buried underground.

637d. *A. hypogaea* L. GROUND-NUT or PEANUT. A much-branched, hairy annual to 50 cm, with leaves with 2 pairs of large leaflets, and stalkless yellow flowers in axillary clusters. Petals 1·5–2 cm. Leaflets 2·5–6 cm, elliptic to obovate. Pod 2–4 cm, netted, constricted between each of the 1–3 large seeds. Native of S. America; cultivated for its edible seeds in GR.BG.

OXALIDACEAE | Wood-Sorrel Family

OXALIS
Flowers yellow
639. *O. pes-caprae* L. (*O. cernua*) BERMUDA BUTTERCUP. A stemless weed of cultivated ground, readily distinguished by its long-stalked umbel (to 20 cm) of large, bright-yellow flowers, and its pale-green clover-like leaves. Petals 2–2·5 cm. An aggressive and persistent weed, spreading largely by means of its bulbils. Native of S. Africa; introduced to the Med. region in GR+Is.CR. ****640. *O. corniculata* L. PROCUMBENT YELLOW SORREL. A perennial spreading to 50 cm and rooting at the nodes, with an umbel-like cluster of 1–7 small yellow flowers with petals 4–7 mm. Leaves trifoliate, alternate. Fruit cyclindrical 1–2·5 cm, hoary. Waste places, disturbed ground. Widespread. **(640).** *O. europaea* Jordan Like 640 but stems erect to 40 cm not rooting at the nodes. Flowers yellow, axillary. Leaves more or less opposite. Fruit 8–12 mm, not hoary. Cultivated ground. Native of N. America; naturalized in YU.AL.BG. **640a.** *O. stricta* L. Differs from (640) in having umbel-like clusters of yellow flowers; deflexed fruit stalks; oblong stipules (stipules absent in (640)). Cultivated ground. Native of N. America; naturalized in YU.AL.

Flowers white
**638. *O. acetosella* L. WOOD-SORREL. Woods, shady places. Widespread on mainland.

Flowers bright red
639e. *O. tetraphylla* Cav. Leaves with 4 leaflets; bulb to 4 cm. Native of Mexico; naturalized in YU.

GERANIACEAE | Geranium Family

GERANIUM | Cranesbill   Distinguished from *Erodium* by leaves which are palmate or palmately lobed, and the beak of fruit splitting into arched or watch-spring-coiled sections. 31 species in our area, including 3 endemic to it.

Annuals or biennials; petals usually less than 1 cm long
Widespread species, particularly on the mainland include:
648. *G. dissectum* L. CUT-LEAVED CRANESBILL; **(648).** *G. columbinum* L. LONG-STALKED CRANESBILL; **649.** *G. rotundifolium* L. ROUND-LEAVED CRANESBILL; **(649).** *G. molle* L. DOVE'S-FOOT CRANESBILL; **(649).** *G. pusillum* L. SMALL-FLOWERED CRANESBILL; **650.** *G. robertianum* L. HERB ROBERT; **(650).** *G. purpureum* Vill.; ****651.** *G. lucidum* L. SHINING CRANESBILL.

Perennials; petals usually more than 1 cm long
1. Petals with a distinct stalk (claw) at
    least a third as long as the blade.   *macrorrhizum, dalmaticum*

1.   Petals with a very short stalk (claw)
    or none.
 2.   Petals entire pointed, toothed, or
     slightly notched.
   3.   Petals spreading horizontally or   **646. *G. phaeum* L.
       deflexed (YU.AL.BG.).   *reflexum, aristatum*
   3.   Petals curving upwards, giving a
       more or less cup-shaped fl.
     4.   Inflorescence compact, usually
         with more than 10 fls.
       5.   Fl. stalk deflexed as fr.
           matures; sepals 11–15 mm
           (YU.GR.BG.).   644. *G. pratense* L.
       5.   Fl. stalk erect as fr. matures;
           sepals 6–12 mm (YU.AL.GR.BG.).   **645. *G. sylvaticum* L.
     4.   Inflorescence lax, spreading,
         usually with less than 10 fls.   *asphodeloides*
         (YU.BG.).   646a. *G. palustre* L.
 2.   Petals two- or three-lobed, or
     distinctly notched.
   6.   Root-stock small and inconspicuous;
       petals usually less than 1 cm.
     7.   Frs. smooth, hairy (Widespread on
         mainland).   **642. *G. pyrenaicum* Burm. fil.
     7.   Frs. rough, hairless (Widespread
         on mainland).   649a. *G. brutium* Gasparr.
   6.   Plant with conspicuous tuber,
       rhizome, or root-stock; petals
       usually more than 1 cm.
     8.   Fls. solitary (Widespread on
         mainland).   **641. *G. sanguineum* L.
     8.   Fls. paired.
       9.   Fl. stalks with long glandular   *macrostylum,*
           hairs.   *peloponesiacum*
       9.   Fl. stalks without long glandular
           hairs.
         10.   Lvs. not more than 4 cm wide,
             usually all basal.   *cinereum*
         10.   Larger lvs. at least 5 cm wide,
             some borne on the stem.
           11.   Fl. stalks with minute almost
               stalkless glands.   *versicolor*
           11.   Fl. stalks without glands.
             12.   Lvs. divided for about three-
                 quarters of the radius;
                 rhizome long slender
                 (YU.).   **(642). *G. nodosum* L.
             12.   Lvs. divided to the base;
                 rhizome short.   *tuberosum*

**647. *G. macrorrhizum* L. ROCK CRANESBILL. Flowers very attractive, large, 2·5 cm
across, blood-red or carmine with wide-spreading or somewhat reflexed petals, and

conspicuous projecting stamens and style. Calyx inflated, reddish, hairy; petals about 15 mm. Leaves divided to about two-thirds into 5–7 broad lanceolate lobes which are pinnately cut or toothed. Shady rocks, in mountains. YU.AL.GR.BG. Pl. 26. **647a.** *G. dalmaticum* (G. Beck) Rech. fil. Like 647 but smaller and more delicate. Blade of basal leaves 2·5–4 cm across, divided almost to the base into 5 wedge-shaped, straight-sided lobes which are three-toothed at the apex; stem leaves absent. Petals about 13 mm. In mountains. YU(south-west).AL.

646c. *G. reflexum* L. Like 646 *G. phaeum* DUSKY CRANESBILL but petals narrower 2·5–4 mm wide (6–10 mm wide in 646), dull lilac or purple, and sharply reflexed. Flower stalks reflexed as fruit ripens. Meadows, mountain woods. YU.AL.GR.BG. **646b.** *G. aristatum* Freyn & Sint. Like 646 *G. phaeum* with spreading or reflexed petals but petals pale lilac with darker veins, 13 mm long (8–10 mm long in 646). Sepals with a long point 4–6 mm (point 1 mm in 646). Mountains. YU(south).AL.GR(north-west). **646d.** *G. asphodeloides* Burm. fil. A tall perennial 30–75 cm, with fleshy spindle-shaped roots arising from a very short rhizome. Flowers few, the petals entire or slightly notched, 15 mm, pinkish lilac with darker veins. Sepals 10 mm, with a point not more than 1·5 mm (roots fibrous; sepals with a point more than 1·5 mm in 646a *G. palustre*). Fruit finely hairy, without ridges. Thickets, woods. AL.GR.BG.TR.

643. *G. tuberosum* L. TUBEROUS CRANESBILL. Flowers rosy-purple, borne in a lax cluster on nearly leafless stems arising direct from a globular, nut-like, underground tuber. Petals 8–13 mm, deeply notched. Stem with a pair of large leaf-like, deeply lobed bracts at the base of the inflorescence; basal leaves 6–8 cm wide, with 5–9 deeply cut lobes which are further cut into 3–6 pairs of narrow, toothed segments. Fields, cultivation. Widespread. Pl. 26. **643b.** *G. macrostylum* Boiss. Distinguished from 643 by the glandular hairs on the flower stalks, and the presence of a pair of deeply cut leaves as well as deeply lobed bracts on the flowering stems. Petals longer 12–17 mm, rosy-purple. Rocks in hills and mountains. AL.GR. **642a.** *G. peloponesiacum* Boiss. Flowers bluish-violet, in a flat-topped cluster, with obovate, notched petals about 15 mm. Sepals glandular-hairy. Leaves 8–12 cm wide, divided for two-thirds of the radius into 3–5 ovate-rhombic, irregularly deeply cut lobes with blunt segments. Shady places in mountains. AL.GR(south).

641a. *G. cinereum* Cav. A low-growing perennial with paired, usually deep reddish-purple but sometimes pale flowers, arising on leafless stalks 5–10 cm directly from a stout vertical root-stock. Petals about 15 mm, shallowly notched, with a very short claw. Leaves circular, 2–3 cm, divided to four-fifths of their radius into 5–7 lobes which are further three-lobed, hairy, and often silvery-grey beneath, but rather variable. Rocky and grassy places in mountains. YU.AL.GR. Page 169. **(642).** *G. versicolor* L. VEINED CRANESBILL. Distinguished from most other species by its few rather large white or pale-lilac flowers with conspicuous dark-violet veins. Petals 15–18 mm, deeply notched. Woods, bushy places. YU.AL.GR.

ERODIUM | Storksbill Distinguished from *Geranium* by its fruits which almost always have spirally twisted beaks derived from the style. Leaves pinnately lobed or pinnate, rarely simple. Important distinguishing features between species are the two apical depressions or pits on the one-seeded units, and the presence or absence of furrows or ridges above or below the pits. 14 rather similar-looking species in our area, including 2 endemics.

Leaves simple, toothed or lobed, but not divided to the mid-vein; annuals or biennials
**652. *E. malacoides* (L.) L'Hér. SOFT STORKSBILL. Distinguished by the glandular pits on

the fruit which have a wide, deep furrow on the lower margin lying between 2 ridges; beak of fruit 1·8–3·5 cm. Petals 5–9 mm, purplish. Leaves ovate, toothed, sometimes lobed. Waysides, dry fields, sandy places. YU.AL.GR+Is.CR.TR.   **652a.** *E. laciniatum* (Cav.) Willd. Distinguished by the shallow, non-glandular pits on the fruits, without a furrow below; beak of fruit 3·5–9 cm. Petals 7–10 mm, purplish. Leaves very variable, undivided to pinnate-lobed. Maritime sands, dry places. AL.GR.CR.TR.   **652b.** *E. chium* (L.) Willd. Pits of fruit covered with glands but without a furrow below; beak of fruit 3–4 cm. Petals 5–9 mm, purplish. Leaves ovate, variably dissected. Sandy places, cultivated ground. GR+Is.CR.TR.

653a. *E. botrys* (Cav.) Bertol. Flowers large, violet, with petals 10–15 mm. Fruit large, 8–11 mm, with deep non-glandular pits with 2 furrows at the base, and a long beak 5–11 cm. Leaves deeply lobed, lobes further toothed or cut. Dry places. YU.GR+Is.CR.BG.TR.   **653g.** *E. hoefftianum* C. A. Meyer Flowers violet, with petals about 8 mm. Fruits 6–7 mm, with more or less non-glandular pits without a furrow below; beak of fruit 5–7·5 cm. Dry places. YU.GR.BG.TR.   ****(652).** *E. gruinum* (L.) L'Hér. LONG-BEAKED STORKSBILL. An erect annual or biennial with oval to oval-lanceolate leaves which are deeply cut and irregularly toothed. Flowers in umbels of 2–6, bracts lanceolate, hairless, whitish. Petals violet 20–5 mm; sepals 15–20 mm. Fruit rough-haired, pits deep, honeycombed, with a wide furrow at base; beak often very long, 6–11 cm. Dry grassland, maritime sands. GR+Is.CR.

Leaves divided to the mid-vein, pinnate or twice-pinnate
(a) Annuals or biennials

653. *E. cicutarium* (L.) L'Hér. COMMON STORKSBILL. Cultivated ground, sandy, stony places. Widespread.   **(654).** *E. ciconium* (L.) L'Hér. Flowers lilac or blue, veined with purple, in umbels of 3–10, on stems 10–70 cm. Petals somewhat unequal, about 8 mm. Fruits 9–11 mm, white-haired, the glandular pits with a furrow below; beak 6–10 cm. Sandy, dry or disturbed ground. Widespread.   **654.** *E. moschatum* (L.) L'Hér. MUSK STORKSBILL. Readily identified by its musk-smelling leaves when crushed; stems 10–50 cm. Flowers in umbels of 5–12, rosy-purple, not blotched; petals about 15 mm. Fruit 5–6 mm, densely rough-hairy, the pits wide, glandular, with a wide deep furrow at the base; beak 2–4·5 cm. Waysides, cultivated ground, waste places. YU.AL.GR+Is.CR.TR.

(b) Perennials

653b. *E. acaule* (L.) Becherer & Thell. A stemless perennial with a stalked umbel of 3–10 lilac flowers. Petals equal, 7–12 mm; hairs on sepals adpressed, not glandular. Fruits with non-glandular pits and a distinct furrow below; beak 2·5–5 cm. Dry places. GR.CR.TR.   **654a.** *E. absinthoides* Willd. Flowers violet, in umbels of 2–8, on stems to 20 cm; petals about 10 mm; sepals 10–13 mm, with spreading glandular hairs. Leaves twice-pinnate, or twice-lobed, the ultimate segments linear-lanceolate. Fruit covered with dense white ascending hairs, pits glandular-hairy, without a furrow at base; beak 4·5–6·5 cm. Rocks, stony ground in mountains. YU(south).GR(north).BG.   **654b.** *E. chrysanthum* DC. Distinguished from all species by its bright-yellow petals 8–10 mm. Leaves twice-pinnate, silvery-haired. Fruit as in 654a. Open stony ground in mountains. GR(central and south). Pl. 26.

ZYGOPHYLLACEAE | Caltrop Family

1. Lvs. alternate, divided to the base
   into linear-pointed lobes.   *Peganum*
1. Lvs. opposite, with 2 or more distinct
   leaflets.
   2. Stipules spiny; fr. not spiny.   *Fagonia*
   2. Stipules not spiny; fr. spiny.   *Tribulus*

PEGANUM 655a. *P. harmala* L. A somewhat fleshy-leaved, glaucous, erect, and much-branched perennial 30–60 cm, with solitary terminal greenish-white flowers 1–2 cm across. Petals 5; sepals 5, linear, persisting; stamens 12–15. Leaves deeply and irregularly cut into many narrow lobes. Fruit a globular, stalked capsule 7–10 mm. Dry waste places, steppes. YU.GR.CR? BG.TR.

FAGONIA 655b. *F. cretica* L. A prostrate, much-branched perennial 10–40 cm, with attractive, solitary, magenta flowers 1–1·5 cm across, and trifoliate leaves with spiny stipules. Sepals and petals 5; stamens 10. Leaflets linear 0·5–1·5 cm, spine-tipped, the leaf resembling the imprint of a bird's foot; stems angled. Fruit of 5 sharply angled carpels, angles ciliate. Dry stony places. GR.CR.

TRIBULUS **655. *T. terrestris* L. MALTESE CROSS. A hairy, creeping annual 10–60 cm, with neat compound leaves which have 5–8 pairs of elliptic leaflets, and small solitary, yellowish, axillary flowers about 1 cm across. Fruit very distinctive, with 5 lobes spreading in a star, each lobe bearing 2 long and 2 short hard spines. Waste places, waysides, cultivated ground. Widespread.

LINACEAE | Flax Family

LINUM | Flax 27 species in our area, including 4 endemic species.

Flowers yellow
(a) *Perennials*
658c. *L. arboreum* L. A shrub to 1 m, with thick persistent oblanceolate leaves, often in dense clusters, at the ends of woody branches. Flowers few, in a rather compact cluster; petals yellow, 1·2–1·8 cm. Limestone rocks. GR.CR+Karpathos, Rhodes. Page 67. **658d.** *L. capitatum* Schultes A herbaceous plant with a woody rhizome often terminating in leaf rosettes. Flowering stems stout, angular, 10–40 cm; flowers yellow, 5–10 or more, in a dense flat-topped head; petals 1·5–2 cm; sepals long-pointed. Rosette leaves oblong-spathulate, stem leaves linear-lanceolate. Rocky mountain slopes. YU.AL.GR.BG. Page 324. **658a.** *L. maritimum* L. A tall erect perennial to 80 cm, with the lower leaves opposite, three-veined, the upper alternate, one-veined. Petals yellow, 0·8–1·5 cm; sepals 3 mm, white, ovate, glandular-hairy; stigmas club-shaped. Damp saline soils. YU.GR.

L. flavum group—YELLOW FLAX. Hairless perennials with woody stock and erect or ascending flowering stems, with yellow flowers. Sepals lanceolate, usually with glandular marginal hairs. A difficult group:

317

1. Inflorescence with 1–9 fls.; stem
usually less than 15 cm. Dwarf plants
of mountains with compact basal leaf
rosettes (YU.AL.GR.BG?).   **658e.** *L. elegans* Boiss.
1. Inflorescence with more than 10 fls.;
stems up to 60 cm; grassy places in
hills.
  2. Inflorescence with 20–40 fls.; stock
    erect or ascending, little-branched.
    3. Sepals 5–8 mm, scarcely longer than
      the capsule. Robust plant with
      erect flowering stems up to 60 cm
      (YU.AL.BG.).   **658.** *L. flavum* L. Pl. 26.
    3. Sepals 8–10 mm, up to twice as long
      as the capsule. Less robust, with
      stem usually 20–40 cm (YU.GR.BG.
      TR.).   **658f.** *L. thracicum* Degen Pl. 26.
  2. Inflorescence usually with 10–20
    fls.; stock much branched with
    flowering stems up to 40 cm. Sepals
    6–8 mm, much longer than the
    capsule (YU.AL.GR.BG.TR.).   **658g.** *L. tauricum* Willd. Pl. 27.

(b) *Annuals*

657. *L. strictum* L. UPRIGHT YELLOW FLAX. Petals yellow, 6–12 mm; sepals 4–6 mm, margin glandular-hairy. Inflorescence dense, spike-like, or flat-topped, or lax. Leaves without glands, margin of leaf rough, minutely toothed. Grassy places in the Med. region. Widespread. **657b.** *L. trigynum* L. Distinguished from 657 by its smaller yellow flowers with petals 4–6 mm; sepals 3–4 mm, and leaves with smooth untoothed margin. Grassy and stone places. Widespread. **(657).** *L. nodiflorum* L. Distinguished by its larger, solitary or few yellow flowers in a lax branched cluster; petals about 2 cm, with a long claw, and sepals 8–13 mm. Upper leaves with a pair of brown glands at the base; stem narrowly winged. Grassy places, olive groves in the Med. region. Widespread.

Flowers pink
(a) *Plants hairy*

662. *L. viscosum* L. STICKY FLAX. A hairy perennial to 60 cm, easily recognized by its glandular sticky lanceolate leaves 3–8 mm wide, its large pink flowers veined with violet about 3 cm across, and its glandular-hairy sepals. Petals 2½–3 times as long as sepals which are 6–9 mm. Grassy places in mountains. YU(Croatia). **663a.** *L. pubescens* Banks & Solander A small finely hairy annual, usually less than 20 cm, with pink petals twice as long as the sepals which are 8–10 mm. Leaves lanceolate, up to 5 mm wide. In herbaceous vegetation in the Med. region. AL.GR.CR.

(b) *Plants hairless*

663. *L. tenuifolium* L. A usually hairless perennial 20–45 cm, with long-stalked pink or pale-lilac flowers, in a lax, rather flat-topped cluster. Petals 2–2½ times as long as sepals which have glandular-hairy margins. Leaves linear, 1 mm wide. Grassy and rocky places, dry hills. YU.AL.GR.BG.TR. Pl. 26.

Flowers blue or violet (about 15 species in our area, not easily distinguished)
(a) *Plants hairy*

662a. *L. hirsutum* L. A hairy perennial with robust many-flowered stems to 45 cm, with rather large pale-blue flowers with petals 2–2·8 cm. Sepals 8–12 mm. Leaves 1–4·5 cm, up to 1 cm wide, ovate or oblong, acute or obtuse, three- to five-veined. Dry grassy places. YU.AL.GR.BG.TR. Pl. 27. **662b.** *L. spathulatum* (Halácsy & Bald.) Halácsy Like 662a but a woody-based, slender, spreading plant up to 25 cm with short non-flowering shoots present at flowering. Flowers few, dull violet; petals 2·5 cm. Mountains. AL.GR(north). Page 127.

(b) Plants hairless or almost so
659. *L. perenne* L. PERENNIAL FLAX. Distinguished by its knob-like globular stigmas (not club-shaped). Petals blue, 1·5–2 cm; sepals oval, unequal, the outer narrower, the inner with a broad papery margin, about half as long as the fruit which is 5–8 mm. Flower stalks erect, straight. Meadows, in mountains. YU.AL.GR.BG. **659a. *L. austriacum* L. Differs from 659 in having reflexed or curved flower stalks. Dry grassy places. Scattered throughout the mainland. **660.** *L. bienne* Miller Distinguished by its slender linear stigmas. Petals blue, 8–12 mm, 2–3 times as long as the sepals. Sepals oval, long-pointed, nearly equal, the inner with papery, ciliate margin, the outer entire. Grassy, stony places. Widespread on mainland. **(660).** *L. usitatissimum* L. CULTIVATED FLAX. Of uncertain origin; widely cultivated and sometimes naturalized in our area.

661a. *L. hologynum* Reichenb. Distinguished from all other European species by its styles which are fused together almost to the apex. Flowers blue, violet, or pinkish-purple; petals 2½ times as long as sepals. Sepals 5–8 mm, lanceolate long-pointed, all with papery margins. Leaves linear or thread-like, 1 mm wide. Mountain meadows. YU.AL.GR.BG. **661b.** *L. aroanium* Boiss. & Orph. A blue-flowered mountain perennial of Greece with petals 2½–3 times as long as the sepals. Petals 12–18 mm; sepals 5–7 mm, lanceolate long-pointed, with narrow papery, ciliate margins. Stigmas club-shaped. Leaves with rough margins, one-veined. Fruit 5–7 mm. GR(central and south). Page 92.

Flowers white
658h. *L. leucanthum* Boiss. & Spruner A densely tufted plant with a woody stock and compact basal leaf rosettes, and short flowering stems usually less than 15 cm bearing one or few white flowers. Petals about 2 cm; sepals narrowly lanceolate, pointed, much longer than the fruit. Leaves spathulate, thick, usually densely hairy. Rocks in mountains. GR(south-east). **664.** *L. catharticum* L. PURGING FLAX. Fields, rocky places. Widespread on mainland.

EUPHORBIACEAE | Spurge Family

CHROZOPHORA ****665.** *C. tinctoria* (L.) A. Juss. TURN-SOLE. A green or greyish, densely hairy Dogs Mercury-like annual to 50 cm, with wedge-shaped leaves, and short clusters of yellowish flowers; the male spike-like erect, the female flower solitary at the base of the spike, long-stalked, drooping. Fruit with 3 rounded segments, covered with conspicuous swellings. Waysides in the Med. region. Widespread. **665a.** *C. obliqua* (Vahl) Sprengel Similar to 665 but plant whitish, with dense hairs, and with cut-off or shallowly heart-shaped bases to the leaves; stamens 4–5 (leaf bases wedge-shaped; stamens 9–11 in 665). Waysides in the Med. region. GR.CR.

MERCURIALIS | Mercury ****666.** *M. perennis* L. DOGS MERCURY. Woods, shady places. YU.AL.GR.BG. **666b.** *M. ovata* Sternb. & Hoppe Like 666 but smaller and with stem leafy throughout (not crowded towards the top and scale-like below as in 666). Leaves ovate to rounded; leaf stalk usually 1–2 mm (5–10 mm in 666). Woods, bushy places. AL.GR.BG.TR. **667.** *M. annua* L. ANNUAL MERCURY. Cultivated ground. Widespread.

RICINUS **668. *R. communis* L. CASTOR OIL PLANT. A robust herb or shrub to 4 m, easily identified by the large palmately five- to seven-lobed leaves up to 60 cm long, and the dense erect clusters of reddish or yellowish flowers. Fruit large, 1–2 cm, often reddish, covered with conical spines, or smooth; seeds large, shining, reddish-brown to blackish-mottled, with a swelling, poisonous. A native of the tropics; cultivated for its oil-bearing seeds; naturalized in YU.AL.GR.CR.BG.

EUPHORBIA | Spurge About 65 species occur in our area, many of them are similar in general appearance and difficult to identify. The leaves below the ultimate branches of the inflorescence, the secondary rays, are called collectively here 'floral bracts'. Distinctive species include:

Woody plants, with or without spines
669. *E. dendroides* L. TREE SPURGE. A robust shrub with numerous stout branches forming dense rounded bushes up to 2 m, with thick, glaucous, oblong-lanceolate blunt leaves 2·5–6·5 cm which fall off in the summer. Glands of involucre irregularly lobed. Fruit smooth, or nearly so. Rocky slopes, usually near the sea. YU.AL.GR+Is.CR. **670. *E. spinosa* L. SPINY SPURGE. A low, twiggy, branched shrublet to 30 cm, with the older branches forming weak woody spines. Leaves small 1 cm, lanceolate; floral bracts wider than stem leaves. Glands oval. Fruit weakly three-angled, covered with long conical swellings. Dry stony places. YU.AL.

(670). *E. acanthothamnos* Boiss. GREEK SPINY SPURGE. Like 670 with branches becoming spiny, but much more branched and forming a low domed shrub with intertwined branches, like wire-netting. Rays of umbels becoming spiny; stem leaves and floral bracts similar. Fruit strongly three-angled, densely covered with short cylindrical swellings. YU?GR+Is.CR. Page 96. **670a *E. glabriflora* Vis. A small spineless shrub 10–20 cm, of stony mountain slopes, with lanceolate to linear-lanceolate leaves and obovate yellowish floral bracts. Rays usually 3–5, each with one 'flower'. Fruit globular, with long cylindrical swellings. YU.AL.GR.

Herbaceous plants, without spines
(a) Glands of involucre oval or rounded
670b. *E. capitulata* Reichenb. A small perennial of mountain screes and rocks with many short stems 1–10 cm, and numerous rounded usually overlapping leaves, each less than 1 cm. Flower cluster solitary (one-rayed), with 8 purple glands. Fruit covered with swellings. YU.AL.GR. Page 178. ****671.** *E. helioscopia* L. SUN SPURGE. A hairless, unbranched annual 10–50 cm, with a broad flat-topped, often golden-yellow inflorescence, and obovate, finely toothed leaves 1·5–3 cm, and similar floral bracts. Fruit without swellings. Cultivated and waste ground. Widespread. ****672.** *E. villosa* Willd. A robust hairless or hairy perennial 30–120 cm, with numerous stems and lanceolate leaves 5–10 cm, which are hairy beneath and finely toothed towards the apex. Floral bracts ovate, yellowish, often toothed, not fused. Fruit smooth, or with swellings. Damp places. YU.AL.GR.BG.

(b) Glands of involucre sickle-shaped or with horn-like projections
(i) Annuals or biennials, with a single stem
**676. *E. lathyris* L. CAPER SPURGE. A robust biennial with a stout erect stem to 1·5 m,

branched above, with 4 very regular ranks of bluish-green linear to lanceolate leaves 3–15 cm. Floral bracts triangular-ovate, not fused. Glands half-moon-shaped, with 2 blunt club-shaped horns. Fruit smooth, hairless. Damp places. YU.GR.BG.   **675a.** *E. taurinensis* All. (incl. *E. graeca*) A hairless annual to 15 cm, with linear-oblanceolate leaves 2–3 cm with short leaf stalks. Glands yellow, with 2 pink horns. Fruit shallowly grooved. Disturbed ground. Widespread.

(ii) Perennials, usually with several stems arising from the base

**678. *E. characias* L. LARGE MEDITERRANEAN SPURGE. A stout, very striking rather woody plant with several stems to 1·5 m or more, and oblong to lanceolate, leathery leaves 3–13 cm, very numerous above, but stem leafless below. Inflorescence cylindrical; floral bracts fused in pairs. Fruit densely woolly-haired. Subsp. *wulfenii* (Koch) A.R.Sm. which occurs in the eastern Med. region has yellowish glands with long horns. Rocky ground. YU.AL.GR+Is.CR.

682. *E. myrsinites* L. BROAD-LEAVED GLAUCOUS SPURGE. A robust procumbent perennial to 40 cm, distinguished by its numerous broad overlapping, fleshy, glaucous leaves borne on stout, fleshy, spreading stems. Glands with long, club-shaped horns, often bright reddish-brown; floral bracts rounded to broadly heart-shaped. Leaves obovate, abruptly pointed. Fruit smooth, hairless. Rocky and grassy places. Widespread.   **683.** *E. rigida* Bieb. (*E. biglandulosa*) NARROW-LEAVED GLAUCOUS SPURGE. Like 682 with stout stems with glaucous fleshy overlapping leaves but leaves narrow-lanceolate and narrowed to a long point; stems erect or ascending 30–50 cm. Glands half-moon-shaped, with short blunt rounded horns. Fruit strongly three-angled, hairless, granular when dry. Dry stony places, rocks. AL.GR+Is.CR? TR. Pl. 28.

684. *E. paralias* L. SEA SPURGE. A glaucous, stiff erect, maritime perennial 30–60 cm, with numerous, often closely overlapping, thick and fleshy ovate to elliptic acute leaves, and broader cup-shaped floral bracts. Glands with short horns. Fruit deeply grooved, hairless, granular. Maritime sands and dunes. Widespread.   **684a. *E. herniariifolia* Willd. A small mat-forming perennial to 20 cm, of limestone mountains with tiny rounded leaves 4–10 mm and similar floral bracts. Rays 2–3; glands with 2 horns. Fruit with 2 wings on each keel, hairless or hairy. AL.GR.CR,Samos. Page 93.

ANDRACHNE   ****685.** *A. telephioides* L. A slender, glaucous, usually spreading, wiry perennial 8–15 cm, with numerous tiny, alternate, oval leaves 3–10 mm. Flowers yellowish-green, solitary, on a short stalk often shorter than the leaves, one-sexed. Fruit globular, hairless, on reflexed stalks. Waste places, waysides. YU.GR+Is.CR.BG.

RUTACEAE | Rue Family

RUTA   Petals and sepals usually 4; petals usually finely toothed or ciliate.

686. *R. graveolens* L. COMMON RUE Very strong-smelling, glaucous-leaved, shrub-like perennial 14–45 cm, and with lax clusters of yellow flowers with undulate petals with finely toothed or entire margins. Leaves 2–3 times pinnately divided into lanceolate or oblong segments. Fruit with blunt spreading lobes. Rocky places. YU.AL.BG.TR.; introd. GR.   **(686). *R. montana* (L.) L. Like 686 but leaf segments linear, fruit stalks shorter than the capsule, and yellow flowers with undulate petals without marginal hairs or teeth. Rocky places in the Med. region. GR.TR.

687. *R. chalepensis* L. FRINGED RUE. A small yellowish-green shrub-like plant 20–60 cm, with lax clusters of yellow flowers, the petals conspicuously fringed with long hairs. Flowers 2 cm across. Leaves strong-smelling, once or twice pinnately divided, with oblong-elliptic segments. Fruit with long-pointed erect lobes. Rocky places in the Med. region. YU.AL.GR+Is.CR.

HAPLOPHYLLUM Distinguished from *Ruta* by the simple undivided, or three-lobed leaves. Sepals and petals 5; stamens 10, filaments hairy below. Fruit five-lobed. 7 rather similar-looking species in our area, including 2 endemic species.

687b. *H. suaveolens* (DC.) G.Don fil. An erect, woody-based, curly-haired perennial 15–30 cm, with lanceolate to oblong-lanceolate, stalkless leaves 0·7–3 cm, and very dense flat-topped clusters of yellow flowers. Flowers 1·5–2 cm across; petals blunt, entire, gland-dotted; sepals hairless to sparsely white-felted. Fruit usually hairless, strongly warted. Dry grassy places, steppes. BG.TR. Pl. 27. **687c.** *H. balcanicum* Vandas Similar to 687b but quite hairless, and with all leaves linear-lanceolate. Rocky places. GR(north-east).BG. **687d.** *H. coronatum* Griseb. An erect perennial to 40 cm, with numerous small narrow leaves 0·8–2·5 cm long, those in the middle of the stem three-lobed, the upper and lower entire. Inflorescence compact, white-felted; petals yellow, about 8 mm. Fruit with warts, and segments crowned by a large tooth-like appendage. Amongst herbaceous vegetation. YU(south).AL.GR. Page 324.

DICTAMNUS **688.** *D. albus* L. BURNING BUSH. An unmistakable, strongly aromatic, bushy perennial 40–80 cm, with a large lax terminal spike of conspicuous pink or white flowers. Flowering stems, bracts, and sepals densely covered with conspicuous dark glands. Petals 2–2·5 cm, streaked and dotted with purple, unequal with 4 erect or spreading and 1 reflexed. Leaves pinnate, with 3–6 pairs of leathery leaflets. Bushy places, woods, stony ground. YU.AL.GR+Is.BG.

CITRUS The following are cultivated largely in the Med. region for their fruit and aromatic oils: **689.** *C. aurantium* L. SEVILLE ORANGE; **690.** *C. sinensis* (L.) Osbeck SWEET ORANGE; **693.** *C. limon* (L.) Burm.fil. LEMON.

SIMAROUBACEAE | Quassia Family

AILANTHUS **694.** *A. altissima* (Miller) Swingle TREE OF HEAVEN. A freely suckering tree to 20 m, with smooth grey bark. Leaves to 90 cm long, with 13–25 large oval, long-pointed, toothed leaflets. Flowers yellowish-green in branched clusters. Fruit conspicuous 3–4 cm, reddish-brown, flattened, with a broad propeller-like wing. Native of China; often grown as an ornamental tree and for soil conservation; naturalized in YU.AL.GR.BG.

MELIACEAE | Mahogany Family

MELIA **695.** *M. azedarach* L. PERSIAN LILAC, INDIAN BEAD TREE. Flowers lilac in branched clusters 10–20 cm; petals about 18 mm. Leaves twice-pinnate; leaflets elliptic, toothed or lobed. Fruit yellow, pea-sized. Native of China; commonly planted as a roadside tree; sometimes naturalized in YU.CR.

POLYGALACEAE | Milkwort Family

POLYGALA | **Milkwort** About 16 similar-looking species are found in our area, 7 of which are restricted to the Balkans. Wings refer to the 2 larger, outer, coloured sepals. Gynophore is the stalk of the capsule.

Annuals
**697. *P. monspeliaca* L. MONTPELLIER MILKWORT. A slender, little-branched erect annual 10–30 cm, with whitish or greenish-white flowers in a lax terminal spike. Wings greenish-white, conspicuously three-veined, 6–8 mm, and much longer than the white petals. Rocky places, open woods. Widespread.

Perennials
The following are widely scattered on the mainland: ****698.** *P. vulgaris* L. COMMON MILKWORT; **700.** *P. comosa* Schkuhr TUFTED MILKWORT; **(700).** *P. nicaeensis* Koch.

699. *P. major* Jacq. LARGE MILKWORT. An erect or ascending woody-based perennial 15–60 cm, with very attractive, dense terminal spikes of many relatively large, rosy-purple, blue, or rarely white flowers. Wings 9–13 mm, enlarging in fruit; corolla tube 9–14 mm, bent upwards. Fruit 5–6 mm oblong, narrowly winged, gynophore 3–4 mm. Meadows. Widespread on mainland. Pl. 27. **699b. *P. anatolica* Boiss. & Heldr. Like 699 but densely tufted and not more than 40 cm. Flowers usually lilac-pink; wings 7–10 mm. Gynophore less than 3 mm in fruit. Grassy places. GR.BG.TR.

699c. *P. supina* Schreber A spreading or erect woody-based perennial with numerous stems 8–30 cm. Flowers blue; wings slightly oblique, rounded to lanceolate, slightly longer than the petals. Filaments of stamens united throughout their length. Style more than twice as long as ovary, slightly curved at apex. Lower leaves obovate to rounded. Stony slopes, mountain rocks. Widespread on mainland. Page 324.

699d. *P. venulosa* Sibth. & Sm. Distinguished by its dull-bluish flowers with corolla tube longer than the wings, which are 7–9 mm, white or lilac with green veins. Fruit stalkless, shorter and wider than the wings. Stony places in the Med. region. GR+Is.CR.BG? Page 324. **699e.** *P. subuniflora* Boiss. & Heldr. A small spreading perennial 5–8 cm, of mountain rocks of Aroania (Chelmos) of Peloponnisos. Flowers blue, 1 or 2 in one-sided clusters; wings longer than petals; filaments fused for most of their length. Leaves numerous, overlapping, obovate. GR(south).

ANACARDIACEAE | Cashew Family

PISTACIA ****703.** *P. lentiscus* L. MASTIC TREE, LENTISC. A dense dark-green evergreen shrub or small tree 1–8 m, with pinnate leaves which have 4–6 pairs of shining elliptic leathery leaflets without a terminal leaflet, and a winged rachis. Flowers yellowish or purple in dense axillary clusters. Fruit aromatic, about 4 mm, red, then black. Dry open scrub, woods. YU.AL.GR+Is.CR. Page 39. ****704.** *P. terebinthus* L. TURPENTINE TREE,

1. *Hypericum triquetrifolium* 768a 2. *Viola parvula* 784c 3. *Viola gracilis* 786c
4. *Polygala venulosa* 699d 5. *Linum capitatum* 658d 6. *Polygala supina* 699c
7. *Haplophyllum coronatum* 687d

TEREBINTH. Distinguished from 703 by its deciduous, pinnate leaves with a terminal leaflet, and without a winged rachis; leaflets 1–4 pairs, ovate, reddish when young. Flowers reddish-purple, in branched clusters appearing with young leaves; fruit aromatic 5–7 mm long, red then brown. Scrub, dry open woods. Widespread.

704a. *P. atlantica* Desf. Like 704 but leaflets lanceolate, blunt (with a fine point in 704). Leaf stalks finely hairy; rachis narrowly winged. Bushy places. GR(northeast)+Rhodes.TR. **704b.** *P. vera* L. PISTACHIO. A small tree with deciduous leaves which usually have 1–2 pairs of ovate or broadly lanceolate, thin, not leathery leaflets, and a winged, hairy rachis. Fruits 1·6–2·9 cm, in branched clusters. Native of Asia; cultivated in Chios, Lesbos, Crete, and elsewhere for its edible nuts.

SCHINUS 705. *S. molle* L. CALIFORNIAN PEPPER-TREE, PERUVIAN MASTIC-TREE. A small slender shrub or tree to 8 m, with pendulous branches and pinnate leaves, with 7–13 pairs of narrow lanceolate leaflets. Flowers white, in branched clusters. Fruit pink, globular 6–7 mm, in pendulous clusters. Native of America; often planted for ornament; sometimes naturalized in GR.

COTINUS ****706.** *C. coggygria* Scop. (*Rhus cotinus*) WIG TREE, or SMOKE TREE. A dense, spreading, hairless, shrub to 3 m, with neat rounded to obovate, stalked leaves 3–8 cm, which are glaucous beneath. Flowers yellowish, in terminal lax pyramidal clusters, with numerous sterile flowers. In fruit the whole inflorescence elongates and becomes grey and plume-like owing to the hairy sterile flower stalks. Dry hills, rocky places, open woods. Widespread, except on smaller islands. Page 35.

RHUS 707. *R. coriaria* L. SUMACH. A shrub or small tree to 3 m, distinguished by its large softly hairy, pinnate leaves with 7–21 oblong to ovate, coarsely toothed leaflets, and a winged rachis. Flowers whitish, in dense, erect, terminal elongated clusters. Fruit fleshy, 3–4 mm, hairy, brownish-purple. Rocky places, waysides. Widespread. **(707).** *R. typhina* L. STAGHORN SUMACH. Like 707 but rachis of leaves not winged. Fruit crimson. A N. American shrub often cultivated for ornament; occasionally naturalized in YU.BG.

ACERACEAE | Maple Family

ACER | Maple   11 species in our area.

Leaves five- to seven-lobed, or pinnate
Widespread on the mainland are:
708. *A. pseudoplatanus* L. SYCAMORE; ****710. *A. platanoides* L. NORWAY MAPLE; **711.** *A. campestre* L. COMMON MAPLE.

709b. *A. hyrcanum* Fischer & C. A. Meyer BALKAN MAPLE. A shrub or small tree with yellow flowers in nearly stalkless clusters opening before the leaves. Leaves to 10 cm, with 5 narrow, parallel-sided acute lobes cut to about half-way, which are hairless or slightly hairy below. Fruit hairless, with straight wings diverging at an acute angle. Woods in mountains. YU.AL.GR.BG. Pl. 27. **709c.** *A. obtusatum* Willd. Like 709b but distinguished by its leaves which are often densely hairy beneath, with 5, or rarely 3, blunt, wide, shallow lobes. Stalks of flower clusters hairy. Woods in mountains. YU.AL.GR.

708a. *A. heldreichii* Boiss. GREEK MAPLE. A tree to 25 m, with quite distinctive leaves which are deeply five-lobed, with the middle lobe free nearly to the base, all lobes pointed and with 2 or 3 large teeth on each side, and usually glaucous beneath. Flowers rather few, yellowish, in more or less erect clusters appearing with the leaves. Fruit hairless, with pointed wings usually diverging at an obtuse angle. Mountains. YU.AL.GR.BG. Pl. 27. **(712).** *A. negundo* L. BOX-ELDER. Leaves pinnate with usually 3–5 ovate-pointed leaflets. Native of N. America; widely planted as a wayside shade tree; and sometimes naturalized in BG.

Leaves three-lobed or entire

712. *A. monspessulanum* L. MONTPELLIER MAPLE. A small tree to 12 m, easily recognized by its leathery three-lobed leaves with the lobes untoothed and diverging almost at right angles to each other. Fruit with parallel or converging wings, becoming hairless. Shady places in hills and mountains. YU.AL.GR.BG.TR? **(712).** *A. tataricum* L. TARTARIAN MAPLE. Readily distinguished from other Maples by its usually unlobed, deciduous leaves, which are oval to heart-shaped and irregularly double-toothed, though leaves sometimes shallowly three-lobed. Flowers greenish-white, in erect clusters. Fruit becoming reddish, usually with parallel wings. Bushy places. YU.AL.GR.BG. **712c.** *A. sempervirens* L. (*A. orientale*) An evergreen or semi-evergreen shrub or small tree to 5 m with small, three-lobed, or less commonly undivided, leathery leaves 2–5 cm, with short leaf stalks. Flowers few, yellowish-green, in erect nearly stalkless clusters. Fruit with nearly parallel or narrowly diverging wings. Bushy places in hills and mountains. GR+Is.CR. Page 70.

SAPINDACEAE | Soapberry Family

CARDIOSPERMUM **712a.** *C. halicacabum* L. HEART-SEED, HEART-PEA, BALLOON-VINE. A tendril-climbing annual to 2 m, with twice-ternate leaves, and small white flowers in long-stalked, axillary clusters. Flowers with 4 sepals and 4 petals 4 mm. Fruit 3 cm, globular or three-lobed, dry, papery, seeds black with a white heart-shaped scar. Native of the tropics; grown for ornament; locally naturalized in YU.GR.

HIPPOCASTANACEAE | Horse-Chestnut Family

AESCULUS ****713.** *A. hippocastanum* L. HORSE-CHESTNUT. A native tree of mountain woods of the central Balkans. YU.AL.GR.BG. Often grown for ornament and shade elsewhere.

BALSAMINACEAE | Balsam Family

IMPATIENS ****714.** *I. noli-tangere* L. TOUCH-ME-NOT. Flowers yellow, 2–3·5 cm, with a curved spur 6–12 mm. Shady woods, by streams. YU.GR.BG. ****716.** *I. glandulifera*

Royle POLICEMAN'S HELMET. A very robust, fetid annual 1–2 m, with terminal clusters of large pink or white flowers each 2·5–4 cm. Native of the Himalaya; naturalized by streams in YU.

AQUIFOLIACEAE | Holly Family

ILEX **717. *I. aquifolium* L. HOLLY. Mountain woods. YU.AL.GR.BG. **717a.** *I. colchica* Pojark. A spreading shrub 1–3 m. Leaves blackish when dry; leaf stalk with a narrow deep groove. Forests, scrub, shady ravines. TR.

CELASTRACEAE | Spindle-Tree Family

EUONYMUS **718. *E. europaeus* L. SPINDLE-TREE. Bushy places, woods. Widespread on mainland. **(718).** *E. latifolius* (L.) Miller Like 718 but with larger leaves 8–16 cm; spindle-shaped pointed buds 7–12 mm, and young branches greyish-brown instead of green. Fruit usually with 5 winged segments (4 in 718). Mountain woods. Widespread on mainland. **718b.** *E. verrucosus* Scop. WARTED SPINDLE-TREE. Young branches green, circular in section (quadrangular in 718), and covered with dark-brown warts. Fruit with 4 rounded segments. Black seeds only partly covered by orange aril. Woods, bushy places. Widespread on mainland.

STAPHYLEACEAE | Bladdernut Family

A small family of trees and shrubs with odd-pinnate leaves. Flowers in branched clusters; sepals, petals and stamens 5. Ovary three-celled; fruit a papery inflated, lobed capsule, splitting at the apex, few-seeded.

STAPHYLEA **718c.** *S. pinnata* L. BLADDERNUT. A shrub or small tree to 6 m, with pinnate leaves with 5–7 leaflets, and a pendulous cluster of small whitish flowers. Petals 6–10 mm, white, about as long as the pale slightly pinkish sepals. Leaflets 5–10 cm ovate-oblong, pointed, finely toothed, hairless. Fruit 2·5–4 cm rounded, green, papery; seeds about 1 cm. Woods. YU.BG. Pl. 27.

BUXACEAE | Box Family

BUXUS
719. *B. sempervirens* L. BOX. Woods, bushy places. YU.AL.GR.TR.

RHAMNACEAE | Buckthorn Family

RHAMNUS | Buckthorn   11 species in our area, including 3 endemic species.

Evergreen shrubs

**720. *R. alaternus* L. MEDITERRANEAN BUCKTHORN. An evergreen, spineless shrub, varying in habit, usually erect to 5 m, but sometimes prostrate in the mountains, with
lanceolate to ovate, leathery shining leaves 2–6 cm. Flower clusters yellowish, small,
dense, axillary; petals absent. Fruit 4–6 mm globular, at first reddish, then black.
Evergreen scrub in the Med. region. YU.AL.GR+Is.CR. Page 38.   **720a** *R. lycioides* L.
A much-branched spiny shrub to 1 m, with leathery, linear to obovate, evergreen or
deciduous leaves 0·5–2 cm, and obovoid, laterally compressed, yellowish or black fruits
when ripe. Flowers yellowish, clustered, usually with 4 sepals and stamens. Dry stony
places, rocks in the Med. region. GR+Is.CR. Subsp. *graecus* (Boiss & Reuter) Tutin has
deciduous obovate leaves 6–18 mm. S.Greece and Aegean region.

Deciduous shrubs or trees (see 720a)
(a) Spineless shrubs; leaves alternate

722. *R. alpinus* L. ALPINE BUCKTHORN. An erect spineless shrub 1–4 m, with large
alternate, elliptic, toothed leaves 4–10 cm, with prominent veins beneath. Subsp. *fallax*
(Boiss.) Maire & Petitmengin is the plant of our area. It has hairless twigs, and
ciliate but otherwise hairless bud scales, and leaves green on both sides with 10–
20 pairs of lateral veins. Bushy places in mountains. YU.AL.GR.BG.   **722a.** *R.
sibthorpianus* Roemer & Schultes SIBTHORP'S BUCKTHORN. An erect, spineless shrub
easily recognized by the dense covering of short cottony hairs on both surfaces of the
ovate to rounded leaves. Leaves 1–9 cm, lateral veins 6–12 pairs. Mountain rocks
GR(south).

(b) Usually spiny small trees or shrubs; lower leaves mostly opposite

**723. *R. catharticus* L. BUCKTHORN. A thorny shrub or small tree 4–6 m, with ovate-
elliptical leaves 3–7 cm, with 2–4 pairs of conspicuous lateral veins. Fruit globular, in
clusters, 6–8 mm, black. Woods. YU.AL.GR? BG.   **(723).** *R. saxatalis* Jacq. ROCK BUCK-
THORN. A prostrate, or erect much-branched, very spiny shrub to 2 m, distinguished from
723 by its smaller narrow leaves 1–5 cm, with short leaf stalks about as long as the
stipules (leaf stalks much longer in 723). Fruit 5–8 mm, black. Rocky places.
YU.AL.GR.BG.

FRANGULA   **724.** *F. alnus* Miller ALDER BUCKTHORN. Damp shady places. Widespread
on mainland.   **724a.** *F. rupestris* (Scop.) Schur *(Rhamnus r.)* An ascending or procumbent shrub to 80 cm of the western Balkans. Flowers yellowish, in small flat-topped,
distinctly stalked clusters (flowers solitary or in stalkless clusters in 724). Petals and
stamens 5. Leaves elliptic to rounded, finely toothed, with 5–8 pairs of lateral veins.
Fruit about 6 mm, red becoming black. Rocks, bushy places in mountains. YU.AL.GR.
Page 179.

ZIZIPHUS   **725.** *Z. jujuba* Miller COMMON JUJUBE. A spiny shrub or small tree to 8 m,
with green twigs, two-ranked leaves, and conspicuous dark-reddish or black, ovoid-
oblong edible fruits, 1·5–3 cm. Flowers yellowish, few, in axillary clusters. Leaves
2–5·5 cm, oblong-blunt, with finely glandular-toothed margin. Native of Asia; cultivated in the Med. region and frequently naturalized in our area.   **725a.** *Z. lotus* (L.)
Lam. Distinguished from 725 by its grey, hairless, zigzag twigs, its very shallowly
glandular-crenate, oval leaves about 1·5 cm, and its globular, deep-yellow, pea-sized
fruits. Dry places. GR.

PALIURUS 726. *P. spina-christi* Miller CHRIST'S THORN. An extremely spiny deciduous shrub 2–3 m, with slender often intricately intertwined branches, two-ranked leaves, and distinctive fruits 'like miniature toadstools'. Flowers tiny, yellow, in axillary clusters. Fruit 2–3 cm across, hemispherical with a flattened disk-like rim, ribbed and undulate on the margin. Sandy arid hills, bushy places. Widespread on the mainland and islands, except CR. Page 34.

VITACEAE | Vine Family

VITIS 727. *V. vinifera* L. COMMON VINE. Cultivated throughout our area in a variety of forms, often grafted on to stock from American species. Subsp. *sylvestris* (C. C. Gmelin) Hegi has small bluish-black acid fruits about 6 mm, with usually 3 seeds. Probably native of south-east Europe. Subsp. *vinifera* has larger fruits 6–22 mm, which are sweet and vary in colour from green, yellow, red, or blackish-purple, with 2 or no seeds. It is the hybrid cultivated form used for fruit, wine-making, etc. Widespread.

TILIACEAE | Lime Tree Family

TILIA A difficult genus owing largely to hybridization and to the cultivation of hybrid forms and varieties for ornament.

Leaves white-woolly beneath, with star-shaped hairs
730. *T. tomentosa* Moench (*T. argentea*) SILVER LINDEN. Tree to 30 m; twigs woolly-haired. Leaves dark green above and contrasting paler beneath. Flowers 6–10, with both infertile and fertile stamens present. Woods and thickets. YU.AL.GR.BG.TR.

Leaves hairless beneath or with simple hairs
(a) Fruit strongly ribbed
729. *T. platyphyllos* Scop. LARGE-LEAVED LIME. Leaves 6–9 cm, saw-toothed. Flowers 2–5 in pendulous clusters; fruit usually 3. Woods, bushy places. YU.AL.GR.BG.TR? **729a.** *T. rubra* DC. Like 729 but teeth of leaves ending in a fine bristle. Woods. YU? AL? GR.BG.

(b) Fruit smooth or slightly ribbed
**731. *T. cordata* Miller SMALL-LEAVED LIME. Tree to 30 m with a spreading crown. Leaves 3–9 cm, hairless except for tufts of reddish-brown hairs on the vein axils beneath. Flower clusters ascending, with 4–15 flowers. Woods, bushy places. YU.AL.GR? BG.TR.

MALVACEAE | Mallow Family

MALOPE Fruits of numerous nutlets in a globular head; epicalyx lobes 3, wider than the sepals.

733. *M. malacoides* L. An erect or ascending, rough hairy perennial 20–50 cm, with oval, irregular-lobed, long-stalked leaves and large rose-coloured flowers veined with purple (recalling 742 *Lavatera trimestris*). Petals 2–4 cm. Fruits strawberry-like but uncoloured, surrounded by 3 short broad epicalyx lobes and longer, narrower sepals. Waste places, thickets in the Med. region. AL.GR.CR+E.Aegean Is.TR. Pl. 28.

KITAIBELA Distinguished from *Malva* by its fruits which are composed of numerous nutlets arranged in about 5 superimposed whorls to form a more or less globular head; epicalyx lobes 6–9.

733d. *K. vitifolia* Willd. A robust leafy plant to 3 m, with lobed leaves and large white, or pale pinkish, mallow-like flowers about 4 cm across. Petals broadly delta-shaped; epicalyx lobes ovate-pointed, slightly longer than sepals. Leaves long-stalked, with blades to 18 cm, with 5–7 triangular, toothed lobes. Fruit dark brown. Amongst herbaceous vegetation, thickets, vineyards. YU. Pl. 28.

MALVELLA Fruit with 9–12 nutlets which are inflated.

733c. *M. sherardiana* (L.) Jaub. & Spach Flowers deep pink; petals about 1 cm. A woody-based procumbent perennial 20–45 cm, with rounded or kidney-shaped toothed leaves 1–3·5 cm. Epicalyx lobes linear, much smaller than the sepals. Cultivated ground, stony places. GR.BG(south).

MALVA | Mallow Epicalyx lobes 2–3, in addition to the 5 sepals.

1. Sepals more than 3 times as long as wide, linear to narrowly triangular (GR+Is.CR.).	**736.** *M. cretica* Cav.
1. Sepals not more than 3 times as long as wide, ovate or triangular.	
2. Epicalyx-lobes ovate to wedge-shaped, not more than 3 times as long as wide.	
3. Ripe frs. smooth or faintly ribbed; lower fls. solitary in lv. axils (YU.BG.).	****735.** *M. alcea* L.
3. Ripe frs. distinctly net-veined; lower fls 2 or more in each lv. axil.	
4. Petals 12–30 mm, 3–4 times as long as sepals; lower surface of sepals with numerous small, star-shaped hairs (Widespread).	**737.** *M. sylvestris* L.
4. Petals 10–12 mm, not more than 2½ times as long as sepals; lower surface of sepals with few star-shaped hairs, or none (YU.GR+Is.CR.TR.).	****738.** *M. nicaeensis* All.
2. Epicalyx lobes linear to narrowly ovate, at least 3 times as long as wide.	
5. Lower fls. solitary in lv. axils, or all fls. in a terminal cluster.	
6. Petals about equalling sepals.	*aegyptia*
6. Petals three times as long as sepals (YU.AL.GR.BG.).	**734.** *M. moschata* L.

5. Lower fls. in groups in each lv.
   axil.
  7. Back of ripe frs. smooth, or only
     faintly ridged.
    8. Petals at least twice as long as
       sepals; calyx not enlarging
       in fr. (Widespread on mainland).   **(738).** *M. neglecta* Wallr.
    8. Petals less than twice as long
       as sepals; calyx much enlarging
       in fr. (introd.YU.GR.).   **739.** *M. verticillata* L.
  7. Back of ripe frs. distinctly net-
     veined.
    9. Petals at least 12 mm, usually
       bright purple or pink
       (Widespread).   **737.** *M. sylvestris* L.
    9. Petals less than 12 mm, pale
       pink or lilac.
      10. Calyx much enlarging in fr.;
         fr. stalks usually less than
         1 cm; angles of frs. winged
         (Widespread).   **738a** *M. parviflora* L.
      10. Calyx only slightly enlarging
         in fr.; fr. stalks usually
         more than 1 cm; angles of frs.
         not winged.
        11. Stamen tube hairy (YU.GR +
           Is.CR.TR.).   ****738.** *M. nicaeenisis* All.
        11. Stamen tube hairless
           (YU.GR.BG.TR.).   **(738).** *M. pusilla* Sm.

736b. *M. aegyptia* L. A bristly annual to 20 cm, with small lilac flowers, mostly in
terminal clusters, with petals about equalling the sepals. Epicalyx lobes usually 2,
linear; sepals 7–11 mm, broadly triangular-ovate. Leaves rounded in outline, deeply cut
into narrow segments. Fruits usually hairless. Arid hills. GR.CR.

LAVATERA Very like *Malva* but usually distinguished by its 3 epicalyx lobes which
are fused together at the base, but some species, notably 741, have lobes almost free
when in full flower.

1. Fls. in clusters in lv. axils.
  2. Epicalyx lobes longer than sepals,
     at least in fr.; stems woody below
     (YU.AL.GR.CR.).   ****740.** *L. arborea* L.
  2. Epicalyx lobes shorter than sepals;
     stem all herbaceous
     (YU.AL.GR.CR.TR.).   **741.** *L. cretica* L.
1. Fls. solitary (rarely in pairs) in lv.
   axils.
  3. Annuals; stems not densely woolly-haired.
    4. Central axis of fr. expanded above
       to a disc that covers and conceals the
       ripe fr.; stem rough-hairy with simple

or few-rayed reflexed hairs.   *trimestris*

4. Central axis of fr. not expanded; stem
    sparsely and minutely covered with
    star-shaped hairs.   *punctata*

3. Perennials; stems densely woolly-haired,
    at least when young.

    5. Shrub; fl. stalks usually not more
    than 1 cm at flowering and 1·3 cm
    in fr.   *bryoniifolia*

    5. Herb; fl. stalks usually more than
    1 cm at flowering and more than 1·3
    cm in fr.   *thuringiaca*

742. *L. trimestris* L. A tall annual to 120 cm, with large showy bright-pink flowers 5–7 cm across, and large baggy epicalyx. Flowers solitary; petals 2–4·5 cm; epicalyx lobes shorter than sepals, enlarging in fruit, united for most of their length. Cultivated ground. YU? GR+Rhodes **742a. *L. punctata* All. A branched erect annual 20–90 cm, often red-flushed, with solitary long-stalked lilac-pink flowers. Petals 1·5–3 cm; epicalyx lobes 6–8 mm, ovate; sepals 8–9 mm, enlarged and converging in fruit. Lower leaves rounded or kidney-shaped, shortly five-lobed, the upper spear-shaped with a long central lobe and short, spreading lateral lobes. Cultivated places, among herbaceous vegetation. AL.GR+Is.CR.TR.

744. *L. thuringiaca* L. A softly-hairy, greyish-green herbaceous perennial to 2 m. Flowers 5–8 cm across, pale pink with darker veining, long-stalked, forming a rather lax terminal cluster. Leaves rounded, five-lobed. Hills, thickets, waysides. Widespread on mainland. Pl. 28. **743b.** *L. bryoniifolia* Miller A coarsely hairy shrub to 2 m, with solitary or paired pinkish-violet flowers in the leaf axils. Petals 1·5–3 cm; epicalyx cup-like, shorter than the sepals. Leaves with 3–5 lobes, the middle and upper stem leaves spear-shaped with the middle lobe longer than the laterals, all toothed and densely covered with star-shaped hairs. Fruit hairless. Rocks, arid ground. GR+Is.CR.

ALTHAEA Distinguished from *Malva* and *Lavatera* by the epicalyx lobes which are 6–9, and united at the base.

Annuals

745. *A. hirsuta* L. HAIRY MALLOW. A bristly hairy annual to 60 cm, with pinkish-lilac flowers with petals about 1·5 cm, borne on flower stalks longer than the leaves. Leaves rounded in outline, becoming increasingly deeply lobed towards the apex of stem, the upper with linear, toothed lobes. Fruit transversely ribbed. Dry places. Widespread.

Perennials

746. *A. cannabina* L. A tall perennial to 2 m, readily distinguished by its deeply cut leaves which have 3–5 linear to lanceolate, toothed or lobed segments. Flowers pink, solitary or in long-stalked axillary clusters; petals 1·5–3 cm. Sepals erect in fruit; fruit hairless. Damp shady places. Widespread on mainland. ****747. *A. officinalis* L. MARSH MALLOW. A tall, densely grey-velvety-haired perennial to 2 m, with upper leaves triangular-ovate, toothed, entire or shallowly three- to five-lobed, and usually with very pale lilac-pink flowers. Petals 1·5–2 cm, rarely deeper pink. Sepals recurved in fruit; fruit hairy. Damp places, field verges. Widespread on mainland.

ALCEA Like *Althaea* but flowers very large, in wand-like spikes; epicalyx lobes usually 6, united at base. Fruit of 18–40 nutlets, each two-celled, the upper empty the lower one-seeded. 4 species in our area, and 1 naturalized species.

748. *A. pallida* (Willd.) Waldst. & Kit. EASTERN HOLLYHOCK. A tall, hairy perennial to 2·5 m, with stiff, erect, nearly leafless densely hairy stems bearing along their length large stalkless pink flowers 6–9 cm across. Petals pale pink or bright pink, usually yellow at the base, notched, not touching at their margins. Fields, disturbed ground. Widespread. Page 164. Pl. 28. **(748). *A rosea* L. HOLLYHOCK. Like 748, but petals touching at the margins; stems hairless or with few bristly reflexed hairs. Widely grown for ornament; naturalized in YU.BG. **748a.** *A. heldreichii* (Boiss.) Boiss. Similar to 748 but less robust, with rather smaller flowers, and epicalyx not more than half as long as the calyx (at least three-quarters as long in 748). Stems with persistent dense star-shaped hairs as in 748. Sepals with conspicuous raised veins on back. Mountain rocks. YU.GR(Macedonia).BG. Pl. 28.

ABUTILON ****749.** *A. theophrasti* Medicus Distinguished by its small yellow flowers, long-stalked heart-shaped leaves, and its fruits with 12–15 hairy carpels each with 2 horns. Petals 7–13 mm. An erect hairy annual 50–100 cm. Waste and damp places. YU.AL.GR.CR.BG.

GOSSYPIUM ****750.** *G. herbaceum* L. LEVANT COTTON. A woody annual to 1 m or more, with large white or pale-yellow flowers with purple centres and large rounded fruits, known as a 'bole', which has greyish-white cottony hairs surrounding the seeds within. Epicalyx lobes with 6–8 triangular teeth usually less than three times as long as wide. Leaves heart-shaped with 3–7 lobes, 11–14 cm. Fruit 2–3·5 cm, rounded. Culti-vated in the Med. region; naturalized in AL.GR.CR. **750a.** *G. arboreum* L. differs from 750 in its more conical fruit, and epicalyx lobes which have only 3–4 teeth. Cultivated locally in CR. **750b.** *G. hirsutum* L. Like 750 but with epicalyx lobes with teeth usually more than 3 times as long as wide; filaments of stamens 4–6 mm (1–2 mm in 750). Cultivated; naturalized in GR.CR.

HIBISCUS **751.** *H. syriacus* L. SYRIAN KETMIA. A deciduous much-branched shrub 2–3 m, with bluish-purple to white flowers 5–8 mm across. Leaves 4–7 cm, rhombic, usually three-lobed, toothed. Native of Asia; often planted for ornament in YU.GR. ****752.** *H. trionum* L. BLADDER KETMIA. An annual 10–50 cm, with pale-yellow, often purple-veined flowers with dark centres, and united sepals which become inflated and bladder-like in fruit with conspicuous veins and hairs. Petals about 2 cm; epicalyx lobes 10–13. Leaves with 3 narrow, pinnately lobed segments. Cultivated ground, waste places. YU.AL.GR.CR.BG.

THYMELAEACEAE | Daphne Family

THYMELAEA Fruit dry, usually enclosed in the remains of the flower.

753. *T. hirsuta* (L.) Endl. A distinctive shrub to 1 m, with white-woolly branches, and numerous small fleshy, rather scale-like, overlapping dark-green leaves which are cottony above. Flowers small, yellowish, in clusters, densely hairy outside. Leaves 3–8 mm, usually blunt. Dry hills, maritime rocks in the Med. region. YU.GR+Is.CR.TR? ****754. *T. tartonraira* (L.) All. A branched, silvery-greyish shrublet 20–50 cm, with numerous oblong, silky leaves 1–2 cm, and small yellowish flowers in clusters amongst the upper leaves. Tube of flowers 5–6 mm, usually silky-haired. Woods and rocks in the Med. region. GR+Is. CR.TR. Page 39. **755.** *T. passerina* (L.) Cosson & Germ. A rigid erect, hairless, flax-like annual 20–50 cm, with narrow lanceolate leaves 8–14 mm. Flowers tiny, green, stalkless, axillary, forming a slender lax spike more than half the length of the plant. Cultivated and stony ground. YU.AL.GR.BG.

DAPHNE   Fruit fleshy, not enclosed in the remains of the flower.

1. All fls. terminal, solitary, or in
   dense more or less stalkless heads
   or clusters.
  2. Lvs. deciduous, not leathery
     (YU.).   **757.** *D. alpina* L. Pl. 29.
  2. Lvs. evergreen, more or less
     leathery.
    3. Fls. solitary or in terminal pairs   *jasminea*
       (YU(Piva gorge)).   **757b.** *D. malyana* Blečić
    3. Fls. in terminal heads of 3 or
       more.
      4. Sepals narrowly triangular,
         long-pointed; inflorescence
         without bracts.   *oleoides*
      4. Sepals ovate or broadly triangular,
         blunt or acute; fls. subtended by
         papery or leaf-like bracts.
        5. Fls. creamy-white.   *blagayana*
        5. Fls. pink or purplish.   *sericea, cneorum*
1. Fls. wholly or partly in axillary clusters,
   or in terminal branched clusters.
    6. Fls. greenish-yellow, hairless.   *laureola, pontica*
    6. Fls. white, cream, or pink, often hairy.
      7. Mature lvs. hairy beneath.   *sericea*
      7. Mature lvs. hairless beneath.
        8. Lvs. deciduous, 8–25 mm wide
           (YU.AL.GR.BG.).   **759.** *D. mezereum* L.
        8. Lvs. evergreen leathery, 3–10
           mm wide.   *gnidium, gnidioides*

Flowers white or cream-coloured
(757). *D. oleoides* Schreber A small shrublet to 50 cm, with leathery evergreen leaves, and sweet-smelling white or cream-coloured flowers in terminal heads of 3–6. Flowers 6–8 mm long. Leaves 1–4·5 cm, obovate to elliptic, hairy when young. Fruit red, finely hairy. Rocks in mountains. YU.AL.GR.CR.BG. Pl. 29.   **(757).** *D. blagayana* Freyer Distinguished by its white or cream-coloured, very sweet-scented flowers, in rounded terminal clusters of 10–15, or more. Flowers 1·5–2 cm long. A low spreading shrublet to 30 cm, with clusters of obovate, evergreen, hairless leaves 3–6 cm. Fruit whitish. Mountains. YU.AL.GR.BG. Pl. 29.   **757a.** *D. jasminea* Sibth. & Sm. A low much-branched often prostrate evergreen shrub to 30 cm, with usually pairs of white or yellowish flowers which are dull purple on the outside. Flowers 10–12 mm long. Leaves 8–11 mm, oblong-obovate, fine-pointed, hairless. Rocky places. GR(south-east). Pl. 29.

758. *D. gnidium* L. MEDITERRANEAN MEZEREON. An erect evergreen shrub to 2 m, of the

Mediterranean region, with linear-pointed leaves, clusters of small cream-coloured flowers and red fruits often present at the same time as the flowers. Flowers 3–4 mm long, with spreading hairs. Leaves 2–5 cm. Bushy places. YU? AL.GR+Is.

Flowers pink

756. *D. cneorum* L. GARLAND FLOWER. A dwarf mountain evergreen shrub like (757) *D. oleoides* but distinguished by its sweet-scented pink flowers in an almost stalkless head of 6–13. Flowers 6–10 mm long, usually with whitish hairs. Leaves 1–2 cm long, stalkless, hairless. Dry stony places. YU.AL.BG. **757c. *D. sericea* Vahl An erect evergreen shrub to 70 cm, distinguished by its leaves which have adpressed hairs beneath and are almost hairless above. Flowers pink, very fragrant, white-hairy, 6–8 mm long, in terminal heads of 5–15, subtended by short silky bracts. Leaves 2–5 cm, oblong-obovate. Rocky places in hills in the Med. region. GR.CR. **758a.** *D. gnidioides* Jaub. & Spach An erect shrub to 2 m with stout branches and pink flowers in terminal heads of 3–8, and with axillary clusters below. Leaves spine-tipped, leathery, glaucous, oblong-lanceolate, 2·5–4 cm; young branches with brown hairs. Rocky places. GR(Skiathos, Euboea)+E.Aegean Is.

Flowers greenish-yellow

760. *D. laureola* L. SPURGE LAUREL. Shady rocks, woods, bushy places in the hills and mountains. YU.AL.GR.BG.TR? **760a. *D. pontica* L. Like 760 but more spreading, with obovate leaves 2½ times as long as wide (at least 3 times as long in 760). Flowers pale yellowish, in pairs (in many-flowered clusters in 760), arising from the axil of bract-like leaves on the current year's growth; tube of flower as long as the spreading lobes. Woods, bushy places. BG(south-east).TR.

ELAEAGNACEAE | Oleaster Family

HIPPOPHAE ****761.** *H. rhamnoides* L. SEA BUCKTHORN. Sands, dunes, river gravels in the hills. YU.BG.

ELAEAGNUS ****762.** *E. angustifolia* L. OLEASTER. A shrub or small tree to 7 m, with silvery twigs and willow-like leaves, and very sweet-scented yellowish flowers which are silvery-scaly on the outside. Flowers 8–10 mm. Leaves oblong to narrow-lanceolate 4–8 cm, with silvery scales beneath; branches spiny or not. Fruit oval, about 2 cm, yellowish, with silvery scales. Native of Asia; often planted and sometimes naturalized in AL.GR+Is.CR.BG.

GUTTIFERAE (HYPERICACEAE) | St. John's Wort Family

HYPERICUM | St. John's Wort A difficult genus with about 43 species found in our area, including 8 which are exclusive to it. In many species the ovary and fruit have glandular swellings and patches: they are referred to as 'oil-glands' (vittae) if flat or slightly swollen, or as 'swellings' (vesicles) if conspicuously swollen. The following are either fairly widespread, or distinctive, or interesting:

Small shrubs or shrublets
(*a*) *Plants without red or black glands on sepals*
(763). *H. calycinum* L. ROSE OF SHARON. Flowers 6–8 mm across. Leaves 4·5–8·5 cm. Shady places. BG(south-east).TR. ****763. *H. androsaemum* L. TUTSAN. Flowers about 2 cm across, in clusters. Damp shady places. YU.BG.TR. **763c.** *H. amblycalyx* Coust. & Gand. Like 763d but leaves in whorls of 4; margins of sepals without glands. Rocky places. CR(east). Page 66.

(*b*) *Black glands present on sepals*
763d. *H. empetrifolium* Willd. A small tufted shrublet to 50 cm, with erect branches, or a procumbent straggling or rooting plant, with narrow leaves in whorls of 3, and usually pyramidal, branched clusters of yellow flowers. Petals 3–4 times as long as sepals which have stalkless black glands on the margins. Leaves 2–12 mm. Rocky places in the Med. region. AL.GR+Is.CR. Pl. 29.

Herbaceous perennials (sometimes woody-based)
(*a*) *Hairless or almost hairless plants*
   (*i*) *Sepals without black glands on the margin*
767. *H. tetrapterum* Fries SQUARE-STEMMED ST. JOHN'S WORT. Damp places. Widespread on mainland+CR. **(767).** *H. maculatum* Crantz IMPERFORATE ST. JOHN'S WORT. YU.AL.GR.BG. ****768.** *H. perforatum* L. COMMON ST. JOHN'S WORT. Dry meadows, shady places amongst rocks. Widespread. **768a.** *H. triquetrifolium* Turra (*H. crispum*) An erect or spreading plant 13–55 cm, with a two-lined stem and numerous spreading lateral branches. Flowers few, at ends of branches, 1–1·5 cm across; yellow petals usually without black dots; sepals entire or finely toothed, without black dots. Leaves linear-oblong to lanceolate-triangular, 3–15 mm, clasping, undulate, sometimes black-dotted. Rocky places, waysides, uncultivated ground. AL.GR+Is.CR.TR. Page 324. **769c.** *H. olympicum* L. A tufted, woody-based perennial with many erect stems and conspicuous rather large yellow flowers 3–6 cm across. Leaves glaucous, lanceolate to narrowly elliptic 0·5–3·6 cm, with glands on the blade and without black glands on margins. Flowers solitary or few; sepals unequal, leafy, and usually without black glands; black glands present on anthers. Dry stony places. YU.GR.BG.TR. Pl. 29.

   (*ii*) *Sepals with black glands on the margin* (*see* 764g)
768b. *H. elegans* Willd. Stems 15–55 cm, two-lined; leaves 1–3 cm, lanceolate to narrow-oblong with several large translucent dots. Yellow petals 10–15 mm, with only marginal black dots; sepals with black glandular teeth. Dry places. YU.BG.TR. **769b.** *H. fragile* Boiss. A small, slender herb 4–11 cm, branching from the base, with ovate glaucous leaves 2–7 mm, with conspicuous pale glands on the margins. Flowers yellow tinged with red, 1–8 in a cluster; sepals with black glandular-ciliate hairs. Crevices in limestone rocks. GR(east). **772a.** *H. perfoliatum* L. A nearly hairless, glaucous perennial 25–75 cm, with 2 raised lines on the stems, with usually clasping, ovate to linear-lanceolate leaves 1–6 cm, with black glands. Yellow petals 9–14 mm, sometimes with superficial black dots and streaks; sepals blunt with dense, irregular black glands or hairs; anthers with black glands. Fruit with oil glands and orange swellings. Bushy places in the Med. region. YU.GR+Is.CR.BG?TR?

772b. *H. montbretii* Spach Differs from 772a in having a narrow pyramidal fruit capsule (ovoid in 772a), and sepals spreading or reflexed in fruit, not erect. Fruit without oil glands, but with numerous orange swellings. Damp or shady, stony places. YU.GR.BG.TR. **772c.** *H. umbellatum* A. Kerner A perennial to 50 cm, with stalkless ovate to ovate-triangular leaves 1·5–4 cm, with conspicuous net veins and numerous superficial translucent or black glands. Yellow petals black dotted and streaked. Sepals acute, with glandular marginal hairs and numerous superficial black dots and streaks. Fruit without black swellings but with interrupted oil glands. Mountain woods. YU(north).BG.

772. *H. richeri* Vill. ALPINE ST. JOHN'S WORT. Leaves ovate to elliptic, with conspicuous net veins, without glands or black dots; stems 10–50 cm. Yellow petals with numerous black dots over the surface. Sepals long-pointed, with black dots and streaks. Fruit without dorsal oil glands, but with numerous round or elongated black swellings. Mountain meadows and woods. YU.AL.GR.BG. **772d. *H. rochelii* Griseb. & Schenk Stems 15–35 cm, with triangular-lanceolate to linear leaves 2–5 cm, usually clasping at the base, with indistinct net veins. Yellow petals sometimes with black dots. Sepals obtuse or acute, with black glandular hairs along the margins. Fruit without dorsal oil glands but with numerous round or elongated orange swellings. Rocky pastures. YU(north-east).BG.

772e. *H. barbatum* Jacq. Leaves 0·6–4 cm, lanceolate to linear-lanceolate or elliptic-oblong, stalkless or nearly so; stems 10–45 cm. Yellow petals usually with black dots over the surface or towards apex only. Sepals with non-glandular hairs along the margin and with numerous superficial black dots and streaks. Orange glands on fruit not prominent. Meadows, stony places. YU.AL.GR.BG. **772f.** *H. rumeliacum* Boiss. Leaves 0·6–3·5 cm, ovate-lanceolate or oblong to linear, stalkless or nearly so; stems 5–40 cm. Yellow petals black dotted all over. Sepals with glandular hairs along the margin and numerous superficial black dots and streaks. Fruit as in 772e. Limestone, stony places. YU.AL.GR.BG. Page 115.

(b) Hairy plants

   (i) Sepals without marginal glands

769d. *H. cerastoides* (Spach) N. K. B. Robson (*H. rhodoppeum*) A spreading or ascending perennial 7–27 cm, with oblong to elliptic or ovate, finely hairy leaves 0·8–3 cm, borne on hairy stems. Flowers usually solitary, rarely few, 2–4·3 cm across. Petals with marginal black glands. Sepals unequal, leafy, broadly overlapping, without glands. Stony places in non-limestone areas. GR.BG.TR. Pl. 29.

   (ii) Sepals with black glandular hairs on margins

764d. *H. annulatum* Moris An erect perennial 20–65 cm, covered with short whitish hairs, with ovate, stalkless leaves 1·5–5·5 cm. Flowers in lax pyramidal, hairless clusters; yellow petals usually without black glands; sepals with black glandular-hairs on margins. Scrub and stony places in mountains. YU.AL.GR.BG. **764e.** *H. delphicum* Boiss. & Heldr. A hairy perennial usually 11–35 cm, with rough-hairy, ovate to oblong-ovate, stalkless leaves 1·2–3·5 cm. Flowers in cylindrical, hairless clusters; petals with marginal black glands toward the apex; sepals with black glandular-hairs on the margins. Woodland and stony places. GR(Euboea, Andros). **764f.** *H. athoum* Boiss. & Orph. Like 764e but smaller in all its parts and covered with soft hairs, and leaves, 0·8–1·5 cm, shortly stalked. Flowers fewer, 1–7. Stony places in shade. GR(N.Aegean region). **764g.** *H. cuisinii* W. Barbey A sprawling shortly hairy, or hairless, perennial 4–28 cm, with elliptic to oblong or obovate, usually stalked leaves, 2–15 mm. Flowers in lax clusters; yellow petals 5–7 mm, with black glands; sepals with black glandular-teeth and black dots and streaks. Limestone rocks. GR(Chios, Karpathos, Kasos).

VIOLACEAE | Violet Family

Calyx appendages are small flaps at the base of each sepal.

VIOLA | **Violet, Pansy** About 58 species are found in our area, 24 of which are restricted to it. Often a difficult genus. The following are mostly widespread on the mainland in the central European climatic region. ****774.** *V. odorata* L. SWEET VIOLET; **775.** *V. hirta* L. HAIRY VIOLET; **776.** *V. canina* L. HEATH DOG VIOLET; ****777.** *V. riviniana* Reichenb. COMMON DOG VIOLET; **(777).** *V. reichenbachiana* Boreau PALE DOG VIOLET; ****783.** *V. tricolor* L. WILD PANSY; **784.** *V. arvensis* Murray FIELD PANSY. Other distinctive species, often restricted to limited habitats or areas where they may be conspicuous, include:

Annuals
784b. *V. kitaibeliana* Schultes An annual 2–10 cm, covered in dense, short hairs. Flowers 4–8 mm; petals creamy-white to yellow, with a yellow centre, shorter than the calyx; spur short, slightly longer than the calyx appendages. Lower leaves rounded 1–3 cm, the others oblong-spathulate, shallowly lobed; stipules pinnately divided. Dry, open habitats. Widespread, except Aegean Islands. **784c.** *V. parvula* Tineo A tiny annual usually 2–3 cm, with long woolly hairs, and pale-yellow flowers about 5 mm, with upper petals flushed violet, found on rocks and screes of high mountains. Petals scarcely longer than calyx; spur shorter than calyx appendages. Leaves almost entire, oblong-rounded 0·5–1·2 cm. YU.GR. Page 324. **784d.** *V. hymettia* Boiss. & Heldr. Distinguished from 784c by the larger yellow, often violet-flushed, or all violet flowers 1–1·5 cm, with petals twice as long as calyx. Spur 3–4 mm stout, slightly longer than calyx appendages. A roughly hairy annual, usually 3–10 cm. Meadows in the Med. region. AL.GR.TR.

Woody-based perennials
781b. *V. scorpiuroides* Cosson A distinctive, shrubby plant 10–20 cm, with small yellow flowers 1–1·5 cm, with a blunt, curved spur. Leaves broadly obovate, entire; stipules linear. Rocky places in the Aegean region. CR+Kithira. **781c.** *V. delphinantha* Boiss. An attractive shrubby perennial 5–10 mm, of limestone rocks in the mountains, with large, beautiful delicate pinkish- or reddish-purple flowers with long slender spurs. Flowers long-stalked; lower petals not notched; spur 16–18 mm, as long as or longer than the petals. Leaves linear to lanceolate, pointed, entire, up to 1·5 cm, stalkless. Limestone rocks in mountains. GR.BG(south). Page 126. Pl. 30. **781d.** *V. kosaninii* (Degen) Hayek Similar to 781c but with lilac-pink flowers, with the lower petal notched and spur 12 mm long. Alpine meadows. YU(Macedonia).AL(north).

Herbaceous perennials
(a) *Flowers with 2 erect petals, 2 lateral petals spreading and directed downwards, and a basal petal, all similar in size and colour* (VIOLETS)
774a. *V. alba* Besser A rhizomatous perennial with a rosette of persistent, heart-shaped leaves and linear-lanceolate stipules, usually producing long, slender stolons. Flowers 1·5–2 cm long, fragrant, white or violet. Lateral petals with hairs (bearded). Woods, scrub. Widespread, except Aegean Islands. **774b.** *V. cretica* Boiss. & Heldr. Similar to 774a but stolons very long, and the leaves covered with rough hairs. Flowers small, violet; lateral petals without hairs at base. Mountains. CR. **774c.** *V. chelmea* Boiss. & Heldr. Flowers 1 cm, pale violet with a rather stout spur shorter than the sepals. A stemless perennial 4–8 cm, with leaf rosettes of triangular, hairy spring leaves, and larger hairless summer leaves; stipules lanceolate, with a glandular fringe. Limestone rocks and mountains. YU(west).GR. Page 93. **778.** *V. rupestris* F. W. Schmidt TEESDALE VIOLET. Rocky pastures and sandy places in mountains. YU.BG.

(b) *Flowers with 2 erect petals, 2 lateral petals directed upwards, and a broader basal petal; petals unequal in size, often multi-coloured* (PANSIES) *(i)* Spur not more than twice as long as the calyx appendages, usually much less

   (x) Stipules deeply divided

783a. *V. aetolica* Boiss. & Heldr. A yellow-flowered perennial 15–40 cm, with flowers 1·5–2 cm long, with a slender, straight spur 5–6 mm, about 1½ times as long as the calyx appendages. Upper petals sometimes violet. Leaves about 2 cm, ovate to lanceolate, coarsely toothed. Stipules pinnately divided, ciliate. Mountain meadows. YU.AL.GR. Pl. 30. **785d.** *V. dacica* Borbás Recognized by the rather large flowers 2–3 cm long, with violet or yellowish petals and a short spur as long as or scarcely longer than the calyx appendages. Leaves ovate or elliptical, with slightly wavy, hairless, or sometimes hairy margins, the blade as long as or longer than the leaf stalk. Stipules coarsely toothed, or pinnately divided to half or a third their width. Grassland and forest margins in mountains. YU.AL.BG. Page 137. **785e.** *V. rhodopeia* W. Becker A hairless, spreading perennial to 20 cm, with flowers 1·5–2 cm long, borne on stalks up to 8 cm. Petals yellow, with a pale violet, straight slender spur 6 mm; sepals lanceolate long-pointed. Leaves linear-lanceolate; lower stipules linear, the upper deeply divided into linear lobes. Damp mountain pastures. BG(Rhodope). Pl. 30.

   (xx) Stipules not deeply divided

782. *V. biflora* L. YELLOW WOOD VIOLET. Shady places, damp rocks in hills and mountains. YU.BG. **785f. *V. grisebachiana* Vis. A slender, stemless perennial 3–8 cm, with violet flowers about 2 cm long; spur blunt 3–4 mm, exceeding the calyx appendages. Leaves ovate-rounded to 3 cm, with blade abruptly contracted to the leaf stalk, which is at least twice as long; stipules like the leaves but smaller. Alpine pastures. YU.AL.BG. Page 136. **785g.** *V. orphanidis* Boiss. A softly hairy, many-stemmed perennial 20–70 cm, with violet or blue flowers 2–3 cm long, and a slender, curved spur 4–5 mm, up to twice as long as the calyx appendages. Leaves 2–4 cm, ovate to rounded, toothed, hairy mainly along the veins and margin; stipules 1–2 cm, obliquely ovate. Mountain meadows, woodland margins. YU(south).AL.GR.BG(south-west).

   (ii) Spur more than twice as long as the calyx appendages

    (x) Stipules usually entire or toothed

786. *V. calcarata* L. LONG-SPURRED PANSY. Flowers large, violet or yellow, 2–4 cm long, by up to 3 cm wide; spur 8–15 mm, equalling petals. Subsp. *zoysii* (Wulfen) Merxm. is the plant of our area. It usually has solitary yellow flowers on short stems. Leaves broadly ovate or rounded; stipules entire or with 1–2 slender teeth. Alpine and mountain meadows. YU.AL.GR? Page 185. **786a. *V. fragrans* Sieber A tufted, often rough-hairy perennial 5–10 cm, with oblong to linear-oblong, entire, blunt leaves about 1·5 cm, with long leaf stalks. Flowers usually about 1 cm long but rather variable, 1–2 per stem, yellow or pale violet, with a stout spur about 3 mm long, 2–3 times as long as the calyx appendages. Stipules like leaves but smaller. Rocks in mountains. GR.CR. Page 70.

786b. *V. poetica* Boiss. & Spruner Like 786a but hairless, with oblong-ovate leaves up to 2·5 cm, and leaf stalks three times as long as the blade. Flowers 1–4 per stem, violet or blue. Rock crevices at about 2000 m in mountains. GR(south-central). Pl. 30. **786f.** *V. magellensis* Strobl A tufted perennial 4–10 cm, of alpine pastures and screes, with flowers violet, pink, or with the upper petal dark reddish-violet, and with a straight or slightly curved spur 8–10 mm. Leaves rounded-ovate to oblong, entire, 0·5–1·5 cm; stipules similar, smaller, sometimes lobed. AL.GR(north). Page 115.

   (xx) Stipules deeply divided

786c. *V. gracilis* Sibth. & Sm. A shortly hairy perennial to 30 cm, with violet or yellow flowers 2–3 cm long, with a straight or slightly curved spur 6–7 mm, 2–3 times as long as the calyx appendages. Leaves rounded-ovate or oblong, weakly toothed; stipules pinnately divided into 4–8 linear segments, the central segment larger and leaf-like. Rocks, and pastures in mountains. YU(south).AL.GR(north-east).BG. Page 324. Pl. 30. **786d.** *V. elegantula* Schott Distinguished by its large unequally pinnately divided stipules which have 2–4 lobes on the inner side, and 4–10 on the outer side and a broad central part. Flowers violet, yellow, or particoloured (rarely pink or white), 2–2·5 cm long; spur 5–8 mm, slender, straight or slightly curved, three times as long as calyx appendages. A hairy or hairless perennial 10–30 cm; lower leaves rounded, the upper lanceolate. Mountain meadows. YU(west).AL.

786e. *V. allchariensis* G. Beck A tufted perennial 8–25 cm, with stem and leaves covered with grey hairs. Flowers violet or yellow 2–2·5 cm long, with upper petal about 1·5 cm wide; spur about 5 mm, blunt, curved upwards, about twice as long as the calyx-appendages. Leaves 2–5 cm, elliptic-oblong, weakly toothed, the upper linear, entire; stipules with 2–5 linear lobes. Rocky hillsides in mountains. YU.AL.GR. Page 168. Pl 30.

CISTACEAE | Rockrose Family

CISTUS Fruit with 5, 6, or 10 valves; sepals all similar in size. 5 species in our area.

Flowers pink
**787. *C. incanus* L. A variable, densely branched, erect shrub to 1 m, with hairy branches and large, pink flowers 4–6 cm across, in a lax terminal cluster of 1–7. Leaves green- or greyish-hairy, oval to elliptic, veins impressed above and prominent beneath. Subsp. *incanus* has larger flat leaves 2·5–5 cm by 1·5–3 cm; it is widespread. Subsp. *creticus* (L.) Heywood has smaller, undulate, wrinkled leaves 1·5–2·5 by 0·8–1·5 cm, and is common in Greece and the Aegean region. Sunny, stony places, uncultivated ground, low thickets. Widespread. Page 34.

787b. *C. parviflorus* Lam. A somewhat spreading shrub to 1 m, with ovate leaves 1–3 cm, covered with dense, short, grey-woolly hairs. Distinguished from 787 by the smaller pink flowers, 2–3 cm across, in clusters of 1–6, and the stalkless stigmas which are shorter than the stamens (style as long as stamens in 787). Scrub, sunny slopes. GR+Is.CR.TR.? Pl. 31.

Flowers white
790. *C. salvifolius* L. SAGE-LEAVED CISTUS. A sage-like shrub to 1 m, with rough wrinkled, stalked leaves and long-stalked, white flowers, often with orange centres, 3–5 cm across. Outer 2 sepals conspicuously larger and investing the inner 3. Leaves ovate to elliptic, 1–4 cm. Scrub, dry sunny slopes, open woods. Widespread. Page 38. ****791. *C. monspeliensis* L. NARROW-LEAVED CISTUS. Distinguished by its linear to linear-lanceolate, very sticky aromatic, stalkless leaves, dark shining green above and grey-haired beneath. Flowers white, small 2–3 cm across, in rather one-sided clusters. Scrub and dry slopes in the Med. region. YU.AL.GR.CR.

TUBERARIA Sepals 5, the 2 outer usually smaller than the 3 inner; fruit with 3 valves. Style absent. Basal leaves in a rosette.

**798. *T. guttata* (L.) Fourr. SPOTTED ROCKROSE. A distinctive annual with slender hairy stems to 30 cm, with a terminal cluster of small yellow flowers usually with striking dark purple-brown centres. Flowers variable in size, 1–2 cm. Leaf rosette persisting at flowering; leaves elliptic, the upper narrower. Fruit stalks reflexed. Open sandy ground, bushy places, pinewoods. Widespread.

HELIANTHEMUM | Rockrose Sepals 5, the outer smaller; fruit with 3 valves; stamens all fertile; style present; leaves opposite. Some species are very variable and have distinctive subspecies.

Shrubs or woody-based shrublets with branched inflorescences (see 801 g)
(a) All leaves with stipules
801a. *H. lavandulifolium* Miller Readily distinguished by its forked inflorescence of yellow flowers, which is at first dense and coiled, and later spreading in fruit with numerous drooping fruits regularly spaced along the lower side of the 3–5 branches. Flowers about 1·5 cm across, erect. Leaves linear-lanceolate, with inrolled margins, greyish-green above, white-woolly beneath. Evergreen thickets in the Med. region. YU.GR.CR.TR.

(b) Lower leaves without stipules (upper leaves with stipules)
801d. *H. cinereum* (Cav.) Pers. Inflorescence lax, branched, nearly leafless, bearing numerous small yellow flowers about 8 mm across. Calyx densely covered with white bristly hairs. Basal leaves ovate to lanceolate, with a rounded or heart-shaped base, grey-woolly beneath, green and often hairless or grey-woolly above, but very variable; upper leaves of flowering shoots with conspicuous leaf-like stipules. Arid sandy and stony places, limestone hills. GR. Pl. 31. **801h.** *H. hymettium* Boiss. & Heldr. Differs from 801d in its leaves which have wedge-shaped bases, and are either covered with short, white hairs on both surfaces, or hairless and greenish above. Flowers small, yellow, in a rather flat-topped dense cluster of 7–9; petals 3–4 mm. Rocky hill slopes in the Med. region. GR(south).CR. Page 88.

Straggling, woody-based shrublets with unbranched inflorescences
(a) All leaves with stipules
**802. *H. nummularium* (L.) Miller COMMON ROCKROSE. An extremely variable species, which can be distinguished by the following combination of characters. All leaves with rather leaf-like stipules longer than the leaf stalk; leaf blades either green on both sides, or grey- or white-woolly below only, or on both sides. Inflorescence simple, one-sided; flowers stalked, golden-yellow, pale yellow, creamy white, orange, or pink; sepals hairy or hairless between the ribs. Capsule equalling the sepals. Rocky places, grassland, mountains. Widespread.

803. *H. apenninum* (L.) Miller WHITE ROCKROSE. Like 802 but leaves green, grey, or white and densely woolly-haired above, with dense star-shaped hairs below, and with inrolled margins. Stipules not leaf-like, linear. Flowers stalked, white with a yellow base; calyx 3–7 mm, shortly hairy all over. Rocky hill slopes in the Med. region. AL.GR.CR. **803c. *H. stipulatum* (Forskål) C. Chr. Identified by the stalkless flowers in a loose cluster of 3–7, with yellow petals 3–4 mm, as long as the sepals. Sepals with star-shaped hairs and marginal hairs. Leaves linear to ovate-lanceolate, with inrolled margins, 8–15 mm, with star-shaped hairs, sparse above and dense beneath. Maritime sands. GR(south).

(b) Lower leaves without stipules
801. *H. canum* (L.) Baumg. HOARY ROCKROSE. A very variable ascending or spreading perennial 4–20 cm, distinguished by the absence of stipules to all leaves except on the upper flowering shoots. Flowers yellow, in dense or lax cluster of 1–5; petals 4–6 mm; sepals with adpressed hairs. Leaves all grey-woolly beneath, green to grey above, with or without star-shaped hairs, elliptic to linear with wedge-shaped bases. Dry sandy, stony,

grassy places. YU.AL.GR.BG.TR. **801g.** *H. oelandicum* (L) DC. Distinguished from 801 by its linear to elliptic leaves which are green and hairless on both surfaces, or with simple or clustered (not star-shaped) hairs. A very variable, lax to densely tufted, dwarf shrub to 20 cm, with simple or branched clusters of yellow flowers; petals 3–10 mm. Rocks, dry sandy places, from the lowlands to the mountains. YU.AL.GR. Page 184.

Annuals
800. *H. salicifolium* (L.) Miller An annual to 30 cm, with yellow unspotted flowers in a rather lax, simple or branched terminal cluster, and with stout flower stalks spreading at right angles and usually up-curved at apex, usually longer than the calyx. Petals 5–12 mm, longer or shorter than the sepals; bracts small, scarcely as long as the flower stalk. Leaves 0·5–3 cm, broadly lanceolate, densely greyish-hairy; stipules linear-lanceolate. Dry grasslands, stony places. Widespread. **(800).** *H. ledifolium* (L.) Miller Like 800, but flower stalks erect, thickened, shorter than the calyx. Flowers yellow with a golden blotch at the base of each petal, which is shorter than the sepals. Leaves 1·5 cm. Dry grassland. YU.GR.CR.BG.

800a. *H. sanguineum* (Lag.) Dunal A glandular-hairy, sticky annual, with a lax three- to six-flowered inflorescence, with stout curved flower stalks which are reflexed in fruit. Petals yellow, shorter than the sepals. Stems 2–10 cm, often tinged with purple; lower leaves elliptic, soon falling, upper leaves oblong-lanceolate. Sandy places. CR. **800b.** *H. aegyptiacum* (L.) Miller Distinguished by its papery sepals (sepals green in 800, (800), 800a). Flowers 3–9, borne on long, slender stalks which become reflexed in fruit; petals yellow, shorter than the sepals which are 6–10 mm. Leaves 1–3 cm, linear-lanceolate, often with inrolled margins, dark green above with dense, short, grey hairs below; stems densely hairy. Dry sandy places in the Med. region. GR.CR+Karpathos. BG (south).

FUMANA Distinguished from *Helianthemum* by the upper leaves which are usually alternate and more or less linear, and the outer stamens which are sterile. Capsule with 3 valves, usually spreading in a star when ripe. 8 very similar-looking species in our area.

Leaves unequally spaced on stem; bracts much smaller than leaves
804. *F. thymifolia* (L.) Webb A small twiggy shrub to 20 cm, with slender heath-like leaves, and lax, three- to nine-flowered clusters of small yellow flowers, each about 1 cm across. Leaves 5–11 mm, opposite, linear to narrowly elliptic, margins inrolled, unequally spaced, with stipules, and having shorter clusters of leaves in their axils; bracts very small. Rocky hill slopes in the Med. region. YU.AL.GR+Is.CR.TR. Page 39. **804b.** *F. laevipes* (L.) Spach Differs from 804 in its narrow, alternate bristle-like green or glaucous leaves 4–8 mm, which are rounded in section. Flowers yellow, 3–8 in a terminal cluster; flower stalks spreading. Sandy hills in the Med. region. YU.GR.CR.

Leaves mostly equally spaced; bracts and leaves similar
804a. *F. arabica* (L.) Spach A much-branched, laxly tufted shrublet to 25 cm, with alternate, evenly spaced, ovate to oblong-lanceolate, glandular-hairy to almost hairless leaves 5–12 mm, with stipules. Flowers yellow, 1–7 in a lax cluster. Rocky places in the Med. region. YU.AL.GR+Is.CR.BG? Pl. 31. ****805.** *F. procumbens* (Dunal) Gren. & Godron Like 804a but a procumbent shrublet with spreading branches to 40 cm, and linear leaves without stipules. Flowers yellow, solitary in the upper leaf axils on stalks about as long as the subtending leaves. Rocky places. YU.AL.GR.CR.BG.TR.

TAMARICACEAE | Tamarisk Family

TAMARIX | Tamarisk 5 very similar-looking species in our area distinguished by different combinations of small botanical characters: a difficult genus. **806a.** *T. parviflora* DC. A shrub or small tree with a dark-brown to purple bark, and pale-pink cylindrical flower clusters 3–5 mm wide. Flowers with 4 or 5 petals and sepals; the petals not more than 2 mm; stamens 4–5. Bracts entirely papery, not exceeding the calyx. Leaves 3–5 mm, with papery margins. Hedges, riverbanks. Widespread, except BG. **806b.** *T. hampeana* Boiss. & Heldr. A hairless tree with brown or reddish bark, readily identified by the large pink flower clusters 2–6 cm or more long and 1–1·2 cm wide, with the bracts not longer than the flower stalks. Petals 3–4 mm; sepals shorter; stamens 6–8. Leaves 2–4 mm. Riverbanks, maritime sands. GR.TR. **806c.** *T. dalmatica* Baum A hairless tree with blackish bark, similar to 806b from which it differs by the narrower clusters of white flowers, 8–10 mm wide, and the bracts longer than the flower stalks and calyx. Coastal marshes, riverbanks. YU.AL.GR.CR.

FRANKENIACEAE | Sea Heath Family

FRANKENIA 809. *F. pulverulenta* L. ANNUAL SEA HEATH. An annual to 30 cm, with numerous radiating, prostrate branches, and small solitary or lax clusters of stalkless pink or pale-violet flowers in the axils of the branches and upper leaves. Petals 4–5 mm. Leaves 1–8 mm long, broadly oval to obovate, hairless above, crisply hairy beneath, often becoming reddish. Maritime sands and shingles, salt-rich soils. Widespread, except TR. **810d.** *F. hirsuta* L. A mat-forming, woody-based, finely hairy perennial spreading to 80 cm, with conspicuous, dense flat-topped clusters of lilac or white flowers terminating the main stems or branches. Petals 4–6 mm. Leaves 2–5 mm, linear, sometimes covered with a white powder. Maritime sands and shingles, salt-rich soils. AL.GR+Is.CR.TR.

ELATINACEAE | Waterwort Family

Aquatic or marsh herbs. Leaves simple, opposite or whorled, with stipules. Flowers regular, solitary or clustered. Ovary superior, two- to five-celled; styles free.

ELATINE 4 species in our area.

810e. *E. alsinastrum* L. An aquatic annual or perennial to 80 cm, with whorled leaves, having up to 18 linear leaves in each whorl in submerged shoots, and as few as 3 broader leaves in terrestrial shoots. Flowers tiny, pale red, stalkless, axillary; petals 4; stamens 8. In still water, or damp mud and sand. YU.GR.CR.BG.

DATISCACEAE | Datisca Family

Flowers one-sexed or bisexual, petals usually absent; male flowers with 4–9 calyx lobes;

stamens 4–25. Female and bisexual flowers with 3–8 calyx lobes; ovary inferior, one-celled with numerous ovules.

DATISCA **810f.** *D. cannabina* L. A robust, hairless perennial to 1 m, resembling 64 *Cannabis sativa* HEMP, with compound leaves, and small stalkless green flowers in long clusters in the axils of the upper leaves. Bracts subtending flowers entire. Leaves pinnate, with lanceolate, long-pointed, coarsely toothed leaflets. Stream banks. CR.

CUCURBITACEAE | Gourd Family

In addition to the few native species there are a considerable number of other species, belonging to this family, from the subtropical and tropical regions, which are often cultivated for their fruit, or have become weeds of cultivation.

Tendrils absent
ECBALLIUM ****811.** *E. elaterium* (L.) A. Richard SQUIRTING CUCUMBER. A stout, spreading, somewhat fleshy, rough-leaved perennial 15–60 cm, with yellow bell-shaped flowers, and long-stalked, inclined, sausage-shaped fruits, which when ripe explode violently when touched. Leaves heart-shaped to triangular, long-stalked. Roadsides, waste places, by the sea. Widespread.

Tendrils present
1. Tendrils unbranched.
  2. Fls. greenish-white; male fls. in
    terminal clusters; fr. 6–10 mm.   *Bryonia*
  2. Fls. deep yellow; male fls. in
    axillary clusters; fr. at least
    2 cm.
    3. Connective of anther prolonged
      beyond the pollen sacs; disk
      present.   *Cucumis*
    3. Connective of anther not
      prolonged beyond the pollen sacs;
      disk absent.   *Citrullus*
1. Tendrils branched.
  4. Fls. whitish; stamens 5, with fused
    filaments; fr. bristly-haired.
    5. Lvs. lobed to about the middle.   *Echinocystis*
    5. Lvs. angled or shallowly lobed.   *Sicyos*
  4. Fls. deep yellow; stamens 3, filaments
    free; fr. not bristly-haired.
    6. Corolla lobed almost to base.   *Citrullus*
    6. Corolla lobed to about half-way. *Cucurbita*

BRYONIA ****815.** *B. cretica* L. WHITE BRYONY. Stigma hairy; fruit red. Bushy places, hedges. YU.AL.GR+Is.CR. **(815).** *B. alba* L. Like 815 but stigma hairless; fruit black. Plant usually one-sexed in our area like 815. Bushy places, hedges. Widespread on mainland.

CITRULLUS   812. *C. lanatus* (Thunb.) Mansfeld WATER MELON. Distinguished by its huge globular or ellipsoid, dark-green, edible fruit with pink or yellow flesh, and usually with black seeds. Flowers solitary, yellow, with corolla lobes about 15 mm. Leaves pinnately lobed, the lobes further divided. Native of S. Africa; often cultivated.   **(812).** *C. colocynthis* (L.) Schrader BITTER APPLE, BITTER CUCUMBER. A spreading perennial with deeply once or twice pinnately lobed leaves, and solitary greenish-yellow flowers with corolla lobes about 5 mm. Fruit dry, globular to about 4 cm, pale yellow. Sandy places by the sea. GR.

CUCURBITA   813. *C. maxima* Duchesne PUMPKIN. Distinguished by its enormous fruits which are usually bright yellow and shallowly segmented, but varicoloured and irregularly swollen fruits are also grown. Flowers deep yellow, 7–10 cm across. Leaves shallowly lobed or almost entire, softly hairy. Native of central America; grown as a vegetable and sometimes for ornament.   **814.** *C. pepo* L. MARROW, ORNAMENTAL GOURD. Distinguished from 813 by its rough bristly-haired leaves which have deeper, more pronounced acute lobes, and stalk of flowers five-angled. Fruits large 15–40 cm across, very variable in shape and colour, dry or fleshy. Native of America; widely cultivated for its fruits and sometimes for ornament.

CUCUMIS   816. *C. sativus* L. CUCUMBER. Fruit oblong or cylindrical, rough, green or yellowish, with white flesh. Flowers 2–3 cm long, lobes acute. Leaves palmately three-to five-lobed, toothed, rough. Native of India; frequently cultivated.   **817.** *C. melo* L. MELON, CANTALOUPE. Leaves palmately five-lobed with rounded, toothed lobes, the middle lobe much broader (lobes acute in 816). Fruit large globular, but very variable in shape and colour, with white, orange, or greenish flesh. Native of Asia and Africa; widely cultivated.

ECHINOCYSTIS   814a. *E. lobata* (Michx) Torry & A. Gray An almost hairless climbing annual 5–8 m, with ovoid fruits 3–5 cm, covered with long slender prickles. Flowers greenish-white; corolla lobes about 5 mm. Leaves three- to seven-lobed, about 5 cm. Native of N. America; naturalized in YU.

SICYOS   814b. *S. angulatus* L. A viscid-hairy climbing annual 5–8 m, with clusters of compressed-ovoid leathery fruits about 1·5 cm, woolly-haired and covered with long bristles. Flowers whitish; corolla lobes about 5 mm. Leaves about 7 cm, five-angled or five-lobed. Native of N. America; naturalized in damp places in YU.

CACTACEAE | Cactus Family

OPUNTIA   **818. *O. ficus-indica* (L.) Miller PRICKLY PEAR, BARBARY FIG. Quite unmistakable with its large thick, spineless or spiny, racket-shaped stem joints 20–50 by 10–20 cm, forming a large tree-like growth to 5 m. Flowers bright yellow, 7–10 cm across, with numerous petals and stamens. Fruit large 5–9 cm, fleshy and edible, red, yellow, or purple. Native of Tropical America; widely naturalized in YU.GR.CR+Rhodes.   **(818).** *O. vulgaris* Miller A low spreading perennial to 50 cm, with oblong, dark-green, spiny stem joints 3–13 by 4–5 cm. Flowers bright yellow, 5–9 cm across. Fruit edible. Native of N. America; sometimes naturalized in YU.GR.CR.

LYTHRACEAE | Loosestrife Family

LYTHRUM   About 9 species occur in our area; often plants of small stature with insignificant flowers.

820. *L. salicaria* L. PURPLE LOOSESTRIFE. Lakes, river-sides, marshes. Widespread on mainland.   ****821. *L. virgatum* L. SLENDER LOOSESTRIFE. Like 820 but less robust, and stems and leaves quite hairless. Flowers smaller; petals about 7 mm; epicalyx segments 1 mm, as long as the sepals. Meadows, marshes, damp woods. YU.AL.GR.BG.   ****822.** *L. hyssopifolia* L. GRASS POLY. An annual with tiny pink flowers with petals 2–3 mm; stamens usually 4–6. Damp fields, stream-sides. Widespread.   **822a.** *L. junceum* Banks & Solander A slender, straggling perennial with long unbranched stems 20–70 cm, bearing regularly placed linear-oblong leaves, each with a solitary pink flower in its axil. Petals 5–6 mm, sometimes cream or white; calyx tubular, spotted with red near the base, with pointed epicalyx lobes alternating with blunt sepals, and shorter than the 12 stamens. Wet places. AL.GR+Is.CR.TR. Pl. 31.

AMMANNIA   Readily distinguished from *Lythrum* by the opposite not alternate leaves. Parts of flowers in fours.

822c. *A. verticillata* (Ard.) Lam. A hairless, erect annual 10–30 cm or more, with linear-oblong to broadly oblanceolate, opposite leaves. Calyx hairy, the epicalyx lobes conspicuous and much longer than the sepals; petals sometimes present, minute, purple. Rice-fields, marshes, and shallow water. Native of S.W. Asia; naturalized in YU.BG?

LAGERSTROEMIA   **822d.** *L. indica* L. CRAPE-MYRTLE. A deciduous shrub or small tree with branched clusters of bright-pink flowers with fringed or crinkled petals. Flowers 3–4 cm across; petals 6; stamens numerous. Leaves elliptic or oblong, entire; stalkless. Native of China and Australia; planted for ornament in the Med. region.

TRAPACEAE | Water Chestnut Family

TRAPA   ****823.** *T. natans* L. WATER CHESTNUT. An aquatic plant with rosettes of floating leaves which are unmistakable; the blades are rhomboid, coarsely toothed at the apex, and the leaf stalks have spindle-shaped hollow swellings below water. Flowers white, 1–2 cm across. Fruit top-shaped, with 2–4 robust spines. Lakes and stagnant waters, usually in non-limestone areas. YU.AL.GR.BG.

MYRTACEAE | Myrtle Family

MYRTUS   ****824.** *M. communis* L. MYRTLE. A dense evergreen shrub to 5 m, with shining leathery leaves and solitary, white, sweet-scented, long-stalked axillary flowers 2–3 cm across, with numerous projecting stamens. Leaves 2–5 cm, oval-lanceolate, gland-dotted, aromatic when crushed. Fruit bluish-black, 7–10 mm. Bushy places, scrub, damp ground. YU.AL.GR+Is.CR.

EUCALYPTUS   Australian trees well suited to the Med. climate have frequently been planted in our area. They include: **825.** *E. globulus* Labill. BLUE GUM; **826.** *E. amygdalinus* Labill.

825f. *E. camaldulensis* Dehnh. A spreading tree to 15 m, with smooth, dull white bark, with umbels of 5–10 white flowers, and with distinctly stalked fruits. Mature leaves lanceolate-pointed, 12–22 cm. Widely planted as a roadside tree in AL.GR.

PUNICACEAE | Pomegranate Family

PUNICA ****827.** *P. granatum* L. POMEGRANATE. Unlike any other shrub in having large scarlet flowers with crumpled petals, and red, fleshy calyx. Leaves deciduous, oblong-lanceolate, shining. Fruit reddish-brown, 5–8 cm, with a tough rind, and purple, yellowish, or white translucent fleshy edible centre. Native of S.W. Asia; widely cultivated and naturalized in our area.

ONAGRACEAE | Willow-Herb Family

LUDWIGIA **828.** *L. palustris* (L.) Elliott Still and flowing waters. Widespread on mainland.

CIRCAEA 3 species in our area. ****829.** *C. lutetiana* L. ENCHANTER'S NIGHTSHADE. Woody and shady places. Widespread on mainland.

OENOTHERA | **Evening Primrose** Natives of America, usually with large attractive yellow flowers.

831. *O. biennis* L. EVENING PRIMROSE. Disturbed ground, waste places, dunes. Naturalized on mainland.

EPILOBIUM | **Willow-Herb** 16 species occur in our area, the majority are widespread European species mostly of the central European climatic region.

Petals more than 1 cm long
835. *E. angustifolium* L. ROSEBAY WILLOW-HERB, FIREWEED. Well known for its long cylindrical, terminal spikes of bright rosy-purple flowers, and its neat, spirally arranged, lanceolate, glaucous leaves. An erect, often massed, perennial to 2 m or more. Woods and shady places. Widespread on mainland. **836. *E. dodonaei* Vill. Distinguished from 835 by the narrower, thick, linear-lanceolate leaves which are green on both sides. Flowers rosy-purple, large, in a short leafy cluster; petals 1·5 cm, not narrowed to a claw. Seeds with fine swellings. Stony places, river banks, railway embankments. YU.AL.GR.BG. ****837.** *E. hirsutum* L. GREAT HAIRY WILLOW-HERB, CODLINS AND CREAM. Ditches, watersides. Widespread on mainland.

Petals less than 1 cm long
838. *E. parviflorum* Schreber SMALL-FLOWERED HAIRY WILLOW-HERB. A tall, robust, very hairy perennial like 837 but with rather small rosy-purple flowers with petals 4–9 mm. Leaves not clasping stem (clasping in 837). Damp places, ditches. Widespread.

347

THELIGONACEAE

Flowers one-sexed, in axillary clusters of 1–3 flowers. Perianth present. Male flowers with 7–20 stamens; female flowers with an inferior, one-celled ovary. Fruit a nut-like drupe; ovule 1, basal.

THELIGONUM 843a. *T. cynocrambe* L. A hairless, somewhat succulent, fetid, leafy annual 5–30 cm, with swollen nodes, and insignificant green flowers. Leaves ovate, entire, stalked, the lower opposite, the upper alternate; stipules papery, sheathing. Fruit ovoid, about 2 mm. Damp or shady rocks and walls. Widespread.

CORNACEAE | Dogwood Family

CORNUS **845. *C. mas* L. CORNELIAN CHERRY. Easily recognized when in flower, in early spring, by its numerous rounded clusters of small yellow flowers appearing on bare branches. Also in autumn, by its scarlet, oblong-elliptic, fleshy, acid-flavoured fruits 1–1·5 cm long. A deciduous shrub or small tree to 8 m, with oval-pointed, strongly veined leaves 4–10 cm. Bushy places, woods; sometimes cultivated. Widespread on mainland. **846. *C. sanguinea* L. DOGWOOD. Flowers white, in flat-topped clusters. Fruit black, 5–8 mm. Woods, bushy places. Widespread on mainland.

ARALIACEAE | Ivy Family

HEDERA **848. *H. helix* L. IVY. Hedges, walls, rocks, woods. Widespread.

UMBELLIFERAE | Umbellifer Family

A large and quite important family in our area, with 84 genera and approximately 265 species. Considerable experience is required to identify many genera and species, and ripe fruit is usually necessary. Space only permits the inclusion of a very few of the most distinctive species.

SANICULA 850. *S. europaea* L. SANICLE. Woods, in mountains. Widespread on mainland.

ASTRANTIA 2 species in our area. **852. *A. major* L. GREAT MASTERWORT. Woods and meadows in mountains. YU.AL.BG.

ERYNGIUM 10 species, including 4 species restricted to our area in Europe. Bracteoles are the small spiny bracts subtending each group of flowers in the flower head; involucral bracts surround the whole flower head.

Some basal leaves undivided (though often deeply toothed)
854. *E. alpinum* L. ALPINE ERYNGO. Alpine meadows, rocky ground. YU(west and central). **(856).** *E. creticum* Lam. SMALL-HEADED BLUE ERYNGO. Distinguished by its numerous small globular, stalked flower heads 0·5–1 cm, in a lax branched, bluish inflorescence. Involucral bracts 5–7, 1–3 cm, spine-tipped and with 1–2 pairs of lateral spines. Basal leaves decaying early, variable, rounded and undivided to three-times cut; leaf stalk unwinged. Fields, arid ground in the Med. region. Widespread. Pl. 31. **(856).** *E. planum* L. Like *E. creticum* with stems 25–100 cm, but basal leaves persistent, elliptic and shortly spiny-toothed, the uppermost stem leaves palmately five-lobed, spiny-toothed. Inflorescence usually bluish, more or less flat-topped; flower heads globular 1–2 cm; involucral bracts 6–8, 1·5–2·6 cm long, sometimes slightly shorter than flower head, with 1–4 pairs of lateral spiny teeth. Bushy and dry places. YU.TR.

All basal leaves divided (see also (856))
(a) Flower heads blue
853. *E. maritimum* L. SEA HOLLY. Sand and shingle beaches. Widespread. ****856. *E. amethystinum* L. BLUE ERYNGO. Distinguished by its basal leaves which have winged leaf stalks and blades twice cut into narrow spiny-toothed lobes. Inflorescence, stems, and bracts becoming bright blue. Flower heads 1–2 cm; involucral bracts 5–9, 2–5 times as long as the flower heads, with 1–4 pairs of lateral spines. Stony ground in the Med. region, and in mountains. YU.AL.GR.CR.BG?

(b) Flower heads green
855. *E. campestre* L. FIELD ERYNGO. Distinguished by its broad domed, pale-green inflorescence of numerous ovoid flower heads 1–1·5 cm across, each with 5–7 involucral bracts 1·5–4·5 cm long, entire or with 1–2 pairs of lateral spines. Bracteoles entire. Dry places. Widespread. **855d. *E. palmatum* Pančić & Vis. Distinguished by the erect green inflorescence 30–75 cm, with up to 10 stalked hemispherical flower heads 1–1·5 cm across. Involucral bracts 5–7, 2–4 cm long, with 1–5 pairs of spiny lateral teeth. Bracteoles three-toothed. Basal leaves persisting, slightly leathery, palmately divided into 5–7 oblanceolate, three-lobed, toothed lobes; leaf stalk unwinged. Dry places, woods. YU.AL.GR(north).BG.

LAGOECIA Flowers in globular umbels with pinnately divided, leafy involucral bracts. Sepals also leafy and pinnately divided. Fruit covered with short brittle club-shaped hairs.

856a. *L. cuminoides* L. An annual 10–30 cm, with pinnately divided upper leaves with narrow fine-pointed lobes, and with a branched inflorescence of dense globular, small, white, silky-haired umbels 0·5–1·5 cm. Dry grassy places, vineyards. GR+Is.CR.BG. Page 353.

ECHINOPHORA ****857.** *E. spinosa* L. A much-branched, domed, very spiny, stiff perennial to 50 cm, with small umbels of white flowers about 3 cm across. Calyx, involucral bracts, and bracteoles spine-tipped. Leaves twice-pinnate, lobes thick, spine-tipped. Maritime sands. YU.AL.GR.

SMYRNIUM Flowers yellowish-green. Fruit broadly ovoid, or rounded, with 3 prominent ridges on each carpel. Foliage characteristically pale yellowish-green.

1. Upper stem lvs. not clasping, the leaf
   blade divided (YU.AL.GR+Is.CR.TR.).   **865.** *S. olusatrum* L.
1. Upper stem lvs. clasping, the blade
   entire or toothed.

   2. Upper lvs. and branches opposite;
      rays of umbel 15–20.
     3. Lobes of basal lvs. ovate-heart-
        shaped; lower stem lvs. ovate-heart-
        shaped, coarsely lobed
        (GR+Is.TR.).   **866a.** *S. orphanidis* Boiss. Page 97
     3. Lobes of basal lvs. oblong-wedge-
        shaped; lower stem lvs. three-lobed
        (GR.CR+E. Aegean Is.).   **866b.** *S. apiifolium* Willd.
   2. Upper lvs. and branches alternate;
      rays of umbel 5–12.
     4. Stems narrowly winged on the angles;
        upper lvs. toothed, sometimes minutely
        so (Widespread).   ****866.** *S. perfoliatum* L.
     4. Stems not winged on angles; upper lvs.
        entire, rarely very finely toothed
        (YU.GR.CR+E.Aegean Is.BG.).   **(866).** *S. rotundifolium* Miller Pl. 31.

CRITHMUM ****871.** *C. maritimum* L. ROCK SAMPHIRE. A fleshy-leaved, somewhat glaucous, densely branched perennial 15–50 cm, of rocky sea shores. Umbels 2–6 cm across, yellow. Leaves 2–3 times pinnate, with cylindrical-pointed segments. Coasts throughout.

FOENICULUM 875. *F. vulgare* Miller FENNEL. A strong-smelling, somewhat glaucous perennial 0·5–1·5 m, with much-divided, rather dark green, feathery leaves and umbels of yellow flowers 4–8 cm across. Ultimate lobes of leaves 1–5 cm long, thread-like. Fruit ovoid-oblong, with stout, prominent ridges. Dry rocky places; also escaped from cultivation in waste places. Widespread.

CACHRYS Leaves 2–4 times cut into linear lobes. Petals yellow. Fruit with thick undulate wings. 3 species in our area.

879a. *C. ferulacea* (L.) Calestani (*Prangos f.*) A robust perennial to 180 cm, with large leaves, several times divided into slender, often thread-like, ultimate lobes 1–4·5 cm, and distinctive fruit with 5 equal broad wings or prominent ridges, but variable. Umbels yellow. Fruit 1–2·5 cm. Among herbaceous vegetation in the mountains. YU.AL.GR.BG.

BUPLEURUM Leaves undivided. Umbels yellow. Fruit usually with prominent ribs. 24 species in our area.

Annuals or perennials
(a) Upper leaves encircling or half-clasping stem
881. *B. rotundifolium* L. HARE'S EAR, THOROW-WAX. Cultivated land, waste places, waysides. Widespread on mainland. **881a.** *B. lancifolium* Hornem. An annual like 881, but clasping leaves usually ovate-lanceolate or oblong-lanceolate; rays of umbel fewer, 2–5 (leaves rounded; rays 5–10 in 881); bracteoles broadly ovate, overlapping, forming a cup round the fruits. Bracts absent. Fruit 3 mm, blackish-brown, smooth. Arable land, dry places. Widespread, except BG. Page 353.

882. *B. longifolium* L. A perennial 30–150 cm, with long-stalked elliptic-spathulate lower leaves and rounded-heart-shaped, half-clasping upper leaves. Rays 5–12; bracts 2–4, ovate to rounded like the bracteoles which are fused slightly at the base. Fruit

4–6 mm, brown or black, ridges prominent. In mountains; open woods, pastures, rocky places. YU.AL.BG.

883. B. falcatum L. SICKLE HARE'S EAR. Grassy places, fallow, woods. YU.AL.GR.BG. **883b. B. ranunculoides** L. A slender hairless perennial to 60 cm, with linear, three- to five-veined basal leaves, and wider, pointed, half-clasping upper leaves. Rays of umbels stout, 5–7; bracts 1–5, ovate; bracteoles variable in shape, usually 5, conspicuously veined. Fruit 2·5–3 mm, with slender winged ridges. Mountain rocks. YU(south-west). Page 179.

(b) Upper leaves not half-clasping or encircling stem
881b. B. flavum Forskål A slender, erect branched annual 20–75 cm. Rays of umbels usually 3–6; bracts more than two-thirds as long as the longest ray, lanceolate-oblong-pointed, yellowish-green semi-translucent, with 3 veins; bracteoles ovate-lanceolate-pointed, semi-translucent. Leaves all linear-lanceolate to linear, the lower stalked. Fruit 1·6–2 mm, with slender ridges. Dry rocky places, especially near the sea. GR+Is.BG.TR.

Shrubs
884. B. fruticosum L. SHRUBBY HARE'S EAR. A bushy evergreen shrub to 2·5 m, with stalkless, shining elliptic-oblong leaves, and yellow umbels 7–10 cm across. Primary rays 5–25; bracts and bracteoles 5–7, soon falling. Fruit 7–8 mm, with narrowly winged ridges. Bushy places in the Med. region; sometimes grown for ornament. GR.

FERULA 3 species in our area.
**897. F. communis L. GIANT FENNEL. A very robust perennial 1–3 m, with a tall inflorescence of many large yellow umbels with 20–40 rays, and dark-green, much-dissected feathery leaves with thread-like lobes. Uppermost leaves reduced to a large sheathing base which at first enclose the young umbels. Fruit about 1·5 cm, elliptic, flattened, with thin lateral wings. Dry hill slopes. YU.AL.GR+Is.CR.TR.

FERULAGO Very like *Ferula* but distinguished by the presence of 5 or more bracts and bracteoles (in *Ferula* bracts and bracteoles absent, or bracteoles soon falling). Fruit often broadly winged on the back as well as being laterally winged. 6 species in our area, including 2 endemic to it.

898. F. campestris (Besser) Grec. (*F. ferulago*) A tall erect perennial 0·5–2 m, with grooved, angular stems and large dark-green leaves, which are triangular in outline and many-times dissected into short thread-like lobes. Umbels yellow, the terminal with 12–20 rays. Fruit 1–2 cm, the lateral wings well developed, the dorsal ridges slender, unwinged. Rocks, among herbaceous vegetation. YU.AL.BG. **898a. F. nodosa** (L.) Boiss. An erect hairless perennial to 60 cm, with feathery leaves and yellow umbels, but readily distinguished from similar-looking species by the conspicuous swollen nodes of the stem. Ultimate lobes of leaves linear. Bracts and bracteoles ovate-lanceolate. Fruit 8–10 mm, with narrow, somewhat undulate, lateral wings and narrow dorsal wings. Rocks and bushy places in the Med. region. AL.GR.CR. Page 96. **898b. F. sylvatica** (Besser) Reichenb. Distinguished by its leaves which are elliptic to nearly linear in outline, and widest at the middle, with linear ultimate lobes (leaves triangular in outline in 898a). A perennial to 125 cm. Fruit 6–10 mm, elliptic, with lateral wings well developed and slender dorsal ridges. Bushy places, woods, in mountains. YU.AL.GR.BG.TR.

OPOPANAX A rather distinctive genus of tall plants with large yellow umbels, and yellowish-green, often once-pinnate leaves with large oval leaflets. Fruit strongly compressed, the lateral ridges united to form a border surrounding the fruit before splitting.

898c. *O. chironium* (L.) Koch A robust perennial to 2 m, branched above only; the lower leaves twice-pinnate with large more or less oval, toothed leaflets 4–12 cm. Upper leaves simple, or reduced to inflated sheaths. Umbels yellow, with 9–25 rays; bracts and bracteoles bristle-like. Fruit whitish, 6–7 mm, border narrow, thickened. Stony and damp places in the Med. region. YU.AL.GR.BG. Page 353. **898d.** *O. hispidus* (Friv.) Griseb. Like 898c but up to 3 m, and fruit larger 7–9 mm, with a wide, thin border. Rays of umbel 6–13. Leaflets smaller, 2–4 cm. Stony ground, olive groves, vineyards, etc. in the Med. region. YU.AL.GR+Is.CR.BG.

TORDYLIUM Fruits distinctive: rounded, strongly compressed, and with a thickened, often white, lobed border. Umbels white or pinkish. 5 species in our area.

**904. *T. apulum* L. A small umbellifer frequently encountered in the Mediterranean phrygana and by waysides, with distinctive flowers and fruits. Umbels white, the outer flowers with the outer petals very much longer and larger than the rest, and deeply lobed. Fruit oval, flattened, with a thick white horny border which appears crinkled on the inner margin. YU.AL.GR+Is.CR.TR.

THAPSIA Leaves two- to three-pinnate. Petals yellow, wedge-shaped. Fruit oblong to ovate, compressed; broadly winged laterally.

906c. *T. garganica* L. Often a tall perennial 30–250 cm, with leaves many times cut into linear-oblong lobes, and with large yellow umbels with 5–20 rays, without bracts or bracteoles. Fruit distinctive, 1·2–2·5 cm, oblong or elliptic, notched at base and apex and with conspicuous lateral wings 3–6 mm broad. Rocky places, fields, sunny slopes. GR+Is.CR.

ORLAYA Fruit ovoid, with spines and bristles on the ribs. Flowers white or pink, the outer petals of the outer flowers much larger than the inner petals. 3 species in our area.

909. *O. grandiflora* (L.) Hoffm. An annual to 40 cm, distinguished by its white umbels with the outer petals of the outer flowers 9–14 mm, up to 8 times as long as the other petals. Upper stem leaves entire or pinnately lobed. Fruit about 8 mm, not compressed, with 2–3 rows of spines on the ribs. Waysides, amongst herbaceous vegetation, dry places. YU.AL.GR.CR.BG.TR.

PYROLACEAE | Wintergreen Family

PYROLA ****912.** *P. minor* L. SMALL WINTERGREEN. Flowers whitish to pinkish, globular, in a dense oval cluster; style 1–2 mm, shorter than petals and stamens. Woods. YU.AL.GR.BG. **(912).** *P. media* Swartz INTERMEDIATE WINTERGREEN. Like 912, with white or pink flowers, but flower cluster rather lax, cylindrical. Style 4–6 mm, straight, longer than petals and stamens, with a disk below the stigma. Mountain woods, moors. YU.BG.

913. *P. rotundifolia* L. ROUND-LEAVED WINTERGREEN. Flowers white, in a lax cluster; petals spreading; style 4–10 mm, down-curved, longer than the up-curved stamens. Sepals lanceolate, 2–3 times shorter than petals. Leaves rounded, 2–4·5 cm. Woods.

1. *Vaccinium arctostaphylos* 936a   **2.** *Lagoecia cuminoides* 856a   **3.** *Bruckenthalia spiculifolia* 934a   **4.** *Bupleurum lancifolium* 881a   **5.** *Opopanax chironium* 898c

YU.BG. **913a.** *P. chlorantha* Swartz GREENISH WINTERGREEN. Like 913 but flowers greenish-white and sepals oval-triangular, almost as broad as long, 3–4 times shorter than the petals. Style 6–7 mm, curved. Leaves smaller 0·5–2 cm. Woods, alpine meadows. YU.GR.BG.

ORTHILIA ***914.** *O. secunda* (L.) House NODDING WINTERGREEN. Flowers 5–6 mm across in a one-sided cluster, white tinged with green. Style about 5 mm, longer than petals; anthers without tubes (anthers opening by pores at the end of short tubes in *Pyrola*). Woods. YU.AL.GR.BG. Page 185.

MONESES ***915.** *M. uniflora* (L.) A. Gray ONE-FLOWERED WINTERGREEN. Flowers white, solitary, 13–20 mm across. Mountain woods. YU.AL.BG.

MONOTROPA ***917.** *M. hypopitys* L. YELLOW BIRD'S NEST. A fleshy, waxy-looking, yellowish or brownish saprophytic plant 5–25 cm, without chlorophyll, with a terminal cluster of drooping yellowish flowers. Leaves scale-like. Damp woods in mountains. YU.AL.GR.BG.

ERICACEAE | Heath Family

RHODODENDRON ***919.** *R. ferrugineum* L. ALPENROSE. A small evergreen mountain shrub 50–120 cm, with dark-green shining leaves which are rust-brown beneath, and terminal clusters of bright pinkish-red, funnel-shaped flowers 1·5 cm long. Leaves 2–4 cm, elliptic-oblong, acute. Mountain pastures, open woods. YU(west). **919a.** *R. myrtifolium* Schott & Kotschy (*R. kotschyi*) Similar to 919 but smaller, to 50 cm; leaves smaller 1–2 cm, blunt, with toothed margins and the undersides often greenish. Petals clear pink; flower stalks hairy (hairless in 919). Shady rocks, stony places in mountains. YU(south).BG.

919b. *R. luteum* Sweet An easily recognized deciduous suckering shrub 2–4 m, bearing crowded clusters of yellow flowers on bare branches before the leaves develop. Flowers 5 cm across; corolla tube 1–1·5 cm, hairy outside. Leaves oblong-lanceolate, 6–12 cm, deciduous, somewhat bluish-green. Pine woods, or wet, peaty ground. YU.GR(Lesbos). Pl. 32. ***920.** *R. ponticum* L RHODODENDRON. A large shrub 2–5 m, with dark evergreen, laurel-like leaves 12–25 cm, and rounded clusters of purple flowers. Flowers 4–6 cm across, hairy within at the base; stamens and style longer than the funnel-shaped corolla. Woods, damp places in mountains. BG(south-east). TR.

RHODOTHAMNUS **920a.** *R. chamaecistus* (L.) Reichenb. A dwarf evergreen shrub to 40 cm, with terminal clusters of pink flowers 2–3 cm across and long projecting stamens and style. Corolla with a short tube and spreading lobes. Leaves elliptic 5–10 mm, hairless but for stiff white marginal hairs. Dry stony slopes, screes in limestone hills. YU(Croatia).

LOISELEURIA ***921.** *L. procumbens* (L.) Desv. CREEPING AZALEA. Rocks, grassy places in mountains. YU(Croatia, Macedonia).

ARBUTUS ***924.** *A. unedo* L. STRAWBERRY TREE. A small evergreen tree or commonly a shrub 1·5–3 m, with large dark-green, elliptic, toothed leaves 4–11 cm, and drooping clusters of nearly globular, cream-coloured flowers appearing in autumn. Twigs mostly glandular-hairy. Fruit unmistakable: globular 1·5–2 cm, red and strawberry-like when ripe, and covered with warts. Evergreen scrub, wood margins, rocky slopes in the Med. region. Widespread, except BG. Page 34. **(924)**. *A. andrachne* L. EASTERN STRAWBERRY TREE. Like 924 but stems smooth, reddish-purple, with bark peeling in papery strips, and leaves grey-green, not toothed, except on sucker shoots; twigs hairless. Flowers white, occurring in spring, in erect, glandular-hairy, branched clusters. Fruit about 8–12 mm, golden-yellow. Evergreen scrub, rocky slopes in the Med. region. AL.GR+Is.CR.TR. Pl. 32.

ARCTOSTAPHYLOS 2 species in our area ****925**. *A. uva-ursi* (L.) Sprengel BEAR-BERRY. A spreading, mat-forming shrub, with shiny leathery, oval leaves, and short clusters of nearly globular pink to greenish-white flowers. Corolla about 5–6 mm. Leaves 1–3 cm, paler beneath. Fruit globular 1 cm, shining red. Woods, rocks in mountains. YU.AL.GR(north).BG. Page 185.

CALLUNA **926**. *C. vulgaris* (L.) Hull LING, HEATHER. Lowland and montane heaths. YU.BG.TR.

ERICA | **Heath, Heather**
927. *E. arborea* L. TREE HEATH. A tall, feathery-looking shrub 1–4 m, or rarely more, with very numerous, very tiny leaves and tiny white or pinkish flowers. Flowers 2–3 mm long, bell-shaped, sweet-scented, in dense pyramidal clusters; anthers brownish, contrasting with, and included within, the corolla, anther appendages flat, ciliate. Young twigs white-hairy; leaves 3–5 mm. Woods, evergreen scrub in the Med. region. Widespread. Page 35. ****933. *E. herbacea* L. (*E. carnea*) SPRING HEATHER. A small spreading under-shrub to 60 cm, with leaves in whorls of 4, and with dark reddish, or rarely white, pendulous flowers in short, one-sided, terminal clusters. Stamens projecting, longer than the corolla which is 4–8 mm. Open woods, rocky places in mountains. YU.AL.GR(north).

(934). *E. manipuliflora* Salisb. (*E. verticillata*) A small spreading or ascending shrub to 50 cm, with leaves in whorls of 3–4, and rounded terminal clusters, or lax lateral clusters of tiny pink flowers with projecting anthers. Corolla 3–4 mm, broadly bell-shaped; sepals 1 mm. Leaves 4–8 mm. Evergreen bushy thickets, stony places in the Med. region. Widespread, except BG? Page 38. ****934**. *E. multiflora* L. Very like (934) but stems erect, rigid to 80 cm or more, and leaves in whorls of 4–5. Distinguished by its projecting anthers which are about 1·5 mm, with parallel lobes (anthers 0·5–1 mm, with diverging lobes in (934)). Flowers pink, usually 4–5 mm; sepals 1·5–2 mm. Leaves 6–11 mm. Rocky places and evergreen thickets in the Med. region. YU(west). AL?

BRUCKENTHALIA Like *Erica* but sepals fused below into a tube about as long as the lobes, and flower stalks without bracteoles.

934a. *B. spiculifolia* (Salisb.) Reichenb. SPIKE-HEATH. An erect heather-like shrub 10–15 cm, of mountain pastures, or woodlands, with downy twigs, numerous bristly-pointed linear leaves, and crowded terminal clusters of magenta-pink flowers. Corolla 3 mm, open bell-shaped with ovate lobes, with anthers included and styles longer; calyx pink. Leaves 4–5 mm, ciliate, margins inrolled, in irregular whorls of 4–5. YU.AL? GR.BG. Page 353.

VACCINIUM | Cowberry, Wortleberry, Cranberry

1. Lvs. with entire, more or less inrolled
   margins.
   2. Lvs. evergreen, glossy; filaments
      hairy; berry red (YU.AL.BG.).   ****935.** *V. vitis-idaea* L.
   2. Lvs. deciduous, blue-green; filaments
      hairless; berry bluish-black (YU.AL.
      BG.).   **(936).** *V. uliginosum* L.
1. Lvs. toothed, flat.
   3. Small shrub; corolla pitcher-shaped;
      calyx scarcely lobed (YU.AL.BG.).   ****936.** *V. myrtillus* L.
   3. Shrub or small tree; corolla bell-
      shaped; calyx distinctly five-lobed.   *arctostaphylos*

936a. *V. arctostaphylos* L. A deciduous shrub or small tree 2–5 m, with pendulous axillary clusters of 5–8 greenish-white, pink-tinged, bell-shaped flowers about 1 cm. Leaves 4–6 cm, oblong-pointed, toothed. Berry purplish-black, globular, 6–8 mm. Wooded ravines. BG(south-east).TR. Page 353.

EMPETRACEA | Crowberry Family

EMPETRUM **938. *A. nigrum* L. CROWBERRY. Heaths and rocky places in mountains. YU.AL.BG.

PRIMULACEAE | Primrose Family

PRIMULA | Primrose, Cowslip About 14 species in our area, 4 of which are restricted to it.

Flowers yellow
(a) Flowers in a terminal stalked umbel
939. *P. veris* L. COWSLIP. Subsp. *columnae* (Ten.) Lüdi is the southern form which is distinguished by its leaves, which are densely white-haired beneath, with twisted and often branched hairs 1 mm, and leaf stalk not or slightly winged. Flowers 1–2·2 cm across, the corolla tube longer than the calyx. Meadows in mountains. YU.AL.GR.BG. Page 168. ****940.** *P. elatior* (L.) Hill OXLIP, PAIGLE. Very like *P. veris* subsp. *columnae* but calyx lobes with slender attentuated tips, and mature capsule equalling or longer than the calyx (calyx lobes acute; mature capsule shorter than calyx in 939). Flowers yellow 1·5–2·5 cm across, sometimes with an orange throat. Meadows in hills and mountains. YU.AL.BG. ****941.** *P. auricula* L. AURICULA, BEAR'S EAR. Limestone rocks in mountains. YU(Croatia,Serbia).

(b) Flowers solitary, arising directly from the root-stock
**942. *P. vulgaris* Hudson PRIMROSE. The widely distributed, pale yellow-flowered subsp. *vulgaris* occurs scattered throughout the mainland. Subsp. *sibthorpii* (Hoffmanns.) W. W. Sm. & Forrest usually has red or purple flowers, and leaves with a wedge-shaped

base, more or less abruptly contracted into the leaf stalk. Woods, thickets. GR(east).BG.TR. Pl. 32.

Flowers pink, purple, or violet (see also 942)
(a) *Leaves with white or yellow mealy powder (farina) beneath*
943. *P. farinosa* L. BIRD'S-EYE PRIMROSE. A delicate perennial with a small rosette of mealy leaves, and a slender stem 3–20 cm bearing an umbel of small pink-lilac flowers with pale-yellow 'eyes'. Flowers 8–16 mm across; petals notched, corolla tube 5–6 mm; calyx cylindrical or urn-shaped with usually blunt lobes. Damp meadows in mountains. YU(Croatia).BG. Page 137. **943a.** *P. frondosa* Janka Like 943 but calyx bell-shaped with acute lobes. Flowers pinkish-lilac to reddish-purple 1–1·5 cm across, in an umbel of 10–30. Leaves usually thickly mealy beneath, but sometimes without mealy powder. Shady cliffs, near melting snow. BG(Stara Planina). Pl. 32. **(943).** *P. halleri* J. F. Gmelin (*P. longiflora*) LONG-FLOWERED PRIMROSE. Like 943 but readily distinguished by the very long violet corolla tube 2–3 cm, 2–3 times as long as the calyx. Flowers rose-violet, in an umbel of 2–12; corolla 1·5–2 cm across. Alpine meadows, rock crevices. YU.AL.BG. Page 184.

(b) *Leaves without mealy powder beneath (see also 943a)*
(944). *P. minima* L. LEAST PRIMROSE. A mat-forming plant with tiny rosettes of leaves bearing much larger solitary, or paired, almost stalkless, rose-coloured flowers 1·5–3 cm across. Leaves wedge-shaped, shining, the longest 0·5–3 cm, toothed only at apex. Alpine pastures by melting snow. YU.AL.BG. Pl. 32. **946b.** *P. deorum* Velen. A very handsome and distinctive species of the Rila Mountains in south-west Bulgaria, with an umbel of large deep reddish to violet-purple flowers, and a sheaf of oblanceolate shining, rather leathery, entire leaves. Flowers 3–18, in an umbel, borne on a sticky stem 5–20 cm, which is purplish towards the apex. Damp peaty soils in mountains. BG. Pl. 32.

946c. *P. glutinosa* Wulfen A small sticky perennial with a cluster of small basal leaves, and a short stem 1·5–8 cm bearing 2–8 deep-violet flowers with notched petals. Bracts 4–12 mm, oblong-elliptic, sticky; calyx 5–9 mm. Leaves oblong-elliptic, toothed, finely glandular, the longest 1·5–8 cm. Wet acid soils. YU(Bosnia-Hercegovina). **946d.** *P. kitaibeliana* Schott Flowers pink to lilac, 1–3, on stems 1·5–8 cm. Bracts linear; calyx 7–13 mm. Leaves sticky, fleshy, lanceolate to obovate, shallowly toothed, covered with three- or five-celled glandular hairs. Rocks and stony ground on limestone. YU-(Bosnia-Hercegovina). Page 191.

ANDROSACE

Annuals
950. *A. maxima* L. An unmistakable, slender annual with a neat basal rosette of usually ovate, toothed leaves, and one or several leafless stems 3–15 cm, bearing umbels of many tiny pink or white flowers about 5 mm. Petals much shorter than the green calyx, and the conspicuous green leafy oval involucral bracts. Fields, dry stony places. YU.GR.BG.TR. **950a.** *A. elongata* L. Distinguished from 950 by its lanceolate or oblanceolate entire or toothed leaves, and its lax umbel of many tiny white flowers each about 3 mm with petals shorter than the calyx. Involucral bracts small, much shorter than the individual flower stalks (bracts equal or longer than flower stalks in 950). Fields, sandy or stony places. YU.BG.

Perennials
(a) *Calyx hairless*
952. *A. lactea* L. A hairless, loosely mat-forming perennial with rosettes of linear leaves, each rosette with 1–4 slender stems 3–15 cm, bearing umbels of 1–6 white flowers with

yellow centres. Flowers to 12 mm across; petals notched: calyx hairless; involucral bracts linear-lanceolate. Rosettes 2–3 cm across, superimposed on each other; leaves parallel-sided. Limestone rocks in mountains. YU(central). **952b.** *A. hedraeantha* Griseb. Like 952a *A. obtusifolia*, but inflorescence in a dense rounded cluster of 5–10 flowers, on shorter stems often only 1–2 cm, and calyx and bracts hairless. Flowers pink or white. Acid rocks in mountains. YU.AL.BG. Pl. 33.

(b) Calyx hairy

952a. *A. obtusifolia* All. A small perennial, like 952, tufted, or with solitary rosettes of lanceolate or oblong-lanceolate, sparsely hairy leaves, but leaves widest above the middle. Flowers white or pink, 6–9 mm across, in 1–7 flowered clusters; calyx densely hairy. Limestone rocks and screes in mountains. YU.BG. ****(954).** *A. villosa* L. Distinguished by its long spreading hairs on the stems and its shaggy-haired, silvery rosettes 5–8 mm across. Plants usually densely tufted, with numerous rosettes; leaves lanceolate. Flowers 3–7, white or pink, on stems 1–7 cm. Limestone rocks and screes. YU.AL.GR(central).BG. Page 184.

CORTUSA ****955.** *C. matthioli* L. ALPINE BELLS. In mountains, damp ravines, wet rocks. YU.BG.

SOLDANELLA | Snowbells

1. Corolla cut into narrow lobes to not more than a third of its length; throat usually naked (YU.BG.).	****(956).** *S. pusilla* Baumg. Page 137
1. Corolla usually cut to half its length or more; throat with scales which are usually two-toothed.	
2. Lv. stalks with stalkless glands, most of which do not persist to maturity.	
3. Lvs. bluish-green beneath with waxy-whitish bloom (GR.).	**956b.** *S. pindicola* Hausskn.
3. Lvs. green beneath, not bloomed (YU.AL.).	**956.** *S. alpina* L. Page 185
2. Lv. stalks with stalked glandular hairs, most of which persist to maturity.	
4. Lvs. green or violet beneath, not bloomed (BG.).	**(956).** *S. montana* Willd. *hungarica*
4. Lvs. bluish-grey beneath, with waxy-whitish bloom.	*dimoniei*

956c. *S. hungarica* Simonkai Distinguished by the persistent stout, dense glandular hairs on the leaf stalks, with stalks of the hairs 2–5 times as long as the head. (**(956)** *S. montana* has glandular hairs with stalks 8–10 times longer than the head.) Flowers 8–15 mm, reddish-violet, in an umbel of 2–8, on a stem 3–25 cm. Leaves undivided, kidney-shaped, 1–3 cm broad or more, often violet beneath. In mountains, by melting snow, woods, alpine meadows. YU.AL.BG. Pl. 33. **956d.** *S. dimoniei* Vierh. Like 956c but distinguished by its leaves which are differently coloured beneath and densely covered with glandular pits. Flowers violet, 2–3, on a stem 3–9 cm. By melting snow in mountains of Macedonia. YU.AL.BG.

HOTTONIA ****957.** *H. palustris* L. WATER VIOLET. Still waters, ditches. YU(Croatia, Serbia).

CYCLAMEN

Plants flowering June–December

958. *C. hederifolium* Aiton (*C. neapolitanum*) SOWBREAD. Flowers appearing in autumn before the leaves, varying from pale pink to white, with a darker blotch at the base of each reflexed petal, and a pentagonal throat formed by the outward-flaring of the petal bases (auricles). Petals to 2 cm. Leaves 3–14 cm, heart-shaped, often angled or lobed, and often purplish beneath, variously silver-grey mottled above. Tubers large, rooting from upper surface. Woods, bushy places. YU.AL.GR+Is.BG.TR? **(958).** *C. graecum* Link GREEK CYCLAMEN. Like 958 flowering in autumn and early winter before the leaves and having auricled flowers. Petals white, or pale pink, with dark-purple blotches, to 2 cm. Leaves 3–7 cm, heart-shaped with horny, toothed margins, richly and variously blotched with grey. Tubers with thick retractile roots from the base. Stony hill slopes. GR+Is.CR.

**959. *C. purpurascens* Miller (*C. europaeum*) COMMON CYCLAMEN. Flowers sweet-scented, petals rose-carmine with darker basal blotch, appearing with the leaves from June to October. Petals 1·5–2 cm. Leaves rounded 2·5–8 cm, not angled, silver blotched above, reddish-purple beneath. Tuber small, rooting all over. Woods, bushy places. YU.

Plants flowering January–May

960. *C. repandum* Sibth. & Sm. Flowers sweet-scented, usually bright rose-coloured, rarely white, appearing in spring with the leaves. Petals 1·5–3 cm, darker coloured towards the rounded throat. Leaves 4–13 cm triangular-heart-shaped, deeply toothed, or with shallow angular lobes, often not blotched. Tuber small, rooting from the base. Woods, thickets, rocky places. YU.GR. Page 89. **960b. *C. creticum* Hildebr. A Cretan endemic with rather small white flowers appearing from March to May; petals 1·6–2·5 cm. Leaves 3·5–9 cm broadly ovate with sharply toothed margin, white-blotched above. Tuber small, rooting from the base. Shady places, thickets. CR. Pl. 33.

960c. *C. persicum* Miller Flowers large, white or pink with darker basal zone, appearing from January to May with the leaves. Petals 2·5–4·5 cm, lanceolate, twisted. Leaves 3–14 cm heart shaped with a thickened, toothed margin, blotched above. Fruiting stalks arched, not coiled like other species. Tuber large 4–15 cm, rooting from the base. Shrubby places. E. Aegean Is. Pl. 33. **959a.** *C. coum* Miller Easily recognized by its small flowers with overlapping petals only 7–15 mm, magenta with prominent dark-violet, pale-centred blotches at their bases. Flowering January–April. Leaves 3–7 cm, rounded, usually not blotched, and usually with entire margins. Tuber small rooting from the base. Shady places, amongst rocks. BG.TR.

LYSIMACHIA | Loosestrife

1. Small creeping or spreading plants.
    2. Lvs. gland-dotted; calyx lobes ovate, overlapping; fls. 8–18 mm across (Widespread on mainland).   ****961.** *L. nummularia* L.
    2. Lvs. not gland-dotted; calyx lobes long-pointed, not overlapping; fls. 6–9 mm across (YU.).   ****962.** *L. nemorum* L.
                                                 serpyllifolia

1. Erect plants.
  3. Fls. yellow.
    4. Fls. with parts in sevens, in dense
      axillary clusters (YU.BG.).   **964. *L. thyrsiflora* L.
    4. Fls. with parts in fives, not in
      dense axillary clusters.
      5. Fls. in a terminal branched
        cluster; petals without glandular
        hairs (Widespread on mainland).   **963. *L. vulgaris* L.
      5. Fls. usually in 2 axillary clusters;
        petals with glandular hairs.
        (Widespread on mainland).   **(963). *L. punctata* L.
  3. Fls. pink or purple.
    6. Fls. purple, stalkless.   *atropurpurea*
    6. Fls. pink, stalked.   *dubia*

962a. *L. serpyllifolia* Schreber Like 962 but a dwarf woody-based shrublet, with ovate-heart-shaped, almost stalkless leaves 6–11 by 3–8 mm. Flowers yellow, 1–1·5 cm across, with flower stalks at least twice as long as the subtending leaf (equalling in 962). Bushy and stony places in mountains. GR(central and south).CR. Page 89. **964a.** *L. atropurpurea* L. An erect perennial 20–65 cm, with a terminal spike of stalkless, dark-purple flowers, and alternate, narrow lanceolate to spathulate, glaucous leaves with toothed and undulate margins. Petals oblong 6–7 mm; stamens longer. Waste ground, damp sandy places. Widespread on mainland. Pl. 33. **964b.** *L. dubia* Aiton Distinguished from 964a by the stalked, pink flowers which are in laxly branched, terminal, spike-like clusters; calyx with white ciliate margins. Petals 4–6 mm; stamens shorter or as long as petals. A hairless annual or biennial 30–60 cm. Ditches, damp places. YU(south).AL.GR.BG.TR.

ASTEROLINON Annuals with 5 petals shorter than calyx. Flowers solitary in axils of the upper leaves. Capsule with 5 valves.

966b. *A. linum-stellatum* (L.) Duby A freely branched, tiny hairless annual 3–18 cm, with narrow pointed leaves and tiny axillary white flowers. Calyx 3–6 mm; petals 1–2 mm. Leaves lanceolate 3–7 mm. Open dry places in the Med. region. Widespread.

ANAGALLIS | Pimpernel
966. *A. tenella* (L.) L. BOG PIMPERNEL. Bogs, damp peaty places. GR.CR. **(966).** *A. minima* (L.) E. H. L. Krause CHAFFWEED. Damp open places, especially on sand. YU.AL.GR.BG. **967.** *A. arvensis* L. SCARLET PIMPERNEL. Flowers red, blue, or pink. Red-flowered forms are common in the central European climatic region and blue-flowered forms commoner in the Med. region. Fields, open sandy places, waste places. Widespread. **(967).** *A. foemina* Miller Very like 967 but flowers always blue and marginal hairs on petals absent or few, usually four-celled (marginal hairs numerous, always three-celled in 967). Cultivation, fields, waste places. Widespread. Pl. 33.

SAMOLUS **969.** *S. valerandi* L. BROOKWEED. Marshes or damp salt-rich places. Widespread.

CORIS **970.** *C. monspeliensis* L. A stiff, branched, thyme-like, woody-based perennial or biennial 10–30 cm, with linear leaves, and dense terminal clusters of small two-lipped, rose-lilac to blue flowers. Corolla 9–12 mm, five-lobed with 3 upper and 2 lower lobes. Calyx tinged purple, bell-shaped, with 2 rows of teeth, the inner 5 triangular, the outer 6–21 spiny. Stony hills, maritime sands. YU.AL.

PLUMBAGINACEAE | Sea Lavender Family

PLUMBAGO **971.** *P. europaea* L. EUROPEAN PLUMBAGO. An erect, much-branched perennial 30–100 cm, with dense clusters of strongly veined, violet or pink flowers, and conspicuously bristly, glandular-hairy calyx. Flowers about 1·5 cm across; corolla tube twice as long as calyx. Leaves 5–8 cm, rough, glandular-toothed, the lower oval, stalked, the upper linear, clasping. Dry rocks, hills, maritime sands. Widespread.

ACANTHOLIMON Spiny, cushion-forming shrubs. Calyx funnel-shaped, ten-ribbed at the base, papery, pleated, five-toothed; petals fused at their bases; stamens free; styles 5.

971a. *A. androsaceum* (Jaub. & Spach) Boiss. (*A. echinus*) A cushion-forming shrub with linear sharply spiny-tipped leaves, and short-stemmed clusters of 3–7 pinkish-purple flowers. Calyx 14–16 mm, the tube about twice as long as the papery limb, which is white with purplish veins. Petals long-stalked, shorter or slightly exceeding calyx. Leaves numerous, 1–2 cm. Mountains. YU(south).AL.GR.CR+Samos. Page 70. Pl. 34.

LIMONIUM | Sea Lavender Flowers grouped in a spikelet with 3 scale-like bracts at its base, the spikelets are further grouped into spikes which are borne at the ends of the inflorescence branches. Identification is often difficult: combinations of small differences in the calyx, corolla, bracts, inflorescence, and leaves distinguish the different species. About 19 species in our area. 6 are endemic. Distinctive and widespread species include:

972. *L. sinuatum* (L.) Miller WINGED SEA LAVENDER. A striking perennial 20–40 cm, with a rosette of leaves with deeply rounded lobes, winged stems, and compact flat-topped clusters of bright blue-mauve, papery everlasting flowers. Calyx blue-mauve, much more conspicuous than the tiny yellowish corolla. Rocks and sands on the littoral. YU.AL.GR+Is.CR. **972b.** *L. ferulaceum* (L.) O. Kuntze An unusual-looking bushy perennial 10–40 cm, with brush-like tufts of numerous flowerless thread-like branches with brown scales. Flowering branches differing, regularly alternate with club-shaped spikes clustered at the ends of the branches; spikelets one-flowered. Corolla pink, 5–6 mm; calyx cylindrical, surrounded by long-pointed bracts. Leaves absent at flowering time. Maritime marshes. YU(Croatia).

972e. *L. latifolium* (Sm.) O. Kuntze A hairy perennial 50–80 cm, readily recognized by the large, long-stalked spathulate or elliptical basal leaves 25–60 cm long, and branched inflorescence to 80 cm, with scale leaves 5–6 mm. Flowers pale violet, in spikes 1·5 cm long, in a dense cluster; calyx 3–4 mm, with short, rounded lobes, half as long as the petals; bracts papery. Dry grassland. BG. **972f.** *L. gmelinii* (Willd.) O. Kuntze A hairless perennial 20–60 cm, somewhat similar to 972e but leaves 7–11 cm; scale leaves to 2·5 cm. Spikes 0·6–1 cm, dense; petals reddish; calyx about 4 mm. Limestone soils. YU.GR?+E.Aegean Is.BG.TR. Page 371.

974. *L. bellidifolium* (Gouan) Dumort. MATTED SEA LAVENDER. A rather small plant 9–30 cm, with a hemispherical, much-branched, zigzag inflorescence with numerous sterile and flowering branches. Flowers pale violet, 4–6 mm, in densely clustered spikes at the ends of the branches, with white papery calyx and bracts. Basal leaves spathulate 1·4–4 cm, usually withered at flowering. Coastal marshes. YU.GR. **(974).** *L. oleifolium* Miller (incl. *Statice virgata* Willd.) Like 974 but flowers larger 8–9 mm, in long lax one-sided spikes to 4 cm, borne on usually wide-spreading branches on stems 5–50 cm. Spikelets curved, compressed; bracts conspicuous, reddish-brown; calyx lobes curved, with broad white papery margins. Maritime cliffs. Coasts throughout, except BG.

974d. *L. echioides* (L.) Miller A slender annual 5–30 cm, with a basal rosette of obovate-spathulate leaves, and widely spaced, curved spikelets arranged along one side of the forked inflorescence branches. Flowers pale pink, short-lived; calyx with 5–10 curved red spines; bracts with minute swellings. Maritime sands and rocks. GR + Is.CR. **974e.** *L. cancellatum* (Bertol.) O. Kuntze A densely hairy perennial 10–18 cm, with spathulate leaves 1·5–2·7 cm forming a lax rosette. Inflorescence a network of branches borne at angles of 90–180° to each other, with numerous non-flowering branches. Spikes to 2·5 cm; corolla 5·5 mm; calyx 4·5 mm. Maritime rocks. YU.GR(west). **974f.** *L. ocymifolium* (Poiret) O. Kuntze A hairless perennial 20–30 cm, with oblong-spathulate leaves 3–4 cm. Inflorescence without non-flowering branches; flowers in pairs in unilateral spikelets forming spikes up to 2·5 cm; calyx about 3·5 mm. Rocks by sea. GR(south).

GONIOLIMON Like *Limonium* but always perennial herbs; stem usually compressed; leaves leathery, few, in a basal rosette; spikelets with pointed-tipped bracts. Stigma knob-like (thread-like in *Limonium*). 6 species in our area, 3 are endemic.

(a) Flowers whitish
974g. *G. heldreichii* Halácsy A hairless perennial 10–20 cm, with slender, flexuous stems. Flowers whitish, in long lax spikes 2–3 cm long, with 3 spikelets per cm; spikelets usually one-flowered; calyx 6 mm, tube hairy. Leaves 5–6 cm, spathulate, white-spotted. Dry grassland, stony hillsides. GR(east).

(b) Flowers pink or reddish-purple
974h. *G. sartorii* Boiss. A hairless perennial with stout, arching stems 10–20 cm, with whitish glands, and a basal tuft of spathulate leaves covered in white glands. Inflorescence branched, with dense spikes 1–2 cm long; spikelets one-flowered; petals pale pink; calyx 7–8 mm, tube hairless. Leaves 3–4 cm, densely whitish-glandular. Maritime rocks. GR(east) + Cyclades. Page 371. **974i.** *G. besseranum* (Reichenb.) Kusn. Flowers in a pyramidal inflorescence with lax spikes 1–2 cm, with reddish-purple corollas. Calyx tube hairless. Leaves 3–8 cm, sparsely whitish-glandular. Steppes, dry grasslands. BG(north-east). Page 148.

ARMERIA | Thrift Flowers in spikelets grouped into dense heads surrounded by bracts. About 7 species in our area.

975. *A. maritima* (Miller) Willd. THRIFT, SEA PINK. A densely tufted perennial with soft, linear, one-veined leaves, and globular heads, 1–3 cm across, of pink or purple flowers, borne on leafless stems up to 20 cm or more. Outer involucral bracts often shining green with narrow papery margins, the inner papery; stem sheath 0·5–1·5 cm. A very variable species with many distinctive subspecies occupying a wide range of habitats from the coasts to the high mountains. YU.AL.BG. **975d. *A. canescens* (Host) Boiss. Distinguished by its blunt leaves which are of two kinds, the outer 2–3 mm or more wide, the inner much longer, 1–2 mm wide. Flower head usually 1·5–2·5 cm across; flowers pinkish to reddish-purple; stem sheath 7–30 mm. Involucral bracts pale brown with narrow or wide papery margins, blunt; calyx 6–10 mm. A very variable species. Mountain meadows and rocks. YU.AL.GR. Pl. 34.

975e. *A. undulata* (Bory) Boiss. Like 975d with leaves of two kinds but flowers usually white, and calyx longer 8–12 mm. Outer involucral bracts with long-pointed tips; stem sheath 8–15 mm. Leaves with somewhat undulate margins. Mountain grasslands. GR. **975f.** *A. rumelica* Boiss. Like 975d but all leaves linear-pointed. Flowers pale lilac or purple, in a head 2–3·5 cm across; flowers shortly stalked; stem sheath 1·5–4·5 cm. Mountain grassland. YU.GR.BG.TR.

EBENACEAE | Ebony Family

Trees and shrubs with simple leaves without stipules. Calyx 4–5, often enlarging in fruit; corolla urn- to bell-shaped. Plants often one-sexed; male flowers with 14–24 stamens; female flowers with 8 sterile stamens, and a superior four- to twelve-celled ovary. Fruit a large berry.

DIOSPYROS 977a. *D. lotus* L. DATE PLUM. A deciduous tree to 14 m, with furrowed bark, and elliptical to oblong-pointed leaves, 6–12 cm, with rounded to broadly wedge-shaped bases. Flowers small 2–10 mm, reddish-green, axillary. Calyx hairy, four-lobed. Fruit globular 1·5 cm, yellow or black. Native of Asia; cultivated for its edible fruit and widely naturalized in our area. **977b.** *D. kaki* L.fil. KAKEE or CHINESE PERSIMMON. A deciduous tree, also sometimes cultivated for its larger edible fleshy orange fruits, 3·5–7 cm across. Male flowers larger, about 1 cm across. Native of Japan and Korea.

STYRACACEAE | Storax Family

STYRAX 977. *S. officinalis* L. STORAX. A shrub or small tree to 7 m, with white woolly twigs, oval deciduous leaves, and short clusters of white bell-shaped flowers. Petals 5–7, lanceolate, fused at base, about 2 cm long. Leaves entire 3–7 cm. Fruit dry white-woolly, with persisting calyx. Woods, thickets, streamsides in the Med. region. YU.AL.GR+Is.CR.TR. Pl. 34.

OLEACEAE | Olive Family

FRAXINUS | Ash
Flowers brownish or greenish, appearing before leaves in axillary clusters
(a) *Buds black*
978. *F. excelsior* L. ASH. Leaves with usually 7–13 leaflets, each leaflet usually with more marginal teeth than lateral veins. Damp woods, riversides. YU.BG.TR?

(b) *Buds dark brown*
978a. *F. angustifolia* Vahl NARROW-LEAVED ASH. Like 978 but a smaller tree to 25 m, with

finely and deeply fissured grey bark, and leaves of mature trees with oblong to linear-lanceolate leaflets which are hairy on the mid-vein beneath. Leaflets 5–13, usually with as many marginal teeth as lateral veins. Fruit 2–4·5 cm, hairless. Woods, riversides. Widespread on mainland.   **978b.** *F. pallisiae* Wilmott Distinguished from 978a by its densely hairy twigs, leaf stalks, and leaf rachis; young leaflets densely hairy on both sides, but becoming hairless above. Fruit finely hairy. Riversides. BG(east).TR.

Flowers white, appearing with leaves in terminal clusters
**979. *F. ornus* L. MANNA or FLOWERING ASH. A small tree to 20 m, with smooth grey bark, compound leaves and dense pyramidal, often terminal clusters of creamy-white flowers. Petals 4, linear, 5–6 mm. Leaflets 5–9, ovate to lanceolate, 3–8 cm, finely toothed, distinctly stalked; buds greyish or brownish. Woods, thickets, rocky places mainly in the Med. region. Widespread on mainland.

SYRINGA   980. *S. vulgaris* L. LILAC. A branched, freely suckering shrub 2–7 m, with large heart-shaped entire leaves, and dense pyramidal clusters 10–20 cm, of sweet-scented, lilac or rarely white flowers. Corolla tube 8–12 mm, lobes 4, spreading; calyx 2 mm. Leaves 4–8 cm. Fruit ovoid-pointed, 8–10 mm long. Bushy and rocky places. YU.AL.GR.BG.

PHILLYREA   (981). *P. latifolia* L. (incl. *P. media*) A small evergreen shrub or tree to 15 m, with smooth grey bark and two kinds of dark-green leaves, and small rounded axillary clusters of greenish-yellow flowers. Juvenile leaves ovate-heart-shaped 2–7 cm, toothed; adult leaves lanceolate to elliptical 1–6 cm, entire or finely toothed, lateral veins 7–11 pairs. Fruit nearly globular 7–10 mm, bluish-black. Evergreen thickets, rocky slopes in the Med. region. Widespread. Page 35.   **981.** *P. angustifolia* L. Distinguished from (981) by its narrower leaves which are all linear to lanceolate, 3–8 cm by 0·3–1·5 cm, and are almost always without teeth, and with 4–6 pairs of lateral veins. Fruit 6–8 mm. Evergreen thickets. YU(Croatia).AL.

LIGUSTRUM   982. *L. vulgare* L. PRIVET. Hedges, bushy places. Widespread on mainland.

JASMINUM   **983. *J. fruticans* L. WILD JASMINE. A dense semi-evergreen shrub to 3 m, with green angular stems, and small few-flowered clusters of scentless, tubular yellow flowers 12–15 mm across. Calyx lobes as long as calyx tube. Leaves oblong, shining, with 1 or 3 leaflets 0·7–2 cm. Berry black, shining. Bushy places. YU.AL.GR.TR.   **983a.** *J. humile* L. An evergreen shrub 1·5 m; leaves usually with 3 ovate to oblong leaflets 2–6 cm. Flowers yellow, about 1 cm across; calyx lobes minute. Native of China; cultivated for ornament and locally naturalized in YU.GR.   ****984.** *J. officinale* L. JASMINE. A climbing plant with opposite pinnate leaves, and clusters of white, very sweet-scented flowers. Native of south-west Asia; grown for ornament and sometimes naturalized in YU.

FORSYTHIA   984a. *F. europaea* Degen & Bald. A deciduous shrub to 2 m, with drooping yellow flowers appearing before the leaves on upright branches. Corolla with tube 4 mm, the lobes 12 mm, narrowly-oblong, spreading. Leaves opposite, ovate-lanceolate, pointed, 4–5 cm. Hills. YU(south).AL(north). Pl. 34.

OLEA   **985. *O. europaea* L. OLIVE. The wild olive is often found as a low spiny bush with four-angled stems and small, opposite, oval juvenile leaves, or with larger elliptic adult leaves, and fruit 8–12 mm long, thus differing markedly from the cultivated plant.

It may also occur as a tree resembling the cultivated olive but with smaller fruits. (Distinguished from similar-looking *Rhamnus* species by the opposite leaves.) The cultivated olive is a small tree with grey-fissured bark, greyish-green lanceolate leaves 2–8 cm. Flower cluster axillary, white. Fruit green then black or brown, 1–3·5 cm. A very important oil-bearing tree; widely cultivated in the Med. region. Widespread, except BG.

GENTIANACEAE | Gentian Family

CENTAURIUM | Centaury A difficult genus with 4 similar-looking, pink-flowered species in our area, and 1 yellow-flowered species.

986. *C. erythraea* Rafn COMMON CENTAURY. A very variable plant. Dry grassland, open woods. Widespread. **(986). *C. pulchellum* (Swartz) Druce Like 986 but the pink flowers forming a lax branched cluster at the ends of short branches. Basal rosette of leaves absent. Damp places. Widespread. ****987.** *C. maritimum* (L.) Fritsch YELLOW CENTAURY. Easily distinguished by its small yellow flowers about 1 cm across, in a few-flowered, branched, and somewhat flat-topped cluster. Calyx three-quarters to nearly as long as the corolla tube. Leaves increasing in length up stem. Sandy or grassy places by the sea. Widespread.

BLACKSTONIA ****988.** *B. perfoliata* (L.) Hudson YELLOW-WORT. A very variable plant. Damp places, stony, grassy places. Widespread.

GENTIANA | Gentian
1. Fls. blue, violet, or blue with yellow
    or white.
  2. Annuals, without non-flowering shoots.   *nivalis, utriculosa*
  2. Perennials, with non-flowering shoots.
    3. Corolla four-lobed.   *cruciata*
    3. Corolla five- or ten-lobed.
      4. Corolla ten-lobed.   *pyrenaica*
      4. Corolla five-lobed.
        5. Basal rosettes absent; stem
        leaves many (YU.AL.BG.).   ****993.** *G. pneumonanthe* L.
                         ascepiadea
        5. Basal rosettes present; stem
        leaves few.
          6. Corolla yellow-white with blue
          stripes and with bluish spots
          in the throat (BG(south-west)).   **993a.** *G. frigida* Haenke
          6. Corolla blue.
            7. Corolla 5–7 cm, tube obconical,
            lobes ascending.   *acaulis, dinarica, clusii*
            7. Corolla 1·5–2·5 cm, tube
            cylindrical, lobes spreading.   *verna*
1. Fls. yellow, or greenish-yellow (*see also* 993a).   *lutea, punctata*

989. *G. cruciata* L. CROSS GENTIAN. A robust, leafy perennial 15–40 cm, with large blunt oval to lanceolate, three-veined, sheathing leaves, and few large, dull-blue flowers, in terminal and axillary clusters. Corolla 2–2·5 cm long, barrel-shaped, with 4 short spreading lobes. Woods and pastures in mountains. YU.AL.GR.BG. Page 184. **990. *G. nivalis* L. SNOW GENTIAN. A slender, rosette annual 1–15 cm, with tiny leaves and small, brilliant deep-blue flowers 15 mm long. Lobes of corolla acute; calyx tube 2–3 mm wide, angled not winged. Rosette leaves oval, about 5 mm. Alpine meadows. YU.BG.

(990). *G. utriculosa* L. BLADDER GENTIAN. Like 990 but flowers larger 1·5–2 cm long, intense blue, lobes oval-blunt. Calyx ovoid, 4–7 mm wide, angled with 5 broad wings, calyx lobes triangular. Grassy and rocky places in mountains. YU.BG. **991. *G. verna* L. SPRING GENTIAN. A rosette-forming perennial 2–20 cm, distinguished by its brilliant deep-blue flowers, often with whitish pleats, and its angled and winged calyx. Corolla 1·5–2·5 cm long, greenish-blue on the outside. Leaves lanceolate to broadly ovate; stem leaves 1–3 pairs. Damp grassy places in mountains. YU.AL.GR(north).BG.

992. *G. acaulis* L. (*G. kochiana*) TRUMPET GENTIAN. Flowers very large 4–6 cm long, with olive-green spots in the throat, and corolla lobes acute or tapering to a rigid point. Calyx lobes ovate, usually less than half as long as the tube. Rosette leaves mostly lanceolate or elliptic. Alpine pastures. YU.BG. **(992).** *G. clusii* Perr. & Song. STEM-LESS TRUMPET GENTIAN. Flowers very large, blue with few or no green spots in the throat; calyx lobes triangular, at least half as long as the calyx tube. Leaves elliptic-lanceolate. Mountain pastures. YU(north). Page 191. **992c.** *G. dinarica* G. Beck Like 992 but leaves broadly elliptic, and flowers unspotted in the throat, and with long-pointed corolla lobes. Calyx lobes narrowly lanceolate, about half as long as the tube. Meadows in limestone mountains. YU(south-west).AL.

994. *G. asclepiadea* L. WILLOW GENTIAN. A robust leafy perennial 15–60 cm, readily distinguished by the row of large stalkless flowers in the axils of the upper leaves. Flowers 1–3 in each axil; corolla 3·5–5·5 cm long, trumpet-shaped, blue and spotted with violet inside and with paler stripes outside. Leaves ovate-lanceolate, strongly three- to five-veined. Hills, and mountains, woods, shady places, damp meadows. Widely scattered on mainland. **995. *G. pyrenaica* L. PYRENEAN GENTIAN. Distinguished from all other species by its medium-sized blue-violet flowers which have 10 nearly equal corolla lobes. A cushion-forming rosette perennial with rather leathery linear-lanceolate leaves, and short stems bearing solitary flowers 2–3 cm long. Mountain pastures. BG(south-west). Pl. 34.

996. *G. lutea* L. GREAT YELLOW GENTIAN. Readily distinguished from other yellow-flowered gentians by the corolla, which is not tubular but has 5–9 narrow spreading lobes. Very robust perennial 50–120 cm, with large, strongly veined leaves, and terminal and axillary clusters of numerous flowers. Marshes and pastures in the hills and mountains. YU.AL.GR.BG. ****997. *G. punctata* L. SPOTTED GENTIAN. Readily recognized by the dense terminal and axillary clusters of large bell-shaped, pale greenish-yellow flowers usually with dark-purple spots. Corolla 1·5–3·5 cm long, lobes short, erect. A stiff erect, rather robust perennial 60 cm, with elliptic, greyish-green, strongly veined, stalked leaves. In mountains, meadows, open woods, rocky places. YU.AL.BG.

GENTIANELLA Distinguished from *Gentiana* by the long hairs in the throat of the corolla, or by the corolla lobes which are fringed with hairs; usually annuals or biennials. About 6 species in our area.

999a. *G. austriaca* (A. & J. Kerner) J. Holub Flowers relatively large, purplish or whitish, 2·4–4·5 cm long, with corolla tube at least half as long as the lobes. Calyx lobes usually distinctly longer than the calyx tube. A biennial 10–20 cm, branched from the base, with a flat-topped inflorescence. Mountain meadows. YU. **999b.** *G. lutescens*

Velen.) J. Holub Like 999a but flowers smaller, 1·8–2·5 cm long, purplish or yellowish. Calyx lobes as long or shorter than the calyx tube. Mountain pastures. YU.BG. **999c.** *G. bulgarica* (Velen.) J. Holub Like 999a but flowers much smaller 1–2 cm long, whitish or pale violet. Calyx lobes much longer than the calyx tube. Alpine meadows. YU.BG. Page 137.

1001. *G. ciliata* (L.) Borkh. FRINGED GENTIAN. Readily distinguished by the large blue flowers with 4 broad, spreading, corolla lobes with margins conspicuously fringed with blue hairs. Flowers 4–5 cm long, few in a lax cluster. Stem leaves linear-lanceolate; biennial 5–30 cm. Calcareous rocks. YU.AL.GR.BG. **1001a.** *G. crispata* (Vis.) J. Holub A small, erect biennial 2–20 cm, with a flat-topped cluster of small violet or whitish flowers. Distinguished from all others by the calyx lobes which have crisped, blackish margins. Flowers 1·2–2 cm long, the corolla tube nearly twice as long as the calyx. Mountain pastures. YU.AL.GR(north).BG. Page 185.

SWERTIA ****1002.** *S. perennis* L. MARSH FELWORT. A gentian-like perennial 15–60 cm, distinguished by its spike of dull violet-purple, or pale-yellow and black spotted flowers, with 5 wide-spreading petals 6–16 mm, each with 2 dark shining nectaries at their base. Leaves pale greenish-yellow, lanceolate, the upper clasping the stem, the lower stalked. Grassy bogs in mountains. YU.BG.

MENYANTHACEAE | Bogbean Family

MENYANTHES ****1003.** *M. trifoliata* L. BOGBEAN. Still waters. YU.AL? BG.

NYMPHOIDES ****1004.** *N. peltata* (S. G. Gmelin) O. Kuntze FRINGED WATER-LILY. Readily distinguished by its golden-yellow flowers 3–4 cm across, with conspicuously fringed, oval petals, borne above the surface of the water. Floating leaves 3–10 cm, rounded, often blotched. Lakes, slow rivers. YU.AL.GR.BG.TR.

APOCYNACEAE | Dogbane Family

Corolla without a corona of lobes.

VINCA | Periwinkle   4 species in our area.

*1005. *V. minor* L. LESSER PERIWINKLE. A creeping or ascending evergreen plant, rooting at the nodes, with blue flowers on stalks often longer than the leaves. Corolla 2–3 cm across, with blunt lobes, corolla tube 9–11 mm; calyx lobes ovate to triangular 3–4 mm. Leaves mostly lanceolate, hairless. Woods, hedges, rocks. YU.GR?BG. ****(1005).** *V. herbacea* Waldst. & Kit. HERBACEOUS PERIWINKLE. Like 1005 but stems and leaves soft and herbaceous and dying down in winter. Flowers blue, 2–3·5 cm across, with acute lobes, corolla tube 10–15 mm; calyx lobes narrowly triangular 4–8 mm. Bushy places. YU.GR.BG.TR.

1006. *V. major* L. GREATER PERIWINKLE. Differs from 1005 in having flower stalks shorter than the leaves, and the blue corolla 3–5 cm across, corolla tube 12–15 mm; calyx lobes

7–17 mm with densely ciliate margins. Leaves ovate, margins ciliate. Woods, hedges, shady rocks, streamsides; sometimes naturalized. YU.; introd.GR.CR.BG.

NERIUM **1007. *N. oleander* L. OLEANDER. A handsome, grey-leaved shrub to 4 m, with clusters of large pink flowers, often found growing in dry river courses and ravines in the Med. region. Flowers 3–5 cm across, with blunt spreading petals, and a frilly throat. Leaves in whorls of 3 or 4, linear-lanceolate, leathery, 6–12 cm. Widespread, except BG.

TRACHOMITUM Corolla bell-shaped, the throat wide and open. Anthers eared, adhering to stigma. Disk of 5 fleshy scales; seeds with a tuft of hairs.

1007a. *T. venetum* (L.) Woodson (*Apocynum* v.) A perennial to 50 cm, with opposite, narrow oblong leaves and a terminal branched cluster of tubular, whitish flowers. Corolla 4–5 mm long, five-lobed. Leaves 2–4 cm, finely toothed, deciduous. Fruit 15 cm, pendulous. Maritime sands. YU.BG.TR. Page 149.

ASCLEPIADACEAE | Milkweed Family

Corolla often with a corona of a single or double row of 5–10 free or fused lobes.

CYNANCHUM **1008. *C. acutum* L. STRANGLEWORT. A slender twining perennial to 3 m, with opposite, triangular-heart-shaped, glaucous leaves, and small axillary stalked umbels of pink or white, sweet-scented flowers. Flowers 8–12 mm across, with corona of 10 projecting lobes. Fruit usually paired, horn-like, each carpel about 8 cm. Hedges scrub, waysides in the Med. region. Widespread, except TR.

PERIPLOCA **1009. *P. graeca* L. SILK-VINE. A twining perennial to 12 m, with dark shining green, oval to lanceolate leaves, and long-stalked, lax, axillary clusters, longer than the leaves, of brownish-purple flowers. Flowers about 2 cm across; petals oblong, fringed with hairs; corona of 5 erect horns. Fruit paired, each carpel 10–15 cm. Woods, thickets, river banks. Widespread on mainland. **1009a.** *P. laevigata* Aiton A thorny, grey-stemmed erect shrub 1·3–3 m, with small leathery, oblong leaves 1–3·5 cm, and dark brownish-purple flowers about 1 cm across, which are green beneath. Fruit paired, each carpel 5–8 cm. Rocky scrub. CR.

VINCETOXICUM Corona of 5 fused, fleshy, not horned lobes.

Flowers dark purple or brownish
1010c. *V. speciosum* Boiss. & Spruner A perennial 60–70 cm with broadly ovate, stalkless, velvety leaves, and nearly stalkless clusters of dark-purple flowers. Corolla 10–12 mm across, lobes lanceolate with straight hairs on upper surface. Individual carpel about 7 cm, spindle-shaped, woolly-haired. Mountain woods, meadows. Widespread on mainland. **1010d.** *V. fuscatum* (Hornem.) Reichenb. fil. Flowers 8–10 mm across, brown on both sides, lobes oblong, hairless or with curved hairs. Individual carpel 4–5 cm, hairless. Dry stony places, scrub, steppes. YU.AL.GR.BG.TR.

Flowers yellow, greenish-yellow, or white
1010e. *V. canescens* (Willd.) Decne Distinguished by its grey-woolly, broadly ovate leaves, and by the lobes of the corona narrowed below to a stalk. Flowers yellow, 7–8 mm

across, lobes triangular, hairy above. Stems more or less twining, 40–70 cm. Rocks, screes. CR+E.Aegean Is. **1010. *V. hirundinaria* Medicus COMMON VINCETOXICUM. A very variable perennial with erect stems, sometimes twining, to 120 cm, broadly ovate to ovate-lanceolate, yellowish-green leaves, and axillary stalked clusters of yellowish or dull-white flowers. Flowers 3–10 mm across; corolla lobes ovate, hairless or with curved hairs. Leaves 6–10 cm, more or less finely hairy. Individual carpel about 6 cm. Wood margins, open places. Widespread on mainland. **1010b.** *V. huteri* Vis. & Ascherson Distinguished from 1010 by its greenish-yellow flowers 5–7 mm across and the corolla lobes with straight hairs on the upper surface. Scrub. YU.AL.

CIONURA Like *Vincetoxicum* but corona with 5 free lobes.

1011a. *C. erecta* (L.) Griseb. (*Marsdenia e., Cynanchum e.*) An erect, sometimes woody-based perennial with numerous stems to 1 m, heart-shaped pointed leaves, and axillary clusters of many white flowers. Corolla 1 cm across, lobes oblong, blunt, hairless. Leaves about 7 cm, stalked. Individual carpel about 8 cm. Rocky places, river gravels, maritime sands in the Med. region. Widespread. Pl. 34.

ASCLEPIAS **1011. *A. syriaca* L. SILKWEED. A robust perennial 1–2 m, with large oblong, opposite leaves, and dense terminal and lateral rounded umbels of sweet-scented pink flowers. Corolla about 1·5 cm across, corona horns erect. Leaves 10–20 cm, white-downy beneath. Individual carpel 8–11 cm, white-woolly, with few short spines. Native of N. America; naturalized YU.BG.

GOMPHOCARPUS **1012. *G. fruticosus* (L.) Aiton fil. BRISTLY-FRUITED SILKWEED. A shrubby perennial 1–2 m, with narrow long-pointed leaves, small white flowers in umbels, and very distinctive large ovoid-pointed, 'boat-shaped' fruits covered with long bristly hairs. Flowers about 5 mm across, with a corona of 5 fleshy, hooded, horns. Carpels 4–5 cm. River-courses, damp places in the Med. region. Native of S. Africa; naturalized in YU.AL.GR.CR.

RUBIACEAE | Madder Family

PUTORIA **1013. *P. calabrica* (L. fil.) DC. A prostrate, mat-forming, strong-smelling shrublet, with leathery leaves and dense heads of long-tubed pink flowers. Flowers 1–1·5 cm long, with 4 short spreading narrow lobes; stamens projecting. Leaves 1–2 cm, opposite, shining, margins inrolled. Rocks, hills, river gravels in the Med. region. YU.AL.GR.CR+Rhodes.

SHERARDIA **1014. *S. arvensis* L. FIELD MADDER. Waste places, fields, stony places. Widespread.

CRUCIANELLA 6 similar-looking species in our area.

1015. *C. angustifolia* L. A slender erect hairless annual to 50 cm, with narrow linear-lanceolate leaves in whorls of 6–8, and long slender four-angled spikes of insignificant yellowish flowers. Corolla 3–5 mm long, little longer than the narrow, densely overlapping, papery-margined bracts. Dry stony and rocky places. Widespread on mainland +CR. **1015a. *C. latifolia* L. Like 1015 but lower leaves obovate-elliptic, and bracts of flower spike usually fused at the base (unfused in 1015). Corolla yellow, the tube 5–8 mm, longer than the bracts; flower spikes more slender. Rocks and dry hills in the Med. region. YU.AL.GR+Is.CR.BG.

ASPERULA About 40 species in our area, including 26 which are endemic; a difficult genus.

Flowers usually in lax, branched clusters, without a conspicuous involucre
(a) Leaves in whorls of 4
1017. *A. cynanchica* L. SQUINANCY WORT. Dry grassy and bare places. YU.AL.GR.CR.BG. **1017c.** *A. aristata* L. fil. Flowers yellowish to reddish, in a freely branched cluster. Corolla tube 6–10 mm long, 2–4 times as long as the narrow lobes. Leaves in whorls of 4, lanceolate to linear, with a short transparent apex; non-flowering shoots green or grey-green. A very variable, often erect, often woody-based perennial 10–60 cm. Rocky places, rocky heaths, mountain grasslands. YU.AL.GR.BG. Page 371. **1017d.** *A. lutea* Sibth. & Sm. A tufted herb, usually woody and hairy at the base, more or less erect or ascending to 50 cm. Flowers yellow, or pink to reddish, in three- to many-flowered clusters, forming a lax spike-like inflorescence. Corolla with a slender tube 1½–3½ times as long as the lobes; anthers blackish. Mountains, among rocks. GR(south and central). Pl. 35.

(b) At least some leaves in whorls of more than 4
(1017). *A. tinctoria* L. DYER'S WOODRUFF. Meadows, rocky places. YU.BG. **1017e.** *A. purpurea* (L.) Ehrend. (*Galium p.*) An erect or ascending perennial 8–50 cm, with rough-margined leaves mostly in whorls of 6–10, and flowers usually purple to yellowish-green in much-branched narrow pyramidal clusters. Corolla with 4 spreading ovate to triangular lobes about 1 mm, hairless or hairy above, and tube very short. Leaves 8–30 mm, narrow-lanceolate to thread-like. Fruit hairless. Dry places. YU.AL.GR.BG. **1017f.** *A. muscosa* Boiss. & Heldr. A dwarf tufted perennial with slender stems 10–16 cm, with pale-yellow flowers which are solitary in the axil of the leaf whorls. Corolla funnel-shaped 3–5 mm long, lobes about 1 mm, ovate, hairless. Leaves 9–12 mm, in whorls of 6–8, lanceolate to linear, long-pointed, ciliate. Mountain coniferous woods. GR(Olympus). Page 127.

Flowers in terminal heads surrounded by a leafy involucre
(a) Annuals; flowers blue
1018. *A. arvensis* L. BLUE WOODRUFF. Fields, waste places. Widespread on mainland.

(b) Perennials; flowers white, yellowish, pinkish, or purplish
1019. *A. taurina* L. Flowers in dense heads, white or pale yellowish. Leaves 3–6 cm, in whorls of 4, lanceolate to ovate; stems 20–50 cm. Deciduous woods, scrub, and shady mountain rocks. YU.AL.GR.BG. **1019a. *A. arcadiensis* Sims A densely tufted, greyish-hairy, woody-based perennial usually 8–15 cm, with 4–8 pale-purplish flowers in stalkless terminal heads. Corolla narrowly funnel-shaped, hairless, the tube 8–12 mm long, lobes 2–3 mm. Leaves 8–10 mm, broadly lanceolate, densely grey-hairy, in whorls of 6–8. Moist limestone rocks. GR(south). Pl. 35. **1019b.** *A. boissieri* Boiss. A small greyish-blue, densely tufted alpine perennial to 16 cm, of rock crevices, with short-stalked clusters of pink flowers in a terminal head surrounded by linear, bristle-pointed involucral bracts as long as the flowers. Corolla usually 3–5 mm long, with lobes one-third as long as the tube. Leaves 4–10 mm, like bracts, and with a curved bristle-like tip. Higher mountains. GR(south). Pl. 35.

GALIUM | **Bedstraw** About 66 species occur in our area; about 22 are restricted to it. A very difficult genus.

1. *Goniolimon sartorii* 974h **2.** *Galium incanum* 1026a **3.** *Limonium gmelinii*
972f **4.** *Convolvulus dorycnium* 1037c **5.** *Asperula aristata* 1017c

Leaves three-veined, in whorls of 4
1020. *G. boreale* L. NORTHERN BEDSTRAW. Damp meadows, bushy places
YU.AL.BG. ****1021.** *G. rotundifolium* L. ROUND-LEAVED BEDSTRAW. Woods in mountains. Widespread on mainland +CR.

Leaves one-veined, usually in whorls of more than 4
(a) Stems with recurved bristles or prickles on the angles
(1022). *G. palustre* L. MARSH BEDSTRAW. Wet places. YU.AL?GR.BG.TR? **1023. *G.*
aparine L. GOOSEGRASS, CLEAVERS. Bushy places, hedges, cultivated ground. Widespread. **(1023).** *G. tricornutum* Dandy (*G. tricorne*) ROUGH CORN BEDSTRAW. Cultivated
ground, waste places. Widespread.

1023a. *G. parisiense* L. WALL BEDSTRAW. A rough, prickly, somewhat scrambling, much-branched annual 5–40 cm, with leaves in whorls of 5–7, and an oblong, narrowly ovoid to
pyramidal inflorescence with partial clusters of 7–16 flowers at 2 nodes below the
central flower. Flowers minute about 1 mm across, greenish inside, reddish outside
with pointed lobes. Fruit about 1 mm, hairless or with curved hairs. Cultivated fields,
roadsides, dry places. YU.AL.BG.

(b) Stems with smooth angles, or stems not angled
1024. *G. verum* L. LADY'S BEDSTRAW. Grassland, bushy places, sand-dunes. Widespread
on mainland. **1025.** *G. mollugo* L. HEDGE BEDSTRAW. Hedges, meadows.
YU.AL.BG? **1025a.** *G. heldreichii* Halácsy Similar to 1025 but without stolons, and
with a dense, narrowly ovoid inflorescence of whitish to greenish or sometimes reddish
flowers. Dry scrub. GR+Is.CR.

1025b. *G. reiseri* Halácsy A bluish-grey, hairless, non-stoloniferous perennial 15–60 cm,
with leathery somewhat fleshy, obovate to broadly elliptical leaves 1–2 cm long, with
margins inrolled. Flowers bright yellow, in a dense narrowly pyramidal cluster with
short erect branches; corolla 2–3 mm across, the lobes conspicuously pointed. Maritime
rocks. GR(N.Sporades). **1025c.** *G. rhodopeum* Velen. A tufted perennial 10–35 cm,
with densely leafy non-flowering shoots. Flowers white or yellowish, few in a lax long
and narrow cluster with short rigid branches; corolla about 4 mm across, the lobes
shortly pointed. Leaves linear, stiff, narrowed to a long colourless tip. Limestone mountain cliffs. YU.GR(north).BG. Pl. 35.

1026a. *G. incanum* Sibth. & Sm. A tufted, more or less woody-based perennial 2–15 cm,
with leaves in whorls of usually 6, linear to linear-lanceolate, with a short transparent
incurved apex, usually hairy, margins inrolled. Flowers white or pink, in an oval to
elongated cluster, few- or many-flowered; corolla 1·5–3 mm long, funnel- or cup-shaped,
lobes lanceolate. Limestone mountain rocks and grassland. GR.CR. Page 371. ****1027.**
G. odoratum (L.) Scop. SWEET WOODRUFF. Deciduous woods. Widely distributed on mainland. ****1028.** *G. glaucum* L. GLAUCOUS BEDSTRAW. Forest margins, dry grassland, stony
places. YU.BG.

CRUCIATA 5 species in our area.

1029. *C. laevipes* Opiz (*Galium cruciata*) CROSSWORT. Grassland, open woods. Widespread on mainland. **(1029). *C. glabra* (L.) Ehrend. (*Galium vernum*) SLENDER CROSS
WORT. Woods, bushy places, meadows. YU.AL.GR.BG. **1029a.** *C. pedemontana* (Bellardi) Ehrend. Like (1029) but an annual 10–35 cm, with spreading hairs and small
prickles on the stem. Flowers yellow, smaller about 0·5–1 mm long (2·5–3·5 mm in
(1029)). Dry grassland, scrub. Widespread on mainland.

VALANTIA

Annuals

1030. *V. hispida* L. Dry places, rocks, sands in the Med. region. GR+Is.CR.　**1030a.** *V. muralis* L. Like 1030, a tiny annual usually not more than 15 cm, with leaves in whorls of 4, and minute yellowish flowers. Differing from it in being hairless or softly hairy only above; and in the hairless fruit with 4 horns covered in hooked bristles (bristles straight and horns absent in 1030). Stony places, dry hills in the Med. region. YU.AL.GR+Is. CR.

Perennials

1030b. *V. aprica* (Sibth. & Sm.) Boiss. & Heldr. A low cushion-like perennial of mountain rocks and screes. Flowers yellowish-white to pinkish, in rounded heads mixed with the leaves and borne on stems 5–20 cm. Flowers with 4 spreading petals, about 3 mm across. Leaves ovate, somewhat fleshy, 3–6 mm. AL.GR(south and west). CR. Page 89.

RUBIA　3 species in our area.　**1031.** *R. peregrina* L. WILD MADDER. Hedges, thickets, rocky ground. Widespread, except BG.　**(1031).** *R. tinctorum* L. DYER'S MADDER. Hedges, thickets, waste places; widely naturalized and perhaps native in our area.

POLEMONIACEAE | Phlox Family

POLEMONIUM　**1032.** *P. caeruleum* L. JACOB'S LADDER. Damp woods and pastures in mountains. YU.

CONVOLVULACEAE | Convolvulus Family

IPOMOEA　The following largely S. American species are now widespread throughout the tropics. They are sometimes cultivated and naturalized in our area:

1033. *I. purpurea* Roth MORNING GLORY. A climbing annual with large, funnel-shaped, violet-purple, pink, blue, or white flowers 4–6 cm long. Sepals lanceolate 10–16 mm, hairy. Leaves 4–16 cm, usually entire, broadly heart-shaped, long-pointed, sparsely hairy. Naturalized in GR.　**1033a.** *I. sagittata* Poiret A creeping or climbing perennial with showy purple or pink flowers 4–7 cm long, and bright shining-green, lanceolate leaves with a heart-shaped base, the upper arrow-shaped. Sepals rounded 5–10 mm, hairless. Ditches and hedges on the littoral. Native in GR.　**1033c.** *I. stolonifera* (Cyr.) J. F. Gmelin Plant creeping and usually rooting at the nodes, with more or less fleshy, usually oblong entire to deeply three- or five-lobed leaves. Flowers usually solitary, white or pale yellow, sometimes with a purple centre, 3·5–5 cm long. Sepals 8–15 mm, acute or obtuse, hairless. Sandy places near the sea. CR.　**1033d.** *I. batatas* (L.) Lam. SWEET POTATO. Flowers white, or violet to purple, 3–5 cm long; sepals 7–12 mm, oblong. Leaves 4–14 cm, ovate entire to deeply three- to five-lobed. Stems creeping; tubers large spindle-shaped. Cultivated in GR.

CONVOLVULUS

1. Dwarf cushion plants with woody stems. — *boissieri, libanoticus*
1. Not cushion plants.
  2. Lvs. very abruptly narrowed into a distinctive lv. stalk.
    3. Lv. blade with a wedge-shaped or cut-off base. — *persicus, siculus*
    3. Lv. blade with a heart-shaped base, or with triangular-pointed basal lobes.
      4. Upper lvs. and bracts very deeply divided. — *althaeoides*
      4. Upper lvs. and bracts not divided except for the basal lobes.
        5. Corolla 3–4·5 cm long. — *betonicifolius, scammonia*
        5. Corolla 0·7–2·5 cm long.
          6. Fls. blue, 7–12 mm long. — *siculus*
          6. Fls. white or pink, 10–25 mm long (Widespread). — **1038.** *C. arvensis* L.
  2. Lvs. usually stalkless or gradually tapered into the lv. stalk.
    7. Mature stems woody and branching divergent throughout; fls. pink. — *dorycnium*
    7. Mature stems herbaceous, at least above, branching not divergent.
      8. Annuals or short-lived perennials, entirely herbaceous; fls. predominantly blue.
        9. Sepals hairy, with a distinct apical and basal region (YU.GR.). — ****1035.** *C. tricolor* L.
        9. Sepals hairless, not divided into 2 distinct regions (YU? GR.TR.). — **(1034).** *C. pentapetaloides* L.
      8. Perennials, shoots woody below, herbaceous above; fls. white, pink, or yellowish.
        10. Inflorescence with all flowers on each main branch or stem crowded into a compact head. — *cneorum, oleifolius*
        10. Inflorescence lax, or with at least some branches distinct and conspicuous.
          11. At least lower parts of stem with mostly spreading hairs. — *cantabrica*
          11. Hairs densely silky, mostly lying flat.
            12. Outer sepals heart-shaped at base and conspicuously pouched on the back. — *holosericeus*
            12. Outer sepals wedge-shaped to rounded at base, not or scarcely pouched on the back. — *lineatus, cneorum, oleifolius*

1034. *C. siculus* L. SMALL BLUE CONVOLVULUS. Flowers blue, conspicuously five-lobed, 7–12 mm long. Leaves 2–3 cm, lanceolate to ovate, stalked. Dry places in the Med. region. GR.CR. **1036b.** *C. boissieri* Steudel A beautiful silvery-leaved, dense cushion-forming plant of the higher mountains, with large almost stalkless, white to red, broadly funnel-shaped flowers 1·5–2 cm long. Calyx with woolly, spreading hairs. Leaves tiny to 2 cm, linear to obovate, folded, densely covered with shining silvery-white hairs. Screes and rocks. YU.AL.GR.BG. Pl. 35.

1036c. *C. libanoticus* Boiss. Like 1036b but less compact, less densely hairy, and leaves hairless above, linear to oblanceolate with only the mid-rib distinct (mid-rib and lateral veins distinct in 1036b). Flowers red, 1–1·5 cm long. Mountain slopes and pastures. GR(south).CR. **1036d.** *C. persicus* L. A densely woolly perennial with a creeping, woody stock and erect, little-branched stems 10–50 cm, with solitary axillary white flowers 3–4·5 cm long. Sepals blunt. Leaves elliptic-oblong, with a rounded or wedge-shaped base, stalked. Maritime sands of the Black Sea. BG.TR.

**1036. *C. cantabrica* L. PINK CONVOLVULUS. An ascending perennial 10–50 cm, branching from the base with stems with spreading hairs, and with a lax spreading inflorescence of axillary clusters of pink flowers, on stalks much longer than the subtending leaves. Corolla 1·5–2·5 cm long, hairy outside. Leaves linear to oblanceolate, with spreading hairs. Bushy, stony places. Widespread on mainland.

1037. *C. lineatus* L. SILVERY-LEAVED PINK CONVOLVULUS. Differs from 1036 in being a low-growing, spreading or ascending perennial 3–25 cm, with stems with silky adpressed hairs and narrow leaves. Flowers pink, in a dense terminal cluster, or clusters axillary on stalks much shorter than the leaves; corolla 1·2–2·5 cm long. Stony places in the Med. region. YU?AL.GR?CR?BG. Pl. 35. **(1037).** *C. cneorum* L. A densely silvery-haired, erect or spreading woody perennial 10–50 cm, with terminal crowded heads of usually white flowers, and linear or oblanceolate silvery-haired leaves. Corolla 15–25 mm long; sepals lanceolate to broadly ovate. Fruit hairy, longer than calyx when ripe. Limestone rocks near the sea. YU.AL.

1037b. *C. oleifolius* Desr. A densely silky perennial like (1037), but with thread-like linear leaves, and usually pink flowers 1·5–2·5 cm long, in a somewhat diffuse cluster. Fruit shorter than calyx when ripe. Limestone rocks by the sea. GR(south). CR+Rhodes. **1037a.** *C. holosericeus* Bieb. Distinguished by its white to yellow, solitary or paired flowers 2–3 cm long. A densely silky-haired woody-based perennial 10–30 cm, with linear to oblanceolate leaves which on the base of the stems are widened and papery at the base. Dry stony slopes. YU(Macedonia).GR(north to central).BG **1037c.** *C. dorycnium* L. A woody-based, widely branched, finely hairy erect shrub 50–100 cm, with small linear to oblanceolate leaves, and axillary clusters of one or a few pink flowers on stalks much longer than the leaves. Corolla 1–2 cm long; sepals blunt or notched. Dry places. GR(south).CR. Page 371.

1039. *C. althaeoides* L. MALLOW-LEAVED BINDWEED. A climbing or trailing perennial to 2 m, with 1–5 large, deep-pink flowers 3–4·5 cm across, borne in the axils of the leaves. Lower leaves triangular-heart-shaped, the upper divided into narrow lobes, with mostly spreading hairs. Subsp. *tenuissimus* (Sibth. & Sm.) Stace (*C. elegantissimus*) differs in having silky adpressed hairs, and almost all leaves more deeply divided into unequal linear lobes. Bushy and stony places by the sea. Widespread. Pl. 35. **1039a. *C. betonicifolius* Miller A slender trailing or climbing perennial to 1 m, with large usually pink flowers 3–4·5 cm long, in axillary clusters of 3 or more, on stalks longer than the leaves. Leaves triangular-ovate with heart-shaped or arrow-shaped base (not lobed), with spreading hairs. Dry, usually acid soils. YU(south).AL.GR.BG.TR. **1039b.** *C.*

scammonia L. A hairless perennial with twining stems to 75 cm, with pale-yellow flowers 2·5–4·5 cm long. Upper leaves deeply lobed. Scrub. E.Aegean Is.

CALYSTEGIA | Bindweed

1040. *C. sepium* (L.) R.Br. BELLBINE, GREATER BINDWEED. Hedges, field verges, damp places. Widespread. **1040a.** *C. silvatica* (Kit.) Griseb. Like 1040 with white flowers, but flowers larger, 5–9 cm long, and with bracteoles 14–32 mm wide and overlapping (bracteoles rarely more than 15 mm wide, not overlapping in 1040). Hedges, open ground. Widespread. **1041.** *C. soldanella* (L.) R.Br. SEA BINDWEED. Flowers pink; leaves kidney-shaped, fleshy; stems creeping. Sands and shingles on the littoral. Coasts throughout.

CRESSA **1042.** *C. cretica* L. A small, dense, grey-leaved annual or perennial to 30 cm, of salt-marshes in the Med. region, with terminal globular clusters of small yellowish or pale-pink flowers. Corolla 3–5 mm across, divided to about half its length into 5 spreading hairy lobes; stamens 5, longer than corolla; calyx lobes blunt. Leaves small, 2–10 mm, lanceolate to heart-shaped, silky-haired. AL.GR+Is.CR.BG.

CUSCUTA | Dodder 14 species occur in our area, including 4 introduced species; all are similar-looking and difficult to distinguish from each other.

1043. *C. europaea* L. LARGE DODDER. Commonly parasitic on nettles and hops, but also on other plants. Widespread on mainland. ****1044.** *C. epithymum* (L.) L. COMMON DODDER. Commonly parasitic on dwarf shrubs, clover, etc. Widespread.

BORAGINACEAE | Borage Family

The following local species present in our area have not been described:
Argusia sibirica (L.) Dandy BG(Black Sea).TR. *Halacsya sendtneri* (Boiss.) Dörfler YU(central).AL.; *Rochelia disperma* (L.fil.) C. Koch GR.

HELIOTROPIUM 6 similar-looking species in our area.

Flowers 2–4 mm across, rarely more

1045. *H. europaeum* L. HELIOTROPE. A softly hairy, often greyish, erect annual 4–40 cm, with tiny white or pale-lilac flowers 2–4 mm across, in leafless forked spikes. Corolla 2–4 mm long; calyx divided almost to the base, spreading in fruit; stigma linear, hairless. Leaves oval. Fruit splitting into 4 nutlets. Cultivated ground, waste places, waysides. Widespread. **1045a. *H. dolosum* De Not. Like 1045 but flowers scented, 3–5 mm long, and distinguished by having a conical, hairy stigma. Calyx at first curving round fruit, later spreading. Coastal areas. GR+Is.CR.BG.TR. **(1045).** *H. supinum* L. Distinguished from other species by its calyx, which is fused for three-quarters of its length, enlarging in fruit and enclosing the single nutlet and falling with it. Flowers white, minute. A white-haired annual. Sandy damp places. Widespread, except TR.

Flowers 4–8 mm across

1045b. *H. suaveolens* Bieb. Differing from 1045 by the larger white, scented flowers, 4–8 mm across, and the shortly conical or hemispherical, entire or shallowly lobed stigma. An annual to 40 cm, with adpressed hairs; leaves ovate to elliptic. Sandy places, cultivation. GR.CR? BG.TR. **1045c.** *H. hirsutissimum* Grauer Like 1045b in habit and vegetative characters, but with dense greyish or yellowish spreading hairs on the stem and leaf stalks. Flowers fragrant, white, 5–8 mm long and 4–8 mm across, with 5 scales at the throat. Cultivated fields, waysides, coastal areas. GR+Is.CR. Pl. 36.

OMPHALODES **1047.** *O. verna* Moench BLUE-EYED MARY. A creeping perennial with slender flowering stems 10–40 cm, bearing a lax cluster of a few sky-blue flowers. Corolla 1 cm across; calyx deeply divided, hairy. Leaves mostly basal, long-stalked, oval with a rounded or heart-shaped base and fine-pointed apex. Woods; sometimes naturalized. YU.; introd.GR. Page 190. **1047b.** *O. luciliae* Boiss. A beautiful tufted plant of rock crevices in high mountains, with clear blue flowers, and glaucous ovate-blunt leaves. Flowers about 1·5 cm across, with a yellow eye; sepals hairless. GR(south and central). Page 127. Pl. 36.

CYNOGLOSSUM | Hound's-Tongue 7 species in our area.

Mature nutlets with a thickened border, flat or slightly concave outside.
1049. *C. officinale* L. HOUND'S TONGUE. Flowers dull purple, 7–10 mm across (a form found in W. Bulgaria has bright reddish flowers 1·2–1·5 cm across). Nutlets 5–8 mm across. Waysides, hedges, woods. YU.AL.GR.BG. (1049). *C. columnae* Ten. Distinguished by its deep-blue or blue-purple flowers in leafless clusters, and its large thick-bordered nutlets 7–10 mm across, with hooked hairs which are shorter on the dorsal surfaces than on the margin. Calyx lobes ovate, 5 mm. An annual 25–45 cm. Dry open places, waysides in the Med. region. YU.AL.GR+Is.CR.

Mature nutlets without a thickened border, convex outside
(1049). *C. creticum* Miller BLUE HOUND'S TONGUE. Flowers blue with conspicuous netted veins, 7–9 mm long, and calyx lobes oblong, hairy, 6–8 mm. A hairy biennial 30–60 cm, with densely hairy oblong-lanceolate leaves which are shortly stalked or clasping. Nutlets 5–7 mm, ovoid, without a distinct border, densely covered with hooked spines. Stony places, roadsides in the Med. region. Widespread. **1049e.** *C. nebrodense* Guss. Like (1049) *C. creticum* but flowers reddish-violet without conspicuous veining to the petals; corolla 4–6 mm long; calyx lobes about 4 mm, hairy. Corolla scales reaching beyond the base of the corolla lobes; fruiting stalks shorter than calyx. Mountain woods in the Med. region. YU? GR. **1049f.** *C. hungaricum* Simonkai Flowers dull red, 4–6 mm. Distinguished by the scales of the corolla almost reaching the bases of the corolla lobes; fruiting stalks longer than the calyx. Very variable and often merging into 1049e. Dry open places. YU.AL.GR.BG.

LAPPULA Distinguished by its flattened nutlets with 1–3 rows of cylindrical, conical, flattened or hooked spines on the margins, and sometimes also on the back. 3 species in our area.

1050. *L. squarrosa* (Retz.) Dumort. (*L. myosotis*) BUR FORGET-ME-NOT. A slender, stiff erect, rough-haired, widely branched greenish annual or biennial 10–70 cm, with small sky-blue flowers 2–4 mm across, in lax leafy spikes. Leaves oblong to narrow-lanceolate. Fruit with spreading sepals and 4 nutlets each with 2–3 rows of hooked spines on the margin, and with corrugated centres. Dry places, cultivation, dunes. Widespread on mainland. **1050a.** *L. barbata* (Bieb.) Gürke (*Echinospermum b.*) Distinguished from 1050 by its greyish, densely adpressed-hairy soft leaves and stems, and larger bright-blue or white flowers usually 5–8 mm across. Leaves linear to lanceolate, the lower rosette leaves stalked. An erect annual or biennial 10–70 cm, branched above. Dry places. AL.BG. Page 379.

ASPERUGO **1051. *A. procumbens* L. MADWORT. A coarse, bristly-haired spreading or scrambling leafy annual to 70 cm, with small, stalkless purplish to violet, funnel shaped flowers, 2–4 mm across, in the axils of the lanceolate leaves. Calyx five-lobed enlarging in fruit into 2 kidney-shaped toothed lobes encircling the nutlets. Nitrogen rich waste ground, fields. Widespread on mainland.

SYMPHYTUM | Comfrey 9 species in our area, including 3 local endemic species.

Scales of corolla included within corolla

**1052. *S. officinale* L. COMFREY. Flowers purple-violet, or dirty pink, or white, 12–18 mm long, corolla lobes recurved. Upper leaves lanceolate, stalkless, blade often decurrent Nutlets black, shining. Damp meadows, marshes, by water. YU.BG.TR. **1052a. *S. tauricum* Willd. Readily recognized by the long-stalked, rough-haired, triangular heart-shaped not decurrent leaves, and the calyx, 4–7 mm, which is divided almost to the base and equalling the length of the corolla tube. Flowers pale yellow, 9–12 mm long corolla lobes not recurved. Woods. BG. Page 148.

****1053. *S. tuberosum* L. TUBEROUS COMFREY. Flowers large, pale yellow, 13–19 mm long, corolla lobes recurved. Lower stem leaves elliptic to lanceolate, the upper stalkless, shortly and narrowly decurrent. Rhizome creeping. Damp meadows, open woods, streamsides. YU.AL.GR? BG.TR. Page 136.

Scales of corolla protruding beyond corolla

**1053a. *S. bulbosum* C. Schimper BULBOUS COMFREY. Flowering stems 15–50 cm, with pale-yellow flowers with cylindrical corolla 8–11 mm long, and with the corolla scales protruding beyond the corolla mouth by 1–4 mm. Calyx lobed to two-thirds or less, 5 mm. Leaves ovate to elliptic, the lower stalked, the upper stalkless. Underground rhizome slender, creeping, producing globular tubers. Woods, shrubby places. YU.AL.GR.BG. Pl. 37. **1053b. *S. ottomanum* Friv. TURKISH COMFREY. Flowers small, pale yellow, corolla 5–7 mm long, distinguished by the narrow scales which protrude beyond the corolla mouth by 2–6 mm. Calyx 3–5 mm long, divided to below the middle, enlarging in fruit to 10 mm. Flowering stems 30–80 cm; upper leaves slightly decurrent, the lower stalked. Roots spindle-shaped, not creeping Woods. Widespread on mainland.

PROCOPIANIA Like *Symphytum* but with the corolla tube contracted at both the base and apex, and corolla lobes longer than the tube; stamens long-protruding. 3 species in our area.

**1053c. *P. cretica* (Willd.) Gusuleac (*Trachystemon c.*) A rough-hairy leafy perennial 10–50 cm, with many-flowered clusters of blue-violet or rarely white flowers, with a short corolla tube and much longer spreading lobes, slightly recurved toward their apices. Corolla lobes 2⅓–4½ times as long as corolla tube; calyx lobes acute. Leaves ovate, the lower stalked, the upper decurrent. Damp woods. GR+Is.CR. Page 379. **1053d. *P. insularis* Pawł. Similar to 1053c but with the corolla lobes 2–3 times as long as the tube and calyx lobes usually blunt. Corolla scales with slender papillae at the base, at least three times as long as wide (1½ times as long as wide in 1053c). Islands of the Aegean region. GR(Is.).CR.

BORAGO **1054. *B. officinalis* L. BORAGE. Readily distinguished by its bright-blue flowers 2–2·5 cm across, with spreading lobes, and its forward-projecting cone of blackish-purple anthers with additional horn-like appendages. A very bristly-haired, stout-stemmed annual 15–70 cm. Waste places, waysides, ditches. YU.GR.CR.TR.

TRACHYSTEMON **(1054). *T. orientalis* (L.) G. Don fil. EASTERN BORAGE. Like 1054

1. *Procopiania cretica* 1053c **2.** *Moltkia petraea* 1060g **3.** *Nonea ventricosa* (1060) **4.** *Lappula barbata* 1050a **5.** *Rindera graeca* 1083j

but flowers bluish-violet, about 1·5 cm across, lobes linear, spreading or somewhat reflexed, scales short. Stamens forward-projecting in a cone, the filaments without an additional appendage. Shady woods. BG(east).TR.

ANCHUSA | Alkanet A difficult genus with 19 species in our area, including 6 restricted to it.

Flowers with a straight corolla tube, with 5 more or less equal spreading lobes.
(a) *Flowers yellow (see also 1055)*
1055a. A. *aegyptiaca* (L.) DC. Distinguished by its pale-yellow flowers in lax leafy clusters, with a straight corolla tube about 4 mm long and spreading lobes 3–5 mm. Calyx divided almost to base. A rough-haired, erect or spreading annual 5–30 cm, with oblong-ovate to lanceolate, toothed leaves 2·5–4 cm. Rocks and sands on the littoral. GR+Is.CR.

(b) *Flowers blue, violet, or white*
1055. A. *officinalis* L. TRUE ALKANET. Flowers violet, reddish, rarely white or yellow, 7–15 mm across. A variable plant. Cultivated ground, grassy places, waysides. Widely scattered on mainland. ****1056. A. *azurea* Miller (*A. italica*) LARGE BLUE ALKANET. Distinguished by its conspicuous bright-blue or violet flowers 1–1·5 cm across, each with a brush-like tuft of white hairs in the throat, borne in a branched many-clustered inflorescence. Corolla tube about as long as calyx which is lobed to the base, 6–8 mm, enlarging to 18 mm in fruit. A thick-stemmed, bristly perennial 20–150 cm, with lanceolate leaves 10–30 cm. Fields, waste places, waysides. Widespread. **1057a.** A. *cespitosa* Lam. A distinctive rough-haired, tufted, stemless, rosette perennial of the mountains of Crete. Flowers deep blue, in almost stalkless clusters of 3–5; corolla 10–12 mm across, the tube 1½ times as long as calyx, which has linear blunt lobes. Leaves blunt, strap-like, covered in bristly hairs with swollen bases. CR. Page 71. Pl. 36.

(1058). A. *undulata* L. (incl. A. *hybrida*) UNDULATE ALKANET. A biennial or perennial 10–50 cm, both with short downy hairs and long bristly hairs mostly without conspicuous swellings at the base, and with small blackish-purple or violet flowers fading to deep blue, 3–8 mm across. Tube of corolla 1½–2 times as long as the calyx which is 5–10 mm, enlarging in fruit to 1·5 cm or more; calyx lobed to half-way, lobes blunt. Leaves often undulate, toothed. Sunny hills, sandy ground, cultivation. AL.GR+Is.CR.BG. TR.

Flowers with corolla tube curved, with 5 unequal oblique lobes
1059. A. *arvensis* (L.) Bieb. (*Lycopsis a.*) A rough-haired annual with blue flowers 4–6 mm across, in leafy clusters. Filaments inserted at or below the middle of the corolla tube. Cornfields. Widespread on mainland. **1059a.** A. *cretica* Miller A rough-haired annual 10–50 cm, with usually a solitary cluster of purplish flowers each 6–9 mm across, becoming blue with white lines. Corolla tube curved, about as long as calyx; filaments inserted above the middle of corolla tube. Calyx enlarging in fruit to 1 cm. Leaves 3–10 cm, linear-lanceolate, entire or shallowly toothed. Cultivated ground, waysides in the Med. region. YU.AL.GR. **1059b.** A. *variegata* (L.) Lehm. Like 1059a but stems spreading not erect, and flowers white or pale purplish, becoming pale blue with reddish markings. Nutlets smooth between the primary network of ridges (densely wrinkled in 1059a). Cultivated ground, waste places, waysides in the Med. region. GR+Is.CR. Page 40.

NONEA Corolla with scales or hairs in the throat, like *Anchusa* but funnel-shaped or tubular.

Flowers white or yellow

(1060). *N. ventricosa* (Sibth. & Sm.) Griseb. (incl. *N. alba.*) A rough-hairy, non-glandular annual 10–40 cm, with pale-yellow or white flowers 4–5 mm across. Calyx about 5 mm in flower, lobes about a third as long as calyx tube, calyx enlarging to 8–12 mm in fruit and becoming strongly inflated and globular. Leaves lanceolate. Nutlets 1·5–2 mm rough, ribbed. Sunny hills, stony ground. YU.GR.BG.TR. Page 379. **1060e.** *N. pallens* Petrović Flowers pale yellow or white 4–5 mm across, and calyx lobes at least three-quarters as long as calyx tube, enlarging in fruit to 1–1·5 cm. A sparsely bristly and glandular-hairy erect annual 10–40 cm. Sunny hills, fields. YU.AL.BG. **1060f.** *N. lutea* (Desr.) DC. Flowers pale yellow 5–15 mm across; calyx lobes about half as long as tube, calyx enlarging in fruit to 1–2 cm. Nutlets longer than wide, rough, netted. Native of Russia; locally naturalized in YU.

Flowers blue or dark reddish-purple

1060. *N. pulla* (L.) DC. Flowers dark reddish to blackish-purple, 5–8 mm across. Nutlets 2–3 mm, ribbed. Grassy and stony places. YU.GR.BG. Pl. 36. **1060d. *N. obtusifolia* (Willd.) DC. Flowers blue, 3–6 mm across. Nutlets about 3 mm, smooth, shiny, blackish. Dry, waste places. GR+Is.BG(south-east).

MOLTKIA Distinguished by its tubular flowers without scales or a ring, and its fruit which usually has only one nutlet which is ovoid and curved, usually smooth and shining. 2 species in our area.

1060g. *M. petraea* (Tratt.) Griseb. A dense much-branched dwarf shrub 20–40 cm, with slender rigid stems covered with adpressed white bristly hairs, and dense rounded clusters of deep violet-blue tubular flowers which are pink in bud. Corolla 6–10 mm long; stamens much longer. Leaves 1–5 cm, linear to oblanceolate, green, with scattered bristly hairs above, whitish with dense bristly hairs beneath. Rock crevices in mountains of central Balkans. YU.AL.GR. Page 379.

ALKANNA Corolla funnel-shaped, with a ring of hairs or tiny transverse swellings alternating with the stamens (scales absent), and spreading lobes; stamens included in corolla. Nutlets usually 2. A difficult genus with 16 species in our area, 14 of which are restricted to it.

Flowers blue

**1061. *A. tinctoria* (L.) Tausch DYER'S ALKANET. A small, spreading, bristle-haired, somewhat shrubby perennial to 30 cm, with bright-blue flowers about 5 mm across, in short, leafy, and often forked spikes. Corolla tube not as long as, or little longer than, the calyx. Nutlets with irregular rough swellings. Maritime sands, uncultivated ground. Widespread.

Flowers yellow or orange

(a). All bracts about twice as long as calyx

1061b. *A. orientalis* (L.) Boiss. A bristly and glandular-hairy perennial 30–50 cm, with lanceolate to oblong-lanceolate leaves with irregularly lobed undulate margins. Inflorescence dense, becoming lax in fruit; bracts up to twice as long as calyx. Flowers yellow hairless, with corolla tube up to 1½ times as long as the calyx, lobes of corolla spreading 9–12 mm. Calyx glandular, 6–8 mm, enlarging to 10–15 mm in fruit. Rocky places. GR(south). Page 96.

1061c. *A. graeca* Boiss. & Spruner (incl. *A. baeotica*) A bristly-haired perennial 15–50 cm, with entire, flat, lanceolate to linear-oblong leaves. Flowers yellow, hairless, tube of corolla from 1–2 times as long as the calyx, corolla lobes spreading 8–10 mm.

Calyx usually without glands, 5–6 mm, enlarging to 10–12 mm in fruit. Mountain rocks. YU.AL.GR. Pl. 36.

(b) Upper bracts about equalling the calyx or less
1061d. *A. primuliflora* Griseb. A bristly, glandular-hairy perennial 10–30 cm, with linear-lanceolate basal leaves and oblong-lanceolate stem leaves, and yellow flowers with corolla tube only slightly longer than calyx. Corolla hairless outside, 5–7 mm across; calyx 4–5 mm in flower, about 7 mm in fruit. Dry rocky slopes on acid rocks. GR(east).BG. **1061e.** *A. calliensis* Boiss. Distinguished by its yellow flowers which are hairy on the outside and corolla tube slightly longer than the densely glandular calyx. Bracts unequal, the lower twice as long as the calyx, the upper half as long. Plant densely glandular. Mountains. GR(south-central). Pl. 36.

1061g. *A. pindicola* Hausskn. Flowers pale yellowish to brownish, sometimes with blue veins; corolla 5–6 mm across, tube hairy, nearly twice as long as the calyx. Calyx about 8 mm in flower, 11 mm in fruit; bracts triangular-heart-shaped, the lower twice as long as the calyx. A rough, bristly, glandular, woody-based perennial 30–50 cm. Bushy places in hills. YU(south).AL.GR. Page 115.

Flowers whitish
1061f. *A. corcyrensis* Hayek A bristly- and glandular-haired perennial 25–40 cm, with oblong leaves, and flowers which are whitish, hairless, with the corolla tube nearly twice as long as the calyx. Calyx 7–8 mm, enlarging to 8–9 mm in fruit, more or less glandular. Bracts as long or slightly longer than calyx. Limestone rocks. AL(south).GR(west).

PULMONARIA | Lungwort About 5 species in our area; a difficult genus.

1063. *P. officinalis* L. LUNGWORT. Woods in mountains. YU.AL.BG. **1063a.** *P. rubra* Schott Distinguished by its dull-red flowers which do not turn blue as they age. Blade of summer leaves 15 by 7 cm, contracted abruptly into the leaf stalk, usually unspotted, rough above; stem leaves shortly decurrent. Woods. YU.AL.BG. Pl. 37. **1064a.** *P. mollis* Hornem. (incl. *P. mollissima*) Distinguished by its very softly-hairy, unspotted summer leaves 60 by 12 cm, which are gradually narrowed into the leaf stalk. Flowers violet to blue-violet and inflorescence densely glandular-sticky. Woods. YU.AL.GR.BG.

MYOSOTIS | Forget-Me-Not About 20 similar-looking species in our area. A difficult genus.

Flowers usually more than 5 mm across; mostly perennials
1066. *M. sylvatica* Hoffm. WOOD FORGET-ME-NOT. Woods, mountain pastures, damp meadows. Widespread on mainland. **1067.** *M. alpestris* F. W. Schmidt ALPINE FORGET-ME-NOT. Rocks and pastures in high mountains. YU.BG. Page 185. **1067b.** *M. suaveolens* Willd. An often bushy, much-branched perennial to 40 cm, similar to 1067, from which it may be distinguished by the numerous hooked hairs on the calyx, its deep-blue corolla, and its nutlets with a complete rim. Mountains. YU.AL.GR. BG.

Flowers usually less than 5 mm across; mostly annuals
Widespread are: **(1068).** *M. ramosissima* Rochel EARLY SCORPION-GRASS; **1068a.** *M. stricta* Roemer & Schultes; **1069.** *M. arvensis* (L.) Hill COMMON SCORPION GRASS.

LITHOSPERMUM **1072. *L. officinale* L. GROMWELL. Flowers yellowish or greenish-white, 4–6 mm long. Fruits shining white. Perennial 20–100 cm. Woods and bushy places. Widespread on mainland.

NEATOSTEMA (1071). *N. apulum* (L.) I. M. Johnston (*Lithospermum a.*) Flowers yellow, 6 mm long. Fruit light brown, covered with swellings. Annual 3–30 cm. Fields, stony places. Widespread.

BUGLOSSOIDES
Perennials
**1073. *B. purpurocaerulea* (L.) I. M. Johnston (*Lithospermum p.*) BLUE GROMWELL. Flowers bright blue, 14–19 mm long. Fruit shining white. Bushy places, wood verges. Widespread on mainland.

Annuals
1071. *B. arvensis* (L.) I. M. Johnston (*Lithospermum a.*) CORN GROMWELL. Flowers white, purplish, or blue, 6–9 mm long. Fruit brownish. Cultivated ground, waste places. Widespread.

LITHODORA **1075a.** *L. hispidula* (Sibth. & Sm.) Griseb. A much-branched, bristly-haired shrublet 10–35 cm, with bluish-violet long-tubed flowers, about 12 mm long, with spreading lobes 1 cm across. Calyx bristly-haired. Leaves oblong-ovate, to 1·5 cm, leathery, dark green above, whitish with adpressed bristly hairs beneath; twigs with white adpressed bristly hairs. Fruit white. Rock crevices, dry banks. CR+Karpathos, Samos, Rhodes. Pl. 37. **1075b.** *L. zahnii* (Halácsy) I. M. Johnston A shrub like 1075a but leaves narrower, linear or oblong-linear 2–4 cm, with strongly inrolled margins. Flowers white or blue; corolla tube funnel-shaped, tube about 12 mm long, the lobes spreading to 1·3–1·5 cm across. Calyx softly hairy. Young branches silvery-haired. Cliffs. GR(Kalamata peninsula). Pl. 37.

MACROTOMIA Calyx lobed to the base; corolla funnel-shaped, without scales or rings; stamens inserted at or above middle of corolla, stamens and style included. Nuts ovoid or three-angled, erect.

1075c. *M. densiflora* (Nordm.) Macbride A unique perennial of Aroania in the Peloponnisos with a dense rounded head of yellow tubular flowers and a rosette of lanceolate leaves. Flowers with a long tube twice as long as the calyx and with spreading lobes 2 cm across. Mountain rocks. GR(Aroania). Pl. 36.

ONOSMA About 19 species in our area, of which 10 are restricted to it. A very difficult genus particularly in the Balkans. Hairs are either simple and with a swollen base, or stellate, with usually 4–20 smaller hairs radiating in a star from the swollen base, the larger central hair usually present but sometimes absent; the microscopic structures of these hairs are often important in distinguishing species, but these are not dealt with here.

Hairs all simple
(a) Corolla 8–15 mm long
1076b. *O. graeca* Boiss. Flowers pale yellow tinged with purple; corolla 14–15 mm long, about 1½ times as long as the calyx. A more or less tufted, much-branched, very bristly-haired biennial 10–30 cm, with a much-branched inflorescence. Leaves linear to linear-lanceolate. Nutlets 6–7 mm, covered with swellings. Dry places in the Med. region. GR+Is.CR. ****(1076).** *O. arenaria* Waldst. & Kit. A stiff erect, very bristly-haired, variable perennial or biennial, with non-flowering rosettes and usually a single flowering stem 15–70 cm, branched above, with terminal clusters of pendulous tubular, pale-yellow flowers. Corolla 12–19 mm long; calyx 6–12 mm in flower, up to 18 mm in fruit.

Leaves pale green or greyish, with mostly simple and sometimes star-shaped hairs. Nutlets smooth, shining. Rocks, grassy and bushy places. YU.AL.BG.

(b) *Corolla 15–25 mm* (*see also* (1076))

1076c. *O. visianii* G. C. Clementi Flowers cream or pale yellow, corolla minutely downy, 1·5–2 cm long, about 1¼ times as long as calyx, which enlarges to 2–2·5 cm in fruit (in 1076b calyx enlarges to 1·5 cm). A much-branched, bristly-haired biennial 50–60 cm. Nutlets 4–6 mm, with minute swellings. Dry grassy places. Widely scattered on mainland. Pl. 37.   **1076d.** *O. rhodopea* Velen. Differs from 1076c in having a pale-yellow, minutely downy corolla scarcely longer than the calyx, which is 1·5–2 cm in flower and enlarging to 2·5–4 cm in fruit. Mountains. GR(north).BG(south).   **1076e.** *O. frutescens* Lam. A spreading bushy perennial with a branched woody stock and many stems 10–25 cm, with usually simple, dense inflorescence of pale-yellow flowers tinged with purple. Corolla hairless, 16–21 mm long, about 1½ times as long as the calyx, which is 10–15 mm in flower and 12–16 mm in fruit. Leaves numerous, densely grey-haired, oblong or lanceolate. Nutlets smooth. Walls, rocks in Med. region. GR(south and central).CR? Page 96.

Star-shaped (stellate) hairs present at least on the lower leaves
(a) *Lower bracts longer than the calyx*
   (i) *Flowers yellow*

1077b. *O. heterophylla* Griseb. Distinguished by its large pale-yellow flowers, 2–3 cm long, tapering from apex to base, and about twice as long as the calyx; and by its narrow-elliptic blunt stem leaves which are enlarged at the base, half clasping the stem, and covered with star-shaped hairs. Inflorescence branched or simple; calyx 8–10 mm in flower, up to 15 mm in fruit. Sandy places, on the littoral. YU.AL.GR.CR.TR. Pl. 37.

1077c. *O. erecta* Sibth. & Sm. A densely tufted greyish perennial, with several short unbranched stems 15–25 cm, bearing dense terminal clusters of large yellow flowers with the corolla about twice as long as the calyx. Corolla 20–4 mm long; calyx 10–14 mm, densely covered with star-shaped hairs or hairless. Leaves densely covered with star-shaped hairs. Mountain rocks, cliffs. GR(south).CR. Pl. 37.   **1077d.** *O. taurica* Willd. Like 1077c but distinguished by the calyx which has many simple hairs. Flowers pale yellow; corolla 2–3 cm long, 2–3 times as long as the calyx, which enlarges to 18 mm in fruit. Lower leaves 4–12 cm, linear-oblong, with star-shaped hairs above and below. Rocks. YU.GR.BG.TR. Pl. 37.

   (ii) *Flowers white*

1077e. *O. spruneri* Boiss. Distinguished by its white, finely hairy, almost stalkless flowers 2–3 cm long, 2–3 times as long as the calyx. A softly-hairy biennial 25–40 cm with unbranched stems. Nutlets dull. Mountain rocks. GR(central and south).

(b) *Lower bracts shorter or as long as calyx*

1077h. *O. echioides* L. Distinguished by its pale-yellow tubular flowers which are twice as long as the calyx. Corolla cylindrical at the apex and tapering towards the base, minutely downy, 18–25 mm long; calyx about 1 cm, up to 1·5 in fruit. Lower leaves linear or linear-oblong, 1–7 mm wide, the upper broader, with sparse or dense star-shaped hairs; bracts shorter or as long as calyx. Dry stony places. A more or less tufted perennial with several erect, unbranched flowering stems 10–30 cm. YU.AL.Pl. 37.

1077f. *O. thracica* Velen. A slender perennial 15–25 cm, endemic to Bulgaria, with pale-yellow flowers about twice as long as the calyx. Corolla 14–18 mm, minutely hairy; calyx 6–8 mm, with star-shaped hairs. Lower leaves 1–2·5 cm, linear 1–3 mm wide, covered with star-shaped hairs; bracts shorter than the calyx. Rocks in hills. BG(south and east). Page 149. **1077g.** *O. leptantha* Heldr. A tufted perennial of the Taiyetos Mountains with several flowering stems 10–20 cm, bearing small hairless pale-yellow flowers. Corolla 13–17 mm, not more than 1¼ times as long as the calyx, which is 9–12 mm in flower, up to 16 mm in fruit. Lower leaves linear-oblong 4–6 mm wide, densely covered with star-shaped hairs; bracts shorter than calyx. Rocks in hills. GR(south). Page 89.

CERINTHE | Honeywort

**1078. *C. minor* L. Distinguished by its small yellow flowers, which are sometimes violet-spotted. Corolla 10–12 mm, with lanceolate-pointed erect lobes almost as long as the tube. Annual, biennial, or perennial 15–60 cm. Bushy and waste places, rocks. YU.AL.GR.CR? BG.TR.

1079. *C. major* L. HONEYWORT. A hairless, glaucous annual 15–60 cm, with broad overlapping clasping stem leaves, and drooping clusters of yellow or cream cylindrical flowers, often with a varying amount of chocolate-brown or red towards the base. Corolla 1·5–3 cm long, the short lobes ovate, recurved. Leaves often blotched with white, with rough swellings. Cultivated ground, waysides, stony places. Widespread, except BG. **(1079). *C. glabra* Miller Distinguished from 1079 by its narrower yellow corolla 8–13 mm by 3–4 mm wide (5–8 mm wide in 1079), with 5 dark-red spots in the throat. A perennial or biennial. Bushy places in hills and mountains. YU.AL.BG(south-west).

1079a. *C. retorta* Sibth. & Sm. Distinguished by its flowers which are pale yellow at the base and violet at the mouth, and the large violet leafy bracts which are as long as the calyx. Corolla curved 1–1·5 cm, the lobes recurved. An annual 10–50 cm, with ovate to oblong-spathulate stalked leaves, the upper stalkless, with some white swollen-based hairs. Rocky places in the Med. region. YU.AL.GR.CR? Pl. 38.

ECHIUM | Bugloss

Stamens included in corolla; flowers usually less than 12 mm long
1080. *E. parviflorum* Moench SMALL-FLOWERED BUGLOSS. Flowers pale or dark blue, 10–13 mm. Calyx 6–8 mm at flowering, up to 15 mm in fruit with lobes 3–6 mm wide at base. Sandy places and rocks on the littoral. YU.AL.GR+Is.CR. **(1080).** *E. arenarium* Guss. Flowers dark blue, 6–11 mm. Calyx 5–7 mm at flowering, up to 10 mm in fruit with lobes 2–3 mm wide at base. Maritime sands. GR+Is.Cr.

At least some stamens protruding beyond the corolla; flowers usually more than 12 mm long.
(a) Flowers 1–1·5 cm long
1081. *E. italicum* L. PALE BUGLOSS. A very hairy, greyish or yellowish biennial with a single stem 40–100 cm, bearing a dense spike or a pyramidal inflorescence of small yellowish or bluish-white flowers. Corolla 10–12 mm long, narrowly funnel-shaped, with 4–5 stamens protruding; filaments of stamens pale. Basal leaves lanceolate, narrowed to the base. Dry uncultivated places, stony ground. Widespread. Page 65. **1081e. *E. russicum* J. F. Gmelin Flowers dark red, 9–12 mm long, with 4–5 long-protruding stamens with red filaments, in a spike-like inflorescence. Stigma club-shaped. Meadows, scrub, uncultivated slopes. YU.AL.BG.

(b) Flowers usually more than 1·5 cm long
1082. *E. vulgare* L. VIPERS BUGLOSS. Corolla bright blue to blue-violet, 10–19 mm, uniformly hairy, with 4–5 long-protruding stamens. Cultivation, waste places, dry stony ground. Widespread on mainland. ****1083. *E. plantagineum* L. (*E. lycopsis*) PURPLE VIPERS BUGLOSS. Distinguished from 1082 by its larger flowers 18–30 mm long, and its oblong to lanceolate stem leaves with the uppermost more or less heart-shaped at the

base. Corolla blue becoming pink through purple, hairless except on the veins and margin, with usually 2 protruding stamens; calyx 7–10 mm, enlarging to 15 mm in fruit. Roadsides, fields, sandy places near the sea. Widespread. **1083h.** *E. angustifolium* Miller A bristly-haired, whitish or greyish perennial 25–40 cm, with several flowering stems and narrow whitish bristly-haired leaves. Corolla reddish, or reddish-purple to purplish-violet, 16–22 mm long, with 4 or more long-protruding stamens. Calyx 7–10 mm, enlarging to 15 mm in fruit, covered with stiff prickly bristles. Upper stem leaves narrow-lanceolate usually only 3–5 mm wide. A very variable species. By the sea. GR+Is. CR.TR.

MATTIASTRUM Flower clusters without bracts. Calyx lobed almost to the base; corolla bell- or funnel-shaped, with scales in the throat; stamens not protruding. Nutlets rounded or ovate, with a broad flat marginal wing.

1083i. *M. lithospermifolium* (Lam.) Brand A tufted, softly-hairy or bristly-hairy perennial 7–35 cm, with dense terminal clusters of tiny blue-violet, bell-shaped flowers, and lanceolate to spathulate, stalked leaves. Corolla 3–5 mm long; calyx 2–4 mm, with linear-oblong lobes. Inflorescence elongating in fruit to 6–10 cm or more; nutlets 7–8 mm. Mountain rocks. CR. Page 70.

RINDERA Like *Mattiastrum* but nutlets flattened, smooth, with a broad wing. Corolla purple, cylindrical or funnel-shaped, without scales in throat, but sometimes with 5 small invaginations.

1083j. *R. graeca* (A.DC.) Boiss. & Heldr. A tufted grey-haired perennial with stems 6–25 cm, linear to oblanceolate woolly leaves, and an umbel-like cluster of purple flowers. Corolla 10–12 mm long, more or less cylindrical, lobed to a third; stamens equalling corolla; calyx 4–8 mm, silvery-haired. Nutlets 11–14 mm. Mountain rocks. GR. Page 379. **1083k.** *R. umbellata* (Waldst. & Kit.) Bunge Differs from 1083j in having reddish-yellow flowers, green leaves, and taller stems 25–70 cm, which are leafy to the top (leaves grey-woolly, stem not leafy above in 1083j). Corolla 11–15 mm long, conical. Dry open places. YU(east).BG.

SOLENANTHUS Corolla cylindrical to funnel-shaped, with 5 sack-like folds in the throat; stamens shorter to longer than corolla; style longer. Nutlets globular, covered with hooked hairs. 4 local species, including 2 endemic species in our area.

1083l. *S. stamineus* (Desf.) Wettst. A perennial with shortly branched velvety-hairy stems 20–60 cm, with grey-velvety lanceolate leaves, and long-branched clusters of reddish-purple tubular flowers with triangular corolla lobes. Corolla 6–7 mm long; stamens protruding, up to twice as long. Mountain rocks. GR(Aroania). Page 93.

VERBENACEAE | Verbena Family

VERBENA 1084. *V. officinalis* L. VERVAIN. An erect perennial with lilac flowers, and fruiting spikes 10–25 cm. Waste ground, waysides. Widespread. **(1084).** *V. supina* L. PROCUMBENT VERVAIN. Differs from 1084 in being a procumbent annual and having leaves mostly twice pinnately cut into oval segments. Flowers lilac, about 3 mm across; corolla about twice as long as calyx, fruiting spikes up to 8 cm. Damp places, sandy ground, waste places. YU.GR.CR.BG.TR.

LANTANA 1085. *L. camara* L. A shrub 1–1·5 m, sometimes prickly stemmed with ovate, toothed, strong-smelling leaves, and dense hemispherical heads of numerous yellow flowers which characteristically turn orange or red. Flower clusters 2·5–5 cm across, pink in bud; corolla tubular, 4–5 mm across, with 4 spreading lobes. Native of tropical America; often grown for ornament in the Med. region.

LIPPIA **1086. *L. nodiflora* (L.) Michx A creeping, rooting perennial with ascending stems 15–30 cm. Flowers white, in dense ovoid heads 5–7 mm across, borne on long leafless stems from the axils of the opposite leaves. Corolla 2 mm across; calyx lobed almost to base. Leaves 1–2·5 cm, elliptic. Wet grassy places, usually near the sea. AL.CR.TR.

VITEX **1087. *V. agnus-castus* L. CHASTE TREE. A grey-felted, aromatic shrub 1–6 m, with slender terminal spikes of usually pale-lilac or rarely pink flowers, and distinctive palmately divided leaves. Corolla 8–10 mm long, two-lipped; stamens projecting. Leaflets 5–7, lanceolate entire, greyish-green above, white-felted beneath. Fruit globular fleshy, reddish-black. Damp places, on the littoral, streamsides. Widespread.

LABIATAE | Mint Family

AJUGA | Bugle Like *Teucrium* but corolla hairy in tube, with upper lip usually short and the lower three-lobed. 9 species in our area. The following are most frequently encountered:

1. Lvs. with 3 linear lobes which are sometimes further three-lobed.	*chamaepitys*
1. Lvs. entire, toothed, or shallowly lobed.	
2. Lvs. 3–6 mm wide, linear to linear-oblong.	*iva*
2. Lvs. 8–40 mm wide, oblong to more or less rounded.	
3. Fls. yellow or cream, 2 at each node.	*laxmannii*
3. Fls. blue or blue-violet, 4 or more at each node.	
4. Stamens included in corolla tube; upper lip of corolla with 2 conspicuous lobes.	*orientalis*
4. Stamens more or less protruding from the corolla tube; upper lip of corolla entire or with 2 short teeth or lobes.	
5. Middle and upper part of the stem hairy on opposite faces, alternating at each node (Widespread on mainland).	**1089.** *A. reptans* L.
5. Middle and upper part of stem more or less equally hairy on all faces.	

6. Stamens conspicuously longer
    than corolla tube, the filaments
    hairy; upper bracts often
    shorter than fls. (Widespread
    on mainland).   **1090. *A. genevensis* L.
6. Stamens only slightly longer
    than the corolla tube, the
    filaments hairless; upper
    bracts longer than fls. (YU.
    AL.BG.).   **1091. *A. pyramidalis* L.

(1091). *A. orientalis* L. EASTERN BUGLE. Distinguished by its densely hairy, oblong, deeply or shallowly lobed leaves, and large oval, purple-tinged bracts of the pyramidal inflorescence. Corolla 12–16 mm long, violet-blue, with tube longer than the calyx and the upper lip two-lobed. Bushy places. AL.GR+E.Aegean Is. Pl. 38. **1092.** *A. laxmannii* (L.) Bentham A pale erect leafy perennial 20–50 cm, with spreading hairs, oblong half-clasping upper leaves, and yellow or cream flowers with purple veins. Corolla 2·5–3·5 cm long; stamens longer. Bracts ovate, like the leaves, much longer than the flowers. Limestone hills. YU.GR.BG.TR. Pl. 38.

1093. *A. iva* (L.) Schreber A very leafy tufted woody-based perennial 5–20 cm, with narrow silvery-haired leaves, smelling of musk, and rosy-purple or yellow flowers. Flowers 2–4 at each node; corolla 12–20 mm long. Bracts linear, longer than the flowers. Leaves linear, usually 3–6 mm wide, entire or with 2–6 short lobes. Dry stony places, olive groves. YU.GR+Is.CR. **1094.** *A. chamaepitys* (L.) Schreber GROUND-PINE. Subsp. *chia* (Schreber) Arcangeli (*A. chia*) is the typical plant of the Balkans. A short-lived perennial 5–30 cm, like 1093 but with leaves deeply cut into 3 linear segments, and with larger yellow flowers 18–25 mm long, with red or purple marking. Fields, uncultivated places. Widespread.

TEUCRIUM | **Germander** Like *Ajuga* in having a corolla with or without a very short upper lip, but lower lip five-lobed, and corolla tube hairless within. 20 species in our area, including 7 restricted to it.

Flowers in dense rounded or flattened, simple or compound heads
1097. *T. polium* L. FELTY GERMANDER. A small branched shrublet 5–45 cm, with densely white-, golden-, or grey-felted stems and leaves, and dense globular heads of pink, white, or rarely yellowish flowers. Flowers numerous, little longer than the felted calyx. Leaves oblong, conspicuously shallowly lobed or 'scalloped', with inrolled margins. A variable species. Dry places, sunny rocks. Widespread. Page 39. **1103.** *T. montanum* L. MOUNTAIN GERMANDER, ALPINE PENNY-ROYAL. A woody-based, shrubby, mat-forming perennial with many spreading white-hairy stems 10–25 cm, narrow entire leaves, and flattened or rounded clusters of small whitish-yellow flowers. Corolla 12–15 mm long; calyx to 10 mm. Leaves 13–30 mm, greenish and often hairless above, white-felted beneath, margins inrolled. Calcareous rocks. YU.AL.GR.BG. Page 178.

Flowers not in dense rounded or flattened heads
(a) Flowers pink, violet, blue, or purple
  (i) Leaves toothed or lobed
1098. *T. scordium* L. WATER GERMANDER. A leafy perennial 10–60 cm, smelling of garlic when crushed, with axillary purplish flowers. Damp places, watersides. Widespread. **1099.** *T. chamaedrys* L. WALL GERMANDER. A small creeping shrublet with

pinkish-purple flowers. Dry places, banks, open woods. Widespread on mainland. **1099a.** *T. divaricatum* Boiss. A very variable, short stout shrublet usually 10–30 cm, with leathery ovate leaves which are shallowly rounded-lobed, hairless or somewhat woolly. Flowers pink or purple, in a lax more or less one-sided inflorescence; calyx often purplish, half as long as corolla. Rocks in the Med. region. GR+Is.CR. Page 67. **1099b.** *T. microphyllum* Desf. A dwarf shrub 5–40 cm, with tiny oblanceolate to ovate leaves about 5 mm long, toothed, and white-woolly beneath. Flowers pink, 10–12 mm long, hairy, in a lax inflorescence; calyx glandular-hairy, 6–7 mm. Rocky places in south Aegean region. CR+Karpathos. **1099c.** *T. halacsyanum* Heldr. A dwarf, woolly-haired shrublet usually up to 15 cm, with violet flowers in a dense elongated cluster. Corolla about 7 mm long, hairy, stamens much longer; calyx distinctly two-lipped, tubular, curved and pouched at the base. Leaves 1–1·5 cm, broadly oval, toothed, woolly-haired. Rocky places. GR(west).

(ii) Leaves entire

1101. *T. fruticans* L. TREE GERMANDER. A handsome evergreen shrub to 2·5 m, with large pale-blue or lilac flowers in lax leafy clusters, with white-felted calyx. Lower lip of corolla about 2 cm long. Twigs four-angled, white-felted; leaves lanceolate to ovate, white-felted beneath. Rocks. YU(Adriatic Is.). **1101a. *T. brevifolium* Schreber A densely branched low shrub to 60 cm, like 1101, but with green or brown (not white-felted) twigs, and smaller blue purple-veined flowers about 1 cm long, in a short leafy cluster. Leaves narrow-lanceolate, grey-felted above and below, with inrolled margins. Dry rocky places near the sea. GR+Is.CR. Page 393. **1101b.** *T. aroanium* Boiss. Like 1101 with blue flowers 1·5–2 cm long, but a creeping much-branched rooting shrublet with white-woolly twigs and leaves 5–15 mm, ovate to obovate. Mountain rocks. GR(south). Page 92. Pl. 38.

(b) Flowers yellow, white, or greenish

1104. *T. flavum* L. YELLOW GERMANDER. A shrubby perennial like 1099a but with yellow flowers and leaf stalks as long as the width of the leaf blade (shorter in 1099a). Corolla about 1·5 cm long; calyx 7–10 mm, hairy or hairless; bracts lanceolate-entire, shorter than the flowers. Leaves 1–4 cm, leathery, hairless or velvety beneath; stems velvety. Rocks in the Med. region. YU.AL.GR+Is.CR.TR. Pl. 38. **1104a.** *T. arduini* L. Distinguished by its very dense, unbranched, cylindrical leafless spike of numerous small, whitish flowers with conspicuous woolly-haired calyx. Corolla 8–9 mm long, glandular-hairy; stamens longer; calyx swollen at base, curved, glandular. Bracts linear-lanceolate. Stems to 30 cm; leaves ovate, conspicuously rounded-toothed, short-stalked. Rocky places. YU(west).AL(north). Page 393.

ROSMARINUS | Rosemary

**1105. *R. officinalis* L. ROSEMARY. A dense, aromatic, usually erect, evergreen shrub to 2 m, readily distinguished by its narrow dark-green leaves which are inrolled and white-felted beneath, and its lilac flowers. Corolla two-lipped; the 2 stamens and style curving outwards well beyond the corolla. Rocks and stony ground in the Med. region. YU.GR+Is.CR. Naturalized BG.

PRASIUM ****1106.** *P. majus* L.

A hairless, much-branched shrub to 1 m, with white or pale-lilac 'dead-nettle' flowers, and oval, conspicuously toothed leaves borne on hairless, square-sectioned twigs. Corolla 17–20 mm long, two-lipped; calyx lobes leafy, bristle-tipped, enlarging to 2·5 cm in fruit. Leaves 2–5 cm, stalked. Nutlets black. Dry places in the Med. region. YU.AL.GR+Is.CR.

389

SCUTELLARIA | Skullcap   11 species in our area, 2 of which are restricted to it.

1. Bracts mostly leaf-like, though
   smaller towards the apex.
   2. Calyx glandular-hairy; lvs. more or
      less lobed at the base (YU.GR? BG.
      TR.).   (1108). *S. hastifolia* L.
   2. Calyx hairless or, if hairy, without
      glandular hairs; lvs. heart-shaped at
      the base (Widespread on mainland).   **1108. *S. galericulata* L.
1. Bracts conspicuously dissimilar to lvs.
   3. Upper lip of corolla yellow or pink.   *orientalis*
   3. Upper lip of corolla purple, blue,
      reddish, or white.
      4. Corolla 18 mm or more long.
         5. Inflorescence four-angled; lvs. not
            more than 3·5 cm, usually blunt
            (YU.AL.GR.BG.).   **1107. *S. alpina* L. Page 190
         5. Inflorescence not four-angled; lvs.
            up to 8 cm, pointed (YU.AL.GR.
            BG.).   **1109. *S. columnae* All.
      4. Corolla less than 18 mm long.
         6. Internodes of inflorescence about
            5 mm.
            7. Lvs. 2–5·5 cm; bracts exceeding
               the calyx (CR.).   1109a. *S. sieberi* Bentham Page 67
            7. Lvs. 1–2 cm; bracts not exceeding
               the calyx (CR.).   1109b. *S. hirta* Sibth. & Sm.
         6. Internodes of inflorescence 8–12 mm.
            8. Bracts 12–20 mm (YU.BG.).   1109c. *S. velenovskyi* Rech. fil.
            8. Bracts 3–12 mm.
               9. Fls. usually white; bracts more
                  than 10 mm (YU.GR.BG.TR.).   1109d. *S. albida* L.
               9. Fls. usually bluish; bracts
                  usually less than 10 mm.   *altissima, rubicunda*

**(1107). *S. orientalis* L. EASTERN ALPINE SKULLCAP. A spreading woody-based perennial with a dense oblong, four-sided head of yellow or rarely pink flowers, and purplish or yellowish-green, broad or narrow bracts longer than the calyx. Corolla 1·5–3 cm, lower lip often reddish. Leaves 1–1·5 cm, ovate, deeply toothed or lobed, grey-woolly beneath. A variable species. Dry rocky places usually on limestone. YU.AL.GR.BG.TR. (1109). *S. altissima* L. A perennial with simple or branched erect stems to 1 m, ovate, toothed leaves, and slender spikes of bluish flowers. Corolla 12–16 mm, the lower lip whitish; calyx without or with few long white non-glandular hairs. Bracts 6–10 mm, ovate, entire, green. Leaves 5–15 cm. Woods. YU.AL.GR.BG. 1109e. *S. rubicunda* Hornem. (*S. peregrina*) Flowers usually purplish to bluish, the lower lip often whitish, rarely all white. Differing from (1109) in its smaller leaves 1–3 cm, and its calyx with numerous long white non-glandular hairs. Rocks in hills and mountains. AL.GR. Pl. 38.

LAVANDULA | Lavender
**1110. *L. stoechas* L. FRENCH LAVENDER. An attractive, small, grey-leaved shrub to 1 m, with dense oval, stalked spikes 2–3 cm of small dark-purple flowers, with a topknot of narrow, pale-purple bracts 1–5 cm long. Bracts subtending flowers purple papery, oval,

strongly veined, woolly-haired. Leaves linear to oblong-lanceolate, usually grey-woolly. Dry stony places, sunny hillsides, pinewoods. GR+Is.CR.TR. Page 38. **1111.** *L. angustifolia* Miller (*L.spica*) COMMON LAVENDER. A very aromatic, greyish-leaved, much-branched small shrub to 1 m or more, with rather dense, cylindrical, long-stalked spikes of blue-purple flowers without a topknot of bracts. Distinguished by the broad, rhomboid-oval bracts 3–8 mm long, subtending each whorl of 6–10 flowers. Corolla 10–12 mm. Leaves entire, at first white-felted, later green. Rocks in the Med. region; often cultivated. YU.GR. **(1111).** *L. latifolia* Medicus Distinguished from 1111 by its linear or lanceolate bracts, its grey-green, more densely felted leaves, and its slender interrupted spike of blue-violet flowers. Corolla 8–10 mm. Rocks in the Med. region. YU(west).

MARRUBIUM | **Horehound**   9 species in our area, including 2 which are exclusive to it.

Calyx with 10 teeth
**1112. *M. vulgare* L. WHITE HOREHOUND..Waste ground. Widespread.

Calyx with 5 teeth
(a) *Calyx longer than corolla; calyx teeth usually at least as long as calyx tube*
1112a. *M. velutinum* Sibth. & Sm. A densely felted, often golden-yellow haired perennial to 40 cm, with several, spaced, globular whorls of many tiny yellow flowers. Calyx 9–11 mm, longer than corolla, with yellowish hairs, and 5 awl-shaped teeth which do not become rigid in fruit. Leaves yellow-felted, ovate to rounded, shallowly toothed, shortly stalked. Rocks in mountains. GR(north and central). Pl. 38. **1112b.** *M.friwaldskyanum* Boiss. A larger perennial 60 cm or more, like 1112a but leaves oblong-ovate to elliptic, irregularly and deeply toothed, densely greyish-hairy beneath, and dark green with adpressed hairs above. Flowers yellow; calyx tube 6–8 mm (calyx tube 4–6 mm in 1112a). Stony ground. BG. Page 393.

(b) *Calyx shorter than corolla; calyx teeth shorter than calyx tube*
1113. *M. peregrinum* L. BRANCHED HOREHOUND. A silvery-white felted perennial to 60 cm, distinguished from other species by its many wide-spreading branches which diverge almost at right angles from the main stem. Flowers whitish, inconspicuous, in few, distinct, many-flowered whorls, or whorls crowded at ends of branches. Calyx adpressed woolly-haired; calyx teeth 1–3 mm. Leaves oblong, toothed, densely white-felted. Stony ground, olive groves, waste places, waysides. YU.AL.GR.BG.TR. Page 165. **1113d. *M. thessalum* Boiss. & Heldr. Distinguished from 1113 by its unbranched inflorescence, and globular whorls of more than 10 flowers. Flowers white, up to 12 mm long, the upper lip conspicuously two-lobed to $\frac{1}{3}-\frac{1}{2}$. Calyx ten-ribbed, with star-shaped hairs; calyx teeth straight, erect, shaggy-haired. Leaves oblong to obovate, toothed, white-woolly beneath, greenish-woolly above. Mountain pastures. GR(east). Page 127. **1113e.** *M. cylleneum* Boiss. & Heldr. Like 1113d but flowers yellowish. Leaves rounded to broadly obovate, with yellowish hairs. Mountain rocks. AL(south).GR (south). Page 92.

SIDERITIS   The genus can be distinguished by the dense whorls of small yellow, white, or brownish flowers, closely invested by broad bracts; by the two-lipped corolla with a flat, very erect, rounded or notched upper lip; by the spiny calyx teeth. A difficult genus. 8 species in our area, including 2 endemics.

Annuals; bracts leaf-like
**(1114). *S. montana* L. A yellowish-green, woolly-haired branched annual to 35 cm, with numerous whorls of yellowish or blackish-brown flowers which go brownish-purple

when dry, and often with a topknot of yellowish leafy bracts to the inflorescence. Corolla about 1 cm, shorter than or equalling calyx. Calyx teeth more or less equal with bristly-tipped apices. Hills, rocky places. Widespread on mainland. **1114.** *S. romana* L. Like (1114) but differing in having calyx more or less two-lipped, and teeth unequal with the upper tooth larger, ovate. Flowers white, yellow, or purple, equalling or longer than the calyx. Stony places in the Med. region. YU.AL.GR. **1114a.** *S. curvidens* Stapf Like 1114 but calyx strongly swollen at base, calyx teeth curved, soft not bristly-tipped, usually ending in a recurved awn. Flowers white, with upper lip pink. Stony places in the Med. region. GR+Is.CR.TR. **1114b.** *S. lanata* L. Plant covered with spreading shaggy hairs and stem glandular-hairy 10–30 cm. Flowers yellow with black lips, and drying black; calyx more or less two-lipped, the upper tooth larger, narrowly lanceolate, rarely more or less equal. Cultivated or undisturbed ground in the Med. region. YU.GR+Is.BG.

Perennials; bracts dissimilar to leaves
(a) *Middle bracts usually equalling or shorter than the flowers; whorls usually distinct*
1115d. *S. syriaca* L. A grey- or white-woolly perennial 10–50 cm, with 5–20 usually widely spaced whorls of yellow flowers subtended by broad rounded or fine-pointed bracts. Corolla 9–15 mm long; calyx 7–12 mm, with teeth half as long to almost as long as the tube. Leaves oblong or linear-lanceolate, the lower leaves finely toothed, the upper entire. A variable species with several distinct local forms. Mountain rocks. YU.AL.GR.CR.BG. Page 126. Pl. 39.

(b) *Middle bracts longer than the flowers; whorls clustered to form a dense spike*
1115e. *S. clandestina* (Bory & Chaub.) Hayek Like 1115d but whorls of yellow flowers crowded, or the lower 1–3 distant. Corolla 1–1·5 cm long; calyx 9–11 mm, the teeth slightly shorter than the tube. Middle bracts with a narrow point 4–10 mm long. Lower leaves oblong-spathulate to oblong-elliptical, entire or toothed, 1·5–5 cm, the upper linear. Mountain rocks. GR(south). Page 89. **1115f.** *S. scardica* Griseb. An attractive woody-based perennial 15–40 cm, with woolly-white leaves, and erect stems bearing a dense terminal spike of whorls of yellow flowers subtended by broad, net-veined, often purple-flushed bracts. Flowers about 1·5 cm long; middle bracts with a narrow point 2–4 mm long. Lower leaves oblong-lanceolate, 4–8 cm. Mountain rocks. YU.AL.GR.BG. Page 393.

NEPETA 13 species, 8 of which are restricted to our area. The following are the most frequently encountered:

Outermost bracteoles equalling or longer than calyx
1116h. *N. scordotis* L. Flowers blue, in a spike-like inflorescence with whorls interrupted only below, and conspicuous bracts which are broadly ovate, acute, and shaggy-haired. Corolla 13–16 mm, longer than calyx which has a straight tube. An ascending woolly- or shaggy-haired perennial 25–60 cm, with ovate-lanceolate leaves, the upper stalkless. Dry places in the Med. region. GR(south).CR.

All bracteoles usually shorter than calyx
(a) *Calyx tube straight, the upper calyx teeth not exceeding the lower*
N. sibthorpii group.
Flowers white; bracts and bracteoles green with more or less papery margins, rarely longer than calyx. Flowers in whorls, usually in a slender spike-like inflorescence with the lower whorls distant; corolla 9–13 mm long; calyx 5–8 mm. Leaves ovate to oblong, with a heart-shaped base, grey-green, all shortly stalked. A difficult group which includes the following 3 species: **1116e.** *N. sibthorpii* Bentham With fine grey hairs which are adpressed at least on the stem above. Bracteoles oblong-lanceolate 6–8 mm.

1. *Nepeta parnassica* 1116f **2.** *Teucrium brevifolium* 1101a **3.** *Teucrium arduini* 1104a **4.** *Marrubium friwaldskyanum* 1112b **5.** *Sideritis scardica* 1115f **6.** *Phlomis cretica* 1124a

Rocks in the Med. region. GR(south). **1116f.** *N. parnassica* Heldr. & Sart. Distinguished by its stems with woolly-shaggy spreading hairs and often sticky. Bracteoles lanceolate, 5–10 mm. Rocks in hills and mountains. AL(south).GR. Page 393. **1116g.** *N. spruneri* Boiss. Hairs on stems adpressed, velvety; leaves grey-green. Distinguished by the narrow outer bracteoles which are less than 1 mm wide (more than 1 mm wide in 1116e and 1116f). Rocks in mountains. AL.GR(north-west,central).

**1117. *N. nuda* L. (incl. *N. pannonica*) HAIRLESS CATMINT. An erect, nearly hairless, much-branched perennial to 1 m or more, with lax branched whorls of pale-violet flowers in slender interrupted spikes. Corolla 6–8 mm long; calyx 4–6 mm, teeth green or blue-tinged; bracteoles 2–3 mm, linear. Leaves ovate to ovate-oblong, toothed, mostly stalkless. Subsp. *albiflora* Gams of Macedonia has white flowers in a compact inflorescence, with white-tinged calyx teeth. Rocks and bushy places in the mountains. YU.AL.GR.BG.

(b) Calyx tube curved, the upper calyx teeth exceeding the lower
1116. *N. cataria* L. CATMINT. Flowers white with small purple spots, 7–10 mm; calyx 5–7 mm. Stony and bushy places. YU.AL.GR.BG. **1117a. *N. melissifolia* Lam. Distinguished by its comparatively large blue flowers 1·2–1·5 cm long with small red spots, in an interrupted spike. Bracts 5–7 mm, lanceolate; bracteoles shorter, linear. A hairy perennial 20–40 cm, with ovate-heart-shaped stalked leaves 1·5–3·5 cm. Bushy places in the Med. region. GR(south).CR.

GLECHOMA ****1118.** *G. hederacea* L. GROUND IVY. Woods, shady places. YU.AL.GR.BG. **1118a.** *G. hirsuta* Waldst. & Kit. Like 1118 but a larger plant and usually with longer and denser hairs. Corolla larger 2–3 cm, pale blue with white spots on the lower lip. Calyx 7–11 mm, the teeth of the upper lip half as long to about equalling the tube (calyx 5–6·5 mm, teeth of upper lip less than half as long as tube in 1118). Woods, shrubby places. YU.AL.GR.BG.

PRUNELLA **1119.** *P. vulgaris* L. COMMON SELF-HEAL. Flowers deep blue-violet 13–15 mm; upper leaves entire or toothed. Grassland, open woods, rocks, waste places. Widespread. ****1120.** *P. laciniata* (L.) L. CUT-LEAVED SELF-HEAL. Flowers whitish-yellow; the upper leaves pinnately lobed. Dry meadows, open woods, rocks. Widespread on mainland+CR. ****(1120).** *P. grandiflora* (L.) Scholler LARGE SELF-HEAL. Flowers deep violet, more than 18 mm; the upper leaves entire or toothed. Woods, stony places. Widespread on mainland.

MELITTIS ****1121.** *M. melissophyllum* L. BASTARD BALM. A strong-smelling, softly hairy perennial 20–70 cm, with large handsome two-lipped tubular, pink, white, purple, or pink-spotted flowers borne in leafy whorls. Flowers 2–6 in each whorl; corolla 2·5–4 cm long; calyx two-lipped. Leaves oblong to ovate, coarsely toothed. Damp shady places, woods in hills and mountains. YU.AL.GR.BG.

PHLOMIS
1. Corolla purple or pink, rarely white.
  2. Root with tubers; upper lip of
      corolla straight.   *tuberosa*
  2. Roots without tubers; upper lip
      of corolla curved.
    3. Stem and bracteoles with glandular
        hairs.   *samia*

3. Stem and bracteoles without
   glandular hairs.   *herba-venti*
1. Corolla yellow or brownish-yellow.
   4. Bracteoles and calyx teeth hooked at
      apex (Karpathos).   **1124c.** *P. floccosa* D. Don
   4. Bracteoles and calyx teeth straight
      at apex.
      5. Bracteoles elliptic-lanceolate to
         ovate.
         6. Shrub to 55 cm; lower leaves 1·5–
            2·8 cm, broadly elliptical to rounded;
            calyx teeth 0·5–1 mm.   *lanata*
         6. Shrub to 130 cm; lower leaves 3–9 cm,
            lanceolate; calyx teeth 1–4 mm. *fruticosa*
      5. Bracteoles awl-shaped, linear or
         narrowly lanceolate.   *cretica*

**(1122). *P. tuberosa* L. Flowers purple or pink in numerous whorls, the upper crowded,
the lower remote. Corolla 1·5–2 cm long, upper lip erect, margin and inner side of lip
with long hairs; calyx 8–13 mm, hairless or rough-haired, spiny-toothed. An erect
variable perennial to 1·5 m, with triangular-heart-shaped, strongly toothed, long-
stalked leaves 5–25 cm. Among herbaceous vegetation. YU.GR.BG. **1122a.** *P. samia*
L. Distinguished by its whorls of large dull-purple flowers with the upper lip helmet-like
and curved over the lower lip. Corolla 3–3·5 cm long; calyx with glandular and star-
shaped hairs 18–25 mm including the bristle-tipped teeth which are up to 12 mm.
Leaves 8–18 cm, leathery, with star-shaped hairs above, whitish beneath with both
star-shaped and glandular hairs, long-stalked. Woods, bushy places. YU(south).GR. Pl.
39. ****1122.** *P. herba-venti* L. Distinguished from 1122a by the smaller pink or purple
flowers 1·5–2·5 cm long, and smaller calyx 8–15 mm. Most widely distributed in our
area is subsp. *pungens* (Willd.) DeFilipps (*P. pungens*) which is distinguished by its
whitish-felted stems, its lanceolate stem leaves with star-shaped hairs above, and its
fewer-flowered whorls of 2–6 flowers. Arid places, steppes. YU.GR.BG.TR.

**1124. *P. fruticosa* L. JERUSALEM SAGE. A handsome grey-leaved shrub with white-felted
branches up to 1·5 m, and 1–3 compact terminal whorls of large orange-yellow flowers.
Corolla 2·3–3·5 cm long, the upper lip hooded, up to 2 cm wide; calyx covered with dense
white star-shaped hairs, teeth spiny-tipped, recurved. Leaves 3–9 cm, elliptic to ovate-
lanceolate, leathery, densely white-woolly beneath, stalked to 4 cm. Rocky and bushy
places in the Med. region. YU.AL.GR+Is.CR. Page 35. **1124a.** *P. cretica* C. Presl Like
1124 with few whorls of orange-yellow flowers and leathery greyish leaves, but a smaller
shrub to 45 cm, and stems with glandular or club-shaped hairs. Flowers 2·5–2·7 cm;
calyx with star-shaped and glandular hairs; bracteoles linear to narrowly lanceolate.
Nutlets hairless. Rocks in the Med. region. GR.CR+Rhodes. Page 393. **1124b.** *P.
lanata* Willd. A handsome small shrub to 50 cm, with small densely-woolly leaves and
one or several whorls of orange-yellow flowers, and young branches often covered with
golden-yellow hairs. Flower clusters of 2–10 flowers; corolla 2–2·3 cm long; calyx
densely woolly, with minute teeth to 1 mm. Leaves 1·5–2·8 cm, oblong to rounded,
stalked. Nutlets hairy. Rocks, bushy places. CR. Pl. 39.

GALEOPSIS | Hemp-Nettle 7 species in our area. The most widely distributed are
the following:

**1126. *G. tetrahit* L. COMMON HEMP-NETTLE. Flowers pink; plant with rigid hairs; stem

395

swollen at the nodes. Woods, fields. YU.AL.BG. **(1126).** *G. speciosa* Miller Flowers yellow with large purple blotch on lip. Woods, bushy places, fields. YU.AL.BG. **1125b.** *G. ladanum* L. Flowers pink with yellow blotches; plant without rigid hairs, not swollen at the nodes. Stony ground, clearings in woods, uncultivated ground. YU.AL.BG.

LAMIUM | Dead-Nettle

Flowers white
1129. *L. album* L. WHITE DEAD-NETTLE. Hedges, waste places. YU.GR.BG. **1129a.** *L. moschatum* Miller Like 1129 with white flowers, but an annual, and corolla with conspicuous triangular lateral lobes (lateral lobes reduced to 2–3 small teeth in 1129); corolla tube shorter than calyx. Leaves and bracts often white-blotched. Shrubby places, rocks, waste ground in the Med. region. GR+Is.CR.BG? TR. Pl. 39.

Flowers pink or purple
(a) Annuals
1127. *L. amplexicaule* L. Readily distinguished by its stalkless, kidney-shaped, strongly toothed bracts which closely encircle the whorls of rosy-purple flowers. Corolla 14–25 mm long, the tube 10–14 mm long, straight. Cultivated ground, waste places. Widespread. **1128. *L. purpureum* L. RED DEAD-NETTLE. Cultivated ground, waste places. Widespread on mainland. **1128a.** *L. bifidum* Cyr. Flowers pink or purple or occasionally white, distinguished by the upper lip of the corolla which is two-lobed (entire in 1128). Corolla 12–25 mm long, the tube 12–14 mm and longer than the calyx; lateral lobes of corolla linear. A hairless or finely hairy annual 10–40 cm, with leaves mostly coarsely and irregularly cut. Waste places, amongst rocks. YU.GR.CR.BG.

(b) Perennials
1130. *L. maculatum* L. SPOTTED DEAD-NETTLE. Distinguished by its hairy anthers, curved corolla tube, and lateral lobe of corolla reduced to one tooth on each side. Flowers 2–3·5 cm long, pinkish-purple, brownish-purple, or rarely white. Leaves often white-blotched. Bushy and waste places, waysides, mountain woods. Widespread on mainland. **1131. *L. garganicum* L. LARGE RED DEAD-NETTLE. Flowers large 2·5–4 cm long but variable in size, pink to purple, rarely white. Distinguished from 1130 by its straight corolla tube much longer than the calyx, short triangular lateral lobes of corolla, and upper lip which is two-lobed or irregularly toothed (entire in 1130). Anthers hairy. Rocks in mountains. Widespread. Pl. 39. ****(1131).** *L. orvala* L. Like 1131 with large pink to dark purple, or rarely white, flowers 2·5–4·5 cm long, but with hairless anthers, and both upper and lower lip of flower irregularly toothed (lower lip notched in 1131). Calyx 12–20 mm. Bushy places. YU(west).

LAMIASTRUM (GALEOBDOLON) ****1132.** *L. galeobdolon* (L.) Ehrend. & Polatschek YELLOW ARCHANGEL. Woods, hedges. YU.AL.GR.BG.TR.

LEONURUS ****1133.** *L. cardiaca* L. MOTHERWORT. Leaves deeply three- to seven-lobed; corolla white or pale pink, 8–12 mm, longer than calyx. Waste places, hedges. YU.AL.GR.BG.TR. **(1133).** *L. marrubiastrum* L. Leaves coarsely toothed; corolla pale pink, 5–7 mm, shorter or scarcely longer than calyx. Damp bushy places, by rivers, waste ground. YU.GR.BG.

BALLOTA | Horehound 5 species in our area, including 1 endemic species.

1134. *B. nigra* L. BLACK HOREHOUND. Hedges, waste places. Widespread. **(1134). *B. acetabulosa* (L.) Bentham Readily distinguished by the much-enlarged, goblet-shaped fruiting calyx which has a wide-spreading, umbrella-like, conspicuously veined wing 1·5–2 cm across. Flowers pink and white, 15–18 mm, in many close whorls; calyx with undulate, smooth, or finely toothed margin. An erect grey-woolly stemmed perennial to 60 cm, with thick, woolly, rounded, toothed leaves 3–5 cm. Rocky places, waste ground in the Med. region. GR+Is.CR.TR? Page 403. **1134b.** *B. pseudodictamnus* (L.) Bentham Like (1134) but stems yellowish-woolly, and fruiting calyx funnel-shaped, the spreading wing 7–8 mm across, with irregularly rounded-toothed margin. Flowers pink and white, 14–15 mm, in separated whorls. Middle stem leaves 1·5–2 cm. Rocks, waste places. GR(Kithira).CR.

STACHYS | Woundwort About 48 species in our area, of which 25 are restricted to it. The following are widespread on the mainland: **1135.** *S. annua* (L.) L. ANNUAL YELLOW WOUNDWORT; ****1136.** *S. recta* L. PERENNIAL YELLOW WOUNDWORT Page 191; **1137.** *S. arvensis* (L.) L. FIELD WOUNDWORT; ****1138.** *S. palustris* L. MARSH WOUNDWORT; **1139.** *S. sylvatica* L. HEDGE WOUNDWORT; **(1139).** *S. alpina* L. ALPINE WOUNDWORT; ****1140.** *S. germanica* L. DOWNY WOUNDWORT; **1141.** *S. officinalis* (L.) Trevisan (*Betonica o.*) BETONY. In addition the following are distinctive:

Flowers white to yellow, sometimes with pink or purple markings
(a) *Plants densely white-woolly*
1136c. *S. chrysantha* Boiss. & Heldr. A white-woolly perennial 10–20 cm, with rounded white-woolly leaves, and crowded whorls of yellow woolly-haired flowers. Corolla 12–14 mm long, the upper lip 4–6 mm, the lower 6–8 mm; calyx 6–9 mm, densely woolly. Cliffs, screes. GR(south). Pl. 39. **1136d.** *S. candida* Bory & Chaub. Like 1136c but flowers white with purple spots, larger 15–18 mm long. Rocks in Taiyetos Mountains. GR(south). Pl. 39. **1136e.** *S. iva* Griseb. Distinguished by its rather large hairy yellow flowers 1·5–2 mm long, and its linear-lanceolate to elliptic leaves, the whole plant being densely white-woolly. Calyx 8–12 mm, white-woolly, teeth spiny-tipped, shorter to as long as the calyx tube. Stony slopes, cliffs on limestone. YU(south).GR(north).

(b) *Plants green, finely hairy, or greyish-hairy*
(i) *Flowers white (or occasionally pink)*
1136f. *S. spruneri* Boiss. A sparsely hairy, densely glandular erect perennial 15–30 cm, with whorls of 4–6 white flowers in a crowded ovoid spike. Corolla 1·6–1·8 cm long, upper lip 5 mm, lower lip 10 mm; calyx 9–12 mm, glandular, hairy, teeth about as long as the calyx tube. Upper leaves linear-lanceolate acute. Cliffs. GR(south-east). Pl. 40. **1136g.** *S. swainsonii* Bentham A procumbent to erect, branched rather variable silky- to woolly-haired perennial 10–20 cm, with crowded whorls of 2–10 white or pale-pink flowers. Corolla 1·4–1·8 cm long, upper lip 2–5 mm, lower lip 5–8 mm; calyx 6–10 mm, shaggy or woolly and glandular-hairy, teeth as long as or slightly shorter than the calyx tube. Leaves elliptic, obtuse. Rock crevices. GR(south and south-central). Page 96. **1136j.** *S. leucoglossa* Griseb. An erect, hairy perennial 30–50 cm, with a lax spike of whorls of usually 6 white or pale-pink flowers with purple markings. Corolla 16–18 mm hairy, upper lip 5 mm, lower lip 8–9 mm; calyx 6–9 mm, glandular, teeth about as long as the calyx tube. Lower leaves elliptic; upper lanceolate acute, usually toothed. Rocks. YU(south).GR.BG. Page 149.

1141a. *S. scardica* (Griseb.) Hayek Flowers white sometimes tinged with pink, in whorls forming a lax spike, distinguished by the star-shaped hairs on the upper part of the stem and the lower surface of the leaves. Flowers 15–18 mm long, upper lip entire; calyx

9–15 mm with teeth about half as long as the calyx tube. A hairy perennial 40–90 cm, with leaves densely grey-woolly beneath, the lower ovate-lanceolate toothed, the upper narrower. Mountain meadows, scrub. YU.AL.GR.BG.

(ii) *Flowers yellow*
1136h. *S. obliqua* Waldst. & Kit. An erect woolly-haired perennial 20–50 cm, with crowded whorls of 6–10 densely hairy, pale-yellow flowers. Corolla about 1·5 cm long, upper lip 6 mm, lower lip 8 mm; calyx about 10 mm with teeth two-thirds as long as the calyx tube. Leaves oblong-lanceolate, toothed, usually green above and grey-green beneath. Meadows, thickets. Widespread on mainland. ****1142.** *S. alopecuros* (L.) Bentham (*Betonica a.*) YELLOW BETONY. Flowers pale yellow in a dense spike-like cluster, the lowest whorls often distant. Upper lip of corolla bilobed; calyx hairy, net-veined, teeth much shorter than the tube. Leaves coarsely toothed, densely hairy. Rocky meadows in mountains. YU.AL.GR. Page 178.

(c) *Plant hairless*
1136i. *S. angustifolia* Bieb. An erect hairless perennial to 30–60 cm, woody at the base, with well-spaced pairs of yellow flowers tinged with pink. Corolla 1·2–1·6 cm long, hairy, upper lip 3–4 mm, lower lip 5–8 mm; calyx 6–8 mm, hairless or sparsely hairy, with teeth half as long as the calyx tube. Lower and middle leaves cut into linear lobes, the upper leaves linear, entire. Dry places. GR.BG.TR.

Flowers pink to purple (see also 1136d, 1136g, 1136j)
1140a. *S. cretica* L. A hairy, often woolly-haired perennial 20–80 cm, with oblong to ovate leaves which are softly hairy, grey-green above and densely woolly-haired, white or greyish beneath. Flowers spotted pink or purple, 1·5–2 cm long, in often spaced whorls; calyx 6–12 mm, densely white-haired, teeth spiny-tipped, about half as long as the calyx tube. Rocks in the Med. region. Widespread. Page 403. **1140b.** *S. spinosa* L. A tussock-forming perennial to 30 cm, with a spiny, silky-haired stem, and narrow densely silky-haired leaves which soon fall. Flowers pale pink, 4–6, in usually 2–3 crowded whorls. Corolla about 15 mm long, silky-haired; calyx 6–9 mm hairy, with stalkless glands. Stony places, cliffs. GR(Cyclades).CR+Karpathos. Page 39.

SALVIA | Sage 27 species, 4 of which are restricted to our area. The following 20 species are most likely to be encountered; the remainder are local.

1. Connective separating the 2 anthers of each stamen shorter than or equal to the filament; arms of connective more or less equal.	
2. Lvs. lobed or pinnate.	*ringens, triloba*
2. Lvs. simple, not lobed or cut.	
3. Calyx enlarged in fr.; bracts exceeding the calyx.	*pomifera*
3. Calyx not enlarged in fr.; bracts shorter than the calyx.	
4. Calyx 5–8 mm; stems with adpressed hairs.	*triloba*
4. Calyx 10–15 mm; stems with spreading hairs.	
5. Lvs. more or less wedge-shaped at the base (YU.AL.GR?; introd. BG.).	**1143.** *S. officinalis* L. Page 179
5. Lvs. heart-shaped or rounded at the base.	*grandiflora*

Connective separating the 2 anthers of
each stamen longer than the filament;
arms of connective very unequal.

5. Short arm of connective awl-shaped;
usually 15–30 fls. in each whorl.   *verticillata*

5. Short arm of connective enlarged at
apex or with a sterile cell; fls. usually
less than 15 in each whorl.

  7. Upper lip of calyx flat, not concave,
in fr.; calyx tubular or bell-shaped.

    8. Bracts longer than the corolla.   *viridis, sclarea*

    8. Bracts shorter than the corolla.

      9. Lvs. finely hairy, or with
bristly hairs.

        10. Fls. pink or blue-violet.   *viridis, forskaohlei*

        10. Fls. yellow with reddish-brown
markings (Widespread on
mainland).   **1146.** *S. glutinosa* L.

      9. Lvs., at least the basal, woolly-
haired, or densely covered with
short curled hairs beneath.

        11. Fls. 10–15 mm, the upper lip
weakly curved.   *aethiopis*

        11. Fls. 15–35 mm, the upper lip
strongly curved.   *candidissima, argentea*

  7. Upper lip of calyx concave and two-
ribbed in fr.; calyx bell-shaped.

    12. Stamens much longer than the
corolla; fls. white to cream (YU.BG.
(north-east)).   **1148d.** *S. austriaca* Jacq. Pl. 40.

    12. Stamens shorter or slightly longer
than the corolla; fls. blue or
violet-blue (rarely white in 1147).

      13. Inflorescence nodding at apex
before flowering; stems leafless
above.   *nutans*

      13. Inflorescence erect before
flowering; stems more or less
leafy above.

        14. Bracts usually violet or
purplish, overlapping in bud.   *amplexicaulis, nemorosa*

        14. Bracts green, not over-
lapping in bud.

          15. Lvs. more or less deeply
divided; fls. usually 6–10
mm (Widespread).   **(1148).** *S. verbenaca* L.

          15. Lvs. entire, undivided; fls.
11–30 mm (Widespread
on mainland).   **1147.** *S. pratensis* L.
     virgata

1143d. *S. grandiflora* Etlinger A small shrub or herb up to 1 m, like **1143.** *S. officinalis* but leaves larger, ovate to oblong, up to 6·5 cm wide, and leaf bases heart-shaped or rounded, long-stalked, and stems with spreading hairs. Flowers to 3·5 cm, lilac, pink, or violet-blue, rarely white, 4–10 in each whorl; calyx 1–1·5 cm, often reddish-purple, glandular-sticky. Dry places. YU.AL.GR.BG.TR. Page 149.   ****(1143).** *S. triloba* L. fil THREE-LOBED SAGE. Readily distinguished by the leaves, some of which have one or two pairs of small oval lobes at the base. Flowers lilac or pink, rarely white, 1·6–2·5 cm, in whorls of 2–6; calyx 5–8 mm, often purple. Leaves oblong-elliptic, wrinkled, greenish above, greyish-white beneath. A strongly aromatic shrub to 120 cm, with adpressed-hairy white stems. Bushy places, rocks in the Med. region. AL.GR + Is.CR. Page 38.

1143e. *S. pomifera* L. A much-branched greyish shrub to 1 m, with whorls of 2–4 large violet-blue flowers, with pale-blue lower lip, and large conspicuous bracts exceeding the calyx. Flowers about 3·5 cm; calyx enlarging in fruit, often reddish-purple, hairy, with stalkless glands. Leaves densely grey-haired, wrinkled, ovate with rounded or heart-shaped base, entire, stalked. Rocky places in the Med. region. GR(south).CR. Pl. 40.   **1143f.** *S. ringens* Sibth. & Sm. A herbaceous perennial to 60 cm, woody-based, with mostly basal leaves, which are wrinkled and deeply cut or pinnately lobed, with 3–6 pairs of small lateral segments and an oval terminal segment. Flowers violet to blue, large, up to 4 cm, in distant whorls of 2–4 flowers, on nearly leafless, usually unbranched stems. Calyx 1–1·7 cm, densely glandular-hairy. Dry places, in mountains. YU.AL.GR.BG. Pl. 40.

1144. *S. sclarea* L. CLARY. Distinguished by its lilac or pale-blue flowers and its conspicuous lilac or white bracts which are longer than the flowers. Corolla 2–3 cm; calyx with spiny teeth, glandular, hairy. A robust, sticky, strong-smelling biennial or perennial to 1 m, with simple ovate-heart-shaped, hairy leaves. Bushy places, rocks. Widespread on mainland. Page 165.   **(1144).** *S. argentea* L. SILVER SAGE. Distinguished by its large white flowers tinged with pink or yellow, with green bracts, borne in a large pyramidal inflorescence, and its white-woolly, irregularly toothed or lobed leaves. Corolla 1·5–3·5 cm; calyx glandular-hairy; bracts about half as long as calyx. A perennial 30–100 cm, with much-branched stems, glandular and nearly leafless above. Dry places. YU.AL.GR.BG.

(1144). *S. aethiopis* L. Like *S. argentea* but with smaller white flowers, 1–1·5 cm, and green or violet-tinged bracts slightly shorter than the calyx which is woolly-haired, not glandular-sticky. Stems not glandular above. Dry places. YU.GR.BG.TR.   **1144b. *S. candidissima* Vahl Like *S. argentea* with largish white flowers 2–3·5 cm, but bracts less than a third as long as the glandular-hairy calyx (⅓–⅔ as long in *S. argentea*). Leaves simple, ovate-heart-shaped, with white-felted hairs; stems little-branched, glandular above. Sandy, dry places in the Med. region. AL.GR.(west). Pl. 40.

1145. *S. verticillata* L. WHORLED CLARY. Distinguished from all other species by the numerous whorls, each of 15–30 blue-violet flowers, in elongated often branched clusters. Corolla 8–15 mm, the tube slightly longer than the calyx. Leaves simple, or with 1–2 pairs of small lateral lobes; stems to 80 cm. Waysides. Widespread on mainland. Pl. 40.   **1146a.** *S. forskaohlei* L. A glandular-bristly, erect, simple or little-branched perennial 25–100 cm, with coarsely lobed or toothed green leaves, and whorls of blue-violet flowers with white or yellow markings. Corolla 2–3 cm, lower lip down-curved, upper lip strongly curved, two-lobed; calyx 8–12 mm, teeth somewhat spiny. Mountain woods. GR.BG.TR.

(1147). *S. nemorosa* L. Distinguished by its long dense, leafless spikes of numerous whorls of small, usually violet-purple flowers with often conspicuous purplish bracts and calyx. Corolla 8–14 mm, rarely pink or white; calyx 6–7 mm, mostly with adpressed hairs. An erect, non-glandular, branched perennial 30–60 cm, with finely adpressed hairy stems and with oblong, toothed leaves which are more or less finely hairy. Stony ground, arid places, waysides. YU.AL.BG.TR? Page 164. **1147b.** *S. amplexicaulis* Lam. Like (1147) but stems with shaggy, spreading hairs. Flowers violet, 8–12 mm, in whorls of 6–8 (2–6 in (1147)); calyx with spreading hairs. Bushy places, waysides. Widespread on mainland. **1147c.** *S. virgata* Jacq. Differs from 1147. *S. pratensis* in having an inflorescence with long ascending lateral branches, and smaller flowers 1–2 cm (inflorescence short-branched; flowers 2–3 cm in 1147). Flowers blue-violet, rarely white, in remote whorls forming elongated clusters; calyx shaggy-haired; bracts green, less than half as long as the calyx. Leaves simple, ovate-heart-shaped, toothed, stem leaves few or many. Grassy places, dry places, rocks. YU(south). AL.GR.BG. TR.

1148c. *S. nutans* L. Distinguished by its slender unbranched leafless, wand-like spikes of violet flowers, nodding before coming into full flower. Corolla 12–16 mm; calyx 5–8 mm, glandular and hairy. Leaves mostly basal, long-stalked, ovate to oblong with heart-shaped base, irregularly toothed. Amongst herbaceous vegetation, dry places. YU? BG.TR. Page 403. ****1149.** *S. viridis* L. (*S. horminum*) RED-TOPPED SAGE. A hairy annual with erect usually unbranched stems to 50 cm, with whorls of pink or violet flowers, and often with a conspicuous crown of bicoloured bracts to the inflorescence which may be violet, blue, pink, or green. Corolla 14–18 mm; calyx 7–10 mm, finely hairy. Leaves ovate to oblong, simple, toothed, stalked. Rocky places in the Med. region. Widespread.

ZIZIPHORA Annuals. Calyx long-tubular with 13 strong veins, upper lip three-toothed, lower lip two-toothed with blunt tips. Corolla tube slender, lips more or less equal, the upper erect, the lower three-lobed. Fertile stamens 2.

1149b. *Z. capitata* L. A small, rough, simple or branched annual to 20 cm, with pink flowers in a terminal rounded head. Corolla 10–15 mm; calyx 6–11 mm, rough-haired; bracts broadly ovate, pointed, ciliate. Leaves 3–35 mm, linear to lanceolate, stalked, adpressed-hairy. Rocks, hills, cultivation. YU.AL.GR.BG.TR.

MELISSA 1150. *M. officinalis* L. BALM. A much-branched, usually hairy perennial 20–125 cm, with pale-yellow flowers becoming white or pinkish, in dense, one-sided, terminal and axillary whorls forming leafy spikes. Corolla 8–15 mm, two-lipped, twice as long as two-lipped calyx. Leaves 2–9 cm, oval, deeply toothed, sweetly lemon-scented or fetid. Shady places, hedges. Widespread.

SATUREJA | Savory 10 species, including 5 restricted to our area. The following are the most distinctive or widely distributed:

1. Calyx teeth very unequal, upper 3 about half as long as the lower pair (BG.). **1152e.** *S. rumelica* Velen.
1. Calyx teeth equal or slightly unequal.
  2. Annuals; at least lower calyx teeth much longer than calyx tube (YU.AL. GR.; introd. BG.). **(1152).** *S. hortensis* L.
  2. Dwarf shrubs or perennials woody at base; calyx teeth shorter than to slightly longer than calyx tube.

3. Bracteoles numerous, about as long
as calyx.   *thymbra*

3. Bracteoles absent or short, rarely
a few almost as long as calyx.

    4. Dwarf shrubs with spiny branches.   *spinosa*

    4. Perennial herbs, woody at base,
without spiny branches.

        5. Stem and lvs. with long rigid
hairs.   *parnassica*

        5. Stem and lvs. hairless, or with
short hairs only (Widespread   **1152. *S. montana* L. Page 178
on mainland).   *coerulea*

1152c. *S. thymbra* L. A much-branched, very aromatic, greyish shrublet 20–35 cm, with dense globular, many-flowered whorls of bright-pink or reddish-purple flowers, and bristly-haired bracts and bracteoles. Corolla 8–12 mm long; calyx 4–7 mm, with long spreading white hairs. Leaves 7–20 mm, oblong to obovate acute, glandular and bristly-haired. Dry hills, stony places in the Med. region. GR+Is.CR. Page 403. **1152d.** *S. spinosa* L. An easily recognized cushion shrublet to 20 cm, with the branches becoming spiny at the apex. Flowers white to pale lilac, in few two-flowered whorls. Corolla 5–8 mm long; calyx 2–4 mm, minutely downy. Leaves to 1 cm, narrowly obovate, rough. Rocks in mountains. CR+Samos.

1152f. *S. parnassica* Boiss. A dwarf plant of rocks, with a thick contorted woody stock, and leafy stems to 10 cm. Flowers in crowded whorls, whitish or pink; corolla 6–7 mm long; calyx 3–4 mm, rough-haired; bracts equalling or longer than whorls. Leaves 5–10 mm, obovate, the upper narrower, acute, rough-haired, densely glandular. GR(south and south-central). Page 114. **1152g.** *S. coerulea* Janka Flowers blue-violet, in a long rather dense spike of two- to four-flowered whorls, with linear-lanceolate bracts shorter or as long as the whorls (bracts exceeding the whorls in 1152 *S. montana*). Corolla 6–10 mm; calyx tube hairy, teeth hairless. Leaves 8–20 mm, linear-oblong, shining, ciliate, leathery. Rocks. BG. Page 148.

MICROMERIA 17 similar-looking species in our area, including 8 which are exclusive to it. Small botanical characters distinguish each species.

1153. *M. juliana* (L.) Reichenb. Flowers tiny purplish, in numerous dense stalkless whorls in long slender interrupted spikes about 5 mm wide. Corolla about 5 mm long; calyx hairless in throat, hairy outside, teeth awl-shaped, erect rigid, hairless. An aromatic, hairy, shrublet 10–40 cm, with many erect stems; upper leaves 3–8 mm, linear-lanceolate, the lower broader. Rocky places in the Med. region. Widespread. Page 38. **(1153).** *M. graeca* (L.)⁻ Reichenb. Like 1153 but whorls of flowers stalked and spreading outward from the axis and forming a much more lax, broader spike 1–1·5 cm wide. Corolla purplish, 6–8 mm or more long; calyx hairy in throat, adpressed-hairy outside. Leaves ovate to lanceolate, 5–12 mm. Rocky places in the Med. region. YU.GR+Is.CR.TR. Page 403.

1153a. *M. nervosa* (Desf.) Bentham Distinguished from (1153) by its calyx which is curved and has long, dense, spreading hairs, in contrast to the rest of the plant, which is shortly and less conspicuously hairy. Flowers purplish, 4–6 mm. Rocky and dry places in the Med. region. GR+Is.CR. **1153b.** *M. cristata* (Hampe) Griseb. A plant 5–20 cm, with many slender erect stems with short spreading hairs, and usually six-flowered whorls of purplish flowers. Corolla 4–6 mm; calyx 3–4 mm, with dense short hairs and teeth from

1. *Satureja thymbra* 1152c   **2.** *Origanum microphyllum* 1161b   **3.** *Micromeria graeca* (1153)   **4.** *Ballota acetabulosa* (1134)   **5.** *Salvia nutans* 1148c   **6.** *Origanum scabrum* 1161d   **7.** *Stachys cretica* 1140a

one-quarter to half as long as the tube. Leaves mostly overlapping, 5–12 mm, oblong or elliptical near the base, linear above, with inrolled margins. Rocky places. YU.AL.GR.BG.   **1153c.** *M. thymifolia* (Scop.) Fritsch A branched, leafy nearly hairless perennial 20–50 cm, with long lax terminal clusters of small two-lipped, white or violet flowers. Whorls short-stalked of 10–30 flowers; corolla 5–9 mm; calyx 3 mm, hairless. Leaves 5–20 mm, elliptic to ovate, usually remotely toothed, gland-dotted beneath. Rocky places. YU.AL. Page 178.

CALAMINTHA | **Calamint** Flowers in stalked clusters. Calyx thirteen-veined, two-lipped, hairy in throat, tube straight. A difficult genus.

1. Calyx 10–16 mm; corolla 25–40 mm long (Widespread on mainland). ****1154.** *C. grandiflora* (L.) Moench
1. Calyx 3–10 mm long; corolla less than 22 mm long.
  2. Calyx with long, dense spreading hairs; lvs. densely grey-felted (CR.). **1156a.** *C. cretica* (L.). Lam.
  2. Calyx with short, adpressed or curled hairs, or with sparse, long, spreading hairs.
    3. Lvs. densely grey-felted on both surfaces (GR(south).CR?). **1156b.** *C. incana* (Sibth. & Sm.) Boiss.
    3. Lvs. green with relatively few spreading hairs.
      4. Lower calyx teeth 2–4 mm, usually with dense, long hairs (YU.AL.BG.TR?). ****1155.** *C. sylvatica* Bromf.
      4. Lower calyx teeth 1–2 mm, without or with a few long hairs (YU.AL.GR.CR.BG.TR.). ****1156.** *C. nepeta* (L.) Savi

ACINOS Like *Calamintha* but flowers in whorls; calyx swollen at base and with a curved tube contracted in the middle, two-lipped, the lower teeth longer than the upper.

1. Fls. 12–20 mm long; perennials.
  2. Lv. blade mostly 1–2 times as long as wide, rounded or broadly wedge-shaped at base, obtuse or acute (Widespread). ****(1157).** *A. alpinus* (L.) Moench Page 18
  2. Lv. blade mostly 2–3 times as long as wide, narrowly wedge-shaped at base, long-pointed, or narrowly acute (YU.AL? GR.BG.TR?). **1157b.** *A. suaveolens* (Sibth. & Sm.) G. Don fil.
1. Fls. 7–12 mm long.
  3. Lv. blade not more than 1½ times as long as wide, very prominently veined beneath, abruptly fine-pointed (YU.GR.CR.BG.TR.). **1157a.** *A. rotundifolius* Pers.
  3. Lv. blade usually at least 1½ times as long as wide, not prominently veined, obtuse or acute.

4. Perennials with branched woody stock;
   fls. 10–12 mm long, usually longer
   than subtending bract (Widespread). **(1157). *A. alpinus* (L.) Moench Page 185
4. Annuals or short-lived perennials;
   fls. 7–12 mm long, not longer than
   subtending bract (Widespread on
   mainland).   1157. *A. arvensis* (Lam.) Dandy

CLINOPODIUM   **1158. *C. vulgare* L. WILD BASIL. Woods and bushy places. Widespread.

HYSSOPUS   **1159. *H. officinalis* L. HYSSOP. Rocky places, hills. YU.AL.BG.

ORIGANUM   Including *Majorana* and *Amaracus*. Bracts differing from leaves, often conspicuous and coloured. Whorls of flowers grouped together in rounded or ovoid spikelets. Calyx two-lipped, with 5 equal teeth, or entire and obliquely cut off without teeth, or one-lipped and deeply slit on ʼone side. Corolla two-lipped.

1. Calyx two-lipped.
   2. Upper lip of calyx entire or nearly so.
      3. Stem densely leafy above; lvs. woolly.   *dictamnus*
      3. Stem more or less leafless above;
         lvs. sparsely hairy (GR(Cyclades).CR.).   1161e. *O. tournefortii* Aiton
   2. Upper lip of calyx three-toothed.
      4. Tufted dwarf shrub; lvs. less
         than 5 mm, densely hairy
         (Karpathos).   1161f. *O. vetteri* Briq. & W. Barbey
      4. Rhizomatous perennial; lvs.
         more than 11 mm, hairless above.
         5. Lvs. heart-shaped at base,
            hairless or with sparse rough
            hairs beneath.   *scabrum*
         5. Lvs. rounded at base, sparsely
            hairy beneath (GR(south)).   1161g. *O. lirium* Halácsy
1. Calyx with 5 equal teeth, or cut off
   and entire at the apex, or one-lipped.
   6. Calyx entire at apex.   *microphyllum*
   6. Calyx with 5 equal teeth, or one-lipped
      and deeply slit.
      7. Calyx one-lipped, with a deep slit on
         one side.
         8. Stem covered with minute pimples,
            shaggy-haired.   *onites*
         8. Stem without pimples, hairless to
            hairy (introd. YU.).   **1161. *O. majorana* L. (*Majorana
                                                      hortensis*)
      7. Calyx with 5 equal teeth.
         9. Bracts 2–3 mm, usually green,
            densely glandular.   *heracleoticum*
         9. Bracts 4–5 mm, usually purple,
            glands sparse or absent (Widespread
            on mainland).   **1160. *O. vulgare* L.

1161a. *O. onites* L. A dwarf shrublet to 60 cm, with white flowers in compact ovoid spikelets 4–10 mm, densely grouped into flat-topped clusters 2–8 cm across. Leaves 5–22 mm, ovate, gland-spotted, woolly-haired. Dry hills, rocks, bushy places in the Med region. YU.GR+Is.CR. **1161b.** *O. microphyllum* (Bentham) Boiss. A multi-stemmed dwarf shrub to 50 cm, with numerous distinctly four-angled, slender purplish branches and few tiny white-woolly leaves. Flowers purple, in globular or obconical, woolly-haired spikelets 4–11 mm long, borne 1–3 at ends of leafless branches. Corolla 5 mm long; calyx 2 mm, hairless. Leaves oval, 4–8 mm, leaf stalk about 1 mm. Mountain rocks. CR. Page 403.

1161c. *O. dictamnus* L. (*Amaracus d.*) DITTANY. A very distinctive plant to 20 cm, with woolly-haired stems, many rounded white-woolly leaves, and oval or oblong spikelets of overlapping purple bracts subtending delicate pink, long-tubed flowers. Spikelets in groups of 3–10, arranged in opposite stalked pairs and forming a lax branched cluster; bracts 7–10 mm, longer than calyx. Leaves 13–25 mm, with conspicuous raised veins. Shady rocks in mountains. CR. Pl. 41. **1161d.** *O. scabrum* Boiss. & Heldr. (*Amaracus s.*) Stems erect to 45 cm, with paired, ovate to rounded, stalkless, nearly hairless leaves, and a nodding rounded spikelet of pink flowers with conspicuous purplish bracts. Flowers twice as long as calyx tube; bracts ovate, 8–10 mm, hairless. Leaves 1–3 cm. Rocks in mountains. GR(south). Page 403.

1160c. *O. heracleoticum* L. (*O. hirtum*) Like 1160 *O. vulgare* MARJORAM but distinguished by its swollen, densely-glandular, usually green bracts 2–3 mm. Flowers 4–5 mm long, usually white, rarely pink, in ovoid or oblong spikelets forming a lax or densely branched inflorescence. Bushy places, stony ground in the Med. region. YU.AL? GR+Is.CR.BG.TR.

THYMUS | Thyme A difficult genus with about 26 species in our area, of which 9 are restricted to it. Hybrids are frequent. Only the commonly encountered or distinctive species are described below:

Calyx tube dorsally flattened, with 20–2 veins
1162. *T. capitatus* (L.) Hoffmanns. & Link (*Coridothymus c.*) A compact, stiff, branched shrublet 20–50 cm, with terminal clusters of pink flowers in oblong cone-like heads, with broad, oval, closely overlapping, greenish and often reddish-tinged, ciliate bracts. Flowers up to 1 cm, upper lip two-lobed. Leaves linear, somewhat three-angled, gland-spotted, margin flat. Dry hills in the Med. region. Widespread, except BG. Page 39.

Calyx tube dorsally convex, with 10–13 veins
(a) Leaf margins distinctly inrolled
1163d. *T. cherlerioides* Vis. A tufted creeping perennial with long branches bearing axillary leaf clusters, and erect flowering stems 1–8 cm. Flowers pink, in globular heads with bracts similar to leaves, sometimes purple; corolla 5–7 mm, tube cylindrical; calyx 3–5 mm, tubular-bell-shaped, usually ciliate and purple. Leaves 4–10 mm, linear to linear-lanceolate, stalkless, hairless or velvety-haired, the margins more or less inrolled, ciliate. Mountains. YU.AL.GR.BG. Pl. 41. **1163e.** *T. teucrioides* Boiss. & Spruner A much-branched shrublet (recalling (1157) *Acinos alpinus*), with woolly, erect branches, axillary clusters of ovate hairless leaves to 1 cm, and whorls of 2–6 purple flowers, not forming a distinct head. Corolla 9–12 mm; calyx 4–6 mm, shorter than corolla tube. Mountains. AL.GR.

(b) Leaf margins not or only slightly inrolled
  (i) Bracts ovate, not similar to the leaves
1163f. *T. atticus* Celak. A shrublet with woody creeping shoots bearing axillary clusters

of leaves, and erect flowering stems to 15 cm, with axillary leaf clusters at least at the base. Flowers white or pink in a dense head; bracts broadly ovate, the lowermost 9 by 3 mm, leathery, straw-coloured; calyx 5–7 mm, with prominent veins and ciliate teeth. Leaves 1–2 cm, linear, ciliate at base. Rocky places in mountains. GR(east).BG.TR. Pl. 41. **1163g.** *T. striatus* Vahl Like 1163f but less robust with purplish bracts, smaller purplish calyx 3–5 mm, and flowers whitish, pink, or purple. Rocky places in mountains. YU.AL.GR.BG.TR? Page 179.

(*ii*) *At least lowermost bracts more or less similar to the leaves, although sometimes wider or coloured*

1164a. *T. sibthorpii* Bentham Flowering stem erect or ascending 10–20 cm, often branched, woody at the base, without creeping non-flowering stems. Flowers pale pink to red; inflorescence often branched. Bracts similar to leaves; calyx 2·5–4 mm, bell-shaped, greenish to straw-coloured, the upper lip longer than the tube. Leaves elliptic-lanceolate, 1–1·5 cm, leathery, with distinct veins, densely gland-dotted, leaf stalk short. In mountains. YU.AL?GR.BG.TR. **1164b.** *T. pannonicus* All. Like 1164a but leaves more herbaceous with indistinct veins, and sparsely gland-dotted; inflorescence rarely branched. Hills among herbaceous vegetation. YU.BG. Page 148.

1164c. *T. longicaulis* C. Presl A plant with long, somewhat woody, creeping branches, either non-flowering or with a terminal inflorescence. Flowering stems to 10 cm, borne in rows, each with a basal cluster of small leaves. Flowers purple; calyx 2·5–3·5 mm, bell-shaped, the tube shorter than the upper lip. Leaves linear-lanceolate to elliptic. A variable species. From the littoral to the mountains. Widespread on mainland. **1164d.** *T. praecox* Opiz Similar to 1164c but the leaves mostly obovate, broadly spathulate to rounded. Flowers purple; calyx 3–5 mm. Very variable in leaf shape and hairiness. Dry places, mountains. YU.AL.GR.BG.TR? **1165.** *T. pulegioides* L. LARGER WILD THYME. Grassy places, uncultivated ground. YU.AL.GR.BG.TR?

LYCOPUS | **Gipsy-wort** **1166. *L. europaeus* L. GIPSY-WORT. Watersides, damp places. Widespread on mainland. **(1166). *L. exaltatus* L. fil. Distinguished from 1166 by its leaves which are deeply cut to the apex of the blade (apex of blade toothed or shallowly lobed in 1166), and bracts longer, 6–9 mm (3–5 mm in 1166), and calyx teeth not spiny. Wet places. YU.AL.GR.BG.

MENTHA | **Mint** A difficult genus with many hybrids and cultivated forms which often become naturalized. About 11 species and distinctive hybrid species occur in our area. The following are fairly widely scattered on the mainland and some islands: **1167. *M. pulegium* L. PENNY-ROYAL; **1168.** *M. arvensis* L. CORN MINT; **1169. *M. aquatica* L. WATER MINT; **1170.** *M. spicata* L. SPEARMINT; **1171. *M. longifolia* (L.) Hudson HORSE-MINT.

1170a. *M. suaveolens* Ehrh. Differs from 1170 in having hairy leaves and stem, and leaves not more than 4·5 cm long by 2–4 cm wide, rounded, and wrinkled above, blunt. Flowers whitish or pink, in crowded whorls except at the base. Damp places, etc. AL.GR.CR.TR. **1170b.** *M. microphylla* C. Koch Differs from 1170a in having narrower ovate to lanceolate usually acute leaves, 5–20 mm wide. Flower whitish or pink, usually in distant whorls. Damp places, etc. Widespread. Pl. 40.

OCIMUM | **Basil** Flowers in whorls forming a terminal spike-like cluster. Calyx reflexed in fruit, two-lipped, with a broad undivided upper lip and a four-lobed lower lip; corolla two-lipped, the upper lip four-lobed, the lower undivided.

1172a. *O. americanum* L. (*O. basilicum*) SWEET BASIL. An erect, much-branched annual with oval, toothed, stalked leaves which are aromatic when crushed. Flowers white, forming a long spike-like cluster to 25 cm; corolla hairy, longer than calyx. Native of subtropical and tropical Asia and Africa; cultivated as a pot herb and for religious purposes.

SOLANACEAE | Nightshade Family

LYCIUM | Tea-Tree

1173. *L. europaeum* L. A spiny-branched shrub 1–4 m, with tubular violet, pink, or white flowers and oblanceolate leaves. Corolla 11–13 mm, lobes 3–4 mm; calyx 2–3 mm. Leaves 2–5 cm by 3–10 mm wide. Fruit a reddish berry. Hedges, rocky places in the Med. region. YU.AL.GR.CR? TR.   **(1173). *L. barbarum* L. Differs from 1173 in having leaves at least 1 cm wide, usually widest below the middle. Flowers purple, becoming brownish; corolla about 9 mm; stamens protruding; calyx 4 mm, two-lipped. Stems arched, with few weak spines. Native of China; planted and naturalized in YU.CR.BG.TR.   **1173b.** *L. intricatum* Boiss. Like 1173 with stout rigid spines, but flowers larger, blue-violet, purple, lilac, pink, or white. Corolla 13–18 mm, lobes 2–3 mm; stamens not protruding. Leaves smaller, 3–15 mm. Fruit orange-red or black. Rocky places. GR.CR.

ATROPA   ****1174.** *A. bella-donna* L. DEADLY NIGHTSHADE. Clearings in woods. Widespread on mainland.

SCOPOLIA   ****1175.** *S. carniolica* Jacq. Woods. YU(Croatia).

HYOSCYAMUS | Henbane   ****1176.** *H. niger* L. HENBANE. Flowers pale yellow with purple veins. All leaves stalkless. Waste places, uncultivated ground. Widespread on mainland.   ****1177.** *H. albus* L. WHITE HENBANE. A leafy, hairy, sticky, biennial or perennial 30–90 cm, with oval or rounded, deeply toothed, stalked leaves, and conspicuously veined yellowish-white flowers with greenish or violet throats, stalkless in a dense spike. Corolla 3 cm long, tubular-bell-shaped, glandular-hairy outside. Waste places, uncultivated ground in the Med. region. Widespread.   **(1177).** *H. aureus* L. GOLDEN HENBANE. Flowers bright golden-yellow with a deep violet throat, with protruding stamens and style, short-stalked in a lax cluster. Corolla 4–5 cm long. Leaves irregularly lobed and toothed, densely glandular-hairy, stalked. Walls in the Med. region. CR+Rhodes.

WITHANIA   Shrubs with entire leaves. Calyx and corolla bell-shaped, five-lobed; calyx enlarging in fruit; stamens 5. Fruit a berry surrounded by an inflated calyx.

1178a. *W. somnifera* (L.) Dunal A shrub 60–120 cm, with dull-green, oval leaves 3–10 cm, and several small, stalked, yellowish bell-shaped flowers in the leaf axils. Corolla about 5 mm; calyx enlarging to 1–2 cm, with star-shaped hairs. Berry shining red. Roadsides, scrub, waste places. GR.CR+Rhodes.

PHYSALIS   ****1178.** *P. alkekengi* L. BLADDER CHERRY. Flowers dull white. Fruit with inflated red or orange calyx encircling a red or orange berry. Woods, bushy places. YU.AL.GR.BG.TR.

NICANDRA **1179. *N. physalodes* (L.) Gaertner APPLE OF PERU. Flowers blue or violet. Fruit a brown berry surrounded by a large green lobed calyx. Waste places, waysides. Native of Peru; introduced to YU.BG.

CAPSICUM (1179). *C. annuum* L. RED PEPPER, CHILI, CAPSICUM. Flowers usually white, blotched. Fruit 5–15 cm, very variable in size and colour from red, to green, to black. Widely cultivated in the south of our area as a vegetable, and sometimes occurring as a casual.

SOLANUM Widespread native mainland species are: **1181. *S. dulcamara* L. BITTERSWEET, WOODY NIGHTSHADE; 1182. *S. nigrum* L. BLACK NIGHTSHADE; (1182). *S. luteum* Miller. Widely cultivated 'vegetables' are: (1181). *S. melongena* L. AUBERGINE, EGGPLANT; 1183. *S. tuberosum* L. POTATO.

1180.*S. sodomeum* L. Distinguished by its pale-violet flowers 2·5–3 cm across, and the large shiny, yellow to brown, globular fleshy fruit 2–3 cm. A shrubby perennial 50–300 cm, with many stout pale-yellow spines on stem and leaves; leaves deeply pinnately lobed. Native of Africa; introduced to YU.AL.GR.BG. **1180c. *S. laciniatum* Aiton A nearly hairless spineless shrub 1–1·5 m, with purplish stems, having both entire leaves and deeply cut leaves with narrow lobes, and an inflorescence of one or several stalkless purple flowers. Corolla 4·5 cm across, appearing ten-lobed. Leaves linear-lanceolate 10–15 cm. Fruit a yellow or orange ovoid berry, 2 by 1 cm. Native of Australia and New Zealand; cultivated for its foliage used as a source of steroid precursors.

1180d. *S. cornutum* Lam. Distinguished by its bright-yellow, broadly funnel-shaped flowers, and its numerous yellow spines on stems, leaves, and particularly on the calyx. Corolla 2–4 cm across, weakly ten-lobed; some anthers purplish; calyx enlarging in fruit. Leaves pinnately lobed, stalked. Fruit a dry, globular berry. Native of America; locally naturalized in GR.BG. Pl. 41. **1180e.** *S. elaeagnifolium* Cav. Distinguished by its clusters of conspicuous purple salver-shaped flowers, and its linear to oblanceolate leaves which are greyish-woolly when mature and usually have few reddish prickles. Corolla 2·5–3·5 cm across, five-lobed; anthers yellow. Fruit yellow, globular, dry, with spreading calyx. A perennial or dwarf shrub 30–50 cm, covered with star-shaped hairs. Native of S. America; locally naturalized in GR. Pl. 41.

LYCOPERSICON | **Tomato 1184.** *L. esculentum* Miller Widely cultivated for its fruit throughout our area. Native of South and Central America.

MANDRAGORA | **Mandrake** The following have been separated and recently described as distinct species. Leaves in a large rosette pressed to the ground, and flowers and fruits borne centrally on short individual stalks; tap-root large, fleshy.

1185b. *M. officinarum* L. (*M. acaulis*, *M. vernalis*) Flowers greenish-white, not more than 2·5 cm long, with narrowly triangular lobes. Fruit globular, yellow, the calyx shorter than the berry. Rocks in the Med. region. YU(west). **1185.** *M. autumnalis* Bertol. Flowers violet, larger 2·5–4 cm long, with broadly triangular lobes. Fruit ellipsoid, yellow to orange, with calyx usually longer. Rocky places in the Med. region. GR+Is.CR.

DATURA **1186. *D. stramonium* L. THORN-APPLE. Waste ground. Widespread on mainland+CR. **1186b.** *D. innoxia* Miller Flowers tubular, large 11–19 cm, white, sometimes violet-tinted. Fruit pendulous. A hairy annual 30–200 cm. Native of Central America; locally naturalized in waste places. Pl. 41.

NICOTIANA | Tobacco ****(1188).** *N. tabacum* L. LARGE TOBACCO. Flowers pink or purple; leaf stalks winged. Widely cultivated in many forms for tobacco. Native of S America. ****1188.** *N. rustica* L. SMALL TOBACCO. Flowers greenish-yellow; leaf stalks no winged. Much less commonly grown for tobacco than formerly. Native of N America. ****1187.** *N. glauca* R. C. Graham A slender, erect, hairless shrub 2–6 m, with blue-grey leaves, and lax terminal clusters of numerous long funnel-shaped yellow flowers. Corolla 3–4 cm long, lobes short. Leaves lanceolate to ovate, 5–25 cm. Native of S. America; often planted for ornament in the Med. region; sometimes naturalized.

SCROPHULARIACEAE | Figwort Family

VERBASCUM | Mullein A large genus with about 70 species in our area, 30 of which are restricted to it. A difficult genus with many similar-looking species; only the more frequently encountered or distinctive species are considered here. Plants of dry, stony waste ground in the lowlands, hills and mountains, except where otherwise described in the following account.

Each bract with a single flower in its axil

(a) Flower stalks absent or very short

1192a. *V. ovalifolium* Sims Distinguished by its stalkless yellow flowers 2–4 cm across the petals with translucent glands; basal leaves ovate to oblong-lanceolate, with wavy margins, hairless or sparsely hairy above, and densely hairy beneath. Bracts rounded, 10–14 mm, with a fine point. Stamens with yellow filament hairs. Inflorescence simple or branched; plant yellowish- or grey-hairy. GR.BG.TR. **1192c.** *V. purpureum* (Janka) Huber-Morath Distinguished from 1192a by its stamens which have kidney-shaped anthers (anthers decurrent in 1192a). A laxly branched biennial with yellow flowers 2–3 cm across, subtended by bracts and bracteoles. Bracts 1–1·5 cm, lanceolate. Stamens with pale-yellow filament hairs. Basal leaves pinnately cut into narrow lobes, or toothed, hairless above, with cobweb hairs beneath. BG(east).TR. Page 148. **1192b.** *V. spinosum* L. A much-branched shrub to 50 cm, distinguished by the branches which end in a spine, and by the rather small solitary yellow flowers 10–18 mm across, with short lilac filament hairs. Leaves greyish-white, oblong-lanceolate, irregularly toothed or lobed, 1·5–5 cm. Mountains, stony hillsides. CR. Page 71.

(b) Flower stalks 5–25 mm or more

1192. *V. blattaria* L. MOTH MULLEIN. Flowers yellow, rarely white. Widespread on mainland. **(1192).** *V. phoeniceum* L. PURPLE MULLEIN. Flowers usually violet. Widespread on mainland. **1196d.** *V. arcturus* L. (*Celsia a.*) A woody-based plant of calcareous cliffs of Crete, with numerous stems, whitish-felted, lobed leaves, and yellow flowers with 4 fertile stamens with violet filament hairs. Flowers 2·5–3 cm across; inflorescence glandular-hairy. Leaves with a much larger terminal lobe and 2–4 smaller lateral lobes. CR. Page 66. **1196e.** *V. acaule* (Bory & Chaub.) O. Kuntze (*Celsia a.*) Quite unlike any other mullein, a stemless, rosette perennial of high mountains. Flowers yellow, 16–22 cm across, in a stemless cluster, borne on flower stalks 2·5–9 cm long. Stamens 4, with violet filament hairs. Leaves oblong-ovate, coarsely toothed. GR(south). Pl. 42.

At least the lower bracts with clusters of several flowers in their axils

(a) Anthers partially or completely elongated down the filament

(i) Stamens with filament hairs white or yellow

1194. *V. phlomoides* L. A greyish or yellowish, densely woolly-haired biennial

30–120 cm, with a rather lax usually unbranched spike of bright-yellow flowers each 2–5·5 cm across, in short-stalked clusters. Bracts 9–15 mm, longer than flower stalks. Filaments of upper stamens with white or yellow hairs, the lower hairless. Upper leaves ovate-lanceolate, long-pointed, toothed, not or very shortly decurrent. Widespread on mainland. **1194a.** *V. densiflorum* Bertol. Like 1194 but stem leaves long-decurrent, and bracts 1·5–4 cm, with a long point, decurrent. Widespread on mainland. **(1194).** *V. longifolium* Ten. Like 1194 in having the filaments of the lower stamens usually hairless, but bracts linear-lanceolate (ovate-lanceolate in 1194). A densely white- or yellowish-woolly plant 50–150 cm, with a usually simple inflorescence, with yellow flowers 2·5–3·5 cm across, on unequal flower stalks up to 16 mm long. Filaments of upper stamens with dense usually white, or yellowish, hairs. Upper stem leaves not or slightly decurrent. YU.AL.GR.BG. Pl. 42.

(ii) Stamens with filament hairs violet
1194b. *V. eriophorum* Godron A very tall whitish, tufted-haired biennial 1–2·5 m, which at length often becomes green and hairless, with a lax simple or branched inflorescence of large yellow flowers 3–4 cm across, with unequal stamens. Inflorescence axis glandular and with densely tufted white hairs; bracts 8–18 mm, linear-lanceolate. Upper stem leaves broadly ovate or rounded, usually crowded. Basal leaves ovate-oblong, entire or toothed, leaf stalk short or absent. YU.GR.BG. Pl. 42. **1194c.** *V. epixanthinum* Boiss & Heldr. Differs from 1194b in upper stem leaves few and small, lanceolate to ovate-lanceolate, and inflorescence axis glandular but finely hairy, becoming hairless. Flowers 2·5–3·5 cm across; 5 stamens with shortly decurrent anthers and violet filament hairs. Bracts 8–18 mm, linear. Basal leaves stalked, oblong-elliptic, entire or weakly toothed, variable in colour and persistence of hairs. Mountains. GR. Pl. 42.

(b) Anthers kidney-shaped, attached at the middle
(i) Basal leaves lobed
1191. *V. sinuatum* L. Distinguished by its distinctive basal rosette leaves which are deeply pinnately lobed, and densely covered with grey or yellowish woolly hairs. Inflorescence 50–100 cm, widely branched with hairy axis, and flowers borne in clusters on twiggy hairless branches. Corolla 1·5–3 cm across; hairs on filaments violet; calyx 2–4 mm. Bracts 3–8 mm, heart-shaped-triangular. Widespread. **1191b.** *V. nobile* Velen. Like 1191 with a lax branched inflorescence, but with hairless axis. Flowers in clusters of 1–3; corolla 2–3·5 cm across; hairs on filaments yellowish. Bracts ovate-lanceolate, 1–2 mm. Basal leaves lobed, grey- or yellowish-woolly at least beneath. GR(north-east).BG(south and east). **(1195).** *V. undulatum* Lam. Differs from 1191 in having usually several simple or little-branched stems with very lax spike-like clusters; larger yellow flowers 2·5–5 cm across; longer bracts 6–12 mm; and hairs on filaments white. Basal leaves entire to pinnately lobed, strongly undulate, stalked, densely white, grey, or yellow woolly-haired. YU.AL.GR+Is. **1195a.** *V. pinnatifidum* Vahl Plant at first woolly-haired but becoming hairless, with several stems 30–50 cm, branched near the base into numerous lax spike-like clusters. Flowers 2·5–3 cm across, stalkless; hairs on filaments white or yellow; bracts 9–15 mm, triangular-ovate. Basal leaves deeply toothed to deeply lobed, the lobes often further toothed or lobed. Maritime rocks and sands; north Aegean region. GR+Is.TR.

(ii) Basal leaves not lobed (see also (1195), 1195a)
(x) Stamens with filament hairs violet
1190. *V. nigrum* L. DARK MULLEIN. Woods, bushy places. YU.AL.GR.BG. **1190a. *V. chaixii* Vill. Like 1190 but basal leaves with blades cut off or very shortly wedge-shaped at base (heart-shaped in 1190), and flower stalks as long as calyx (2–3 times as long in 1190). Inflorescence 50–100 cm, much-branched, with wide-spreading branches; flowers

15–22 mm across; bracts 2–5 mm, linear-lanceolate. Mountain woods, thickets. YU.GR.BG. **1190b.** *V. lanatum* Schrader Basal leaves usually heart-shaped at base, singly or doubly coarsely toothed, green and sparsely hairy above, grey-woolly beneath. Inflorescence unbranched; flowers 16–28 mm across; bracts 6–15 mm, linear, hairless; calyx hairless. Mountain woods. YU.BG. **1190c.** *V. glabratum* Friv. Like 1190b with bracts and calyx hairless, but inflorescence branched. Basal leaves more or less densely white-woolly beneath, long-stalked with a white-woolly leaf stalk 15–22 cm. Woods. YU.AL.GR(north and central).BG. Pl. 42.

(xx) Stamens with filament hairs white or yellow
1195. *V. lychnitis* L. WHITE MULLEIN. Flowers white or yellow. YU.AL.BG. **(1195).** *V. speciosum* Schrader A perennial to 2 m, with freely branched inflorescence, and entire, short-stalked, greyish, white, or yellowish thickly hairy basal leaves. Flowers yellow, 1·3–3 cm across; lower bracts ovate-lanceolate 8–15 mm, the upper lanceolate. Leaves entire, oblong-lanceolate to obovate 12–40 cm. Widespread on mainland. Pl. 42. **1195b.** *V. graecum* Boiss. A densely white-haired biennial 40–150 cm, which has tufts of soft white hairs but becoming nearly hairless above, with a lax branched inflorescence of yellow flowers each 1·5–3 cm across. Bracts mostly 2–4 mm, linear to linear-lanceolate; calyx 2–5 mm. Leaves ovate to lanceolate, entire or toothed, short- or long-stalked. YU(south).AL? GR.TR. Pl. 42. **1195c.** *V. delpicum* Boiss. & Heldr. Inflorescence usually branched, 50–150 cm, white-woolly; flowers 1·5–2 cm across; filament hairs white. Bracts 4–10 mm, lanceolate. Basal leaves ovate, with a rounded or heart-shaped base, usually toothed, sparsely hairy above, white-woolly beneath. GR (east). Pl. 42. **1195d.** *V. mallophorum* Boiss. & Heldr. A white-woolly biennial with a much-branched rather dense inflorescence of yellow flowers 1·5–3 cm across, with violet filament hairs and kidney-shaped anthers. Bracts linear 4–7 mm. Basal leaves entire oblong-ovate, 15–45 cm; stem leaves not decurrent. Hills, mountains. AL.GR (central and south). Pl. 42.

ANTIRRHINUM | Snapdragon
**1197. *A. majus* L. SNAPDRAGON. Native of S.W. Europe. Widely grown for ornament; naturalized on walls and cliffs in YU.AL.GR.CR+Rhodes.TR.

MISOPATES ****1199.** *M. orontium* (L.) Rafin. (*Antirrhinum o.*) WEASEL'S SNOUT. Cultivated ground, vineyards, waysides. Widespread.

LINARIA | Toadflax A difficult genus with about 15 species in our area, including 3 endemic species.

Flowers predominantly purple
(a) Spur less than half length of corolla
1201. *L. arvensis* (L.) Desf. Fields, sandy places. YU.AL.GR.BG. **1201l.** *L. micrantha* (Cav.) Hoffmanns. & Link Like 1201 but lilac-blue flowers smaller 3–5 mm, including the straight or slightly curved spur which is 1 mm or less (spur strongly curved in 1201). Waste places, cultivation. YU.GR.CR.

(b) Spur more than half length of corolla
1203. *L. pelisseriana* (L.) Miller Flowers bright violet with a white throat-boss, and a straight slender violet spur about as long as the rest of the corolla. Corolla including spur 1·5–2 cm; calyx hairless, lobes linear acute, white-margined; bracts oval, short. A hairless glaucous, unbranched erect annual 15–50 cm. Fields, stony and grassy places. Widespread. Pl. 43. ****1204.** *L. alpina* (L.) Miller ALPINE TOADFLAX. Flowers violet with

a yellow throat-boss; usually a spreading plant 5–25 cm. Rocks and screes in mountains. YU.AL.GR.

Flowers predominantly yellow

1201k. *L. simplex* (Willd.) DC. A glaucous erect annual 10–50 cm, like 1201 with a dense glandular-hairy inflorescence, but flowers pale yellow, often veined with violet. Corolla including spur 5–9 mm; spur 2–4 mm, straight. Leaves linear to linear-lanceolate. Dry open places. Widespread. **1205.** *L. vulgaris* Miller COMMON TOADFLAX. Corolla including spur 2·5–3·3 cm, spur 10–13 mm. Seeds disk-like. Hedges, banks, rocks, waste places. Widespread on mainland. **(1205).** *L. angustissima* (Loisel.) Borbás ITALIAN TOADFLAX. Differs from 1205 in having dense slender spikes of smaller pale lemon-yellow flowers with the spur at least as long as the lower lip (shorter in 1205). Corolla including spur 1·5–2 cm, spur 7–10 mm. Leaves usually 1–2 mm wide, very glaucous. Rocky places. YU.AL.GR? BG.

(1205). *L. genistifolia* (L.) Miller (incl. *L. dalmatica*) A hairless perennial 20–100 cm, with lax or dense, usually branched spike-like clusters of yellow flowers with leaves reaching to the base of the clusters. Corolla including spur 1·5–5 cm, spur 4–25 mm; calyx 2–12 mm. Leaves linear to ovate, often more or less clasping, alternate, glaucous. Seeds not disk-like. A variable species. Sandy, stony places. Widespread on mainland. Pl. 43. **1205b. *L. peloponnesiaca* Boiss. & Heldr. Distinguished from (1205) *L. genistifolia* by the flowering stems which are more or less leafless some distance below the short, very dense inflorescence, and by the leaves which are not more than 3 mm wide. Flowers pale yellow, 1·3–2 cm including spur; spur 7–8 mm; calyx 2–4 mm. An erect, sparingly branched perennial 25–70 cm, with usually a glandular-hairy inflorescence, and linear leaves. Seeds not disk-like. Rocks, in mountains. YU.AL.GR. Pl. 43.

Flowers white or conspicuously multicoloured

1207. *L. chalepensis* (L.) Miller WHITE-TOADFLAX. Distinguished from all other species by its small, pure-white flowers with a long slender curved spur. Corolla including spur 12–16 mm, spur 8–11 mm. A slender erect, usually unbranched, hairless annual 20–40 cm, with a slender, lax inflorescence, and linear leaves. Grassy places, stony ground in the Med. region. Widespread. ****1208.** *L. triphylla* (L.) Miller THREE-LEAVED TOADFLAX. Unmistakable with its compact terminal heads of large tricoloured flowers, and its glaucous oval leaves often in whorls of 3. Corolla including spur 2–3 cm, white with an orange throat-boss, and a curved pointed, violet spur 8–11 mm. Erect usually unbranched annual 10–45 cm. Fields, cultivated ground, vineyards in the Med. region. YU.GR+Cyclades.CR.

KICKXIA | **Fluellen** 3 species in our area.

CYMBALARIA ****1210.** *C. muralis* P. Gaertner, B. Meyer & Scherb. IVY-LEAVED TOADFLAX. Shady rocks, walls, YU(west); introd. GR.CR.BG. Page 190. **1210b.** *C. longipes* (Boiss. & Heldr.) A. Cheval. Distinguished from 1210 by its longer spur 4–5 mm which is much longer than the calyx (spur 2–3 mm, about as long as calyx in 1210), and its glandular-hairy fruit. Corolla 9–15 mm, lilac or violet with a yellow throat-boss. A slender, trailing plant with stem and leaves hairless at maturity, and small rounded leaves not more than 1·5 cm long by 2 cm wide. Rocks and stony places by the sea. GR+Is.CR. Pl. 43. **1210c.** *C. microcalyx* (Boiss.) Wettst. Distinguished from 1210 by the soft shaggy hairs on the leaves and stems. Flowers 9–13 mm long; violet with a yellow throat-boss; spur 1–3 mm; calyx 1–2 mm. Leaves 1 by 1·5 cm, kidney-shaped to semicircular, entire or usually with 3 shallow rounded lobes. Fruit hairy. Rocks in mountains. YU.AL.GR.CR+Rhodes, Karpathos.

413

CHAENORHINUM   1211e. *C. minus* (L.) Lange (*Linaria m.*) SMALL TOADFLAX. Stony, sandy places. Widespread on mainland.   **1211f.** *C. rubrifolium* (DC.) Fourr. An erect annual to 20 cm, with a rosette of ovate basal leaves which are red beneath, smaller green stem leaves, and blue or violet short-spurred flowers. Corolla including spur 1–1·5 cm, spur 1–4 mm, throat-boss yellow; calyx 4–7 mm. Stony places in the Med. region. GR.

SCROPHULARIA | Figwort   14 similar-looking species, 4 of which are restricted to our area.

Leaves undivided, margin entire or toothed
(a) Calyx lobes without a membraneous margin
1212. *S. peregrina* L. NETTLE-LEAVED FIGWORT. Corolla dark red, two-lipped. Cultivated ground, bushy places. YU.AL.GR+Is.CR. Pl. 43.   **1213.** *S. vernalis* L. YELLOW FIGWORT. Corolla yellow, not two-lipped, lobes equal. Woods, shady bushy places. YU.

(b Calyx lobes with a membraneous margin
   (i) Stems winged
1214. *S. nodosa* L. FIGWORT. Damp woods, hedgebanks. YU.AL.GR.BG.TR?   **(1214).** *S. umbrosa* Dumort. (*S. alata*) Riversides, damp places. Widespread on mainland.

   (ii) Stems not winged
1214a. *S. scopolii* Hoppe Flowers greenish 7–12 mm, with a purplish-brown upper lip, in stalked clusters of 4–7; calyx lobes with a broad papery margin. A more or less hairy perennial to 1 m, with ovate to oblong, doubly toothed leaves. Woods and damp places, usually in mountains. Widespread on mainland +E. Aegean Is.

Leaves once or twice cut into narrow, toothed or lobed segments
(a) Sterile stamen linear, lanceolate, or absent
1216. *S. canina* L. Subsp. *bicolor* (Sibth. & Sm.) W. Greuter which is widespread in our area has conspicuous whitish margins to the blackish-purple corolla lobes. Subsp. *canina* has uniformly dark reddish-purple flowers. Dry stony and sandy places. Widespread.

(b) Sterile stamen rounded or kidney-shaped
(1216). *S. lucida* L. Rocky places. GR+Is.CR.   **1216b. *S. heterophylla* Willd. A hairless or glandular, often glaucous perennial 10–70 cm, with dark reddish-purple to greenish flowers, like 1216 *S. canina* but the sterile stamen-flap rounded to kidneyshaped (linear-lanceolate or absent in 1216). Corolla 6–9 mm long. Leaves up to 15 cm, deeply toothed to twice-divided into often fleshy rounded lobes. Rocks, walls in the Med. region. Widespread, except TR. Page 416.   **1216c.** *S. myriophylla* Boiss. & Heldr. A biennial like (1216) *S. lucida* with greenish-brown flowers 4–9 mm, but leaves 4 by 2 cm, all twice-cut into very small linear lobes (leaves 9 by 5 cm, lobes ovate in (1216)). Mountain rocks. GR(south). Page 92.

MIMULUS   ****1217.** *M. guttatus* DC. MONKEY-FLOWER. Native of N. America. Grown for ornament; naturalized by water in YU.BG.

GRATIOLA   ****1218.** *G. officinalis* L. GRATIOLE. Wet meadows, riversides. Widespread on mainland.

LINDERNIA   Hairless annuals with solitary, irregular flowers in the leaf axils. Corolla two-lipped, the upper lip small, flat, erect, two-lobed, the lower spreading, three-lobed. Stamens 4. **1218b.** *L. procumbens* (Krocker) Philcox (*L. pyxidaria*) A small spreading annual 5–18 cm, with entire oval-oblong leaves to 2 cm, and small solitary, pale-violet or pink flowers on slender stalks usually longer than the subtending leaves. Corolla usually 2–4 mm long, closed at the mouth, and shorter than calyx, rarely longer. Muddy river banks, wet sands. YU.BG.

VERONICA | Speedwell About 45 species occur in our area, including 8 species which are endemic. The following are distinctive or frequently encountered:

Flowers solitary in axils of leaves
Widely distributed annuals in the central European climatic region are: **1222.** *V. cymbalaria* Bodard PALE SPEEDWELL. Pl. 43; **(1222).** *V. hederifolia* L. IVY SPEEDWELL; ****1223.** *V. persica* Poiret BUXBAUM'S SPEEDWELL; **(1223).** *V. polita* Fries GREY SPEEDWELL.

Flowers in terminal clusters
(a) Perennials
1219. *V. spicata* L. SPIKED SPEEDWELL. Distinguished by its very numerous blue flowers in a dense cylindrical, terminal spike up to 30 cm long. Corolla 4–8 mm across; calyx lobes oval-elliptic, blunt; flower stalks usually less than 1 mm, much shorter than the bracts. Leaves variable, linear to ovate-lanceolate, margin entire or rounded-toothed, densely white-woolly or sparsely hairy. Dry grasslands, rocky slopes. YU.AL.BG.TR. Page 190. **(1219).** *V. paniculata* L. (*V. spuria*) Like 1219 but a taller and more robust perennial to 80 cm, the leaves with saw-toothed margins. Flowering spikes often laxer, branched below; flowers larger 7–14 mm across, blue; flower stalks often as long or longer than the calyx and the bracts. Grassy, rocky places. YU.BG.

1219a. *V. bellidioides* L. Distinguished by its deep violet-blue flowers borne in short rounded terminal clusters, on erect, sparsely leafy stems. Corolla 9–10 mm across; flower stalks 2–6 mm. Inflorescence elongating in fruit. A tufted, greyish-haired perennial 5–20 cm, with oblong-obovate, mostly toothed basal leaves in a rosette; stock creeping. Dry alpine pastures. YU.BG. Page 416. **1220.** *V. serpyllifolia* L. THYME-LEAVED SPEEDWELL. Flowers whitish-lilac to bright blue, 6–10 mm across, borne in a spike. Stems creeping; leaves oval 1–2·5 cm. Grasslands, in mountains. Widespread on mainland.

1220d. *V. saturejoides* Vis. A creeping woody-based perennial of rocks and mountain screes, with numerous crowded, rather fleshy rounded leaves 6–9 mm, and a terminal head of bright-blue flowers. Corolla about 7 mm across; stamens longer; inflorescence with shaggy hairs; calyx five-lobed, with one very short lobe. YU.AL.BG. **1220e.** *V. erinoides* Boiss. & Spruner Distinguished from 1220d by its lanceolate, conspicuously toothed leaves 4–8 mm; its lilac flowers in a head of 3–5 (flowers 6–12 in 1220d), with stamens not longer than the corolla. Inflorescence glandular-hairy. Mountain screes, stony slopes. GR(south). Pl. 43.

(b) Annuals
Widespread on the mainland, particularly in the central European climatic zone are: **(1220).** *V. arvensis* L. WALL SPEEDWELL; ****1221.** *V. triphyllos* L. FINGERED SPEEDWELL.

1221a. *V. acinifolia* L. A small erect, glandular-hairy annual 5–15 cm, with tiny pale-blue flowers 2–3 mm across, in a terminal dense, then lax, cluster. Leafy bracts shorter than or equalling the flower stalks. Leaves 4–10 mm, ovate. Fruit glandular-hairy. Cultivated ground, damp grassland. Widespread on mainland+CR. **1221b.** *V. glauca* Sibth. & Sm. (incl. *V. peloponnesiaca*) Flowers deep blue, in long terminal clusters, with flower stalks very slender, more or less spreading and much longer than the bracts, and

1. *Veronica bellidioides* 1219a **2.** *Melampyrum barbatum* 1258a **3.** *Scrophularia heterophylla* 1216b **4.** *Odontites linkii* 1240c **5.** *Digitalis viridiflora* 1233b

at least 4 times as long as the calyx. Flowers 1–1·5 cm across. An erect, branched annual to 20 cm, with crisped hairs, and short-stalked, triangular-ovate leaves. Fruit usually hairless. Cultivated ground, dry grassland in the Med. region. AL.GR.CR?

Flowers in axillary clusters, the main stem usually terminating in leaves
Widely distributed on the mainland are the following: ****1225.** *V. beccabunga* L. BROOK-LIME; **(1225).** *V. anagallis-aquatica* L. WATER-SPEEDWELL; **1227.** *V. officinalis* L. COMMON SPEEDWELL; **(1227).** *V. montana* L. WOOD SPEEDWELL; **1228.** *V. chamaedrys* L. GERMANDER SPEEDWELL.

**1229. *V. austriaca* L. (*V. latifolia*) An attractive but very variable perennial, with usually dense lateral clusters of bright-blue flowers, borne on leafless stalks from the axils of the upper leaves. Flowers 1–1·5 cm across; calyx distinctive in having usually 5 very unequal narrow lobes or less commonly with only 4 lobes. Leaves very variable, from entire to deeply once- or twice-cut into narrow segments. Subsp. *austriaca* has leaves once or twice cut into linear segments, and long flower clusters. Widespread on mainland. Subsp. *vahlii* (Gaudin) D. A. Webb has entire lanceolate to oval-oblong, toothed leaves 1·5–3 cm, and short flower clusters. YU. Page 169. Subsp. *teucrium* (L.) D. A. Webb is a robust plant 30–100 cm, with larger toothed leaves 2–7 cm, half-clasping the stem, and with long flower clusters. Dry places, meadows, mountain grasslands. Widespread on mainland.

1229a. *V. urticifolia* Jacq. Like 1229 but calyx with 4 elliptic-oblong, more or less equal lobes, and flowers pale blue or rose-lilac, smaller about 7 mm across. Individual flower stalks longer than leafy bracts. Leaves 4–8 cm, triangular-ovate, saw-toothed; stem hairy all round. Fruit 3–4 mm; fruit stalk sharply upturned below fruit. Damp rocks in woods in mountains. YU.AL.GR.BG. **1229b.** *V. prostrata* L. Like 1229 but non-flowering stems procumbent; calyx hairless, smaller 2–4 mm (4–6 mm, hairless or hairy in 1229); corolla pale blue, 6–8 mm across. Flowering stems to 25 cm, ascending; leaves toothed, densely hairy, oblong-lanceolate to ovate, margins often inrolled. Dry grass-lands. Widespread on mainland.

1229c. *V. rhodopaea* (Velen.) Stoj. & Stefanov A creeping perennial with elliptic to oblong stalkless leaves 8–13 mm, entire or with minute rounded lobes, and with inrolled margins. Flowers deep blue, 1–4 in dense rounded clusters; corolla 9 mm across; calyx lobes 4, ciliate, unequal. Mountains, grasslands. BG(south). **1229d.** *V. thymifolia* Sibth. & Sm. A creeping matted, woody-based, crisp-haired perennial of mountain rocks and screes. Flowering clusters 1–2, axillary, few-flowered; flowers blue, lilac, or pink, 7–8 cm across; calyx with 4 blunt lobes. Leaves 4–6 mm, oblong entire, with inrolled margins. Fruit hairy. GR(south).CR. Page 70.

PAEDEROTA Like *Veronica* but corolla tubular, two-lipped, the upper lip entire, the lower three-lobed. Leaves all opposite.

1229e. *P. lutea* Scop. (*Veronica l.*), Flowers yellow, in a dense leafy spike-like cluster; corolla 10–13 mm, corolla tube longer than the lips. A leafy perennial 7–20 cm or more, with narrowly ovate to lanceolate, sharp-toothed leaves. Limestone crevices. YU(west).

WULFENIA Flowers tubular-bell-shaped, shortly two-lipped, the upper lip entire or two-lobed, the lower shortly three-lobed; stamens 2. Leaves alternate, or basal. Fruit with 4 valves, many-seeded. 2 species in our area.

1229f. *W. carinthiaca* Jacq. Flowers dull blue, in a dense terminal one-sided cluster 6–10 cm long, borne on an erect almost leafless stem 20–40 cm. Corolla 12–15 mm long.

Leaves nearly all basal, 8–17 cm, oblanceolate to obovate, toothed. Damp grassland in mountains. YU(Montenegro).AL?

DIGITALIS | Foxglove

Middle lobe of lower lip much longer than other lobes
(a) Axis of inflorescence hairless and without glands
1230. *D. ferruginea* L. RUSTY FOXGLOVE. A tall wand-like plant to 120 cm, with a dense many-flowered spike of almost globular yellowish- or reddish-brown flowers, with a network of darker brown veins, and a long protruding lip. Corolla 1·5–3·5 cm; calyx lobes blunt, with wide papery margins. Leaves oblong to lanceolate, hairless or slightly hairy beneath. Woods, bushy places. Widespread on mainland. Pl. 44. **1231. *D. laevigata* Waldst. & Kit. Differs from 1230 in having acute calyx lobes with very narrow papery margins, or papery margin absent. Flowers yellowish with brown or purple veins or markings; corolla variable, from 1·5–3·5 cm long, with lip 5–15 mm. Bushy places, woods. YU.AL.GR.BG. Pl. 44.

(b) Axis of inflorescence glandular-hairy
1231a. *D. lanata* Ehrh. Distinguished by its long very dense inflorescence which has a densely glandular-hairy, often reddish-purple axis, and by its white to yellowish-white flowers with brown or violet veins. Corolla 2–3 cm long, middle lobe of lower lip 8–13 mm, oblong, whitish; calyx glandular-hairy, lobes acute. Bracts lanceolate, Leaves oblong to lanceolate, mostly hairless. Woods, bushy places. Widespread on mainland. Pl. 44. **1231b.** *D. leucophaea* Sibth. & Sm. Like 1231a but bracts linear; corolla 1–2 cm long, the middle lobe of the lower lip rounded, 4–7 mm, whitish with purple veins. Woods, dry grassland. GR(Athos, Thasos).

Middle lobe of lower lip only slightly longer than other lobes
1232. *D. grandiflora* Miller LARGE YELLOW FOXGLOVE. Flowers yellow, 4–5 cm long, stout-tubular; leaves usually hairless and shining green above, finely toothed. In mountains, rocks, open woods, clearings. Widespread on mainland. **1233b. *D. viridiflora* Lindley A shortly hairy perennial to 80 cm, with a dense, many-flowered spike of small dull greenish-yellow, and conspicuously veined flowers. Corolla narrow-cylindrical, 1–2 cm long, the lower lip scarcely longer than the upper. Woods. Widespread on mainland. Page 416.

BARTSIA ****1236.** *B. alpina* L. ALPINE BARTSIA. Damp places in mountains. YU.AL? BG(south-west).

BELLARDIA ****1237.** *B. trixago* (L.) All. Distinguished by its conspicuous pinkish or yellowish, two-lipped flowers clustered into a short dense, four-sided, leafy spike. Corolla 2–2·5 cm long; upper lip hooded, the lower much longer, three-lobed. A glandular-hairy, erect, leafy, unbranched annual to 70 cm. Stony, grassy places. Widespread.

PARENTUCELLIA ****1238.** *P. viscosa* (L.) Caruel YELLOW BARTSIA. Flowers yellow, sometimes white, 16–24 mm; calyx teeth nearly as long as calyx tube. Damp, grassy, or sandy places in the Med. region. YU.AL.GR+Is.CR.TR. ****1239.** *P. latifolia* (L.) Caruel SOUTHERN RED BARTSIA. Distinguished by its small reddish-purple, rarely white, persistent flowers 8–10 mm long, with calyx teeth about half as long as calyx tube. A small, erect unbranched annual 5–30 cm, with triangular-lanceolate, very deeply toothed leaves. Stony, sandy places. Widespread.

ODONTITES

Flowers yellow

1240. *O. lutea* (L.) Clairv. Flowers 5–8 mm; calyx 3–4 mm. Dry grassland and scrub, in mountains. YU.AL? BG. **1240b.** *O. glutinosa* (Bieb.) Bentham A finely glandular-hairy annual 12–30 cm, with a dense inflorescence of yellow flowers, becoming lax. Flowers 1·2–1·5 cm long, corolla tube 9–12 mm, yellow sometimes tinged brownish-purple, finely hairy; calyx 6–8 mm. Leaves linear, entire, 1–2 cm. Stony mountain slopes. YU.AL.GR.BG. **1240c.** *O. linkii* Boiss. (incl. *O. frutescens*) A hairless or sparsely hairy dwarf shrub to 50 cm, with a rather lax inflorescence 2–4 cm long of pale-yellow flowers 7 mm long, with corolla tube about 5 mm. Leaves linear to 4 cm, 4 mm wide or less. Mountain rocks. GR.CR. Page 416.

Flowers reddish-pink or rarely white

1241. *O. verna* (Bellardi) Dumort. RED BARTSIA. Cultivated ground, cornfields, waste places. Widespread on mainland.

EUPHRASIA A very difficult genus with about 11 species in our area.

RHINANTHUS | **Yellow-Rattle** A difficult genus with about 17 similar-looking species in our area.

PEDICULARIS | **Lousewort** About 14 species in our area, including 8 which are exclusive to it.

Flowers predominantly yellow to white

(a) Upper lip of flower rounded at apex, neither beaked nor toothed

1249a. *P. oederi* Vahl Flowers yellow with a crimson tip to the upper lip, at first in a dense cluster but becoming lax. Corolla 1·2–2 cm long, hairless; calyx shaggy-haired, lobes unequal, ciliate; bracts lanceolate, hairy, deeply toothed, shorter than the flowers. Leaves pinnately lobed, with ovate-oblong, deeply toothed lobes, hairless, shorter than stem. Damp grassland in mountains. YU.BG. Pl. 45

(b) Upper lip of flower terminating in 2 narrow teeth, sometimes shortly beaked

1. Calyx split on lower side, the two lateral lobes on each side more or less united (YU(central and east)).	**1249b.** *P. heterodonta* Pančić
1. Calyx with 5 (or 4) distinct lobes, not split on lower side.	
2. Calyx lobes wider than long, usually blunt (YU.AL.BG.).	**1249.** *P. comosa* L.
2. Calyx lobes at least as long as wide, acute.	
3. Calyx lobes about as long as wide (YU.AL.GR.BG.).	**1249c.** *P. brachyodonta* Schlosser & Vuk. Pl. 44
3. Calyx lobes 2–3 times as long as wide.	
4. Inflorescence sparsely hairy; calyx lobes toothed near apex only (YU.AL.BG.).	**1249d.** *P. leucodon* Griseb.
4. Inflorescence woolly-haired; calyx lobes uniformly toothed.	
5. Calyx glandular shaggy-haired; upper bracts toothed (AL.GR.).	**1249e.** *P. graeca* Bunge Pl. 45.

5. Calyx non-glandular hairy; upper   **1249f.** *P. friderici-augusti*
bracts entire (YU.AL.).   Tommasini Page 168

Flowers predominantly pink to reddish-purple
(a) Upper lip of flowers rounded at apex, neither beaked nor toothed
1253. *P. verticillata* L. WHORLED LOUSEWORT. Flowers reddish-purple, 12–18 mm; calyx
only hairy on the veins, lobes short, entire. Upper stem leaves and bracts in whorls of
3–4. Damp mountain pastures. YU.AL.GR? BG. Pl. 45.   **1253a.** *P. orthantha* Griseb.
Flowers reddish-purple in a dense rounded cluster, with corolla tube about twice as long
as calyx. Corolla 1–1·5 cm, hairless, upper lip straight below and curved above. Calyx
white-woolly, papery, the lobes blunt, toothed. Leaves alternate, twice-cut, with lanceolate, deeply toothed segments; stems short, 8–12 cm, blackish above. Stony mountain
pastures. YU(south).BG. Page 136.

(b) Upper lip of flowers terminating in 2 narrow teeth sometimes shortly beaked
1256. *P. palustris* L. RED-RATTLE. Marshes, wet meadows, heaths. YU.BG.   **1256a.** *P.
petiolaris* Ten. Flowers deep pinkish-red, borne in a white-woolly rounded cluster.
Corolla up to 2·2 cm, the lower lip hairless, beak very short; calyx woolly-haired; bracts
shaggy-haired, exceeding the calyx. Leaves twice-cut with ovate to lanceolate toothed
segments. A hairy-stemmed perennial 7–25 cm. Mountain meadows, stony hillsides.
YU.AL.GR? BG.

MELAMPYRUM | Cow-Wheat   A very difficult genus with much variation within
the main species, which is associated with season and habitat. About 15 species in our
area.

Flowers in dense spikes
**1257. *M. cristatum* L. CRESTED COW-WHEAT. Flowers 12–16 mm, pale yellow, tinged
purple on lower lip; bracts yellowish-green, folded along mid-rib, with apex reflexed. Dry
pastures, rocky slopes, wood margins. YU.AL.BG.   ****1258.** *M. arvense* L. FIELD COW-
WHEAT. Flowers 2–2·5 cm, purplish-pink, with a yellow closed throat; bracts green,
whitish, or reddish-pink, flat. Fields. YU.AL.BG.TR.   **1258a.** *M. barbatum* Willd. Spike
cylindrical; flowers pale yellow, white, or purplish, with green, whitish, or reddish-
purple bracts which have long spreading teeth. Corolla 1·5–3 cm long, lower lip with
reflexed margin (lower lip with up-turned margin in 1258), throat open. Calyx 8–12 mm,
with shaggy-haired tube, and bristle-like lobes. Leaves lanceolate, entire or toothed.
Meadows, bushy places. YU.AL.GR? Page 416.

Flowers paired, in lax, often one-sided leafy clusters
1259a. *M. scardicum* Wettst. Flowers yellow, the tube curved downwards and with
throat closed; bracts lanceolate, sharply toothed, violet-blue. Corolla 1–1·7 cm long:
calyx shortly hairy about 8 mm, with up-curving lobes about three times as long as the
tube. A leafy plant to 40 cm; leaves ovate-lanceolate 5–15 mm wide. Meadows in hills
and mountains. YU.AL? BG.   **1259b.** *M. hoermannianum* K. Malý Flowers large
16–22 mm long, yellow, with a half-closed throat, in a lax one-sided head. Bracts usually
violet-blue, ovate-heart-shaped, toothed, rarely green. Calyx teeth 8–10 mm; calyx tube
with a tuft of hairs on the upper side near the base. Meadows. YU. Pl. 45.

1260. *M. pratense* L. COMMON COW-WHEAT. Corolla 10–18 mm, whitish to bright yellow,
throat more or less closed; bracts green, entire or toothed. Meadows, woods, and heaths.
YU.BG.   **(1260).** *M.sylvaticum* L. WOOD COW-WHEAT. Corolla 8–10 mm, usually deep
golden-yellow, throat open; bracts green, mostly entire. Mountain woods. YU.BG.

TOZZIA   ****1261.** *T. alpina* L. Damp places in mountains and hills. YU.BG.

LATHRAEA | Toothwort This genus is now included in *Scrophulariaceae* in *Flora Europaea*. ****1269.** *L. squamaria* L. TOOTHWORT. Parasitic on hazel, poplar, elm, and other trees. YU.BG.TR. **1269b.** *L. rhodopea* Dingler Like 1269 but a larger yellowish-brown plant to 50 cm (plant white or cream below, tinged lilac-pink above in 1269). Flowers in dense cylindrical, not one-sided, clusters. Corolla about 1 cm long; calyx 5–7 mm (corolla 14–17 mm; calyx about 1 cm in 1269). Parasitic on various trees. GR(north-east).BG(south). Pl. 45.

RHYNCHOCORYS Distinguished by its corolla which has a short bell-shaped tube, a beaked upper lip, and a spreading three-lobed lower lip; calyx with an entire upper lip and a longer two-lobed lower lip. **1261a.** *R. elephas* (L.) Griseb. A leafy hairless perennial to 50 cm, with obovate to elliptic, shallowly toothed leaves, 1–3 cm, and a leafy cluster of yellow flowers with a conspicuous straight, forward-projecting beak. Corolla 1–1·5 cm; beak 6–10 mm; lower bracts leaf-like. Mountain grasslands. GR.BG. Pl. 44.

SIPHONOSTEGIA Corolla with a long tube, two-lipped, the upper lip cut off, the lower three-lobed; calyx tubular, two-lipped, the upper lip three-lobed, the lower two-lobed.

1261b. *S. syriaca* (Boiss. & Reuter) Boiss. A glandular-hairy, erect rigid perennial to 60 cm, with a terminal leafy cluster of purple flowers. Corolla about 3 cm; calyx 17–20 mm, with narrow lobes. Leaves oblong-lanceolate, entire; bracts narrow elliptical. Scrub in mountains. GR(east). Pl. 45.

GLOBULARIACEAE | Globularia Family

GLOBULARIA

Shrubs or creeping woody-based shrublets
**1262. *G.alypum* L. SHRUBBY GLOBULARIA. A much-branched, bristly-leaved, erect ever-green shrub 30–100 cm, with stems covered with lime-secretions, and with globular, lilac-blue flower heads 1–2·5 cm. Involucral bracts broadly ovate, overlapping, brown-margined, ciliate. Leaves very leathery, with a spiny tip sometimes or with 3 spiny teeth. Dry bushy places, rocks in the Med. region. Widespread, except BG. Page 35.

1263. *G. cordifolia* L. MATTED GLOBULARIA. A low mat-forming perennial with spread-ing woody, rooting branches bearing leaf rosettes, and short, almost leafless stems 1–10 cm bearing globular blue flower heads 1–2 cm. Involucral bracts ovate-lanceolate. Basal leaves to 2·5 cm, fleshy, hairless, broadly spathulate with a rounded, notched or short-pointed apex, or with 3 teeth. In mountains, rocks, screes. YU.AL.BG. Page 185. **1263b. *G. meridionalis* (Podp.) O. Schwarz Like 1263 but a more robust plant with larger, narrower, lanceolate to oblanceolate acute leaves 2–9 cm, which are rarely notched. Flower head blue 1–2 cm; involucral bracts ovate, abruptly narrowed to a point. Mountains. YU.AL.GR.BG. Pl. 45. **1263c.** *G. stygia* Boiss. Like 1263 with procumbent rooting woody stems, but distinguished by the presence of subterranean stolons, by its regular (not two-lipped) calyx, and its rounded rosette leaves. Flower heads blue, almost stalkless, with numerous involucral bracts. Mountains. GR(Peloponnisos). Page 92.

Perennials with herbaceous stems
1264b. *G. punctata* Lapeyr. A herbaceous perennial without stolons, but with basal leaf

rosettes, and stems up to 30 cm bearing blue flower heads about 1·5 cm across. Involucral bracts numerous, lanceolate, long-pointed. Rosette leaves evergreen, stalked, obovate to spathulate, with rounded or weakly notched apex; stem leaves narrower, stalkless. Dry grassy places, rocky ground. YU.GR.BG.TR. **1264f.** *G. trichosantha* Fischer & C. A. Meyer Differs from 1264b in having stolons up to 30 cm long; flowering stems to 20 cm with blue flower heads up to 2·5 cm across. Calyx lobes 3–4 times as long as calyx tube (about as long in 1264b). Dry hills. BG.TR.

BIGNONIACEAE | Bignonia Family

CATALPA ****1265.** *C. bignonioides* Walter CATALPA, INDIAN BEAN. A deciduous tree with very large, pale-green, heart-shaped leaves, and broad pyramidal clusters of white, bell-shaped flowers. Corolla 3–4 cm across, with 2 yellow streaks and purple spots within. Fruit linear to about 40 cm, pendulous. Native of N. America; often planted as an ornamental tree.

ACANTHACEAE | Acanthus Family

ACANTHUS

Basal leaves not spiny
1266. *A. mollis* L. BEAR'S BREECH. Shady places, roadsides. YU(Croatia). **1266b. *A. balcanicus* Heywood & I. B. K. Richardson (*A. longifolius*) Like 1266 with non-spiny deeply lobed basal leaves, but leaf lobes narrowed at the base (not narrowed at base in 1266); upper leaves spiny-toothed, stalked. Lower lip of calyx hairy at apex (hairless in 1266). Flowers whitish with purple veins, 3·5–5 cm long, in dense glandular spikes with large spiny bracts; stems to 1 m. Woodland, scrub, stony hillsides. YU.AL.GR.BG.TR? Pl. 45.

All leaves spiny
1267. *A. spinosus* L. SPINY BEAR'S BREECH. An erect thistle-like perennial 20–80 cm, with once- to twice-cut leaves with numerous long white spiny teeth. Flower spike dense cylindrical, of many whitish one-lipped flowers, borne in the axils of long-spiny toothed and often coloured bracts. Corolla 3·5–5 cm long; calyx four-lobed, the upper and lower lobes larger. Basal leaves stalked, oblong in outline. Woodlands, meadows. Widespread, except TR.

PEDALIACEAE | Sesame Family

Herbs with mucilage glands. Flowers solitary, or in small axillary clusters. Calyx five-lobed; corolla five-lobed, weakly two-lipped; stamens 4, the fifth sterile. Ovary superior, four-celled; fruit a capsule.

SESAMUM **1267a.** *S. indicum* L. SESAME. An erect, hairy annual to 60 cm, with solitary axillary, whitish, bell-shaped two-lipped flowers about 3 cm long, often with purplish or yellow markings. Leaves about 10 cm, stalked, the lower usually opposite, lobed, the upper alternate, narrow entire. Fruit oblong 2·5 cm, splitting into 2 valves. Probably native of S.E.Asia; cultivated for its oil-rich seeds. Locally naturalized in YU.GR.BG.TR?

GESNERIACEAE | Gloxinia Family

1.   Anthers free, at least as long as the
    filaments; corolla almost symmetrical,
    with tube equalling or shorter than
    the lobes.
  2.   Corolla tube much shorter than the
      lobes.   *Ramonda*
  2.   Corolla tube almost as long as the
      lobes.   *Jankaea*
1.   Anthers joined in pairs, much shorter
    than the filaments; corolla two-lipped,
    with tube longer than the lobes.   *Haberlea*

RAMONDA **1268a.** *R. nathaliae* Pančić & Petrović A rosette plant of shady rock crevices with glandular-hairy, leafless stems 6–8 cm, bearing 1–3 large lilac to violet flowers with orange-yellow centres. Corolla 3–3·5 cm across, with usually 4 spreading lobes; anthers yellow. Leaves stalked, ovate to rounded, with short white hairs above and long brown shaggy hairs beneath. YU(south).GR(north).   **1268b.** *R. serbica* Pančić Like 1268a but leaves narrowly obovate and tapering to a wedge-shaped base (base cut off (truncate) in 1268a), blade sometimes deeply toothed. Flowers cup-shaped, usually with 5 lobes which are semi-erect, not spreading flat; anthers dark violet-blue. Shady crevices in limestone rocks. YU(south).AL.GR(north-west).BG(north-west). Pl. 46.

JANKAEA **1268c.** *J. heldreichii* (Boiss.) Boiss. A beautiful silvery-silky rosette plant of cliffs and shady rocks of Mount Olympus. Flowers 1–2, pale lilac, bell-shaped, with usually 4 spreading lobes about as long as the tube, on short leafless stems about 4 cm. Leaves 2–4 cm, obovate entire, with brown shaggy hairs beneath. GR(Olympus). Page 126. Pl. 46.

HABERLEA **1268d.** *H. rhodopensis* Friv. A rosette plant of rocks with short-stemmed clusters of 1–5 pale bluish-violet, cylindrical, irregularly two-lipped flowers. Corolla 1·5–2·5 cm long, hairy within. Leaves 3–8 cm, in a lax rosette, obovate to oblong, blunt, coarsely toothed, softly hairy, tapering to a short broad stalk. Rocks and cliffs in mountains. GR(north-east).BG(south and central). Pl. 46.

OROBANCHACEAE | Broomrape Family

CISTANCHE Differs from *Orobanche* in having a more or less symmetrical five-lobed corolla, and a five-lobed calyx with equal blunt lobes. Parasitic on woody members of the *Chenopodiaceae*.

1269a. *C. phelypaea* (L.) Coutinho (*Phelypaea tinctoria, P. lutea*) A yellowish, hairless plant 20–100 cm, with a stout erect stem with numerous oval-lanceolate, papery-margined scales, and a dense oblong-cylindrical spike of large tubular, shiny yellow flowers. Corolla 3–5 cm long, usually curved, with a wide five-lobed mouth; calyx bell-shaped. Coasts. CR.

OROBANCHE | **Broomrape** About 30 species occur in our area. A very difficult genus. A key to the commoner ones is given below as a guide to identification of many plants encountered. The identity of the host plant is important, but often difficult to determine in the field. (Note: measurements in brackets indicate the maximum or minimum dimensions recorded.)

1. Each fl. subtended by 2 bracteoles (more
   or less fused to the calyx) as well as
   by a bract.
   2. Anthers more or less densely hairy.   *arenaria, lavandulacea*
   2. Anthers hairless, or sparsely hairy
      at the base.
      3. Stems usually branched; corolla
         10–20(22)mm.   *ramosa, oxyloba*
      3. Stems usually simple; corolla
         (18)20–5 mm.   *purpurea*
1. Bracteoles absent.
   4. Stigma purple, orange, or dark red at
      flowering.
      5. Lower lip of corolla glandular-
         ciliate.   *caryophyllacea, alba*
      5. Lower lip of corolla not glandular-
         ciliate.
         6. Corolla shining dark red inside,
            mostly dark red or purple outside.   *sanguinea*
         6. Corolla not shining dark red inside,
            usually white or pale yellow outside,
            at least towards the base.
            7. Corolla with shaggy hairs.   *pubescens*
            7. Corolla hairless or finely hairy.   *reticulata, minor* group
   4. Stigma yellow or white (rarely purple)
      at flowering.
      8. Corolla shining dark red inside,
         usually dark purplish-red or bright
         yellow outside.   *gracilis*
      8. Corolla not shining dark red inside,
         usually pale yellow, white, or bluish
         outside, at least towards the base.
         9. Corolla predominantly white,
            cream, bluish or violet.

10. Corolla conspicuously inflated,
    papery and shining at base.   *cernua*
10. Corolla not as above.
    11. Corolla 20–30 mm, with large
        strongly divergent lips;
        bracts shaggy-haired.   *crenata*
    11. Corolla 10–23 mm, with small
        lips; bracts glandular-hairy.   *minor* group
9. Corolla predominantly yellow or
   reddish.
   12. Lower lip of corolla glandular-
       ciliate; parasitic on Leguminosae.   *lutea*
   12. Lower lip of corolla not ciliate.
       13. Corolla at least 2 cm.
           14. Lvs. 2–3 cm; calyx 10–17
               mm; parasitic on
               Leguminosae.   *lutea*
           14. Lvs. 1–2 cm; calyx 6–11 mm;
               parasitic on Compositae.   *elatior*
       13. Corolla usually not more than
           2 cm.
           15. Calyx lobes fused at base.   *elatior*
           15. Calyx lobes free at base.   *minor* group

1270. *O. ramosa* L. (*Phelypaea r.*) BRANCHED BROOMRAPE. A very variable species growing on a wide variety of hosts. Flowers blue, violet, or cream, yellowish-white at the base, 10–22 mm long, glandular-hairy; calyx 6–8 mm. Stems with a swollen base, often branched; leaves ovate-lanceolate 3–10 mm. Widespread. **1270a.** *O. oxyloba* (Reuter) G. Beck (incl. *O. dalmatica*) Parasitic on *Anthemis chia* and other herbs and differs from 1270 in having the lower lip of the corolla with acute to long-pointed lobes (blunt in 1270). Flowers 15–20 mm, lilac with a white base. YU.GR.CR.BG.TR. **1270b.** *O. lavandulacea* Reichenb. Flowers 16–22 mm, glandular-hairy, bright blue with a white base, narrowly bell-shaped with strongly diverging lips. Calyx 6–8 mm, usually blue, lobes as long as or shorter than tube. Parasitic on *Psoralea bituminosa* and other herbaceous species in the Med. region. YU.GR. Pl. 47.

1271. *O. purpurea* Jacq. Parasitic on Compositae particularly on *Achillea*. Flowers narrowly bell-shaped, white below with bluish-violet tips and deep-violet veins, 18–25(30) mm long. Anthers and filaments hairless or nearly so, attached about 7 mm above the base of the corolla; stigma white or pale blue. Stem 15–70 cm, minutely glandular-hairy, often greyish-tinged. YU.GR.CR? BG.TR. Pl. 47. **(1271).** *O. arenaria* Borkh. Parasitic on *Artemisia* species. Flowers large 25–35 mm long, bluish-violet, in a dense glandular-hairy inflorescence, with lanceolate bracts 15–20 mm and linear bracteoles. Anthers hairy, filaments hairless or nearly so; stigma white. YU.BG.

1271a. *O. gracilis* Sm. Flowers fragrant, glandular-hairy, yellow outside, usually with red veins and reddish towards the lips, shining dark red inside, 15–25 mm long. Anthers hairless, filaments hairy; stigma yellow. Stems 15–60 cm, glandular-hairy, yellow or reddish. Parasitic on Leguminosae, rarely on *Cistus*. YU.AL.GR.CR.BG. **1271c.** *O. sanguinea* C. Presl Parasitic on *Lotus* in the Med. region. Flowers small 10–15 mm, hairless, dark red or purple, yellow at base, shining dark red inside, not fragrant. Anthers hairless, filaments hairless above, attached 1–2 mm above base of corolla; stigma purple. YU.GR.BG.

1273. *O. elatior* Sutton TALL BROOMRAPE. Parasitic on *Centaurea* and other Compositae, and *Thalictrum*. Flowers yellow, often tinged with pink, glandular-hairy, usually 18–25 mm long, narrowly bell-shaped and uniformly curved, in a dense spike 6–20 cm. Anthers hairless, filaments hairy, attached 3–6 mm above base of corolla; stigma yellow. YU.AL.GR.BG. **1273a.** *O. lutea* Baumg. Parasitic on *Trifolium*, *Medicago*, and other Leguminosae. Flowers 20–30 mm long, yellowish or reddish-brown; stigma yellow or whitish. Stems swollen at base. YU.AL.GR.BG. ****1274.** *O crenata* Forskål A pest of bean fields, where it may be seen in great numbers; it also attacks the garden sweet-pea and other Leguminosae. Flowers 20–30 mm, white with often violet veins to lips, fragrant. Widespread. **1274a.** *O. cernua* Loefl. Flowers 12–20 mm long, violet-blue; stamens hairless; stigma whitish. Stems yellowish, up to 40 cm. Parasitic on *Artemisia* and *Helianthus annuus*, or rarely on other herbaceous species. YU.GR.CR.TR.

1275. *O. caryophyllacea* Sm. (*O. vulgaris*) CLOVE-SCENTED BROOMRAPE. Parasitic on Rubiaceae. Flowers large 20–32 mm long, yellowish or pink, variously tinged with dull purple, fragrant, regularly curved in profile, densely glandular-hairy, in a lax spike. Anthers hairless; filaments hairy; stigma purple. YU.AL.GR.BG. **1276. *O. alba* Willd. THYME BROOMRAPE. Parasitic on Labiatae and distinguished from 1275 by the hairs on the corolla many of which are dark at base or apex (colourless or pale yellow in 1275). Flowers 15–25 mm, purplish-red, yellow, or whitish, fragrant, slightly curved. Filaments densely hairy; stigma red or purple. Widespread. Pl. 47.

1276a. *O. pubescens* D'Urv. Flowers 10–20 mm, pale yellow tinged with violet above, shaggy-haired. Filaments of stamens hairy below; stigma violet. Stem 15–50 cm, pale yellow often tinged with pink, glandular-hairy. Parasitic on Compositae and Umbelliferae and perhaps others. YU.GR+Is.CR.BG.TR. **1276b.** *O. reticulata* Wallr. Distinguished from 1276 *O. alba* by being usually parasitic on *Cirsium*, *Carduus*, and *Knautia*; by its hairless or almost hairless filaments; and by the lower lip of the corolla with equal and not ciliate lobes (lower lip glandular-ciliate with middle lobe much larger in 1276). Often a taller more robust plant. YU.AL.GR.BG. Pl. 47.

O. minor group

1. Bracts 7–15 mm; stamens inserted 2–3
   mm above base of corolla.   *minor*
1. Bracts 10–22 mm; stamens inserted 3–5
   mm above base of corolla.
   2. Corolla rather sharply curved inwards
      near the base, the upper lip deeply
      two-lobed.   *amethystea*
   2. Corolla not sharply curved inwards near
      the base, the upper lip notched or
      slightly two-lobed.   *loricata*

1277. *O. minor* Sm. LESSER BROOMRAPE. Flowers pale yellow usually tinged with violet towards the tip. Corolla 10–18 mm long; stigma purple, rarely yellow. Most frequently parasitic on *Trifolium* but also on many other species. Widespread on mainland. **1278.** *O. loricata* Reichenb. (incl. *O. picridis*) Flowers white or pale yellow, tinged and veined with violet. Corolla 14–22 mm long; stigma purple. Parasitic on Compositae and Umbelliferae. YU.AL.GR.CR.BG. **1278a.** *O. amethystea* Thuill. Flowers white or cream, usually tinged with violet, pink, or brown towards the tip. Corolla 15–25 mm long; stigma purple or yellow. Parasitic on various herbaceous plants. YU.AL.GR.BG.

LENTIBULARIACEAE | Butterwort Family

PINGUICULA | Butterwort
1279. *P. alpina* L. ALPINE BUTTERWORT. Distinguished by its whitish flowers with 1–2 yellow spots in the throat. In mountains: springs, damp meadows, damp rocks. YU(Croatia). **1279b. *P. hirtiflora* Ten. Flowers 16–25 mm or more long, with pink to pale-blue unequal lips, white and yellow throat, and an awl-shaped, straight or slightly curved spur 6–13 mm. Lobes of lower lip of corolla notched; lobes of upper lip obovate. Flowering stems 1–3, slender 5–14 cm long. Plant overwintering as a rosette. Wet rocks in mountains. YU.AL.GR. Pl. 46.

1280. *P. vulgaris* L. COMMON BUTTERWORT. Recognized by its bright-violet or lilac flowers 15–30 mm long, with a broad white patch in the throat, and a straight purple spur 3–6 mm or more. Lobes of lower lip of corolla oblong, divergent, not overlapping or touching. Bogs, wet heath, damp rocks. YU. **1280c. *P. balcanica* Casper Like 1280 with blue-lilac flowers, but distinguished by the lower lip of the corolla which has rounded, overlapping or touching lobes, with the middle lobe often larger, and white-spotted at its base; lobes of upper lip strap-shaped. Corolla 14–19 mm long, less commonly to 23 mm; spur 3–7 mm, straight. Plant overwintering as a bud. Wet places in mountains. YU.AL.GR.BG. Pl. 46.

UTRICULARIA **1281. *U. vulgaris* L. GREATER BLADDERWORT. A submerged aquatic plant with deeply cut leaves, bearing small animal-catching bladders, and carrying above water a cluster of deep-yellow, spurred flowers. Corolla 15–18 mm long, lower lip with reflexed margin and a large throat-boss; spur conical at base, rather abruptly narrowed to apex. Still waters. YU.AL.GR.CR? BG. **1281a.** *U. australis* R.Br. Like 1281 but distinguished by the lemon-yellow flowers with the lower lip of the corolla which is more or less flat with an undulate margin, and spur gradually tapering from base to blunt apex. Flower stalks 3–5 times as long as the subtending bract (2–3 times as long in 1281). Still waters. YU.CR? BG.TR. **(1282).** *U. minor* L. LESSER BLADDERWORT. Flowers only 6–8 mm; spur very short, sac-like. Bogs, marshes, lake verges. YU.AL.GR.BG.

PLANTAGINACEAE | Plantain Family

PLANTAGO | Plantain 23 similar-looking species are found in our area.

Leaves opposite; stems branched, leafy
1283. *P. arenaria* Waldst. & Kit. (*P. indica*, *P. psyllium*) BRANCHED PLANTAIN. Readily distinguished by its hairy branched non-glandular, leafy stems bearing a number of ovoid spikes about 1 cm across. Upper and lower bracts of flower heads dissimilar in shape, the lower with lateral veins. Poor fields, sandy places, waysides. Widespread. **1283a. *P. afra* L. Very like 1283 but stems and bracts conspicuously glandular-hairy, and bracts of flower spikes all similar in shape, without lateral veins. Dry places. Widespread.

Leaves in a rosette
(a) *Leaves linear to oblong, more than three times as long as broad*
 (i) *Flowering stems smooth, not grooved or ribbed*
**1285. *P. coronopus* L. BUCK'S HORN PLANTAIN. Leaves toothed or deeply once or twice cut

into linear lobes. Dry sandy places, waysides, on the littoral. Widespread. **1286.** *P. subulata* L. (incl. *P. holosteum*) A densely tufted perennial with short, stout branches densely covered with bases of old leaves and ending in a rosette of leaves. Leaves fleshy, narrow-linear, triangular-sectioned, stiff-tipped, and hairless. Flower spike dense, long-cylindrical 2–5 cm. Rocky places, dry pastures in the Med. region. YU.AL.GR.BG.TR.

1288. *P. bellardii* All. SILKY PLANTAIN. An annual with rosettes with linear-lanceolate leaves 1–5 mm broad, densely covered with shaggy hairs. Flowering stems usually 1–7, densely hairy, not thickened in fruit. Flower spikes dense, usually 1–2 cm. Sandy, stony places in the Med. region. Widespread. **1288b. *P. cretica* L. Differs from 1288 in having often more than 10 flowering stems which thicken up in fruit and curve inwards, the whole plant becoming a ball which is blown about over the ground. Flower spikes to 1 cm; corolla lobes rounded-ovate. Dry places. GR(Is.).CR. **1288c.** *P. atrata* Hoppe MOUNTAIN PLANTAIN Leaves linear-lanceolate, untoothed, hairless or nearly so, in rosettes. Flowering stems as long or longer than leaves; flower spikes oblong-ovoid, 1·5–3 cm. Bracts with brown or colourless margin, ciliate; sepals ciliate; corolla hairless. Mountain pastures. YU.AL.GR.BG. Pl. 46.

(ii) Flowering stems grooved or finely ribbed
1290. *P. lanceolata* L. RIBWORT PLANTAIN. Meadows, waysides, waste ground. Widespread. **1290a.** *P. argentea* Chaix Differs from 1290 in having leaves which are densely covered with adpressed sometimes silky hairs; flowering stems with more than 5 grooves; and usually solitary rosettes. Flowering spikes 0·5–2 cm; anthers white, not yellowish. Roots more than 1 mm thick. Dry grassy and stony places. YU.GR(north).BG. Page 179. **1290b.** *P. lagopus* L. Similar to 1290 but bracts and sepals with dense shaggy hairs so that the shortly cylindrical flower heads appear silky. Dry, stony ground in the Med. region. Widespread, except BG.

(b) Leaves ovate to elliptic, less than three times as long as broad
1292. *P. media* L. HOARY PLANTAIN. Meadows, waysides. Widespread on mainland. **1292a. *P. gentianoides* Sibth. & Sm. A mountain perennial of the central Balkans distinguished by its rosette of ovate, usually hairless, stalked leaves. Flowering stems much longer than leaves, finely grooved; flower spikes 1–3·5 cm, up to 4 cm in fruit. Bracts hairless; sepals usually purplish-brown, with papery margins. Corolla tube hairless; anthers yellowish, filaments whitish. Damp places, by melting snow. YU.AL.GR(north).BG. **1293.** *P. major* L. GREAT PLANTAIN. Waysides, cultivated ground. Widespread.

CAPRIFOLIACEAE | Honeysuckle Family

SAMBUCUS ****1295.** *S. ebulus* L. DANEWORT. Roadsides, waste places. Widespread, except on most islands. **1296.** *S. nigra* L. ELDER. Hedges. Widespread on mainland; often cultivated. ****1297.** *S. racemosa* L. ALPINE or RED ELDER. Montane woods. YU.AL.BG.

VIBURNUM ****1298.** *V. opulus* L. GUELDER ROSE. Damp woods, thickets, hedges, mountainsides. YU.AL.BG. ****1299.** *V. lanata* L. WAYFARING TREE. Open woods, scrub, hedges. YU.AL.GR.CR.BG. ****1300.** *V. tinus* L. LAURUSTINUS. An evergreen shrub with dark-green, oval leathery leaves, and dense flat-topped clusters of white flowers, pink in bud. Fruit metallic blue-black, poisonous. Bushy places, open woods in hills in the Med. region. YU.AL.GR.

LONICERA | Honeysuckle

1. Erect, non-climbing shrubs.
  2. Corolla regular or nearly so
    (YU.AL.BG.).   **(1303). *L. caerulea* L.
  2. Corolla conspicuously two-lipped.
    3. Bracts shorter than ovary; berries
      bluish-black (YU.GR.BG.).   1303. *L. nigra* L.
    3. Bracts at least as long as ovary;
      berries red or yellowish.
      4. Twigs and lvs. hairless or almost
        so (YU.AL.GR.).   **(1303). *L. alpigena* L. Page 191
      4. Twigs and lvs., at least on the
        lower surface, finely hairy.
        5. Stalks of fl. clusters absent
          or very short.   *nummulariifolia*
        5. Stalks of fl. clusters 1–2 cm.
          6. Stalks of fl. clusters not
            glandular, minutely hairy.   *xylosteum*
          6. Stalks of fl. clusters densely
            glandular-hairy.
            7. Lvs. glandular-hairy on both
              surfaces; ovaries and berries
              fused in pairs (YU.(west)).   1302b. *L glutinosa* Vis. Page 178
            7. Lvs. velvety-haired beneath;
              ovaries and berries free or
              almost so (GR(south)TR?).   1302c. *L. hellenica* Boiss.
1. Woody climbers.
  8. Uppermost pair of lvs. below fls.,
    not fused together (YU.AL.GR.).   **1304. *L. periclymenum* L.
  8. Uppermost pair of lvs. below fls.
    fused together.
    9. Fl. clusters stalked.   *etrusca*
    9. Fl. clusters stalkless.   *caprifolium, implexa*

**1302. *L. xylosteum* L. FLY HONEYSUCKLE. A branched upright shrub to 2 m, with grey-green opposite, oval, stalked leaves, and paired white to yellowish, sometimes pinkish-tinged, two-lipped flowers. Flowers 8–12 mm; flower stalk 12–20 mm. Fruit paired but not fused, red. Bushy places. YU.AL.GR.BG.   1302a. *L. nummulariifolia* Jaub. & Spach Like 1302 but differs in having flower stalk very short or absent. Flowers flesh-coloured to whitish-pink, hairy. Leaves 2–4 cm (3–7 cm in 1302). Rocky places in mountains. GR(south).CR. Pl. 47.

1305. *L. etrusca* G. Santi A deciduous-leaved climber distinguished by its long-stalked clusters of whitish-yellow, often pink-flushed flowers with the uppermost leaf-pair fused round the stem below each cluster. Flowers 3·5–4·5 cm. Leaves glaucous or whitish-green. Fruit red. Bushy places. Widespread.   1306. *L. caprifolium* L. PERFOLIATE HONEY-SUCKLE. A deciduous-leaved climber like 1305 but differing in having stalkless clusters of white or yellowish, often pink-flushed, flowers surrounded by the fused cup-like uppermost pair of leaves. Flowers 3–5 cm. Leaves elliptic, dark green above, glaucous beneath. Fruit red or orange. Bushy places. YU.AL.GR? TR.   **(1306). *L. implexa* Aiton Like 1306 with stalkless clusters of whitish-yellow, pink-flushed flowers, surrounded by cup-like fused leaves, but leaves evergreen ovate to oblong, dark green, shining above, glaucous beneath. Flowers 2·5–4·5 cm. Upper leaves borne on hairless glaucous twigs, often fused round stem. Fruit red. Bushy places in the Med. region. YU.AL.GR. Pl. 47.

ADOXACEAE | Moschatel Family

ADOXA ****1308.** *A. moschatellina* L. MOSCHATEL. Woods, damp shady places in mountains. YU.AL.BG.

VALERIANACEAE | Valerian Family

VALERIANELLA | **Lamb's Lettuce**   About 17 similar-looking, pale pink- or bluish-flowered, erect, dichotomously branched annuals; distinguished largely by their fruit and calyx. Corolla tube less than twice as long as corolla lobes.

Fruit with conspicuous curved spines
1310. *V. echinata* (L.) DC. SPINY LAMB'S LETTUCE. The fruit has the appearance of a cluster of spines borne at the end of conspicuously swollen stalks. Cultivated ground. Widespread in the Med. region. Page 40.   ****1311.** *V. coronata* (L) DC. CROWNED LAMB'S LETTUCE. Fruits in dense globular clusters, each fruit topped with a membraneous crown with 6 crooked spines. Dry arid places. Widespread.   **(1311).** *V. discoidea* (L.) Loisel. Fruits in dense globular clusters, each fruit with crown-like calyx with 8–15 unequal teeth having barbed apices. Cultivated ground, waste places in the Med. region. YU.GR+Is.CR.TR.   **1311a.** *V. obtusiloba* Boiss. Fruit in dense globular clusters, each with crown-like calyx with 6 teeth, each tooth with 3 or more barbed spines. Grassy places in the Med. region. GR(south)+Is.CR.

Fruit inflated, globular
1311b. *V. vesicaria* (L.) Moench Fruits in dense clusters, each fruit conspicuously inflated, netted, the calyx with a circular opening with 6 tiny incurved teeth. Cultivated places, hills in the Med. region. GR+Is.CR.TR.

FEDIA   Slender annuals distinguished from *Valerianella* by the long corolla tube, more than twice as long as the corolla lobes.

**1312. *F. cornucopiae* (L.) Gaertner Fields, waste places in the Med. region. GR.CR.

VALERIANA | **Valerian**   A difficult genus with 11 species, including 3 restricted to our area.

Basal leaves pinnate or pinnately lobed (see also 1313e)
1313. *V. officinalis* L. VALERIAN. Bushy places, rocks, woods. YU.AL.GR? BG.TR.

Basal leaves entire or toothed
(a) Rhizome with tubers
(1313). *V. tuberosa* L. Flowers pink, in a dense, simple or somewhat branched rounded

cluster borne on a single hairless stem 10–40 cm. Basal leaves ovate entire, the upper cut into linear leaflets. Fruit about twice as long as wide, hairy on both surfaces. Dry grassland. YU.AL.GR.BG. **1313e.** *V. dioscoridis* Sibth. & Sm. Differs from (1313) in its fruits which are more than twice as long as wide, and hairy on one surface. Flowers pink or white, in a compound inflorescence with dense subsidiary clusters. Stem solitary, 25–75 cm. Basal leaves elliptic, entire or pinnately lobed; stem leaves pinnate, with shallowly toothed leaflets. Rock crevices, rocky woods, damp grasslands. YU.AL.GR+Is.BG.TR.

(b) *Rhizome without tubers*
1315. *V. tripteris* L. THREE-LEAVED VALERIAN. Middle and upper stem leaves mostly three-lobed, or with an additional pair of basal lobes. Flowers pink or white, in a somewhat flat-topped lax cluster borne on stems 10–40 cm. Woods, scrub, rocky ground, usually on alkaline soils. YU.GR.BG. **(1315).** *V. montana* L. Like 1315 and not infrequently showing intermediate characters but distinguished by the stem leaves which are usually entire or toothed. Flowers lilac, pink, or white; corolla tube 3–5 mm; flowering stem 12–50 cm. Scrub and rocky places, mainly in mountains. YU.AL.BG. Page 178. **1315d.** *V. crinii* Boiss. Distinguished from (1315) by the short flowering stem 5–12 cm, and leaves which are simple and entire or very shallowly indented. Corolla tube 2–2·5 mm. Mountain cliffs. AL.GR. Page 93. **1315e.** *V. olenaea* Boiss. & Heldr. A local plant of rocks in mountains (from 1500 to 2200 m) of the north Peloponnisos with long-stalked, ovate or obovate basal leaves and simple or three-lobed stem leaves with elliptic or ovate lobes. Flowers pink; corolla tube 4–5 mm; flowering stems several, hairless, 6–25 cm. Fruit square in section. GR(south). Page 93.

CENTRANTHUS (KENTRANTHUS)

Perennials
1316. *C. ruber* (L.) DC. RED VALERIAN. Rocks, walls in the Med. region; naturalized elsewhere. Widespread, except BG. **1316b. *C. longiflorus* Steven A glaucous perennial 40–200 cm, of mountain rocks, with linear to ovate leaves 4–10 cm by 2–35 mm. Inflorescence compound with oblong clusters of pink or lilac flowers with conspicuously long corolla tubes 12–18 mm, and spurs 10–14 mm. Subsp. *junceus* (Boiss. & Heldr.) I. B. K. Richardson has stems 40–150 cm and leaves mostly about 2 mm wide. GR. Page 127. Subsp. *kellereri* (Stoj., Stefanov & Georgiev) I. B. K. Richardson is a very robust plant to 2 m, with broader ovate-lanceolate leaves 1–3·5 cm wide. Calcareous screes. BG(west). **1316c.** *C. nevadensis* Boiss. Subsp. *sieberi* (Heldr.) I. B. K. Richardson is like 1316b in having long corolla tubes 12–14 mm and spurs 13–17 mm, but it is a tufted plant with stems less than 40 cm, with 2–6 smaller elliptical stem leaves, 2–4 cm long. Mountain rocks. CR.

Annuals
1317. *C. calcitrapae* (L.) Dufresne Rocks, dry hills. YU.GR+Is.CR.

DIPSACACEAE | Scabious Family

Flowers small, clustered into a head, or *capitulum*, which is surrounded at the base by usually green leafy *involucral bracts*. Among the flowers on the receptacle are usually scaly, bract-like *receptacular scales*. The individual flowers are surrounded at the base by an additional structure, or epicalyx, known as the *involucel*, and this may be enlarged into a crown-like structure or *corona*.

DIPSACUS | Teasel

Stem leaves shortly stalked; flower heads globular
1319. *D. pilosus* L. SMALL TEASEL. An erect, branched, weakly prickly-stemmed biennial
to 120 cm, with small white globular flower heads 1·5–2·5 cm across, with violet
anthers. Involucral bracts with long apical spine, white-hoary. Basal leaves ovate, in a
rosette, long-stalked, upper leaves short-stalked, lobed. Woods, bushy places. YU.BG.

Stem leaves stalkless; flower heads ovoid to cylindrical
1318. *D. fullonum* L. TEASEL. A stiff erect biennial to 2 m, with prickly stems and
mid-veins of leaves, and ovoid-cylindrical, bristly flower heads with pinkish-purple
florets. Involucral bracts linear, curved upwards, the longest longer than the flower
head. Stem leaves entire or toothed, fused at base round stem. Waysides, marshes.
Widespread on mainland.   ****(1318).** *D. laciniatus* L. Differs from 1318 in having stem
leaves deeply cut into narrow blunt lobes and stems covered with slender prickles.
Involucral bracts lanceolate, unequal, curved upwards but not longer than pinkish
flower head. Receptacular scales longer than florets. Marshes, waysides. Widespread on
mainland.   **(1318).** *D. sativus* (L.) Honckeny FULLER'S TEASEL. Cultivated in the past for
the fruiting heads used in preparing cloth. Distinguished by its receptacular scales
which have rigid recurved apical spines, and its more or less spreading involucral bracts.
Naturalized in YU.BG.

MORINA   Perennials with thistle-like, spiny leaves, and flowers in whorled clusters.
Epicalyx funnel-shaped, spiny. Corolla tube very long, curved, two-lipped.

1319a. *M. persica* L. A robust very spiny, thistle-like perennial 30–90 cm, with an erect
spike-like inflorescence, of several whorls of long-tubed, curved flowers which are at
first yellow but soon turn brick-red. Corolla about 3 cm long, two-lipped; calyx not spiny,
deeply two-lipped; bracts with long marginal spines. Leaves lanceolate, pinnately lobed,
the lobes with long pale terminal and marginal spines. Rocks in mountains.
YU(south).AL.GR.BG.TR. Page 168. Pl. 47.

CEPHALARIA   9 species, including 3 which are endemic to our area. A difficult
genus with several very similar-looking yellow-flowered species.

Shrubs
1320b. *C. squamiflora* (Sieber) W. Greuter A shrub to 90 cm, with leathery lanceolate to
elliptic, entire or toothed leaves, and white or yellow flower heads. Corolla 9–12 mm
long; involucral bracts 4–6 mm, ovate, with adpressed hairs; receptacular scales
6–7 mm. Involucel with 4 teeth on the angles and 4 very short intermediate teeth. Rocks
in the Med. region. CR+Karpathos.

Annuals or herbaceous perennials
(a) Annuals; flowers mostly blue or violet
**1320. *C. transylvanica* (L.) Roemer & Schultes Flower heads pale violet or yellow
1–1·5 cm across, with lanceolate involucral bracts. Receptacular scales 7–10 mm,
ovate-lanceolate, with a short spine and purple vein at apex. Involucel of fruit eight-
angled, with 8 equal bristles. A slender annual to 120 cm, with deeply cut leaves
5–12 cm, with entire, toothed, or cut lobes. Vineyards, waysides, uncultivated ground.
Widespread on mainland.   **1320a.** *C. syriaca* (L.) Roemer & Schultes Distinguished
from 1320 by the receptacular scales which have a long spine, as long as or longer than
the limb. Flowers blue or lilac, 8–14 mm long. Involucel of fruit with 4 long and 4 shorter
bristles. Leaves entire or toothed, rarely weakly lobed. Fields. Native of Anatolia,
naturalized in GR.BG.

(b) Perennials; flowers yellow or white

(1320). C. leucantha (L.) Roemer & Schultes Flowers white or pale yellow, with white anthers, densely hairy, in globular heads 2–3 cm across. Corolla 10–15 mm. Involucral bracts and receptacular scales similar, hairy, oval-blunt, papery, pale with a dark tip. Involucel with a papery, ciliate, entire or toothed crown. An erect, many-stemmed perennial to 1 m, with leaves deeply cut into narrow toothed or cut lobes. Rocks, dry stony places in the Med. region. YU.AL.GR. Page 434. **1320c. C. uralensis** (Murray) Roemer & Schultes A perennial to 1 m. Flower heads yellow; corolla 8–14 mm long; involucral bracts and receptacular scales dissimilar, the former ovate-blunt, the latter lanceolate-pointed. Basal leaves shallowly lobed or pinnately cut, with 2–4 pairs of oblong side lobes and a larger entire or weakly lobed terminal lobe; stem leaves with narrow oblong lobes. Involucel of fruit with 4 long and 4 short teeth. Grassland, dry places. YU.GR? BG.

1320d. C. ambrosioides (Sibth. & Sm.) Roemer & Schultes A robust perennial herb to 1·5 m, with yellow flower heads. Flowers about 12 mm long; involucral bracts ovate-lanceolate 4–9 mm. Distinguished by its long receptacular scales 12–15 mm, which are lanceolate with a short terminal spine. Leaves shallowly lobed, or pinnately cut, with ovate to lanceolate, somewhat silky-haired, toothed lobes. Involucel of fruit with 4 short teeth. Mountains, sunny grasslands, bushy places YU(south).AL.GR(north and central). **1320e. C. flava** (Sibth. & Sm.) Szabó A perennial to 90 cm, somewhat woody at the base. Flower heads yellow; corolla about 12 mm long; involucral bracts ovate-blunt 3–5 mm; receptacular scales 8–11 mm, oblong-lanceolate acute. Leaves shallowly lobed or pinnately cut, with ovate to lanceolate, toothed lobes, more or less silky-haired. Grassland, stony ground. YU.GR(north).BG.

SUCCISA 1321. *S. pratensis* Moench DEVIL'S-BIT SCABIOUS. Damp meadows, marshes, fens, damp woods. YU.AL.GR.BG.

SUCCISELLA Like *Succisa* but involucel urn-shaped; calyx four-lobed, without bristles. 2 species in our area.

1321c. *S. inflexa* (Kluk) G. Beck A scabious-like nearly hairless plant 60–80 cm, with pale lilac-blue flower heads 1–1·5 cm across, on branched stems which are hairless below. Receptacular scales shorter than the fruits. Basal leaves obovate, though sometimes absent; stem leaves lanceolate-blunt, almost entire, with winged leaf stalks. Involucel about 4 mm, hairless. Wet places. YU.AL.

KNAUTIA Receptacle hairy, scales absent. Involucel four-angled; corona absent; calyx cup-shaped or saucer-shaped with 8–16 minute awns or teeth. A very difficult genus with frequent hybridization and with many intermediate populations. About 23 species in our area, including 13 endemic to it.

Annuals, with a slender root

1324. *K. integrifolia* (L.) Bertol. A slender, erect, branched annual 20–80 cm, with violet flower heads 1·5–3 cm across. Involucral bracts mostly lanceolate, densely grey-haired. Calyx cup-shaped, with 12–24 teeth. Basal leaves in a rosette, toothed or lobed, hairy or hairless; upper leaves entire, linear-lanceolate, somewhat clasping stem. Fields, bushy places, disturbed ground in the Med. region. YU.AL.GR+Is.CR.BG.TR. **1324d. *K. orientalis* L. Readily distinguished by its almost cylindrical involucre and its relatively few (5–10) reddish-purple flowers in each head. Flower heads about 2 cm across; involucral bracts lanceolate, with bristly and glandular hairs. An erect annual

1. *Knautia orientalis* 1324d   **2.** *Scabiosa hymettia* 1328e   **3.** *Pterocephalus papposus* 1324f   **4.** *Tremastelma palaestinum* 1329c   **5.** *Cephalaria leucantha* (1320)

20–60 cm, much branched above; leaves all entire, the upper lanceolate, or some lower leaves pinnately lobed. Bushy places, disturbed ground. GR(north-east).BG.TR. Page 434.

Perennials or biennials with root-stock or thick tap-root
(a) Leaves all undivided
**1323. *K. drymeia* Heuffel (*Scabiosa sylvatica*) A very variable perennial 30–100 cm, generally distinguished by its broad, thin, undivided, toothed, hairy leaves, its erect slender branched stems, and its purple to pink flower heads 1·5–3 cm across. Leaf rosette terminal, flowering stem arising laterally; upper stem leaves more or less heart-shaped, stalkless. Calyx with 8–16 awns. Bushy places, wood verges. YU.AL.GR.BG.   1323a. *K. midzorensis* Form. Flower heads few, pale yellow, pink, or light purple, 3·5–5 cm across. Lower stem leaves broadly lanceolate (4–6 times as long as wide), the upper often clasping, ovate; lower internodes hairless, upper glandular-hairy; stems 60–120 cm. Mountain pastures. YU.AL?GR?BG.   1323b. *K. longifolia* (Waldst. & Kit.) Koch Differs from 1323a in having upper stem leaves with a wedge-shaped or rounded, not clasping base, and longer lower stem leaves (6–9 times as long as wide). Flower heads pinkish-purple, 3·5–5 cm across. Mountain pastures, wood verges. YU.AL.GR(north).

(b) Some leaves deeply divided
**1322. *K. arvensis* (L.) Coulter A very variable perennial or biennial 25–75 cm, which hybridizes with many other species, with lilac to bluish-violet or rarely pink or purple flower heads 2–4 cm across. Basal leaves thin, green, undivided or pinnately lobed; stem leaves mostly on lower half of stem, pinnately cut with 4–12 ovate-lanceolate lateral lobes. Calyx cup-shaped, usually eight-awned. Pastures, open woodland. YU.AL? BG.   1322a. *K. ambigua* Boiss. & Orph. Like 1322 but with pale-yellow or pale-pink flower heads 2–3 cm across. Stems of flower heads usually glandular. Lower stem leaves usually undivided, often grey-hairy, the upper pinnately cut. Bushy places, wood margins. YU(south).GR(north).BG.   1322b. *K. macedonica* Griseb. Distinguished from 1322 by the dark-red flower heads, or flower heads sometimes lilac or pink, 1·5–3 cm across. Leaves evenly distributed up stem, the basal withering at flowering, the upper pinnately cut with an ovate, toothed terminal lobe. Pastures, scrub, open woodland. YU.AL.GR? BG.

PTEROCEPHALUS Distinguished by the calyx which has a short limb and 5–24 feathery-haired bristles; involucel deeply grooved, with terminal awn, minute teeth, or a short crown. Corolla equally five-lobed. 3 species in our area.

**1324f. *P. papposus* (L.) Coulter (*P. plumosus*) An erect, glandular-hairy, scabious-like annual to 60 cm, with hemispherical pink or purple flower heads surrounded by equal or longer involucral bracts. Soon distinguished by its 11–12 delicate feathery, purplish, spreading bristles, topping the conspicuously grooved fruits. Leaves pinnately lobed, the terminal lobe larger, lobes coarsely toothed. Dry hills in the Med. region. Widespread. Page 434.

**1324e. *P. perennis* Coulter A creeping, mat-forming rather woody perennial with usually silvery-grey leaves, and handsome, short-stalked, pink or pale-purplish flower heads about 3 cm across, the outer most flowers larger and spreading. Corolla 12–20 mm long; calyx with 13–16 long, hairy, purplish bristles, very distinctive in fruit; involucel silky-haired, with a crown of short feathery bristles. Leaves 2–5 cm, ovate, toothed, undivided or deeply lobed with terminal lobe broadest, sometimes green and glandular. Mountain rocks. AL.GR. Pl. 48.

SCABIOSA | Scabious Scales of receptacle present. Involucel a cylindrical tube encircling the base of the flower, eight-ribbed and enlarged above into a rounded or funnel-shaped papery corona, with many veins. Calyx usually with 5 long bristles. About 30 species in our area, 9 of which are restricted to it. A difficult genus with many intermediate forms.

Involucel tube with 8 conspicuous grooves from base to apex
(a) Corona with 8 veins
**1326. S. *atropurpurea* L. MOURNFUL WIDOW, SWEET SCABIOUS. Flowers lilac to dark purple, in heads 2–3 cm across, becoming oblong in fruit. Corona broadly funnel-shaped, about as long as the involucel tube; calyx bristles stalked, 3–5 times as long as corona. A biennial 20–60 cm, with most leaves deeply lobed. Dry places. Widespread.

(b) Corona with 20–4 veins
  (i) Flowers lilac-blue, purple, or reddish
1325c. S. *tenuis* Boiss. A slender, usually branched annual 10–70 cm, with purple flower heads 2·5–3 cm across. Involucel tube 2–3 mm, corona about 1 mm; calyx bristles 5–12 mm, stalked. Basal leaves oblanceolate, toothed, the upper leaves twice-cut into narrow segments to 1 mm wide. Rocky places. AL.GR(north and central). **1325d.** S. *taygetea* Boiss. & Heldr. A densely yellow-haired or greenish long-hairy perennial with reddish flower heads 2–4 cm across, of limestone rocks. Corona shorter than involucel tube, about 24-veined; calyx bristles up to 6 times as long as corona. Basal leaves obovate, simple or lobed, long-stalked; stem leaves lobed and with a very large, rounded to elliptical, terminal lobe. YU(south).GR(south and central). Page 88.

1325e. S. *silenifolia* Waldst. & Kit. Flower heads lilac-blue, 1·5–2·5 cm across, with involucral bracts ovate-lanceolate as long as or shorter than the flowers. Marginal flowers 9–12 mm, longer than the central florets. Calyx bristles 2–3 times as long as corona. Leaves of non-flowering rosettes and lower stem leaves spathulate entire, ciliate; upper stem leaves deeply cut with lanceolate segments, the terminal widest. Rocky pastures in mountains. YU.AL. Page 185.

  (ii) Flowers yellow or whitish
**1327. S. *ochroleuca* L. A perennial with finely hairy stems, and yellow or white flower heads 1·5–2·5 cm across, with the outer flowers distinctly longer than the inner. Involucral bracts finely hairy. Calyx bristles 2–3 times as long as corona. Non-flowering rosette leaves obovate-lanceolate, toothed; lower stem leaves entire or broadly lobed, finely hairy; upper stem leaves pinnately lobed. Dry meadows, stony places. YU.AL.BG.TR.

1327a. S. *webbiana* D. Don Similar to 1327 but differing in having leaves densely covered with woolly hairs. Involucral bracts densely white-woolly-haired; flowers pale yellow. Lower leaves entire, toothed, or lobed; upper stem leaves once- or twice-lobed. Dry stony places, alpine meadows. Widespread on mainland. **1327b.** S. *triniifolia* Friv. Similar to the previous two species but differing in having lower stem leaves and non-flowering rosette leaves 2–3 times cut into narrow, linear slightly hairy lobes. Flower heads very pale yellow. Dry stony places. Widespread on mainland.

Involucel tube rounded, but with 8 pits towards the apex
(a) Calyx bristles shorter than the corona
  (i) Leaves at least some with 3–9 lobes, the early ones hairless
1328e. S. *hymettia* Boiss. & Spruner A tufted perennial with woody stems covered with old leaf scars, and terminal clusters of simple, or three- to five-lobed, densely silvery-

haired leaves. Flower heads solitary, pale blue-violet, 2·5–4 cm across, on long stems to 25 cm. Outer flowers twice as large as inner; involucral bracts ovate, blunt. Rock crevices. GR(central and south). Page 434.

(ii) Leaves entire, the early ones hairy

1328f. *S. albocincta* W. Greuter An endemic of rock crevices in Crete. A densely tufted, woody-based perennial, with a white-woolly stem and elliptical, silvery-hairy leaves. Flower heads lilac, 3·5–5 cm across, solitary, borne on a long stem 30–40 cm, much longer than the leaves. Involucel tube densely hairy, pits about 2 mm long; corona hairy. CR. **1328g.** *S. minoana* (P. H. Davis) W. Greuter Like 1328f but with broader elliptic-obovate leaves 1½–2¾ times as long as wide (3½–5 times as long as wide in 1328f), and leaves with adpressed-silky hairs and short curved hairs (long straight hairs in 1328f). Calcareous rocks. CR. Page 67.

(b) Calyx bristles as long as or longer than corona
(i) All leaves entire or toothed

1328. *S. graminifolia* L. GRASS-LEAVED SCABIOUS. Distinguished by its grass-like, linear, densely silky-haired leaves 1–3 mm wide, borne on woody-based stems. Flower heads lilac, 3–4 cm across, on stems 20–30 cm, which are leafy in the lower half. Outer flowers about 2 cm, twice as long as inner; involucral bracts triangular-ovate about half as long as flowers. Rocky and stony places in the Med. region. YU.AL.GR. **1328h. *S. rhodopensis* Stoj. & Stefanov Differs from 1328 in having pale-yellow flower heads 1·5–3 cm across, and narrower leaves mostly 1–2 mm wide. Limestone rocks in Rhodope Mountains. GR.BG.

(ii) Some leaves once or twice pinnately lobed

1329. *S. argentea* L. (*S. ucranica*) Flower heads yellowish-white, yellow, or pinkish-yellow, 1·5–2·5 cm across. Outer flowers 12–15 mm, distinctly larger than the central flowers; involucral bracts narrow-lanceolate. A variable perennial or biennial with branched stems 30–70 cm, and upper leaves linear, entire. Stony hills. Throughout the mainland. **1329a.** *S. crenata* Cyr. A tufted woody-based perennial with pinkish-lilac, usually solitary flower heads 2–4 cm across, with marginal flowers larger than central flowers. Leaves oblong-ovate, the lower spathulate, toothed, the upper once or twice pinnately lobed with broadly elliptic or narrow-lanceolate lobes which are hairy or hairless. Involucral bracts elliptic-ovate, densely white-woolly; calyx bristles 2–3 times as long as the corona. Rocky places in the Med. region. YU.AL.GR. Page 88. **1329b.** *S. sphaciotica* Roemer & Schultes A low tufted, woolly-haired perennial of mountain screes of Crete. Flower heads small 1–1·5 cm across, with 7–8 lilac-pink flowers, borne on usually leafless stems 3–10 cm. Involucral bracts ovate-lanceolate, white-woolly, blunt. Involucel tube less than 2 mm long; corona with 25–30 veins, half as long as calyx bristles. CR. Page 71.

TREMASTELMA Like *Scabiosa*, but calyx shortly stalked, and with 10 feathery-haired bristles.

1329c. *T. palaestinum* (L.) Janchen A slender, little-branched annual to 50 cm, with violet, hemispherical, long-stalked flower heads about 2 cm across. Outer flowers spreading, hairy outside; involucral bracts lanceolate, covered with long, bristly white hairs. Fruit very distinctive and beautiful: top-shaped with 8 ribs and 8 pits above, with a delicate papery cup-shaped corona, and surmounted by a short-stalked calyx with 10 long radiating bristles. Stony, grassy places in the Med. region. Widespread. Page 434.

CAMPANULACEAE | Bellflower Family

CAMPANULA | Bellflower A large and often difficult genus with about 90 species in our area, of which more than 50 are endemic.

Fruit splitting by almost apical or lateral pores, or valves
(a) Perennials or biennials
1334. *C. rapunculus* L. RAMPION. Meadows, bushy places. Widespread on mainland. **1334a.** *C. spathulata* Sibth. & Sm. Flowers 1–5, blue, on long almost leafless stems 20–30 cm, forming a lax inflorescence. Corolla broadly funnel-shaped, 1–2 cm long, longer than the awl-shaped calyx lobes. Upper leaves lanceolate, long-pointed, stalkless; lower oval or oblong, stalked; root turnip-shaped. Fruit obconical, grooved. Mountain meadows, scrub, screes. YU.AL.GR.CR.BG. Page 88. ****1335.** *C. patula* L. SPREADING BELLFLOWER. Woods and meadows in hills and mountains. YU.AL.GR.BG.

**1336. *C. persicifolia* L. NARROW-LEAVED BELLFLOWER. Distinguished by its large open-bell-shaped, blue-violet flowers borne in a slender, unbranched, spike-like cluster of 2–8 flowers. Corolla 3–4 cm across, shallowly lobed; calyx lobes lanceolate-pointed, half as long as corolla. A hairless unbranched perennial to 70 cm, with linear-lanceolate stem leaves. Woods, bushy places. Widespread on mainland.

(b) Annuals
1335b. *C. ramosissima* Sibth. & Sm. A simple or branched annual 20–40 cm, distinguished by its many long-stalked, salver-shaped, violet-blue flowers with paler centres. Corolla 1–3 cm long, with spreading lobes as long as or longer than the tube; calyx lobes narrowly lanceolate, long-pointed, three-veined, hairy, mostly shorter than the corolla. Leaves lanceolate to spathulate, toothed, the upper stalkless, leaves and stems bristly-haired. Fruit top-shaped, strongly grooved with bristles on the ridges. Rocks, stony places, olive groves, in mountains. YU.AL.GR. Pl. 49. **1335c.** *C. phrygia* Jaub. & Spach Differs from 1335b in having calyx lobes one-veined, and a smaller corolla 6–7 mm long, with spreading lobes twice as long as the calyx. Grassy places. YU.AL.GR.BG.

1335d. *C. sparsa* Friv. An erect, branched, hairy annual 20–40 cm, with slender flower stalks, bearing deeply lobed, bell-shaped, violet flowers. Corolla lobed to one-third, variable, about 3 cm long, or smaller-flowered and 12–23 mm long. Calyx lobes about as long as the corolla, awl-shaped, finely toothed at base, longer than the ovary. Basal leaves oblong-lanceolate, toothed, the upper linear-pointed. Woods, thickets. Widespread on mainland.

Fruit splitting by basal pores or valves
(a) Fruit mostly with 5 cells; stigmas mostly 5
  (i) Calyx with very small appendages between the calyx lobes
1331d. *C. celsii* A.DC. A velvety-haired, branched, ascending biennial 20–30 cm, with terminal and axillary erect, tubular, velvety lilac or blue-lilac flowers 1·8–3 cm long. Calyx lobes one-quarter as long as the corolla, triangular to lanceolate; calyx appendages very small, tooth-like. Basal leaves variable, usually not more than 5 cm, irregularly lobed, with ovate-acute, toothed terminal lobe, the upper leaves obovate, toothed, stalkless. Rocks. GR(south-east). Pl. 49. **1331e.** *C. anchusiflora* Sibth. & Sm. A shortly hairy biennial with a robust, long erect central stem and spreading lateral stems, and a large basal rosette of shallowly lobed, toothed, and stalked leaves more than 6 cm, with terminal heart-shaped lobes. Flowers blue, 12–15 mm long, tubular, in a branched inflorescence; calyx appendages very small. Limestone rocks. GR(east).

  (ii) Calyx with large appendages between the calyx lobes
1331f. *C. tubulosa* Lam. A biennial of rock fissures in west Crete with numerous slender

branches, each with one to several blue-lilac, tubular, velvety flowers about 2 cm long. Calyx appendages conspicuous, ovate, becoming inflated and papery in fruit; calyx lobes triangular-acute, bristly-haired, Lower leaves oblong-ovate, toothed, long-stalked, the upper oblong-lanceolate, saw-toothed, stalkless, all softly hairy. Damp rock crevices. CR. Page 67. **1331g.** *C. pelviformis* Lam. An ascending, often unbranched biennial 20–30 cm, with large very broad urn-shaped, blue-lilac, or rarely white flowers about 3 cm, which are as broad as long. Calyx appendages deflexed, almost as long as calyx lobes. Leaves rough-hairy, the basal ovate-acute, saw-toothed, stalked, the upper stalkless. Stony slopes, thickets. CR(east and central).

1331h. *C. laciniata* L. A hairy erect perennial 20–60 cm, with a many-flowered inflorescence of very large, broadly bell-shaped flowers 4–5 cm across, with broadly ovate-acute, ascending lobes. Calyx lobes triangular; calyx appendages ovate, deflexed. Leaves to 30 cm, ovate, deeply cut into narrow toothed lobes. Limestone rocks. GR(south).CR. **1331i.** *C. topaliana* Beauverd Stems numerous, spreading from a basal rosette of silvery or woolly-haired lobed leaves, with terminal and axillary blue flowers varying from 8 to 15 mm long. Corolla tubular, hairy, lobes erect or spreading; calyx lobes ovate long-pointed, the calyx appendages about as long as the ovary. Upper stem leaves ovate to elliptic, toothed. Limestone rocks. GR(south). Page 97.

(b) Fruit with 3 cells; stigmas 3
  (i) Calyx with additional appendages between the calyx lobes
1330a. *C. alpina* Jacq. A low-growing perennial 10–20 cm, of mountain pastures, with a rosette of narrow leaves, and a pyramidal cluster of usually numerous, blue, bell-shaped flowers 1·5–2 cm long. Calyx lobes long-pointed, shaggy-haired, shorter than corolla; calyx appendages ovate-acute, woolly-haired. Stem leaves strap-shaped. Mountain pastures. YU.AL.BG. Pl. 49. **1330b.** *C. oreadum* Boiss. & Heldr. A rather delicate creeping, greyish-hairy plant of rock crevices and fissures, endemic to Mount Olympus. Flowers deep blue, usually solitary, long-stalked; corolla tubular-bell-shaped, 2–3·5 cm long. Calyx lobes one-third as long as corolla, densely hairy; calyx appendages very short. Basal leaves oblong-spathulate, stalked, adpressed-hairy, the upper oblong-linear, stalkless. Alpine rocks. GR(Olympus). Page 126.

1330c. *C. rupicola* Boiss. & Spruner A creeping greyish, fragile-stemmed perennial to 10 cm, of rock crevices in the mountains, with 1–3 blue-purple narrowly bell-shaped flowers about 3 cm long. Calyx lobes broadly ovate-oblong blunt, ciliate, finely toothed; calyx appendages very short. Basal leaves long-stalked, ovate with wedge-shaped base, the uppermost smaller, linear-lanceolate nearly stalkless. Mountain rocks. GR(south-central). Pl. 48. **1330d.** *C. lingulata* Waldst. & Kit. A bristly-hairy biennial 20–30 cm, with one or several stems bearing dense terminal heads of violet, tubular-funnel-shaped flowers 2–2·5 cm long. Flower heads encircled by lanceolate involucral leaves. Calyx bristly-haired, the lobes oblong-blunt, the appendages ovate. Basal leaves oblong-spathulate, narrow to the base, toothed, the upper leaves narrower. Grassy places, scrub. Widespread on mainland. Pl. 48.

1330e. *C. formanekiana* Degen & Dörfler A handsome softly hairy biennial 10–20 cm, with very large white or blue-lilac, broad bell-shaped flowers 5–6 cm long, borne in a branched pyramidal cluster. Flowers long-stalked, solitary and terminal on lateral branches; corolla sparsely hairy. Calyx lobes large, triangular-ovate, finely toothed; appendages triangular, reflexed. Leaves ovate-spathulate, toothed, narrowed to the winged leaf stalk, finely hairy. Rock fissures in Macedonia. YU.GR. **1330f.** *C. sibirica* L. A slender erect, branched, rough-haired biennial 20–50 cm, with a lax pyramidal cluster of blue-lilac, tubular-bell-shaped flowers 1·5–4 cm long. Calyx bristly-haired, lobes long-pointed; calyx appendages lanceolate, little shorter than the calyx lobes. Lower leaves obovate, toothed, stalked, the upper lanceolate, stalkless. A very variable species. Dry grasslands, rocks. YU.AL.BG.

1330g. *C. incurva* A.DC. A handsome erect biennial with branched stems bearing large blue-lilac bell-shaped flowers up to 4 cm long. Calyx lobes broadly triangular ovate; calyx appendages ovate, as long as ovary. Lower leaves ovate-heart-shaped, stalked, the upper ovate, short-stalked, the uppermost stalkless, all finely hairy, toothed. Scrub, rocky places. GR(east). Pl. 48.

(ii) Calyx without additional appendages between the lobes
  (x) Flowers in dense terminal heads
1333. *C. glomerata* L. CLUSTERED BELLFLOWER. Meadows, scrub, forest margins. YU.AL.GR.BG. **(1333). *C. cervicaria* L. Woods, damp meadows. YU.AL.GR.BG. **1333a.** *C. foliosa* Ten. Like 1333 with a terminal cluster of rather large funnel-shaped violet flowers 2–3·5 cm long, on stems 30–50 cm, but with large ovate, toothed involucral leaves. Calyx lobes narrow-linear, ciliate, shorter than corolla tube. Basal leaves broadly ovate, with a heart-shaped or rounded base, stem leaves elliptic-ovate, toothed, all stalked except the uppermost. Meadows in mountains, forest verges. YU.AL.GR.BG? **1333b.** *C. moesiaca* Velen. A hairy biennial to 40 cm, with a dense terminal cluster of blue–lilac flowers. Corolla about 3 cm, more than twice as long as the triangular calyx lobes. Basal leaves oblong, toothed, stalked, stem leaves lanceolate, shortly stalked, the uppermost stalkless, lanceolate-long-pointed, with heart-shaped base. Mountain meadows. YU.AL.BG. Pl. 48.

(xx) Flowers in short lax clusters, or solitary
1337. *C. erinus* L. ANNUAL BELLFLOWER. A slender, branched, rough-haired annual usually 3–10 cm, with small pale-blue, lilac, reddish, or white tubular flowers 3–5 mm long. Flower stalks very short; calyx lobes lanceolate, shorter than corolla, spreading in a star in fruit. Leaves 1–2 cm, oval-wedge-shaped, toothed, stalkless. Grassy, stony places, walls in the Med. region. Widespread, except BG. **1337a.** *C. drabifolia* Sibth. & Sm. Like 1337 but corolla blue, bell-shaped, larger 8–16 mm long, twice as long as calyx lobes, which are usually spreading in fruit. Dry ground, rocks in the Med. region. GR(south)+Is.CR. Page 41.

1338. *C. cochleariifolia* Lam. Rocks, screes, stony places mainly in mountains. YU.AL.BG. Page 191. **(1338).** *C. scheuchzeri* Vill. Mountain meadows. YU. AL.BG. **1338a.** *C. waldsteiniana* Schultes A delicate hairless, many-stemmed perennial 20–30 cm, of rock fissures in Yugoslavia, with lanceolate, toothed upper stem leaves, and a short terminal cluster of small blue-violet flowers each 1·5–2 cm across. Corolla with wide-spreading lobes about as long as the tube; calyx lobes awl-shaped, much shorter than corolla; flower stalks spreading widely. Limestone rocks in mountains. YU(west). Page 443. **1338b.** *C. fenestrellata* Feer A small ascending or pendent perennial 15–20 cm, of limestone crevices, distinguished by its oval-heart-shaped, stalked leaves which are conspicuously double-toothed. Flowers numerous, 1–2 cm across, blue-violet, stalked, in short terminal clusters; corolla with wide-spreading lobes; pollen blue. Calyx lobes linear-lanceolate. Plants hairless or woolly-haired. Rocks, crevices. YU.AL.

1338c. *C. hawkinsiana* Hausskn. & Heldr. A perennial with numerous slender, flexuous, spreading or ascending leafy stems 10–20 cm, with blue-violet flowers on long slender stalks. Corolla 10–12 mm long, with spreading ovate lobes 2–3 times as long as the lanceolate calyx. Leaves hairless, rounded, entire or toothed, the lower stalked, the uppermost stalkless. Crevices in serpentine rocks in mountains. AL.GR(north). Page 114.

(xxx) Flowers in elongated spikes, branched or not
1340. *C. trachelium* L. BATS-IN-THE-BELFRY. Woods, bushy places. Widespread on mainland. **1341. *C. latifolia* L. LARGE BELLFLOWER. Woods. YU.BG. **1342.** *C. rapunculoides* L. CREEPING BELLFLOWER. Grassy places, fields, walls. YU.AL.BG. **1343.** *C. bononiensis* L. Distinguished by its small blue-lilac, pendulous flowers 1–2·5 cm long, borne in a long, dense spike-like cluster on stems to 70 cm. Corolla with spreading lobes, hairy within; calyx lobes triangular-lanceolate spreading, rough, much shorter than corolla. Basal leaves ovate-pointed, toothed, stalked, shortly hairy above, greyish-woolly beneath, the uppermost leaves stalkless and half clasping the stem. Fruit pendulous. Meadows, bushy places. YU.AL.BG.

1343a. *C. versicolor* Andrews A usually hairless but very variable perennial with stout stems 20–40 cm, and a spike-like or branched inflorescence of clusters of pale-blue or lilac flowers with darker purple centres. Corolla 1·5–2·5 cm, about 3 cm across, salver-shaped, with spreading triangular lobes; calyx lobes narrowly lanceolate. Basal leaves leathery, stalked, ovate or ovate-heart-shaped, toothed, the uppermost almost stalkless with a wedge-shaped base. Fruit erect. Rocky places in hills and mountains. YU.AL.GR.BG. Pl. 48. **1343b.** *C. pyramidalis* L. Like 1343a but leaves with glandular-tipped teeth. Inflorescence pyramidal, many-flowered; corolla broadly bell-shaped, to 5 cm across, pale blue-lilac. Flowering stems to 1·5 m; stem leaves ovate-lanceolate, stalkless. Rocks, walls. YU.AL.

SYMPHYANDRA Like *Campanula* but anthers, at flowering, joined to form a tube round the style. Fruit splitting by 3 valves near the base.

Calyx without an additional reflexed appendage between each calyx lobe
1344a. *S. cretica* A.DC. A hairless perennial up to 45 cm, with large heart-shaped, toothed, stalked lower leaves, and pendulous blue or white flowers in a one-sided cluster. Flowers bell-shaped, about 3 cm long. Calyx lobes 1·5–2 cm, linear to lanceolate, erect or ascending. Rocks, walls. GR(Samothrace, Sporadhes).CR. Page 66. **1344b.** *S. wanneri* (Rochel) Heuffel Distinguished from 1344a by its elliptic to lanceolate lower leaves which are gradually narrowed at the base to a winged stalk. Flowers violet, 2–3·5 cm long, in a branched, one-sided cluster. A hairy perennial to 40 cm. Mountain rocks. YU(east).BG.

Calyx with an additional reflexed appendage between each calyx lobe
1344c. *S. hofmannii* Pant. A hairy biennial with pendent yellowish-white bell-shaped flowers in a one-sided, much-branched inflorescence. Corolla 2–3 cm long; calyx with ovate lobes 1·5–2 cm, and conspicuous deflexed appendages. Leaves ovate to lanceolate, gradually narrowed at the base, coarsely toothed, the lower with a winged stalk. Rocky places. YU(Bosnia). Pl. 49.

ADENOPHORA Distinguished from *Campanula* by the tubular disk surrounding the base of the style.

1345. *A. lilifolia* (L.) A.DC. Flowers pale blue or whitish, funnel-shaped 1·2–2 cm, with protruding styles. An erect perennial to 1 m, with a branched inflorescence, and stem leaves lanceolate, toothed, the lower short-stalked, the upper stalkless. Woods, damp fields. YU(north and central). Page 443.

LEGOUSIA (SPECULARIA) Erect annuals, with narrow leaves; fruit cylindrical.
Corolla shorter than calyx
1346. *L. hybrida* (L.) Delarbre Flowers reddish-purple to lilac, few, stalkless in small terminal clusters. Fields, stony places. YU.AL.GR.CR.TR. **(1346).** *L. falcata* (Ten.)

Fritsch Flowers violet, paired or solitary in leaf axils, forming a lax spike at least half total length of stem. Thickets, rocky slopes in the Med. region. YU.GR + Is.CR. Page 443.

Corolla longer than calyx

1347. *L. speculum-veneris* (L.) Chaix VENUS'S LOOKING-GLASS. Flowers deep violet-purple, about 1 cm long, at least as long as the calyx lobes, which are shorter than or about as long as the ovary at flowering. Fruit 1–1·5 cm. Fields, cultivated ground. Widespread. **(1347). *L. pentagonia* (L.) Druce Like 1347 but flowers larger with corolla 1·5–1·8 cm long, and calyx lobes ⅓–½ as long as the ovary at flowering. Fruit 2–3 cm. Fields and cultivated ground in the Med. region. GR + Is.CR.BG.TR. Page 41.

TRACHELIUM (incl. **DIOSPHAERA**) Flowers in a flat-topped cluster. Corolla tubular, with 5 lobes; style much longer than corolla, thickened towards apex. Fruit with 2–3 pores near the base.

1348a. *T. jacquinii* (Sieber) Boiss. (incl. *T. rumelianum*) A plant of rock crevices, with a stout stock, many leafy stems, and dense terminal, more or less flat-topped clusters of pale bluish-lilac flowers, with long protruding stamens and styles. Corolla with a slender tube about 5 mm long and spreading lobes. Leaves leathery elliptic, toothed, mostly stalkless 2·5–5 cm. GR + Sporadhes.CR.BG(south). Pl. 49. **1348b.** *T. asperuloides* Boiss. & Orph. A cushion plant, with tiny crowded, rounded shining stalkless leaves, and 1–5 pink flowers in the axils of the upper leaves. Corolla tube about 6 mm, twice as long as lobes. Leaves to 5 mm. Rock crevices. GR(south).

PETROMARULA Flowers funnel-shaped, divided nearly to the base into 5 linear lobes. Stamens with filaments swollen at base; anthers free. Fruit with 3 pores opening at the middle.

1348c. *P. pinnata* (L.) A.DC. A robust perennial to 50 cm or more, endemic to Crete, with pinnately lobed leaves (unique in this family), and with a long unbranched spike of pale-blue flowers in clusters. Corolla about 1 cm long, lobes linear; style much longer, with a knob-like stigma. Leaves to 30 cm, leaf lobes further toothed or cut. Rocks. CR. Page 66.

ASYNEUMA Like *Phyteuma* but flowers solitary, or in axillary clusters in a long interrupted inflorescence. Corolla deeply divided into narrow lobes which are fused only at the base. Fruit cylindrical, with apical pores. 4 similar-looking species in our area.

1349. *A. limonifolium* (L.) Janchen An erect often unbranched perennial 10–100 cm, with leaves mostly basal, and a long, interrupted spike-like cluster of blue-lilac flowers. Flowers in clusters of 1–4, in axils of tiny triangular bracts; lobes of corolla narrow 8–9 mm; calyx lobes 1–2 mm. Rosette leaves linear to oblanceolate, undulate, entire or toothed, 3–6 cm. A variable species. Stony slopes, rocky ground on limestone. Widespread on mainland. Pl. 49.

1349b. *A. anthericoides* (Janka) Bornm. Similar to 1349 with a basal rosette of leaves and a long cluster of blue flowers with narrow petals. Differing in the stalked flowers (stalk 1–4 mm), longer calyx lobes 4–5 mm, and larger flowers 10–12 mm, in a laxer cluster. Dry stony places. YU(south-east).AL? BG. Page 148. **1349a.** *A canescens* (Waldst. & Kit.) Griseb. & Schenk An erect perennial with more or less uniformly leafy stems 40–70 cm, with greyish, woolly-haired, narrow-elliptic, toothed leaves, and a long interrupted spike of blue flowers. Flowers in clusters of 2–6; corolla lobes linear 7–9 mm. Stem leaves variable. Steppes, mountain grassland. YU.AL.GR.BG. Page 443.

1. *Jasione laevis* 1355d **2.** *Legousia falcata* (1346) **3.** *Campanula waldsteiniana* 1338a **4.** *Adenophora lilifolia* 1345 **5.** *Edraianthus serbicus* 1354e **6.** *Asyneuma canescens* 1349a

PHYTEUMA | Rampion 6 species in our area, mostly in the north and west.

Flowers in elongated heads, twice as long as broad, or more
1350. *P. spicatum* L. SPIKED RAMPION. Flowers whitish to greenish-yellow or bluish; bracts ovate. Meadows, woods. YU. **1351. *P. ovatum* Honckeny (*P. halleri*) Flowers blackish-violet; bracts linear. Mountain meadows. YU.

Flowers in globular heads, as broad as or broader than long
1352. *P. orbiculare* L. ROUND-HEADED RAMPION. Mountain meadows. YU.AL. Page 191. **1352b. *P. pseudorbiculare* Pant. Like 1352 but basal leaves more or less rounded, toothed (linear-lanceolate to rounded-heart-shaped in 1352). Stem leaves diminishing in size above, more or less stalked. Flowers intense blue-violet; bracts broadly ovate, toothed. Mountain meadows, rocks. YU(west).AL. **1352c.** *P. confusum* A. Kerner Flower heads dark blue-violet or rarely white, on stems 1–15 cm; bracts rounded to ovate, entire or with a few teeth. Leaves linear to oblong-spathulate, toothed. Acid rocks, and stony pastures. YU.AL.BG. Pl. 49.

EDRAIANTHUS Like *Campanula* but fruit splitting irregularly at the apex. Flowers in terminal heads, closely surrounded by a leafy involucre. 9 similar-looking species in our area, 8 of which are restricted to it.

Leaves not ciliate, irregularly toothed; basal and upper leaves dissimilar
1354c. *E. parnassicus* (Boiss. & Spruner) Halácsy. Distinguished from all other species by the leaves having no ciliate hairs on the margin, and being irregularly rounded-toothed. Flowers violet, 1–2 cm long, 3–4 in globular, short-stalked, clusters. A tufted, woody-based perennial with basal leaves oblong-lanceolate to spathulate, stalked, the upper broad or narrow lanceolate with rounded base, stalkless. Mountains. GR.

Leaves ciliate, not toothed; basal and upper leaves similar
(a) Leaves spathulate
1354d. *E. serpyllifolius* (Vis.) A.DC. A tufted plant distinguished by its spathulate, blunt or notched leaves, and several stems 2–5 cm, with solitary dark-violet flowers 1·5–2 cm long, with blunt bracts shorter than the flowers. Mountain rocks. YU(west).AL(north). Pl. 50.

(b) Leaves linear to linear-lanceolate
1354. *E. graminifolius* (L.) A.DC. FALSE BELLFLOWER. A low tufted, rosette-forming perennial with narrow leaves and several erect stems 5–20 cm, bearing a globular cluster of blue-violet or rarely white flowers. Corolla 1–2 cm long, or rarely more, funnel-shaped, usually hairless; bracts ovate long-pointed, usually shorter than the flowers. Rosette leaves linear, 1–4 cm by 0·5–4 mm, ciliate only at base, entire. A very variable plant. Rocks in mountains. YU.AL.GR. Pl. 49. **1354a.** *E. tenuifolius* (Waldst. & Kit.) A.DC. Like 1354 but leaves very narrow 0·4–1·5 mm wide, ciliate to apex. Leafy bracts broadly ovate, abruptly narrowed to a long point as long as or longer than the flowers. Flowers blue-violet, about 2 cm long, in a head of up to 15; calyx lobes linear. Ovary hairy. Rocks in mountains. YU.AL.GR. Page 178.

1354b. *E. dalmaticus* (A.DC.) A.DC. Flowers blue-violet, several, surrounded by broadly ovate, slender-pointed bracts, the outer up to twice as long as the flowers. Calyx lobes broadly triangular, shorter than the calyx tube. Basal leaves linear-lanceolate, mostly 2–3 mm wide, long ciliate at the base, hairless above; stem leaves few. Stems hairless, 3–7 cm. Mountain rocks. YU(west). **1354e.** *E. serbicus* Petrović Like 1354b but outer bracts shorter than or as long as the flowers, and with a short abrupt tip (not long and gradually tapering). Stems hairy, 12–18 cm. Mountains. YU.BG. Page 443.

1354f. *E. pumilio* (Portenschl.) A.DC. A very dwarf perennial with densely leafy stems 1–3 cm, and solitary blue-violet flowers usually 14–18 mm long. Distinguished by its densely grey-haired linear leaves 1 mm wide, with inrolled margins. Mountains. YU(Croatia). **1354g.** *E. dinaricus* (A. Kerner) Wettst. Differs from 1354f in having taller stems, usually 2–6 cm, which are sparsely leafy above, and wider leaves 1·5–2·5 mm. Flowers blue-violet, 12–15 mm. Mountains. YU(Croatia).

JASIONE | Sheep's Bit

1355. *J. montana* L. SHEEP'S BIT. Sandy places, woods. YU.BG. **1355d. *J. laevis* Lam. (*J. perennis*) A densely tufted perennial 5–15 cm, with a stout stock, narrow nearly entire leaves without undulate margins (margine undulate in 1355), and numerous non-flowering shoots. Flower heads blue, 1–2 cm across, with numerous usually green involucral bracts which are ovate to triangular, deeply toothed, the teeth bristly-tipped. Calyx lobes hairless. Alpine meadows. YU.AL.GR.BG. Page 443. **1355g.** *J. bulgarica* Stoj. & Stefanov Like 1355d but stems, leaves, and bracts quite hairless (sparsely or densely hairy in 1355d). Leaves oblanceolate. Flowers bluish-lilac. Mountain pastures, pine scrub. BG. Page 137. **1355h.** *J. heldreichii* Boiss. & Orph. Like 1355d but outer involucral bracts lanceolate to linear-lanceolate, very deeply toothed, the teeth bristly-tipped. Flowers bluish-lilac. Leaves hairy. Often a biennial. Rocks in mountains. YU.AL.GR.BG.

LAURENTIA Flowers solitary, tubular, two-lipped. Stamens fused by their anthers.

1356a. *L. gasparrinii* (Tineo) Strobl (incl. *L. tenella*) A slender hairless annual or perennial with tiny long-stalked, blue, lilac, or white tubular five-lobed flowers 4–11 mm. Leaves obovate to oblong, rounded-toothed or nearly entire, often in a rosette; stems to 25 cm. Marshes, springs. GR.CR.TR.

COMPOSITAE | Daisy Family

A very important family with about 100 genera of native plants in our area, and more than 850 species. Many of these are widespread in Europe, and only the most distinctive species of our area are described briefly, or are keyed.

Subfamily **ASTEROIDEAE** Flower heads with disk-florets, and with or without ray-florets.
Genera not included: *Bellium* 1 sp.; *Filago* 7 spp.; *Logfia* (part of *Filago*) 3 spp.; *Bombycilaena* 2 spp.; *Omalotheca* (part of *Gnaphalium*) 6 spp.; *Filaginella* (part of *Gnaphalium*) 1 sp.; *Gnaphalium* 1 sp.; *Galinsoga* 2 spp.; *Leucanthemopsis* 1 sp.; *Chlamydophora* 1 sp.; *Ligularia* 1 sp.; *Saussurea* 1 sp.; *Wagenitzia* 1 sp.

EUPATORIUM ****1357.** *E. cannabinum* L. HEMP AGRIMONY. Damp places, marshes, clearings in woods, scrub. Widespread on mainland.

SOLIDAGO ****1358.** *S. virgaurea* L. GOLDEN-ROD. Woods, alpine meadows. Widespread on mainland. **(1359).** *S. gigantea* Aiton Native of N. America; introduced to YU.BG.

BELLIS 4 species in our area. **1360.** *B. perennis* L. DAISY. Meadows, damp grassy places. Widespread.**(1360).** *B. sylvestris* Cyr. SOUTHERN DAISY. Distinguished from 1360 by the oblong-spathulate leaves up to 25 mm wide which are gradually narrowed to the short leaf stalk (abruptly narrowed in 1360), leaf blade three-veined, and flower heads larger 2–3·5 cm across. Ray-florets white above, pink or purple beneath. Grassy places

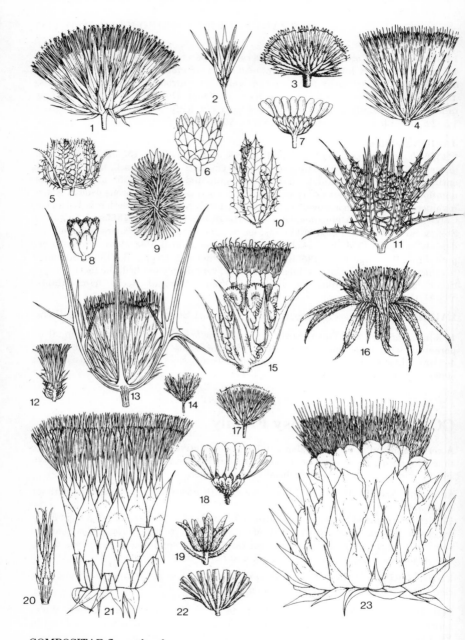

COMPOSITAE flower-heads
1. *Jurinea mollis* 1476   **2.** *Crupina vulgaris* 1496   **3.** *Inula verbascifolia* 1388b
4. *Galactites tomentosa* 1493   **5.** *Atractylis cancellata* 1470   **6.** *Xeranthemum
inapertum* (1464)   **7.** *Anacyclus radiatus* (1414)   **8.** *Helichrysum stoechas* 1385
9. *Xanthium strumarium* 1401   **10.** *Cnicus benedictus* 1509   **11.** *Carthamus lanatus*
1508   **12.** *Cardopatum corymbosum* 1463   **13.** *Notobasis syriaca* 1481

in the Med. region. Widespread. **1361. *B. annua* L. ANNUAL DAISY. An annual, hardly forming rosettes, and with stems 5–12 cm, usually branched below and bearing a few leaves. Flower heads white, small 5–15 mm across. Grassy and rocky places in the Med. region. Widespread.

ASTER 9 native species in our area, and 4 which have escaped from cultivation and are locally naturalized.

Stems unbranched; flower heads solitary
**1363. *A. alpinus* L. ALPINE ASTER. Rocks, mountain pastures. YU.AL.GR.BG. Page 93. **1363a. *A. bellidiastrum* (L.) Scop. A small daisy-like perennial 10–30 cm, of the mountains, with rounded to obovate, toothed or entire, stalked leaves in a basal rosette, and leafless hairy stems (leafy in 1363) bearing solitary flower heads. Ray-florets white or pinkish, up to 50, about 1 cm long; involucral bracts in two rows, linear-pointed, white-margined. Fruit with pappus. Rocks, woods. YU.AL. Page 184.

Stems branched; flower heads several or many
**1364. *A. amellus* L. EUROPEAN MICHAELMAS DAISY. Flower heads conspicuous, 3–5 cm across, with usually bluish-lilac ray-florets and yellow disk-florets, usually rather few in a lax, flat-topped cluster. Lower leaves oval-elliptic, stalked, the upper lanceolate, stalkless; stems 10–70 cm. A very variable plant. Rocky places. YU.AL.GR? BG. **1365. *A. tripolium* L. SEA ASTER. Distinguished by its narrow, fleshy, hairless leaves and its lax branched clusters of bluish-purple flower heads each 1–3 cm across. Coastal salt-marshes, sea cliffs, salt-rich soils. Widespread.

1366. *A. sedifolius* L. Usually a perennial 25–120 cm, distinguished by its linear-lanceolate to elliptic, entire leaves which are often strongly gland-dotted and sometimes with grey cobweb hairs. Flower heads about 1·5 cm across, numerous in a dense cluster; ray-florets lilac or blue, usually 5–10, sometimes absent. A very variable species. Hills, dry uncultivated ground. YU.AL.BG. **1366a. *A. linosyris* (L.) Bernh. (*Linosyris vulgaris*) Flower heads small, about 1 cm long, narrowly funnel-shaped, golden yellow, without ray-florets, numerous in a dense flat-topped cluster. A densely leafy, spreading or ascending perennial 10–70 cm, with narrow stalkless, often gland-dotted leaves, with one vein. Rocks, sunny hills. Widespread on mainland.

ERIGERON | Fleabane 7 species in our area, mostly similar-looking mountain species which show many intermediate and local variants.

**1369. *E. acer* L. BLUE FLEABANE. Cultivated ground, rocks, walls, waste places. Widespread on mainland. **1369a. *E. atticus* Vill. Distinguished from 1369 by the dense glandular hairs covering the whole plant (1369 without glandular hairs). Flower heads 3–10, in a branched cluster; ray-florets purple; involucral bracts purplish towards apex. A robust perennial 15–50 cm. In mountains. YU.BG. **1370. *E. alpinus* L. ALPINE FLEABANE. Flower heads usually 1–3, 2–3 cm across, with numerous spreading violet or pink ray-florets in several rows. Basal leaves narrowly elliptic to spathulate, stalked, the upper narrower and stalkless, all hairy and ciliate. A very variable, erect, perennial usually more than 10 cm. In mountains. YU.AL.GR.BG.

---

14. *Dittrichia viscosa* 1392   15. *Carduncellus caeruleus* (1508)   16. *Pallenis spinosa* 1395   17. *Phagnalon graecum* 1386a   18. *Chamaemelum mixtum* 1413b
19. *Asteriscus aquaticus* 1399   20. *Echinops spinosissimus* 1461b   21. *Onopordum illyricum* 1495   22. *Calendula arvensis* 1460   23. *Cynara cardunculus* 1491

1370b. *E. epiroticus* (Vierh.) Halácsy Like 1370 but with solitary flower heads, purplish ray-florets, and strongly hairy involucral bracts. Plant usually less than 10 cm. In mountains. YU.AL.GR. Page 457. **1370c.** *E. glabratus* Bluff & Fingerh. Distinguished from 1370 by the involucral bracts of the flower heads which are sparsely hairy or almost hairless, less than 1 mm wide, usually green with a brown centre (bracts somewhat downy in 1370). Flower heads usually 1–2; ray-florets lilac, rarely white, without thread-like florets between the ray- and disk-florets as in 1370b. Leaves sparsely ciliate but otherwise almost hairless. In mountains. YU.AL.GR.BG. Page 114.

CONYZA **1373.** *C. canadensis* (L.) Cronq. (*Erigeron c.*) CANADIAN FLEABANE. Female florets 25–45; involucre hairless or nearly so. Cultivated and waste ground. Native of N. America; widely naturalized. **(1373).** *C. bonariensis* (L.) Cronq. (*C. ambigua*, *Erigeron crispus*) Female florets 50–120; involucre usually hairy. Cultivated places. Native of tropical America; widely naturalized.

EVAX 3 species in our area. ****1374.** *E. pygmaea* (L.) Brot. A tiny annual with a rosette of narrow woolly-haired leaves 5–15 mm, and an almost stalkless cluster 5–35 mm across of tiny yellowish flower heads nestling in the rosette. Dry places, stony ground in the Med. region. Widespread, except BG.

ANTENNARIA ****1378.** *A. dioica* (L.) Gaertner CAT'S FOOT. Dry grasslands, stony places. YU.AL.BG.

LEONTOPODIUM ****1379.** *L. alpinum* Cass. EDELWEISS. Alpine meadows, screes. YU.AL.BG. Subsp. *nivale* (Ten.) Tutin has all leaves densely white-woolly with spreading hairs, and the leaves subtending the flower cluster broader, spathulate, and about as long as the flower heads. YU(Peć).BG(Pirin).

HELICHRYSUM 9 species in our area, including 4 endemic species.

Involucral bracts white, much longer than the florets and spreading
1384c. *H. sibthorpii* Rouy (*H. virgineum*) A greyish-white, woolly-haired, rounded cushion plant of Mount Athos. Flower heads white, 1–3, about 1·5 cm across, on white-woolly stems 5–10 cm. Rosette leaves and lower stem leaves narrow, spathulate 1·5–6 cm. Rocks. GR(Athos). **1384d.** *H. amorginum* Boiss. & Orph. Flower heads white, 1–1·5 cm across, stalked, at least 5 in a flat-topped cluster. A loosely tufted perennial 10–30 cm, with white-woolly leaves and stems. Rocks. GR(Amorgos). **1384e.** *H. doerfleri* Rech. fil. A densely tufted, white-woolly perennial, with a dense cluster of 2–4 white flower heads on short stems not more than 8 cm. Rosette leaves with blades 5–10 mm long, stem leaves numerous, longer 2–3 cm, the upper narrowly linear. Rocks in mountains. CR(east).

Involucral bracts yellow, usually equalling the florets and remaining erect
(a) Leaves mostly flat, their margins not inrolled
1384a. *H. orientale* (L.) Gaertner Distinguished by its relatively large yellow flower heads 7–10 mm across, in a dense terminal flat-topped cluster 2–8 cm across. Involucral bracts shining lemon-yellow, the outer ovate. Leaves and stems with adpressed white-woolly hairs; lower leaves oblong-spathulate blunt 2–6 cm, the upper linear, stalkless. Rocks in the Med. region. GR+Is.CR. Page 67. **1384b.** *H. plicatum* DC. Like 1384a but flower heads smaller, about 5 mm across, and leaves and stems green, densely glandular-sticky, only finely or sparsely hairy. Mountain pastures. YU(south).AL.GR. Pl. 50. ****1384.** *H. arenarium* (L.) Moench Distinguished by its numerous golden-yellow, shining ovoid flower heads which are 4–5 mm across, borne in an inflorescence 2–5 cm across. Leaves and stems not glandular, with adpressed densely white-woolly

hairs, the upper linear; herbaceous perennial 8–30 cm. Sandy places. YU.BG.

(b) Leaves with inrolled margins

1385. *H. stoechas* (L.) Moench A loosely tufted woody-based perennial with narrow leaves with inrolled margins and many unbranched stems bearing dense flat-topped clusters 1·5–3 cm across, of small yellow flower heads. Flower heads globular, 4–6 mm across; involucral bracts oval-blunt, straw-coloured, the inner 2–3 times as long as the outer. Leaves linear, greenish to white-woolly; stems white-woolly 10–50 cm. Plant strongly aromatic, or not aromatic. Rocks in the Med. region. YU.AL.GR+Is.CR.TR. **(1385). *H. italicum* (Roth) G. Don fil. Very like 1385 but distinguished by its cylindrical flower heads which are longer than wide, and the inner involucral bracts with minute reddish glands, 5–8 times as long as the outer oval bracts. Flower heads 2–4 mm across. An aromatic, curry-smelling perennial. Dry places, rocks in the Med. region. YU.GR+Is.CR. Page 38.

PHAGNALON ****1386.** *P. rupestre* (L.) DC. Dry places. YU(Croatia). **(1386).** *P. saxatile* (L.) Cass. A small branched shrublet to 60 cm, with numerous erect stems with narrow leaves, and solitary terminal brownish-yellow flowers heads about 1 cm across. Distinguished from the similar 1386 by the undulate margin of the middle involucral bracts (margin of bracts flat in 1386) which are linear-lanceolate, acute, inner bracts linear. Leaves 2·5–3·5 cm, linear to linear-oblanceolate, greenish above, densely felted beneath. Rocks, walls in the Med. region. GR. **1386a.** *P. graecum* Boiss. & Heldr. Like (1386) but outer involucral bracts brownish, narrowly triangular to lanceolate, acute; margin of middle bracts flat. Leaves smaller 1·5–2·5 cm, oblanceolate to obovate, margin undulate, lobed or toothed. A dwarf shrub to 30 cm. Rocky places in the Med. region. AL.GR+Is.CR. Page 97. **1386b.** *P. pumilum* (Sibth. & Sm) DC. A tufted Cretan perennial to 15 cm, of rock crevices, with solitary yellowish flower heads. Leaves hairless or densely hairy, oblong-spathulate, flat, entire 1·5–2·5 cm. Involucral bracts pale, lanceolate, long-pointed, entire. Rock crevices in mountains. CR.

INULA Flower heads all yellow. 14 species in our area, including 2 endemic species.

Ray-florets more than 12 mm long, more than 1½ times as long as involucre
(a) Fruit hairless

1388. *I. helenium* L. ELECAMPANE. Damp hedges, meadows, woods. YU.GR.BG. **1389. *I. salicina* L. WILLOW-LEAVED INULA. Distinguished by its upper leaves which have a heart-shaped clasping base. Ray-florets 1·5–2·5 cm; involucral bracts with ciliate margin otherwise hairless. Leaves net-veined. Rocky and woody slopes, marshes. Widespread on mainland. ****(1389).** *I. ensifolia* L. NARROW-LEAVED INULA. Distinguished by its linear-lanceolate stalkless stem leaves, with usually 3–7 prominent parallel veins. Ray-florets 1·5–2·2 cm; involucral bracts silky-haired at base. Grassy places, stony ground. Widespread on mainland. **1389a.** *I. hirta* L. An erect hairy perennial 15–50 cm, with leaves which are prominently net-veined on both surfaces, and medium to large flower heads with ray florets 1·5–3 cm. Involucre hemispherical 10–13 mm; bracts linear to lanceolate, the outer as long as the inner, more or less densely hairy, 1–1·2 cm. Grassy and bushy places. YU.BG. Page 191.

(b) Fruit hairy

1390. *I. crithmoides* L. GOLDEN SAMPHIRE. A shrubby coastal plant to 1 m, distinguished by its fleshy linear to linear-lanceolate leaves, entire or three-toothed at the apex. Ray-florets 1·4–2·5 cm; involucral bracts linear. On the littoral, marshes, cliffs. YU.AL.GR.CR+Rhodes. **(1391).** *I. britannica* L. Stem leaves elliptic, more or less heart-shaped at base and clasping stem; upper surface of leaves without prominent netted veins, hairy. Ray-florets 1·5–2·5 cm; outer involucral bracts 7–12 mm, spreading or reflexed, the inner shorter 5–8 mm. Damp places, meadows. YU.AL.GR.BG.TR. **1391a.** *I. oculus-christi* L. Like (1391) but outer involucral bracts 5–7 mm, silvery-woolly, usually erect; inner longer 10–12 mm. Flower heads 2·5–3 cm across; ray-floret 1·6–2 cm. Leaves silvery-woolly, the upper more or less clasping the stem. Stony and dry places. YU.AL.GR.BG.TR. Pl. 50.

Ray-florets not more than 12 mm long, not more than 1½ times as long as the involucre
(*a*) *Leaves hairless to sparsely grey-woolly*
**1387. *I. conyza* DC. PLOUGHMAN'S SPIKENARD. Ray-florets shorter than the involucre. Woods, thickets. Widespread on mainland.

(*b*) *Leaves white-felted*
1388b. *I. verbascifolia* (Willd.) Hausskn. A very distinctive white woolly-haired perennial to 50 cm, of rocks and cliffs, with a branched cluster of yellow flower heads with snowy-white involucral bracts. Flower heads very variable in size, short- or long-stalked, and ray-florets shorter or longer than the involucre. Leaves variable, oval-lanceolate, toothed or entire, often acute, with dense or sparse woolly hairs, the veins usually prominent beneath. In the Med. region. YU.AL.GR+Is.CR.BG. Page 97. **1388c.** *I candida* (L.) Cass. Like 1388b but a smaller plant to 30 cm, covered with dense adpressed silky hairs, and with simple or sparsely branched inflorescences. Distinguished by the basal leaves which are usually 4–9 cm, lanceolate and gradually narrowed to a leaf stalk, blunt, without teeth, and with no prominent veins beneath. Involucre 8–9 mm long, with bracts erect or recurved and longer than the ray-florets. Cliffs, rocks. GR+Is.CR. Page 66.

DITTRICHIA Included in *Inula* by many botanists, but distinguished by the cylindrical achenes which are abruptly contracted below the pappus; pappus hairs fused near the base.

1392. *D. viscosa* (L.) W. Greuter (*Inula v.*) AROMATIC INULA. A very strongly resinous-smelling shrubby perennial 40–130 cm, with terminal leafy clusters of yellow flower heads each about 1·5 cm across. Ray-florets 1–1·2 cm, longer than the involucre; involucre 6–8 mm, bracts linear-lanceolate. Leaves lanceolate, sticky, glandular-hairy, entire or toothed, the upper half-clasping the stem. Rocky places, pine woods, olive groves, roadsides in the Med. region. Widespread. Pl. 50. **1392a.** *D. graveolens* (L.) W. Greuter (*Inula g.*) A densely glandular, much-branched annual 20–50 cm, smelling of camphor, with small flower heads 6–8 mm across, in a dense cluster. Ray-florets 4–7 mm, about as long as the involucre. Lower leaves lanceolate entire, the upper half-clasping the stem. Rocks, waysides in the Med. region. Widespread.

PULICARIA | Fleabane   4 species in our area.

1393. *P. dysenterica* (L.) Bernh. FLEABANE. Flower heads yellow, 1·5–3 cm across. Damp meadows, marshes, ditches, saline soils. Widespread. **(1393). *P. odora* (L.) Reichenb. Distinguished from 1393 by the upper stem leaves which are half-clasping and basal leaves green at flowering time (upper leaves heart-shaped, clasping, basal leaves withered at flowering 1393). Flower heads few, yellow, 2–3 cm across, on thickened stems. A woolly-haired perennial 20–70 cm, without stolons. Olive groves, pine woods, bushy places in the Med. region. YU.AL.GR+Is.CR.TR. **(1393).** *P. vulgaris* Gaertner SMALL FLEABANE. Like 1393 but an annual 7–30 cm, with smaller flower heads 0·8–1 cm across, and ray-florets erect and little longer than the involucre. Leaves lanceolate to elliptic, with undulate margins. Ditches, wet places. YU.AL.GR.CR? BG.TR.

CARPESIUM 1394. *C. cernuum* L. FALSE BUR-MARIGOLD. Distinguished by its solitary nodding, globular, yellow flower heads 1·5–2·5 cm across, each surrounded by an involucre of spreading or recurved leafy bracts (recalling *Bidens*, Bur-Marigold). Ray-florets absent. An annual or biennial 20–80 cm, with elliptic glandular leaves. Wood margins, thickets. YU.BG. **1394a.** *C. abrotanoides* L. Like 1394 but flower heads about 0·5 cm across, stalkless (stalked in 1394), and outer involucral bracts not leaf-like. Woods. YU(Croatia).

PALLENIS ****1395.** *P. spinosa* (L.) Cass. Flower heads bright yellow with stiff spiny-tipped, spreading bracts much longer than the ray-florets. A softly hairy annual or biennial to 60 cm. Dry uncultivated places in the Med. region. Widespread.

BUPHTHALMUM ****1396.** *B. salicifolium* L. YELLOW OX-EYE. Woods, bushy and grassy places. YU(central).

TELEKIA ****1397.** *T. speciosa* (Schreber) Baumg. LARGE YELLOW OX-EYE. A robust, broad-leaved perennial to 2 m, with few or solitary, large orange-yellow flower heads 5–6 cm across. Disk-florets brownish-yellow; ray florets 1–1·5 cm. Lower leaves large, to 30 cm, broadly triangular-heart-shaped, stalked, the upper rhomboid or oblong, stalkless, all strongly toothed. Wood margins, streamsides, in mountains. YU.AL.BG.

ASTERISCUS ****1398.** *A. maritimus* (L.) Less. Maritime rocks in the Med. region. GR(south and west). ****1399.** *A. aquaticus* (L.) Less. A dichotomously branched hairy annual to 50 cm, with yellow flower heads and leafy outer involucral bracts 1–2 cm, spreading beyond the florets. Ray-florets little longer than disk-florets. Leaves oblanceolate, entire, blunt. Maritime rocks, on the littoral in the Med. region. Widespread.

AMBROSIA Flower heads one-sexed, inconspicuous. Fruit enclosed in a nut-like involucre; pappus absent.

1400a. *A. maritima* L. A densely grey-haired, aromatic, leafy annual to 120 cm, with deeply twice-cut leaves, and terminal spikes of tiny, almost stalkless yellow flower heads. Male flower heads 3 mm across, terminal, with several florets; female heads in axillary clusters, female involucres with 5 conical spines. Leaves green above, densely grey-hairy beneath, lobes oblong, toothed or lobed. Maritime sands, waste places in the Med. region. YU.AL.GR.CR(+Rhodes).

XANTHIUM ****1401.** *X. strumarium* L. COCKLEBUR. A non-spiny annual 20–120 cm, with long-stalked, triangular-heart-shaped, grey-green leaves. Fruit cylindrical 12–15 mm, covered with spines. Disturbed ground, waste places, damp ground. Widespread. ****1402.** *X. spinosum* L. SPINY COCKLEBUR. A spiny, much-branched annual with narrow rhomboid, three- to five-lobed almost stalkless leaves, dark green above, white-felted beneath, and with three-pronged spines at base. Fruit 6–10 mm, with hooked spines. Waysides, waste places, disturbed ground. Native of N. America; widely naturalized.

HELIANTHUS | Sunflower 1404. *H. annuus* L. SUNFLOWER. Often cultivated for its oil-bearing seeds. Native of N. America; locally naturalized in YU.AL.GR.BG.TR. **1405.** *H. tuberosus* L. JERUSALEM ARTICHOKE. Sometimes cultivated for its tubers. Native of N. America; locally naturalized in YU.AL.

BIDENS | Bur-Marigold ****1406.** *B. cernua* L. NODDING BUR-MARIGOLD. Leaves undivided. Flower heads yellow, nodding. Damp places. YU.AL.GR.BG. **(1406).** *B. tripar-*

tita L. TRIPARTITE BUR-MARIGOLD. Leaves usually three-lobed. Flower heads yellow, mostly erect. Marshes, damp places. Widespread on mainland.

SANTOLINA 1408. *S. chamaecyparissus* L. LAVENDER COTTON. Rocks by the sea. YU(Croatia).

ANTHEMIS | **Chamomile**   A very difficult genus requiring ripe fruit for identification. Small characters of the fruit, involucral bracts, receptacular scales are important in distinguishing species. About 43 species in our area, including 22 which are restricted to it.

Dwarf shrubs, or perennials, with non-flowering shoots present
1409. *A. tinctoria* L. YELLOW CHAMOMILE. Distinguished by its yellow flower heads 2·5–4·5 cm across, with yellow ray-florets and central dome of yellow disk-florets, but ray-florets sometimes absent. Outer involucral bracts triangular-acute; receptacular scales stiff, bristle-like. Leaves often grey-woolly, 1–5 cm or more, twice-cut into oblong to linear lobes. Fruit more or less compressed, corona very short. A very variable species. Dry places, sunny slopes. Widespread on mainland.   **1409b. *A. cretica* L. A cushion-forming perennial of mountains, with solitary, rather large flower heads 2·5–4·5 cm across, with white ray-florets and yellow disk-florets, borne on long stems 12–30 cm, which are leafless in the upper half. Involucral bracts with a wide pale-brown margin; receptacular scales oblong, abruptly pointed. Leaves at first densely white-hairy, becoming hairless, the lower with usually 3–4 pairs of narrower lobes, gland-dotted. Fruits not compressed. A very variable species. Rocks. Widespread on mainland. Page 89.

1409c. *A. tenuiloba* (DC.) R. Fernandes (*Lyonnetia t.*) A tufted perennial 6–30 cm, with densely adpressed-hairy greyish-white leaves, with solitary flower heads up to 2·5 cm across, with or without white ray-florets. Leaves once- or twice-cut into linear to oblong lobes 0·5–2 mm wide. Involucral bracts grey-haired, with or without papery margins, or margins sometimes brownish, acute or blunt; receptacular scales acute. Dry places, stony and grassy places in mountains. YU.AL.GR.BG.TR. Page 457.   **1409d.** *A. orientalis* (L.) Degen (*A. pectinata*) A tufted woody-based, branched perennial 11–27 cm, with yellow, usually rayless flower heads 6–12 mm across, and hairless involucral bracts with wide papery margins and tips. Receptacle conical, scales papery, blunt. Leaves oblong to linear, with numerous regular comb-like linear lobes. Pine-woods, scrub in mountains. GR.TR. Pl. 50.

Annuals or biennials, usually branched and without non-flowering shoots
1410. *A. altissima* L. (*A. cota*) A tall, nearly hairless annual to 1 m or more, with white flower heads 2·5–4 cm across, on stalks which become thickened in fruit. Ray-florets white, up to 2 cm long. Inner involucral bracts with pale-papery or brown-papery margins and tips, the outer without papery margins. Receptacular scales with a long rigid point, longer than the florets. Leaves ovate, 2–3 times cut into narrow, bristle-pointed lobes. Cultivated and waste ground. Widespread. **(1410).** *A. chia* L. A nearly hairless annual 5–40 cm, with many stems bearing long-stalked flower heads up to 4·5 cm across, with long white ray-florets, longer than the width of the disk. Outer involucral bracts triangular-lanceolate, acute, with wide brown papery margins. Receptacular scales oblong to lanceolate, not fine-pointed. Leaves once- to twice-cut into triangular or ovate lobes. Fruit 2–2·5 mm, cylindrical with 7–10 ribs, the outer fruits with an ear-like projection, the inner with a short crown. Waste places, cultivated ground in the Med. region. YU.AL.GR+Is.CR.TR. Page 41.

(1410). *A tomentosa* L. Usually a woolly-haired, branched annual 2–30 cm, with stems swollen below the flower heads which are 1·5–3·7 cm across. Ray-florets white, shorter than the width of the disk. Involucral bracts grey-hairy, the outer acute, the inner with papery margins. Receptacular scales oblong. Fruit 1·5–2 mm, obconical ribbed. Sandy places by the sea, dry hills in the Med. region. GR+Is.TR. **1410a.** *A. rigida* (Sibth. & Sm.) Boiss. & Heldr. A low-growing annual with usually numerous, spreading, unbranched stems to 15 cm, which later become rigid and somewhat swollen and curved above. Flower heads small 3–9 mm across, solitary, yellow, sometimes pink-flushed, usually without ray-florets. Involucre obconical; bracts triangular-lanceolate, thickened at base. Receptacle shortly conical, 3–4 mm across. Leaves oblong, once- or twice-cut, lobes linear to wedge-shaped. Fruits obconical. Maritime sands, dry places. GR+Is.CR.TR. Pl. 50.

1411. *A. arvensis* L. CORN CHAMOMILE. Fields, uncultivated ground. Widespread. **1412.** *A. cotula* L. STINKING MAYWEED. A stinking, branched annual 20–70 cm, with a more or less flat-topped cluster of long-stalked flower heads 1–3 cm across; rays white 5–14 mm. Involucral bracts with pale-brown papery margins. Receptacular scales linear. Leaves 2–3 times cut into many narrowly linear lobes. Waste places, cultivated ground. Widespread.

CHAMAEMELUM Like *Anthemis* but disk-florets with the base of the corolla tube enlarged and sac-like.

1413b. *C. mixtum* (L.) All. (*Anthemis m.*, *Ormenis m.*) A widely branched, hairy annual 10–60 cm, with deeply cut leaves, and solitary flower heads with white ray-florets about 1 cm long. Involucre 3–4 mm long, the bracts oblong blunt, greenish with a wide papery margin. Lower leaves once- or twice-cut into linear-lanceolate, toothed lobes, the upper once-cut or toothed. Cultivated ground, roadsides, maritime sands in the Med. region. AL.GR.CR.

ANACYCLUS Outer fruits flattened, winged.

1414. *A. clavatus* (Desf.) Pers. (*A. tomentosus*) Flower heads solitary 1·5–2 cm across, with white erect or recurved ray-florets and yellow disk-florets, and the stem becoming conspicuously thickened below the flower heads. Inner involucral bracts without a papery appendage. Leaves twice-cut into linear, fine-pointed lobes. A widely branched, hairy annual to 50 cm. Maritime rocks, disturbed ground in the Med. region. YU.GR.TR. **(1414).** *A. radiatus* Loisel. Like 1414 but ray-florets yellow, spreading and inner involucral bracts with a large rounded, papery, fringed appendage. Sandy places, rocks in the Med. region. YU.GR.

ACHILLEA A difficult genus with much hybridization, making the identification of individual plants very difficult. About 35 species occur in our area, 9 of which are restricted to it.

Flower heads with white, pink, or mauve ray-florets (rarely pale yellow)
(a) Flower heads solitary

1419a. *A. ageratifolia* (Sibth. & Sm.) Boiss. A tufted, white-felted, creeping perennial of the mountains, with linear to lanceolate woolly leaves, and erect stems to 30 cm bearing a solitary white flower head 1·5–2·5 cm across. Ray-florets many, ovate 7–9 mm, longer than the involucre; involucral bracts silky-haired, with a brown-papery margin. Leaves 2–4 cm, entire or toothed, or pinnately cut at base. Rocks in mountains. YU.AL.GR.BG. Page 126.

(b) Flower heads mostly in clusters

    (i) Large, usually lowland, perennials 30 cm or more

1417. *A. millefolium* L. YARROW. This and 7 related species can be distinguished by their stem leaves which are lanceolate or linear in outline, usually more or less rounded in profile, and with more than 15 pairs of primary leaf segments; also by their tiny flower heads. Ray-florets white, pink, or mauve, 1–2 mm; involucre 3–5 mm across. They are widely distributed from mountain pastures to the lowlands on the mainland. **(1417).** *A. nobilis* L. Distinguished by its small leaves 1–3 cm long, elliptic to ovate in outline, usually flattened, with not more than 10 pairs of primary leaf segments each 6–8 mm. Flower heads many in a flat-topped cluster. Ray-florets white or pale yellow, about 1 mm long; involucre 2–3 mm across; bracts hairy. Perennial 15–60 cm, without stolons. Sunny hill slopes, stony ground. YU.AL.GR.BG. Pl. 51.

1417b. *A. crithmifolia* Waldst. & Kit. Flower heads many; ray-florets about 2 mm, white to pale yellow. Involucre 3–5 mm by 2–3 mm with shiny, hairless bracts. Basal leaves much dissected into linear ultimate lobes; middle stem leaves 4–6 cm, ovate to lanceolate in outline, flat, deeply cut, with primary leaf segments 5–10 mm, lanceolate or linear in outline, mostly with a few lanceolate teeth. Stems simple erect, 20–60 cm. Mountain meadows and scrub. YU.AL.GR.BG.TR.

    (ii) Small alpine perennials, usually less than 30 cm

1420a. *A. umbellata* Sibth. & Sm. A low more or less tufted white-woolly perennial with simple stems to 15 cm, and with flat-topped clusters of usually 3–6 flower heads with white, oblong ray-florets 3–5 mm long. Involucre 4–6 mm across; bracts ovate-blunt with wide brown margins. Leaves mostly basal 1–2 cm, ovate in outline, cut into spathulate entire blunt lobes. Mountain rocks. GR(central and south). Pl. 50. **1420b.** *A. clavennae* L. A silvery-woolly perennial with flat-topped clusters of 6–25 flower heads, with white oval ray-florets 4–6 mm long, borne on stems to 40 cm. Involucre 4–6 mm across; bracts ovate with wide brown margins. Leaves pinnately cut into few, mostly entire blunt lobes, the lower leaves to 8 cm stalked, the uppermost stalkless, about 2 cm. Mountain rocks. YU.AL.GR(west). Page 114.

1420c. *A. ambrosiaca* (Boiss. & Heldr.) Boiss. A small grey-woolly alpine endemic of Mount Olympus. Flower heads 4–10, with white, rounded ray-florets about 2 mm. Involucre 5–6 mm across; bracts ovate with wide brown margins. Leaves 1–2 cm, mostly twice-cut into numerous lanceolate-acute lobes; stems to 20 cm. Rocks, stony ground. GR(Olympus). **1420d.** *A abrotanoides* (Vis.) Vis. Distinguished from the previous two species by the adpressed hairs on the leaves. Stems to 40 cm, with 12–30 flower heads; ray-florets obovate, about 4 mm; involucre 5–7 mm across; bracts ovate with narrow pale-brown margins. Mountain rocks and screes. YU(west).AL.GR(west). Page 178.

Flower heads with bright-yellow ray-florets

    (a) Perennials of mountain rocks and pastures

1421b. *A. holosericea* Sibth. & Sm. Distinguished by its more or less silvery-haired, once-cut, mostly basal leaves, with entire or toothed, ovate to lanceolate lobes. Stems 15–60 cm, hairy, bearing 10 to many flower heads with rounded, yellow ray-florets 1–3 mm. Involucre about 5 mm across; bracts ovate-blunt with narrow brown margins. Mountain rocks. YU.AL.GR. Page 127. Pl. 51. **1421c.** *A chrysocoma* Friv. A woolly-haired perennial to 40 cm, with leaves twice-cut into very numerous linear, bristle-tipped lobes, and with numerous yellow flower heads in a flat-topped cluster. Ray-florets about 3 mm; involucre 3–5 mm across; bracts 3–4 mm with dark-brown margins. Mountain pastures. YU(south).AL.GR(north).BG.

(b) Perennials of lowlands and dry hills
1421d. *A. clypeolata* Sibth. & Sm. A shortly woolly-haired perennial to 60 cm, with flat, deeply cut leaves and dense flat-topped clusters of many tiny flower heads on woolly stalks. Ray-florets yellow, about 1 mm; involucre about 3 mm across; bracts 1–2 mm. Lower leaves stalked, to 15 cm, once-cut into ovate toothed or lobed segments, the uppermost stalkless, 1–2 cm. Rocks in hills. YU.AL.GR.BG.TR. Pl. 51. **1421e.** *A. coarctata* Poiret Differs from 1421d in having basal leaves to 30 cm, once-cut into lobes which are further once- or twice-cut and with more or less lanceolate-blunt teeth (1421d lobes saw-toothed or deeply cut with ovate-acute teeth). Ray-florets yellow, 0·5–1 mm; involucre 3–4 mm across; bracts with brown hairs, or silky. Dry hills, sandy soils. YU.AL.GR.BG.TR. Page 165. Pl. 51.

OTANTHUS **1423.** *O. maritimus* (L.) Hoffmanns. & Link. *(Diotis m.)* COTTON-WEED. An unmistakable snowy-white, woolly-haired, shrubby plant to 50 cm, of sand-dunes and seashores. Flower heads globular, yellow, without ray-florets; involucre woolly-white, like the stems and leaves. Leaves fleshy, oblong 5–17 mm. YU.AL.GR+Is.CR. TR.; introd. BG.

CHRYSANTHEMUM Yellow-flowered annuals. Fruits of two kinds, the outer three-angled, the inner cylindrical, ribbed.

1424. *C. segetum* L. CORN MARIGOLD. Leaves entire or deeply toothed, glaucous; stems little branched. Flower heads golden-yellow, 3·5–5·5 cm across. Cultivated or waste ground; often naturalized. YU.GR+Is.CR.TR. ****1425.** *C. coronarium* L. CROWN DAISY. Leaves mostly twice-cut into linear lobes; stems much branched above. Flower heads 3–6 cm across, with ray-florets yellow, or white with yellow base. Cultivated and waste ground in the Med. region. YU.GR+Is.CR.TR.

COLEOSTEPHUS Leaves simple. Fruit cylindrical, with 8–10 white ribs and a basal thickening, and a prominent membraneous crown. **(1424).** *C. myconis* (L.) Reichenb. fil. *(Chrysanthemum m.)* An erect, simple or branched annual 14–45 cm, with conspicuous flower heads 2 cm across, with yellow, whitish, or variegated ray-florets 6–15 mm. Involucral bracts brownish with narrow papery margins. Basal leaves obovate, toothed, stalked, the middle stem leaves half-clasping. Cultivated ground, waste places. YU.GR.CR.

TANACETUM Like *Chrysanthemum* but fruits all similar, usually 3–10 ribbed, with a crown-like pappus. Flower heads without or with ray-florets which are either strap-shaped or tubular, white or yellow.

Ray-florets absent
**1426. *T. vulgare* L. *(Chrysanthemum v.)* TANSY. Flower heads small, 5–8 mm across, numerous, in dense flat-topped clusters borne on stems 30–100 cm. Leaves ferny, deeply twice-cut into narrow lobes, covered with glandular dots, and very aromatic when crushed. Roadsides, waste places, river gravels. Widespread on mainland.

Ray-florets present
(a) Ray-florets white
**1429. *T. parthenium* (L.) Schultz Bip. *(Chrysanthemum p., Pyrethrum p.)* FEVERFEW. A strongly aromatic perennial 25–60 cm, with pale yellowish-green leaves, deeply cut into oblong to ovate segments, which are often further divided. Flower heads short-stalked, numerous in lax flat-topped clusters; ray-florets white, rays 3–7 mm; involucre 6–8 mm across. Scrub, rocky places in mountains; often naturalized. Throughout the mainland

+CR. **(1429).** *T. corymbosum* (L.) Schultz Bip. (*Chrysanthemum c.*) Distinguished by its long-stalked flower heads forming a regular, loose, flat-topped cluster. Flower heads usually 3–15; involucre 8–14 mm across; bracts with brown margins; ray-florets white, rays 10–16 mm. Leaves 3–4 cm, once-cut into 16–40 oblong, toothed lobes, shining green, hairless or finely hairy, the upper leaves similar, smaller, stalkless; stems 30–120 cm. Bushy places, woods. Throughout the mainland.

(1429). *T. macrophyllum* (Waldst. & Kit.) Schultz Bip. Distinguished from 1429 *T. parthenium* by its more numerous (40–100), smaller flower heads in very dense clusters, and ray-florets white, rays 2–5 mm. Involucre 4–7 mm across. Leaves variable, usually hairless above, densely hairy below, pinnately cut with toothed lobes. Mountain woods. YU.AL.GR.BG. Page 457. **1429a.** *T. cinerariifolium* (Trev.) Schultz Bip. (*Pyrethrum c.*) A tufted, silvery-grey silky perennial with solitary white flower heads on unbranched, grey-haired stems 15–45 cm. Flower heads 3–4 cm across; involucre 12–18 mm across; ray-florets white, rays 8–16 mm. Leaves gland-dotted, silvery-haired, deeply twice-cut into narrow lanceolate lobes. Rocky ground; cultivated as an insecticide. Native in YU(west).AL. Page 179.

(b) Ray-florets yellow

1426a. *T. millefolium* (L.) Tzvelev (*Chrysanthemum millefoliatum.*) Flower heads 2–10, about 1 cm across, in a lax irregular flat-topped cluster, with tiny ray-florets with rays 2–3 mm which are shorter than the disk-florets. Leaves twice-cut, grey-haired when young; stems 20–50 cm. Dry grassy places. BG. Pl. 51.

LEUCANTHEMELLA Differs from *Tanacetum* in its simple leaves and fruit without a pappus or crown. Ray-florets in one row, sterile; corolla tube compressed; receptacle convex.

1429b. *L. serotina* (L.) Tzvelev (*Tanacetum s.*) Distinguished by its lanceolate, coarsely toothed leaves and its erect, densely leafy stem bearing 1–8 large white or pinkish flower heads 4–6 cm across, in a lax flat-topped cluster. Ray-florets 1–2·5 cm. Leaves stalkless, the lower with 2–4 lobes at the base. Fruit with white ribs. Wet places. YU(east-central).BG. Page 457.

LEUCANTHEMUM Like *Chrysanthemum* but fruits of one kind, ten-ribbed, with secretory canals between the ribs, and mucilage cells present in the epidermis. Ray-florets usually white. One widespread species in our area, and 3 local species in Yugoslavia, including 1 endemic species.

****1427.** *L. vulgare* Lam. (*Chrysanthemum leucanthemum*) MARGUERITE, MOON-DAISY, OX-EYE DAISY. A very variable species with many local forms often named as separate species, subspecies, or varieties. Flower heads solitary or few, 2·5–4 cm across or more; involucral bracts with dark papery margins. Grasslands, waysides, in mountains. Widespread on mainland.

CHAMOMILLA Fruits with an oblique, lateral attachment-scar, weakly three- to five-ribbed, without apical resin glands.

1431. *C. recutita* (L.) Rauschert (*Matricaria chamomilla*) WILD CHAMOMILE. Cultivated fields, wet places, and saline soils. Widespread. ****1432.** *C. suaveolens*(Pursh) Rydb. (*Matricaria matricarioides*, *M. discoidea*) PINEAPPLEWEED, RAYLESS MAYWEED. Native of Asia; naturalized in cultivated ground, waste places in YU.BG.

MATRICARIA Differs from *Chamomilla*, which is often included in it, by the fruits

1. *Tanacetum macrophyllum* (1429)   2. *Erigeron epiroticus* 1370b   3. *Senecio thonnae* 1452c   4. *Anthemis tenuiloba* 1409c   5. *Leucanthemella serotina* 1429b

which have a transverse basal attachment-scar, and are strongly three-ribbed, with apical resin glands. 6 species in our area, including 2 endemic species.

1432a. *M. trichophylla* (Boiss.) Boiss. (*Tripleurospermum tenuifolium*) A hairless, glandular, much-branched annual or biennial 50–150 cm, with many (10–60) flower heads 3–5 cm across, with white ray-florets longer than the width of the disk. Involucral bracts hairless, with wide colourless or pale-brown papery margins. Leaves 2–3 times cut into thread-like pointed lobes. Fruit usually without crown or pappus. Hedges, roadsides, cultivated ground. Widespread on mainland.

1432b. *M. perforata* Mérat (*Tripleurospermum inodorum*) Differs from 1432a in its narrow triangular involucral bracts with narrow colourless or dark-brown papery margins; and its fruits with a short crown and well-separated ribs and rounded resin glands. An annual 30–80 cm. Cultivated ground, waste places, salt-rich soils. Widespread on mainland. **1432c.** *M. rosella* (Boiss. & Orph.) Nyman A pink-flowered perennial 15–40 cm, of mountain slopes, endemic to Mount Parnon (Greece), with usually solitary flower heads about 2·5 cm across, with pink ray-florets with rays about 7 mm. GR(south).

ARTEMISIA | Wormwood About 18 similar-looking species in our area.

1435. *A. vulgaris* L. MUGWORT. Waste places. Widespread on mainland. ****1436.** *A. absinthium* L. WORMWOOD. Thickets. YU.AL.GR.BG. **(1436).** *A. dracunculus* L. TARRAGON. Widely cultivated as a pot herb; naturalized in YU. ****1437.** *A campestris* L. FIELD SOUTHERNWOOD. Meadows, fallow fields, rocky places. Widespread on mainland. **1438.** *A. arborescens* L. SHRUBBY WORMWOOD. Rocky places. YU.GR+ Is.CR. **(1438).** *A. alba* Turra CAMPHOR WORMWOOD. Rocky places in lowlands and mountains. YU.AL.GR.BG. **1434a.** *A. pedemontana* Balbis (*A. lanata, A. assoana*) A whitish woolly-haired perennial to 30 cm, of rocks, with small rounded yellowish flower heads 4–6 mm across; involucral bracts densely hairy. Leaves twice-cut into linear acute lobes, the upper pinnately lobed. BG. Page 148.

TUSSILAGO ****1439.** *T. farfara* L. COLTSFOOT. Damp stony places, riversides, waysides. Widespread on mainland.

PETASITES 5 species in our area, including 1 endemic species.

Involucral bracts hairless except for a few hairs at base
**1440. *P. hybridus* (L.) P. Gaertner, B. Meyer & Scherb. BUTTERBUR. Riversides, damp meadows, marshes. Widespread on mainland.

Involucral bracts minutely hairy
1441. *P. albus* (L.) Gaertner WHITE BUTTERBUR Flowers whitish-yellow; corolla lobes 2–4 mm. In hills and mountains, damp woods, streamsides, springs. YU.AL.BG. **(1441). *P. paradoxus* (Retz.) Baumg. (*P. niveus*) Distinguished by its white or reddish-pink flowers with purplish involucral bracts (bracts pale green in 1441). Leaves triangular-heart-shaped, or rarely with acute divergent lobes, toothed, densely white-felted beneath. Gravel by streams in mountains. YU(central). **1441a.** *P. kablikianus* Berchtold (*P. glabratus*) Distinguished by its triangular-heart-shaped leaves which are hairless beneath except on the veins, and bracts of involucre pale green. Flowers white or pale yellow, all tubular; corolla lobes 1–2 mm. Gravel, stream banks, wooded gorges. YU.AL.BG.

HOMOGYNE ****1443.** *H. alpina* (L.) Cass. ALPINE COLTSFOOT. Leaves rounded-toothed, becoming hairless beneath. Flower heads solitary, purplish-red; pappus pure white. Damp, shady places in mountains. YU.AL.BG. **1443a.** *H. discolor* (Jacq.) Cass. Like 1443 but basal leaves white-felted beneath. Flower heads bright purple; pappus dirty white. Stony slopes, screes in mountains. YU(central). **1443b.** *H. sylvestris* Cass. Like 1443 but basal leaves with 5–9 shallow lobes which are toothed, green, sparsely hairy on veins beneath. Stems often branched with more than one flower head 10–12 mm long. Woods, scrub in mountains. YU(west and central).

ADENOSTYLES 1444. *A. alpina* (L.) Bluff & Fingerh. (*A. glabra*) Woods, stream margins, damp slopes, usually in mountains. YU(Croatia). ****1445.** *A. alliariae* (Gouan) A. Kerner Woods, streamsides, scrub, damp rocky places, usually in mountains. YU.AL.GR(north).BG.

ARNICA **1446. *A. montana* L. ARNICA. Mountain pastures. YU(central).

DORONICUM | Leopard's Bane Distinguished from *Arnica* by the alternate stem leaves.

1. Basal leaves narrowed to the base.
   2. Frs. of outer florets without a pappus;
      plant glandular-hairy. *hungaricum*
   2. Frs. of all florets with a pappus;
      plant hairy, not glandular except
      flower head stalks (YU(west).AL.). ****1448.** *D. grandiflorum* Lam.
1. Basal leaves heart-shaped at base, not
   or scarcely narrowed.
   3. Frs. of all florets with a pappus
      (YU(west).AL.). ****1448.** *D. grandiflorum* Lam.
   3. Frs. of outer florets without a
      pappus.
      4. Inflorescence with 4–12 or more
         flower heads (YU.AL.GR.(north).BG.). **(1447).** *D. austriacum* Jacq.
      4. Flower heads solitary. *columnae, orientale*

1448a. *D. columnae* Ten. A hairless or sparsely hairy stoloniferous perennial to 60 cm, with solitary flower heads 2–6 cm across, on glandular-hairy stalks. Basal leaves ovate-orbicular or heart-shaped, toothed, with long leaf stalks, stem leaves all clasping. Shady rocks in mountains. YU.AL.GR.BG. Pl. 51. **1448b.** *D. orientale* Hoffm. (*D. caucasicum*) Like 1448a with solitary flower heads 2·5–5 cm across, but stem leaves usually 1–2, weakly rounded-toothed, the lower distinctly stalked, not clasping, and rhizome with conspicuous tufts of silky hairs. Woods, and shady rocks in mountains. YU.AL.GR.BG.TR. **1448c.** *D. hungaricum* Reichenb. fil. HUNGARIAN LEOPARD'S BANE. A glandular-hairy perennial to 80 cm, recognized by the oblong to narrowly elliptical basal leaves 2–3·5 cm which are narrowed to an often indistinct leaf stalk. Lower stem leaves clasping, the upper lanceolate. Flower heads usually solitary, 3–5 cm across, with glandular-hairy stalks. Mountains. YU.GR? BG.

SENECIO Involucral bracts in one row, often with shorter additional bracts at the base of the involucre; receptacle without scales. A large and difficult genus with 29 species in our area, including 3 which are endemic.

Some or all leaves deeply lobed
(a) *Ray-florets absent, or inconspicuous and inrolled*

Widespread on the mainland are: **1449.** *S. vulgaris* L. GROUNDSEL; **1450.** *S. viscosus* L. STINKING GROUNDSEL.

(*b*) *Ray-florets conspicuous, spreading*
 (*i*) *Green, nearly hairless annuals or perennials*
Widely scattered on the mainland are: **1451.** *S. jacobaea* L. RAGWORT; **(1451).** *S. erucifolius* L. HOARY RAGWORT; **(1451).** *S. aquaticus* Hill MARSH RAGWORT; ****1452.** *S. vernalis* Waldst. & Kit. SPRING GROUNDSEL.

1452a. *S. squalidus* L. An annual to short-lived perennial with a branched inflorescence to 60 cm, bearing few to numerous flower heads 1·5–2·5 cm across, in a lax irregular cluster. Ray-florets about 13, rays 10–12 mm long, bright yellow, spreading; involucral bracts about 20, additional bracts 5–13, all black-tipped. Leaves pinnately lobed, with a winged leaf stalk, nearly hairless, the upper clasping the stem. Open stony and sandy places. YU.AL.GR.CR.BG. Page 115. **1452b.** *S. pancicii* Degen Flower heads golden-yellow, 3–4 cm across, with spreading rays, 2–8 in a lax flat-topped cluster. Ray-florets 12–15, rays 7–10 mm. Involucral bracts linear, purple-flushed, additional bracts 5–8. Leaves with sparse cobweb hairs, soon nearly hairless; lower leaves long-stalked, ovate, coarsely toothed, the upper smaller, lanceolate, deeply cut into narrow teeth, clasping; stems 25–50 cm. Alpine meadows. YU(Serbia).BG. Page 137.

1452c. *S. othonnae* Bieb. A hairless perennial to 2 m, with long stolons; the basal and lower stem leaves pinnately cut into large oblong-lanceolate, coarsely toothed lobes. Flower heads many, golden-yellow, small, 5–15 mm across, forming a large, dense, compound cluster. Ray florets 1–2, rays 8–12 mm. Involucre 5–7 mm with often black-tipped bracts, and 1–4 additional bracts. Woods in hills. YU.AL.GR(north).BG.TR. Page 457. **1452d.** *S. abrotanifolius* L. A hairless, erect perennial to 30 cm, with a creeping rhizome; the lower leaves 2–3 times cut into linear lobes. Flower heads large, solitary or few, golden-yellow, 2·5–4 cm across. Involucral bracts and ray-florets often purple-flushed, additional bracts few. Subsp. *carpathicus* (Herbich) Nyman is a dwarf plant to 15 cm, of the Balkan Mountains, with solitary flower heads, and uppermost stem leaves simple, bract-like. Mountain meadows, screes. YU.AL.BG. Page 169.

 (*ii*) *White-felted, at least when young; dwarf shrubs*
1453. *S. bicolor* (Willd.) Tod. (incl. *S. cineraria*) CINERARIA. A dwarf silvery-white shrub to 50 cm or more, with densely white-woolly stems, usually pinnate leaves, and dense clusters of many small yellow flower heads, each 1–1·5 cm across with 10–12 ray-florets. Involucral bracts densely white-felted. Leaves ovate to lanceolate, toothed or pinnate, greenish above, white-felted beneath. Limestone cliffs in the Med. region; also grown for ornament. YU.GR+E.Aegean Is. **1453c.** *S. ambiguus* (Biv.) DC. (incl. *S. taygeteus*) A dwarf shrub differing from 1453 in having stems with a few slender branches at the base and more or less equally leafy throughout. (Stems with many stout branches from the base, and leaves crowded towards the base and sparse above in 1453.) Flower heads many, 10–12 mm across; ray-florets about 10, rays 2–5 mm. Leaves deeply lobed, the lobes toothed or lobed. Rocky and sandy places. GR(south).

All leaves entire or toothed
(*a*) *Ray-florets absent or inconspicuous*
1453d. *S. thapsoides* DC. Distinguished by its white-woolly stem and undivided oblong-lanceolate often toothed leaves, which are densely white-woolly beneath and sparsely hairy above, the upper narrower and clasping stem. Flower heads cylindrical, 8 mm across, without ray-florets, in an elongate branched cluster. Involucre 1–1·5 cm; bracts hairless or white-woolly. A stout woody-based perennial 20–70 cm. Rocks in mountains. YU(west).AL.GR. Pl. 51.

(b) Ray-florets conspicuous
(i) Dwarf shrubs

1453e. *S. gnaphalodes* Sieber A dwarf shrub to 50 cm, with distinctive linear leaves which are white-woolly beneath and greenish above, and with white-woolly stems. Flower heads many, 12–15 mm across with 10–15 yellow ray-florets, borne in a branched cluster; involucre reddish-tinged, hairless or with cobweb hairs. Some leaves with a few small lobes. Rocky, stony ground. CR+Karpathos. Pl. 51.

(ii) Herbaceous perennials

1455e. *S. macedonicus* Griseb. A nearly hairless leafy perennial 50–80 cm, with erect stems and elliptic to lanceolate leaves which rapidly decrease in size towards the inflorescence, the uppermost half-clasping. Flower heads 2·5–4 cm across with 13–14 yellow ray-florets, 4–12 in a terminal flat-topped cluster. Mountain woods and rocky slopes. YU? GR.BG.TR. **1455f.** *S. ovirensis* (Koch) DC. Flower heads yellow, 3–4 cm across with 18–21 ray-florets, 3–15 in a dense flat-topped cluster. Leaves at first with white cobweb hairs which are soon rubbed off; lowest leaves oblong or ovate, with a winged stalk; middle leaves with a rounded base, half-clasping; uppermost leaves linear. A perennial 20–80 cm. Woods, bushy places in mountains. YU.AL. Pl. 51. ****(1456).** *S. nemorensis* L. (incl. *S. fuchsii*) A tall, leafy perennial to 2 m, of mountains, with a terminal domed cluster of numerous yellow flower heads each 2–3·5 cm across, with usually 5–6 spreading ray-florets. Involucre 5–9 mm long, hairless or nearly so, with 3–6 additional bracts. Leaves 5–20 cm, ovate to lanceolate, toothed, the upper stalkless and clasping the stem, or shortly stalked, hairless above. Woods, ravines, shady places. YU.AL.BG.TR.

CALENDULA | Marigold ****1460.** *C. arvensis* . MARIGOLD. Cultivation, grassy places in the Med. region. Widespread, except BG. **1460b.** *C. suffruticosa* Vahl A perennial to 50 cm or more, often woody for some distance above the base, with green, glandular-hairy, lanceolate leaves. Flower heads 3–4 cm across, with yellow or orange ray-florets usually more than twice as long as the involucral bracts (less than twice as long in 1460). Outer fruits spreading or deflexed, up to 3 cm long, the inner shorter, boat-shaped. Rocks and sands on the littoral. GR.TR.

ECHINOPS | Globe-Thistle The globular flower clusters comprise numerous small individual cylindrical flower heads, each consisting of one floret surrounded by 3–4 rows of scale-like involucral bracts, and outside them a brush of white bristles. Careful comparisons of involucral bracts and bristles are required for identification.

1. Inner involucral bracts fused for at least the basal third.
  2. Stem without glandular hairs but with cobweb hairs, or hairless.   *graecus*
  2. Stem with glandular hairs, and with dense cobweb-hairs (TR.).   **1461d.** *E. orientalis* Trautv.
1. Inner involucral bracts free to the base.   *spinosissimus*
  3. Involucral bristles up to one-fifth as long as the individual fl. head, not more than half as long as outer involucral bracts.
    4. Stems densely glandular-hairy; fl.. clusters purplish. (YU(south).BG.).   **1462b.** *E. oxyodontus* Bornm. & Diels
    4. Stems with cobweb hairs, also with few glandular hairs; fl. clusters blue.   *microcephalus*

461

3. Involucral bristles at least a third as long as the individual fl. head, slightly shorter to much longer than outer involucral bracts.

    5. Fl. clusters bluish.

       6. Lvs. flat, lobes with a slender apical spine 2–4 mm (YU.AL.GR.BG.).   **1462c. *E. bannaticus* Schrader

       6. Lvs. with inrolled margins, lobes with a stouter apical spine 3–15 mm (YU.AL.GR.BG.TR.).   **1462. *E. ritro* L.

    5. Fl. clusters white or greyish.

       7. Lvs. with glandular hairs above.   *sphaerocephalus*

       7. Lvs. with rigid, non-glandular hairs above (YU.BG.).   **1461c. *E. exaltatus* Schrader

1461a. *E. graecus* Miller Flower clusters shiny silver-white, 3–4 cm across, on branched hairless or cobweb-hairy stems 25–70 cm. Corolla blue. Leaves hairless or with cobweb hairs above, white-woolly beneath, twice cut into short linear-lanceolate lobes with short slender spines. Dry rocky ground. GR(east)+Cyclades. Pl. 52. **1461b.** *E. spinosissimus* Turra (*E. viscosus*) Flower clusters greyish to greenish, or greenish-blue, 3·5–7 cm across, on branched stems which are densely glandular-hairy, and cobweb-hairy. Corolla white or pale blue. Leaves twice-cut, glandular-hairy above, white-woolly beneath, lobes spiny. Dry rocky places. YU.AL.GR+Is.CR. Pl. 52.

1462a. *E. microcephalus* Sibth. & Sm. Flower clusters blue, 1·5–4·5 cm across, on branched stems to 60 cm, which are covered with white cobweb hairs. Corolla blue. Leaves once-cut, hairless above, white-woolly beneath, lobes triangular, with short slender spines. Dry rocky places. YU(south).AL.GR.BG.TR. **1461.** *E. sphaerocephalus* L. (incl. *E. albidus*, *E. taygeteus*) PALE GLOBE-THISTLE. Flower clusters solitary, or several in a branched cluster, greyish or whitish, 3–6 cm across, borne on woolly-cobweb-hairy stems up to 1·5 m or more. Corolla white or greyish. Leaves once- or twice-cut into spiny lobes, clasping, glandular-hairy above, white-woolly beneath. Dry rocky places. Widespread on mainland.

CARDOPATUM **1463.** *C. corymbosum* (L.) Pers. A much-branched, erect perennial 8–25 cm, with numerous blue flower heads in a dense, very spiny, broadly domed cluster, surrounded by many spiny leaves. Flower heads 5–10 mm across; involucral bracts with recurved spiny apex and spiny margin, with sparse cobweb hairs. Leaves pinnately cut with segments further cut into spiny lobes; stem with a spiny wing above. Dry places by the sea. GR+Is.CR.TR. Pl. 52.

XERANTHEMUM ***1464.** *X. annuum* L. PINK EVERLASTING. Readily distinguished by the spreading pink, shiny everlasting, papery inner involucral bracts of the globular flower heads. Flower heads 3–5 cm across, outer involucral bracts pointed, silvery-brown, hairless. A slender, little-branched, whitish woolly-haired annual 25–75 cm, with linear to oblong leaves. Stony places, waysides. YU.AL.GR.BG. Page 164. **(1464)** *X. inapertum* (L.) Miller Like 1464 but flower heads smaller, 1–2 cm across, oblong-cylindrical with pale-pink inner involucral bracts, erect, not spreading. Stony places, dry hills. YU.AL.GR.CR.TR. **1465.** *X. cylindraceum* Sibth. & Sm. (*X. foetidum*) Distinguished by its small pink flower heads 8–15 mm across, with the outer involucral bracts blunt or notched and with a whitish patch of adpressed hairs on the outer surface, the inner bracts pink, erect, not spreading. Leaves linear to oblong; stem 15–65 cm. Cultivated ground, open dry places. Widespread on mainland.

AMPHORICARPOS Involucral bracts overlapping in several rows, with papery margins; receptacle with scales. Outermost florets female, the remainder two-sexed. Fruit with a pappus of about 10 linear scales. Leaves entire not spiny.

1465a. *A. neumayeri* Vis. A plant of mountain rocks with solitary white or pink flower heads 2–3 cm across, borne on almost leafless stems 20–40 cm. Involucral bracts ovate, hairless, the inner about 13 mm. Basal leaves 5–18 cm by 4–25 mm, linear or oblong, green above and white woolly beneath. YU.AL.GR(north-west). Page 467.

CARLINA | Carline Thistle   10 species, including 4 that are endemic and very local.
1. Fl. heads usually solitary, stemless.
   2. Inner involucral bracts silvery-
     white or pinkish; pappus about 13 mm
     (YU.AL.GR(north)).   ****1467.** *C. acaulis* L.
   2. Inner involucral bracts yellowish;
     pappus 18–25 mm (YU.AL.GR(north).
     BG.).   **1466.** *C. acanthifolia* All.
1. Fl. heads usually several on branched
   stems.
   3. Inner involucral bracts bright
     yellow or brownish-yellow.   *corymbosa*
   3. Inner involucral bracts reddish-purple,
     pink, white, or pale yellow.
     4. Inner involucral bracts reddish-purple
       above, at least towards the apex.   *lanata*
     4. Inner involucral bracts white, pale
       pink, or pale yellow.
       5. Inner involucral bracts at least
         3 cm (YU.AL.GR(north)).   ****1467.** *C. acaulis* L.
       5. Inner involucral bracts not more
         than 2 cm.
         6. Outer involucral bracts 2–2·5 cm
           wide, pinnately cut with spiny
           lobes (YU.AL.GR.).   **1468a.** *C. frigida* Boiss. & Heldr.
         6. Outer involucral bracts 4–9 mm
           wide, spiny-toothed (Widespread
           on mainland).   **1468.** *C. vulgaris* L.

1469. *C. corymbosa* L. (incl. *C. graeca*) FLAT-TOPPED CARLINE THISTLE. Distinguished by its branched, more or less flat-topped cluster of bright golden-yellow flower heads each 1·2–2 cm across. Leaves with spiny margins; stems white-felted, 20–50 cm. Very variable. Dry stony places. Widespread.   **(1469). *C. lanata* L. PURPLE CARLINE THISTLE. Readily distinguished by the reddish-purple inner involucral bracts and the pinkish-purple florets. Flower heads 1·5–4 cm across, sometimes solitary. Leaves pinnately lobed, spiny, woolly-haired beneath. An annual 5–40 cm. Rocks in the Med. region. YU.AL.GR + Is.CR.BG(south).TR.

ATRACTYLIS   Like *Carlina*, but involucre bell-shaped or globular, the outer bracts leafy and deeply cut into spiny teeth, the innermost bracts papery-tipped, not brightly coloured.

1470. *A. cancellata* L. A slender annual with small purple flower heads 1·5–2 cm across, which are loosely encircled by several longer, shiny, comb-like upper leaves, recalling a miniature Chinese lantern. Leaves oblong to spathulate, spiny; stems white-felted. 3–30 cm. Dry rocky places, tracksides, hills in the Med. region. YU.GR+Is.CR. **1471** *A. gummifera* L. A spiny, stout thistle-like rosette perennial, bearing at the centre a large, solitary stalkless purple flower head 3–7 cm across. Involucre with cobweb hairs middle bracts with 3 spreading apical spines and much shorter lateral spines. Leaves 15–40 cm, pinnately cut, with lobes further cut and spiny. Dry places, field verges waysides in the Med. region. GR+Is.CR. Pl. 54.

ARCTIUM | Burdock   4 species in our area, but there are many intermediates.

1472. *A. tomentosum* Miller WOOLLY BURDOCK. Distinguished by the densely cobweb-haired involucre. Alluvium, waysides, habitations. YU.GR.BG.   **1473. *A. lappa* L. (*A. majus*) GREAT BURDOCK. Waste places, tracksides, uncultivated ground. Widespread or mainland.   **(1473).** *A. minus* Bernh. LESSER BURDOCK. Waste places, roadsides. Wide-spread on mainland.

STAEHELINA   Small shrubs. Flower heads narrow-cylindrical, in clusters; ray florets absent; involucral bracts not spiny; receptacle with scales. Fruit usually hairless, pappus a single row of long, white branched hairs.

Leaves green, hairless on both sides
1474b. *S. fruticosa* (L.) L. A tufted shrub to 150 cm, with downy-glandular branches bearing terminal rosettes of leaves, and short-stemmed clusters of whitish, cylindrical flower heads. Involucre 10–12 by 4–5 mm, bracts light brown, pointed, hairless. Stem leaves 3·5–5·5 cm, lanceolate acute, rosette leaves oblong-spathulate blunt. Fruit shaggy-haired. Limestone cliffs. GR(south).CR+Karpathos. Page 66.

Leaves silvery-silky or white-woolly beneath
1474c. *S. arborea* Schreber (*S. arborescens*) A shrub to 1 m, with branches covered with silvery-silky hairs, and rosettes of ovate leaves 5–8 cm, which are dark green above silvery-silky-haired beneath. Flower heads pink, in simple or compound clusters, involucre 15–20 by 5–7 mm, bracts hairless to silky-haired. Fruit hairless. Limestone gorges. CR. Page 67.   **1474d.** *S. uniflosculosa* Sibth. & Sm. A small shrub to 50 cm with branches with white-woolly hairs, and ovate-pointed, toothed leaves 1·5–4 cm, which are dark green above and with white-woolly hairs beneath, not in terminal rosettes. Flower heads with only 1–2 pink florets, borne in simple or compound clusters; involucre 8–10 by 2 mm, bracts purple, the lower shortly woolly-haired, the upper hairless. Fruit hairless. Mountain rocks, bushy places. YU.AL.GR. Pl. 52.

JURINEA   Like *Carduus* but fruits four- to five-angled; pappus hairs in several rows and with a membraneous crown round the base of the pappus hairs. Leaves not spiny. 8 species in our area, the majority of which are eastern plants extending only as far west as Bulgaria.

1476. *J. mollis* (L.) Reichenb. (incl. *J. anatolica*) Flower heads rosy-purple, often solitary, globular or hemispherical 2–5 cm across, on long, nearly leafless, simple stems 30–70 cm (though stems much branched in subsp. *moschata* (DC.) Nyman). Involucral bracts lanceolate, recurved, purple towards the tips, with cobweb hairs, the inner much longer than the outer. Leaves variably pinnately cut, the lobes lanceolate to ovate usually greyish; stem leaves often entire. Dry grassy places. Widespread on mainland Page 96.   **1476c.** *J. glycacantha* (Sibth. & Sm.) DC. Very like 1476 but distinguished by the fruits which are not ribbed and have minute swellings especially in the angles (fruits ribbed in 1476). Flower heads larger 4·5–7·5 cm across, on almost leafless stems

Involucral bracts linear-lanceolate, the outer strongly recurved and slightly hooked, densely cobweb-hairy, the inner crested, hairless. Rocks, dry hills in the Med. region. YU.AL.GR.BG.

1476d. *J. taygetea* Halácsy A stemless or nearly stemless perennial, with purple globular flower heads 2–2·5 cm across, and involucral bracts straight and adpressed, with purple tips. Leaves all pinnately cut. Rocky places in mountains. GR.BG.

1476e. *J. albicaulis* Bunge Flower heads 1–3, shortly stalked, cylindrical or globular 18–25 mm long, with purple florets, borne on simple stems 30–50 cm. Involucral bracts linear-lanceolate long-pointed, with cobweb-hairs. Leaves linear, the lower weakly pinnate, grey-haired above, white-woolly beneath. Sandy and grassy places by the sea. GR(Thrace).BG.TR. Page 149.

CARDUUS | Thistle Involucral bracts many, overlapping, with spiny tips. Ray-florets absent; receptacle with scales. Fruit ovoid, hairless; pappus of several rows of bristles. A difficult genus with about 24 species in our area, 4 of which are endemic.

Flower heads small; involucre oblong or cylindrical, 1–2 cm wide
1477a. *C. acicularis* Bertol. A rather slender annual to 70 cm, with a very narrowly winged stem, leafless above, and with rosy-purple flower heads. Flower heads oblong 1·5–2 cm by 1–1·5 cm, long-stalked; involucral bracts cobweb-hairy, narrow with spiny apex. Leaves ovate to lanceolate in outline, with 2–5 pairs of triangular spine-tipped lobes, cobweb-hairy beneath, sparsely so above. Waste places in the Med. region. YU.GR.BG(south).TR. **1477b.** *C. pycnocephalus* L. Distinguished from 1477a by the involucral bracts which are widened in basal third and at least 1·5 mm wide above (1477a has narrower bracts, widened in basal quarter and not more than 0·5 mm wide above). Flower heads stalked or stalkless, solitary or two- to three-clustered. A very variable, erect simple or branched annual to 80 cm, with narrow discontinuous spiny wings to the stems. Waste places. Widespread.

Flower heads large; involucre globular or bell-shaped, often more than 2 cm wide
**1478. *C. nutans* group. MUSK THISTLES. Flower heads large, solitary, reddish-purple, 2·5–5·5 cm across, more or less drooping and borne on spiny-winged stems which are leafless below the heads. Several closely related species occur in this group; they are identified largely by their involucral bracts and are not easily distinguished. They include *C. macrocephalus* Desf. YU.GR.; *C. taygeteus* Boiss. & Heldr. GR.; *C. thoermeri* Weinm. YU.AL.GR.BG.TR.; *C. micropterus* (Borbás) Teyber YU.AL.; *C. nutans* L. YU. They occur in waste places, tracksides, pastures, cultivation.

1479a. *C. tmoleus* Boiss. A perennial to 1 m, with usually congested clusters of several rosy-purple flower heads borne on spiny-winged, nearly hairless stems. Flower heads globular or broadly bell-shaped, 1·5–3 cm wide. Leaves spiny, deeply cut, nearly hairless above and below. In mountains. YU(south).AL.GR.BG. Page 467. **1479b.** *C. cronius* Boiss. & Heldr. Differs from 1479a in the greyish or whitish cobweb-hairy stems (stems nearly hairless in 1479a) and wings up to 1·5 cm wide, and greyish- or white-hairy spiny leaves. Flower heads usually in stalkless clusters of 2–5, rounded to bell-shaped, 2–2·5 cm wide. Bare mountain slopes. AL.GR.

NOTOBASIS ****1481.** *N. syriaca* (L.) Cass. SYRIAN THISTLE. A tall erect annual thistle to 60 cm or more, distinguished by the purple-flushed, very spiny, rigid uppermost leaves which encircle and spread beyond the purple, globular, clustered flower heads. Leaves white-veined above, spiny. Field margins, wayside ditches. Widespread in the Med. region.

CIRSIUM | Thistle Distinguished from *Carduus* by its pappus which has feathery-haired bristles. A difficult genus with many intermediates; about 28 species are found in our area, and 7 are restricted to it.

Flower heads whitish to pale yellow
****1482.** *C. oleraceum* (L.) Scop. CABBAGE THISTLE Distinguished by its dense cluster of 2–6 yellow flower heads surrounded by much longer, broad, pale-yellowish, bract-like leaves. Perennial 50–150 cm. Damp meadows. YU.BG.   **1484.** *C. erisithales* (Jacq.) Scop. YELLOW MELANCHOLY THISTLE. Flower heads lemon-yellow, 2–3 cm across, usually solitary, nodding and borne on long almost leafless stems, or sometimes in a dense cluster of 2–5. Involucre 15–20 by 16–22 mm; bracts adpressed. Leaves deeply pinnately cut into oblong bristly-toothed lobes, the uppermost bract-like; stems 60–100 cm. Meadows and woods in mountains. YU.AL.GR.

****(1484).** *C. candelabrum* Griseb. CANDELABRA THISTLE. A tall pyramidal biennial to 2 m, with many lateral branches bearing terminal clusters of 4–12 pale whitish-yellow flower heads. Involucre 14–19 by 7–15 mm; bracts adpressed, spiny-tipped, pale. Leaves lobed, the lobes with numerous stiff pale spines, upper leaves stalkless. Stony, bushy places, disturbed ground, waysides. YU.AL.GR.BG. Page 165.   **1484a.** *C. hypopsilum* Boiss. & Heldr. Flower heads white, numerous in a much-branched dense narrow cluster, with several subtending leaves longer than the flower heads. Involucre 10–27 by 17–30 mm, hairy; bracts spine-tipped. A biennial to 1 m, with a much-branched stem. In mountains. GR(central and south).

Other white-flowered mountain species include: *C. morinifolium* Boiss. & Heldr. CR; *C. heldreichii* Halácsy GR(central).

Flower heads purple or pink (occasional individual plants white-flowered)
(a) Leaves with rigid prickles on upper surface of blade
  (i) Stems winged
(1485). *C. vulgare* (Savi) Ten. SPEAR THISTLE. Flower heads purple, in a branched cluster; involucre 3–4 by 2–4 cm, usually woolly-haired; bracts spine-tipped. Lower and middle stem leaves decurrent; stems winged; biennial 50–150 cm. Pastures, waste places. Widespread on mainland.

  (ii) Stems not winged
****1485.** *C. eriophorum* (L.) Scop. (incl. *C. vandasii*) WOOLLY THISTLE. A very variable often robust biennial to 1·5 m or more, with few large purple flower heads in a lax cluster. Involucre globular 3–5 by 4–7 cm, densely covered with cobweb hairs; bracts mostly spine-tipped, spreading or recurved. Leaves surrounding flower heads shorter to much longer than heads. Roadsides, waste places. YU.AL.GR.   **1485f.** *C. ligulare* Boiss. Like 1485 in having large purple flower heads in a loose cluster but differing in the middle to inner involucral bracts which are gradually widened towards the apex to a spoon-like papery appendage 2–5 mm wide, fringed with papery teeth and with an apical spine (bracts with appendage usually not more than 1·5 mm wide in 1485). A very variable biennial 20–150 cm. Stony places, in mountains. YU.AL.GR.BG. Page 467.

(b) Leaves without rigid prickles on the upper surface of the blade
  (i) Stems leafy to apex
1486. *C. palustre* (L.) Scop. MARSH THISTLE. Damp meadows, marshes. YU.AL.   **1486a** *C. creticum* (Lam.) D'Urv. A perennial 50–100 cm, much branched above with spiny-winged branches, and solitary or clustered purple flower heads at the ends of the branches. Involucre 12–17 by 7–10 mm; bracts with spreading spiny tips. Leaves leathery, undulate, pinnately cut, the lobes with very stout spines 5–15 mm, and with

1. *Cirsium ligulare* 1485f   **2.** *Carduus tmoleus* 1479a   **3.** *Tyrimnus leucographus*
1493a   **4.** *Amphoricarpos neumayeri* 1465a   **5.** *Lamyropsis cynaroides* 1488f

spreading hairs and sparse or dense cobweb hairs. Wet meadows, marshes. Widespread. Pl. 52. **1487.** *C. arvense* (L.) Scop. CREEPING THISTLE. Cultivated ground. Widespread on mainland. **1487a.** *C. appendiculatum* Grisb. A robust, usually unbranched perennial to nearly 2 m, with a terminal cluster of 6–8 purple flower heads on sparsely leafy stems. Involucre 15–20 by 15–20 mm; bracts oval-lanceolate, spiny-pointed. Leaves green, smooth, nearly hairless, decurrent, pinnately cut into triangular lobes with stout yellow spines 3–13 mm long. Mountain streamsides, damp places. YU.AL.GR.BG.TR. Pl. 52.

(*ii*) *Stems without leaves towards apex*
(1487). *C. rivulare* (Jacq.) All. Flower heads purple, in a terminal cluster of 2–5, or solitary, borne on stems which are leafless above the middle. Involucre 15–20 by 15–20 mm; bracts not spine-tipped. Leaves soft hairy, with bristly margins, usually deeply lobed, green on both sides. A perennial 40–100 cm. Damp places. YU.AL.

PTILOSTEMON Placed either in *Cirsium* or in *Chamaepeuce* by some botanists. Distinguished from *Cirsium* by minute characters of the fruit and pollen. Fruit obliquely obovoid, usually scarcely compressed, smooth, woody, with a cut-off apex; pappus hairs in several rows, feathery, fused at base, soon falling.

Spiny annuals to perennials
1488a. *P. afer* (Jacq.) W. Greuter (*Cirsium afrum*) An extremely spiny biennial thistle to 75 cm or more, with 10–16 purple flower heads in a dense domed or cylindrical cluster. Involucre 2–4 cm long; bracts purple, long needle-like as long as or longer than the flower head. Stems usually white-woolly; leaves lanceolate, pinnately cut, the lobes further cut into 2–3 spiny lobes with long marginal spines in clusters of 3, green above with pale veins, white-woolly beneath. Rocks in hills and mountains. YU.AL.GR.BG(west). Pl. 52. **1488b.** *P. strictus* (Ten.) W. Greuter Distinguished from 1488a by its smaller purple flower heads, with involucre 17–24 mm long, in dense branched clusters, with nearly hairless flat, spiny-tipped involucral bracts (bracts grooved and cobweb-hairy in 1488a). Stems 60–100 cm, narrowly winged; leaves decurrent; perennial. Shady places in mountains. YU(west).AL.GR. **1488c.** *P. stellatus* (L.) W. Greuter An annual 15–30 cm, distinguished by its linear-lanceolate entire leaves, with 1–3 stout basal spines on each side and a short apical spine. Flower heads purple, short-stalked, solitary or few, in the axils of the upper leaves; involucre 15–25 mm. Waste places, stony ground. YU(Vis).AL.GR+Ionian Is.CR.

Spineless shrubs
1488d. *P. chamaepeuce* (L.) Less. A shrubby plant of rocks and cliffs. Leaves numerous, linear, spineless with inrolled margins, green and hairless above and white-woolly beneath. Flower heads purple, 1–9 in stalked rather flat-topped clusters. Involucre 14–18 mm long; bracts without a spiny tip, ascending or deflexed, hairy or hairless. Limestone rock crevices. GR+Is.CR+Rhodes. Page 66 Pl. 52. **1488e.** *P. gnaphaloides* (Cyr.) Soják Like 1488d but leaves on flowering branches shorter, bristly-tipped, and leaf bases with 1–2 narrow lobes on each side. Rocks. GR(south)+Corfu.CR.

LAMYROPSIS Like *Ptilostemon* but differing in its fruits, which are leathery, oblong, compressed, with a cut-off apex with a raised margin surrounding a cylindrical central boss; pappus of several rows of feathery hairs.

1488f. *L. cynaroides* (Lam.) Dittrich (*Cirsium c.*) A perennial 20–50 cm, with white cobweb-hairy stems, bearing solitary or few rosy-purple stalked flower heads, with spreading or deflexed, long-tapering spiny-pointed involucral bracts. Florets 2·6–3·4 cm long; involucre 2·7–3·5 cm long. Leaves pinnately cut, each segment with 3–5 triangular spiny lobes, white-woolly beneath. Pappus of fruit 17–21 mm. Dry places. GR(south)+Is.CR. Page 467.

PICNOMON   Like *Cirsium* but the involucral bracts without resin tubes and with a recurved, pinnate, spiny apical appendage.

**1488. *P. acarna* (L.) Cass. (*Cirsium a.*) An extremely spiny, branched annual 20–50 cm, with greyish- to white-woolly winged stems and leaves, and a dense cluster of purple flower heads encircled by longer very spiny upper leaves. Involucre cylindrical 22–30 mm long. Leaves decurrent, oblong, remotely lobed, with slender marginal spines. Cultivated ground, dry waste places in the Med. region. Widespread. Page 165.

CYNARA | **Cardoon, Artichoke**   Flower heads very large; involucral bracts conspicuous, stout, leathery, blunt or pointed, in many ranks. Ray-florets absent; receptacle fleshy, scales present.

1491. *C. cardunculus* L. CARDOON. A robust perennial 20–100 cm, with large blue to whitish globular flower heads 4–5 cm, with conspicuous leathery often purplish involucral bracts, each with a long stout yellow spine. Leaves white-cottony beneath, all deeply once- or twice-cut into narrow lobes, with rigid yellow spines 1·5–3·5 cm, in clusters. Stony or waste places, dry grassland in the Med. region; occasionally cultivated for its edible leaves. GR.CR. Pl. 53.   ****(1491).** *C. scolymus* L. GLOBE ARTICHOKE Flower heads 8–15 cm across; involucral bracts blunt, fleshy. Cultivated for its edible, young flower heads in YU.GR.CR.   **1491e.** *C. cornigera* Lindley (*C. sibthorpiana*) Flower heads yellowish, globular 4–5 cm, solitary. Outer involucral bracts with a slender apical spine about 5 mm, middle bracts with a shell-like appendage abruptly narrowing into a spine. Leaves mostly basal, to 40 cm, leathery, pinnately cut into triangular lobes with terminal yellow spines, bright green with pale veins above, white-hairy with prominent veins beneath; stems short to 30 cm, woolly-haired. Coasts. GR(south).CR+Karpathos.

SILYBUM   ****1492.** *S. marianum* (L.) Gaertner MILK-THISTLE, HOLY THISTLE. Flower heads solitary 2·5–4 cm across, rosy-purple, with involucral bracts with long spreading and recurved spines 2–5 cm. Leaves pinnately cut, white-veined or variegated, hairless, the upper with spines; stems 20–150 cm. Waste ground, waysides, uncultivated places in the Med. region. Widespread, largely on mainland, Page 164.

GALACTITES   ****1493.** *G. tomentosa* Moench A rather slender thistle-like plant 15–100 cm, with white-cottony stems and undersides of leaves, and few stalked ovoid, rosy-purple, lilac, or rarely white flower heads in a lax cluster. Outer florets larger, spreading; involucre 10–15 mm long; bracts with cobweb hairs, spine-tipped. Leaves pinnately cut, spiny, white-veined. Uncultivated ground, waysides, waste places in the Med. region. YU.GR.CR.

TYRIMNUS   Like *Carduus* but filaments of stamens fused, and fruits four-angled.

1493a. *T. leucographus* (L.) Cass. A cottony, thistle-like plant with spiny-winged lower stems, spiny leaves, and a solitary long-stalked, purplish-pink globular flower head 1·5 cm across. Involucral bracts numerous, adpressed, weakly spine-tipped. Leaves oblong-lanceolate, spiny-toothed or lobed, white-veined and green above, cottony beneath. Annual or biennial 20–60 cm. Waste places, stony and sandy ground in the Med. region. Widespread. Page 467.

ONOPORDUM   Stem with spiny wings; leaves spiny-lobed. Flower heads globular, with several rows of leathery, spine-tipped involucral bracts. Receptacle hairless, pitted; disk-florets only present. 10 species in our area, 2 of which are endemic.

Key to the most frequently encountered species:
1. Stems and mature lvs. not densely woolly-
   haired, usually greenish, with
   multicellular hairs.
   2. Involucral bracts erect and closely
      overlapping (GR(south)+Is.).   **1494c.** *O. laconicum* Rouy
   2. Involucral bracts with more or less
      diverging spiny tips.
      3. Lvs. with at least 10 pairs of
         lobes; fl. heads with more or less
         dense cobweb hairs (GR(south)).   **1494d.** *O. argolicum* Boiss.
      3. Lvs. with not more than 8 pairs of
         lobes; fl. heads hairless, or with
         very sparse cobweb hairs.
         4. Fl. heads 3·5–5 cm across; stem
            wings with spines up to 15 mm
            (YU(central)).   **1494e.** *O. corymbosum* Willk.
         4. Fl. heads 5·5–7 cm across; stem
            wings with spines up to 5 mm.   *tauricum*
1. Stems and lvs. densely white- or grey-
   woolly, with unicellular hairs.
   5. Involucral bracts linear or awl-shaped;
      corolla lobes without glands (YU.AL.
      GR.BG.).   ****1494.** *O. acanthium* L.
   5. Involucral bracts lanceolate to ovate,
      often with spiny apices; corolla lobes
      glandular.
      6. Longest involucral bracts, longer
         than the florets (GR+Is.TR.).   **1495c.** *O. caulescens* D'Urv.
      6. Involucral bracts not longer than
         the florets.
         7. Fl. heads hairless, 5–7 cm
            across.   *bracteatum*
         7. Fl. heads with cobweb hairs
            below, 4–6 cm across.   *illyricum*

(1494). *O. tauricum* Willd. Flower heads rosy-purple, large globular 5·5–7 cm, borne on a stout yellowish-brown spiny-winged stem to 2 m. Involucral bracts 4–7 mm wide at base, tapering to a rigid spine; middle bracts ascending, the outer usually reflexed. Leaves to 25 cm, pinnately lobed, dark green, very sparsely hairy above, densely hairy beneath; wings of stem 1·5 cm wide with spines to 5 mm. Stony, rocky places, waysides, dry open ground. GR+Is.CR.BG.TR. Page 165.

1495. *O. illyricum* L. Usually a densely grey- or white-felted biennial, with densely felted, spiny-winged stems to 130 cm, and large globular to ovoid, purple flower heads 4–6 cm across, with cobweb hairs. Corolla 3–3·5 cm long, lobes glandular-hairy. Involucral bracts 5–7 mm wide, overlapping and more or less adpressed to the florets, or middle and outer bracts spreading and recurved towards apex, the outer bracts shortest. Leaves to 55 cm, lobed, usually densely grey-felted, spiny. Dry open ground in the Med. region. YU.AL.GR+Is.CR.BG. **1495b. *O. bracteatum* Boiss. & Heldr. Distinguished by its hairless, broadly lanceolate spiny-tipped involucral bracts, and by its larger purple flower heads 5–7 cm across. Corolla 3–4 cm long, glandular. Stems to 180 cm, winged, with palmately arranged spines 1–1·5 cm long. Mainly in mountains, stony ground in the Med. region. GR.CR+Karpathos, Rhodes.BG.

CRUPINA 1496. *C. vulgaris* Cass. FALSE SAW-WORT. Flower heads purple, with 3–5 florets; involucre 8–15 mm long; bracts lanceolate, closely overlapping, spineless, green and sometimes purple-flushed. An erect, branched annual 20–50 cm, leafy to the branches; leaves deeply pinnately cut into linear toothed but spineless lobes. Dry grassy and stony ground. Widespread on mainland. **(1496).** *C. crupinastrum* (Moris) Vis. Like 1496 but stems leafy only to the lower third or half. Flower heads with 9–15 florets. Dry places in the Med. region. Widespread. Page 41.

SERRATULA 5 species in our area. ****1497.** *S. tinctoria* L. SAW-WORT. A very variable plant. Damp meadows, marshes, wood clearings. YU.AL.GR.BG. **1497b.** *S. radiata* (Waldst. & Kit.) Bieb. A slender erect; rough-haired perennial to 60 cm, with solitary or few, purplish bell-shaped flower heads usually 2·5–3 cm long. Outer involucral bracts with short rigid spiny apex and with tufts of soft woolly hairs. Leaves deeply pinnately cut with numerous parallel lobes. Grassy and rocky places. YU.AL.BG.

CENTAUREA | Knapweed, etc. Important distinguishing characters are those of the middle involucral bracts. These may have a papery appendage which is often cut into a comb-like fringe; this appendage may run as a narrow margin down each side of the involucral bracts. Involucral bracts may have one or several terminal spines, often with smaller lateral spines on the margins. A difficult genus with 126 species in our area, of which about 56 are endemic.

Involucral bracts with a terminal spine, usually with smaller lateral spines
(a) Flower heads yellow, orange, or white
1499. *C. solstitialis* L. ST. BARNABY'S THISTLE. Florets yellow, not glandular. Waste ground, cultivation, Widespread. Page 164. **1499c *C. idaea* Boiss. & Heldr. Like 1499 but stems numerous, the central stem short unbranched (stem single, erect wide-branched in 1499). Florets yellow, glandular; involucral bracts hairy, yellowish, with apical spines 1·5–3 cm. A biennial 10–15 cm, with greyish-hairy rosette leaves cut into numerous small lobes, the terminal lobe 3 times as large as the laterals. Dry places. CR. **(1499).** *C. melitensis* L. MALTESE STAR THISTLE. Distinguished from 1499 in having an erect, sparingly branched stem to 80 cm, with rough-haired, green leaves. Flower heads one or several; involucral bracts hairless, with apical spine 5–8 mm; florets yellow, glandular, the outer spreading. Dry places, disturbed ground. YU.GR+Is.CR.

1503. *C. salonitana* Vis. YELLOW KNAPWEED. An erect, sparingly branched perennial to 1 m, with conspicuous yellow flower heads. Involucre 1·5–3 cm across; bracts with pale papery borders and an apical spine varying from 3 mm or up to 4 cm. Leaves usually hairless, rough, mostly pinnately cut into entire or toothed lobes. Grassy places, rocks, bare ground. YU.AL.GR.CR?+Rhodes.BG. Page 164. Pl. 53. **1503f. *C. macedonica* Boiss. Flower heads pale yellow, the outer florets longer and spreading. Involucre about 12 mm across; bracts with a reddish-brown, fringed margin, and with a slender apical spine usually 1–6 mm, sometimes longer. Leaves sparsely cobweb-hairy, pinnately cut into narrow lobes which are about 2 mm wide. Rocky places. AL.GR.

1503e. *C. ragusina* L. A white-woolly perennial with yellow flower heads, and involucral bracts with short recurved spines, found on maritime rocks and walls of Yugoslavia. Involucre globular 2–2·5 cm; bracts white-woolly, with brownish triangular margin, and apical spine 4 mm. Leaves mostly basal, pinnately cut into entire or lobed segments; stems 30–60 cm. YU(west). **1503g.** *C. pelia* DC. Flower heads small, yellow, with ovoid to cylindrical involucre 10 mm by 4–5 mm, and bracts with a brown appendage and a short, somewhat recurved apical spine 1–3 mm. Stems 30–50 cm; leaves rough, with cobweb hairs, the lower leaves pinnately lobed. Dry rocky places. GR(central and east). Page 96.

(b) Flower heads pink to purple

1500. *C. calcitrapa*. L. STAR THISTLE. Flower heads pale purple, closely surrounded by upper leaves. Involucre 6–8 mm across; bracts leathery with papery margins and a spreading terminal, thickened spine 1–1·8 cm long. A much-branched, erect biennial 20–100 cm. Waste places, disturbed ground. Widespread, except TR. **1500c. *C. raphanina* Sibth & Sm. (incl. *C. mixta*) Flower heads pink or purple, 2–4, usually stalkless and clustered in the middle of a flattened rosette of deeply and irregularly lobed oblong leaves. Involucre 1·2–2 cm across; bracts with bristly teeth on margin and with an apical spine up to 2·5 cm. A variable perennial, with leaves hairless and shiny, or rough and dull. Dry stony ground. GR(south and east)+Is.CR. Page 71.

1500d. *C. graeca* Griseb. An erect perennial 50–180 cm, branched above, with leaves mostly in a basal rosette, and pinkish-purple flower heads with outer florets spreading. Involucre 1·8–2 cm long; bracts with a finely toothed, straw-coloured margin, and an apical spine 3–20 mm. Leaves usually pinnate, with cobweb hairs, leaflets oblong. Young leaves white-woolly. Rocky places in hills and mountains. AL.GR(north and central). Pl. 53. **1500e.** *C. redempta* Heldr. A handsome plant of cliffs and gorges in Crete, with very large solitary, dark-purple, long-stemmed flower heads about 5 cm across. Involucre of broad green bracts with a broad black, frilled, horseshoe-shaped margin, and usually a long stout brown terminal spine 1·5–3 cm. Leaves greyish-white, pinnately cut; stems white-cottony. CR. **1500f.** *C. diffusa* Lam. A much-branched annual or biennial 10–50 cm, with usually many small pink flower heads about 1·5 cm across. Involucre cylindrical-ovoid, 7–10 mm by 4–5 mm; bracts with a short erect or spreading spine 2–3 mm long. Lower leaves twice-cut, green, with cobweb hairs. Waste places, maritime sands. YU.GR.BG.TR.

1500g. *C. attica* Nyman A branched, usually erect, perennial 5–30 cm, and with usually white-woolly leaves, and small pink or purple solitary flower heads. Involucre 10–16 mm by 5–10 mm; bracts with 3 veins, a black margin, and a long apical spine to 7 mm. Lower leaves cut into narrow lobes 1 mm or more wide, the upper undivided. Mountain rocks. GR(north and east). Pl. 53. **1500h.** *C. laconica* Boiss. Distinctive are the globular pinkish-purple flower heads with involucral bracts with long apical spines 1–2 cm. A little-branched perennial 30–50 cm, with twice-cut leaves with narrow segments. Rocky places GR(south). Pl. 53.

Involucral bracts without apical spines, often with a papery appendage fringed with teeth or bristles

(a) Flower heads yellow, orange, or white

1504c. *C. orientalis* L. An erect, sparsely branched perennial 80–120 cm, with pale-yellow flower heads (recalling 1503 *C. salonitana*) but involucral bracts quite different. Involucre 2–2·5 cm across, globular; bracts broadly ovate and with an ovate, papery, fringed appendage with a central brown spot. Outer florets longer than inner, spreading. Leaves pinnately cut into narrow lobes. Dry pastures. YU.BG. Pl. 53. **1504d.** *C. argentea* L. A white-leaved cliff perennial of Crete and Kithira, with solitary yellow flower heads with ovoid involucres 8–10 mm long. Involucral bracts with pale-brown, triangular to half-moon-shaped fringed appendages, without terminal spines. Leaves deeply cut with oblong lobes. Mountain rocks. CR+Kithira. Page 67.

1501e. *C. pindicola* Griseb. A dwarf alpine perennial with a rosette of leaves and usually a single stem to 15 cm, bearing a solitary cream-coloured flower head. Involucre ovoid 16–18 mm across; bracts green, with a broad black margin fringed with silvery-white bristles. Stem leaves entire or toothed, the basal leaves lobed. Pastures in mountains. YU.AL.GR. Page 126. **1501f.** *C. baldaccii* Bald. A dwarf alpine perennial of Crete, with cream-coloured flower heads on stout stems. It differs from 1501e, having a smaller ovoid involucre 6–8 mm across and leaves which are linear-lanceolate in outline (oblong in 1501e). CR(west).

(b) Flower heads blue or violet
1501. *C. cyanus* L. CORNFLOWER, BLUEBOTTLE. Dry open places, cornfields. Widespread on mainland. **1501d. *C. triumfetti* All. A very variable perennial, with entire leaves and erect stems bearing solitary flower heads with larger spreading blue outer florets, and smaller violet central florets. Involucre ovoid or cylindrical 7–25 mm across; bracts with a conspicuous brown or black margin and with usually longer, marginal bristles which are brown, white, or silvery. Widespread on mainland. Subsp. *cana* (Sibth. & Sm.) Dostál Basal leaves white-woolly, usually deeply lobed, stem leaves linear-lanceolate. Flowering stems short 3–20 cm, unbranched. Involucre about 1·5 cm across; bracts with dark-brown margin and silvery bristles. Throughout the Balkan Peninsula. Page 168. Pl. 53. **1501g.** *C. napulifera* Rochel Like 1501d but roots spindle-shaped or turnip-shaped and flower heads usually bicoloured with inner florets purple or lilac, and outer florets dark blue, purple, pink, cream, or white. Involucre ovoid 8–14 mm across; bracts with a dark-brown or black margin and longer silvery-tipped bristles. Stems 3–35 cm, simple or little-branched; leaves variable. Grassy slopes in mountains. YU.AL.GR.BG. TR. Page 136.

**1504. *C. uniflora* Turra (incl. *C. nervosa*) Flower heads usually purplish-violet with outer florets spreading, usually solitary, on erect unbranched stems usually 10–15 cm. Involucre 12–20 mm across; middle bracts with a blackish-brown margin, and with a long slender arched tip with numerous brown comb-like bristles arranged along its length. Leaves green, oblong to elliptic, not lobed but toothed, the upper narrow lanceolate. Dry grasslands in mountains. YU.GR(north).BG.

(c) Flower heads pink to purple
(i) Involucral bracts with a papery fringed margin (see also 1501g, 1504)
1502a. *C. affinis* Friv. An erect perennial with white- or greyish-hairy rough leaves, and small solitary pink flower heads on branched stems 30–80 cm. Involucre ovoid-globular 12–15 mm long; bracts usually with a dark-brown or black triangular margin with 5–9 bristles on each side, or a rounded and scarcely fringed appendage. Leaves pinnately cut, with narrow oblong lobes 3–5 mm wide, or entire. Rocky places, mainly in mountains. Widely distributed on mainland.

(ii) Involucral bracts with an enlarged papery terminal appendage
1505. *C. jacea* L. BROWN-RAYED KNAPWEED. Flower heads purple, rarely white, in a flat-topped cluster. Involucre ovoid 15–18 mm long; bracts with a pale-brown, rounded appendage with a papery fringe usually covering the bracts. Leaves entire, toothed, or pinnately cut. Meadows, thickets, shady places. Widespread on mainland. **1505c.** *C. alba* L. A very variable biennial or perennial, with up to 20 subspecies recognized, with erect branched stems bearing at least 10 pink or purple flower heads with conspicuous rounded papery appendages hiding the bracts. Involucre variable, globular to cylindrical, 8–22 mm long; bracts with entire, toothed, or fringed margin, and rounded

appendages with white, green, or black to brown centres. Leaves very variable, white-woolly to green, once- or twice-cut into linear to ovate lobes. Dry places. YU.AL.GR.BG. Pl. 53. **1505d.** *C. zuccariniana* DC. Readily distinguished by its tiny pink-purple flower heads with pale-brown or yellowish papery involucral bracts which are recurved and cut into feathery tassels. Flowers 2–4 in each head; involucre 12–14 mm. An erect biennial 20–30 cm, with pinnately cut leaves. Dry scrub. AL.GR. Pl. 53.

CARTHAMUS   5 species in our area.

Flower heads yellow or orange-yellow
1508. *C. lanatus* L. A greyish-haired, thistle-like, glandular annual 30–60 cm, with yellow ovoid flower heads 2–3 cm across. Involucre of numerous, green leafy, spiny, outer involucral bracts as long as the florets, and papery inner bracts with a lanceolate, finely fringed apex. Leaves pinnately cut with spiny lobes. Pappus greyish, of narrow ciliate scales. Waste ground, cultivation. Widespread. **(1508). *C. tinctorius* L. SAF-FLOWER. Leaves usually entire, oblong to elliptic with marginal bristles. Flower heads orange-yellow; involucral bracts shorter than the florets. Pappus usually absent. Native of W. Asia. Cultivated for its oil-bearing seeds; sometimes naturalized in YU.GR?

Flower heads violet or pinkish-purple
1508c. *C. dentatus* (Forskål) Vahl A cobweb-haired, glandular annual with pale-violet to pinkish-purple, oblong-ovoid flower heads with spreading, spiny outer involucral bracts. Inner involucral bracts with a distinct ovate-lanceolate, toothed, papery apical appendage. Stem leaves lanceolate to ovate-lanceolate, pinnately cut, lobes with spiny margins. Rocky places. YU.GR+Is.CR.BG.TR. Pl. 54. **1508d.** *C. boissieri* Halácsy Distinguished from 1508c by the inner involucral bracts with a simple spine (not a papery appendage). Florets violet to pinkish-purple; outer involucral bracts 4–5 cm, spreading, with spines 7–9 mm long. Basal leaves usually with more than 10 pairs of lobes, stem leaves with dense cobweb hairs; stems pale brown, glandular and with dense cobweb hairs. Rocky places, cultivation. GR(south). CR+Karpathos, Rhodes. Page 480.

CARDUNCELLUS   **(1508)**. *C. caeruleus* (L.) C. Presl (*Carthamnus c.,Kentrophyllum c.*) Flower heads clustered, blue, with cobweb-hairy involucre with leafy bristle-toothed outer involucral bracts a little longer than the florets. Leaves shiny, simple and toothed, or pinnately cut into 6–10 pairs of lobes with spiny margins and apex; stems 30–60 cm, usually cobweb-hairy. A variable species. Bushy places, cultivation in the Med. region. GR.CR+Karpathos.

CNICUS   ****1509.** *C. benedictus* L. BLESSED THISTLE. Cultivated ground, sandy fields. Widespread mostly on mainland.

Subfamily **CICHORIOIDEAE**   Flower heads with all florets having strap-shaped corollas, known as ray-florets or ligulate-florets. Many are yellow-flowered and are widespread in Europe. Only the most distinctive or interesting species in this subfamily are included; the following genera are not described: *Aposeris* 1 sp.; *Arnoseris* 1 sp.; *Aetheorhiza* 1 sp.; *Cephalorrhynchus* 1 sp.; *Calycocorsus* 1 sp.; *Taraxacum* about 10 species groups; *Hieracium* over 100 spp.

SCOLYMUS   **1510.** *S. hispanicus* L. SPANISH OYSTER PLANT. Flower head yellow, about 3 cm long, closely encircled by longer, leafy, spiny bracts. Involucre 1·5–2 cm wide;

---

12. *Tragopogon porrifolius* 1528   **13.** *Leontodon crispus* 1524a   **14.** *Sonchus tenerrimus* (1539)   **15.** *Urospermum picroides* 1526   **16.** *Crepis rubra* (1546) **17.** *Scorzonera purpurea* 1529   **18.** *Prenanthes purpurea* 1545   **19.** *Hymenonema graecum* 1510c   **20.** *Andryala integrifolia* 1533   **21.** *Picris echioides* (1525)

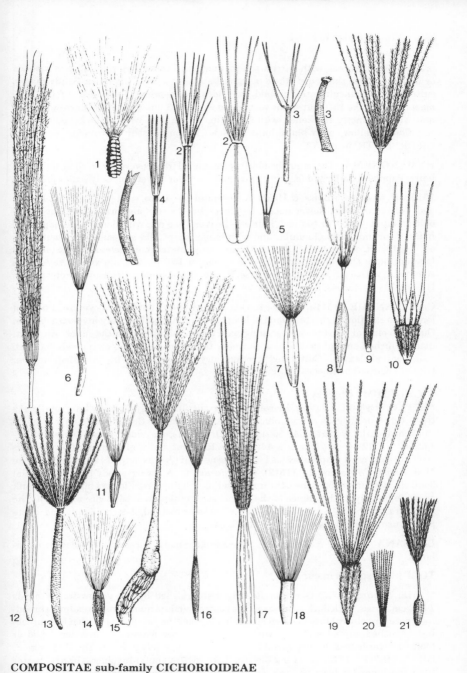

COMPOSITAE sub-family CICHORIOIDEAE

1. *Reichardia picroides* 1536 **2.** *Hyoseris radiata* 1516 **3.** *Rhagadiolus stellatus* 1518 **4.** *Hedypnois cretica* 1519 **5.** *Tolpis barbata* 1515 **6.** *Chondrilla juncea* 1534 **7.** *Cicerbita pancicii* 1537a **8.** *Lactuca graeca* 1541b **9.** *Hypochoeris maculata* 1522 **10.** *Catananche lutea* 1511a **11.** *Mycelis muralis* 1540

bracts mostly hairless, with narrow papery margins. Leaves and wings of stem without or with a slightly thickened white margin. Pappus of a few rigid hairs. Sandy places, cultivated and uncultivated ground, waysides. Widespread. Page 165 **(1510). *S. maculatus* L. Like 1510 but stems with broad wings 2–5 mm wide at the narrowest part, and with distinctly thickened white margin; leaves also with a thickened margin. Ray-florets yellow, with black hairs outside. Pappus absent. Cornfields, waysides. GR.CR? BG.TR. introd.YU.

HYMENONEMA Distinguished by its five-angled fruit covered with rigid adpressed hairs. Pappus of both rigid feathery hairs and scales, or sometimes only scales present.

1510b. *H. laconicum* Boiss. & Heldr. A perennial 20–70 cm, with pinnately cut leaves, and 1–3 large yellow flower heads. Involucre 1·5–2·4 cm long; bracts ovate or ovate-oblong, blunt, entire or toothed, hairless. Leaves 7–25 cm, with dense rigid adpressed hairs; stems minutely glandular and with longer hairs. Dry grassland. GR(south). Pl. 54. **1510c.** *H. graecum* (L.) DC. Like 1510b but pinnately cut leaves with the terminal lobe not more than 1 cm wide (1·5–3 cm wide in 1510b), and the yellow ray-florets sometimes with a purple spot at the base. Pappus of scales only. Roadsides, stony places. GR(south)+Cyclades.CR.

CATANANCHE **1511a.** *C. lutea* L. YELLOW CUPIDONE. Flower heads yellow, 1·5–2 cm long, with ray-florets shorter than the long narrow inner, pale papery involucral bracts. Outer involucral bracts ovate, shining pale brown, shorter than the inner. An erect, little-branched annual to 40 cm, with linear to lanceolate, entire or sparsely toothed, rough-haired leaves. Cultivated ground, dry rocky places in the Med. region. GR.CR+Karpathos,Rhodes.TR. Page 480.

CICHORIUM | Chicory
1512. *C. intybus* L. CHICORY. Flower heads bright blue, rarely pink or white; ray-florets three times as long as the involucre. Stalks of terminal flower heads only slightly thickened. A leafy perennial 30–120 cm. Dry fields, waysides, waste places. Widespread. **(1512).** *C. endivia* L. ENDIVE. Like 1512 but stalks of terminal flower heads strongly thickened. Pappus scales ⅓–½ as long as fruit (1/10–¼ as long in 1512). Rocks in the Med. region. Widespread. **(1512).** *C. spinosum* L. A dwarf woody-based perennial 3–18 cm, with many branches, the upper ending in spiny tips. Flowers blue, rarely pink or white, mostly in the angles of the branches. Ray-florets about twice as long as the involucre which is 5–8 mm. Leaves fleshy, lobed or toothed, 2–9 cm. Sandy and waste places, usually by the sea. GR+Is.CR. Pl. 54.

LAPSANA **1513.** *L. communis.* L. NIPPLEWORT. Shady places. Widespread on mainland.

TOLPIS 3 species in our area.

1515. *T. barbata* (L.) Gaertner Readily distinguished by the yellow flower heads 1·5–3 cm across, with dark reddish-purple centres, and its numerous spreading, thread-like, outer involucral bracts, often spreading beyond the ray-florets. A slender, spreading, branched annual 6–90 cm, with lanceolate upper leaves and broader, lobed or toothed lower leaves. A very variable species. Grassy, sandy places in the Med. region. GR+Is.CR.BG. **1515b.** *T. virgata* Bertol. Like 1515 but outer involucral bracts shorter than the inner (at least as long as inner in 1515). Flower heads several; involucre 6–8 mm long, bracts adpressed,shorter than florets. A perennial or biennial 30–100 cm. Grassy, stony places in the Med. region. AL.GR+Is.CR.TR.

HYOSERIS **1516.** *H. radiata* L. A dandelion-like perennial with leaves all basal and deeply cut into backward-pointing, triangular, toothed lobes, and erect unbranched stems 6–35 cm, bearing solitary yellow flower heads about 3 cm across. Involucre 1–1·5 cm long; bracts blunt, much shorter than the showy florets, spreading in fruit. Fruit brown, with a pappus of both rigid hairs and scales. Rocks, dry places, cultivated ground in the Med. region. YU.GR.CR. **(1516).** *H. scabra* L. Like 1516 but an annual with spreading or ascending flowering stems 0·5–7 cm, as long as the leaves, hollow and often conspicuously swollen below the yellow flower heads. Florets little longer than the involucral bracts; involucre 7–10 mm long; bracts erect in fruit. Dry open places in the Med. region. YU.AL? GR+Is.CR.TR.

RHAGADIOLUS 1518. *R. stellatus* (L.) Gaertner STAR HAWKBIT. Readily distinguished from other small yellow-flowered members of the family Compositae by its outer fruits, which enlarge and spread outwards to form a persistent five-rayed star, while the erect inner cylindrical fruits soon fall. A hairy, erect branched annual 7–40 cm. Rocky places, cultivated and uncultivated ground in the Med. region. Widespread.

HEDYPNOIS ****1519.** *H. cretica* (L.) Dum.-Courset (*H. rhagadioloides*) A small variable rosette annual, with spreading or ascending branched stems 3–45 cm, bearing yellow flower heads about 1·5 cm across, the branches conspicuously swollen below the flower heads. Fruiting heads distinctive, crown-like, with stiff incurved involucral bracts surrounding the fruits. Leaves elliptic, entire or lobed. Dry stony ground, grassy places in the Med. region. Widespread.

HYPOCHOERIS 7 similar-looking species in our area.

Widespread are: **1520.** *H. radicata* L. CAT'S EAR; **(1520).** *H. glabra* L. SMOOTH CAT'S-EAR. ****1522.** *H. maculata* L. SPOTTED CAT'S-EAR. A rather robust rosette perennial with simple or sparingly branched stems with one or several large pale-yellow flower heads 4–5 cm across. Involucral bracts 12–25 mm, blackish-green, bristly-haired. Leaves elliptic to ovate, toothed, often with brownish-violet streaks or spots. Stem leaves few or absent; stems 15–75 cm. Meadows, among herbaceous vegetation. YU.AL.GR.BG.

LEONTODON | Hawkbit A difficult genus with about 12 similar-looking species in our area.

1523. *L. autumnalis* L. AUTUMNAL HAWKBIT. Pastures, meadows, waysides. YU.GR.BG. ****1524.** *L. hispidus* L. ROUGH HAWKBIT. Meadows, pastures, rocky places. YU.AL.GR.BG. **1524a.** *L. crispus* Vill. (incl. *L. asperrimus*, *L. graecus*) A perennial with slender tubers. Leaves in a rosette, oblong-lanceolate, toothed to pinnately lobed, with numerous forked to four-branched hairs. Stems 7–40 cm, often several, simple or branched, thickened below flower heads, bearing 1 to 3 yellow flower heads with ray-florets sometimes reddish-purple beneath. Involucre 10–25 mm long; bracts oblong-lanceolate, often with comb-like margins, with simple or branched hairs. Fruit with a beak as long as the nut; pappus of brownish-white feathery hairs. Mountain rocks. Widespread on mainland. Pl. 54.

1524b. *L. tuberosus* L. TUBEROUS HAWKBIT. Like 1524a with a rosette of basal leaves and solitary yellow flower heads, but outer ray-florets with a greenish stripe beneath. Involucre 9–15 mm long; bracts oblong, narrowed to a blunt apex. Leaves oblong, wavy-margined or lobed, with sparse bristly two- to three-branched hairs. Fruit of two kinds; the outer curved, beakless, or shortly beaked, with pappus of very short hairs, the inner straight, beaked, with pappus hairs feathery. A perennial with long spindle-shaped tubers. Mountains, cultivated ground, stony places in the Med. region. YU.AL.GR+Is.CR.TR.

PICRIS | Ox-tongue 6 similar-looking species in our area.

1525. *P. hieracioides* L. HAWKWEED OX-TONGUE. Grassland, rocky places, waysides. Widespread on mainland. **(1525).** *P. echioides* L. BRISTLY OX-TONGUE. Grassland, cultivated fields, waste places. Widespread. **1525a.** *P. pauciflora* Willd. An annual 10–50 cm, with few yellow flower heads on long stalks which are thickened after flowering. Involucre 10–12 mm long; bracts shorter than florets, linear-lanceolate, mostly pointed, with short star-shaped and longer rigid hairs mainly on the median line, the outer bracts spreading. Basal and lower leaves usually elliptic to oblanceolate, entire to wavy-toothed. Fruit dark brown, transversely wrinkled, more or less curved, with a short beak. Fields and roadsides. YU.AL.GR+Is.CR.BG. **1525b.** *P. sprengerana* (L.) Poiret Similar to 1525a but with numerous flower heads on short stalks that do not thicken after flowering. Involucre 8–12 mm long; bracts linear-lanceolate, with dense rigid hairs. Fruit dark brown, more or less curved, without a beak. Fields, waysides. Widespread.

UROSPERMUM ****1526.** *U. picroides* (L.) F. W. Schmidt Flower heads pale yellow, 5 cm across. Involucral bracts ovate-lanceolate, long-pointed, bristly-haired. Pappus white. Cultivated ground, dry grassland, waste places in the Med. region. YU.AL.GR+Is.CR.TR. ****(1526).** *U. dalechampii* (L.) F. W. Schmidt Flower heads up to 4 cm across on swollen stems. Involucral bracts lanceolate, softly hairy. Pappus pale reddish-brown. Cultivated ground. YU(Croatia).

TRAGOPOGON A difficult genus, with about 14 similar-looking species in our area, including 3 endemic species; distinguished largely by their fruits and involucral bracts.

Flower heads yellow

1527. *T. pratensis* L. GOATSBEARD, JACK-GO-TO-BED. Meadows, waysides. Widespread on mainland.

Flower heads with at least outer florets violet-purple

1528. *T. porrifolius* L. SALSIFY. Readily distinguished by its solitary violet-purple flower heads on stems which are thickened above, and its narrow grass-like, often wavy-margined leaves. Flower heads variable in size; ray-florets often as long as involucral bracts and spreading to 4 cm. Biennial 20–125 cm. Bushy and grassy places in the Med. region. Widespread, except AL. **1528a.** *T. balcanicus* Velen. Like 1528 but distinguished by the swollen club-shaped beak of the fruit which is constricted below the pappus (beak not swollen in 1528). Flower head purplish-violet; stem not thickened below head. Involucral bracts 4–5 (about 8 in 1528), about as long as the florets at flowering. Leaves narrow-linear, widened at the base. Fruit dark brown. Rocky places. YU.AL.GR.BG.TR. Page 480.

1528b. *T. crocifolius* L. Flower heads with outer florets violet with a yellow base, and inner florets yellow, or rarely all florets yellow or violet, on stems which are not thickened. Involucral bracts 5–12. Annual or biennial with simple or branched stems 10–80 cm; leaves narrowly linear. Fruit with a stout club-shaped beak. Rocky places in mountains. YU.GR.BG. **(1528).** *T. hybridus* L. (*Geropogon glaber*) Flower heads pink to rosy-violet, with 8 narrow acute involucral bracts much longer than the florets. Distinguished by the outer row of fruits with a beak of 1·5–2 cm and with 5 unequal bristles 1–2 cm, the inner fruits with a shorter beak and feathery hairs. A hairless annual, with linear leaves, and stems 20–50 cm. Fields, cultivated ground in the Med. region. YU.AL.GR+Is.CR.TR.

SCORZONERA A difficult genus, with about 14 species in our area, 4 of which are restricted to it.

Flower heads pink, violet, or purple
1529. *S. purpurea* L. PURPLE VIPERGRASS. Flower heads pale lilac to pale purplish, with florets 1½–2 times as long as the involucre which is 1·5–2·5 cm. Leaves grass-like, straight, channelled above and keeled below, 1–3 mm wide, the stem leaves clasping; stems usually branched, 5–70 cm. Dry places, alpine meadows. YU.AL.GR.BG. Subsp. *rosea* (Waldst. & Kit.) Nyman (*S. rosea*) PINK VIPERGRASS has simple unbranched stems, flat leaves 3–7 mm wide, and pale pinkish-purple florets. Alpine meadows. YU.AL.GR.BG. Pl. 54.

Flower heads yellow
(a) Leaves entire, linear to lanceolate
  (i) Fruits hairless
1530. *S. austriaca* Willd. AUSTRIAN VIPERGRASS. Stony and grassy places, rocks. YU.AL.BG. **(1530). *S. mollis* Bieb. Distinguished by its cottony white hairs on leaves and stems, its swollen rhizome, and its hairless fruits. Involucre 18–27 mm long at flowering, usually little shorter than the yellow florets which are reddish beneath; involucre enlarging to 3·5 cm in fruit. Stems solitary or few 5–50 cm, leafless above. Rocky places. YU.AL.GR.CR? BG. Page 480. **1531.** *S. hispanica* L. Flower heads large with yellow florets often twice as long as the involucre, which is 2–3 cm long in flower and up to 4 cm in fruit. Stems 25–100 cm, leafy below, often branched; lower leaves linear-lanceolate to ovate-lanceolate, long-pointed but very variable. Meadows, bushy places. YU.AL.GR.BG.

  (ii) Fruits hairy
1531d. *S. cretica* Willd. Distinguished by its densely hairy fruits with pale reddish- or whitish-brown feathery pappus hairs 1½–2 times as long as the nut. Stems one or few, rigid, branched below, or simple, with yellow flower heads often reddish outside. Involucre 1–3 cm long, up to 4 cm in fruit, hairless to densely hairy; florets 1¼–1¾ as long as involucre. Leaves linear, long-pointed, nearly hairless or with dense hairs. Rocks, dry places. GR(south).CR.

(b) Leaves pinnately lobed
1532. *S. laciniata* L. (*Podospermum l.*) Flower heads yellow with florets less than 1½ times as long as the involucral bracts, on prostrate or ascending leafy branches. Involucre 7–20 mm long, up to 4 cm in fruit. Leaves pinnately cut with linear or lanceolate lobes. Nut linear, up to 17 mm, swollen and hollow at the base; pappus feathery, as long as the nut. Grassy places. Widespread on mainland. **1532a.** *S. cana* (C. A. Meyer) O. Hoffm. (*Podospermum c.*) Flower heads with pale-yellow florets which are reddish or purplish outside (yellow outside in 1532), and more than 1½ times as long as the involucre which is 1·2–2 cm long, and up to 2·5 cm in fruit. Stems several 5–60 cm, usually branched; basal leaves with linear acute lobes, with cobweb hairs. Nut 6–10 mm, pappus as long as or up to twice as long. Salt-rich soils, grassy places, waysides. Widespread on mainland +E.Aegean Is.

ANDRYALA ****1533.** *A. integrifolia* L. Maritime rocks. GR+Is.

CHONDRILLA 3 species in our area.

1. *Chondrilla juncea* 1534 **2.** *Carthamus boissieri* 1508d **3.** *Catananche lutea* 1511a **4.** *Scorzonera mollis* (1530) **5.** *Tragopogon balcanicus* 1528a

1534. *C. juncea* L. A glaucous, stiff broom-like perennial 50–100 cm, with green, almost leafless branches and small yellow flower heads in clusters of 2–5 ranged along the upper branches. Flower heads numerous, about 1 cm across; involucre 9–12 mm long. Upper leaves entire, linear to lanceolate, basal leaves soon withering, deeply lobed. Sandy and stony places, cultivated and uncultivated ground, waysides. Widespread. Page 480.

REICHARDIA (PICRIDIUM)   3 species in our area.

1536. *R. picroides* (L.) Roth A perennial 10–45 cm, with yellow flower heads with florets 1½–2 times as long as the involucre. Inner fruits smooth. Fields, waysides, banks. Widespread.   **1536a.** *R. tingitana* (L.) Roth A hairless Hawkbit-like annual to perennial 4–35 cm, covered with small white projections, and solitary or up to 4 golden-yellow, purplish-centred flower heads 2–2·5 cm across. Involucre urn-shaped 1–1·5 cm; bracts ovate with wide papery margins, in several rows. Flower stalks swollen below flower heads. Basal leaves glaucous, in a rosette, entire or shallowly cut into broad finely toothed lobes, stem leaves clasping, entire or toothed. All fruits transversely wrinkled. Sandy and grassy places, cultivation, dry hills. GR.CR.

CICERBITA   **1537.** *C. alpina* (L.) Wallr. (*Mulgedium a.*) BLUE SOW-THISTLE. A robust perennial to 2·5 m, distinguished by its dense reddish glandular hairs on the inflorescence. Flower heads about 2 cm across, blue-violet, in a dense cylindrical cluster. Involucre 1–1·5 cm long, brownish, glandular. Leaves large 8–25 cm, deeply lobed. Grassy places and streamsides in mountains. YU.AL.BG.   **1537a.** *C. pancicii* (Vis.) Beauverd Like 1537 but differing in having the flowering stem and branches hairless, not glandular. Leaves with lateral lobes shorter and not as wide as the terminal lobe, all lobes with prominent nipple-shaped teeth. Woods, meadows and gorges in mountains. YU.AL.BG.   **(1537).** *C. plumieri* (L.) Kirschleger Like 1537a in being hairless but differing in having the mid-rib between the leaf segments with a broad wing, and fruit flat in section (mid-rib narrowly winged, fruit triangular-sectioned in 1537a). In mountains. YU? BG(south-west).

SONCHUS | Sow-thistle   6 species in our area.

Widespread are: **1538.** *S. arvensis* L. FIELD SOW-THISTLE; **1539.** *S. asper* (L.) Hill PRICKLY SOW-THISTLE; **(1539).** *S. oleraceus* L. COMMON SOW-THISTLE.

(1539). *S. tenerrimus* L. Distinguished from (1539) *S. oleraceus* by its leaves which are cut into rhomboid to narrowly linear lobes which are entire, toothed, or rarely lobed, the terminal and lateral lobes about equal (lobes ovate, the terminal lobe much larger in *S. oleraceus*). Upper leaves with long-pointed or blunt basal lobes clasping the stem. Flower heads yellow; involucre hairless above, usually white-woolly-haired at the base and on the upper part of the stem. Rocks, waste places in the Med. region. YU.GR+Is.CR? TR?

MYCELIS   **1540.** *M. muralis* (L.) Dumort. WALL LETTUCE. Woods, bushy places, rocks. Widespread on mainland.

LACTUCA | Lettuce   12 species in our area, including one endemic.

Flower heads blue or violet
1541. *L. perennis* L. BLUE LETTUCE. Flower heads blue or violet, 3–4 cm across, with numerous florets, in a loose spreading branched cluster. Florets 12–20; involucre

1·2–2 cm long at flowering. Stems 30–80 cm, branched in upper part only; leaves pinnately cut into narrow lobes, glaucous, hairless. Fruit black, long-beaked; pappus white. Rocky places. YU.AL.BG. **1541b.** *L. graeca* Boiss. Differing from 1541 in having stems branching from the base; smaller blue flower heads with 6–15 florets; and involucre about 1 cm long at flowering. Leaves with crisped hairs beneath. Fruit greyish; pappus white. Rocks and screes in mountains. AL.GR(north and central). **1541c.** *L. tatarica* (L.) C. A. Meyer Flower heads blue-lilac, with 16–23 florets; involucral bracts reddish with a white margin. Lower leaves pinnately cut, the upper lanceolate entire, half-clasping. Fruit yellowish to black, with a very short beak of the same colour. Seashores, river banks, waste places. BG.TR. Page 148.

Flower heads yellow
(a) Leaves with prickles on margins and underside of mid-vein
1542. *L. serriola* L. (*L. scariola*) PRICKLY LETTUCE. Readily distinguished as a 'compass' plant by its upper leaves with the blade held vertically and orientated in a north–south direction. Flower heads pale yellow, numerous, 11–13 mm across, in a long pyramidal or spike-like cluster. Leaves stiff, glaucous, oblong-obovate, entire to deeply lobed, the upper clasping, narrower; stem to 180 cm. Fruit greyish-green, beaked. Rocks, sandy places, waste ground. Widespread. **(1542).** *L. virosa* L. Like 1542 but leaf blades held horizontally; fruit blackish. Flower heads pale yellow, about 1 cm across, in a long pyramidal cluster. Rocks. YU.GR.TR.

(b) Leaves without prickles on margins, and mid-vein
1543. *L. saligna* L. LEAST LETTUCE Flower heads pale yellow, in groups of 1–3, in a narrow elongated spike-like cluster; florets 6–15. Upper leaves linear, entire, with arrow-shaped basal lobes clasping the stem, lower leaves often lobed. Fruit pale brown with beak 1½–3 times as long as nut. Annual or biennial 30–100 cm. Poor pastures, rocks, waysides. Widespread, but largely on mainland. ****1544.** *L. viminea* (L.) J. & C. Presl PLIANT LETTUCE. Readily distinguished by its narrow glaucous stem leaves which have their basal lobes running down the white stem as conspicuous green ribs. Flower heads pale yellow, with 4–8 florets, in stalkless groups of 3, forming a variously branched, or long slender spike-like cluster. Lower leaves deeply lobed, in a rosette; stems to 1 m. Fruit black with beak shorter than or as long as nut. Rocks, bushy places. Widespread.

STEPTORHAMPHUS Like *Latuca* but fruit compressed, with very long slender beak, and pappus of 2 unequal rows of simple hairs, the outer few forming a very short fringe.

1544a. *S. tuberosus* (Jacq.) Grossh. (*Lactuca cretica*) Flower heads yellow, 1–2; involucral bracts up to 4 cm long, lanceolate to ovate, hairless or nearly so, and usually purplish-tinged. Lower leaves entire to coarsely lobed with backward-pointing lobes, hairy, clasping, uppermost leaves lanceolate entire; stem 40–60 cm or more, usually unbranched. Thickets, cultivation. GR+Is.CR.BG.

PRENANTHES ****1545.** *P. purpurea* L. Woods, shady places. YU.AL.GR.BG.

CREPIS | Hawksbeard A large and difficult genus with about 49 species in our area, which are mostly yellow-flowered, including: **(1546).** *C. setosa* Haller fil. BRISTLY HAWKSBEARD. Widespread on mainland. **1547.** *C. biennis* L. ROUGH HAWKSBEARD. YU.AL.GR.BG.

Pink-, pale-lilac-, or white-flowered plants
(1546). *C. rubra* L. PINK HAWKSBEARD. Readily distinguished by its solitary or paired pink

flower heads up to 4 cm across, drooping in bud and borne on long nearly leafless stems 4–40 cm. Involucre 11–15 mm long; outer bracts pale or papery, usually hairless, half as long as the inner glandular-hairy bracts. Lower leaves toothed or deeply lobed, the stem leaves few, narrow bract-like. Fruit dark brown, the outer short-beaked, the inner long-beaked. (May be confused with pink-flowered *Scorzonera*, but the latter have all leaves entire.) Bushy or grassy places, olive groves in the Med. region. YU.AL.GR.CR. Pl. 54.

1546a. *C. incana* Sibth. & Sm. A dwarf pink-flowered perennial 3–15 cm, of alpine rocks. Florets few or numerous; involucre 1 cm long, densely woolly-haired; flower stems glandular-hairy. Leaves cobweb-hairy or hairless, lobed, the uppermost bract-like. GR(south). **1546b.** *C. praemorsa* (L.) Tausch subsp. *dinarica* (G. Beck) P. D. Sell has pink or white flower heads in a lax often more or less flat-topped cluster. Involucre 7–12 mm long, hairless or woolly-haired. Leaves all basal, mostly obovate entire; stems 15–75 cm. Fruit pale brown, 4–5 mm. In mountains. YU (west).

MONOCOTYLEDONES

ALISMATACEAE | Water-Plantain Family

BALDELLIA **1558.** *B. ranunculoides* (L.) Parl. LESSER WATER-PLANTAIN. Watersides, ditches. YU.GR(west).

ALISMA 3 species in our area. **1560.** *A. plantago-aquatica* L. WATER-PLANTAIN. Muddy places, damp ground, shallow waters. Throughout. **1561.** *A. lanceolatum* With. NARROW-LEAVED WATER-PLANTAIN. Damp places, shallow waters. Throughout.

DAMASONIUM ****1562.** *D. alisma* Miller THRUMWORT. Still waters. GR.

SAGITTARIA **1563.** *S. sagittifolia* L. ARROWHEAD. Still or slow-flowing shallow waters. YU.GR.BG.TR.

CALDESIA Distinguished from *Sagittaria* by the bisexual flowers usually borne in whorls of 3 on a leafless stem. Petals and sepals 3; stamens 6. Fruits free, crowded round the disk-like receptacle.

1563a. *C. parnassifolia* (L.) Parl. (*Alisma p.*) Aquatic plant with floating, triangular-heart-shaped blunt leaves. Flowers white; petals 5–7 mm. Fruits 5–10, ribbed. YU.BG.

BUTOMACEAE | Flowering Rush Family

BUTOMUS ****1564.** *B. umbellatus* L. FLOWERING RUSH. Still and slow-flowing waters. YU.AL.GR.BG.TR.

HYDROCHARITACEAE | Frog-Bit Family

STRATIOTES ****1565.** *S. aloides* L. WATER SOLDIER. Still waters. YU.BG.TR.

HYDROCHARIS ****1566.** *H. morsus-ranae* L. FROG-BIT. Still waters. YU.GR.BG.TR.

VALLISNERIA ****1567.** *V. spiralis* L. Still waters. YU.GR.BG.TR.

ELODEA **1568.** *E. canadensis* Michx CANADIAN PONDWEED. Still and slow-flowing waters. Native of N. America; naturalized in YU.BG.

JUNCAGINACEAE | Arrow-Grass Family

TRIGLOCHIN ****1569.** *T. palustris* L. MARSH ARROW-GRASS. Marshes, damp places. YU.AL.BG. **(1569).** *T. maritima* L. SEA ARROW-GRASS. Coasts. YU.BG. **1569a.** *T. bulbosa* L. Root-stock bulbous, fibrous-coated. Damp saline places. YU.AL.GR.CR.

POTAMOGETONACEAE | Pondweed Family

POTAMOGETON | **Pondweed** About 20 species in our area. The following are widely distributed on the mainland: ****1570.** *P. natans* L. BROAD-LEAVED PONDWEED; ****1572.** *P. lucens* L. SHINING PONDWEED; **1573.** *P. perfoliatus* L. PERFOLIATE PONDWEED; **1574.** *P. crispus* L. CURLED PONDWEED; ****1576.** *P. pectinatus* L. FENNEL-LEAVED PONDWEED; **1576a.** *P. pusillus* L. LESSER PONDWEED; **1576b.** *P. berchtoldii* Fieber; **1576c.** *P. nodosus* Poiret

RUPPIACEAE | Ditch Grass Family

RUPPIA 1578. *R. cirrhosa* (Petagna) Grande TASSEL PONDWEED. YU.AL.GR.CR.BG. **1578a.** *R. maritima* L. (*R. rostellata*) Widespread, except AL.

ZOSTERACEAE | Eel-Grass Family

ZOSTERA 3 species in our area. **1579.** *Z. marina* L. EEL-GRASS, GRASS-WRACK. Shallow coastal waters. YU.GR.CR.BG.TR? **1579a.** *Z. noltii* Hornem. (*Z. minor, Z. nana*) Shallow coastal waters. YU.AL?GR.BG.TR.

POSIDONIACEAE

POSIDONIA ****1580.** *P. oceanica* (L.) Delile Coastal waters. YU.AL.GR.CR.TR?

ZANNICHELLIACEAE | Horned Pondweed Family

CYMODOCEA 1581a. *C. nodosa* (Ucria) Ascherson Shallow coastal waters. YU.AL.GR.CR.TR.

ZANNICHELLIA **1581. *Z. palustris* L. HORNED PONDWEED. Still or flowing, fresh or brackish water. YU.GR.CR.BG.TR.

NAJADACEAE | Najas Family

NAJAS   3 species in our area. **1582. *N. marina* L. A dark-green, brittle, submerged aquatic annual. Leaves conspicuously toothed, about 2 cm long 1–6 mm wide, sheathing at the base; stem spiny-toothed or smooth. Still fresh or brackish waters. YU.GR.BG.   **1582a.** *N. minor* All. Like 1582 but leaves narrower 0·5 mm; stem smooth. Still waters. YU.BG.TR.

LILIACEAE | Lily Family

The perianth segments, referred to here simply as petals, are the outermost and second outermost parts of the flower, which are usually similar and brightly coloured.

TOFIELDIA **(1583). *T. calyculata* (L.) Wahlenb. Damp meadows, rocks. YU.

VERATRUM **1586. *V. album* L. WHITE FALSE HELEBORINE. A robust perennial to 1 m or more, with whorls of broad oval, pleated leaves, and terminal clusters of numerous white to green or yellowish flowers each about 1·5 cm across. Petals 7–15 mm. Leaves hairy beneath. Mountain pastures. YU.AL.GR.BG. (1586). *V. nigrum* L. BLACK FALSE HELEBORINE. Differs from 1586 in having blackish-purple flowers, flower stalks shorter than the bracts and narrower linear-lanceolate upper leaves which are hairless beneath. Mountain meadows, scrub, wood margins. YU.AL.GR?BG.

MERENDERA 1587b. *M. sobolifera* C. A. Meyer A bulbous plant like an Autumn Crocus flowering in winter or spring with pale rosy-purple or white flowers with very narrow petals 3–4 mm wide which spread widely apart at flowering. Limb of petals with 2 basal appendages. Leaves 3, appearing with the flowers. Distinguished by its corm which has 2–3 stout horizontal stolons, and from the swollen end of one stolon the flowering shoot arises. Grassland. YU.GR?BG.TR. Page 497. **1587c.** *M. attica* (Tommasini) Boiss. & Spruner (*M. rhodopaea*) A coastal plant without horizontal stolons flowering in autumn. Flowers slightly larger with narrower petals 2–4 mm wide, limb of petals without basal appendages. Leaves 3–4, narrowly linear, grooved. Dry hills. GR(south).BG(south).

COLCHICUM | Autumn Crocus   About 19 species, including 6 endemic in our area. A very difficult genus. The petals are often conspicuously marked with rectangles or squares of darker colour and are described as *tesselated* or chequered.

Leaves partially or well developed at flowering
(a) Autumn-flowering
1588d. *C. pusillum* Sieber (incl. *C. cretense*) Flowers 1–4, pinkish-lilac to white, with
small blunt petals 1–2 cm by 1·5–2 mm. Leaves 3–8, usually developed at flowering,
thread-like or narrowly linear, usually 1–4 mm wide (at high altitudes in Crete
not developed at flowering). Rocky places. GR(central & south).CR. Page 71. Pl.
55. **1588f.** *C. cupanii* Guss. A common plant of rocky ground, from sea level to the lower
mountains of Greece, with small pinkish-purple flowers appearing with usually 2
leaves. Flowers 1–5; petals to 2·5 cm, blunt or acute; anthers dark-purplish. Leaves
usually 10–18 mm wide. AL? GR.CR? Pl. 55.

(b) Late winter- to spring-flowering
1588e. *C. triphyllum* G. Kunze A small purplish-pink-flowered plant, flowering in the
new year and spring, with usually 3 partially developed lanceolate leaves. Petals
1·5–3 cm by 6–12 mm, blunt; anthers dark-purplish. Leaves with rough margins,
4–10 mm wide at maturity. Corm with a papery, short-lived covering. Open stony
places. GR.BG.TR. Pl. 55. **1588g.** *C. hungaricum* Janka (incl. *C. doerfleri*) A winter- to
spring-flowering species of stony places in the mountains, with 3–6 purplish-pink, pink,
or white flowers and usually 2 leaves partially developed at flowering. Petals 3 cm by
6–7 mm wide, acute; anthers dark-purplish. Leaves with ciliate margins, sometimes
hairy above, up to 20 cm by 1–2 cm at maturity. YU.AL.GR(north).BG.

Leaves undeveloped at flowering (rarely beginning to appear as flowers fade) (see also
1588d)
(a) Flowers one-coloured, not tesselated or chequered
**1588. *C. autumnale* L. MEADOW SAFFRON, AUTUMN CROCUS. Flowers 1–6, rosy-purple or
rarely white, appearing leafless in the autumn from colourless basal sheaths. Petals
4–6 cm by 1–1·5 cm, blunt; anthers yellow. Leaves usually 4, broadly lanceolate, appearing in the spring, up to 35 cm by 2–5 cm. Damp meadows. YU.AL.GR.BG. **1588j.** *C.
lingulatum* Boiss. & Spruner Like 1588 but smaller-flowered, found in open stony places
in hills. Flowers 2–5, pink-lilac; petals 2–4 cm by 3–10 mm, blunt, occasionally obscurely tesselated. Leaves 4–6, up to 15 cm by 1–2 cm undulate, spreading, often
pressed to the ground. Corm scales leathery. Autumn-flowering. AL.GR(south-east).

1588n. *C. parnassicum* Boiss. An autumn-flowering species from Greece, differing from
1588j in its thinner papery scales round the corm, its larger leaves up to 23 cm by 5·5 cm,
arched, not undulate at the margins, and its larger and fewer pink to lilac-purple, or
whitish flowers. Damp places, rocks in hills. GR(south and central). **1588h.** *C. boissieri*
Orph. A local uncommon autumn- to early-winter-flowering species, readily identified
by its horizontal 'rhizome-like' corm. Flowers usually solitary, bright rosy-lilac, appearing leafless. Petals 2–4 cm by 5–11 mm; anthers yellow. Leaves usually 2, linear, blunt.
Stony ground, scrub. GR(south).

1588i. *C. arenarium* Waldst. & Kit. Pink-flowered species appearing without leaves
from September to October in sandy fields in Yugoslavia. Petals pink 2·5–4 cm by
3–10 mm, acute or obtuse; anthers yellow. Leaves 3–5, strap-shaped blunt, 8–20 cm by
4–17 mm. YU(north-east). **1588k.** *C. turcicum* Janka Distinguished by its deep
reddish-purple flowers which may be obscurely tesselated, and its glaucous leaves with
undulate ciliate margins, which appear after flowering. Flowers 3–8; petals 3–4 cm by
4–13 mm. Leaves 5–9. Flowering August to October. Fields, open places.
GR(east).BG.TR. Pl. 55.

(b) Flowers conspicuously tesselated or chequered
1588l. *C. bivonae* Guss. (*C. latifolium*) Distinguished by its 1–6 large, strongly che-

quered, deep pinkish-purple flowers which appear in autumn before the leaves. Petals 5·5–6·5 cm by 1–2 cm, acute or obtuse; anthers dark purplish. Leaves 5–9, up to 25 cm by 8–13 mm, hairless. In mountains. YU.GR.BG.TR. Pl. 55. **1588m.** *C. variegatum* L. Distinguished by its funnel-shaped, strongly chequered flowers with petals which are gradually tapering and often twisted at the apex, deep red or violet-purple, sometimes paler at the base. Flowers 1–3, petals 4·5–7 cm by 8–15 mm. Leaves 3–4, 10–15 cm by 7–18 mm, with a firm whitish undulate margin, developing after flowering and spreading over the ground. Flowering October to December. Sunny, grassy places in the Med. region. GR(south)+Is. Pl. 55.

ASPHODELUS | Asphodel

Robust perennials; leaves flat or keeled
(a) Fruits 8–20 mm long
1590. *A. albus* Miller WHITE ASPHODEL. A very distinctive species with stout, erect, usually unbranched stems to 1 m or more, bearing a dense spike of large white or pinkish flowers, each 3–5 cm across, and subtended by long dark-brown bracts. Petals narrowly elliptic 1·5–2 cm. Leaves all basal, flat, usually 1–2 cm wide. Fruit globular 16–20 mm. Dry hills, meadows in mountains. YU.AL.GR.BG. Pl. 55. **1590a.** *A. ramosus* L. (*A. messeniacus.*) Usually a much-branched perennial to 1·5 m with lateral flowering branches almost as long as the terminal. Flowers white or pinkish; petals 1·5–2 cm, blunt. Fruit ovoid 8–14 mm long. Dry uncultivated ground in the Med. region. YU.GR(south-west).

(b) Fruit 5–7 mm long
**1591. *A. aestivus* Brot. (*A. microcarpus*) ASPHODEL. Distinguished by its stout, much-branched, pyramidal inflorescence of pinkish-white flowers to 2 m, with ascending lateral flowering branches shorter than the terminal. Petals 10–14 mm. Leaves keeled, usually 1–2 cm wide. Fruit 5–7 mm. Common in dry stony places, hills, rocky ground in the Med. region. YU.AL.GR+Is.CR.

Annuals or short-lived perennials; leaves hollow, semicircular in section
**1592. *A. fistulosus* L. HOLLOW-STEMMED ASPHODEL. A rather slender perennial 15–70 cm, with a simple, or lax and sparsely branched inflorescence, bearing pale-pinkish flowers about 2 cm across. Petals 5–12 mm, oblong, blunt. Leaves about 4 mm wide, more or less hollow. Fruit globular 5–7 mm. Dry sunny places, roadsides. YU.GR+Is.CR.TR.

ASPHODELINE
Flowers yellow
1593. *A. lutea* (L.) Reichenb. KING'S SPEAR, YELLOW ASPHODEL. Unmistakable with its large dense, spike-like cluster of yellow flowers 10–15 cm long, borne on a stout leafy stem to 1 m. Petals 2–2·5 cm, spreading; bracts 2–3 cm. Leaves numerous, densely clustered, linear, triangular in section, stiff-pointed, smooth. Rocky and stony places. Widespread. **1594. *A. liburnica* (Scop.) Reichenb. A more slender perennial than 1593, with rough-margined, very narrow leaves, about 1–2 mm wide. Flowers yellow; bracts to 15 mm; inflorescence more lax, and flowering stem only leafy below (leafy throughout in 1593). Bushy places, field verges. Throughout mainland +CR. Rhodes. Pl. 56.

Flowers white
1594a. *A. taurica* (Bieb.) Kunth Flowers white, in a dense spike-like cluster 15–25 cm, with large papery ovate-acute bracts partially hiding the flowers. Stem 30–65 cm, leafy

its whole length; leaves narrow linear to 2 mm wide, triangular in section, rough-margined. Rocks in hills and mountains. YU?AL.GR(north).BG. Page 497.

ANTHERICUM **1596. *A. liliago* L. ST. BERNARD'S LILY. Flowers white, with spreading petals 16–22 mm and longer than stamens by 6–10 mm. Inflorescence usually unbranched. Fruit 8–10 mm, ovoid. Dry pastures, stony places, open woods. YU.AL.GR.BG. **(1596).** *A. ramosum* L. Like 1596 but petals 10–14 mm, longer by 2 mm or less than the stamens. Inflorescence usually branched, pyramidal. Fruit 5 mm, globular. Dry sunny places, scrub. YU.AL.BG.TR. Pl. 56.

ALOE Flowers in a simple or branched spike; corolla tubular, toothed or lobed. Leaves fleshy, often spiny, in rosettes. **1598.** *A. vera* (L.) Burm. fil. (*A. vulgaris*) Leaves in numerous, usually stemless rosettes, glaucous sometimes red-tinged, spineless, 35–60 cm. Flowers yellow, 2·5–3 cm long, drooping, in a simple or one- to two-branched leafless spike 30–50 cm. Maritime sands and rocks. Native of S. Africa; known since classical times in GR+Is.CR.

GAGEA A very difficult genus. Basal leaves are those arising directly from the bulb; stem leaves subtend the first branch of the inflorescence; those above the first branch are referred to as bracts. Flowers all with yellow petals which are greenish outside, except (1601). About 14 species in our area.

Basal leaves on most plants usually solitary
(a) Bulbs 3, one with a scale
1599a. *G. pratensis* (Pers.) Dumort. Flowers 2–6, on hairless stalks; petals 15–20 mm, blunt or almost acute. Stem leaves 2, opposite, ciliate. Grassland, disturbed ground. YU.GR.BG.

(b) Bulbs 1, or 2 enveloped in a single scale
1599b. *G. minima* (L.) Ker-Gawler Flowers 1–7; petals 10–15 mm, fine-pointed; flower stalks hairless or sparsely and finely hairy. Stem leaves 1–2, opposite, long-pointed. Bulbs 2. Woods, grassland. YU.AL? GR.BG. **1599c.** *G. pusilla* (F. W. Schmidt) Schultes & Schultes fil. Flowers 1–3; petals about 13 mm, bluntish, flower stalks hairless. Stem leaves 2–3, opposite, usually hairless. Bulb usually one. Dry grassland, stony ground. YU.AL.GR.BG.TR? **1599.** *G. lutea* (L.) Ker-Gawler Flowers usually 1–7; petals 15–18 mm, oblong-linear blunt; flower stalks hairless or hairy. Stem leaves 2, opposite, ciliate. Bulb 1. Woods, scrub. YU.GR.BG.TR. **1599d.** *G. reticulata* (Pallas) Schultes & Schultes fil. Flower usually 1, with lanceolate long-pointed petals 14–20 mm. Basal leaf linear, rounded in section, stem leaves alternate, crowded, narrow linear. Plants often in clusters; bulb with a long collar. Dry grassy slopes. GR.CR.BG.

Basal leaves on most plants usually 2 or more
(1601). *G. graeca* (L.) A. Terracc. (*Lloydia g.*) Distinguished from all others by the 3–7 white flowers usually with purple veins. Basal leaves 2–4. Rocky places. GR+Is.CR. **1601.** *G. fistulosa* (DC.) Ker-Gawler Flowers 3–5, with hairy flower stalks; petals 13–17 mm, elliptic-lanceolate blunt. Basal leaves usually 2, rounded, hollow; stem leaves 2, broadly lanceolate. In mountains. GR.BG. **1600.** *G. arvensis* (Pers.) Dumort. Flowers 5–12; petals 13–15 mm, lanceolate, hairless or sparsely hairy outside, the tips often reflexed. Basal leaves grooved; stem leaves sometimes with bulbils; stems hairy above. Dry places. YU.GR.BG.TR.

1600a. *G. peduncularis* (J. & C. Presl) Pascher Flowers 1–3; petals 15–20 mm, linear-oblong; flower stalks finely hairy. Basal leaves 2, thread-like, grooved; stem leaves lanceolate-spathulate long-pointed, ciliate or hairless. Bulbs more or less covered with recurved fibrous roots. Dry hills, stony places. YU.GR+Is.CR.BG.TR. **1600b.** *G. bohemica* (Zauschner) Schultes & Schultes fil. Flowers 1–3 borne on stems about 2 cm; petals 13–17 mm, blunt. Basal leaves thread-like, often curved and wavy; stem leaves lanceolate long-pointed. Dry grassland. YU.GR.CR+E.Aegean Is.BG.

ALLIUM | Onion, Garlic, Leek The spathe is the papery sheath which encircles the umbel of flowers and later splits into valves or lobes. About 60 species in our area, many of which are probably endemic to our area.

Leaves linear
(a) Leaves hollow, either cylindrical, or half-cylindrical, or flat in section
1603. *A. schoenoprasum* L. CHIVES. Flowers pink or violet, in very dense globular umbels 1·5–5 cm across, borne on a hollow stem 5–50 cm. Leaves 1–2, hollow. Rocky pastures, damp mountain meadows. YU.GR.BG. **1604. *A. vineale* L. CROW GARLIC. Flowers pale purple to dark red, or greenish-white, in a rather lax umbel 2–5 cm across, with numerous stalkless bulbils, or umbels with bulbils only present. Stem slender, 30–120 cm, leafy to about the middle. Spathe one, soon falling. Leaves 2–4, linear, about 2 mm wide, hollow, grooved above. Sandy ground, dry places, cultivation. Widespread on mainland.

**(1604). *A. sphaerocephalon* L. ROUND-HEADED LEEK. Like 1604 but flowers usually dark reddish-purple, in a dense globular umbel 2–2·5 cm across, usually without bulbils. Stamens longer than the blunt petals, the inner petals with minute swellings on keel and both surfaces. Spathe valves 2, persistent. Stony and rocky places, cultivation. YU.AL.GR.BG.TR. Subsp. *arvense* (Guss.) Arcangeli has petals white with green keels, without minute swellings. Rocky hills. AL.GR(south)+Cyclades.

1604a. *A. guttatum* Steven (incl. *A. margaritaceum*, *A. dalmaticum*) Flower head dense, globular to ovoid 1·5–3·5 cm across, with tiny whitish or pale- to deep-purplish flowers borne on very unequal flower stalks. Central flower stalks up to 2 cm, erect in fruit, the outer shorter and deflexed. Petals 2·5–4·5 mm; stamens longer. Leaves 2–4, thread-like, hollow; stem 20–60 cm. Dry slopes. Widespread on mainland. Pl. 56. **1604c.** *A. amethystinum* Tausch Flower heads dark purple with numerous cylindrical flowers with protruding stamens on more or less equal stalks. Flower heads 2·5–6 cm across; spathe 2–7 cm long-pointed, soon falling. Leaves 2–3 linear, hollow, grooved, usually dead at flowering. Cultivation, rocky places YU.AL.GR.CR.BG. Pl. 56.

(b) Leaves solid, either cylindrical, or flat, or grooved in section
  (i) Stamens without lateral projections at the apex of the filament
    (x) Flowers yellow, white, or straw-coloured (see also 1607b, 1611, 1612, 1615)
1606. *A. flavum* L. YELLOW ONION. Flowers pale to golden-yellow, bell-shaped, in a loose cluster of 9–60, unequally long-stalked, the outer drooping in flower, becoming erect in fruit. Spathe valves 2, very long and slender. Stamens longer than petals. Leaves 2–3, linear 2 mm wide, solid, grooved; stem 8–50 cm. Rocks, dry slopes. YU.AL.GR.BG.TR. **1608. *A. subhirsutum* L. A rather slender plant with pure-white flowers in a lax spreading umbel borne on a cylindrical stem 7–30 cm. Petals spreading in a star, ovate-lanceolate 7–9 mm long; stamens shorter; spathe short. Leaves 2–3, linear-acute 2–10 mm wide, flat, ciliate. Stony hills, dry waste places in the Med. region. YU.GR+Is.CR. Pl. 56.

1609. *A. neapolitanum* Cyr. A larger plant than 1608 with numerous pure shining-white, cup-shaped flowers 1–2 cm across, in an umbel 5–8 cm across, borne on a triangular-sectioned stem 20–50 cm. Petals spreading, blunt, 7–12 mm long; spathe, broad, persistent. Leaves usually 2, hairless, flat, keeled beneath, 0·5–2 cm wide. Grassy places, cultivation in the Med. region. YU.GR+Is.CR.TR. **1611b. *A. chamaemoly* L. Flower head 1·5–2·5 cm across, appearing stalkless among the leaves, with 2–20 white star-shaped flowers with green or purplish mid-veins to the petals. Petals 5–9 mm long; stamens shorter; spathe two- to four-lobed. Leaves 2–5, linear, ciliate, usually spread flat over the ground. Open places, sandy slopes in the Med. region. YU.AL.GR.

1611c. *A. circinnatum* Sieber A tiny plant with a hairy stem 5–18 cm, with an umbel of 3–5 small, star-shaped whitish, pink-striped flowers. Leaves 2–4, very narrow, very hairy, spirally coiled above, lying on the ground. Dry stony hills. CR. **1611d.** *A. callimischon* Link Flower head with 8–25 very unequal-stalked, tubular flowers with white petals with reddish-brown mid-veins, often turning pink with age. Petals 5–7 mm; stamens shorter, young anthers red. Stem 9–38 cm, leafy almost up to the flower head; leaves 3–5, thread-like, hairless, Hills, mountains. GR(south). CR. Page 93.

(xx) Flowers pink, violet, or purple
(1607). *A. carinatum* L. (incl. *A. pulchellum*) Flower head reddish-purple, usually with bulbils, or with bulbils only, with unequal flower stalks, the outer curving downwards. Petals 4–6 mm, blunt; stamens much longer; spathe valves 2, with a long beak, much longer than the flower head. Leaves 2–4, 1–2 mm wide, hairless, weakly grooved above, ribbed below; stem 30–60 cm. Meadows, rocks. YU.AL.GR.BG.TR. Pl. 56. **1607a. *A. parnassicum* (Boiss.) Halácsy A slender plant 10–20 cm, with a lax umbel of 7–13 pale-purplish, funnel-shaped flowers. Petals about 7 mm, blunt; flower stalks unequal; spathe valves 2, long-pointed, little longer than flower head. Leaves 3, more or less cylindrical, hairless. Rocks in mountains. GR(south).

1607b. *A. paniculatum* L. A very variable species with pale-purplish, whitish, brownish, or yellowish flowers on very unequal flower stalks in a lax many-flowered head. Flowers bell-shaped; petals 4–5 mm, variously streaked with red or green: stamens shorter or slightly longer, anthers yellow; spathe valves 2, very unequal. Leaves 3–5, usually 2 mm wide, grooved above; stem rounded in section 10–70 cm. Grassy and rocky places. Widespread. **1607c.** *A. melanantherum* Pančić Flower pale purplish with red mid-veins, in a many-flowered umbel with a few bulbils, and unequal flower stalks. Flowers cylindrical; petals 9–10 mm; stamens shorter, anthers blackish; spathe valves 2, unequal. Leaves 1–3, thread-like 2 mm wide, with rough margins; stem 13–40 cm, leafy nearly to the middle. Rocks and grassy places in mountains. YU.GR.BG.

1611. *A. roseum* L. ROSE GARLIC Flowers rose or flushed pink, rarely white, numerous in a rounded head to 6 cm across, usually without bulbils, less commonly with bulbils and fewer flowers. Flowers bell-shaped; petals 7–12 mm, acute; spathe three- to four-lobed, persistent. Leaves basal 2–4, linear 1–14 mm wide; stem 10–65 cm. A common and very variable species. Grassland, gravelly places by the sea. Widespread, except BG. Pl. 56. **1612. *A. nigrum* L. A robust plant 60–90 cm, with a dense globular flower head up to 10 cm across of pale-lilac or white flowers with green mid-veins. Petals 6–9 mm, blunt, spreading then reflexed; spathe two- to four-lobed, shorter than flower stalks. Leaves 3–4, broadly linear 1·5–7 cm wide. Cultivated ground, waste places. Widespread. **1612a.** *A. atropurpureum* Waldst. & Kit. Rather similar to 1612 but flower heads smaller 3–7 cm across, and flowers dark purple. Petals spreading, then reflexed, 7–9 mm, linear acute; spathe valves 2. Leaves 3–7, 1–4 cm wide; stem 40–100 cm. Dry hills, cultivation. YU.BG.TR.

(ii) Stamens, at least the inner, with lateral projections at the apex of the filament
1614. *A. scorodoprasum* L. subsp. *rotundum* (L.) Stearn SAND LEEK is the plant of our

area. Flowers many, purple, in a lax rounded umbel on a stem 25–90 cm (subsp. *scorodoprasum* has purplish bulbils and few or no flowers). Flowers ovoid, on unequal stalks; petals 4–7 mm, the outer darker purple; stamens shorter, the 3 inner with 2 long projections longer than the anthers; spathe 1·5 cm, soon falling. Leaves 2–5, linear 2 cm wide, flat or grooved. Fields, grassy places. YU.GR+Is.BG.TR. Pl. 56. **1615.** *A. ampeloprasum* L. WILD LEEK Flower heads large globular 5–9 cm, very dense with numerous red, pale-purplish to whitish flowers borne on a stout cylindrical stem to 180 cm, with withered leaves at flowering. Flowers cup- or bell-shaped; petals 4–6 mm; stamens slightly to distinctly longer; spathe to 12 cm, soon falling. Leaves 4–10, flat, grooved, with rough margins, 5–40 mm wide. Dry places, cultivation, particularly in the Med. region. Widespread. **1615a.** *A. atroviolaceum* Boiss. Differs from 1615 in having smaller dark-purple globular flower heads 3–6 cm across, with stamens distinctly longer than the petals. Leaves flat with finely toothed margins, 4–10 mm wide. Dry grassy slopes, fields. YU.GR.BG.TR.

Leaves elliptic or lanceolate, stalked; blade flat
1616. *A. ursinum* L. RAMSONS. Flowers white, with spreading petals, in a somewhat spreading, lax, flat-topped umbel. Petals 1 cm, acute; stamens shorter. Stem 10–45 cm, two-angled or sometimes three-angled, with 2–3 oval-elliptic, stalked leaves near the base. Damp woods. YU.GR(north).BG. **1617.** *A. victorialis* L. ALPINE LEEK. Flowers greenish-white, becoming yellowish, in a dense globular or hemispherical cluster. Petals 4–5 mm; stamens longer. Stems 30–60 cm, stiff, two-edged above, leafy to the middle. Leaves 2–3, oblong-elliptic, short-stalked. Rocks and stony places in mountains. YU.BG.

NECTAROSCORDUM Distinguished from *Allium* by the flower stalks which are swollen into a disk at the apex.

1617a. *N. siculum* (Ucria) Lindley (*N. bulgaricum*) Flowers large bell-shaped, to 1·5 cm long, dull greenish-white tinged pale pink outside with a green mid-rib. Umbel lax with many flowers borne on long unequal pendent flower stalks, which later become erect. Leaves linear, 1–5 cm wide, keeled; stem to 125 cm, leafy at base. Woods, damp places. BG.TR. Pl. 56.

LILIUM | Lily
Flowers with petals recurved from the base
(a) *Lower leaves whorled; flowers pink or purplish*
1618. *L. martagon* L. MARTAGON LILY. Flowers pendulous, pale pink to dark purple, in a lax several-flowered cluster. Lower leaves in whorls of 4–8, the upper leaves alternate. Mountain woods. YU.AL.GR.BG.TR.

(b) *Lower leaves alternate; flowers yellow, orange, or scarlet*
1618a. *L. chalcedonicum* L. (incl. *L. heldreichii*) SCARLET MARTAGON LILY. Stem to 1·2 m, bearing up to 12 brilliant orange-red, unspotted, pendulous flowers with thick waxy, strongly reflexed petals, and red anthers. Petals 6–7 cm. Leaves lanceolate to ovate, silver-edged, the lower spreading, abruptly changing to the upper leaves which are pressed to the stem. Mountain woods. AL.GR+Euboea,Zakinthos. Pl. 57. **(1619).** *L. carniolicum* Koch (incl. *L. albanicum*, *L. bosniacum*, *L. jankae*) Distinguished by its often solitary or up to 6 nodding, bright vermilion, orange, or golden-yellow flowers with strongly recurved spotted petals, and orange to orange-red anthers. Petals 3–6·5 cm. Leaves lanceolate, changing gradually from spreading to erect up stem. A very variable species with distinctive forms in the Balkans often described as distinct species. Mountain meadows. YU.AL.GR.BG. Page 137. Pl 57. **1619a.**

L. rhodopaeum Delip. An endemic species of Bulgaria with large yellow, unspotted flowers. Petals 8–12 cm, recurved; anthers red. Meadows, rocky places. BG(Rhodope Mountains).

Flower tubular to bell-shaped with petals only curved at apex
1620. *L. bulbiferum* L. ORANGE LILY. Recognized by the 1–3 erect flowers which are bright orange to red with black spots and swellings within, and which are surrounded by a whorl of 3–5 oval to lanceolate leaves. Stem 40–60 cm. Mountain pastures, woods, rocky places. YU. **1621. *L. candidum* L. MADONNA LILY. Unmistakable with its 5–6 large white, funnel-shaped flowers to 8 cm, which are held more or less horizontally. Leaves numerous, lanceolate. Stem 90–120 cm. Rocky slopes and scrub. YU(south). GR(west and south); sometimes naturalized elsewhere. Pl. 57.

FRITILLARIA | Fritillary A difficult genus with about 18 species in our area, about 13 of which are endemic and local.

Flowers narrowly bell-shaped or conical-bell-shaped; nectaries at the base of the petals within
(a) *Flowers yellow*
1623c. *F. conica* Boiss. Flowers yellow; petals 1·2–2 cm; nectaries 1 mm, greenish-yellow; style trifid. Leaves 5–7, shining green, oblong-lanceolate, mostly alternate; stem 7–25 cm. Fruit not winged. Rocky hills. GR(W. Peloponnisos).

(b) *Flowers blackish, purplish, or greenish, usually glaucous outside*
1623d. *F. obliqua* Ker-Gawler Flowers usually 1–2, conical-bell-shaped; petals equal, 2–3 cm, rather pointed, dark purplish-maroon overlaid with a silvery-blue sheen. Nectaries linear 4 mm, green; style trifid. Leaves 8–11, glaucous, twisted, lanceolate, the lower often opposite, the upper alternate bract-like; stem 10–20 cm. Fruit not winged. Scrub. GR(south).

Flowers broadly bell-shaped; nectaries above the base of the petals within
(a) *Petals evenly chequered, without a stripe; nectaries 1 cm above the base of the petals*
**1622. *F. meleagris* L. SNAKE'S HEAD FRITILLARY. Damp meadows. YU(central).

(b) *Petals green, brown, or blackish, usually with a stripe, chequered or not; nectaries about 5 mm above the base of the petals*
1622a. *F. messanensis* Rafin. (incl. *F. gracilis*) A variable species with a solitary, pendent, broadly bell-shaped flower, yellow or brownish outside, usually chequered or marked with purplish-brown, sometimes with a well-marked green stripe on the back of each petal. Petals 2·2–3·2 cm; nectaries 6–10 mm. Leaves 7–10, linear 3–7 mm wide, the uppermost sometimes in whorls of 3; stem 15–30 cm. Fruit not winged. Open woods, scrub, stony places. YU.AL.GR.CR. Pl. 57. **1622b.** *F. pontica* Wahlenb. Flowers 1–3, broadly bell-shaped, green tinged with reddish-brown, not chequered, petals always with a broad greenish stripe. Petals 2·5–4·5 cm; nectaries 3 mm, circular. Leaves usually 8, lanceolate, upper 3 in a whorl overtopping the flowers. Fruit winged. Open woods, scrub. YU.AL.GR.BG.TR. Pl. 57.

1622c. *F. graeca* Boiss. & Spruner Flowers 1–2, broadly bell-shaped, dark purple or blackish, strongly or lightly chequered, and usually with a definite green stripe down the centre of each petal. Petals 1·8–3·5 cm; nectaries 4–6 mm. Leaves 5–12, sometimes glaucous, lower ovate to lanceolate, upper narrower lanceolate. Fruit not winged. Open woods, scrub. YU.AL.GR.CR. Pl. 57. Subsp. *thessala* (Boiss.) Rix (*F. ionica*) Petals green, usually lightly chequered, 2·8–3·8 cm. Lowest leaves up to 2·5 cm wide, opposite. YU(south).AL.GR(north-west). Page 115.

TULIPA | Tulip About 7 native species in our area, and 3 introduced from cultivation.

Filaments of stamens hairy at the base
1625. *T. sylvestris* L. subsp. *australis* (Link) Pamp. Flowers yellow tinged with pink or crimson on the outside of the outer petals; buds nodding. Petals usually 2–3·6 cm; filaments 5–8 mm. Leaves 2–3, the lowermost less than 1·2 cm wide. Mountain meadows, stony places. YU.AL.GR.BG. **1627a. *T. saxatilis* Sprengel (incl. *T. bakeri*) Flowers pink to lilac-pink with a yellowish basal blotch within; petals white-edged, 3·8–5·5 cm. Anthers 5–7 mm. Leaves 2–3, shiny green above, flattish, the basal leaf more than 2·5 cm wide. Field verges, hillsides, rocky slopes. GR? CR.

1627b. *T. cretica* Boiss. & Heldr. A small plant of mountains in Crete, differing from 1627a in its smaller flowers 1·5–3·2 cm long, which are white tinged with pink or purple, greenish outside, and with a dull-yellow basal zone inside. Anthers 1–3 mm. Leaves 2–3, deep shining green above; stem 7–11 cm. Rocky, stony places. CR. Page 71. Pl. 58. **1627c.** *T. orphanidea* Heldr. (*T. hageri, T. thracica*) Flowers bright orange or orange-brown sometimes with a darker basal blotch within. Petals 3–5 cm, the outer usually tipped with green outside. Anthers 5–10 mm. Leaves 3–7, deep dull green, the basal leaf less than 2 cm wide. Stony ground, cultivation. GR.CR.BG.TR. Pl. 58.

Filaments of stamens hairless at the base
(a) Flowers red to orange
1627d. *T. boeotica* Boiss. & Heldr. (*T. scardica*) Flowers crimson-scarlet with a yellow-edged black blotch inside each petal at the base. Petals 2·8–6·9 cm, long-pointed; anthers 7–16 mm. Leaves 3–4, the lower 2 with undulate margins. Cultivation, stony places in the Med. region. YU.GR.TR. Pl. 58. **1627e.** *T. praecox* Ten. Flowers orange, tinged with green outside, each petal having a brownish-green basal blotch edged with yellow inside, and inner petals with a yellowish median stripe. Outer petals 4–8 cm acute, the inner shorter, blunt. Leaves 3–4, more or less flat, glaucous. Bulb with scales with dense felty hairs inside. A native of Asia; sometimes naturalized in cultivated areas in GR.TR.

(b) Flowers white, pink-flushed outside
1627. *T. clusiana* DC. LADY TULIP. Flowers predominantly white but flushed with pinkish-crimson on the outside of the outer petals, and inner petals with a small purplish basal blotch within. Leaves 3–5. Bulb with a tuft of hairs protruding from the neck. Native of W. Asia; naturalized in GR.

ERYTHRONIUM ****1628.** *E. dens-canis* L. DOG'S TOOTH VIOLET. Alpine pastures. YU.AL.BG.TR.

URGINEA ****1630.** *U. maritima* (L.) Baker SEA SQUILL. Unmistakable with its tall, quite leafless long-stemmed spike of numerous white flowers, arising in autumn from the bare ground. Leaves broad, persisting through winter to the following summer; bulbs very large. Sandy and rocky hillsides, particularly by the coast. YU.AL. GR+Is.CR. Pl. 59.

SCILLA | Squill Petals free, not fused at the base into a tube. 6 species native in our area.

Flowers 15 or more, in an elongated cluster
1631. *S. hyacinthoides* L. A robust plant 30–80 cm, with a conical-cylindrical spike 15–40 cm of very numerous (40–150) long-stalked, blue-violet flowers. Petals 5–7 mm; flower stalks and anthers violet. Leaves 8–12, 2–3 cm wide, acute, ciliate. Grassy and

rocky places in the Med. region. YU(Croatia).GR. Pl. 58. **1631a.** *S. litardierei* Breistr. (*S. pratensis*) Flowers brilliant blue, 15–35 in a dense cluster which later elongates, borne on a stem 20–30 cm. Petals spreading, 3–5 mm long; flower stalks 2–3 times as long as flower. Leaves 3–6, linear 4–8 mm wide. Meadows. YU. Pl. 58.

Flowers few, usually less than 15, in a more or less oval cluster

1635. *S. bifolia* L. A small plant 5–20 cm, distinguished by its usually paired shining, channelled leaves, and its lax cluster of 1–5, long-stalked, blue flowers with wide-spreading petals 6–10 mm. Anthers violet; bracts absent; flower stalks unequal. Leaves occasionally 3–5. Meadows, thickets, woods to the alpine zone; often by melting snow. Widespread on mainland. **1635b. *S. messeniaca* Boiss. Differs from 1635 in having angled stems; 5–7 leaves 12–20 mm wide; and 7–15 or more pale-blue flowers subtended by tiny bracts. Rocky grasslands. GR(Peloponnisos). Page 89. ****1636.** *S. autumnalis* L. Unmistakable with its dull lilac to pinkish-blue flowers in a more or less flat-topped cluster, which later elongates, appearing in the autumn before the leaves. Petals 3–5 mm; bracts absent. Leaves 5–10, narrowly linear 1–2 mm wide, channelled; stem 5–20 cm. Hills, grassy, rocky places. Widespread.

ORNITHOGALUM About 26 often similar-looking species in our area, 6 of which are endemic. A difficult genus.

Flowers in a spreading or more or less flat-topped cluster, often with very unequal flower stalks

(a) *Leaves with a white stripe*

1639. *O. umbellatum* L. STAR OF BETHLEHEM. Distinguished by its leaves which have a broad white band in the channel on the upper surface and by its numerous offsets with tufts of leaves surrounding the bulb. Flowers 8–20, in a flat-topped cluster 20–30 cm high, with lower flower stalks up to 10 cm and longer than the bracts. Petals 15–22 mm, white with a green band outside, spreading in the sun. Fruit stalks spreading. Grassy places. Widespread on mainland. **1639f. *O. collinum* Guss. (*O. tenuifolium*) A slender plant to 20 cm, with a stemless or short-stemmed cluster of few white flowers with a green band on the outside of the petals. Petals 10–17 mm, spreading. Leaves 4–15, 2–8 mm wide with a white stripe on upper surface. Fruit six-sided, keeled, on spreading or ascending fruiting stalks. Grassy and stony places in the Med. region. Widespread except BG. Pl. 59. **1639g.** *O. sibthorpii* W. Greuter (*O. nanum*) A dwarf plant 5–15 cm, with 1–10 stemless white flowers with a broad green stripe on the outside of the petals. Petals 13–26 mm. Leaves 5–9, 1–4 mm wide, with a central white band. Fruit ovoid, winged above, on fruit stalks that are curved downwards near the base, then up-curved. Sandy hills, rocks, mountains. YU.GR.CR.BG.TR.

(b) *Leaves without a white stripe*

(i) *Leaves hairless*

1639h. *O. oligophyllum* E. D. Clarke A slender plant 9–24 cm, with a rather dense cluster of 1–5 white flowers with a broad green band on the outside of the petals. Petals 11–15 mm; flower stalks ascending, shorter than or as long as the flowers; bracts usually longer. Leaves 2–3, linear to lanceolate 5–20 mm wide. Fruit pendent, usually with broad wings. Rocks, grassy places in mountains. YU.AL? GR.BG.TR. **1639e.** *O. montanum* Cyr. Like 1639h but fruit not winged, fruit stalks erect or horizontal or lying on ground. Petals 10–25 mm, spreading; bracts shorter than flower stalks. Leaves 3–6. Grassy and rocky places in mountains. YU.GR.BG.TR. **1639i.** *O. arabicum* L. A tall plant 30–80 cm, with a rather flat-topped cluster of 6–25 large white or cream-coloured, shallowly bell-shaped scented flowers, 5–6·5 cm across. Petals 1·5–3 cm long, with a green stripe on the back; ovary blackish-violet. Leaves 7–8, flat or concave, without a white stripe, 1–3 cm broad. Fruit cylindrical, three-sided. Bulb with many offsets. Rocky places in the Med. region. YU.GR.

(ii) Leaves with ciliate margins, or hairy all over

1639j. *O. comosum* L. Flowers 8–30, in an oval cluster, the lower flower stalks not much longer than the upper. Petals 1–2 cm spreading, white with a narrow green band beneath. Leaves 4–6, 3–6 mm wide, without a white band, minutely ciliate on margin; stem to 30 cm. Fruit cylindrical with more or less rounded sides. Grassy, stony places. YU.GR.BG. **1639k.** *O. fimbriatum* Willd. Distinguished from other species by the leaves which usually have silvery hairs beneath, but hairiness of leaves and flower stalks variable. Flowers stemless or with a short stem; petals spreading, 1–1·5 cm, white, green-banded beneath. Leaves 2–6, flat or concave, 2–15 mm wide, green. Fruit ovoid, six-sided. Dry rocky places. GR.BG.TR. Page 497.

Flowers in a narrow cluster or an elongated spike, with flower stalks more or less equal (see also 1639j)

(a) Flowers usually 20–80

1640. *O. pyrenaicum* L. BATH ASPARAGUS. Flowers numerous in a lax cylindrical spike, faintly sweet-scented, with pale-yellowish petals within, greenish with a darker greenish stripe on the outside. Petals 11–13 mm. Leaves 5–8, often withered at flowering; stem 30–80 cm. Fruit with 3 rounded sides, borne on erect stalks usually pressed to the stem. Woods, scrub, meadows. YU.GR? BG. **1640a. *O. sphaerocarpum* A. Kerner Differs from 1640 in its colourless or pale greenish-white petals, and ovoid to globular ovary (ovary ovoid-lanceolate 2–3 mm long in 1640). Meadows, scrub. Widespread on mainland. ****1641.** *O narbonense* L. (*O. pyramidale*) Like 1640 but flowers milky-white within, with a greenish band on the outside of the petals, scentless. Petals 12–16 mm, spreading; bracts projecting beyond the young flower buds. Leaves 6–7, persisting during flowering. Grassy, bushy places, fields, meadows. Widespread.

(b) Flowers less than 20

1642. *O. nutans* L. DROOPING STAR-OF-BETHLEHEM. Very distinctive with its large drooping, silvery-greenish bell-shaped flowers in a lax, one-sided, narrow cluster 20–60 cm long. Petals 1·5–3 cm, white within, with a broad green band outside, not spreading; bracts much longer than the flower stalks, which become arched in fruit. Leaves 4–6, 1–1·5 cm wide, with a white band, channelled. Grassy places, mountain rocks. GR.BG.TR; naturalized YU. Pl. 59.

Key to **MUSCARI–HYACINTHUS** complex, with petals (perianth segments) fused at the base into a corolla tube (perianth tube). When the petals are fused for more than ⅓ their length the segments are referred to as lobes or teeth.

1. Corolla at flowering constricted at mouth,
   urn-shaped or globular.   *Muscari* Page 498
1. Corolla at flowering not constricted at
   mouth, tubular or funnel-shaped.
   2. Filaments of stamens with 2 appendages
      at apex; bracts with a down-pointing
      cylindrical spur.   *Strangweia* Page 496
   2. Filaments of stamens without apical
      appendages; bracts without spur.
      3. Filaments of stamens winged.   *Chionodoxa* Page 496
      3. Filaments of stamens not winged, or
         sometimes slightly winged at base.

4. Corolla remaining attached throughout the fruiting stage; leaves with
   prominent fibre strands.   *Hyacinthella* Page 496
4. Corolla coming off as a cap from the
   ripening fruit; leaves without
   prominent fibre strands.
   5. Anthers longer than the corolla
      tube but not longer than the petals.   *Bellevalia* Page 498
   5. Anthers not longer than the   *Hyacinthus* Page 496
      corolla tube.   *Brimeura* Page 496

CHIONODOXA | Glory of the Snow Distinguished from *Scilla* by its petals which
are shortly fused below; filaments of stamens winged.

1643b. *C. cretica* Boiss. & Heldr. Flowers 1–5, blue, borne in a lax cluster on erect flower
stalks, without bracts. Stem slender 8–18 cm; leaves 2, broadly linear 4–10 mm wide,
flat, flaccid, apex hooded. Bushy places in the mountains. CR. Page 70. Pl. 58.   **1643c.**
C. nana (Schultes & Schultes fil.) Boiss & Heldr. Distinguished by its usually solitary
smaller white flowers which are sometimes lilac-tipped. Petals 9–12 mm, tube 2–3 mm
(petals 12–19 mm, tube 3·5–5 mm in 1643b). Stony places in mountains. CR.

HYACINTHUS ****1643.** *H. orientalis* L. HYACINTH. Native of the Eastern Med. region.
Grown in gardens and sometimes naturalized in YU.GR.

BRIMEURA Distinguished from *Hyacinthus* by its small bulbs with tight scales;
petals joined at their bases (joined for half their length in *Hyacinthus*); bracts as long as
the flower stalks (minute in *Hyacinthus*).

1643a. *B. fastigiata* (Viv.) Chouard (*Hyacinthus f.*) A small bulbous plant to 5 cm, from
the Taiyetos Mountains, with almost flat-topped clusters of 1–6, lilac, lilac-pink, or
white bell-shaped flowers, and several longer leaves. Flowers 10–13 mm across, with
spreading lobes longer than the tube. Dry stony places. GR(Peloponnisos).

HYACINTHELLA Like *Hyacinthus*, and often included in it, with the petals fused to
the middle but with 6 ovate or oblong, straight (not curved) lobes which remain surrounding the fruit. Leaves with raised fibre strands; bulbs with a whitish powder
between the scales.

1643d. *H. leucophaea* (C. Koch) Schur Flowers pale blue, rarely white or deep blue, often
becoming green, in a lax spike-like cluster of 10–20 or more, on stems to 15 cm. Flowers
4–5 mm long, lobes about a third as long as the tube. Leaves usually 2, erect, flat with
prominent raised veins. Scrub and rocky places. YU.GR.BG. Pl. 59.   **1643e.** *H. dalmatica* (Baker) Chouard Like 1643d, but leaves spreading or recurved, and veins not
prominent. Flowers pale blue, the lobes slightly shorter than the tube. Grassy places
amongst rocks. YU(west). Pl. 59.

STRANGWEIA Distinguished from *Hyacinthus* by the filaments of the stamens
which are flattened at the base and joined into a cup surrounding the ovary, and
filaments winged with 2–3 teeth at the apex; bracts with a cylindrical spur.

1643f. *S. spicata* (Sibth. & Sm.) Boiss. (*Hyacinthus s.*) A tiny bulbous plant with a dense
spike of 5–10 deep purplish-blue, bell-shaped flowers, and several linear leaves with
ciliate margins. Flowers about 7–10 mm; perianth tube short, lobes spreading and twice

1. *Stangweia spicata* 1643f **2.** *Romulea ramiflora* 1681d **3.** *Asphodeline taurica* 1594a **4.** *Ornithogalum fimbriatum* 1639k **5.** *Merendera sobolifera* 1587b
6. *Bellevalia trifoliata* 1644b

as long; anthers dark blue. Leaves 4–8, longer than flowering spike which is up to 15 cm. Rocky hills. GR(south and west). Page 497.

BELLEVALIA Like *Muscari* but corolla tubular or funnel-shaped and deeply divided into more or less spreading lobes (not constricted at the mouth); flowers white, blue, or violet, becoming greenish or yellowish when mature. Anthers held at the mouth of the corolla and level with the lobes. 5 species in our area.

Flowering spike conical
(1644). *B. ciliata* (Cyr.) Nees Flowering spike with 30–50 bell-shaped, lilac flowers with greenish lobes on stems 6–50 cm. Flowers 9–11 mm, lobes ½–⅓ as long as tube; lower flower stalks 2–3·5 cm, much longer than upper; fruit stalks horizontal and rigid. Leaves 3–5, shorter than flowering spike, margins ciliate. Fields, cultivation. GR.TR.

Flowering spike cylindrical or oblong
1644b. *B. trifoliata* (Ten.) Kunth Flowering spike cylindrical with 10–40 violet, tubular-bell-shaped flowers which later become brownish with greenish lobes, borne on stems to 60 cm. Flowers 12–14 mm long; flower stalks 4–8 mm. Leaves 2–4, about as long as flowering spike, often finely ciliate. Fields, grassy places in the Med. region. GR+Is.CR?TR. Page 497. **1644c.** *B. dubia* (Guss.) Reichenb. Distinguished by its greenish-violet bell-shaped flowers with whitish lobes and green veins, and its hairless leaves which are procumbent and longer than the flowering spikes. Flowers 5–8 mm long; flowering and fruiting spikes cylindrical. Cultivation, grassy places in the Med. region. YU.GR.CR? ****1644.** *B. romana* (L.) Reichenb. Distinguished by its oblong spike of top-shaped or obconical flowers which are whitish, sometimes tinged with blue at the base, becoming dingy brown. Flowers 8–10 mm long, lobes as long as or longer than the tube. Leaves 3–6, hairless, longer than flowering spike. Fruiting spike cylindrical. Fields, meadows in the Med. region. YU.AL.GR.

MUSCARI | Grape-hyacinth
Lower fertile flowers with brownish, yellow, or greenish corolla tube when mature
(a) Tube of corolla yellow; sterile flowers usually absent
1645a. *M. macrocarpum* Sweet Flowers in a dense cluster, becoming lax, at first purplish then bright yellow when mature, sweet-scented. Corolla egg-shaped, teeth minute, spreading, brownish. Leaves 10–20 cm, longer than flowering stem. Limestone cliffs. GR(Cyclades,Fournoi,Simi). Pl. 59.

(b) Tube of corolla brownish or greenish; sterile flowers usually many
  (i) Teeth of corolla blackish
1645b. *M. tenuiflorum* Tausch Flowers in lax cylindrical cluster 6–30 cm. Fertile flowers light greyish-brown, teeth blackish, recurved; sterile flowers bright violet, borne on fleshy bright violet, erect or ascending stalks 3–16 mm. Flowering stem 20–60 cm, longer than the 3–7 linear, channelled leaves. Dry places. YU.AL.GR.BG.TR.

  (ii) Teeth of corolla whitish to pale-brown, or yellow
1645. *M. comosum* (L.) Miller TASSEL HYACINTH A very variable plant but very distinctive with its erect terminal tuft of bright blue-violet sterile flowers, and its lower spreading, brownish-green mature fertile flowers. Teeth of fertile flowers creamy or pale yellowish-brown; stalks of sterile flowers violet, ascending, 2–25 mm or more. Leaves 3–5, linear 5–17 mm wide; stems 15–50 cm. Rocky ground, cultivation. Widespread. **1645c. *M. cycladicum* P. H. Davis & Stuart Differs from 1645 in having stalkless or very short-stalked (less than 2 mm) fertile flowers with bright brownish-yellow teeth to the greenish-brown corolla. Stalks of sterile flowers spreading. GR(Cyclades).CR.

1645d. *M. spreitzenhoferi* (Heldr.) Vierh. (*M. creticum*) Flower spikes dense conical, becoming lax; fertile flowers deep brown with yellow spreading or recurved teeth, sterile flowers usually absent. Flower stalks ascending. Leaves 3–5, longer than flowering stem which is 5–15 cm. Mountains. CR. **1645e.** *M. weissii* Freyn Differs from 1645d in having a lax cylindrical flower spike with spreading flower stalks, and numerous purple, short-stalked sterile flowers. Fertile flowers oblong, brown with bright brownish-yellow teeth. Rocks. GR(Cyclades, E.Aegean Is.).CR.

Lower fertile flowers with pale-blue to blackish-blue corolla tube when mature
(a) Autumn-flowering
1646a. *M. paviflorum* Desf. The only autumn-flowering species, with a very lax cylindrical cluster of pale-blue fertile flowers which have paler recurved teeth with a darker blue stripe. Fertile flowers 3–5 mm, sterile flowers few and minute, or absent. Leaves narrowly linear, shorter than flowering stem which is 15–35 cm. Rocks, dry hills. YU.GR.CR.

(b) Spring-flowering
 (i) Teeth and tube of corolla blackish-blue
**1646. *M. commutatum* Guss. Readily distinguished by its deep blackish-blue fertile flowers with tiny black incurved teeth, and smaller paler sterile flowers. Flower cluster dense ovoid; corolla 4–7 mm. Leaves 2–5, channelled, about as long as the flowering stem which is 6–30 cm. Grassy places, dry hills. YU.GR.CR.

 (ii) Teeth of corolla pale blue or white, tube darker blue
**1648. *M. botryoides* (L.) Miller SMALL GRAPE-HYACINTH Distinguished by its tiny globular bright-blue flowers with white recurved teeth, borne in a dense cluster. Corolla 3–4 mm. Sterile flowers few, paler. Leaves ribbed, glaucous, often hooded at apex; stem 7–30 cm. Meadows, fields, woods. YU.AL.GR.BG.TR.

1647. *M. neglectum* Ten. (incl. *M. pulchellum*, *M. atlanticum*, *M. racemosum*) A very variable species which can readily be distinguished by the white teeth to the oblong-ovate, dark-blue to blackish-blue corolla of the fertile flowers. Corolla 4–8 mm; sterile flowers smaller, paler. Flower heads usually dense, on stems 4–30 cm, often as long as the 3–6 channelled leaves. Dry hills, grassy places, mountains. Widespread. **1647a. *M. armeniacum* Baker Distinguished from 1647 by its bright azure-blue sometimes purple-tinged flowers, with paler or white teeth, borne in a dense oval to cylindrical cluster. Sterile flowers few, smaller, similar-coloured or paler than fertile flowers. Leaves usually 3–5; stem 10–40 cm. Very variable. Grassy places in mountains. YU.GR.BG.TR. Pl. 59.

ASPARAGUS True leaves reduced to papery scales or spines; branchlets (cladodes) green, needle-like, usually in whorls. About 8 similar-looking species in our area.

Stems herbaceous, not spiny; fruit red
1649. *A. officinalis* L. ASPARAGUS. Woods by streams, bushy places. YU.AL? GR.BG.TR. **1649a.** *A. maritimus* (L.) Miller Differs from 1649 in having stem and branches which are rough to the touch and needle-like cladodes usually in whorls of 4–7, with rough teeth (whorls with 3–8 smooth cladodes in 1649). Flowers greenish. Fruit blood-red, 6–12 mm. By the sea. YU.AL.GR.BG? ****(1649).** *A. tenuifolius* Lam. Needle-like cladodes smooth, very slender, 1–2·5 cm long, in whorls of 15–40. Flowers greenish. Stems and branches smooth. Fruit red, 10–16 mm. Bushy places. YU.GR?BG.TR.

Stems woody, at least at base, usually spiny; fruit black
(a) Cladodes in whorls or clusters
1650. *A. acutifolius* L. A much-branched, climbing shrub to 2 m, with whitish rough woody stems, and rigid clusters of 10–30 sharp-pointed, spreading, green, more or less equal cladodes 2–8 mm long. Old stems with stout spines. Flowers 3–4 mm, yellowish-green. Bushy places. Widespread in the Med. region + BG. Page 39. **(1650).** *A. aphyllus* L. Differs from 1650 in having smooth green stems and branches, and stout angled unequal, spine-tipped cladodes usually 1–2 cm long, in clusters of 3–10. Flowers green, 3–4 mm, with petals unequal, the inner shorter. Sunny hills, by the sea. GR+Is.CR.TR.

(b) Cladodes solitary
1650a. *A. stipularis* Forskål (*A. horridus*) Stems woody, much-branched, smooth, spiny below; cladodes stout, angled, spiny-pointed, 1·5–3 cm long, solitary except sometimes at the tips of the branches. Flowers 3–4 mm, shortly stalked. Stony places in the Med. region. GR(Cyclades).CR.

RUSCUS | Butcher's Broom Lateral branches (cladodes) flattened, green, leaf-like; flower clusters and fruit borne on one side of the cladodes.

1651. *R. aculeatus* L. BUTCHER'S BROOM. Bushy places, woods. Widespread. **(1651).** *R. hypoglossum* L. LARGE BUTCHER'S BROOM. Cladodes elliptic or lanceolate, 5–9 cm (cladodes to 2·5 cm, rigid or spiny-tipped in 1651). Flowers 3–5, borne on the upper surface; bract green. Fruit scarlet, about 2 cm. Woods, bushy places. YU.GR.BG.TR.

MAIANTHEMUM **1652.** *M. bifolium* (L.) F. W. Schmidt MAY LILY. Woods. YU.

STREPTOPUS **1653.** *S. amplexifolius* (L.) DC. Mountain woods. YU.BG(south-west).

POLYGONATUM | Solomon's Seal

All leaves alternate, elliptic or ovate; fruit bluish-black
(a) Leaves hairless
1654. *P. odoratum* (Miller) Druce (incl. *P. pruinosum*) SWEET-SCENTED SOLOMON'S SEAL. Flowers usually 1–2 from each leaf axil, sweet-scented; corolla 12–30 mm long. Stem 15–65 cm, angled, hairless. Woods, bushy and rocky places. YU.AL.GR.BG. **1655.** *P. multiflorum* (L.) All. SOLOMON'S SEAL. Flowers 2–6 from the lower leaf axils, odourless; corolla 9–20 mm long, somewhat constricted in the middle. Stem 30–80 cm, rounded, hairless. Woods, bushy places. Throughout the mainland.

(b) Leaves minutely downy on the veins beneath
(1655). *P. latifolium* Desf. BROAD-LEAVED SOLOMON'S SEAL. Flowers 1–5 from each leaf axil; corolla 10–18 mm long; flower stalks finely downy. Stem 20–100 cm, angled, sparsely hairy above. Woods. YU.AL? GR(north).BG.TR.

Upper leaves in whorls, linear-lanceolate; fruit reddish-violet
1656. *P. verticillatum* (L.) All. WHORLED SOLOMON'S SEAL. Mountain woods. YU.AL.BG. Page 185.

CONVALLARIA **1657.** *C. majalis* L. LILY-OF-THE-VALLEY. Woods, thickets, YU.AL.GR.BG.

PARIS **1658.** *P. quadrifolia* L. HERB PARIS. Damp woods. YU.GR.BG.

SMILAX **1659.** *S. aspera* L. (*S. nigra*) A creeping or climbing shrub to 15 m, with angled, prickly or smooth stems, usually spiny leathery leaves, and with tendrils. Flowers greenish-yellow, 2–4 mm, in umbels or clusters of 5–30 flowers on branched stems 2–15 cm. Fruit red or black. Bushy places. Widespread in the Med. region. Page 34. **1659a.** *S. excelsa* L. Similar to 1659 but leaves with a rounded base (heart-shaped, spear- or arrow-shaped base in 1959), often without spines. Flowers 4–12, in a simple unbranched long-stemmed umbel. Fruit red. Bushy places. GR.BG.TR.

AGAVACEAE | Agave Family

AGAVE **1660.** *A. americana* L. CENTURY PLANT. Readily identified by its enormous rosettes, 2–4 m across, of massive spiny spear-shaped leaves 1–2 m long. Flowering stem produced after at least 10 years, tree-like to 10 m; flowers greenish-yellow, including ovary about 9 cm. Fruit an oblong, three-angled capsule. Native of Mexico; widely naturalized in the Med. region.

AMARYLLIDACEAE | Daffodil Family

The coloured perianth segments are described as petals in this family.

LEUCOJUM **1661.** *L. vernum* L. SPRING SNOWFLAKE. Flowers usually solitary, rarely 2, nodding, white with a yellow or green spot below the tip of each petal. Flowering stem usually longer than leaves. Seeds whitish. Damp woods, meadows. YU(central).

1662. *L. aestivum* L. SUMMER SNOWFLAKE. Like 1661 but a taller plant 35–60 cm, with usually 2–5 flowers on a stem usually shorter than the leaves. Longest flower stalk as long as or longer than the spathe. Leaves linear 5–20 mm wide. Seeds black. Damp meadows, ditches. YU.AL.GR.BG.TR.

GALANTHUS 3 species in our area. **1663.** *G. nivalis* L. SNOWDROP. Flowering January to April with leaves appearing before the flowers open. Outer petals usually 12–25 mm long. Woods, meadows. YU.AL.GR.BG.TR. Subsp. *reginae-olgae* (Orph.) Gottl.-Tann. Flowering from October to December with leaves generally appearing after flowering; outer petals 20–35 mm. Woods. GR(south-west). **1663a.** *G. elwesii* Hooker fil. Distinguished from 1663 by having leaves with incurved margins in bud or at base in sheath (not flat as in 1663). Inner petals with a green patch at base as well as at the apex; outer petals white, 15–28 mm. Spring-flowering. Woods, scrub, rock pastures. YU.GR(N. Aegean).BG.

STERNBERGIA **1664.** *S. lutea* (L.) Sprengel COMMON STERNBERGIA. Flowers solitary bright yellow, crocus-like, 3–5 cm long, appearing in the autumn with the young leaves, borne on a short stalk 4–10 cm, with a papery spathe about half as long as the blunt petals. Leaves broadly linear 5–15 mm wide. Stony places, sunny hillsides in the Med. region. YU.AL.GR+Is.CR. Subsp. *sicula* D. A. Webb differs in having narrow leaves 3–5 mm wide. Flowers smaller, with acute petals 4–8 mm wide (7–15 mm wide in subsp. *lutea*). S. Greece and Aegean region. Pl. 59.   **(1664)**. *S. colchiciflora* Waldst. & Kit. SLENDER STERNBERGIA. Flowers pale yellow, appearing stalkless before the leaves in autumn. Petals 2·5–5 cm long, narrow, tube as long as or longer than petals. Leaves dark green, linear 2–5 mm wide, often twisted. Stony, rocky ground. YU.GR.BG.

NARCISSUS | Daffodil, Narcissus
Flowers several
1669. *N. tazetta* L. Flowers in a cluster of 3–15, fragrant, with spreading or reflexed white, cream, yellow, or orange variable-sized petals, and a golden-yellow corona. Flowers 2–3 cm across. Leaves 5–25 mm wide; stem 20–40 cm. Spring-flowering. Fields, vineyards, damp grassy places in the Med. region; often naturalized from cultivation. YU.AL.GR+Is.CR. Subsp. *aureus* (Loisel.) Baker has golden-yellow petals and an orange corona. Subsp. *italicus* (Ker-Gawler) Baker has cream-coloured petals and a yellow corona. **(1667).** *N. papyraceus* Ker-Gawler PAPER-WHITE NARCISSUS. Like 1669 but flowers pure white, including the corona, in a cluster of up to 20. Leaves glaucous, 7–15 mm wide, grooved. Sunny hills, vineyards, cultivated ground in the Med. region. YU.GR.

Flowers solitary, rarely 2
(a) Spring-flowering
1671. *N. poeticus* L. PHEASANT'S-EYE NARCISSUS. Flowers large 4–6 cm across, with broad white or pale-cream petals and a short yellow corona with a crimped red margin. Leaves glaucous, 5–13 mm wide; stem to 50 cm, compressed. Damp meadows. YU.AL.GR.   **1665. *N. pseudonarcissus* L. DAFFODIL. Flowers with a long, deep-yellow trumpet-shaped corona 2–4 cm, as long as or longer than the free, pale-yellow petals. Cultivated and naturalized in YU.AL.BG.

(b) Autumn-flowering
**1672. *N. serotinus* L. Flowers sweet-scented, usually solitary, appearing before the leaves in autumn; petals white, 9–12 mm long, corona golden-yellow, very short. Leaves 1 or 2, thread-like, appearing in spring; stem slender 10–20 cm. Dry hills in the Med. region. YU.GR+Is.CR.

PANCRATIUM   **1673. *P. maritimum* L. SEA DAFFODIL. Flowers pure white, very large to 15 cm long, in an umbel of 3–15 on a stout stem to 40 cm. Leaves glaucous, broadly linear about 2 cm wide. Maritime sands. YU.AL.GR+Is.CR.BG.TR.

DIOSCOREACEAE | Yam Family

TAMUS   1674. *T. communis* L. BLACK BRYONY. Hedges, thickets, woods. Widespread.

IRIDACEAE | Iris Family

The coloured perianth segments of this family are described as petals here.

CROCUS | Crocus   About 27 species in our area; many are very local; 14 are endemic.

Scales of corms with parallel fibres
(a) *Flowers yellow or cream*
1675. *C. flavus* Weston (*C. aureus, C. maesiacus*). Flowers pale yellow to deep orange-yellow, appearing with the leaves from March to April. Style obscurely divided into 3 short branches. Corm with a persistent brown neck of old scales. Grassy hillsides, open woodlands of E. and C. Balkans. YU.GR(north).BG.TR. Pl. 60. **1675a.** *C. olivieri* Gay Like 1675 with yellow or orange flowers appearing from February to April but found in the S. Balkan and Aegean region. Style divided into 6 slender branches. Corm without a neck of old scales. Grasslands, open woods. YU(south).AL.GR+Is.CR?BG.TR.

1675b. *C. boryi* Gay An autumn-flowering species, from September to December, of the S. Aegean region, with creamy-white flowers which are sometimes slightly purple-flushed. Style deeply divided into many slender branches, yellow or orange; filaments of stamens finely hairy. Leaves very narrow 1–3·5 mm, appearing with the flowers. Like 1680c but corm scales papery, splitting into parallel vertical fibres. Stony hillsides, scrub, olive groves. GR(south and west).CR. Pl. 60.

(b) *Flowers lilac*
1675c. *C. tournefortii* Gay An autumn-flowering species of the S. Aegean region, with corm scales like 1675b, but flowers pale lilac with a yellow or rarely white throat like 1680c. Scrub, stony hillsides. GR.(Cyclades,Idhra)+Karpathos.

Scales of corm with a coarse or fine network of fibres
(a) *Spring- or early summer-flowering*
  (i) *Flowers orange-yellow*
1675d. *C. cvijicii* Košanin Flowers cream to deep orange-yellow with orange-yellow anthers and orange deeply three-cleft style, flowering May to June. Leaves 2–4, appearing with the flowers. Throat of perianth finely hairy. Corm with finely fibrous netted scales. By snow-patches in mountains. YU(Macedonia).AL.GR(Macedonia). Pl. 60. **1675e.** *C. scardicus* Košanin Flowers orange-yellow, usually purplish towards the base of the petals and tube, flowering May to July. Anthers yellow; style shortly divided in 3–5 shallow lobed, flattened stigmas. Leaves 3–4, without a median white stripe. Corm with finely fibrous netted scales forming a persistent neck at the apex. By snow-patches in mountains. YU(south-west). Pl. 60.

  (ii) *Flowers lilac to purple, rarely white*
1677. *C. reticulatus* Adam (*C. variegatus*) Flowers white or lilac and strongly striped with purple outside, appearing from February to March, with the very narrow leaves about 1 mm wide. Petals acute; anthers yellow; style orange or scarlet, deeply three-lobed. Corm with a very coarse network of fibres. Meadows, open woods. YU.BG. **1677c.** *C. tommasinianus* Herbert Flowers rather slender with narrow pointed petals, clear lilac to purple with a white tube, flowering usually with well-developed leaves from March to May. Anthers yellow; style yellow to orange with indistinctly frilled divisions. Leaves narrow 2–3 mm wide, dark green with a very pronounced mid-rib; bracts densely spotted and veined greyish and greenish. Woods, shady banks. YU.BG(north-west).

1677d. *C. sieberi* Gay Flowers pale lilac to deep lilac-purple with a deep-yellow throat, sometimes with a white band separating the purple and yellow, or (in Crete), petals white with a yellow throat and purple markings. Anthers yellow; style yellow to orange-red, large and frilled at the apex with 3 rather obscure lobes. Flowering usually from March to June together with the broad 2–6 mm wide, dark-green leaves, or flowering shortly before the leaves. Mountain grassland, woods, scrub. YU(south). AL.GR+Euboea,Samos.CR. Page 71. Pl. 60. **1677e.** *C. veluchensis* Herbert Like

1677d but flowers solitary, pale to deep lilac-blue, or rarely white, stained with purple-blue at the base, throat whitish finely hairy. Flowering April to June by snow-patches. Corm with very fine fibres. Rocky ground in mountains. YU(south).AL.GR.BG Pl. 60.

(b) Autumn-flowering

1678c. *C. hadriaticus* Herbert Flowers white or rarely flushed pale lilac, throat hairy yellow or rarely white inside; flowering from October to December, usually with the leaves. Anthers yellow; style three-branched, scarlet, 1–1·6 cm long, less than half as long as the petals which are 2–4·5 cm. Stony hillsides. GR(west and south)+Cyclades. Pl. 60. **1678d.** *C. cartwrightianus* Herbert Closely related to 1678c but style 1–2·7 cm, more than half as long as the petals which are 1·4–3·2 cm. Flowers deep lilac-purple and strongly dark-veined, or white, or white with a purple base; throat white or lilac. Rocky hillsides. GR(south)+Is. CR.

1678e. *C. cancellatus* Herbert Flowers white to deep purple often with darker veins, throat yellowish. Petals 2·5–5 cm, oblanceolate or obovate; tube white tinged with purple 5–15 cm; anthers deep yellow; style yellow or orange, divided into slender branches. Leaves 4–5; scales of corm coarsely netted. Flowering September to November. Rocky hillsides, open woods. YU(south).GR. Pl. 60.

Scales of corm breaking off in concentric rings, or splitting into triangular teeth
(a) Spring-flowering

1679. *C. chrysanthus* Herbert Flowers pale to deep yellow, appearing from January to May at the same time as the leaves, honey-scented. Anthers usually yellow, sometimes blackish or with black markings at the base; style orange-red, with 3 slender or broad branches. Stony hill slopes. YU.AL.GR.BG.TR. Page 169. **1680.** *C. biflorus* Miller Flowers white or blue-lilac, often strongly veined or flushed with deeper purple; throat white, lilac, or yellow. Anthers yellow sometimes with blackish basal lobes; style yellow to scarlet. Flowering January to April. Sunny hill slopes, grassy and stony places. YU.GR.BG.TR.

(b) Autumn-flowering

1680a. *C. crewei* Hooker fil. Very like 1680 but flowering from October to January and found in the S. Aegean region. Flowers white outside usually with strong purple veining, sometimes stippled greyish, bluish, or rarely yellow; throat yellow. Anthers usually blackish. Hills. GR(south)+Cyclades. **1680b.** *C. pulchellus* Herbert Flowers pale lilac-blue with darker veining, and a deep-yellow throat, appearing from September to October without leaves. Petals 2·3–4 cm, blunt; anthers white, filaments yellow, densely hairy; style yellow to orange, much divided, much longer than stamens. Open woods. YU(south).GR(north).BG.TR. Pl. 60. **1680c.** *C. laevigatus* Bory & Chaub. Flowers white or lilac, often strongly veined or flushed purple, and throat yellow, hairless; appearing from October to December with the leaves. Anthers creamy-white, filaments hairless; style orange or yellow, deeply cut into many slender branches. Leaves 3–8, narrow 1–3 mm wide. Corm with smooth, hard, leathery scales which split at the base into long narrowly triangular teeth. Scrub, stony grassland. GR+Is.CR.

ROMULEA Like *Crocus* but flowers stalked, corolla tube very short; leaves without a white stripe above. A very difficult genus, with many colour variants and many minor forms described as distinct species by some botanists.

Throat of corolla yellow or white
1681. *R. bulbocodium* (L.) Sebastiani & Mauri Flowers 1–6, usually lilac or violet, often greenish or lined outside, and with yellow throat and tube, but very variable, rarely

wholly white or yellow. Flowers funnel-shaped; petals 2–3·5 cm or more; stigmas usually longer than the stamens, divided to the base into 2 whitish lobes. Basal leaves 2, rush-like, grooved, 1–2 mm broad, stem leaves up to 5. Rocky, sandy places and scrub in the Med. region. YU.AL.GR.CR.BG(south). Pl. 61.

(1681) *R. columnae* Sebastiani & Mauri Flowers small, 1–3, pale violet to pale lilac with purple veins, usually with a yellow throat. Petals 9–19 mm, acute; stigmas shorter than anthers which reach to half the length of the petals. Bracteole almost entirely papery. Basal leaves 2, up to 1 mm broad. Grassy and stony places in the Med. region. GR.CR.TR. **1681d.** *R. ramiflora* Ten. Flowers 1–4, small, blue-lilac sometimes with darker veins and with a white or yellow throat, sometimes pinkish or yellowish outside. Petals 1–3 cm; anthers reaching to two-thirds length of petals; fruit stalks elongating to 10 cm. Bracteole mostly green, conspicuously veined. Maritime sands, hills in the Med. region. GR.CR.TR Page 497.

Throat of corolla violet-purple
1681c. *R. linaresii* Parl. A very small plant less than 5 cm, with 1–2 uniform-coloured dark violet-purple flowers. Petals 1·3–2 cm. Bracteole mostly papery, spotted reddish-brown. Short turf, stony places in the Med. region. GR+Is.CR?TR.

HERMODACTYLUS Distinguished from *Iris* by the ovary which is one-celled, and by its very short style. Leaves tetragonal in section.

**1683. *H. tuberosus* (L.) Miller SNAKE'S HEAD IRIS, WIDOW IRIS. Flower solitary, greenish-yellow with dark-brownish or blackish-purple falls, fragrant. Spathe green, slender, to 20 cm. Leaves 1·5–3 mm broad, rush-like; stems to 30 cm. Hills, stony places, rocks in the Med. region. YU.AL.GR+Is.CR.

GYNANDRIRIS Like *Iris* but with a corm; stamens closely pressed to style-branches; petals attached to the sterile beak of the ovary.

**1684. *G. sisyrinchium* (L.) Parl. (*Iris s.*) BARBARY NUT. Flowers variable in size 1–3 cm across, with blue, rounded spreading falls with greater or lesser amount of white or pale yellow at the base, short-lived. Spathe papery, brownish. Leaves rush-like, grooved with inrolled margins when dry, sheathing at base. Dry places in the Med. region. GR+Is.CR.TR.

IRIS Flowers with 3 outer coloured spreading or recurved petals, the *falls*, each with a terminal *blade* and narrow basal *shaft*, and 3 inner coloured petals, the *standards*, which are usually erect. The falls may have a tuft of hairs, the *beard*, on the upper surface. The spathe, which is green or papery, encircles the flower buds and later the ovary. About 16 species in our area.

Falls hairless (or minutely hairy), thus without a beard
(a) *Flowers blue-violet (rarely white)*
1686. *I. graminea* L. GRASS-LEAVED IRIS. Distinguished by its flattened two-winged stem, and its paired fragrant, blue-violet white-veined flowers, with dull-pinkish styles and oblanceolate standards. Spathes both leafy and papery. Leaves grass-like, 5–15 mm wide; stems 20–40 cm. Grassland, scrub. YU.BG. **1686a. *I. sintenesii* Janka Like 1686 but stem cylindrical or slightly flattened. Flower blue-violet, not fragrant. Spathes all similar, green, leafy. Basal leaves 3–6 mm wide. Dry grasslands, scrub. YU.AL. GR.BG.TR. Pl. 61. **1687.** *I. sibirica* L. Flowers blue-violet, rarely white, about 6 cm across, 1–5 on hollow stems up to 1 m or more, and with distinctive brown spathes at flowering time. Falls 3–5 cm, with a rounded blade; standards with a narrow erect limb.

Leaves 4–10 mm wide. Damp grassland. YU.BG. **1687a.** *I. unguicularis* Poiret (incl. *I. cretensis*) Flowers solitary short-stalked, fragrant, with whitish falls with violet veins and apex, and lilac standards. Tube of perianth slender 6–20 cm; style-branches yellow; spathes mostly green. Leaves grass-like, 1–5 mm wide. Stony ground, banks. GR(south and west).CR+Rhodes,Karpathos. Pl. 61.

(*b*) *Flowers white or yellow*
1688. *I. spuria* L. subsp. *ochroleuca* (L.) Dykes has white flowers with a yellow centre to the falls which are 6–8 cm long. Basal leaves 1–2 cm wide; stems 30–90 cm. Wet places. GR(Thrace)+Lesbos. ****1690.** *I. pseudacorus* L. YELLOW FLAG. Flowers yellow. Marshes, swamps, ditches. Widespread.

Falls bearded
(*a*) *Stems simple or absent*
****1691.** *I, pumila* L. Flowers solitary, almost stalkless, yellow, blue, or purple and with rounded spathes which are papery in the upper part. Falls 3·5–6 cm, broadly spathulate, with a dense beard; standards oblong, stalked, a little longer than the falls; perianth tube 4–9 cm. Leaves straight or nearly so, up to 15 cm long by 1·5 cm wide. Subsp. *attica* (Boiss & Heldr.) K. Richter has leaves strongly curved, 8 cm long by 9 mm wide. İt is found in the southern part of the species range. Dry grassy and rocky places. YU.AL.GR.BG. **1691a.** *I. reichenbachii* Heuffel Differs from 1691 in having both spathes green, boat-shaped, and strongly keeled. Flowers violet to brownish-purple, or greenish-yellow, sometimes with violet veins, 1 or 2, on slender stems 4–30 cm. Falls and standards 4–6 cm; perianth tube 1·5–2·5 cm, much shorter than petals. Leaves 5–15 mm wide. Rocky, grassy places, mostly in mountains. YU.GR.BG. Pl. 61. **1691b.** *I. suaveolens* Boiss. & Reuter (incl. *I. rubromarginata*) Like 1691a but stems shorter, usually 1–6 cm; perianth tube 3–5·5 cm, about as long as the petals. Falls 3–5·5 cm, violet to brownish-purple, or greenish-yellow sometimes violet-veined or -spotted. Leaves up to 1 cm wide. Dry grassy and rocky places. YU.GR.BG.TR.

(*b*) *Stems branched*
****1693.** *I. germanica* L. (incl. 1694 *I. florentina*) COMMON IRIS. Flowers 3–5, bluish-violet, or white tinged with blue (var. *florentina*), on a tall stout branched stem 40–90 cm. Flowers about 10 cm across; falls and standards 5·5–9 cm; spathes papery in upper half, often purple-tinged. Leaves somewhat glaucous, 2–3·5 cm wide. Widely grown for ornament and for perfume which is extracted from the rhizome. Origin unknown; naturalized in waste places in YU.AL.GR+Is.CR.BG. **1693a.** *I. albicans* Lange Differs from 1693 in having pure-white flowers. Originally planted in Muslim cemeteries, naturalized in YU? CR. **1693b.** *I. pallida* Lam. Flowers 3–6, lilac to violet, borne usually on branched stems 15–120 cm. Leaves glaucous, 1–4 cm wide. Differs from 1693 in its spathes which are silvery-white and entirely papery at flowering. Rocky places in the Med. region; often grown for ornament and naturalized elsewhere. YU(west). **1693c.** *I. variegata* L. Flowers 3–6, whitish-yellow with conspicuous deep-violet to brownish-red veins on the falls, and standards and style-branches yellow. Spathes green inflated, sometimes purple-tinged, with only the extreme tip papery. Leaves curved, deep green, 7–28 mm wide; stems 15–40 cm, usually branched above. Bushy, rocky places. YU.BG.

GLADIOLUS Flowers in a spike, curved funnel-shaped, asymmetric, with a very short tube. Species are similar-looking and hybridization is not uncommon, making identification sometimes difficult.

Anthers longer than filaments; seeds not winged

1695. *G. italicus* Miller (*G. segetum*) A common weed of crops, particularly in the Med. region. Flowers 6–16, 4–6 cm long, reddish-purple to light pink, in a lax weakly two-ranked spike on a stem 50–100 cm. Anthers sometimes aborted, then shorter than filaments. Lower bracts leafy. Leaves 3–5, 5–10 mm wide, with irregular veins. Widespread. **1695a. *G. atroviolaceus* Boiss. Like 1695 but flowers dark violet, and leaves with prominent equidistant veins. Native of south-west Asia; a cornfield weed in GR(north-east).TR.

Anthers equalling or shorter than filament; seeds winged
(a) Spike dense; lowest leaf with blunt apex
1696a. *G. imbricatus* L. Flowers usually violet-purple, 4–12, in a dense but strongly one-sided spike. Perianth tube strongly curved, petals more or less equal and bunched together. Leaves usually 3. Water meadows, scrub. YU.GR.BG.

(b) Spike lax; lowest leaf narrowing to an acute apex
1696. *G. communis* L. COMMON GLADIOLUS. Flowers pink, red, or purplish-red, 10–20 in a lax often branched spike, on stems 50–100 cm. Petals 3–4·5 by 1–2·5 cm, lower petals frequently blotched, or with white or dark-red lines. Leaves 4–5, 5–22 mm wide. Grassy places, cultivation in the Med. region. YU.AL.GR.BG. **1697.** *G. illyricus* Koch Very like 1696 and not easily distinguished from it. A smaller plant 25–50 cm, with fewer (3–10) red to purplish-red flowers, with petals 2·5–4 by 0·6–1·6 cm. Maquis, open woods. YU.AL.GR.BG. Pl. 61. **1697b.** *G. palustris* Gaudin Like 1697 but spikes not more than six-flowered and strongly one-sided, without side brances. Flowers red to purplish-red; petals 2·5–4 cm, the lowermost longest. Leaves usually 3, the stem leaf bract-like. Wet meadows, scrub. YU.AL.BG.

JUNCACEAE | Rush Family

A difficult but not very important family in our area, with about 30 species of *Juncus*, RUSH, and 12 species of *Luzula*, WOOD RUSH. The majority of species are widespread in Europe; those present in the Balkans are usually to be found in similar habitats to the rest of Europe, viz. marshes, salt-marshes, damp places, thickets, and mountain pastures.

PALMAE | Palm Family

PHOENIX ****1718.** *P. canariensis* Chabaud CANARY PALM. Trunk about 6–8 m; leaves glossy-green, arching upwards, to 7 m, with numerous pairs of flexible, not rigid, narrow leaflets. Fruit dry, the size of an olive. Planted in the Med. region. **(1718).** *P. dactylifera* L. DATE PALM. Trunk slender to 20 m. Leaves smaller, arched or drooping, grey-green, the leaflets more rigid than in 1718. Fruit about 3 cm, borne in long hanging clusters. Planted in hotter parts of the Med. region. **1718a.** *P. theophrasti* W. Greuter CRETAN PALM. Considered to be native of East Crete. Very like (1718) but differing in its small, dry, inedible fruits 1·5 cm long, and its short-stalked flower clusters. CR.

507

Several palms in addition to those described above are planted and have become a conspicuous feature of the landscape in the E. Mediterranean region. They include: *Trachycarpus fortunei* (Hooker) H. Wendt (*T. excelsa*) CHUSAN PALM; *Jubaea chilensis* (Molina) Baillon WINE PALM; *Washingtonia filifera* (J. A. Linden) H. A. Wendl.

GRAMINEAE | Grass Family

A very large and important family with about 100 genera and 400 species in our area. Often requiring considerable experience and sometimes specialist knowledge for their identification. Space does not allow for their inclusion here.

ARACEAE | Arum Family

ACORUS **1816. *A. calamus* L. SWEET FLAG. Still and slow-flowing waters; introduced to YU.BG.

ARUM

Spathe blackish to brownish-purple
1818b. *A. petteri* Schott (*A. nigrum*) Leaves appearing in autumn; flowering in spring. Spathe purplish-black, 15–20 cm; spadix stout, purplish-grey, about 5 cm shorter than spathe. Stony ground, bushy places. YU(west).GR(north). **1818d.** *A. dioscoridis* Sibth. Spathe purplish strikingly black-spotted; spadix blackish-purple, shorter. By paths and tracks. E.Aegean Is. Pl. 61. **1818e.** *A. elongatum* Steven Spathe brownish-purple inside, to 25 cm; spadix red, about as long as the spathe. Leaves appearing in autumn; flowering in spring. Woods, scrub. GR(east).BG.TR?

Spathe white or greenish, often purple-tinged or -spotted inside
(a) Leaves appearing in spring
1818. *A. maculatum* L. LORDS-AND-LADIES. Leaves appearing in spring; usually shiny, often with blackish spots. Spathe 10–20 cm, pale yellowish-green, but often purple-spotted, or with purple margin; spadix purple, rarely yellow. Tuber horizontal with shoot at one end. Woods, hedges. YU.AL? GR.CR.BG.TR? **(1818).** *A. orientale* Bieb. Very like 1818 and differing only in having a flattened disk-like tuber with the shoot arising from the centre. Leaves appearing in spring, not spotted; flowering late spring or early summer. Spathe 7–15 cm, pale green, purple-flushed, not spotted; spadix dull purple. Shady places. YU.AL? GR.BG.TR?

(b) Leaves appearing in autumn
****(1818).** *A. italicum* Miller ITALIAN ARUM. Spathe pale greenish-yellow, 15–40 cm, drooping early at apex; spadix stout, yellow, less than half as long as spathe. Leaves appearing in autumn or early winter, with or without white veins. Waste ground, hedges. Widespread. Subsp. *byzantinum* (Blume) Nyman has spathe pale green tinged with purple. It occurs in the east Balkans and CR. Pl. 61. **1818c.** *A. creticum* Boiss. & Heldr. Spathe white or pale green, rarely bright yellow, 7–12 cm; spadix dark purple, rarely yellow. Leaves dark shiny green, with basal lobes pointing outwards, appearing in autumn. Sterile flowers few or absent. Stony mountainsides. CR+Karpathos, Samos. Page 71.

DRACUNCULUS **1819.** *D. vulgaris* Schott DRAGON ARUM. Spathe very large to 40 cm, deep reddish-purple inside; spadix large fleshy, deep reddish-purple, stinking. Leaves palmate, with 9–15 elliptic leaflets, purple-spotted; stem stout, spotted. Shady places in the Med. region. YU.AL?GR+Is.CR.BG.

BIARUM **1820.** *B. tenuifolium* (L.) Schott A small plant with a long brownish-purple tongue-like spathe 8–20 cm, and a longer cylindrical purple spadix, often appearing direct from the soil without leaves. Leaves linear to lanceolate, usually appearing after flowering. Winter- or spring-flowering. Hedges, waysides in the Med. region. YU.AL.GR.CR. **1820c.** *B. spruneri* Boiss. Like 1820 but spathe 10–15 cm, about equalling the spadix; flowering in late spring with the leaves. Sterile flowers present only between male and female flowers (not also above the male flowers as in 1820). Dry hillsides. GR(south). **1820d.** *B. davisii* Turrill A rare plant with a greenish-white and spotted pinkish-brown flask-shaped spathe 3–5 cm with a forward-curved lip, and a shorter dark-red spadix. Flowering in late autumn without leaves, which are ovate, stalked, with a blade 1·5–3 cm. Stony hillsides. CR. Pl. 61.

ARISARUM **1821.** *A. vulgare* Targ-Tozz. FRIAR'S COWL. Unmistakable with its brown-and-green-striped flask-shaped spathe with a brownish upper lip, with a slightly longer forward-curved, brown-purple spadix. Leaves oval to arrow-shaped 6–12 cm, long-stalked. Grassy places, open ground, cultivation in the Med. region. YU.AL.GR+Is.CR.

LEMNACEAE | Duckweed Family

The following are found on the surface of ponds, lakes, and still waters, largely in the central European climatic region:
1822. *Spirodela polyrhiza* (L.) Schleiden (*Lemna p.*) GREAT DUCKWEED; **1823.** *Lemna minor* L. DUCKWEED; **(1823).** *L. gibba* L. GIBBOUS DUCKWEED; **1824.** *L. trisulca* L. IVY-LEAVED DUCKWEED; **(1824).** *Wolffia arrhiza* (L.) Wimmer

SPARGANIACEAE | Bur-Reed Family

SPARGANIUM | **Bur-Reed** Plants of still, slow-flowing waters or swamps.

1825. *S. erectum* L. BUR-REED. Widespread. **1826.** *S. emersum* Rehman (*S. simplex*) UNBRANCHED BUR-REED. YU.AL.BG. **(1826).** *S. angustifolium* Michx (*S. affine*) FLOATING BUR-REED. YU.BG. **(1826).** *S. minimum* Wallr. SMALL BUR-REED. YU.BG.

TYPHACEAE | Reedmace Family

TYPHA | **Reedmace** Plants of still, slow-flowing waters or swamps. 6 species in our area.

Leaf sheath usually tapering gradually into leaf blade (not jointed)
(a) Female flowers without scales
1827. *T. latifolia* L. GREAT REEDMACE. Leaves 0·8–2 cm wide, flat; plant to 2 m. Lower female and upper male part of spike almost equal in length and usually contiguous, female part 8–15 cm, brown when mature. Widespread on mainland. **1827a. *T. shuttleworthii* Koch & Sonder Distinguished from 1827 by the female part of the inflorescence being silvery-grey when mature and distinctly longer than the upper male part. Leaves usually 5–10 mm wide; stem robust, to 1·5 m. In more acid waters than the previous species. YU.AL.BG.

(b) Female flowers with scales
1828a. *T. domingensis* (Pers.) Steudel (*T. angustata*) Often confused with 1828 but differing in the presence of scales to the female flowers which are light brown, translucent; scales of male flowers cut into narrow teeth. Leaves 5–12 mm wide; stems to 3 m. Widespread.

Leaf sheath jointed at junction with leaf blade
1828. *T. angustifolia* L. LESSER REEDMACE. Male and female parts of inflorescence separate by 3–8 cm, on stems to 2 m, shorter than the leaves. Female part cylindrical, 8–20 cm, dark reddish-brown, becoming mottled with age; scales of female flowers dark brown, opaque; scales of male flowers entire or forked. Leaves 3–6 mm wide. YU.AL?GR?BG.TR. **1828b.** *T. laxmannii* Lepechin (*T. stenophylla*) Distinguished by the male part of the inflorescence being 2–4 times as long as the female and separated by 1–6 cm. Female part oblong to ovoid, 4–9 cm, pale brown, female flowers without scales. Leaves usually 2–4 mm wide; plant 80–120 cm. AL?GR.BG.

CYPERACEAE | Sedge Family

A difficult family requiring special study; the majority of species in our region are widespread in Europe. The genera and the approximate numbers of species found in our area are: *Cyperus* 3 spp.; *Fuirena* 1 sp.; *Fimbristylis* 2 spp.; *Eriophorum* 5 spp.; *Scirpus* 15 spp.; *Blysmus* 1 sp.; *Eleocharis* 9 spp.; *Schoenus* 2 spp.; *Rhynchospora* 1 sp.; *Cladium* 1 sp.; *Carex c.* 90 spp.; *Kobresia* 1 sp.

MUSACEAE | Banana Family

MUSA *M. cavendishii* Paxton BANANA is cultivated in the warmest parts of the Med. region particularly in Crete.

ORCHIDACEAE | Orchid Family

CYPRIPEDIUM **1872.** *C. calceolus* L. LADY'S SLIPPER ORCHID. Bushy places. YU.BG.

OPHRYS An interesting but difficult genus with a considerable amount of variation in the colour and markings of the flowers; most species hybridize to add to the confusion. The 1–2 stamens are fused to the stigma to form a *column*, bearing detachable pollen masses, or *pollinia*; the tip of the column may be blunt and rounded, or acute, or long-pointed (beaked). The perianth segments comprise an outer whorl of 3 (sepals), and an inner whorl of 2 smaller segments (petals) and a larger lip. The majority of species are found in grassy, rocky places largely in the Med. region.

1. Column blunt at the tip.
   2. Inner 2 perianth segments white,
      yellowish, or green.
      3. Lip with a flat yellow margin.   *lutea*
      3. Lip with a reflexed brown velvety margin.   *fusca*
   2. Inner 2 perianth segments purplish, at
      least at base.
      4. Outer 3 perianth segments green or
         yellowish.
         5. Inner 2 perianth segments linear
            (YU.).   **1877.** *O. insectifera* L.
         5. Inner 2 perianth segments ovate to   *speculum,*
            lanceolate.   *bombyliflora*
      4. Outer 3 perianth segments pink to
         purplish; lip usually entire.   *tenthredinifera*
1. Column with an acute or long-pointed tip
   (beaked).
   6. Lateral lobes of lip each with a basal
      swelling or hump.
      7. Outer 3 perianth segments green.
         8. Middle lobe of lip entire.   *sphegodes, fuciflora*
         8. Middle lobe of lip three-lobed.   *sphegodes, spruneri,*
                                    cretica, carmelii
      7. Outer 3 perianth segments pink or
         purplish; lip usually with apical
         appendage.
         9. Middle lobe of lip entire or
            nearly so.   *fuciflora*
         9. Middle lobe of lip three-lobed
            (YU.AL.GR+Is.CR.TR.).   **1878.** *O. apifera* Hudson
                                    cretica, scolopax
        10. Lip entire.
           11. Outer 3 perianth segments
              green.   *sphegodes, ferrum-equinum*
           11. Outer 3 perianth segments
              pinkish or purplish
              (YU(west).AL.)   **1881.** *O. bertolonii* Moretti
                                    ferrum-equinum, argolica,
                                      fuciflora
        10. Lip three-lobed.   *ferrum-equinum, reinholdii,*
                                      argolica, spruneri, sphegodes
   6. Lateral lobes of lip without a
      basal swelling or hump.

**1873. *O. fusca* Link BROWN BEE ORCHID. Flowers with greenish-yellow perianth segments (rarely pink), and a long three-lobed, dark-chocolate velvety-haired lip with a narrow yellow margin and bluish or slaty paired basal reflective patches. Flowers 1–9; outer perianth segments ovate, inner linear, at least half as long. Widespread in the Med. region. YU(west).AL.GR+Is.CR.TR. Subsp. *iricolor* (Desf.) O. Schwarz differs in having fewer flowers with a larger lip, 2·5 by 2·1 cm, without a yellow margin and with paired shining-blue reflective patches. Med. region. Subsp. *omegaifera* (Fleischm.) E. Nelson OMEGA ORCHID. Readily distinguished by the white or yellowish wavy band across the lip which is strongly bent or down-curved. Inner 2 perianth segments brownish. GR.CR+Rhodes.

1874. *O. lutea* (Gouan) Cav. YELLOW BEE ORCHID. Distinguished by its broad rounded, shallowly three-lobed lip 12–18 mm, with a broad yellow or greenish margin, and a dark-brown, raised hairy central boss with bluish reflective patches. Outer 3 perianth segments green, blunt, the inner 2 green or yellowish. Med. region. YU(west). GR+Is.CR.TR. Subsp. *melena* Renz has a smaller blackish-purple lip 9–12 mm, with blackish-purple hairs on the margin. GR(south). **1875. *O. speculum* Link MIRROR ORCHID, MIRROR-OF-VENUS. Distinguished by its broad rounded, brown shaggy-haired lip which has a large brilliant metallic-blue reflective patch in the centre, and swollen lateral bumps at the base. Outer 3 perianth segments usually greenish or yellowish, the inner 2 dark purple, densely hairy. Med. region. GR(south)+Is.CR? TR. Pl. 63. Subsp. *regis-ferdinandii* Acht. & Kellerer differs markedly from 1875 in that the lip is slender, not rounded, and the lateral lobes slender and arm-like. Rhodes, Chios.

1876. *O. scolopax* Cav. WOODCOCK ORCHID. Flowers with a large cylindrical-ovoid, brown hairy lip, strongly marked with a variable pattern of white and yellow curves or lines encircling blue or violet reflective patches, and with a conspicuous yellowish knob-like apex, and humped hairy basal sidelobes. Perianth segments pink or purplish (rarely green or white), the inner hairy, much shorter than outer. Widespread. Subsp. *cornuta* (Steven) Camus differs in the lip having 2 forward-projecting basal horns to 10 mm, sometimes as long as the lip. Widespread. Subsp. *heldreichii* (Schlechter) E. Nelson has larger flowers with lip 13–15 mm (lip 8–12 mm in other subspecies) as long as or longer than wide and with a wide hairless margin and basal lobes to 5 mm. GR(south)+Is.CR. Pl. 62. ****1879. *O. bombyliflora* Link BUMBLE BEE ORCHID. Readily distinguished by its small rounded, velvety-brown lip 7–9 mm, which is shorter than the 3 oval, concave greenish outer perianth segments. Lip with short lateral lobes with hairy humps, and central lobe with a dull blue-violet central patch. Plant small 7–25 cm. Med. region. YU.AL? GR.CR.

1880. *O. sphegodes* Miller EARLY SPIDER ORCHID. A very variable species with several distinctive subspecies considered by some as distinct species. The colour and size of the perianth segments and the shape, size, and markings on the lip are critical. Subsp. *sphegodes* has the outer 3 perianth segments green, rarely flushed purple, the inner 2 about three-quarters as long, often wavy-margined. Lip 10–12 mm, usually ovoid, entire, velvety-brown with paler hairless bluish markings either of parallel lines, or in an H- or X-shaped pattern. Widespread. Subsp. *aesculapii* (Renz) Soó AESCULAPIUS' ORCHID has a round to quadrangular lip, wider than long, dark velvety-brown, with a yellow or brownish margin 4 mm wide, and an H-shaped hairless reflective patch, without basal humps. Perianth segments pale olive green. GR. Pl. 62. Subsp. *atrata* (Lindley) E. Mayer (*O. atrata*) has a blackish- or purplish-brown lip with 2 large basal humps densely covered with long brown hairs, and with an H-shaped or U-shaped hairless reflective patch; inner 2 perianth segments wide. Med. region.

Pl. 62. Subsp. *mammosa* (Desf.) E. Nelson has a rounded to oblong lip 13–17 mm, with 2 pronounced forward-projecting basal humps, and with an H-shaped hairless reflective patch. Outer 3 perianth segments often greenish to purplish, inner 2 narrow. GR+Is.CR.TR.

1880c. *O. spruneri* Nyman Flowers with a dark blackish or blackish-purple central lip 10–13 mm, and arm-like deflexed lateral lobes. Central lobe of lip rounded or kidney-shaped with a blue-violet pale margined reflective patch of 2 parallel lines, or H-shaped. Perianth segments usually green or greenish-purple. GR+Is.CR. **1880d.** *O. cretica* (Vierh.) E. Nelson CRETAN ORCHID Distinguished by its dark blackish-purple, pear-shaped three-lobed lip, conspicuously but variably marked with a white or bluish H-shaped, shield-like, or spotted pattern all with white margins; and with 2 hairy, spreading arm-like side lobes. Lip 11–14 mm. Perianth segments usually green to brownish. GR(south)+Cyclades.CR. Pl. 62. **1880g.** *O. carmelii* Fleischm. & Bornm. (*O. attica*) Lip oval, brown with a blue or brownish-violet shield-like pattern with a yellow margin, and a reflexed point, and brown-hairy, arm-like side lobes with basal swellings. Lip 6–10 mm. Perianth segments usually green, the middle segment incurved. GR+Rhodes.TR. Pl. 62.

1880f. *O. argolica* Fleischm. Lip brown, rounded to ovate with or without lateral lobes, and with a small apical flap. Markings on lip violet with white margins, variable, semicircular, H-shaped, horseshoe-shaped or a single line or spot. Perianth segments usually pinkish or purple, the inner velvety purple or lilac. GR(south)+Is.CR+Karpathos. **1880e.** *O. reinholdii* Fleischm. Differs from 1880f in having lip with 2 thick comma-shaped lines, or 2 separated or connected spots, white or pale violet. Lip three-lobed, 11–13 mm. Outer perianth segments lilac, whitish to green-purple, the inner green, brown, or lilac. GR+N.Aegean Is. Pl. 62. **1881a.** *O. ferrum-equinum* Desf. HORSESHOE ORCHID. Lip rounded or squarish 10–12 mm, entire or indistinctly three-lobed, velvety, deep purple or brown-purple, with a metallic-blue horseshoe-shaped reflective patch in the centre, sometimes reduced to 2 patches, with a knob-like apex and without humps at base. Perianth segments usually pink or purple, the inner hairless. AL? GR+Is. Pl. 62. Subsp. *gottfriediana* (Renz) E. Nelson differs in having the outer perianth segments greenish, greenish-purple, or whitish, and the lip usually three-lobed. GR(Ionian Is.)

1882. *O. tenthredinifera* Willd. SAWFLY ORCHID. In general, distinguished by its bright-pink or purplish perianth segments, and its broad squarish or obovate lip with a wide conspicuous, yellowish hairy margin and dark-brown velvety centre. Base of lip with a reddish-brown patch, and narrow collar-like blue reflective patch, basal humps often present, apex of lip often with a deeply notched greenish appendage. GR+Is.CR. TR. **1883. *O. fuciflora* (F. W. Schmidt). Moench (*O. arachnites*) LATE SPIDER ORCHID. Perianth segments white or pink, the 2 inner very short. Lip ovate 9–13 mm, dark velvety-brown, rather broad, usually unlobed, with a bold variable pattern of yellowish lines or curves surrounding the violet-blue reflective patch. Lip with basal humps present, and with a yellowish, often three-lobed upcurved apical appendage. YU.AL.GR+Is.CR.TR. Pl. 62.

ORCHIS | **Orchid** Distinguished from *Ophrys* by the presence of a spur, and 5 more or less equal perianth segments. These 5 segments either come together in a helmet above the lip, or the 2 outer segments are spreading or deflexed. Tubers 2–3, rounded, not lobed.

1. Spur ⅓–½ as long as ovary, sac-like 1–3 mm
   (YU.AL.GR.BG.).   **1888. *O. ustulata* L.
1. Spur usually at least half as long as
   the ovary, linear to conical.
   2. All 5 perianth segments coming
      together in a helmet.
      3. Lip not lobed, but fan-like and toothed.   *papilionacea*
      3. Lip three- to four-lobed.
         4. Spur horizontal or somewhat up-
            curved; lip about as wide as long, or
            somewhat wider.   *morio, boryi*
         4. Spur directed downwards; lip longer
            than wide.
            5. Middle lobe of lip entire, or finely
               toothed, without purple spots.   *coriophora, sancta*
            5. Middle lobe of lip deeply notched, or
               bilobed, usually with purple spots.
               6. Bracts somewhat shorter, or
                  about as long as the ovary; notch
                  in middle lobe of lip, without
                  additional tooth.   *tridentata, lactea*
               6. Bracts much shorter than ovary;
                  notch in middle lobe of lip with
                  a tooth
                  7. Middle lobe of lip with linear
                     lobes.   *italica, simia*
                  7. Middle lobe of lip with ovate or
                     oblong lobes (YU.BG.TR.).   **1891. *O. militaris* L.
                                                        purpurea, punctulata
   2. 3 middle perianth segments sometimes
      coming together in a helmet, at least
      the 2 outer segments spreading or
      deflexed.
      8. Flowers pale yellow.   *pallens, provincialis*
      8. Flowers not yellow, usually reddish-
         purple.
         9. Spur slender, thread-like, or narrow-   *anatolica, quadripunctata*
            conical narrowed to apex.   *boryi, laxiflora*
         9. Spur cylindrical or sac-like.
            10. Spur directed downwards.   *saccata, spitzelii*
            10. Spur horizontal or directed up-
                wards (Widespread on mainland).   1894. *O. mascula* L.

5 perianth segments more or less coming together in a helmet
**1884. *O. papilionacea* L. PINK BUTTERFLY ORCHID. Perhaps the most beautiful of our
orchids, with rosy-pink flowers strongly veined with a deeper pink, and with a broad
fan-shaped lip with a toothed margin. Perianth segments also conspicuously veined
pink. Dry grassy places. Widespread. **1885.** *O. morio* L. GREEN-WINGED ORCHID. Grassy
meadows, woodlands. Widespread on mainland and some islands. **1885c.** *O. boryi*
Reichenb. fil. Distinguished from 1885 by the topmost flower of the spike opening first
(the lowest flower opens first in 1885) and by its slender, thread-like spur. Flowers

violet, with a rounded, shallowly three-lobed lip, in a short dense spike; outer perianth segments often more or less spreading. Bushy places. GR(south).CR. Pl. 63.

1886. *O. coriophora* L. BUG ORCHID. Helmet violet-brown, hairy; lip purplish-green, three-lobed; spur conical down-curved, half as long as lip. Flowers fetid. Damp meadows. Widespread. Subsp. *fragrans (Pollini)* Sudre is usually sweetly vanilla-scented. Central lobe of lip much longer than lateral lobes, and spur as long as lip. Widespread. Pl. 64. **1886b. *O. sancta* L. HOLY ORCHID. Distinguished from 1886 by the much larger pale-lilac flowers to 2 cm, the lip with a longer central lobe and lateral lobes with teeth. Spur hook-like, incurved, narrowed to the apex; perianth segments 9–12 mm, long-pointed. Stems only with sheaths, leaves all in a rosette. Grassy, sandy places. Aegean Is.CR.

1889. *O. tridentata* Scop. TOOTHED ORCHID. Distinguished by its usually dense head of conspicuously spotted pale pinkish-violet flowers, but spike may be lax. Lip four-lobed, the large wedge-shaped central lobe being bilobed often with a central tooth between, and with 2 arm-like lateral lobes. Perianth segments in a helmet, usually darker veined; spur stout, directed downward. Bracts of inflorescence small papery. Plant 15–45 cm. Grassy places, thickets, woods. Widespread. **1889a. *O. lactea* Poiret Like 1889 but flowers pale greenish-pink or white, and outer perianth segments more pointed. Lip with a central, broadly fan-shaped lobe, which is usually not notched; spur sometimes longer than ovary. Inflorescence cylindrical, rather lax; plant 7–20 cm. Grassy and sandy places. YU.GR(south).CR.BG.TR. Pl. 63.

1890. *O. simia* Lam. MONKEY ORCHID. Distinguished by its pink, spotted, monkey-like five-lobed lip comprising a slender 'body', curved 'arms' and 'legs', and a central 'tail'. Flowers in a dense ovoid cluster; perianth segments in a helmet, pale greyish-pink, faintly veined. Leaves flat, shining. Grassy places, woods. Widespread on mainland +CR. **(1890). *O. italica* Poiret (*O. longicruris*) Like 1890 but distinguished by its perianth segments, which are pink and conspicuously striped with darker veins. Lip like that of 1890 but with 2 triangular plates at its base and flat, acute lobes. Leaves with undulate margins. Grassy and stony places, pine woods. YU.AL?GR+Is.CR.

1892. *O. purpurea* Hudson LADY ORCHID. Distinguished by its very dark brownish-purple often spotted helmet, and its whitish four-lobed lip covered with raised purple dots and swellings. Lip wider than long with oblong side lobes, and a central tooth between the two broad middle lobes. Bracts usually much shorter than ovary. Banks, thickets. YU.AL.GR.BG.TR. **1892a. *O. punctulata* Lindley subsp. *sepulchralis* (Boiss. & Heldr.) Soó Helmet of pink outer and narrow yellowish-green inner perianth segments; lip yellowish-green often with purple spots, the middle lobe narrowed at the base and abruptly broadened at apex into 2 oblong lobes with a tooth between. Spur broadly cylindrical, directed downwards. Bushy places. GR.BG?TR. Pl. 63.

2 lateral perianth segments spreading or deflexed, 3 inner sometimes in a helmet, or spreading (see also 1885c)
1893. *O. laxiflora* Lam. JERSEY ORCHID. Flowers large, unspotted, dark pinkish-purple with pale centres, borne in a long lax cluster, the whole inflorescence purple-flushed. Lip with 2 rounded lateral lobes which become conspicuously turned downward and almost touching, middle lobe entire or shallowly two-lobed. Perianth segments blunt, 3 erect and 2 lateral deflexed. Spur horizontal or upward-pointing, slender, swollen and notched at apex. Marshy meadows and ditches. Widespread. Subsp. *elegans* (Heuffel) Soó has flowers with a larger almost entire lip, and broader leaves 1·5–2·5 cm wide. YU.BG. Pl. 63. **1894b. *O. saccata* Ten. Distinguished by its undivided fan-shaped lip,

which is red to dark violet, and unspotted or unstreaked. Perianth segments greenish-brown to reddish-brown, the lateral spreading segments often green. Spur stout sac-like, whitish, about half as long as ovary. Bracts purple-violet, somewhat longer than ovary. Leaves usually spotted, in a rosette. Grassy and rocky places. AL.GR(south).CR. Rhodes.TR. Pl. 63.

1894c. *O. spitzelii* Koch Like 1894 *O. mascula* but lip deeply three-lobed, spotted, base of lip with 2 conspicuous ridges. Flowers purple, in a dense ovoid or cylindrical spike. Spur conical-cylindrical, longer than lip. Leaves basal. Meadows in mountains. YU.AL.GR.CR.BG. Pl. 63.   ****1895.** *O. quadripunctata* Ten. FOUR-SPOTTED ORCHID. A slender plant 10–30 cm, with a narrow cluster of pale-pink or purple-violet flowers with whitish centres and 2–4 conspicuous purple spots at the base of the three-lobed lip. Spur slender thread-like, downward-pointing. Bracts shorter than ovary; leaves usually spotted, in a rosette. Grassy banks, hillsides in the Med. region. YU(south). AL.GR+Is.CR.   **1895a.** *O. anatolica* Boiss. A slender plant like 1895 but with larger pink or pale-purple flowers with pale centres and with a three-lobed lip with many purple spots or blotches. Spur long slender, widened to the base, usually upward-pointing. Leaves densely spotted, in a rosette. Thickets, pine woods. S.Aegean Is.CR. Pl. 63.

1896. *O. provincialis* Balbis PROVENCE ORCHID (incl. *O. pauciflora*). Flowers pale yellow, the lip often with a darker centre and with brownish spots, borne in a lax cylindrical spike. Lip three-lobed with 2 rounded side lobes, and a small square-cut often notched middle lobe, with a smaller lobe in the notch. Spur horizontal or upward-pointing, as long as or longer than the ovary. Leaves lanceolate, usually spotted, crowded at base of stem. Bushy and grassy places, woods, mountains. Widespread, except TR. Pl. 63.   **(1896).** *O. pallens* L. PALE-FLOWERED ORCHID. A mountain plant, like 1896 but pale-yellow flowers sweet-scented (recalling elder flowers), in a dense spike, rarely purplish. Lip rounded, three-lobed, unspotted. Spur cylindrical, horizontal, slightly shorter than ovary. Leaves oblong to oblong-ovate, unspotted, spreading half way up stem. Woods, pastures in mountains. YU.AL.GR.BG. Pl. 63.

DACTYLORHIZA   Like *Orchis* but tubers lobed; perianth segments usually not coming together in a helmet. A difficult genus with about 10 species in our area.

Flowers yellow or yellow flushed purple, less commonly purple
1899. *D. sambucina* (L.) Soó ELDER-FLOWERED ORCHID. A mountain plant with pale-yellow flowers, or flowers less commonly purple with a yellowish patch at the base of the lip. Lip shallowly three-lobed, indistinctly spotted; spur 8–15 mm, very stout conical, blunt, horizontal or downward-pointing. Stem with 4–5 separated, unspotted obovate to narrowly oblanceolate leaves in the lower part. Meadows, usually in mountains. YU.AL.GR.BG.   **1899a. *D. sulphurea* (Link) Franco. Differs from 1899 in its narrow linear-oblong leaves which form a basal rosette of up to 10 leaves. Flowers yellow or purple, middle lobe of lip rounded or almost square; spur 12–25 mm, horizontal. Thickets, woods, on acid soils. Widespread.

Flowers purple, pink, red, rarely white
1897. *D. incarnata* (L.) Soó MARSH ORCHID. Flowers numerous in a dense cylindrical cluster, pinkish to purple, or rarely white, with spotted lip and usually unspotted erect leaves which are narrow and gradually widened to the base. Lip 5–8 mm, rounded to three-lobed; outer perianth segments 5–6 mm, the lateral spreading, the inner convergent; spur downward-pointing, half as long as ovary. Stem hollow, to 70 cm, with 4–5 leaves. Wet meadows, marshes. YU.AL.GR?BG.   **1898b.** *D. cordigera* (Fries) Soó

HEART-FLOWERED ORCHID. Recognizable by its usually dense oval or cylindrical cluster of purple or violet flowers each with a large spreading, dark-purple heart-shaped, rarely three-lobed lip. Spur 6–8 mm, stout, downward-pointing, shorter than ovary. Leaves broad, usually dark-spotted; bracts much longer than ovary. Damp meadows in mountains. YU.AL.GR.BG. Pl. 64.

1899b. *D. iberica* (Bieb.) Soó Distinguished by its acute perianth segments all forming a helmet, the outer segments three-veined. Flowers pink; lip purple-spotted, three-lobed, the median lobe narrowest; spur curved, to almost half as long as ovary, white towards base. Leaves long and narrow, unspotted; stem with stolons. Damp meadows in mountains. GR.TR? Pl. 64. **1900a.** *D. saccifera* (Brough). Soó Like (1900) *D. fuchsii* SPOTTED ORCHID but distinguished by its conspicuous bracts which are longer than the flowers (sometimes up to twice as long), and its stout spur 7–13 mm, which is sac-like. Flowers pink or lilac, rarely white; perianth segments to 10 mm, the 2 outer spreading, the inner coming together in a helmet; lip deeply three-lobed, lobes more or less equal. Leaves not spotted; plant 30–80 cm. Damp meadows, in clearings, mountains. YU.AL.GR.BG.TR? Pl. 64.

TRAUNSTEINERA 1887. *T. globosa* (L.) Reichenb. (*Orchis g.*) ROUND-HEADED ORCHID. Mountain pastures. YU.AL?BG.

NIGRITELLA ****1901.** *N. nigra* (L.) Reichenb. fil. BLACK VANILLA ORCHID. Alpine meadows. YU.AL.GR.BG.

SERAPIS | Tongue Orchid The lip has 2 indistinct rounded lateral lobes arising from towards the base (often hidden by the helmet) and a conspicuous tongue-like, undivided middle lobe. Perianth segments coming together in a helmet; spur absent.

Middle lobe of lip heart-shaped, as wide as basal part of lip (see also 1904)
1903. *S. cordigera* L. HEART-FLOWERED SERAPIS. Lip black-purple, hairy, twice as long as the reddish-violet or wine-coloured helmet. Flower spike dense ovoid, with 2–10 flowers; bracts pale pinkish, little shorter than flowers; stem 15–50 cm with purplish, usually spotted sheaths. Pine woods, damp sandy places. YU.AL.GR.TR. Pl. 64. ****(1903).** *S. neglecta* De Not. Distinguished from 1903 by its green stem with unspotted basal sheaths. Lip yellow to pinkish-orange, about three times as long as the lilac perianth segments; bracts as long as or longer than flowers. Coastal grasslands, sandy soils in the Med. region. YU(south-west).GR(Ionian Is.).

Middle lobe of lip triangular-lanceolate, narrowed towards base, narrower than basal part of lip
(a) Lip with one black hump at its base
1902. *S. lingua* L. TONGUE ORCHID. Flowers with a long reddish to violet (rarely yellow or white), slightly hairy lip about twice as long as the similar-coloured helmet. The single swelling at the base of the lip (only seen when the lip is removed) distinguishes it from all other species. Bracts usually purple, as long as the flowers. Grassy or bushy places, olive groves in the Med. region. YU.AL.GR+Is.CR.

(b) Lip with 2 humps at its base
**1904. *S. vomeracea* (Burm.) Briq. LONG-LIPPED SERAPIS. Often a tall plant 20–55 cm, with narrow channelled, pointed leaves, and with a spike of pale-red flowers with pale-red darker-veined bracts longer than the flowers. Helmet pale red with darker veins; lip 1½ times as long as helmet, brick-red to reddish-brown, hairy, the lateral lobes reddish and tipped with black, partly hidden by the helmet. Damp or dry grassy

places. Widespread. **(1904)**. *S. parviflora* Parl. SMALL-FLOWERED SERAPIAS. Distinguished by its smaller flowers 1·5–2 cm (flowers 1·5–3 cm in 1904), with a rusty-red lip little longer than the helmet. Outer perianth segments lilac, the inner greenish or reddish. Lip with 2 dark parallel humps at base. Bracts red to greenish, usually purple-veined, about as long as the flowers. Grasslands, sandy places by the sea in the Med. region. YU.AL.GR+Is.CR.TR.

ACERAS ****1905.** *A. anthropophorum* (L.) Aiton fil. MAN ORCHID. Woods, grassy and bushy places. YU.GR+Is.CR.

NEOTINEA **1905a.** *N. maculata* (Desf.) Stearn (*N. intacta*) DENSE-FLOWERED ORCHID. A slender plant 10–25 cm, with a spike of numerous small, densely clustered pinkish flowers smelling of vanilla. Lip three-lobed, the lateral lobes narrow, the mid lobe longer, oblong, often 2- or 3-lobed at apex, perianth segments 3–4 mm in a helmet, pink or white; spur short, stout. Bracts shorter than ovary. Thickets, grassy places, sometimes among scrub or trees. YU.GR+Is.CR.

BARLIA ****1906.** *B. robertiana* (Loisel.) W. Greuter (*Himantoglossum longibracteatum*) A very robust orchid 30–60 cm, with greenish-yellowish flowers more or less flushed with pink or dull purple, in a dense cylindrical spike. Lip large, deeply four-lobed, often wavy-margined. Spur short, stout. Thickets, grassy banks. YU.GR+Is.CR.

HIMANTOGLOSSUM ****1907.** *H. hircinum* (L.) Sprengel LIZARD ORCHID. A very distinctive, robust, strong-smelling orchid to 1 m, with a lax cylindrical cluster of greenish-yellow flowers with very long, strap-shaped, irregular twisted or coiled lips. Lip brown-spotted, with lateral lobes 5–10 mm; helmet greenish; spur conical, about 4 mm. Subsp. *calcaratum* (G. Beck) Soó is the typical plant of the Balkans with larger flowers more flushed with purple. Lateral lobes of lip 12–20 mm. Spur cylindrical 7–12 mm. Woods, thickets, grassy banks. YU.GR.CR.BG.TR. Pl. 64.

ANACAMPTIS ****1908.** *A. pyramidalis* (L.) L. C. M. Richard PYRAMIDAL ORCHID. Woods, thickets, grassy places, meadows. Widespread.

HERMINIUM ****1909.** *H. monorchis* (L.) R.Br. MUSK ORCHID. Dry and damp meadows, marshes. YU.BG.

COELOGLOSSUM **1911.** *C. viride* (L.) Hartman FROG ORCHID. Grassy places, mountain meadows. YU.AL.BG.

GYMNADENIA ****1912.** *G. conopsea* (L.) R.Br. FRAGRANT ORCHID. Spur 11–18 mm, nearly twice as long as ovary; lip much broader than long. Meadows. YU.AL.GR.BG. **(1912).** *G. odoratissima* (L.) L. C. M. Richard SHORT-SPURRED FRAGRANT ORCHID. Like 1912 but spur 4–5 mm, thicker and shorter than ovary; lip little broader than long. Alpine meadows and rocks. YU(central).

PSEUDORCHIS ****1913.** *P. albida* (L.) Á. & D. Löve (*Leucorchis a.*) SMALL WHITE ORCHID. Flowers numerous, greenish-white, very small, half-drooping, in a dense narrow cylindrical, one-sided spike 3–6 cm long. Perianth segments in a helmet, the three-lobed lip scarcely longer; spur short thick, less than half as long as ovary. Bracts three-veined. Leaves 4–6; stem 12–30 cm. Mountain pastures. YU.BG. **1913a.** *P. frivaldi* (Griseb.) P. F. Hunt Like 1913 but flowers white, rarely pink, in a shorter more globular or pyramidal cluster. Lip shallowly three-lobed, spur thread-like. Bracts one-veined. Leaves 3–4, narrower. Damp meadows in mountains. YU(south).AL? GR(north).BG. Pl. 64.

PLATANTHERA Lip undivided, strap-shaped; spur usually long and slender.

1914. *P. bifolia* (L.) L. C. M. Richard LESSER BUTTERFLY ORCHID. Flowers whitish, 11–18 mm across, with a slender lip and a long curved greenish acute spur 1·5–2 cm. Distinguished by its 2 yellow anthers (placed under the helmet) lying parallel to each other. Woodlands, grassy places, in mountains. Widespread on mainland. ****1915. *P. chlorantha* (Custer) Reichenb. Like 1914 but flowers larger 18–23 mm across; spur swollen at apex, 2–3 cm; the 2 yellow anthers conspicuously diverging below. Woodlands, grassy places, in mountains. Widespread on mainland.

EPIPACTIS | **Helleborine** Lip divided and jointed into a hard basal cup-shaped *hypochile*, and an apical triangular-heart-shaped *epichile*. The *rostellum* is a beak-like process formed from the third stigma. A difficult genus.

Marsh plants with creeping rhizomes; lip jointed and movable
**1916. *E. palustris* (L.) Crantz MARSH HELLEBORINE. Distinguished by its brownish or greenish-purple flowers borne in a lax, hairy-stemmed spike to 50 cm. Outer perianth segments greenish or brownish, the inner whitish with purple veins. Lip with a basal pinkish-white, yellow-spotted, and purple-lined hypochile, and a jointed movable epichile which is white with red veins and is undulate-margined. Marshes, damp meadows in mountains. Widespread on mainland.

Plants usually of woods and bushy places, with short rhizomes; lip jointed but not movable
(a) Flower stem, flower stalks, and ovary hairless or rough to the touch
1917. *E. helleborine* (L.) Crantz BROAD HELLEBORINE. Flowers greenish or dull purple, drooping, numerous in a long one-sided spike 10–40 cm, with the lowest leafy bracts equalling the flowers. Flowers about 1 cm across; lip shorter than perianth segments; hypochile cup-shaped, dark reddish-brown inside, epichile broader than long, apex recurved, purplish-rose or greenish-white with 2–3 more or less smooth bosses. Woods, shady places in mountains. Widespread on mainland. **1917d.** *E. purpurata* Sm. Distinguished by its greyish or purplish leaves, arranged spirally, and its many-flowered spike 15–25 cm long, of greenish flowers flushed with purple. Perianth segments 10–12 mm, green outside, whitish within; lip 8–10 mm, hypochile greenish outside, mottled violet within, epichile as long as or longer than broad, whitish, with 2–3 smooth basal bosses. Woods. YU.GR?BG.

(b) Flower stem, flower stalks, and ovary densely hairy
1918. *E. atrorubens* (Hoffm.) Besser DARK RED HELLEBORINE. Unmistakable with its dull reddish-purple, vanilla-scented flowers. Perianth segments 6–7 mm, deep purple; lip shorter, hypochile green with red margin, epichile deep reddish-violet, with a small recurved tip and with 2 rough basal bosses. Leaves two-ranked, 4–10 cm. Woods, bushy and rocky places. Widespread on mainland. **1918a.** *E. microphylla* (Ehrh.) Swartz Distinguished from 1918 by its smaller leaves 1–2·5 cm, arranged spirally. Flowers greenish, tinged reddish; perianth segments whitish-green, violet-flushed; lip with a greenish-brown hypochile, and white often red-flushed epichile with 2 ribbed and crisped bosses. Woods and bushy places. YU.AL.GR.CR. BG.

519

CEPHALANTHERA

Flowers white or pinkish
(a) Spur absent
1919. *C. damasonium* (Miller) Druce WHITE HELLEBORINE. Leaves oval-lanceolate; outer perianth segments blunt. Woods, shady and bushy places. Widespread on mainland. ****(1919).** *C. longifolia* (L.) Fritsch LONG-LEAVED HELLEBORINE. Leaves lanceolate, the upper linear; outer perianth segments acute. Woods, shady places. Widespread on mainland.

(b) Spur present, short
1919a. *C. epipactoides* Fischer & C. A. Meyer Like (1919) but lip with a short spur. Flowers white; outer perianth segments 25–36 mm; bracts longer than the ovary. Leaves short; plant tall 30–80 cm. Bushy places. GR(north-east)+E.Aegean Is.TR. **1919b.** *E. cucullata* Boiss. & Heldr. Flowers spurred like 1919a but spur 1–2 mm and flowers pinkish; outer perianth segments 14–20 mm. Plant more slender 10–30 cm. Woods, in mountains. CR. Pl. 64.

Flowers red
**1920. *C. rubra* (L.) L. C. M. Richard RED HELLEBORINE. Readily distinguished by its bright-red or reddish-violet flowers, with perianth segments in a helmet. Lip shorter than helmet, erect, epichile with a red-violet margin, and crested yellowish ridges. Lower leaves oblong, the upper lanceolate; stem 20–60 cm. Woods and bushy places. YU.AL.GR.CR.BG.TR.

LIMODORUM ****1921.** *L. abortivum* (L.) Swartz LIMODORE. Unlike any other orchid in having a long slender spike of stalkless violet flowers each about 4 cm across, and violet-flushed scale leaves on the stem below (green leaves absent). Lip shorter than perianth segments, violet with some yellow; spur stout. Woods, bushy places. Widespread, except TR.

SPIRANTHES ****1922.** *S. spiralis* (L.) Chevall. AUTUMN LADY'S TRESSES. Flowers white; autumn-flowering. Flowering stems with scales only; leaves about 2·5 cm, all basal. Woods, stony, grassy places, heaths. Widely scattered. **(1922).** *S. aestivalis* (Poiret) L. C. M. Richard SUMMER LADY'S TRESSES. Flowers white; summer-flowering. Stem leafy; leaves 5–15 cm. Damp meadows, particularly in mountains. YU.GR?

EPIPOGIUM Leafless saprophytes with branched underground rhizomes. Flowers inverted with lip and spur uppermost.

1922a. *E. aphyllum* Swartz SPURRED CORAL-ROOT. A strange, rare, quite leafless orchid with pale-yellow stems and a few large pinkish flowers, growing in damp shady woods. Spur pointing upwards. Woods in mountains. YU.GR(north-west).BG. Pl. 64.

LISTERA **1923.** *L. ovata* (L.) R.Br. TWAYBLADE. Flowers greenish-yellow; plant large 20–60 cm; leaves 5–20 cm. Woods, pastures, damp meadows, in mountains. Widespread, except TR. ****(1923).** *L. cordata* (L.) R.Br. LESSER TWAYBLADE. Flowers reddish-green; plant small 6–20 cm; leaves 1–2·5 cm. Bogs and damp meadows, in mountains. YU.BG.

NEOTTIA ****1924.** *N. nidus-avis* (L.) L. C. M. Richard. BIRD'S-NEST ORCHID. Shady woods, particularly beech woods. Widespread on mainland.

GOODYERA **1925. *G. repens* (L.) R.Br. CREEPING LADY'S TRESSES. Mossy coniferous woods, in mountains. YU.BG.

CORALLORHIZA **1926. *C. trifida* Chatel. CORAL-ROOT ORCHID. Woods in mountains. YU.AL.GR.BG.

List of Indexes

GENUS	ENGLISH	YUGOSLAV	BULGARIAN	GREEK
Alkanna	Alkanna	Vučji jezik, Krvavica	Ajvaživa	Vaforriza
Alliaria	Garlic Mustard	Češnjača, Lučac	Lážičina	Agrióskordo, Agrioprasso
Allium		Luk	Luk	Kremmídi
cepa	Onion	Crni luk	Kromid	
porrum	Leek	Praziluk, Porluk		
sativum	Garlic	Bijeli luk, Češnjak	Česán	Skórdo
schoenoprasum	Chives	Vlašac		
ursinum	Ramsons	Srijemuš	Levurda	
Alnus	Alder	Joha	Elša	Sklithra, Sklithro
Althaea		Sljez	Ruža	Dendromolócha
officinalis	Marsh Mallow	Pitomi sljez, Bijeli sljez		Neromolócha
Amelanchier	Snowy Mespilus	Divlja mušmula, Merala	Jrga, Diva djula	Ipsos
Anagallis	Scarlet Pimpernel	Vidovčica	Ognivče	
Anagyris	Bean Trefoil	Smrdež, Smrduša		Anágyros, Vromoklári
Anchusa	Alkanet	Volovski jezik, Pačje gnezdo	Vinče, Pače gnezdo	Voidóglossa
Anemone	Anemone	Breberina	Sásánka	Anemóna
nemorosa	Wood Anemone	B. bijela, šumarica	Bjala sásánka	
ranunculoides		B. žuta, Žuta šumarica	Ljutikova sásánka	
Anethum	Dill	Kopar, Mirodija	Kopár Merudia	Ánitho
Anthemis	Chamomile	Žablja trava, Prstenak	Podrumiče	Ágria margarita
Anthyllis		Ranjenika	Bjala ráda, Ramenka	
hermanniae				
vulneraria	Kidney-vetch	Bjelodun, Ranjenica	Obiknovena Ramenka	
Antirrhinum	Snapdragon	Zijevalica, Žabica	Zejka	Alogothímaro
Aquilegia	Columbine	Kandilka, Pakujac	Kandilka	

Arbutus	Strawberry Tree	Magrinja, Ranka		Koumariá
andrachne				Agriokoumariá, Andrákla
Arctium	Burdock	Čičak, Repuh	Repej	
Arctostaphylos	Bearberry	Mečije grozdje	Mečo grozde	
Arisarum	Friar's Cowl	Mali kozlac, Kozlić		Lychnaráki
Aristolochia	Birthwort	Vučja jabuka, Vučja stopa	Válča jabálka	Bekroladóchero
Armeria	Thrift	Babina svila	Lážičniče	
Armoracia	Horse-radish	Hren, Ren	Hren	
Arnica	Arnica	Brdjanka, Moravka		
Artemisia		Pelin	Pelin	
abrotanum	Southernwood	Božje drvce		
absinthium	Wormwood	Pelin, Osjenač	Obknoven Pelin	Apsithiá
dracunculus	Tarragon	Zmijavičica, Zmijina trava	Taros	
vulgaris	Mugwort	Komonika	Div pelin	
Arum	Lords-and-Ladies	Kozlac	Zmijarnik, Zmiiski lapad	Drakontiá, Fidóchorto
Arundo	Giant Reed	Trs, Rozga	Ispanska trástika	Kalámi
Asarum	Asarabacca	Kopitnjak	Kopitnik	
Asparagus	Asparagus	Špargla	Zajča sjanka, Sparža	Spárángi
Asperugo	Madwort	Broćanica, Prilep	Ostrec	Kollitsída
Asphodelus	Asphodel	Čepljez, Čapljan	Bárdun	Asphodíli
Aster		Zvijezdan	Zvezdel, Dimitrovče	
tripolium	Sea Aster			
Astragalus	Milk-vetch	Kozinac	Klin, Zgrabiče	Tetrágatho (spiny spp.)
Atriplex	Orache	Loboda	Loboda	Alimia

Caltha	Kingcup, Marsh Marigold	Kaljužnica, Kopitac	Blatnjak	
Calystegia	Large Bindweed	Slatkovina, Slak	Čadárče	
Campanula	Bellflower, Campanula	Zvonce, Zvončić	Kambanka, Zvánče	Kampanoúla
Cannabis	Hemp	Konoplja	Konop	Kánnavis
Capparis	Caper	Kapara, Capra		Kápari
Capsella	Shepherd's Purse	Tarčužak, Rusomača	Ovčarska torbička	
Capsicum	Chili, Pepper	Paprika	Piper	Piperiá
Cardamine	Bitter-Cress	Režuha	Gorva	
Carduus	Thistle	Striček	Magareški bodil	Gaidourágatho
Carlina	Carline Thistle	Kravljak, Kraljevac	Rešetka	
Carpinus	Hornbeam	Grab	Gabár.	Gávros
Carthamus lanatus		Bodolist, Bodalj	Aspurt	Atraktíli
tinctorius	Safflower	Šafranika		
Castanea	Sweet Chestnut	Kesten pitomi, Maron	Kesten	Kastaniá
Celtis	Nettle Tree	Koprivić, Košćela	Koprivka	Melikoukkiá
Centaurea	Knapweed	Različak	Metličina	
Centaurium	Centaury	Kičica, Kitica	Červen kantarion	Thermovótano
Centranthus	Red Valerian	Divlji veslidjen	Kentrautus	
Ceratonia	Carob, Locust Tree	Rogač	Ceratonia	Charoupiá, Xilokeratiá
Cercis	Judas Tree	Judino drvo	Div rožkov	Koutsoupiá, Kótsikas
Cerinthe	Honeywort	Pepeljuša	Medenik	
Cheiranthus	Wallflower	Šeboj	Siboj	Agrioviolétta
Chelidonium	Greater Celandine	Rosopas	Zmijsko mljako	Chelidóni
Chenopodium	Goosefoot	Pepeljuga,	Sladka treva	Vromóchorto

GENUS	ENGLISH	YUGOSLAV	BULGARIAN	GREEK
Datura	Thorn Apple	Kužnjak, Tatula	Tatul	Stramónio, Tátulas
Delphinium	Delphinium	Kokotić, Žavornjak	Ralica	Psarovótano
staphisagria	Stavesacre	Vašljivka		Agriostafida
Dianthus	Pink	Karanfil	Karamfil	Garifaliá
Dictamnus	Burning Bush	Jasenak	Rosen	Rakovótano
Digitalis	Foxglove	Pustikara, Naprstak	Besen buren, Naprastnik	Chelidonóchorto, Korakovótano
Dipsacus	Teasel	Češljuga	Lugačka	Konizós, Kóniza
Dittrichia	Aromatic Inula	Bušina	Zvezdiče	
Dorycnium		Bjeloglavica		
Dracunculus	Dragon Arum	Zmijino zelje	Zmiiska hurka	Drakondiá, Fidóchorto
Drosera	Sundew	Rosika, Rosulja	Rosjanka	
Ecballium	Squirting Cucumber	Štrkavac	Cărkalo, Luda krastavica	Pikrangouriá
Echinops	Globe Thistle	Glavoč	Čeljadnik	Kephalágatho, Achinágatho
Echium	Viper's Bugloss	Lisičina	Usojniče, Lisiča opaška	Voidóglossa, Skylóglossa
Elaeagnus	Oleaster	Dafina	Mirizliva vărba	
Erica	Tree Heath	Crnjuša	Piren, Gariga	Ríki
arborea		Drveni vrijes		Chamoríki (Khamoríki)
manipuliflora		Mali vrijes		
Eriobotrya	Loquat	Japanska mušmula		Mousmouliá
Erodium	Storksbill	Čaplja	Časovniče	Moscholáchano, Mironi
Eryngium		Vetrovalj	Vetrogon	Moschángatho
maritimum	Sea Holly	Morski sikavac		Galanóchorto

Erythronium	Dog's Tooth Violet	Košutac, Pasji zub	Samodivsko cvete	
Eucalyptus	Gum, Eucalyptus	Eukalipt	Čaškodrjan	Eukályptos
Euonymus	Spindle-Tree	Kurika	Div konop	Evónimos
Eupatorium	Hemp Agrimony	Konopljuša	Mlečka	
Euphorbia	Spurge	Mlječika		Galastiví, Koukoulafána
acanthothamnos				
Fagopyrum	Buckwheat	Heljda	Grečiha, Elda	
Fagus	Beech	Bukva	Buk	Oxiá
Ferula	Giant Fennel	Devesilj		Nárthikas
Ficus	Fig	Smokva	Smokinja	Sykiá
Filipendula	Meadow-sweet	Suručica, Medunika	Orehče, Otrvatniče	
Foeniculum	Fennel	Komorač	Morač, Rezene	
Fragaria	Wild Strawberry	Jagoda	Jagoda	Agriofráoula
Frangula	Alder Buckthorn	Krušina	Boja, Zárnastec	Vourvouliá
alnus				
Fraxinus	Ash	Jasen	Jasen	Meliós, Mélegos
ornus	Manna Ash	Jasen crni		Kapniá, Kapnóchorto
Fumaria	Fumitory	Dimnjača	Rosopas	
Galanthus	Snowdrop	Visibaba	Kokiče	Skoularíkia
Galega	Goat's Rue, French Lilac	Ždraljevina	Žablek, Konski rebra	
Galium	Bedstraw	Broć, Broćac	Enjovče, Dragajka	
Genista	Greenwood	Žutilovka	Žáltuga	
acanthoclada				
Gentiana	Gentian	Lincura, Srčanik	Tintjava, Gorčivka	Afána
lutea	Great Yellow Gentian	Zdravac, Iglica	Zálta Tintjava	Agriokapnós
Geranium	Cranesbill	Zečija stopa	Zdravec	
Geum	Herb Bennet, Avens	Sabljičica, Mačika	Omajniče	
Gladiolus	Gladiolus		Petljovo pero	Spathóchorto

		Bogorodičina trava	kantarion	Chelonóchorto
androsaemum	Tutsan	Krvavica		
Hyssopus	Hyssop	Isop, Miloduh	Isop	
Iberis	Candytuft	Ognjivac	Iberis	Lióprino, Arkoudopoúrnaro
Ilex	Holly	Božikovina, Zelenika	Džel, Košličor	
Impatiens	Balsam, Touch-me-not	Nedirak, Netek	Slabonoga	
Inula			Oman	
crithmoides	Golden Samphire	Morski koprc		
helenium	Elecampane	Oman		
Iris	Iris, Flag	Perunika	Perunika	Krínos
Isatis	Woad	Sinj	Sárpica	
Jasminum	Jasmine	Jasmin	Smin, Hrištel	Yasemí
Juglans	Walnut	Orah	Oreh	Karydiá
Juncus	Rush	Sita	Dzuka, Šavar	Voúrla
Juniperus	Juniper	Venja, Smreka	Smrika, Hvojna	Milókedros
foetidissima	Stinking Juniper	Pitoma foja		Agriokyparíssi
phoenicea	Phoenician Juniper	Gluvi smrič		Kédros
Knautia	Field Scabious	Udovica	Červenoglavče	
Laburnum	Laburnum	Zanovijet negnjila	Zlaten dážd	
Lactuca	Lettuce	Salata	Salata	
Lamium	Dead-nettle	Mrtva kopriva	Mártua kopriva	
Lapsana	Nipplewort	Repunjača	Sgárbun	
Lathyrus	Vetchling	Grahor, Grahorina	Sekirče	
cicera	Red Vetchling	Grašnjak	Nahutovo Sekirče	Lathoúri
sativus	Chickling Pea	Grahor, Grah-poljak	Posevno Sekirče	
Laurus	Laurel	Lovorika, Lovor	Lavár, Dafinov list	Dáfni

Pancratium	Sea Daffodil	Balučka	Pjasačna lilija	Krínos tís Thálassas
Papaver	Poppy	Mak	Mak, Bulica, Kadanka	Paparoúna
Parietaria	Pellitory-of-the-Wall	Prilip, Vijošnica		Perdikoúli
Paris	Herb Paris	Vranino oko, Petrov krst	Vransko oko	
Pastinaca	Wild Parsnip	Pastrnak	Paštarnak	
Peganum		Pegan	Zárneš	
Periploca	Silk-vine	Brkva	Gárbač	
Petasites	Butterbur	Lopuh, Repuh	Čobanka	
Petroselinum	Parsley	Peršun	Merudija, Magdanoz	Maintanós
Phaseolus	Bean	Grah	Bob, Fasul, Gripa	Fasóli
Phillyrea		Zelenika	Zelenika, Runica	Fillýki
Phlomis fruticosa	Jerusalem Sage	Veliki pelin		Afáka, Asfáka
Phoenix	Palm	Datula, Urma		Fínikas
Phragmites	Reed	Trska	Trâstika	
Picea	Spruce	Smrča, Smreka	Smârč	
Pinguicula	Butterwort	Tučnica	Petluga	
Pinus	Pine	Bor	Bor	Pévko
cembra	Arolla Pine	Limba		
halepensis	Aleppo Pine	Bili bor, Alepski bor		
leucodermis	Bosnian Pine	Munika	Černa mûra	Rómbolo
mugo	Dwarf Mountain Pine	Krivulj	Klek	
nigra	Black Pine	Crni bor	Čeren bor	Agriópevko, Moschoelato
peuce	Macedonian Pine	Molika	Bela mûra	
pinaster	Maritime Pine	Primorski bor	Morski Bor	
pinea	Stone or Umbrella Pine	Pinjol		Koukounariá
silvestris	Scots Pine	Bijeli bor	Bel Bor	

GENUS	ENGLISH	YUGOSLAV	BULGARIAN	GREEK
Pistacia			Kukuč	
lentiscus	Mastic Tree, Lentisc	Trišlja		Skínos, Schínos
terebinthus	Turpentine Tree, Terebinth	Smrdljika		Kokkorevithiá
Pisum	Pea	Divlji grašak	Grah	Bizéli, Arakás
Plantago	Plantain	Bokvica, Trputac	Živovlek	
Platanus	Plane	Platan, Činar	Činar	Plátanos, Platáni
Plumbago	Leadwort	Vranjemil	Sarkofaj	
Polygala	Milkwort	Krstušac	Telčarka	
Polygonatum	Solomon's Seal	Pečatnik, Salamunov pečat	Momkova sálza	
Polygonum	Knotgrass	Troskot	Piperiče	
persicaria	Persicaria, Redleg	Lisac, Ljutača	Ljutivče	
Populus	Poplar	Topola	Topola	Lévki, Lévka
alba	White Poplar	Topola bijela	Kavak, Bjala topola	
nigra	Black Poplar	Topola crna, Jagnjed	Černa topola	
tremula	Aspen	Jasika, Trepetljika	Trepetlika	
Portulaca	Purslane	Prkos	Tučenica	
Potamogeton	Pondweed	Talasinje, Mrijesnjak	Žegal, Ráždavec	
Potentilla	Cinquefoil	Petoprstnica	Prozorče, Očibolec	
Primula	Primrose, Cowslip	Jagorčevina, Jaglac	Iglika	
Prunella	Self-Heal	Crnjevac	Živeniče	
Prunus				
armeniaca	Apricot	Kajsija, Marelica	Zarzala, Kaisia	Verikokkiá
avium	Wild Cherry	Trešnja	Čereša	Kerasiá
domestica	Plum	Šljiva	Sliva	Damaskiniá
laurocerasus	Common or Cherry-laurel	Lovorvišnja, Zeleniče	Lavrovišnja	

padus	Bird-cherry	Sremza	Pesākinja	
persica	Peach	Breskva	Praskova	
spinosa	Blackthorn, Sloe	Trnjina	Trānka	Tsapourniá
Psoralea	Pitch Trefoil	Djetelnjak		Vromóchorto
Pulicaria	Fleabane	Businjak, Buhača	Blāšnica, Šumkavče	
Pulsatilla	Pasque Flower	Sasa	Sāsānka, Kotence	
Punica	Pomegranate	Nar, Mogranj	Nar	Rodiá, Roidiá
Pyracantha	Pyracantha	Divlja trnovina	Pirakanta	Pyrákantha
Pyrus	Pear	Kruška	Kruša	Agriapidiá
amygdaliformis		Kruška trnovača, Slanopadja		Gortsiá, Agrapidiá
Quercus	Oak	Hrast	Dāb	
cerris	Turkey Oak	Cer	Cer	
coccifera	Kermes or Holly Oak	Prnar	Pārnar	Pournári, Prinári
ilex	Holm Oak	Česmina		Ariá, Areós
macrolepis	Valonia Oak			Valanidiá
petraea	Durmast Oak	Kitnjak	Skalen dāb	Valanidiá
pubescens	White or Downy Oak	Medunac	Kosmat dāb	Valanidiá
robur	Common Oak	Lužnjak		
suber	Cork Oak	Plutnjak	Korkov dāb	
Ranunculus	Buttercup, Crowfoot	Ljutič	Ljutiče	
Raphanus	Radish	Rotkva	Diva repica, Ognica	Rapánia
Reseda	Mignonette	Rezeda, Žutica	Rezeda, Bojka, Boja, Žarnastec	
Rhamnus				
alaternus	Mediterranean Buckthorn	Pasdren	Monjen	Kitrinoxylo
catharticus	Buckthorn	Pasdren	Pasdren	
Rhinanthus	Yellow Rattle	Šuškavac	Klopačka	
Rhododendron	Rhododendron	Sleč, Pjenišnik		Leukágatho

Trollius	Globe Flower	Jablan, Planinčica	Planinski božur Ablen	
Tulipa	Tulip	Tulipan, Lala	Lale	Toulípa
Tussilago	Coltsfoot	Podbjel	Podbel	
Typha	Reedmace, Bulrush	Rogoz botur	Papur, Rogoz	Psathí
Ulmus	Elm	Brijest	Brjast	Fteliá
Umbilicus	Pennywort, Navelwort	Pupčić	Videliče	
Urginea	Sea Squill	Morski luk		Skilokremýda, Askeletoúra
Urtica	Nettle	Kopriva	Kopriva	Tsouknída
Vaccinium myrtillus	Bilberry, Whortleberry	Borovnica	Borovinka, Sinja borovinka	
Valeriana	Valerian	Odoljen	Diljanka	
Valerianella	Corn Salad	Matovilac	Motovilka	
Veratrum	False Helleborine	Čemerika	Čemerika	
Verbascum	Mullein	Divizma	Lopen, Ovča opaška	Flómos
Verbena	Vervain	Sporiš	Vărbinka	
Veronica	Speedwell	Čestoslavica	Velikdenče	
Viburnum		Udika, Udikovina	Kalina	
lanata	Wayfaring Tree	Sibikovina	Tutuñiga, Bubuljak	Klimatsída
opulus	Guelder Rose	Lemprika	Obiknovena Kalina	
tinus	Laurustinus			Pseudodáphni
Vicia	Vetch, Tare	Grahorica	Glušina	
faba	Broad Bean	Bob	Bakla	Koukiá
sativa	Common Vetch	Grahor, Grahorica	Fii, Urov	Víkos
Vinca	Periwinkle	Pavenka	Zimzelen	
Viola	Violet, Pansy	Ljubica, Ljubičica	Temenuga	Menexés
Viscum	Mistletoe	Imela	Bjal imel	Meliós, Ixós
Vitex	Chaste Tree	Konopljika	Avrañovo Darvo	Lygariá
Vitis	Grape Vine	Vinova loza	Loza	Ampéli

GENUS	ENGLISH	YUGOSLAV	BULGARIAN	GREEK
Xanthium				
strumarium	Cocklebur	Dikica, Boca	Rogačica, Kazaški bodil	Agriopampáki
Xeranthemum	Everlasting	Nevenka, Poljska metta	Bezsmártniče	
Zea	Maize	Kukuruz	Carevica, Kukurz	
Zelkova				Ampelitsá
Ziziphus	Jujube	Čičimak	Hinap	
Zostera	Eel-grass, Grass-wrack	Svilina	Morska treva	

Index of place names

LISTING PLACES OF BOTANICAL INTEREST

Index of plant names

Numbers in bold type are running numbers, those preceded by pl. refer to Plates, those preceded by p. refer to page numbers, and those preceded by d. refer to drawings. Scientific names in bold type are synonyms.

Crowfoot
  Common Water, (250)
  Pond, 250
  Scarlet, 246a
  Snakestongue, (243)
Crucianella, p. 369
  angustifolia, 1015
  latifolia, 1015a
Cruciata, p. 372
  glabra, (1029)
  laevipes, 1029
  pedemontana, 1029a
Cruciferae, pp. 247–59
Crupina, p. 471
  crupinastrum, (1496), d. 41
  vulgaris, 1496
Cucubalus, p. 225
  baccifer, 177
Cucumber, 816
  Bitter, (812)
  Squirting, 811
Cucumis, p. 345
  melo, 817
  sativus, 816
Cucurbita, p. 345
  maxima, 813
  pepo, 814
Cucurbitaceae, pp. 344–5
Cupidone
  Yellow, 1511a, d. 480
Cupressaceae, pp. 203–4
Cupressus, p. 203
  sempervirens, 11
Currant
  Black, 415
  Mountain, (414)
  Rock Red, (415)
Cuscuta, p. 376
  epithymum, 1044
  europaea, 1043
Cyclamen, p. 359
  coum, 959a
  creticum, 960b, pl. 33
  europaeum, 959
  graecum, (958)
  hederifolium, 958
  neapolitanum, 958
  persicum, 960c, pl. 33
  purpurescens, 959
  repandum, 960, d. 89
Cyclamen
  Common, 959
  Greek, (958)
Cydonia, p. 276
  oblonga, 463
Cymodocea, p. 485
  nodosa, 1581a
Cymbalaria, p. 413
  longipes, 1210b, pl. 43
  microcalyx, 1210c
  muralis, 1210, d. 190
Cynanchum, p. 368
  acutum, 1008

  erectum, 1011a, pl. 34
Cynara, p. 469
  cardunculus, 1491, pl. 53
  cornigera, 1491e
  scolymus, (1491)
  sibthorpiana, 1491e
Cynoglossum, p. 377
  columnae, (1049)
  creticum, (1049)
  hungaricum, 1049f
  nebrodense, 1049e
  officinale, 1049
  sphacioticum, p. 68
Cyperaceae, p. 510
Cyperus, p. 510
Cypress
  Funeral, 11
Cypress Family, pp. 203–4
Cypripedium, p. 510
  calceolus, 1872
Cytinus, p. 215
  hypocistis, 78
  ruber, (78)
Cytisus, p. 283
  decumbens, 500
  procumbens, 500a
  scoparius, 505
  villosus, 501

Dactylorhiza, p. 516
  cordigera, 1898b, pl. 64
  fuchsii, 1898
  iberica, 1899b pl. 64
  incarnata, 1897
  saccifera, 1900a pl. 64
  sambucina, 1899
  sulphurea, 1899a
Daffodil Family, pp. 501–2
Daffodil, 1665
  Sea, 1673
Daisy, 1360
  Annual, 1361
  Crown, 1425
  European Michaelmas,
    1364
  Moon, 1427
  Oxe-eye, 1427
  Southern, (1360)
Daisy Family, pp. 445–83
Damasonium, p. 483
  alisma, 1562
Dame's Violet, 296
  Cut-leaved, (296)
Damson, 477
Danewort, 1295
Daphne, p. 334
  alpina, 757, pl. 29
  blagayana, (757), pl. 29
  cneorum, 756
  gnidioides, 758a
  gnidium, 758
  jasminea, 757a, pl. 29
  laureola, 760

  malyana, 757b
  mezereum, 759
  oleoides, (757), pl. 29
  pontica, 760a
  sericea, 757c
Daphne Family, pp. 333–5
Date Plum, 977a
Datisca, p. 344
  cannabina, 810f
Datisca Family, pp. 343–4
Datiscaceae, pp. 343–4
Datura, p. 409
  innoxia, 1186b, pl. 41
  stramonium, 1186
Dead-nettle
  Large Red, 1131, pl. 39
  Red, 1128
  Spotted, 1130
  White, 1129
Degenia, p. 253
  velebitica, 323e, d. 190
Delphinium, p. 233
  albiflorum, 211j
  balcanicum, 211i, d. 238
  elatum, 211
  fissum, 211k
  hellenicum, 211h
  peregrinum, (211)
  staphisagria, 212, pl. 6
Descurainia, p. 247
Dewberry, (429)
Dianthus, p. 228
  arboreus, 186a, d. 224
  armeria, 186
  barbatus, 185
  biflorus, 193f
  carthusianorum, 187
  caryophyllus, (194)
  corymbosus, 193g, d. 224
  cruentus, 187c, d. 89
  degenii, 193e
  deltoides, 193
  fruticosus, 186b
  giganteus, 187b, pl. 4
  gracilis, 193d
  haematocalyx, 193e,
    d. 127, pl. 4
  subsp. *pindicola*, 193e
  juniperinus, 186c
  microlepis, 194g, pl. 4
  minutiflorus, 194d, d. 168
  moesiacus, p. 150, 153, 183
  myrtinervius, 193b, d. 224
  nardiformis, p. 151
  petraeus, 189a
  subsp. *simonkaianus*,
    p. 155
  pinifolius, 187a, pl. 4
  pontederae, p. 151
  prolifera, 184
  pseudarmeria, 186d, d. 148
  '*pulviniformis*', p. 75
  scardicus, 194f, pl. 4

Bibliography

(I = illustrated comprehensively)

Reference Floras

BOISSIER, E., *Flora orientalis sive enumeratio plantarum in Oriente a Graecia et Aegypto ad Indiae fines hucusque observatarum*, 5 vols. and suppl., Genève & Bâle, Lyon, 1867–88.

DAVIS, P. H. (ed.), *Flora of Turkey and the East Aegean islands*, vols. 1–5, Edinburgh, 1965–75. Standard flora of Turkey and eastern Aegean Islands; to be completed; in English.

DIAPOULIS, H. A., *Ellēnikē hlōris*, 3 vols., Athēnai, 1939–49; in Greek.

HALÁCSY, E. von, *Conspectus florae graecae*, 5 vols. and suppl., Leipzig, 1900–8.; in Latin.

HAYEK, A. von, 'Prodomus florae peninsulae balcanicae', *Repert. Spec. Nov. Regni Veg. Beih.* 30(1–3), 1924–33. Standard flora of Balkans; in Latin.

JORDANOV, D. (ed.), *Flora na Narodna Republika Balgarija*, vols. 1–5, Sofia, 1963–73. Standard flora of Bulgaria; to be completed; in Bulgarian. I

STOYANOV, N., STEFANOV, B., and KITANOV, B., *Flora na Balgarija*, 4th edn., vols. 1–2, Sofia, 1966–7. I.

TUTIN, T. G., *et al.* (eds.), *Flora Europaea*, vols. 1–4, Cambridge, 1964–76. Standard flora of Europe; to be completed; in English.

Mapping

JALAS, J., and SOUMINEN, J., *Atlas Florae Europaeae*, vols. 1–3, Helsinki, 1972–6. Distribution maps of European species; to be completed; in English.

Floristic Accounts

(For comprehensive bibliographies see Horvat, Glavač, and Ellenberg, 1974; Rechinger, 1943)

ADAMOVIC, L., *Die Pflanzenwelt der Adrialänder*, Jena, 1929.

DAVIS, P. H., 'Cliff Vegetation of the Eastern Mediterranean', *J. Ecol.* 39, 1951.

—— 'Notes on the Summer Flora of the Aegean', *Notes Roy. Bot. Gard. Edinburgh*, 21, 1953.

GREUTER, W., 'Zur Paläogeographie und Florengeschichte der südlichen Agäis', *Feddes Repert.* 81, 1970.

HORVAT, I., GLAVA, Č, V., and ELLENBERG, H., *Vegetation Südosteuropas*, Stuttgart, 1974. Descriptions of the types of vegetation in the Balkans with sample species lists; well illustrated, with good bibliography; in German. I.

RECHINGER, K. H., *Flora Aegaea*, Wien, 1943 (repr. 1973). Lists and distribution of all

Aegean species, including lower plants, with keys to larger genera; 25 plates of drawing and photographs; in Latin and German.

—— 'Phytogeographia Aegaea', *Denkschr. Akad. Wiss. Wien*, 105, 1951.

—— 'Der Endemismus in der griechischen Flora', *Rev. Roumaine Biol. Sér. Bot.* 10, 1965.

RIKLI, M., *Das Pflanzenkleid der Mittelmeerländer*, Berne, 1943–8.

RUNEMARK, H., 'Distribution Patterns in the Aegean', in *Plant Life of South-West Asia*, Edinburgh, 1971.

STOYANOV, N., and JORDANOV, D., 'Über die Vegetationsverhältnisse des Olymps', *God. Univ. Sofia*, 34, 1938.

TURRILL, W. B., *The plantlife of the Balkan peninsula. A phytogeographical study*, Oxford, 1929.

Popular Accounts, Guides, Geography, etc. of the Balkan area

DAVIS, P. H., 'On the Rocks', *Bull. Alp. Gard. Soc.* 15, 1947.

NAVAL INTELLIGENCE DIVISION, *Yugoslavia*, vols. 1–2, 1944.

—— *Greece*, vols. 1–3, 1944–5.

NEWBIGIN, M. I., *Southern Europe*, London, 1932.

OGILVIE-GRANT, M., 'Interesting Plants of the Eastern Mediterranean', *Bull. Alp. Gard. Soc.* 29, 1961.

OSBORNE, R. H., *East-Central Europe*, London, 1967.

PHILIPPSON, A., *Das Klima Griechenlands*, Bonn, 1948.

POLUNIN, O., *Flowers of Europe*, London, 1969. I.

—— *Trees and Bushes of Europe*, London, 1976. I.

—— and HUXLEY, A., *Flowers of the Mediterranean*, London, 1965. I.

ROGER-SMITH, H., 'Botanising on an Hellenic Traveller's Tour', *Bull. Alp. Gard. Soc.* 9, 1941.

SCHACHT, W., 'Balkan Memories', *Bull. Alp. Gard. Soc.* 16, 1948.

SUNDERMANN, H., *Europäische und Mediterrane Orchideen*, Hanover, 1970. I.

TREVAN, D. J., and WHITEHEAD, M. J., 'Plant Collecting in Macedonia', *Bull. Alp. Gard. Soc.* 42, 1974.

WALKER, D. S., *The Mediterranean Lands*, London, 1960.

Greece (mainland)

BEUERMANN, A., 'Waldverhältnisse im Peloponnes unter besonderer Berücksichtigung der Entwaldung und Afforstung', *Erdkunde*, 10, 1956.

BOTTEMA, S., *Late Quaternary Vegetation History of Northwestern Greece*, Groningen, 1974.

DEBAZAC, E. F., and MAVROMMATIS, G., 'Les Grandes Divisions écologiques de la végétation forestière en Grèce continentale', *Bull. Soc. Bot. France*, 1971.

DEPARTMENT OF PHARMACY, UNIVERSITY THESSALONKI, *Pharmacognostic Map of Greece*, 1961.

DIAPOULIS, H. A., 'Beitrag zur Kenntnis der Waldvegetation des Olymps und des Pierriagebirges', *Repert. Spec. Nov. Regni Veg.* 40, 1936.

GAMS, H., 'Nachtrag zur Flora und Vegetation des Olymp', *Österr. Bot. Z.* 107, 1960.

GANIATSAS, K. A., 'Botanikai ereunai epi tou orous Bermiou', *Epist. Epet. Shol. Fus. Math. Epist. Panepist. Thessalonikes*, 5, 1939.

—— 'Sumbolē eis tēn gnōsin tēs hlōridos tou orous Tumfrēstou', *Epist. Epet. Shol. Fus. Math. Epist. Panepist. Thessalonikēs*, 6, 1940.

—— 'Ē hlōris tōn oreinōn boskōn tou Bermiou (futokoinonikē meletē)', *Epist. Epet. Fus. Math. Shol. Aristot. Panepist. Thessalonikēs*, 7, 1955.

—— *Ē blastēsis kai ē hlōris tēs hersonēsou tou Agiou Orous*, Thessalonike, 1963.

—— *Ē hlōris kai ē blastēsis tēs limnēs tōn Iōanninōn*, Ēpeirōt, Estia, 1970 (repr.).

—— *Botanikai ereunai epi tēs haradras tou Bikou*, Ēpeirōt, Estia, 1971 (repr.).

GOULIMIS, C., 'Report on Species of Plants requiring Protection in Greece and Measures for securing their Protection', *Int. Union for Conservation of Nature*, 5, 1959.

HARISTOS, P. A., *Ē farmakeutikē hlōris tēs Halkidikēs*, Thessalonikē, 1969.

KITANOV, B., 'Die Vegetation des Boz-Dagh-Gebirges in Ostmazedonien', *Jahrb. Univ. 111 Climent v. Ochrid in Sofia*, 1943.

—— 'Novi materiali za florata na planinata Bozǎ-Dagǎ vǎ iztočna Makedonija', *God. Sofijsk. Univ. Fiz. Mat. Fak. Kn. 3 Estestv. Istorija*, 41, 1945.

KROCHMAL, A., and LAVRENTIADES, G. I., 'Poisonous Plants of Greece', *Economic Botany*, 9, 1955.

LAVRENTIADES, G. I., 'On the Vegetation of the Keramoti Coast', *Bull. Ist. Bot. Univ. Catania*, 4, 1963.

—— 'The Ammophilous Vegetation of the Western Peloponnesos Coast', *Vegetatio*, 12, 1964.

—— 'Studies on the Flora and Vegetation of the Ormos Archangelou in Rhodos Island', *Vegetatio*, 19, 1969.

MACRIS, C. G., *Planta Medica Graeciae*, University of Thessaloniki.

MAVROMMATIS, G., *Recherches phytosociologiques et écologiques dans le massif de l'Ossa (Grèce) en vue de sa gestion forestière*, Montpellier, 1971.

OBERDORFER, E., 'Nordägäische Kraut- und Zwergstrauchfluren im Vergleich mit den entsprechenden Vegetationseinheiten des westlichen Mittelmeergebietes', *Vegetatio*, 5–6, 1954.

PHITOS, D., 'Futogeōgrafikai paratērēseis epi tou oreinou sugkrotēmatos Tumfrēstou-Oxuas', *Das. Hron.* 25, 1960.

—— 'Beitrag zur Kenntnis der Flora von Nord-Pindos', *Mitt. Bot. Staatssamml. München*, 4, 1962.

—— 'Florula Sporadum', *Phyton (Horn)*, 12, 1967.

POLITIS, J., 'Contribution à l'étude de la flore de la Chalcidique', *Pragm. Akad. Athēnon*, 19, 1953.

QUÉZEL, P., 'Végétation des hautes montagnes de la Grèce méridionale', *Vegetatio*, 12, 1964.

—— 'La Végètation des hauts sommets du Pinde et de l'Olympe de Thessalie', *Vegetatio*, 14, 1967.

—— 'La Végétation du massif de Bela Voda (Macédoine nord-occidentale)', *Biol. Gallo-Hellen.* 2, 1969.

—— 'Contribution à l'étude de la végétation du Vardoussia', *Biol. Gallo-Hellen.* 5, 1973.

—— and CONTANDRIOPOULOS, J., 'Contribution à l'étude de la flore des hautes montagnes de la Grèce', *Naturalia Monspel. Sér. Bot.* 16, 1965.

—— 'Contribution à l'étude de la flore du Pinde central et septentrional et de l'Olympe de Thessalie', *Candollea*, 20, 1965.

—— 'Contribution à l'étude de la flore de la Macédoine grecque', *Candollea*, 23, 1968.

QUÉZEL, P., and KATRABASSA, M., 'Premier aperçu sur la végétation du Chelmos (Péloponèse)', *Rev. Biol. Écol. Medit.* 1, 1974.

RECHINGER, K. H., 'Ergebnisse einer botanischen Sommerreise nach dem ägäischen Archipel und Ostgriechenland', *Beih. Bot. Centralbl.* 54B, 1936.

—— 'Zur Flora von Ostmazedonien und Westthrazien', *Bot. Jahrb. Syst.* 69, 1939.
—— 'Florae aegaeae supplementum', *Phyton (Horn)*, 1, 1949.
—— 'Zur Flora der Kykladen', *Österr. Akad. Wiss. Math.-Naturwiss. Kl. Anz.* 92, 1955.
—— 'Der Endemismus in der griechischen Flora', *Rev. Roumaine Biol. Ser. Bot.* 10, 1965.
REGEL, C. de, 'La Végétation du Pinde et du Taygète', *Bull. Soc. Bot. Genève*, 27, 1937.
—— 'La Végétation du mont Oeta en Grèce', *Boissiera*, 7, 1943.
—— 'Florula montis Oeta in Graecia', *Feddes Repert.* 54, 1944.
—— 'Studien über die Florenelemente in Griechenland. Die Florenelemente des Oeta', *God. Biol. Inst. u Sarujevu*, 5, 1953.
SAMUELSSON, G., 'Symbolae ad floram graecam', *Ark. Bot.* 26A(5), 1933.
STOYANOV, N., and JORDANOV, D., 'Botanische Studien auf dem thessalischen Olymp', *God. Sofijsk. Univ.* 34, 1938.
TURRILL, W, B., 'A Contribution to the Botany of Athos Peninsula', *Kew Bull*, 1937.
VOLIOTIS, D. T., *Ereunai epi tēs blastēseōs kai hlōridos tou Holomōntos kai idia, tēs aromatikēs, farmakeutikēs kai melissotrofikēs toiautēs*, Thessalonikē, 1967.
ZAGANIARIS, D., *Ta xerofuta tēs Attikēs*, Athēnai, 1932.
—— *Ē hlōris tēs Manēs*, Athēnai, 1934.
—— *Ē hlōris tēs periohēs Lauriou*, Athēnai (n.v.), 1935.
—— 'Ta zizania tēs eparhias Thessalonikēs', *Epist. Epet. Geopon. Dasal. Shol. Panepsit. Thessalonikēs*, 1, 1939.
ZERLENDIS, K. K., *Sumbolē eis tēn futogeōgrafian tōn Kukladōn*, Athēnai, 1952.
—— 'Ta futa tou Lukabēttou', *Delt. Agrot. Trapezēs*, 109, 1959.
—— *Sumbolē eis tēn hlōrida tou Umēttou*, Athēnai, 1965.

Greece (mainland)—Popular Accounts

ATCHLEY, S. C., *Wild Flowers of Attica*, Oxford, 1938. I.
ARGYROPOULO, K. A. (ed.), *Wild Flowers of Greece*, Athens, 1965. I.
DIANNELIDIS, T. D., *To Pēlion apo futologikēs apopseōs*, Bouno, 1935.
DIAPOULIS, H. A., *Apo tēn hlōrida tou Olumpou*, Bouno, 1935.
—— *Apo tēn hlōrida tēs Parnēthos*, Bouno, 1958.
—— 'Endēmika futa tōn Kukladōn nēsōn, *Epet. Etair. Kuklad. Melet.* 1, 1961.
GOULIMY, C. N., 'I chloris tou Chelmou', *J. Greek Alp. Soc.* 48, 1948–9.
—— *The Richness of the Greek Flora,* Eklogi, 1953.
—— *From the Flora of Mt. Gamila*, Bouno, 1954.
—— 'Some Alpines of Greece', *Bull. Alp. Gard. Soc.* 23, 1955.
—— 'The Colchicums of Greece and the Distribution of this Genus', *Pharmaceutical Archives*, 1956.
—— 'Some Notable Greek Plants', *Bull. Alp. Gard. Soc.* 25, 1957.
—— 'A Letter from Greece', *Bull. Alp. Gard. Soc.* 25, 1957.
—— *To Oros Boyrinos kai i chloris tou*, Bouno, 1960.
GREY-WILSON, C., 'The Heights of Olympus', *Bull. Alp. Gard. Soc.* 37, 1969.
HARITONIDOU, P., *Wild Flowers of Greece*, Athens, 1965. I.
HERMJAKOB, G., *Orchids of Greece and Cyprus. The genus Ophrys*, Kiffisia, 1974. I.
HUXLEY, A. J., 'An Ascent of Parnassus', *Bull. Alp. Gard. Soc.* 29, 1961.
—— and TAYLOR, W., *Flowers of Greece and the Aegean*, London, 1977. I.
MATHEW, B., 'Greek Crocuses', *Bull. Alp. Gard. Soc.* 43, 1975.
NIEBUHR, A. D., *Herbs of Greece*, Athens, 1970. I.
SCHACHT, W., 'A Botanical Excursion to the Athos Peninsula', *New Flora and Silva*, 9, 1937.

SFIKAS, G., *Flowers of Mount Olympus*, Athens, 1974. I.
THOMSON, H. P., 'Plant Hunting in South Greece', *Bull. Alp. Gard. Soc.* 8, 1940.
TREVAN, D. J. and WHITEHEAD, M. J., 'Plant Hunting in the Pindus', *Bull. Alp. Gard. Soc.* 43, 1975.
TURRILL, W. B., 'The Genus Fritillaria in the Balkan Peninsula and Asia Minor', *J. Roy. Hort. Soc.*, 1937.

Crete

BAKER, G. P., 'Plant Hunting in Crete', *J. Roy. Hort. Soc.*, 1929.
DAVIS, P. H., 'A Collector in Crete', *Bull. Alp. Gard. Soc.* 5, 1937.
FERNS, F. E. B., 'A Land called Crete', *Bull. Alp. Gard. Soc.* 44, 1976.
GREUTER, W., 'Additions to the Flora of Crete, 1938–1972', *Ann. Mus. Goulandris*, i, 1973.
—— 'Die Insel Kreta—eine geobotanische Skizze', *Veröff. Geobot. Inst. Zürich*, 1975.
—— 'Floristic Report on the Cretan Area', *Mem. Soc. Brot.* 24, 1975.
RAVEN, J., 'Crete in March', *Bull. Alp. Gard. Soc.* 40, 1972.
RECHINGER, K. H., 'Neue Beiträge zur Flora von Kreta', *Akad. Wiss. Wien*, 105, 1944.
RIKLI, M., and RÜBEL, E., 'Über Flora und Vegetation von Kreta und Griechenland', *Vierteljahrsschr. Naturf. Ges. Zürich*, 68, 1923.
ZOHARY, M., and ORSHAN, G., 'An Outline of the Geobotany of Crete', *Israel J. Botany*, 14, 1966.

Greek islands

Aiyina (Agina)

HELDREICH, T., *Flora de l'île d'Égine*, Genève, 1898.
PODLECH, D., *Pflanzenliste der Pfingstexkursion 1967 des Instituts für systematische Botanik der Universität München nach Griechenland*, München, 1967.

Ayios evstratios

RAUH, W., 'Klimatologie und Vegetationsverhältnisse der Athos-Halbinsel und der ostägäischen Inseln Lemnos, Evstratios, Mytiline und Chios', *Sitzungsber. Heidelberger Akad. Wiss.*, 1949.

Chios

MEIKLE, R. D., 'A Survey of the Flora of Chios', *Kew Bull.*, 1954.
RAUH, W., 'Klimatologie und Vegetationsverhältnisse der Athos-Halbinsel und der ostägäischen Inceln, Lemnos, Evstratios, Mytiline und Chios', *Sitzungsber. Heidelberger Akad. Wiss.*, 1949.

Elaphonesos

YANNITSAROS, A. G., 'Notes on the Flora of Elaphonesos Island (Laconia, Greece)', *Biol. Gallo-Hellen.* 3, 1971.

Evvoia (Euboea)

PHITOS, D., *Futogeografikē ereuna tēs kentrikēs Euboias*, Athēnai, 1960.
RECHINGER, K. H., 'Die Flora von Euböa', *Bot. Jahrb. Syst.* 80, 1961.
SKROUMBIS, B. G., and PINATZIS, K. L., *Sumbolē eis tēn meletēn kai axiopoiēsin tēs aromatikēs hlōridos tēs Euboias*, Athēnai, 1963.

Ikaria

FORSYTHE MAJOR, C. F., and BARBEY, W., 'Ikaria. Étude botanique', *Bull. de l'Herbier Boissier*, 5, 1897.

Ithaki (Ithaca)

FORSYTH MAJOR, C. F., and BARBEY, W., 'Kalymnos. Étude botanique', *Bull. de l'Herbier Boissier*, 4, 1896.

MAILLEFER, A., 'Herborisation's pendant une croisière dans l'Adriatique et autour de la Grèce en 1938', *Bull. Soc. Vaud. Sci. Nat.* 61, 1940.

Karpathos

STEFANI, C., FORSYTH MAJOR, C. J., and BARBEY. W., *Karpathos. Étude geologique, paleontologique et botanique*, Lausanne, 1895.

DAVIS, P. H., 'Notes on the summer flora of the Aegean', *Notes Roy. Bot. Gard. Edinburgh*, 21, 1953.

GREUTER, W., 'Floristic report on the Cretan area', *Mem. Soc. Brot.* 24, 1975.

Kassandra

LAVRENTIADES, G. I., 'Hlōristikē, futogeōgrafikē kai futokoinōniologikē ereuna tēs hersonēsou tēs Kassandras', *Epist. Epet. Fus.-Math. Shol. Aristot. Panepist. Thessalonikēs*, 8, 1961.

Kefallinia (Cephalonia)

BORNMÜLLER, J., 'Ergebnis einer botanischen Reise nach Griechenland im Jahre 1926 (Zante, Cephalonia, Achaia, Phokis, Aetolien)', *Repert. Spec. Nov. Regni Veg.* 25, 1928.

CUFODONTIS, G., 'Die von Dr. Th. Just in Jahre 1929 auf den Ionischen Inseln und im nordwestlichen Peloponnes gesammelten Farn- und Blütenpflanzen', *Repert. Spec. Nov. Regni Veg.* 39, 1936.

HELDREICH, T., *Flora de l'île de Cephalonie*, Lausanne, 1882.

KNAPP, R., *Die Vegetation von Kephallinia, Griechenland. Geobotanische Untersuchung eines mediterranen Gebietes und einige ihrer Anwendungsmöglichkeiten in Wirtschaft und Landesplanung*, Giessen, 1965.

Kerira (Corfu)

CUFODONTIS, G., 'Die von Dr. Th. Just im Jahre 1929 auf den Ionischen Inseln und im nordwestlichen Peloponnes gesammelten Farn- und Blütenpflanzen', *Repert. Spec. Nov. Regni Veg.* 39, 1936.

MAILLEFER, A., 'Herborisations pendant une croisière dans l'Adriatique et autour de la Grèce en 1938', *Bull. Soc. Vaud. Sci. Nat.* 61, 1940.

RIKLI, M. and RÜBEL, E., 'Uber Flora und Vegetation von Kreta und Griechenland', *Vierteljahrsschr. Naturf. Ges. Zürich*, 68, 1923.

Kithera (Cerigo)

GREUTER, W., and RECHINGER, K. H., 'Flora der Insel Kythera, gleichzeitig Beginn einer nomenklatorischen Überprüfung der griechischen Gefässpflanzenarten', *Boissiera*, 13, 1967.

YANNITSAROS, A. G., *Sumbolē eis tēn gnōsin tēr hlōridos kai blastēseōs tēs nēsou tōn Kuthērōn*, Athēnai, 1969.

Lesvos (Lesbos)

CANDARGY, P. C., 'Flore de l'île de Lesbos', *Bull. Soc. Bot. de France*, 45, 1898.

Levkas (St. Maura)

HOFMANN, U., 'Untersuchungen an Flora und Vegetation der Ionischen Insel Levkas,' *Vierteljahrsschr. Naturf. Ges. Zürich*, 113, 1968.

CUFODONTIS, G., 'Die von Dr. Th. Just im Jahre 1929 auf den Ionischen Inseln und im nordwestlichen Peloponnes gesammelten Farm- und Blütenpflanzen', *Repert. Spec. Nov. Regni Veg.* 39, 1936.

Lemnos

RAUH, W., 'Klimatologie und Vegetationsverhältnisse der Athos-Halbinsel und der ostägäischen Inseln Lemnos, Evstratios, Mytiline und Chios', *Sitzungsber. Heidelberger Akad. Wiss.*, 1949.

Naxos

MAILLEFER, A., 'Herborisations pendant une croisière dans l'Adriatique et autour de la Grèce en 1938', *Bull. Soc. Vaud. Sci. Nat.* 61, 1940.

Nisiros

PAPATSOU, S. H., *E hlōris kai ē blastēsis tēs N. Nisyrou kai tōn peri autēn nēsidōn*, Patrai, 1975.

Poros

ZAGANIARIS, D., 'La Flore de Poros', *Actes Inst. Bot. Univ. Athènes*, 1, 1940.

Psara

GREUTER, W., 'The Flora of Psara—an Annotated Catalogue', *Candollea*, 31, 1976.

Rhodos (Rhodes)

FINKL, A., 'Beiträge zur Kenntnis der Flora der Insel Rhodos', *Acta Albertina Ratisb.* 24, 1962.

FIORI, A., *La Flora dell'isola di Rodi*, Instito Agrario Coloniale Italiano, Firenze, 1924.

HANSEN, A. and SNOGERUP, S., 'Beiträge zur Kenntnis der Flora der Insel Rhodos', *Osterr. Akad. Wiss.* 75, 1966.

LAVRENTIADES, G. I., 'Studies on the Flora and Vegetation of the Ormos Archangelou in Rhodos Island', *Vegetatio*, 19, 1969.

Samos

STEFANI, C., FORSYTHE MAJOR C. J., and BARBEY, W., *Samos. Étude géologique, paleontologique et botanique*, Lausanne, 1892.

Samothraki (Samothrace)

ADE, A., and RECHINGER, K. H., 'Samothrake', *Feddes Repert.* 100, 1938.

DEGEN, A., 'Botanische Reise nach der Insel Samothrake', *Österr. Bot.* 41, 1891.

STOYANOV, N., and KITANOV, B., 'Prinos kămă izučavaneto na florata i rastitelnitě săotnošenija na ostrov Samotraki', *God. Sofijsk. Univ. Fiz.-Mat. Fak. Kn. 3 Estestv. Istorija*, 40, 1944.

Siros

MAILLEFER, A., 'Herborisations pendant une croisière dans l'Adriatique et autour de la Grèce en 1938', *Bull. Soc. Vaud. Sci. Nat.* 61, 1940.

Skiathos

ECONOMIDOU, E., *Geobotanikē ereuna nēsou Skiathou*, Athēnai, 1969.

Skopelos

ECONOMIDOU, E., 'Contribution à l'étude de la flore et de la phytogéographie de l'île de Skopelos', *Biol. Gallo-Helen.* 5, 1973.

Spetsai

PHOUPHAS, C., 'Contribution à l'étude de la flore de l'île "Spetsai" ', *Bull. Soc. Bot. France*, 115, 1968.

Thasos

STOYANOV, N., and KITANOV, B., 'Florata na ostrov Tasos', *God. Sofijsk. Univ.* 41, 42, 1945–6.
—— 'Rastitelnitě otnošenija na ostrov Tasos', *Izv. Bot. Inst. (Sofia)*, 1, 1950.

Thira (Santorin)

HANSEN, A., 'Flora der Inselgruppe Santorin', *Candollea*, 26, 1971.
MAILLEFER, A., 'Herborisations pendant une croisière dans l'Adratique et autor de la Grèce en 1938', *Bull. Soc. Vaud. Sci. Nat.* 61, 1940.

Zakinthos (Zanthe)

RONNIGER, K., 'Flora der Insel Zante', *Verh. Zool.-Bot. Ges. Wien*, 88–9, 1940.
BORNMÜLLER, J., 'Ergebnis einer botanischen Reise nach Griechenland im Jahre 1926 (Zante, Cephalonia, Achaia, Phokis, Aetolien)', *Repert. Spec. Nov. Regni Veg.* 25, 1928.

Yugoslavia

ADAMOVIC, L., *Die Pflanzenwelt Dalmatiens*, Leipzig, 1908–11.
BECK, G., *Flora Bosne, Hercegovine i Novipazarskog Sandzaka*, vols. 1–2, Sarejevo, Beograd, 1903–27.
—— and MALY, K., *Flora Bosnae et Hercegovinae*, Sarejevo, 1950.
—— and BJELCIC, Z., *Flora Bosnae et Hercegovinae*, Sarejevo, 1967–74.
BORNMÜLLER, J., *Beiträge zur Flora Mazedoniens*, Leipzig, 1925–8.
DEGEN, A., *Flora Velebetica*, vols. 1–4, Budapest, 1936–8.
DOMAC, R., *Mala Flora Hrvatske*, Zagreb, 1973.
FREYN, J., 'Die Flora von Süd-Istrien', *Abh. Zool.-Bot. Ges. Wien*, 1877.
GREBENSHCHIKOV, O. S., *Vegetation of the High-Mountain areas of Yugoslavian Macedonia (The Vardar Basin)*, Moscow, 1960.
HORVAT, I., *Biljni svijet Hrvatske*, Zagreb, 1942.
HORVATIC, S., *Illustrirani Bilinar*, Zagreb, 1954.
—— *et al., Analiticka Flora Jugoslavije*, vols. 1–2, Zagreb, 1967–74.
JOSFOVIƇ, M. (ed.), *Flora Serbije*, vols. 1–4, Belgrade, 1970–4. Very useful line drawings; to be completed.
MARTINČIČ, A., SUSNIK, F., *et al., Mala flora Slovenije*, Ljubljana, 1969.
MATVEJEV, S. D., *Biogeographical map of Yugoslavia*, Belgrade, 1961.
MAYER, E., *Selnam Praprotnic Incvetnic Slovenskega Ozemlja*, Ljubljana, 1952.
PISKERNIK, A., *Kijuc za Dolocanje Cvetnic in Praprotnic*, 2nd edn., Ljubljana, 1951.
ROHLENA, J., *Conspectus Florae Montenegrinae*, Prague, 1941–2.

Yugoslavia—Popular Accounts

BARRETT, G. E., 'Land of the Black Mountain (Durmitor)', *Bull. Alp. Gard. Soc.* 37, 1969.
BERTOVIC, P. M., *Prilog Planinarskoj Karti Risnjak-Snjeznik*, Zagreb, 1952.
BERTOVIC, S., KAMENAROVIC, M., and KEVO, R., *The Protection of Nature in Croatia*, Zagreb, 1961.
KAMENAROVIC, M., *Nacionalni Park Risnjak*, Zagreb, 1970.
KEVO, R., *Mljet—Zeleni Otok*, Split, 1962.
MATHEW, B., and GREY-WILSON, C., 'Some Flowers of Yugoslavia', *Bull. Alp. Gard. Soc.* 39, 1971.
RADOS, D., *Paklenica National Park*, Starigrad, 1972.
ŠILIĆ, Č., *Atlas Drvećai i Grmlja*, Sarajevo, 1973. I (colour photographs).
—— *Sumzke Zeljaste Biljke*, Beograd, 1977. I (colour photographs).
THOMPSON, H. P., 'A Yugoslav Memory', *Bull. Alp. Gard. Soc.* 11, 1943.
WOJTÉRSKI, T., *National Parks of Yugoslavia*, Poznan, 1971; in Polish.

Bulgaria

GANTCHEV, I., (ed.) *The vegetation of the meadows and pastures in Bulgaria*, Sofia, 1964; in Bulgarian.
GEORGIEV, T., 'Phytogeographische Skizze des Vitoscha-Gebirges, *God. Univ. Sofia, Agronom. Fak.* 6, 1928; in Bulgarian.
—— *Determinant of the wild and cultivated wood and shrub species in our country*, Sofia, 1956; in Bulgarian.
GRAMATIKOV, D., *Identification of wild and cultivated trees and shrubs in Bulgaria*, Sofia, 1974; in Bulgarian. I.
JORDANOV, D., 'Über die Phytogeographie des Westbalkans', *God. Univ. Sofia*, 20, 1924; in Bulgarian.
—— 'Pflanzengeographische Studien der Sümpfe Bulgariens in ihrer Beziehung zur höhren vegetation', *God. Univ. Sofia, Phys.-Mathem. Fak.* 27, 1931; in Bulgarian.
—— 'Die Vegetationsverhältnisse im bulgarischen Teile des Strandja-Gebirges, *God. Univ. Sofia, Phys.-Mathem. Fak.* 34, 3, 1938, 1939; in Bulgarian.
KITANOV, B., and PENEV, I. V., *Flora of Vitosa*, Sofia, 1963. Local flora with descriptions and keys.
PODPERA, J., 'Ein Beitrag zu den Vegetationsverhältissen in Südbulgarien', *Verh. Bot. Ges. Wien*, 1902.
REICHINGER, K. H., 'Vegetationsskizzen aus Bulgarien', *Feddes Repert.* 33, 1933; in German.
STEFANOV, B., *The origins and development of the types of vegetation in the Rodopes*, Sofia, 1927; in Bulgarian.
—— *Phytogeographische Elemente in Bulgarien*, Sofia, 1943; in Bulgarian.
—— and JORDANOV, D., 'Topographische Flora von Bulgarien', *Englers botanisches Jahrbuch*. 64, 1932; in German.
STOYANOV, N., 'Über die Vegetation des Ali-Botŭs-Gebirges (Slavjanka mountains), *God. Univ. Sofia, Phys.-Mathem. Fak.* 28, 1921; in Bulgarian.
—— 'Versuch einer Analyse des relikten Elements in der Flora der Balkanhalbinsel', *Ebenda*, 63, 1930.
—— 'Caractère phytogéographique du massiv de Rila, des Rhodopes et de Pirin', *Comptes rendus du IVe Congrés des géographes et etnographes slaves*, Sofia, 1938; in French.

—— 'Zur Frage über Herkunft des arktischen Elements in der Flora der Balkanhalbin-sel', *God. Univ. Sofia, Phys.-Mathem. Fak.* 36, 1940; in German.

—— 'Versuch einer phytozönologischen Characteristik Bulgariens', *God. Univ. Sofia, Phys.-Mathem. Fak.* 37, 1941; in Bulgarian.

—— *Phytogeography*, 1950.

—— 'Phytogeographical sketch of Bulgaria', *J. Bot.*, 41, 8, 1958; in Russian.

—— and STEFANOV, B., 'Phytogeographische und floristische characteristik des Pirin-Gebirges', *God. Univ. Sofia, Phys.-Mathem. Fak.* 18, 1922; in Bulgarian.

STRANSKI, I., 'Die Vegetationsverhältnisse in dem Mittel-Rhodopen', *Sbornik, Balg. Akad. Nauk.* 16, 1921; in Bulgarian.

TCHERNJAVSKI, P., *et al.*, *Trees and Bushes in the Forests in Bulgaria*, Sofia, 1959.

VALEV, S., *et al.*, *Excursion flora of Bulgaria*, Sofia, 1960; in Bulgarian.

Bulgaria—Popular Accounts

BARRETT, G. E., 'Moussalla-Rila Mountains', *Bull. Alp. Gard. Soc.* 40, 1972.

—— 'Vitosa and the Rila Mountains', *Bull. Alp. Gard. Soc.* 42. 1974.

HALDA, J., 'Plant Hunting in the Bulgarian Mountains', *American Rock Gard. Soc.* 31, 1973.

INGWERSEN, W. E. T., 'Notes on Bulgarian Plants', *Bull. Alp. Gard. Soc.* 4, 1936.

—— 'Plant Hunting in the Pirin Mountains', *Bull. Alp. Gard. Soc.* 10, 1942.

IVANOV, D., and NIKOLOV, P., *Fito terachiya* (*Medicinal Plants*), Sofia, 1969; in Bulgarian. I.

JANEV, A., *Ornamental plants in the flora of Bulgaria*, Sofia, 1959; in Bulgarian.

—— and VALEV, S., *Early vernal plants*, Sofia, 1965.

JORDANOV, D., *et al.*, *Phytotherapy*, 3rd edn., Sofia, 1969. Descriptions of the medicinal plants in Bulgaria; many prescriptions; in Bulgarian. I.

SCHACHT, W., 'Bulgarian Plants worth Cultivating', *New Flora and Silva*, 38, 1938.

STANEV, S., *The stars fade in the mountain*, Sofia, 1976. Stories about Bulgarian rare plants; in Bulgarian.

STOYANOV, N., and KITANOV, B., *Plants of the high mountains of Bulgaria*, Sofia, 1966. Descriptions; colour plates of species found over 2000 m; in Bulgarian. I.

STOYANOV, N., *et al.*, *Our Reserves and Natural Sites*, 3 vols., Sofia, 1968–74; in Bulgarian.

TOSHKOV, M., and VIHODZEVSKI, N., *Protected Natural Objects,* Sofia, 1971; in Bulgarian.

TOSHKOV, M., *et al.*, *In Protection of the Natural Environment*, Sofia, 1972; in Bulgarian.

VIHODZEVSKI, N., *Our autumn plants*, Sofia, 1965; in Bulgarian. I.

VODENITCHAROV, D., *The vegetation of our water basins*, Sofia, 1959; in Bulgarian.

Albania

ALSTON, A. H. G., and SANDWITH, N., 'Results of two Botanical Expeditions to South Albania', *J. Bot.*, 1940.

JAVORKA, A., *Additamenta ad floram Albaniae*, Budapest, 1926.

LEMBERG, F., 'Northern Albania', *New Flora and Silva*, 7, 1935.

MARKGRAF, F., *Pflanzen aus Albanien*, Wien, 1931.

—— 'Pflanzengeographie von Albanien', *Bibl. Bot.* 105, 1932.

Turkey-in-Europe

WEBB, D. A., 'The Flora of European Turkey', *Proc. Roy. Irish Ac.* 65, 1966.

icea omorika × 1/40

14a *Juniperus foetidissima* × 1/30

1

Arceuthobium oxycedri × 1/6

6a *Pinus brutia* × 1/4

Aristolochia longa × 1/2

75a *Aristolochia cretica* × 1/2

132c *Minuartia stellata* × $\frac{1}{10}$

101a *Beta trigyna* × $\frac{1}{12}$

2

140b *Cerastium candidissimum* × $\frac{1}{3}$

132d *Minuartia baldaccii* × $\frac{1}{3}$

128b *Arenaria filicaulis* × $\frac{1}{4}$

150 *Paronychia kapela* × $\frac{1}{3}$

7a *Drypis spinosa* × ¼

158 *Lychnis coronaria* × ½

3

b *Silene auriculata* × ⅓

168a *Silene roemeri* × ⅛

a *Silene asterias* × 1

176b *Silene compacta* × ⅓

(180) *Saponaria bellidifolia* × ½

181a *Saponaria glutinosa* × ¼

187b *Dianthus giganteus* × ⅘

4

194f *Dianthus scardicus* × ½

187a *Dianthus pinifolius* × ½

194g *Dianthus microlepis* × ⅗

193e *Dianthus haematocalyx* × ⅘

194e *Dianthus serratifolius* × ½

00b *Helleborus orientalis* × ¼

200c *Helleborus odorus* × ¼

5

13) *Consolida orientalis* × ⅛

210f *Aconitum 'pentheri' (divergens)* × 1/10

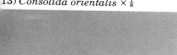

)e *Helleborus dumetorum* × ⅛

200d *Helleborus multifidus* subsp. *serbicus* × ⅔

(214) *Anemone blanda* × ¾

214a *Anemone trifolia* × ⅛

6

216 *Anemone coronaria* × ⅓

(216) *Anemone pavonina* × ½

222 *Pulsatilla pratensis* × ⅕

212 *Delphinium staphisagria* × ⅛

(211) *Delphinium peregrinum* × ⅜

239d *Ranunculus creticus* × ⅓

234b *Ranunculus brevifolius* × ½

7

246a *Ranunculus asiaticus* × ¼

246a *Ranunculus asiaticus* × ½

46a *Ranunculus asiaticus* × ⅔

(246) *Ranunculus platanifolius* × ⅕

253d *Aquilegia amaliae* × ⅖

228 *Clematis recta* × ⅙

8

253i *Aquilegia aurea* × ⅓

226a *Clematis viticella* × ⅓

260b *Leontice leontopetalum* × ⅓

260c *Bongardia chrysogonum* × ¼

8b *Paeonia clusii* × ⅕

258a *Paeonia rhodia* × ⅙

59) *Paeonia tenuifolia* × ⅓

(279) *Corydalis bulbosa* × ½

6b *Papaver nigrotinctum* × ⅖

278 *Corydalis lutea* × ¼

298f *Malcolmia angulifolia* × ⅛

298a *Malcolmia macrocalyx* × ½

10

318a *Arabis bryoides* × ⅓

294e *Erysimum pectinatum* × ¼

320b *Aubrieta intermedia* × ⅛

323a *Alyssoides cretica* × ⅖

)a *Aubrieta gracilis* × 1/15

331e *Draba parnassica* × 1/2

11

5d *Thlaspi microphyllum* × 1/2

329c *Bornmuellera tymphaea* × 1/8

' *Iberis sempervirens* × 1/8

324d *Alyssum handelii* × 1/4

(348) *Iberis umbellata* × $\frac{1}{10}$

12

373 *Reseda alba* × $\frac{1}{10}$

368c *Crambe tataria* × $\frac{2}{5}$

e *Sempervivum zeleborii* × ⅓

379 *Umbilicus rupestris* × ⅛

13

d *Sempervivum kindingeri* × ⅖

385a *Jovibarba hueffelii* × ¼

b *Sedum laconicum* × ⅔

396 *Sedum cepaea* × ¼

389a *Sedum sartorianum* × ½

387 *Sedum ochroleucum* × ½

14

(387) *Sedum sediforme* × ⅔

396d *Sedum creticum* × ⅖

396c *Sedum magellense* × ⅖

393 *Sedum hispanicum* × ⅓

392a *Sedum rubens* × ½

d *Saxifraga juniperifolia* × ½

411j *Saxifraga sibthorpii* × ½

409a *Saxifraga glabella* × ⅕

15

i *Saxifraga stribrnyi* × ½

410g *Saxifraga sempervivum* × ¼

403a *Saxifraga chrysosplenifolia* × ⅕

d *Saxifraga spruneri* × ⅕

(411) *Saxifraga exarata* × ⅓

398e *Saxifraga scardica* × ⅙

419e *Sibiraea altaiensis* × ¼   419d *Spiraea media* × ⅛

16

435a *Rosa glauca* × ⅗   433a *Rosa glutinosa* × ⅕

437 *Aremonia*
agrimonoides × ⅕   443a *Geum bulgaricum* × ½   433b *Rosa turcica* × ⅖

2e *Geum rhodopeum* × ¼   442c *Geum coccineum* × ⅗

17

7f *Potentilla clusiana* × ⅓   455 *Potentilla aurea* × ⅓

e *Potentilla deorum* × 1/10   453 *Potentilla recta* × ⅖

474 *Cotoneaster nebrodensis* × ½

473a *Cotoneaster niger* × ¼

18

464b *Pyrus amygdaliformis* × ⅖

460 *Mespilus germanica* × ¾

475 *Pyracantha coccinea* × ¼

464f *Pyrus elaeagrifolia* × ¼

9a *Sorbus umbellata* $\times \frac{1}{50}$ 462b *Crataegus heldreichii* $\times \frac{2}{5}$

19

2a *Crataegus laciniata* $\times \frac{1}{2}$ 462c *Crataegus pycnoloba* $\times \frac{1}{5}$

7a *Prunus cocomilia* $\times \frac{1}{2}$ 479a *Prunus tenella* $\times \frac{1}{10}$

497a *Podocytisus caramanicus* × ⅓   489 *Acacia cyanophylla* × ½

20

507h *Chamaecytisus heuffelii* × 1/12   507 *Chamaecytisus supinus* × ⅔

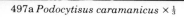

506b *Chamaecytisus subidaeus* × ⅓

509a *Genista januensis* × ⅓   508i *Genista sericea* × ⅓

1g *Genista parnassica* × ¼

507g *Chamaecytisus banaticus* × ¼

21

0a *Lupinus varius* × ⅕

528h *Astragalus thracicus* × ⅓

527f *Astragalus ponticus* × ¼

2) *Astragalus onobrychis* × ⅓

534a *Oxytropis prenja* × ½

545e *Vicia pinctorum* × $\frac{3}{5}$

(547) *Vicia melanops* × $\frac{4}{5}$

22

547 *Vicia hybrida* × $\frac{1}{2}$

550 *Vicia narbonensis* × $\frac{1}{2}$

(536) *Glycyrrhiza echinata* × $\frac{1}{4}$

561a *Lathyrus articulatus* × 1

65f *Lathyrus digitatus* × ¼

559a *Lathyrus grandiflorus* × ½

23

65a *Lathyrus laevigatus* × ⅖

567d *Ononis pubescens* × ½

98a *Trifolium uniflorum* × 1

606a *Trifolium pannonicum* × ⅓

596a *Trifolium rubens* × ½   (596) *Trifolium purpureum* × ¼

24

598b *Trifolium pilulare* × 1

604b *Trifolium physodes* × ½   604a *Trifolium resupinatum* × ⅔

596b *Trifolium alpestre* × ⅓   592b *Trifolium aurantiacum* × ⅓

20 *Anthyllis barba-jovis* × ¼

621c *Anthyllis aurea* × ⅖

25

2 *Anthyllis vulneraria* subsp. *praepropera* × ⅔

636g *Onobrychis ebenoides* × ⅓

5a *Lotus cytisoides* × ⅔

637b *Ebenus cretica* × 1/10

647 *Geranium macrorrhizum* × ⅖

643 *Geranium tuberosum* × 1

26

654b *Erodium chrysanthum* × ⅗

658 *Linum flavum* × ⅕

658f *Linum thracicum* × ⅓

663 *Linum tenuifolium* × ⅓

a *Linum hirsutum* × ½

658g *Linum tauricum* × ⅙

b *Haplophyllum*
veolens × ⅙

699 *Polygala major* × ⅘

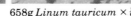

a *Acer heldreichii* × ¼

709b *Acer hyrcanum* × ¼

718c *Staphylea*
pinnata × ½

748 *Alcea pallida* × ¹⁄₁₀

733d *Kitaibela vitifolia* × ⅓

28

744 *Lavatera thuringiaca* × ⅔

748a *Alcea heldreichii* × ½

683 *Euphorbia rigida* × ⅛

733 *Malope malacoides* × ½

Daphne alpina × ⅓

(757) *Daphne blagayana* × ⅛

29

ᵃ *Daphne jasminea* × 1½

(757) *Daphne oleoides* × 1/10

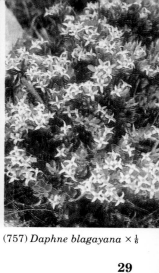

ᵈ *Hypericum empetrifolium* × ½ 769c *Hypericum olympicum* × ⅕

769d *Hypericum cerastoides* × ⅓

781c *Viola delphinantha* × ⅔

786b *Viola poetica* × ½

30

783a *Viola aetolica* × ⅓

785e *Viola rhodopeia* × ⅖

786e *Viola allchariensis* × ⅗

786c *Viola gracilis* × ⅓

37b *Cistus parviflorus* × ⅖

801d *Helianthemum cinereum* × ½

31

4a *Fumana arabica* × ¾

822a *Lythrum junceum* × ¾

56) *Smyrnium rotundifolium* × ⅓

(856) *Eryngium creticum* × ⅔

919b *Rhododendron luteum* $\times \frac{1}{12}$

(924) *Arbutus andrachne* $\times \frac{1}{10}$

32

946b *Primula deorum* $\times \frac{1}{4}$

943a *Primula frondosa* $\times \frac{1}{4}$

(944) *Primula minima* $\times \frac{1}{5}$

942 *Primula vulgaris* subsp. *sibthorpii* $\times \frac{4}{5}$

c *Soldanella hungarica* × 1

952b *Androsace hedraeantha* × ½

33

b *Cyclamen creticum* × ½

960c *Cyclamen persicum* × ⅕

a *Lysimachia atropurpurea* × ⅛

(967) *Anagallis foemina* × ¾

977 *Styrax officinalis* × ½

984a *Forsythia europaea* × ⅔

34

995 *Gentiana pyrenaica* × ⅔

975d *Armeria canescens* × ½

971a *Acantholimon
androsaceum* × ¼

1011a *Cionura erecta* × 1/10

19a *Asperula arcadiensis* × ⅘   1019b *Asperula boissieri* × 1   1017d *Asperula lutea* × ½

37 *Convolvulus lineatus* × ⅓   1025c *Galium rhodopeum* × ⅓

9 *Convolvulus althaeoides* subsp. *issimus* × ⅓   1036b *Convolvulus boissieri* × ⅓

1045c *Heliotropium hirsutissimum* × ⅔   1047b *Omphalodes luciliae* × ⅔

36

1061c *Alkanna graeca* × ½   1061e *Alkanna calliensis* × ½

1060 *Nonea pulla* × ⅓   1075c *Macrotomia densiflora* × ¼   1057a *Anchusa cespitosa* × ⅖

63a *Pulmonaria rubra* × ⅖

1053a *Symphytum bulbosum* × ⅖

1075b *Lithodora zahnii* × 1

37

75a *Lithodora hispidula* × ½

1077c *Onosma erecta* × ⅕

1077d *Onosma taurica* × ⅖

77h *Onosma echioides* × ⅛

1076c *Onosma visianii* × ⅓

1077b *Onosma heterophylla* × ⅕

1079a *Cerinthe retorta* × ⅙

1092 *Ajuga laxmannii* × ¾

38

1101b *Teucrium aroanium* × ½

(1091) *Ajuga orientalis* × ⅖

1104 *Teucrium flavum* × 1/12

˙1109e *Scutellaria rubicunda* × ⅗

1112a *Marrubium velutinum* × ⅛

15d *Sideritis syriaca* × ½

1122a *Phlomis samia* × ⅓

39

24b *Phlomis lanata* × ½

1131 *Lamium garganicum* × 1

29a *Lamium moschatum* × ⅓

1136d *Stachys candida* × ⅓

1136c *Stachys chrysantha* × ⅓

1143f *Salvia ringens* × ⅖   1170b *Mentha microphylla* × ⅓   1143e *Salvia pomifera* × ¼

40

1136f *Stachys spruneri* × ⅓   1144b *Salvia candidissima* × ⅛

1148d *Salvia austriaca* × ⅖ 1145 *Salvia verticillata* × ⅖

51c *Origanum dictamnus* × ⅓

1163f *Thymus atticus* × ½

53d *Thymus cherlerioides* × ½   1180e *Solanum elaeagnifolium* × ½

50d *Solanum cornutum* × ⅓   1186b *Datura innoxia* × ¹⁄₁₀

1195c *Verbascum delphicum* × $\frac{1}{20}$

1194b *Verbascum eriophorum* × $\frac{1}{18}$ 1195b *Verbascum graecum* × $\frac{1}{15}$

42

1194c *Verbascum epixanthinum* × $\frac{1}{2}$

1195d *Verbascum mallophorum* × $\frac{1}{20}$

(1194) *Verbascum longifolium* × $\frac{1}{15}$

(1195) *Verbascum speciosum* × $\frac{1}{25}$

1196e *Verbascum acaule* × $\frac{1}{2}$

1190c *Verbascum glabratum* ×

05) *Linaria genistifolia* × ⅕

1203 *Linaria pelisseriana* × 1

1205b *Linaria peloponnesiaca* × ¼

43

10b *Cymbalaria longipes* × 1½

1212 *Scrophularia peregrina* × ⅓

20e *Veronica erinoides* × ⅘

1222 *Veronica cymbalaria* × ½

1231a *Digitalis lanata* × ½

1231 *Digitalis laevigata* × ½

1230 *Digitalis ferruginea* × ½

44

1261a *Rhynchocorys elephas* × ⅔

1249c *Pedicularis brachyodonta* × ⅖

1269b *Lathraea rhodopea* × ¼

3 *Pedicularis verticillata* × ½

1249e *Pedicularis graeca* × ⅓

45

1b *Siphonostegia syriaca* × ⅖

1259b *Melampyrum hoermannianum* × ½

1249a *Pedicularis oederi* × ⅔

3b *Globularia meridionalis* × ⅕

1266b *Acanthus balcanicus* × ¼

1268d *Haberlea rhodopensis* × ½

1268c *Jankaea heldreichii* × ½

46

1268b *Ramonda serbica* × ⅖

1279b *Pinguicula hirtiflora* × ⅙

1288c *Plantago atrata* × ½

1280c *Pinguicula balcanica* × ⅓

270b *Orobanche lavandulacea* × ½

1276b *Orobanche reticulata* × ½

1276 *Orobanche alba* × ⅓

47

271 *Orobanche purpurea* × ⅘

1302a *Lonicera nummulariifolia* × $\frac{1}{20}$

306) *Lonicera implexa* × ¼

1319a *Morina persica* × ⅛

1324e *Pterocephalus perennis* × ½

1333b *Campanula moesiaca* × ⅓

48

1330d *Campanula lingulata* × ¼

1330c *Campanula rupicola* × ½

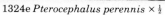

1330g *Campanula incurva* × ⅓

1343a *Campanula versicolor* × ⅓

5b *Campanula*
iosissima × ⅘

1331d *Campanula celsii* × ¾

1330a *Campanula*
alpina × ½

49

4c *Symphyandra hofmannii* × ¼

1348a *Trachelium jacquinii* × ½

9 *Asyneuma limonifolium* × ⅖

1352c *Phyteuma confusum* × ½

1354 *Edraianthus*
graminifolius × ¼

1354d *Edraianthus serpyllifolius* × ½

1384b *Helichrysum plicatum* × ¼

50

1391a *Inula oculus-christi* × ⅓

1392 *Dittrichia viscosa* × ⅕

1410a *Anthemis rigida* × ⅘

1409d *Anthemis orientalis* × ⅕

1420a *Achillea umbellata* × ½

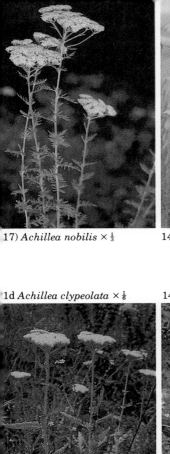

17) *Achillea nobilis* × ⅓

1421e *Achillea coarctata* × ¼

1421b *Achillea holosericea* × ¼

51

1d *Achillea clypeolata* × ⅙

1426a *Tanacetum millefolium* × ¼

1448a *Doronicum columnae* × ⅕

3e *Senecio gnaphalodes* × ⅓

1453d *Senecio thapsoides* × ⅛

1455f *Senecio ovirensis* × ⅓

1461a *Echinops graecus* × ⅔

1461b *Echinops spinosissimus* × ⅓

1486a *Cirsium creticum* × ¼

52

1487a *Cirsium appendiculatum* × ⅓

1474d *Staehelina uniflosculosa* × ¹⁄₁₀

1488d *Ptilostemon chamaepeuce* × ⅛

˙1463 *Cardopatum corymbosum* × ½

1488a *Ptilostemon afer* × ⅓

1491 *Cynara cardunculus* × ⅕

1505c *Centaurea alba* × ¼

1500g *Centaurea attica* × ⅛

53

1501d *Centaurea triumfetti* bsp. *cana* × ½

1500h *Centaurea laconica* × ⅕

1500d *Centaurea graeca* × ⅔

1505d *Centaurea zuccariniana* × ½

1504c *Centaurea orientalis* × ⅓

1503 *Centaurea salonitana* × 1/10

1471 *Atractylis gummifera* × ½

(1546) *Crepis rubra* × ⅔

54

(1512) *Cichorium spinosum* × ½

1510b *Hymenonema laconicum* ×

1524a *Leontodon crispus* × ⅕

1529 *Scorzonera purpurea* subsp. *rosea* × ⅔

1508c *Carthamus dentatus* × 1

881 *Colchicum bivonae* × $\frac{3}{5}$

88k *Colchicum turcicum* × $\frac{3}{4}$

1588m *Colchicum variegatum* × $\frac{2}{5}$

1588f *Colchicum cupanii* × $\frac{1}{4}$

590 *Asphodelus albus* × $\frac{1}{15}$

1588d *Colchicum pusillum* × $\frac{2}{5}$

1594 *Asphodeline
liburnica* × ½

(1596) *Anthericum ramosum* × ½

1614 *Allium scorodoprasum
subsp. rotundum* × ⅔

56

1608 *Allium
subhirsutum* × ⅓

1611 *Allium roseum* × ⅔

1604a *Allium guttatum* × ⅘

1604c *Allium
amethystinum* ×

1617a *Nectaroscordum
siculum* × ⅖

(1607) *Allium carinatum* × ⅔

618a *Lilium chalcedonicum* × ⅖

1621 *Lilium candidum* × ⅕

57

619) *Lilium carniolicum* × ¼

1622b *Fritillaria pontica* × ¼

22a *Fritillaria messanensis* × ⅘

1622c *Fritillaria graeca* × ⅓

1627d *Tulipa boeotica* × ½   1627b *Tulipa cretica* × ⅙

58

1627c *Tulipa orphanidea* × ½   1631 *Scilla hyacinthoides* × ⅙

1643b *Chionodoxa cretica* × ½   1631a *Scilla litardierei* × ⅔

542 *Ornithogalum nutans* × ⅓

639f *Ornithogalum collinum* × ⅓

1645a *Muscari macrocarpum* × ⅖

1643d *Hyacinthella leucophaea* × ⅓

43e *Hyacinthella dalmatica* × ½

1647a *Muscari armeniacum* × ⅓

1664 *Sternbergia lutea* subsp. *sicula* × ¾

1677e *Crocus veluchensis* × ⅓

1680b *Crocus pulchellus* × ⅓

1675b *Crocus boryi* × ⅓

60

1677d *Crocus sieberi* × ⅛

1675 *Crocus flavus* × ⅓

1675d *Crocus cvijicii* × ⅘

1678e *Crocus cancellatus* × ¼ 1675e *Crocus scardicus* × ⅔

1678c *Crocus hadriaticus* × ⅓

687a *Iris unguicularis* × 1

1691a *Iris reichenbachii* × $\frac{2}{5}$

61

686a *Iris sintenesii* × $\frac{3}{5}$

1681 *Romulea bulbocodium* × $\frac{1}{2}$

1697 *Gladiolus illyricus* × $\frac{2}{5}$

18d *Arum dioscoridis* × $\frac{1}{4}$

(1818) *Arum italicum* subsp. *byzantinum* × $\frac{1}{4}$

1820d *Biarum davisii* × 1

1876 *Ophrys scolopax* subsp. *heldreichii* × 1

1880 *Ophrys sphegodes* subsp. *aesculapii* × 2

1880 *Ophrys sphegodes* subsp. *atrata* × 1

62

1881a *Ophrys ferrum-equinum* subsp. *gottfriediana* × 2¼

1881a *Ophrys ferrum-equinum* × 2

1880d *Ophrys cretica* × 2½

1880e *Ophrys reinholdii* × 2½

1880g *Ophrys carmelii* × 2

1883 *Ophrys fuciflora* × ⅘

85c *Orchis boryi* × ¾

1889a *Orchis lactea* × ¾

1893 *Orchis laxiflora*
subsp. *elegans* × 1

63

92a *Orchis punctulata*
bsp. *sepulchralis* × ½

1894b *Orchis saccata (Ophrys
speculum)* × ⅗

1894c *Orchis spitzelii* × 1

95a *Orchis anatolica* × 1¼

1896 *Orchis provincialis* × ¼

(1896) *Orchis
pallens* × ⅕

1886 *Orchis coriophora* subsp. *fragrans* × 1

1898b *Dactylorhiza cordigera* × ½

1899b *Dactylorhiza iberica* × 1¼

64

1900a *Dactylorhiza saccifera* × ¾

1903 *Serapias cordigera* × ¾

1907 *Himantoglossum hircinum* subsp. *calcaratum* × ½

1913a *Pseudorchis frivaldi* × 1

1919b *Cephalanthera cucullata* × ⅓

1922a *Epipogium aphyllum* × ⅓

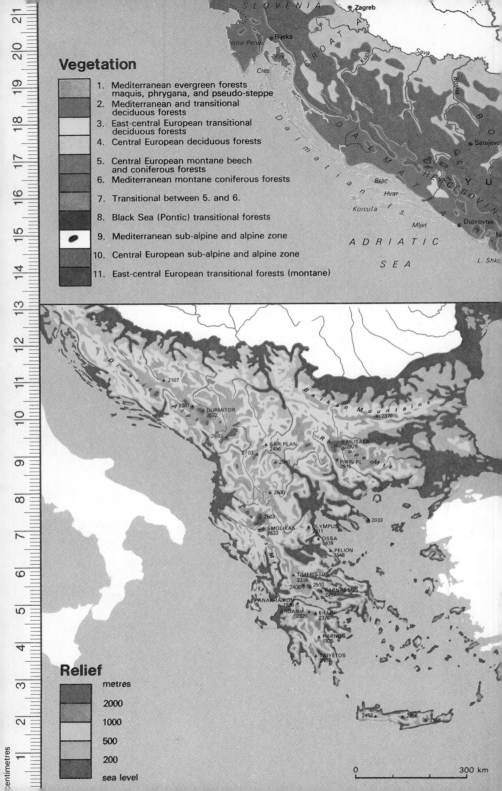

Vegetation

1. Mediterranean evergreen forests maquis, phrygana, and pseudo-steppe
2. Mediterranean and transitional deciduous forests
3. East-central European transitional deciduous forests
4. Central European deciduous forests
5. Central European montane beech and coniferous forests
6. Mediterranean montane coniferous forests
7. Transitional between 5. and 6.
8. Black Sea (Pontic) transitional forests
9. Mediterranean sub-alpine and alpine zone
10. Central European sub-alpine and alpine zone
11. East-central European transitional forests (montane)

Relief

metres

2000

1000

500

200

sea level

0   300 km